Proceedings

1997 Rapid Excavation and Tunneling Conference
Las Vegas, Nevada, June 22-25, 1997

Editors
J.E. Carlson
T.H. Budd

Sponsored by
Society for Mining, Metallurgy, and Exploration, Inc.
American Society of Civil Engineers

Society for Mining, Metallurgy, and Exploration, Inc.
Littleton, Colorado • 1997

Printed in the United States of America
by Cushing-Malloy, Inc., Ann Arbor, Michigan

Library of Congress Catalog Card Number 96-72547
ISBN 0-87335-148-7

Preface

The 1997 Rapid Excavation and Tunneling Conference in Las Vegas will be a premier showcase of contributors from the United States and around the world with papers on a wide variety of quality subjects with specific details on means and methods employed on these innovative and challenging underground projects.

The Program Chairmen, the Executive Committee, and the SME Staff welcomes all attendees to this year's RETC, which will again offer the participants a unique opportunity to inform themselves on the new techniques and creative solutions now being employed on projects throughout the industry. This 1997 conference will encompass 58 presentations (52 papers included here) in 15 sessions and will offer participants details on achievements and advancements in a number of categories — including tunneling, shaft sinking, ground support, design and planning, geotechnical and new technology which have resulted in new design solutions, alternative work practices and increased productivity within the industry. Of special interest are presentations involving the Deer Island Boston Harbor Project and Metro West Boston Water Supply Project, the Inland Feeder Project in Southern California, the Yucca Mountain Project in Nevada, the New York Water Tunnel Project, the Tri-Met Project at Portland, TARP Projects in Chicago, and international projects from Canada, Great Britain, Egypt, Japan, Australia, New Zealand, and Argentina.

The 1997 RETC Session Chairmen and Authors this year represent a platinum list of contractors, engineers, owners, and educational institutions all dedicated to the continued advancement of this organization and the development of new ideas within the industry. The Executive Committee extends their highest appreciation to these individuals for their commitment of time and effort in the preparation and presentation of these briefs. An additional key element for our continued survival is the presence and support of our loyal exhibitors, whom we recognize not only for their participation, but also for their contributions to the underground business through advancements in technology, supply, and equipment innovations. A partial listing of these organizations is contained in the back of the proceedings. When you visit their exhibits, take the time to let them know how valuable they are to the industry.

In every organization there are a few individuals who often represent the difference between success and failure. In our case, it is the behind-the-scenes work and results of the dedicated staff members from SME who have provided the expertise and guidance to the Executive Committee to enable them to perform their responsibilities with ease. We extend our special thanks and greatest appreciation to these people for their contributions and efforts in developing the new CD-ROM RETC Proceedings from 1972 to 1997.

The Chairman, Co-chairman, and Executive Committee extend our sincere welcome to all participants at this 1997 RETC meeting being held in one of the most exciting cities in the world. We anticipate that each of you will leave with new insights and ideas which can be put to further use for you and the underground industry in general.

June 1997

Jay E. Carlson, SME
Theodore H. Budd, ASCE

1997 RETC Executive Committee

Conference Chairman
Patrick B. Kenny
Kenny Construction Co.
Wheeling, IL

SME

W.D. (Toby) Wightman
Kiewit/PB
Las Vegas, NV

I. Michael Gowring
Dillingham Construction N.A., Inc.
Pleasanton, CA

Robert Reseigh
Guy F. Atkinson Co.
Lakewood, CO

Jay E. Carlson
S.A. Healy Co.
Plano, TX

Donald E. Hilton
Frontier-Kemper Construction Inc.
Evansville, IN

ASCE

Theodore H. Budd
Kenny Construction Co.
Wheeling, IL

James E. Monsees
Parsons Brinckerhoff, Inc.
Villa Park, CA

George E. Williamson
Traylor Bros., Inc.
Pleasanton, CA

Lance W. deStwolinski
Kiewit Construction, Co.
Omaha, NE

Joette E. Cross
SME, Inc.
Assistant Conference Manager

Tara K. Davis
SME, Inc.
Publicity & Programming Manager

Franklin Margiotta
ASCE
Exhibits Manager

Jacqueline Williams
ASCE
Director, Conferences
& Conventions

1997 RETC PROGRAM COMMITTEE

Conference Program Chairmen

Jay E. Carlson
S.A. Healy Company
Plano, TX

Theodore H. Budd
Kenny Construction Co.
Wheeling, IL

Committee Members

Patrick B. Kenny
Kenny Construction Co.
Wheeling, IL

W.D. (Toby) Wightman
Kiewit/PB
Las Vegas, NV

Session Chairmen

A.C. Abrams
Atkinson Construction Company
Lakewood, CO

C. Daugherty
Parsons Brinckerhoff Quade & Douglas
Boston, MA

D.B. Desai
DMJM
Los Angeles, CA

P. Freeland
Granite Construction Co.
Watsonville, CA

M. Gay
Kiewit Construction Co.
Omaha, NE

D. Heinz
Kenny COnstruction CO.
Wheeling, IL

B. Kaub
Granite Construction Co.
Watsonville, CA

M. Kobler
UCM
Santa Rosa, CA

T.P. Kwiatkowski
Jenny Engineering Corp.
Springfield, NJ

L.M. Lenahan
McNally Tunneling Corp.
Westlake, OH

C. Linden
Obayashi Corp.
No. Hollywood, CA

W. Mariucci
Kiewit Construction Co.
Omaha, NE

J. McDonald
Traylor/Obayashi Corp. J.V.
San Ysidro, CA

L. Ozdemir
Colorado School of Mines
Golden, CO

D. Rogstad
Frontier-Kemper
Evansville, IN

R.L. Shufflebotham
S.A. Healy Co.
Plano, TX

T.P. Smirnoff
Parsons Brinckerhoff Quade & Douglas
Los Angeles, CA

TABLE OF CONTENTS

SECTION 4. Conventional Excavation

SECTION 5. New Technology

SECTION 6. TBM Case Histories II

SECTION 7. Difficult Ground

SECTION 8. Geotechnical

SECTION 9. Shafts and Raises

SECTION 10. Future/Special Projects

SECTION 11. International Projects I

SECTION 12. NATM

SECTION 13. International Projects II

SECTION 14. Soft Ground Case Histories

1

Design and Planning

Chairman
T.P. Kwiatkowski
Jenny Engineering Corp.
Springfield, NJ

Chapter 1

SECOND MANAPOURI TAILRACE TUNNEL DESIGN AND CONSTRUCTION CONSIDERATIONS

Brian Heer,[1] Randall Essex,[2] and David Young[3]
[1]Assistant Project Director, ECNZ; [2]Project Manager, Woodward-Clyde;
[3]Project Engineer, Woodward-Clyde

ABSTRACT

The Manapouri Power Station is New Zealand's largest hydroelectric facility located on the South Island of New Zealand. Constructed in the 1960's, it includes a 700 MW underground power station and a 9.8 km (6.1 mile), 9.2 m (30 ft) diameter tailrace tunnel. The output of the power station is currently limited due to excessive head losses in the tailrace tunnel. To remedy this situation, the Electricity Corporation of New Zealand (ECNZ) retained Woodward-Clyde to design a second tailrace tunnel and appurtenant facilities and prepare Contract Documents for construction. The Project will involve the construction of a 9.8 km (6.1 miles), 10 m (33 ft) diameter, TBM-mined tunnel, hydraulic control structures at the upstream and downstream ends of the tunnel, an underground dewatering shaft and station, and tailrace channel improvements. This paper summarizes the key design and construction issues addressed during the planning and design phases, outlines the risk management provisions incorporated in the Contract, and overviews the Interactive Tender Process (ITP) that was implemented during the pre-tendering and tendering periods.

INTRODUCTION/BACKGROUND

The Manapouri Power Station is an existing hydroelectric facility located near the southwest tip of the South Island of New Zealand, as shown on Figure 1. ECNZ is planning to construct a second tailrace tunnel at the station to supplement the flow capacity of the existing tailrace tunnel and improve the overall efficiency of the station. The second tailrace tunnel (Tunnel No. 2) will be approximately 9,830 m (6.1 miles) long and have an excavated diameter of 10 m (33 ft). Tunnel No. 2 will be excavated using a tunnel boring machine (TBM) and will be aligned approximately parallel to the existing tailrace tunnel (Tunnel No. 1).

Design Process and Schedule

The Team of Woodward-Clyde and DesignPower (WC) was retained by ECNZ to provide engineering and design services for the Project. One of the first challenges presented by the Project was that ECNZ mandated that the design effort be completed within 27 weeks. In order to meet this aggressive time schedule, WC established a project office two blocks from ECNZ's Wellington headquarters. The initial five weeks comprised the development of a Project Manual addressing all formats, guidelines, and protocols, a detailed Scope of Work, intensive research of the existing project records and documents provided by ECNZ, development of preliminary design concepts and construction approaches, and the development of key design and construction questions whose answers would shape the project geometrics and component details. In order to answer these key questions in as efficient a timeframe as

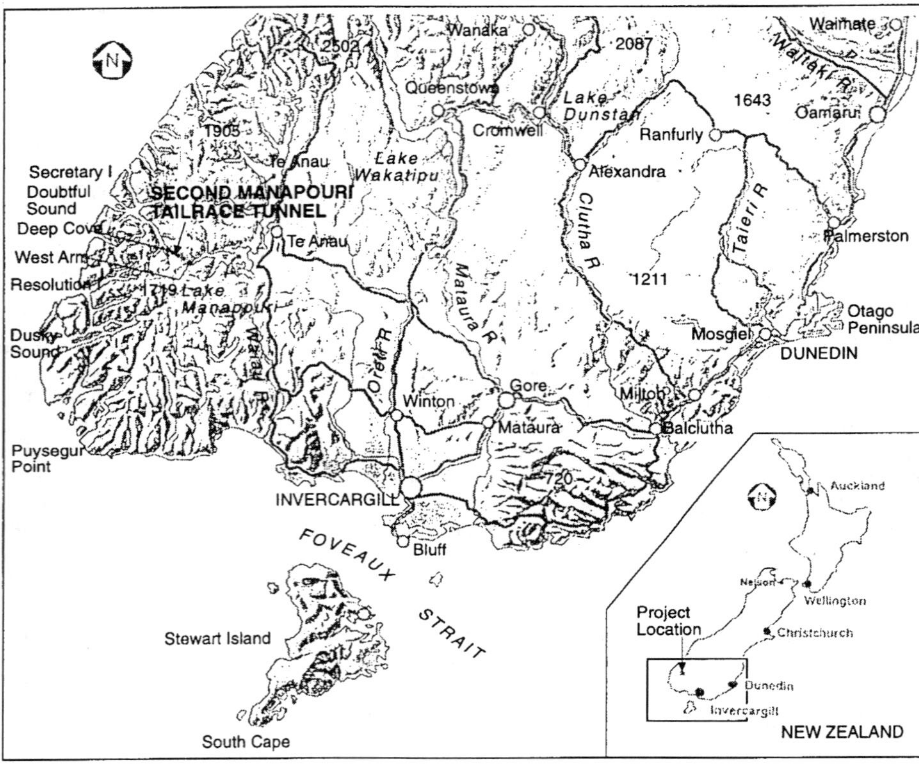

Figure 1. Project Location

possible, WC carried out a Technical Action Team (TAT) meeting, which brought together 28 ECNZ personnel, WC staff, and selected consultants in an intensive, 10-day workshop. The TAT members were provided with briefing documents in advance of the meeting, and were guided through a series of parallel working sessions that addressed civil, mechanical, electrical, tunnel, and contractual issues. Group Interaction and consensus opinion were able to eliminate undesirable alternatives and bring focus to the key issues quickly. Where there was clear consensus, the design would be refined and fixed. The inability to reach consensus was an indication that there was no single correct answer, and that a greater degree of flexibility should be afforded in the design and Contract Documents, in order to engage tenderers' ideas and preferred approaches. In WC's estimation, the TAT process eliminated 3 to 4 months of preliminary design and review activities.

Based on the results of the TAT Meeting, 30% design documents were completed by the 15th week, and 80% documents were submitted by the end of the 29th week. Shortly thereafter, ECNZ's need for an accelerated schedule was eliminated, the design schedule was relaxed, and the design effort refocussed to a more detailed re-visitation of site conditions and detailed value-engineering review of the project hydraulics and component designs. The Interactive Tender Process (ITP) followed during the pre-tendering and tendering periods is described later in this paper.

Existing Facilities

The Manapouri Power Project develops the hydropower potential of the portion of the Waiau River watershed located upstream from the Lake Manapouri control structure including Lake Te Anau. Power is developed by diverting the outflow from Lake Te Anau and the inflow into Lake Manapouri through a surface intake structure located on Lake Manapouri at West Arm and directing it to an underground power station. The water is conveyed through seven vertical penstocks to seven 100 MW (nominal) turbine generators, which are located about 178 m (585 ft) below the level of Lake Manapouri. Flow from the turbines passes through a series of draft tubes and a draft tube manifold to Tunnel No. 1, where the water is conveyed approximately 9.8 km (6.1 miles) to Deep Cove. Deep Cove connects to the Tasman Sea through Doubtful Sound. The existing facilities are illustrated on Figure 2.

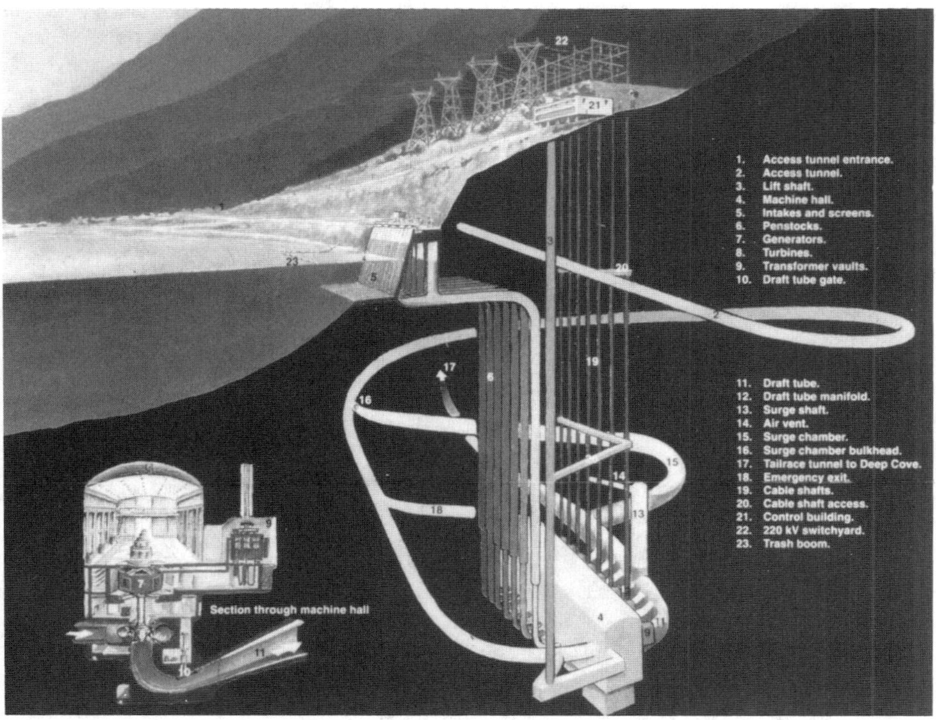

Figure 2. Existing Power Station Facilities

Tunnel No. 1 is 9.2 m (30 ft) wide, and has a horseshoe-shaped concrete lining. The tunnel slopes downward from an invert elevation of approximately El -3 m (-10 ft) at the power station to a low point of El -40 m (-131 ft) about 2.3 km (1.4 mile) from the (downstream) outlet portal, and then upward at a 1.67% grade to the portal. The soffit of the outlet portal is at approximately sea level (see Figure 3).

The existing facilities were designed by Bechtel-Pacific Corporation. Construction was carried out by Utah Construction between 1963 and 1969. Drill and blast methods were used to excavate the underground power station, access tunnel, and tailrace tunnel. The rock encountered during construction consisted of strong metamorphic rock, primarily gneiss, with lesser amounts of intrusive igneous rocks. Groundwater inflows to the tailrace tunnel during construction were significant, with measured point source flows as high as 91 liters per second (1,440 gpm), and cumulative inflows measured over the length of the tunnel as high as 900 liters per second (14,370 gpm). Spalling occurred in reaches of the tunnel with high ground cover. Tunnel No. 1 was excavated full face over most of its length and supported with either rock bolts or steel sets. Concrete lining was placed during mining.

The Project has not been able to operate at its installed capacity for two principal reasons. One reason relates to frictional head losses in the system. The hydraulic design was based on an assumed Manning roughness factor, "n", of 0.0125 for the concrete-lined tailrace tunnel. However, operational experience has indicated a Manning roughness factor of approximately 0.0145. The second reason regards the level of Lake Manapouri. The original design of the project intended that the level of Lake Manapouri would be raised by 8.2 m (27 ft) to increase storage in the lake for improved flow regulation, but this proposal was not acceptable for environmental reasons.

Figure 3. Tunnel No. 1 Outlet Portal

Over the years several options to reduce the head loss in the tailrace system were studied, but in general the economics were not favorable due to the long station outages required, and the resulting significant loss in generation. In 1993 studies were initiated to investigate the use of a Tunnel Boring Machine to construct a second tunnel parallel to the existing tunnel. This option required only a short duration station outage of 21 days to break-through into the existing tailrace manifold as compared with outages of up to 3.5 years for some of the other options previously considered. These studies culminated in the publication of a feasibility report (Brown et al, 1993).

Because the economics of this option appeared favorable, the Manapouri Tailrace Investigations Project was established in January 1994 to progress the findings of the feasibility report and in November 1994 Woodward-Clyde were engaged by ECNZ to undertake detailed design.

Project Objectives

The fundamental objective of adding the second tailrace tunnel is to increase the overall cross-sectional area of flow, thereby reducing the flow velocities and associated frictional head losses. Other important objectives are to minimize the construction costs by minimizing lengths of tunnel to receive a cast-in-place concrete lining, to construct the new facilities with the least impact to the operating power station, and to complete the construction without impacting the pristine and natural beauty of Doubtful Sound (see Figure 4).

PROJECT DESCRIPTION

Tailrace Tunnel

An isometric view of Tunnel No. 2 is shown on Figure 5. The new tunnel will be excavated to a diameter of 10 m (33 ft) using a tunnel boring machine (TBM). The Contract Documents anticipate the tunnel to be unlined for approximately 56% of its 9830 m (6.1 miles) length. A tunnel lining of either shotcrete or concrete will be installed where warranted by the ground conditions. Tunnel No. 2 will be connected to the downstream end of the existing draft tube manifold during a limited station outage period. From its upstream end at the draft tube manifold, the tunnel will slope downward at a 10% gradient for approximately 100 m (328 ft), then downward at a 3% gradient for approximately 640 m (2100 ft), followed by a transition to a relatively flat downward grade of 0.1% for approximately 8.7 km (5.4 miles), to a low point in the tunnel profile. At this point the tunnel will trend upward at a 12.5% gradient for a distance of approximately 300 m (984 ft) until reaching the outlet portal. To reduce the groundwater control requirements in alluvial and glacial soils outside the outlet portal, the portal invert will be constructed at an elevation of El -5 m (16 ft), rather than at El -10 m (33 ft), which is the invert elevation for Tunnel No. 1. To maintain hydraulic efficiency with this geometry, the outlet of the tunnel will "flare" from its circular cross section into a 19 meter-wide horseshoe-shaped section with a flat invert.

New Appurtenant Facilities

In addition to Tunnel No. 2, the project will include tunnel headworks, outlet works, and dewatering facilities. The headworks will be constructed at the upstream end of the new tunnel near the connection with the draft tube manifold (referred to herein as the bifurcation). An isometric view of the headworks is shown on Figure 5. The headworks will include: a maintenance adit, which branches off from the existing power station access tunnel; a maintenance shaft, used to access Tunnel No. 2 for construction and future maintenance purposes; and a stop log adit, shaft and stop log structure which will be used to seal off flows through Tunnel No. 2.

The outlet works, to be constructed at the portal, are shown on Figure 6. These facilities will consist of: a concrete structure that will facilitate a smooth hydraulic transition from the

Figure 4. Doubtful Sound with Deep Cove in Foreground, Fiordland National Park

tailrace tunnel to the tailrace channel; a stop log structure which can seal off waters in the outlet channel from the tailrace tunnel during maintenance periods; a maintenance ramp that transitions from the dewatering adit portal downward into the tailrace tunnel portal flare; and a bridge to facilitate operation of the stop log structure and maintain access to the upstream portion of Tailrace Channel No. 1.

The dewatering facilities are shown in an isometric view on Figure 5. These facilities will consist of: a dewatering adit which extends eastward from a portal located south of the Tunnel No. 2 portal; a dewatering shaft which connects the eastern limit of the dewatering adit to tailrace tunnel level; and dewatering intake lines that connect the bottom of the dewatering shaft to the low point of Tunnel No. 2 through cased boreholes or pipes embedded in a concrete backfilled adit. These facilities will permit the pumping of groundwater inflows from the tunnel during construction, and will provide a long-term tunnel dewatering capability to facilitate future tunnel inspections and maintenance work. During dewatering for maintenance, water will flow from Tunnel No. 2 through valved intake lines to the dewatering pumps located at the bottom of the dewatering shaft. Water will be pumped up the dewatering shaft, and conveyed through a series of cased drill holes (dewatering discharge lines) to Tunnel No. 1.

Environmental Sensitivities

The entire area of the Project is in the Fiordland National Park. This is the largest national park in New Zealand and is a World Heritage Area, registered with the United Nations

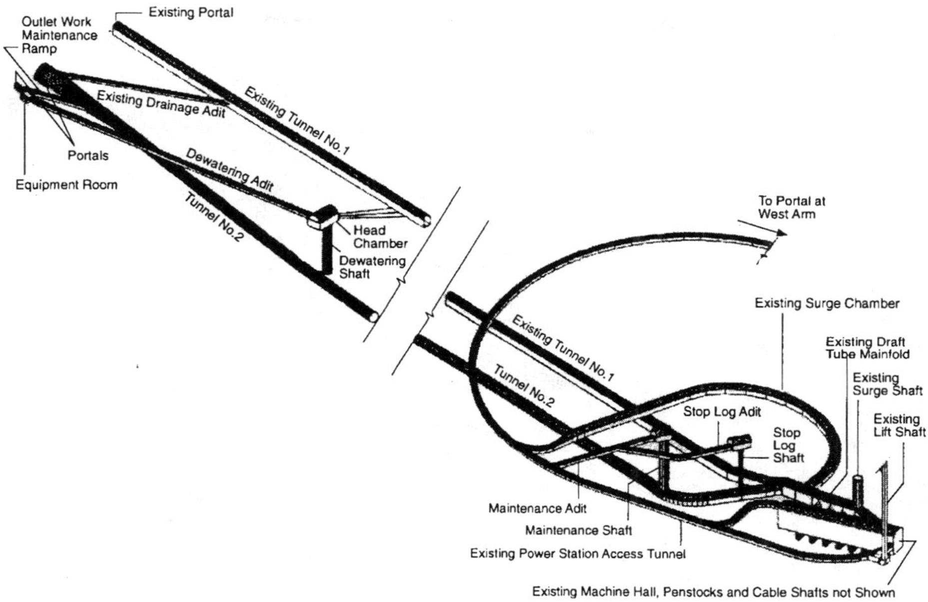

Figure 5. Project Isometric

as one of the world's natural wonders. The unique black coral in Doubtful Sound has been the subject of two Jacques Cousteau expeditions. The Manapouri project first came to the forefront of New Zealand public environmental awareness during the late 1960's when plans included damming the natural lake outlet and raising the level of Lake Manapouri. In what has been called the birth of the environmental movement in New Zealand, public concern halted plans for raising the lake. Operating rules for the power station have required that Lake Manapouri be maintained within its range of natural levels ever since.

ECNZ has undertaken the planning of the project through a comprehensive environmental consultation process with governing agencies and other interested and affected parties. Due to the pristine conditions that prevail at the site, strict environmental controls have been agreed and will be imposed on all construction activities at Deep Cove and West Arm. One such restriction pertains to discharge water quality standards. Particles larger than silt size are not permitted to be discharged from the work area, and the level of turbidity and impurities in discharge water are limited by specifications. There is also a restriction on all equipment and materials entering the National Park to prevent the spread of non-native vegetation in the park. As a result, all equipment must be steam cleaned before entering the park.

The work areas at Deep Cove and West Arm will generally lie within the area disturbed by the original construction. Spoil dumps at Deep Cove will be contoured and vegetated progressively as construction proceeds to hide the impacts of construction. The planned site restoration will improve the present appearance where impacts of the original construc-

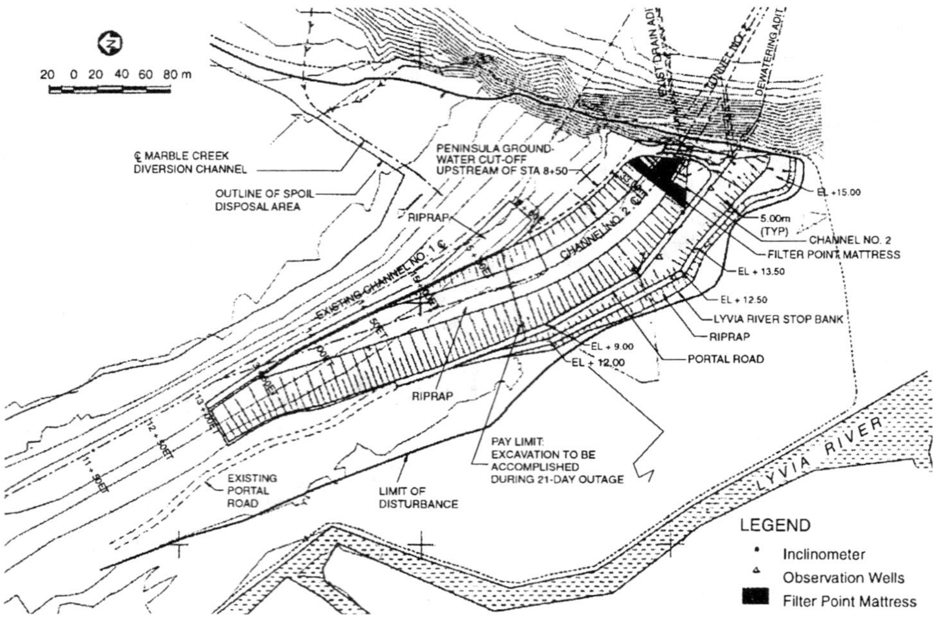

Figure 6. Tunnel No. 2 Outlet Works

tion are still evident.

EVALUATION OF EXISTING CONDITIONS

Analysis of Existing Information

Information from the design of the original project as well as information from other sources collected following the original construction were reviewed. The most important of these data were as-built geologic logs of the original tailrace tunnel construction. These geologic logs contained information on geology, groundwater inflows, grouting, blasting, and ground support. Interviews of personnel who were on site for the original construction were conducted by WC. These interviews provided valuable information about construction activities that were not apparent from the as-built records.

Field Investigations and Laboratory Testing

Field investigations were completed to collect additional data to supplement the information discussed above. Field investigations and laboratory testing consisted of: underground mapping; sampling of test pits excavated in the Tunnel No. 1 spoils; geomechanical testing and petrographic analyses for estimation of TBM performance and concrete aggregate suitability; pumping tests in the vicinity of the Tunnel No. 2 portal, probe hole drilling to identify the subsurface bedrock topography in the vicinity of the outlet works stop log structure; and water sample testing for assessment of corrosion potential.

The geomechanical testing for TBM performance supported performance assessment methods developed by the Robbins Company, the University of Trondheim/NTH, the University of Illinois, and others. Histograms of the test results were presented in the GBR for each of the different rock types. Baseline values for the various parameters were also included in the GBR.

Tunnel No. 2 Groundwater Inflows

There are a number of factors that will influence whether the magnitude of inflows observed in Tunnel No. 1 will be encountered in Tunnel No. 2, as follows:

1. Effectiveness of formation grouting in Tunnel No. 1.

2. Depressurization "shadow" created by steady state flow gradients toward Tunnel No. 1.

3. Relatively quicker advance rate in Tunnel No. 2 compared to Tunnel No. 1.

4. Tunnel No. 2 will be driven at a greater initial downward gradient than that of Tunnel No. 1. The operation of Tunnel No. 1 could provide a source of groundwater recharge in the rock mass that did not exist during the excavation of Tunnel No. 1.

Because of the impact such inflows could have on project schedule and cost, a somewhat conservative position was reflected in the baseline inflows presented in the GBR (discussed later).

Few water pressure measurements were made during the construction of Tunnel No. 1. The maximum pressure measured was 738 m (2421 ft) head. The pressure data, together with hydrogeological inference and modeling, was used to develop probable initial head conditions along the alignment of Tunnel No. 1. The head data and flow data measured at the portal, were used to back-calculate hydraulic conductivity values for reaches where an increase in flow was measured. Permeability calculations assumed both dispersed and discrete inflow to the reach so as to provide a range of values for use in modeling inflows to Tunnel No. 2.

Modeling was undertaken to assess the effect of different lengths, orientations, and diameters of drain hole drilled in advance of the tunnel face. These analyses were used to guide the drain hole drilling requirements included in the specifications.

Portal Area Groundwater Conditions

The groundwater level in the portal area is generally at about El. + 2 m (6.6 ft). This level is higher than the water level in Channel No. 1 but is below the level of the adjacent reach of the Lyvia River. In contrast to Channel No. 1, which appears to have been excavated in lower permeability deltaic deposits, the upper reaches of Channel No. 2 and excavations near the Tunnel No. 2 portal will be in more permeable alluvial and glacial/fluvioglacial deposits.

DESIGN CONSIDERATIONS

A number of aspects were key to the development of the various Project geometries and components. The more important design considerations are discussed below.

System Hydraulics

Overall system hydraulics and hydrologic considerations were the dominant considerations that affected project layout and component design. The more important of these analyses are summarized below.

System Head Losses. A series of analyses were carried out to estimate the frictional and geometric head loss for each component of the completed two tailrace tunnel system. Components of the total system head loss include: the bifurcation head loss, head loss through the headworks stop log structure; head loss associated with each of the typical stabilization and lining systems along the length of the tunnel, head loss at the portal flare, outlet, and transition to the tailrace channel, and channel losses.

Using the assumption that one meter of hydraulic head loss through the power station represents approximately $10 million NZ in present worth of the improvements (when discharging approximately 450 cubic meters per second (7.1 million gpm)), the above estimates were utilized in value engineering analyses of different design alternatives. This process was utilized to optimize the configuration of project geometries and component details.

Hydrodynamic Surge Analyses. Results from the above analyses were used as input into transient flow hydrodynamic analyses carried out for the existing and proposed facilities. A surge model, developed by Dr. John Undrill of the General Electric Company, was used to estimate maximum upsurges and downsurges in the system to determine elevation requirements for new facilities, to estimate spinning reserve available with the new system (operation limitations), and to evaluate the potential for hunting problems associated with short-term water surface oscillations between the surge facilities and turbine governors.

A program of commissioning tests will be carried out to confirm the hydraulic characteristics in the tailrace system during surges associated with load acceptance and load rejection, as well as confirm estimates of hydraulic characteristics during steady-state flow conditions. These tests will be used to develop operating rules for the power station.

Tailrace Tunnel Design

Benefits of using a TBM. Based on a review of the favorable rock mass conditions encountered during the construction of Tunnel No. 1, a number of TBM related benefits were identified: the TBM can produce a hydraulically acceptable rock surface that would preclude the need for lining; faster advance rates can be achieved resulting in a shorter overall construction schedule; stabilization requirements will be reduced; and where lining is needed for erosion protection, less overbreak around the perimeter of the tunnel excavation will result in reduced volumes of shotcrete and concrete lining materials

Tunnel Stabilization. Ground loads were developed by back-calculating the load carrying capacity of existing tunnel stabilization systems based on a review of the geologic conditions encountered and ground support measures installed during the excavation of Tunnel No. 1. These ground loads were factored down recognizing the decreased rock mass disturbance with TBM mining before being used to design a system of tunnel stabilization measures for

Tunnel No. 2. A preliminary version of the proposed stabilization system was subjected to critical review at the TAT Meeting, and was further refined during the design effort to ensure its compatibility with TBM operations. The stabilization approach consists of four types of stabilization as a function of rock mass quality, as summarized on Table 1. These stabilization types are to be augmented as required with separately measured and priced supplementary stabilization measures such as additional rock reinforcement, channels, and shotcrete. As explained earlier in this paper, the maximization of unlined tunnel is desirable from a cost standpoint. It is anticipated that approximately 45% of the tunnel length will be unlined, 45% will be shotcrete lined, and 10% will need to be concrete-lined. The need for a lining will depend upon the rock mass quality and the need to protect the exposed rock surfaces. The need to place shotcrete versus a concrete lining will depend upon several considerations, including the ability to control groundwater inflows during lining placement, and the need to install steel set stabilization during mining; a shotcreted lining in steel set ground would not be hydraulically acceptable, and would be difficult to carry out effectively.

Plan Alignment. The plan alignment for Tunnel No. 2 was controlled by existing facilities at the headworks, compatibility with existing conditions and facilities at the outlet works, a desired offset from Tunnel No. 1, and the desire to maintain a straight tunnel drive. Tunnel No. 2 begins at the bifurcation with the existing draft tube manifold. Except for a short tangent reach through the headworks stop log structure and a 40 m (131 ft) radius curve to the headworks maintenance shaft, the tunnel follows a tangent alignment in a north-westerly direction for its 9.7 km (6.0 miles) length.

The location of the Tunnel No. 2 portal was limited by the offsets required from the existing Tailrace Channel No. 1 to the north and the line of “mature bush” to the south, beyond which no construction-related disturbance is permitted. The orientation of the outlet works stop log structure was designed to conform with the rock contours at the Tunnel portal, and the bearing of the tunnel at that location.

Offset from Tunnel No. 1. The distance between Tunnel No. 1 and Tunnel No. 2 ranges from 75 m (246 ft) at Deep Cove to 65 m (213 ft) at the headworks maintenance shaft, and then tapers to zero at the bifurcation. The distance between the tunnels was the subject of numerous discussions and hydrologic analyses. The preferred separation distance to prevent hydraulic communication of Tunnel No. 1 water through crosscutting discontinuities such as joints, shears, and faults was considered to be 50 to 100 m (164 to 328 ft). This distance is significantly more than the 2 tunnel diameters needed to mitigate concerns related to high in-situ stresses. In contrast, however, there was a hydrogeologic reason to keep Tunnel No. 2 reasonably close to Tunnel No. 1.

Groundwater inflow data recorded during the construction of Tunnel No. 1 were reviewed and back-analyzed to estimate regional hydrogeologic parameters. These parameters were then incorporated into a finite element program that estimated the steady-state drawdown likely to exist around the operating Tunnel No. 1. Groundwater pressures monitored during the original construction were as high as 7 MPa (1,000 psi), indicating a fully saturated groundmass above the tunnel. However, because Tunnel No. 1 carries water under no more than 30 to 40 m (98 to 131 ft) of internal pressure, there is a net flow gradient toward the tunnel. This gradient was estimated to create a 200 m (656 ft) wide “shadow” of reduced hydrostatic pressure centered along the Tunnel No. 1 centerline. It was considered that if tunneling were maintained within this shadow, the groundwater pressures and inflow rates would be lower than encountered during the construction of Tunnel No. 1.

Profile Alignment. The overall profile for the tunnel and tailrace channel is controlled by

Table 1. TBM Tunnel Stabilization Systems

Stabilization Type	Applicable Ground Conditions	Rock Dowel/ Steel Rib Type	Length and Spacing Circumferential (C) Longitudinal (L)	Coverage	Welded Wire Fabric	Lining	Timing of Installation
Type T1	Moderately Blocky	32 mm (1.25 in) dia. deformed bar	3 m (10 ft) long, 1.5 m (5 ft) (C) by 1.8 m (6 ft) (L)	Upper 120°	N/A	N/A	Install dowels within 7 m (23 ft) of heading.
Type T1A	Spalling Ground	32 mm (1.25 in) dia. deformed bar	3 m (10 ft) long, 1.8 m (6 ft) (C) by 1.2 m (4 ft) (L)	Upper 240°	Type 661/0 (6 x 6 - w8 x w8) galv. in Shotcrete areas only	Shotcrete only to fill overbreak areas	Install 3 m (10 ft) dowels within 7 m (23 ft) of heading; install 1.2 m (4 ft) dowels within 10 m (33 ft) behind grippers.
			1.2 m (4 ft) long, 1.5 m (5 ft) (C) by 1.2 m (4 ft) (L)	Lower 120°			
Type T2	Blocky and Seamy; Faults and Shears less than 3 m (10 ft) wide	32 mm (1.25 in) dia. deformed bar	3 m (10 ft) long, 1.5 m (5 ft) (C) by 1.2 m (4 ft) (L)	Upper 210°	Type 661/0 (6 x 6 - w8 x w8) galv.	75 mm (3 in) Shotcrete	Install 3 m (10 ft) dowels and welded wire fabric within 7 m (23 ft) of heading; install 1.2 m (4 ft) dowels and welded wire fabric within 10 m (33 ft) behind grippers.
			1.2 m (4 ft) long, 1.5 m (5 ft) (C) by 1.2 m (4 ft) (L)	Lower 150°			
Type T3	Crushed Ground	200UC 46.2 (w8 x 31)	1.2 m (4 ft) spacing (L)	Full Perimeter	Type 364/0 (3 x 6 - w12 x w12) galv.	375 mm (15 in) C-I-P Concrete	Install within finger shield (within 5 m (16 ft) of heading; expand immediately following finger shield.

hydraulic design considerations relating to downsurge events, the elevation of the existing draft tube manifold at the bifurcation, and the range in tide levels at the outlet works. Modeling of the system hydraulics indicated that it was desirable to match the profiles of Tunnel Nos. 1 and 2. The profile of Tunnel No. 2 was established by matching the crown elevations of the two tailrace tunnels at 2 key points along the profile: the upstream limit of the 3% gradient; and the transition between the 3% and 0.10% gradients. From the bifurcation, Tunnel No. 2 slopes downward at a gradient of approximately 11% for 105 m (344 ft), and then at 3% for 668 m (2192 ft). The tunnel then slopes at a downward gradient of 0.1% for the next 8.7 km (5.4 miles), to the tunnel lowpoint, where the dewatering shaft is located. From this point, the tunnel slopes upward at a 12.5 % gradient to the outlet portal.

The above alignment description follows the direction of water flow. However, the tunnel will be excavated in the opposite direction, from Deep Cove toward the power station. The 12.5% and 0.1% gradients were selected primarily on the basis of constructability considerations. Construction records from Tunnel No. 1 indicate that groundwater inflows during construction were significant. Under these circumstances, it was considered a priority to be able to drive the tunnel uphill so that inflows will drain away from the heading, The tunnel lowpoint needed to be established as early in the construction as possible, to both maximize the length of uphill mining and minimize the tunneling associated with the dewatering facilities. The 12.5% gradient was considered the steepest gradient achievable considering haulage limitations, and is readily achievable using drill and blast methods. At this gradient, the tunnel lowpoint is achieved within 350 m (1148 ft) of the tunnel portal, a tunnel length that was relatively dry during the excavation of Tunnel No. 1. The 0.1% gradient is considered the flattest grade desirable (and achievable) for a tunnel of this length, that will encourage water to flow away from the tunnel heading and toward the tunnel lowpoint. The location of the tunnel lowpoint and its elevation were defined by the intersection of the 12.5% decline and the 0.1% upward gradient for the main drive. The 12.5% decline was shown on the Drawings, however the Contractor was given the option to propose a flatter gradient, subject to ECNZ's approval. Tenderers were warned, though, that a flatter decline would extend the low point farther from the portal, which would increase the potential for encountering groundwater inflows in the downward heading. This would also increase the length of the dewatering adit, and would nominally increase the depth of the dewatering shaft. During the Interactive Tender Process, none of the four shortlisted tenderers questioned the 12.5% decline, and none proposed a flatter gradient.

Headworks Facilities

The headworks facilities, shown on Figure 5, include: a stop log shaft and structure; a maintenance shaft; an access adit that provides access to both shafts from the existing power station access tunnel; and a transition tunnel connecting the TBM tunnel through to the bifurcation with the existing tailrace tunnel. The stop log structure is required for isolating Tunnel No. 2 for inspection and maintenance activities, while maintaining flows through Tunnel No. 1. The maintenance shaft allows for access to Tunnel No. 2 downstream from the stop log shaft, as well as construction access for construction of the transition tunnel, stop log structure and plug excavation. The layout of the headworks facilities was controlled by two sets of criteria; one set applies at the Tunnel No. 1 level and one at the access adit level.

The layout of the tailrace tunnel through this area is designed to maximize the length of tunnel to be driven by TBM on a straight alignment, as well as minimize the length and complexity of drill and blast tunneling. Drill and blast tunneling is required to construct the tunnel transition starting at the maintenance shaft and ending with the removal of a "plug" located between the stop log shaft and the bifurcation. The layout of the access adit is designed to provide a straight alignment from the intersection with the power station access tunnel to the maintenance shaft (which will be the main mining shaft for headworks development), and a branch adit that will access the stop log chamber and shaft (see Figure 5).

Maintenance Adit. The headworks maintenance adit is a horseshoe shaped, 5.0 m x 5.0 m (16 ft x 16 ft) tunnel which branches off from the power station access tunnel. The maintenance adit grades upward at a +10% gradient to a landing at elevation El + 46.2 m (152 ft) at the Maintenance Chamber. Another branching adit that provides access to the stop log shaft also grades upward to a landing at El +46.2 m (152 ft). The elevation of the landings in both chambers was set above the maximum upsurge level, so that workers servicing the hoists and stored stop logs will not be inundated during an extreme surge event.

Maintenance Chamber and Shaft. The maintenance shaft has two purposes; it will serve as the main mining shaft for excavation of the Transition Tunnel, and it will provide permanent access for maintenance equipment to Tunnel No. 2. The shaft will be 9.0 m (30 ft) in diameter, will be excavated using drill and blast techniques, and will be stabilized with rock reinforcement and shotcrete as excavation proceeds. The shaft is sized to allow a drilling jumbo or front end loader up to 8 m (26 ft) long and 20 tonnes (22 short tons) in weight to be lowered down to Tunnel No. 2.

Stop Log Chamber and Shaft. The stop log shaft is a slot excavation which measures 10 m (33 ft) by 3 m (10 ft) in plan. It could be excavated either from the top using drill and blast or alternatively by raise-boring two 3 m (10 ft) holes and slashing between the two. The stop log chamber at the top of the stop log shaft will contain the stop log handling gear and will be used to store the stop logs. Surging of water levels in the access shaft and stop log shaft is expected to create air gusts within the stop log chamber and access adit. However, the wind velocities to be generated, on the order of 7 to 9 m/s (15 to 20 mph), were considered to be of acceptable magnitude. The design includes safety handrails and removable grillage panels over the stop log shaft opening.

The stop logs will be fabricated from high strength steel and will be nominally a meter deep at mid span. The individual stop log sections are nominally 8.4 m long and 2.5 m high and each weighs approximately 12 tonnes (13.2 short tons). They are designed to resist the high stresses that can develop as a result of extreme surge in Tunnel No 1. When lowered into place to close off Tunnel No. 2, each section will seal on the upstream or pressure side of the stop logs, making the shaft a "dry" shaft.

Transition Tunnel. The transition tunnel is the drill and blast section of Tunnel No. 2 which connects the TBM excavated Tunnel No 2 to Tunnel No 1 through the stop log structure and bifurcation. The 10 m (33 ft) diameter horseshoe shape will be shotcrete lined for hydraulic efficiency. A concrete slab will be provided in the invert, and will be anchored with rock dowels to prevent uplift during dewatered periods.

The transition tunnel consists of a 41 m (135 ft) long, 40 m (131 ft) radius curve from the maintenance shaft and a 67 m (220 ft) tangent length leading to the bifurcation. The curve, easily constructable using drill and blast methods, was considered the tightest radius possible without enduring significant head losses. The stop log structure will be located at the base of the stop log shaft close to the bifurcation (see Figure 5). A 30 m (98 ft) section of tunnel will transition the horseshoe-shaped transition tunnel into the circular TBM excavated tunnel. The layout is designed to maximize hydraulic efficiency taking into account the practical problems of sequencing and carrying out construction activities in a confined space.

It is anticipated that the transition tunnel, together with the two shafts and stop log structure, will be completed as soon as practicable in the construction schedule to allow the early removal of the tunnel plug at the bifurcation at a time stipulated by ECNZ.

Tunnel Plug. The tunnel plug is an approximately 20 m (66 ft) reach of 7.5 m (25 ft) wide horseshoe shape tunnel that will be removed during the 21 day power station outage. Removal of the plug will create the connection between Tunnel No. 2 and the existing draft tube manifold. This plug, shown in plan on Figure 7, will be removed according to a specified and controlled sequence of small excavations. This will minimize blasting-related damage to the rock and the completed stop log shaft and reduce the lining requirements in the bifurcation area. The length of the tunnel plug was sized: (1) to provide a stable mass of rock between drill and blast excavation operations and the operating Tunnel No. 1; and (2) to minimize the length of time required to complete the plug excavation and secure the connecting tunnels during the station outage. Stabilization of the bifurcation will consist of rock reinforcement and shotcrete designed to achieve maximum placement rates, high early strength, and high durability.

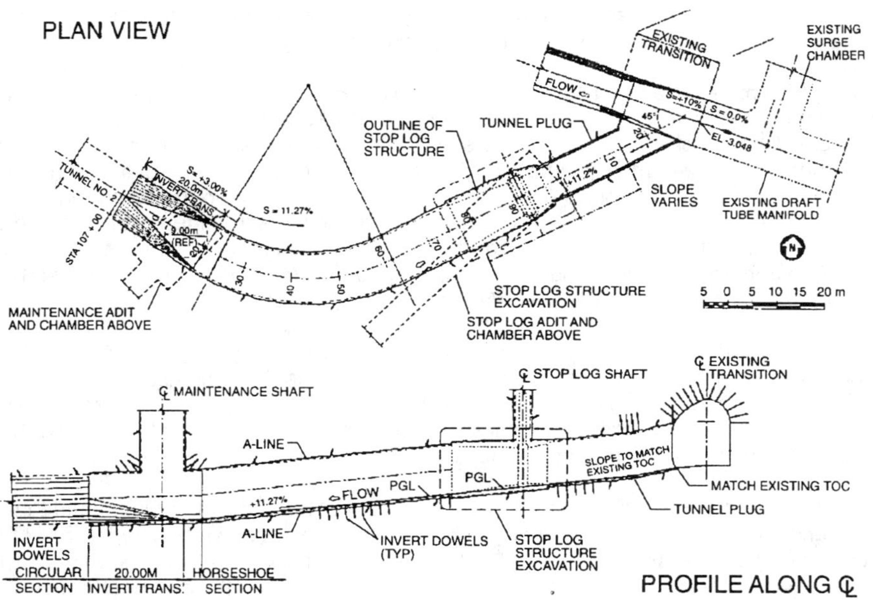

Figure 7. Tailrace Transition Tunnel and Bifurcation

Blasting in Proximity to Operating Power Station. Of a general concern with all headworks excavations was the need to carry out drill and blast excavation within close proximity to the operating power station. Highly sensitive power generating equipment and controls such as Mercury switches could not withstand the vibrations induced by blasting. Discussions were held with power station staff to identify those sensitive devices and facilities that were in closest proximity to the construction areas. Using scaled distance blasting relationships, it was concluded that charge weights per delay would be limited by the need to limit blast damage to the rock openings, and not the vibration levels that would be detected in the power station. Nevertheless, maximum allowable particle velocities of 200 mm/sec (8 in/sec) measured 15 m (50 ft) from a blast, and 50 mm/sec (2 in/sec) in the Machine Hall were included in the Specifications, to emphasize the importance of controlling blasting vibrations during this important element of the construction.

Dewatering Facilities

The dewatering facilities consist of three components, as illustrated on Figure 5: a dewatering adit, a dewatering head chamber, and a dewatering shaft. The dewatering shaft is connected to the Tunnel No. 2 lowpoint via several bored holes and a rock trap/sump. The dewatering facilities are designed to be able to dewater a completely filled Tunnel No. 2 within approximately 5 days, as well as provide for groundwater control pumping during construction.

Dewatering Adit. The 340 m (1115 ft) long dewatering adit commences from its portal, which is located to the south (or true left) of the Tailrace Tunnel No. 2 portal. The access adit, which is nominally a 5 m (16 ft) wide horseshoe shape, follows a south-easterly alignment that crosses over Tunnel No. 2 to reach the head chamber. The adit slopes upward toward the head chamber at a gradient of 0.14%. Any groundwater inflows or leakage from discharge pipework will drain away from the head chamber toward the portal.

Head Chamber. The head chamber is located at the end of the access adit and over the dewatering shaft. The chamber is nominally 25 m by 15 m (82 ft by 49 ft) in plan. This chamber will house electrical transformers for the six submersible (and removable) pumps, pump controls, and a gantry-type traveling crane to hoist the pumps in and out of the dewatering shaft. During construction, construction water will need to be pumped through temporary pipes that convey the water along the dewatering adit to a treatment plant located outside the portal. During the dewatering of Tunnel No. 2, tunnel water will be able to be discharged from the head chamber through three 500 mm (20 in) diameter pipes installed in three separate holes drilled obliquely from the Head Chamber to Tunnel No. 1.

Dewatering Shaft. The dewatering shaft extends down approximately 54 m (177 ft) from the head chamber. The shaft will be 9 m (30 ft) in diameter, the minimum area for locating the six dewatering pumps on the same level in the shaft invert. Cables, valves and pipework required in the shaft to operate the pumps were designed to be removed from the shaft and stored by ECNZ when the pumps are not in use.

The pumps have been designed to meet a number of operational requirements that will occur during the dewatering of Tunnel No. 2. The pump chamber will be fitted with six Flygt (or equivalent) submersible pumps that will act in a "dry" well or "wet" well mode depending on the stage of dewatering. The types of pumps to be used and their manner of operation were subjected to extensive study and value engineering review, because of the uniqueness of the application.

The Contractor will be allowed to use the permanent dewatering equipment during construction, provided that all used equipment is refurbished to an "as new" condition prior to completion of the Project.

Outlet Works Facilities

The outlet works stop log structure is designed to be founded on rock as close as possible to the portal rock headwall to avoid damage from falling trees and rock. The current portal area is concealed beneath approximately 7 m (23 ft) of cover so the rock surface topography had to be investigated by exploratory drilling. With the structure tight against the rock face, access to the dewatered tunnel is provided by the maintenance ramp that connects between the Dewatering Adit portal and the Tunnel No. 2 portal flare. Numerous construction sequence and layout scenarios were evaluated for TBM launching and stop log structure design ranging from constructing the stop log structure prior to launching the TBM to constructing the structure following completion of mining. The objective was to find a way to get the stop log structure work off the critical path at the end of the construction period. To minimize the work required at the end of the job, it was decided to minimize the number of central piers and to allow the Contractor to launch the TBM at a higher level than final invert. This allows flexibility in portal groundwater control and in timing of stop log structure construction. The sequence of construction at the tunnel portal and outlet works is ultimately the Contractor's responsibility, but the design was intended to provide the Contractor the most flexibility in coming up with an innovative solution while satisfying minimum requirements for groundwater control and flood protection.

CONSTRUCTION CONSIDERATIONS

Tunnel Boring Machine Requirements

The TBM has been specified as new, open-type TBM with side grippers and a finger shield, having the following capabilities and attributes:

- The ability to advance drain holes ahead of the face. This is desired in order to help drain the rock mass in advance of the excavation, and to better control inflows that do occur. In the event that pre-draining the rock mass ahead of the face does not prove effective in maintaining reasonable advance rates through very wet ground, the TBM will also be required to have the capability of drilling a 20-hole fan grout curtain out ahead of the machine to reduce groundwater inflows.

- The ability to install spilling pre-reinforcement to reduce the potential for collapse of the tunnel face.

- A cutterhead drive system capable of rotating in both directions to free the cutterhead if seized by loosened rock, and a two-speed electric drive with clutches or variable frequency drive to provide high torque at low revolutions per minute (RPMs).

- Cutters mounted in a recessed position to protect cutter housings and buckets from damage in blocky rock conditions.

- A retractable slotted (finger) shield to cover a forward roof drill and steel set erector station behind the cutterhead. The steel set erector and roof drills will facilitate the installation of stabilization measures within 5 m to 7 m (16 ft to 23 ft) of the heading.

- Grippers designed to "step over" steel sets installed ahead of the grippers.

- Roof drills mounted behind the grippers to install dowels in the tunnel sidewalls.

Staging Considerations

Because of the extreme environmental sensitivity of the site, limitations were agreed by ECNZ as to where staging could occur, what personnel would be permitted within Park Boundaries on a temporary and ongoing basis and where the construction camp could be located.

Access to the project will be generally by ferry or barge approximately 25 km (16 miles) across Lake Manapouri from Pearl Harbor to West Arm. Manapouri Township, which is at Pearl Harbor, connects to the State Highway system and is approximately 2 1/2 hours travel by road from Invercargill and 3 1/2 hours from Dunedin as shown on Figure 1.

The construction of the headworks will be carried out in the power station with access via the existing power station access tunnel. This route will have to be shared with the power station traffic and tour buses that visit the power station.

Access to Deep Cove can be achieved from West Arm by using the Wilmot Pass Road. Particularly large or heavy loads, such as TBM components and transformers, can be shipped to the existing wharf at Deep Cove. This wharf was originally used for the heavy items delivered for the original construction. Accommodation for approximately 160 personnel will be located at West Arm.

Site offices for contract administration recording work time, etc. and engineering support services will be located off site at Manapouri Township. Staff from these offices will travel to the site for the day as required, returning in the evening.

Construction Sequence

The construction is anticipated to begin in the first half of 1997 with the TBM order being placed at that time. In the 12 to 15 months before the TBM is assembled on site, the headworks and portions of the dewatering facilities will be constructed. With the headworks area constructed and the headworks stop logs in place, the plug will be excavated and the bifurcation will be completed. The power station will then resume normal operation while the TBM tunnel is being mined. Once the TBM arrives at the headworks, it could be dismantled and removed out the headworks maintenance shaft and power station access tunnel, or backed out through the tunnel at the contractor's option. The work to be carried out during the 21-day power station outage is discussed below.

Power Station Outage

Completion of the bifurcation and excavation in Channel No. 1 and the junction of the two channels requires a temporary interruption in tailrace discharge and a corresponding outage in power generation at the power station. The period within which the outage may occur will be scheduled according to ECNZ operational requirements. Time windows within which the Contractor will be permitted to carry out this work are specified in the Contract. The Contractor will be required to complete the outage works as early in the construction schedule as practical. The power station outage work will be carried out at the earliest window available to ECNZ following completion and testing of the headworks stop log structure.

The work anticipated to be completed while the power station is shutdown will comprise: construction of temporary cofferdams in the tailrace channel; partial dewatering of Tunnel No. 1 for excavation of the bifurcation plug; excavation, stabilization and lining of the plug; and excavation in Channel No. 1 and at the junction between Channel No. 1 and Channel No. 2.

Outlet Works Dewatering

The predicted volume and inflows of groundwater in the vicinity of the stop log structure, TBM assembly area, and Tailrace Channel No. 2 excavation have influenced the proposed design and layout of these works. Site investigations have indicated high permeability ground in this area. It has been concluded that excavating and dewatering for a portal invert at El.

-10 m (-33 ft) similar to Tunnel No. 1 would be both difficult and expensive. Groundwater pump tests and modeling analyses confirmed these concerns.

As a consequence, El. -5 m (-16 ft) was set as the invert level for the facilities in this area. For partial excavations no deeper than El. -2 m (-7 ft), the Contractor will be permitted to control groundwater using a combination of deep wells, well-pointing, and sump-pumps. For excavations below El. -2 m (-7 ft), the Contractor will be required to install a full perimeter cut-off to El. - 10 m (-33 ft), in addition to dewatering methods to control groundwater inflows through the subgrade of the excavation.

Tailrace Tunnel Dewatering

Construction water together with natural groundwater inflows to the tunnel, will constitute the water to be collected and discharged from the tunnel during construction. The water originating at or being used at or near the face of the TBM will be relatively "dirty" water. This water will be carrying the "flour" from the rock pulverized by the disc cutters, and any fuel, oil, or hydraulic fluid inadvertently spilled near the TBM. Behind the heading, groundwater inflows will continue to occur through discontinuities in the rock mass. This water will also become contaminated by fuel, oil, hydraulic fluid and dirty water in the invert of the tunnel, which will inevitably occur due to tunnel stabilization work, the advancement of utilities, and the transport of men and materials to and from the tunnel heading. Though these sources of water may be cleaner than those emanating from the heading area, they will nevertheless be required to meet stringent standards for discharge.

Considering the high groundwater inflow rates that could occur during tunnel construction, other considerations relating to positioning of a muck conveyor and the rail for man and material transport need to be considered. One approach would be for the conveyor and rail to be mounted on the side haunches of the circular invert, with all water allowed to flow along the invert until it reached the low point at the base of the decline. It would then be channeled into the dewatering chamber and pumped to treatment facilities established adjacent to the existing tunnel portal on the north bank of Tailrace Channel No. 1. After treatment to the required standard, the water will then be discharged to the tailrace channel.

An alternative approach would be to control clean water inflows, collect and protect them and be able to discharge them directly to tunnel No. 1 from the dewatering shaft. The relative percentages of "dirty" water requiring full treatment outside the portal would be reduced, but the complexity of dual water conveyance systems would be increased. The Contractor will be responsible for designing and implementing the water conveyance, treatment, and discharge plan.

Potential Lyvia River Flooding

Flooding of the Portal Area. The Lyvia River discharges to Deep Cove in close proximity to the Tunnel No. 2 portal. Hydrologic and hydraulic evaluations of the Lyvia River were made using the U.S. Army Corps of Engineers' HEC-1 rainfall-runoff model and HEC-2 water surface profile model. Peak discharges for 50-, 100-, and 1,000-year return frequency storm events were calculated and compared to results from previous studies and regional flood estimation techniques. Results of the studies formed the bases for the design of temporary flood protection works required during the construction period to protect the portal and construction against flooding.

The specifications contain requirements for portal protection against flooding, a groundwater cut-off around the perimeter of the portal excavation, and construction of a stop bank between the Lyvia River and Channel No. 2 whose crest height and length will provide protection against a water level caused by a flood with a return period of 1,000 years. For portal protection against flooding the Contractor is required to develop a river level monitoring system and have provisions for a bulkhead or weir system to seal the portal and tunnel should the portal excavation become flooded. The extent of the cut- off around the perimeter of the excavation will depend on the contractor's excavation sequence at the portal.

Risk Management Provisions

Provisions included in the Contract Documents which are intended to avoid disputes, and provide efficient means for the resolution of disputes that may develop during the construction are a Geotechnical Baseline Report, a Disputes Review Board, and Escrowing of Bid Documents. In addition, a Partnering Charter will be offered to the Contractor, and implemented if agreed.

Payment provisions in the contract were developed to provide flexibility during construction to adapt to variable conditions. Where a project component was clearly defined, payment is by lump sum. Where precise quantities for items are not known, payment is by unit price. Where the scope of work is not defined, such as work to address environmental concerns, provisional sums have been included in the contract, and the Tenderers are required to submit "dayworks" rates for labor and equipment for carrying out work paid by provisional sum.

INTERACTIVE TENDER PROCESS

It was generally acknowledged during the TAT Meeting that TBM tunnels have been driven through groundwater inflow volumes of the order to be encountered for this Project, and that 10 m (33 ft) diameter tunnels have previously been excavated with TBMs. However, it was acknowledged that a TBM of this size has not been advanced through groundwater inflows of the magnitudes anticipated. This consideration, compounded with a number of other unique Project aspects including a remote site, extreme environmental sensitivities that placed severe limitations on construction staging, the portal invert level 5 m (16 ft) below sea level amidst difficult geologic conditions, and the need to carry out the construction while avoiding impacts to an operating hydroelectric power station, demonstrated to the TAT Meeting members that this Project presented certain construction challenges that had not been addressed on previous tunneling projects.

The combined challenges presented by this Project clearly indicated that a standard approach to tendering and "hard money" construction contracting would not be in the best interest of either ECNZ or the tendering parties. ECNZ was willing to consider any new approach that could lead toward the best possible project for a fair price. ECNZ and WC concluded that a working team approach with the successful Contractor was a necessary ingredient for success. Because ECNZ was not required to selected the lowest price proposed, the attitude and demonstrated teamsmanship of the tenderers would also be an important consideration in the selection process.

It was concluded that the best approach for this Project was to commence an informal form of "partnering" with the tenderers long before the construction Contract would be signed. To make this as effective as possible, ECNZ preferred to solicit tenders from a limite d number of joint venture parties. WC prepared a detailed qualifications questionnaire, and ECNZ evaluated the submittals and selected four for further consideration. WC developed an

Interactive Tender Process (ITP) with the following objectives in mind:

- permit the shortlisted parties to review a set of draft Tender Documents;
- invite the parties to the project site for a site visit and Project Briefing;
- at the Project Briefing, provide background information regarding what was included in the project design and why, and what was NOT included and why - in this manner, all parties would have a better appreciation for the rationale behind what was included in the draft Tender Documents, in addition to explanations provided in documents such as the Geotechnical Baseline Report;
- allow the parties to comment on the draft documents in public and in private; this would permit them to raise questions regarding any ambiguities in the documents, and to request clarifications that might permit them to be more responsive in their Tenders, and possibly offer innovations in approach or component design, for consideration by ECNZ and WC during the subsequent Tender Period;
- make modifications as necessary to the draft Tender Documents, to alleviate the ambiguities, make corrections as required, and generally incorporate, to the extent possible, feedback received from the parties;
- respect those comments received in private as proprietary, in order to maintain each party's competitive advantages; and
- generally allow the teams to demonstrate their ability to "communicate and work" with ECNZ and the Design Team, as well as demonstrate ECNZ's willingness to incorporate tenderer comments and suggestions.

The ITP was begun in August, 1996. In general, all of the above objectives were achieved. The responses and degree of interaction with the parties were excellent. Fears that the parties would distrust the overall process and would "clam up" were dispelled early in the site visit. ECNZ and WC have each received informal feedback from the parties that the process was extremely worthwhile, that it provided them with a more in-depth understanding of the project specifics and constraints, and that it helped them develop their most competitive tenders.

At the time that this manuscript has been submitted, ECNZ is evaluating tenders. Pending approval by ECNZ in February, 1997, construction is anticipated to commence in March or April, 1997.

ACKNOWLEDGMENTS

The authors wish to commend ECNZ's willingness to consider recent advances and new approaches in risk management and tendering procedures, specifically Kieran Devine, General Manager of Generation, Mark France, Project Manager, and Murray Gillon, Technical Specialist. Hopefully this open attitude and team approach will continue through to successful completion of the construction. Also, the initial fast track stages of the project, the schedule demanded long hours, seven-day work weeks, and solid teamwork among many of the Design Team, including John Bischoff, Tom Freeman, David Hughes, Tom MacDonald, Mike Piepenburg, Ted Feldsher, Ken Susilo, and Mike Zusi from WC-US, Nick Eldred, Bryan Grover, Wayne Russell, and Shirley Graham from WC-NZ, and Allan Brown and Bryce Kennington from DesignPower. The efforts of Jim Bell and Ron Flemming (WC-NZ) later if the design schedule are also acknowledged. Without these individuals' and others' dedicated efforts and commitment, the design of the Second Manapouri Tailrace Tunnel Project would not have been possible.

Chapter 2

DESIGN OF THE METROWEST PRESSURE TUNNEL*

J.N. Eskilsson,[1] V.S. Romero,[2] W.P. Chen,[3] and J.L. Kaplin[4]
[1]Senior Associate, Jacobs Associates; [2]Associate, Jacobs Associates;
[3]Senior Engineer, Sverdrup Civil, Inc.;
[4]Design Manager, Massachusetts Water Resources Authority

ABSTRACT

The Massachusetts Water Resources Authority is currently improving its primary water supply system for the Boston Metropolitan area. The improvements planned by the Authority, which provides water for 46 communities and over 2.5 million people in eastern Massachusetts, include the MetroWest Water Supply Tunnel (MWWST) to provide redundancy in the transmission system, a large capacity water treatment plant and extensive covered storage facilities.

The 28.3 km (17.6 miles) long, mostly 4.3 m (14 ft.) I.D. MetroWest Water Supply Tunnel will allow repair of the aging Hultman Aqueduct, which for 56 years has served as the transmission main from the water supply reservoirs to Boston. On completion of the tunnel, the old pipeline aqueduct will be repaired and brought back into service as a parallel, redundant transmission main.

This paper describes design considerations and the design process for pressure tunnels in general and the MWWST in particular. It provides an overview of how the design of pressure tunnel and shaft components progresses.

INTRODUCTION

Planning, design and construction of a major water transmission main facility in an urban/suburban area is a complex undertaking which may take 10 years, or more, to implement. The planning and design efforts may take five or more years of work by a wide variety of engineering and non-engineering disciplines to define, obtain required governmental approval and gain public acceptance of the required facility. Long range planning to identify the need for a facility early enough to have it completed when needed is consequently an ongoing challenge for metropolitan water supply agencies.

*This article represents the opinions and conclusions of the authors and not necessarily those of the MWRA. This article shall not be used as evidence of design intent, design parameters or other conclusions which are contrary to express provisions in contract documents for the MWRA MetroWest Water Supply Tunnel.

The design of major facilities such as the MWWST facility proceeds through logical, progressive steps of development, and reviews. Considerations are given to a wide variety of planning and design issues. These issues include definition of serviceability requirements, rights-of-way, community relations, environmental, regulatory, geological/geotechnical, construction costs and financial planning. Each major step of development culminates in production of reports and increasingly detailed plans and specifications, finally leading to production of construction contract documents, bidding and construction of the facility. At selected milestones, major reviews are made by independent expertise from specific design disciplines or formally assembled review boards.

In 1989 the Massachusetts Water Resources Authority (Authority) issued a planning and design contract for a second transmission main to provide redundancy for the Hultman Aqueduct. As originally conceived, the project consisted of a tunnel combined with reconstruction of an existing, unused gravity aqueduct, the Sudbury Aqueduct. During feasibility studies, it was recognized that costs and environmental and community impact issues related to reconstruction of the Sudbury Aqueduct through an urban/suburban area compared unfavorably with a full length tunnel in rock, deep under existing structures and facilities. The alignment of the tunnel, which is currently under construction, generally coincides with the existing Hultman Aqueduct and is constructed in permanent underground easements below several hundred private properties.

The full length, unreinforced concrete lined, pressure tunnel design concept was selected and the facility was named the MetroWest Water Supply Tunnel. The extensive planning and environmental studies required for this $540 million project have been described in two earlier papers presented by design team members, Caspe, et. al., 1994a and 1994b.

The final horizontal and vertical alignments for the MWWST facility are shown in Figure 1 and Figure 2, respectively. Figure 1 also shows the existing Hultman Aqueduct, Wachusett Reservoir and tunnels in the Authority's water transmission system.

TUNNEL DESIGN

The essential design criteria which must be incorporated into design of pressure tunnels constructed in rock may be categorized as follows:

- Serviceability Criteria
- Geotechnical Conditions
- Constructability Considerations
- Pressure Tunnel Design Principles

These categories of design criteria are discussed in general, as well as specifically for the MetroWest Water Supply Tunnel, in the following paragraphs.

Serviceability Criteria

A number of essential serviceability criteria requires early definition during design of a pressure tunnel to serve as a water transmission main. These are:

- Tunnel Termination and Connection Points and Intermittent Connections
- Static Hydraulic Gradeline (HGL)
- Design Transient Pressure
- Tunnel and Shaft Conduit Diameters
- Maximum Operating Flow Velocities
- Leakage Design Considerations

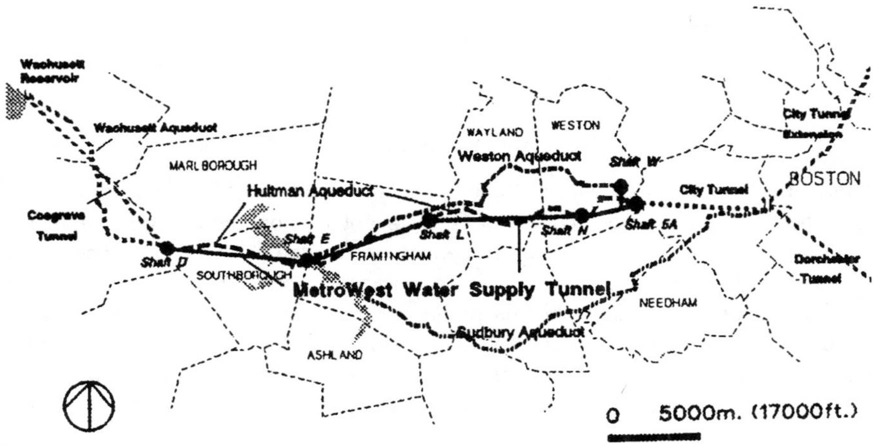

Figure 1. Existing transmission system and proposed MetroWest Water Supply Tunnel

Tunnel Termination and Connection Points

Termination point criteria for a specific water tunnel facility is often simple and straight forward. In the case of the MWWST, it was determined that the tunnel would be redundant to the existing Hultman Aqueduct by connecting Shaft C on the existing Cosgrove Tunnel in the west with Shaft 5 on the existing City Tunnel, and with the Weston Aqueduct Terminal Chamber Area in the east end. These criteria resulted in terminal connections at Shafts D, 5A and W, respectively, and definition of the general alignment of the tunnel facility. Additional major connections to existing Authority transmission facilities resulted in Shaft E, Shaft L, Shaft NW and Shaft NE. Connections to serve community clients resulted in the five small diameter riser shafts to supply water to customers along the tunnel alignment. The locations of these existing transmission system connection points and the resulting horizontal alignment of the facility are shown on Figure 1.

Static Hydraulic Gradeline (HGL)

Definition of the design static HGL is one of the most critical design elements of a pressure tunnel, because of its importance in determining the vertical alignment (depth) of the

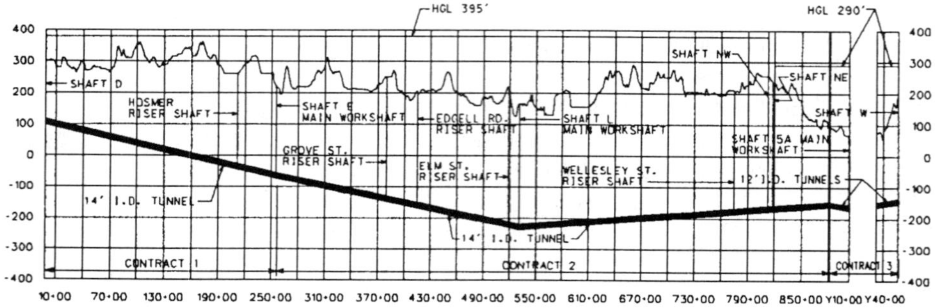

Figure 2. MetroWest Water Supply Tunnel — Profile

constructed tunnel. It is essential that long term planning of a water transmission system be considered in the selection of the HGL so as to prevent development of undesirable "bottlenecks" which may restrict future improvements to the system. This may be considered of particular importance in "upstream" portions of a transmission main system. It is suggested that the HGL should be set at the highest probable level which may be considered for operation in a 100 year (or longer) period of planning, as the useful life of a deep pressure tunnel constructed in rock may be extended for very long periods by periodically performing limited repairs/upgrades of specific components of the overall system.

For the MWWST facility, an HGL corresponding to the maximum water surface elevation of the Wachusett Reservoir, feeding the proposed facility by gravity from an overflow elevation of 120.4 m (395 ft. Boston City Base Datum, BCB), was selected by the Authority. A 120.4 m (395 ft.) HGL was used for design of the tunnel from Shaft D to Shaft NW at the Norumbega Reservoir where water will continue to be stored for distribution to the Boston area. The remaining downstream portion of the MWWST facility, Shaft NE to Shaft 5A and Shaft W, is designed for an HGL of El. 88.4 m (290 ft. BCB). Both HGLs used for design are shown in Figure 2.

Design Transient Pressure

The design transient pressure is determined by hydraulic analysis of the overall future transmission system, including accounting for characteristics of control mechanisms, e.g., shutting times for major valves or catastrophic type changes in flow rate. The resulting design transient pressure for the MWWST facility was established to correspond to 15.2 m (50 ft.) of head increase, or approximately 137.9 KPa (20 psi). Transient pressure is used for design of impermeable lining system components, i.e., localized steel linings in the shafts and tunnels.

Conduit Diameters

The final shaft and tunnel conduit sizes (Internal Diameter or I.D.) were also selected largely from the results of the hydraulic analysis. In determination of conduit sizes, long term planning type flow rates are commonly used to avoid creation of future bottlenecks due to under-sizing of conduits in the system. The design flow rate used in hydraulic analysis for the MWWST facility was 1.893 x 10^6 m^3 per day (500 MGD).

In addition to meeting the fundamental hydraulic requirements of operation, such as flow capacities and operating hydraulic characteristics, additional important operations and maintenance provisions require identification and incorporation in the design, including provisions for future reentry into the tunnels and associated unwatering and safe future access for men and some essential equipment. Considerations of filling and unwatering (e.g., air release/vacuum valves, filling/unwatering procedures and requirements for pumping), as well as the need and means for metering of flows during operation, also must be incorporated into the design. Future access to inspect and maintain the tunnels must be accommodated through the shaft conduits making sizing of manholes and access hatches, invert gradients (the tunnel must generally slope toward the access shafts) and locations of metering devices important design considerations.

The selected I.D. of the MWWST tunnel conduit is 4.3 m (14 ft.) from Shaft D to Norumbega Reservoir and 3.7 m (12 ft.) for the remainder to Shaft 5A and Shaft W. The shaft conduits vary from 3.7 m (12 ft.) to 2.4 m (8 ft.), depending on the location, metering, hydraulic and future access requirements. The five small diameter community riser shafts were selected to be 0.51 m (30 in.) I.D.

Maximum Flow Velocities

Maximum flow velocities are calculated to assess if these are acceptable for the various final lining materials incorporated in the facility. For the MWWST tunnel conduits, all projected flow velocities are less than 3 m (10 ft.) per second and thus acceptable for all envisioned types of lining systems.

Leakage Design Considerations

Issues related to leakage of groundwater into completed gravity flow or low pressure water tunnels, or leakage from the completed pressure tunnel, may affect tunnel final lining criteria and, in effect, the overall design approach.

The MWWST facility is designed to be lined with a cast-in-place, unreinforced concrete final lining, which, in principle, is considered a permeable lining not specifically intended to prevent or control leakage from the tunnel into the surrounding rock mass (exfiltration) during operation of the facility (see Moore, 1989). For the MWWST, leakage is controlled by the rock mass which serves as an integral part of the lining system. For a specific tunnel, the loss of water is largely controlled by the ability of the host rock mass to contain pressurized water, and implementation of localized, special lining treatment where required. Total leakage rates cannot be accurately estimated prior to commissioning of a concrete lined or unlined pressure tunnel.

Geotechnical Conditions

As mentioned earlier, systematically concrete lined and unlined pressure tunnels utilize the host rock mass as the main lining system component to contain the pressurized water within the tunnel and shaft conduits. Several geotechnical characteristics of the host rock mass are critical for design and operation of such facilities:

- In situ stress
- Topographical characteristics
- In situ hydraulic conductivity (permeability)
- Location of groundwater table
- Modulus of deformation

Additional geotechnical characteristics are important for other design aspects of design of tunnels constructed in rock, e.g., rock boreability, rock mass quality and durability. However, the scope of this paper is limited to the discussion of the aspects listed above and their applicability for design of pressure tunnels and the MWWST facility.

In Situ Stress

The prevailing state of minimum principal stress in the rock mass is critical to design of concrete lined and unlined pressure tunnels, because of the long-term exposure of the rock mass to water pressures equal to the internal static pressure. In flat terrain, the vertical stress is generally directly proportional to depth below ground surface and corresponds to the weight of the overburden. Although horizontal stresses will often be less than the vertical stresses, horizontal stresses may be higher due to tectonic conditions, i.e., the vertical stress may be the minor principal stress. This has been reported to be the case in the northeastern United States (Zoback and Zoback, 1980). However, site specific geotechnical exploration is required to verify in situ stresses along the alignment of a pressure tunnel.

The issue of in situ stress in regards to design of a concrete lined or unlined pressure tunnels is associated with the phenomenon of "hydraulic jacking," i.e., opening of existing rock mass joints by the action of pressurized water introduced into the joints. Hydraulic jacking phenomena result in dramatically increased permeability and leakage out of a concrete lined or unlined pressure tunnel through the rock mass. In addition to increased leakage, secondary, undesirable environmental effects may also result.

As mentioned above, hydraulic jacking of joints occurs when the water pressure in a joint exceeds in situ stress normal to the joints in the rock mass opening these joints. Critical stress across joints which may be subject to hydraulic jacking may be different from the minor principal stress due to differences in orientation of principal stresses and joints. Complete exploration of in situ principal stresses (i.e., magnitudes and directions) at different depths in numerous boreholes along a tunnel alignment is expensive, impractical and unnecessary. However, assessment of "hydraulic jacking pressure" at different depths, and at various locations along a proposed pressure tunnel is required for design.

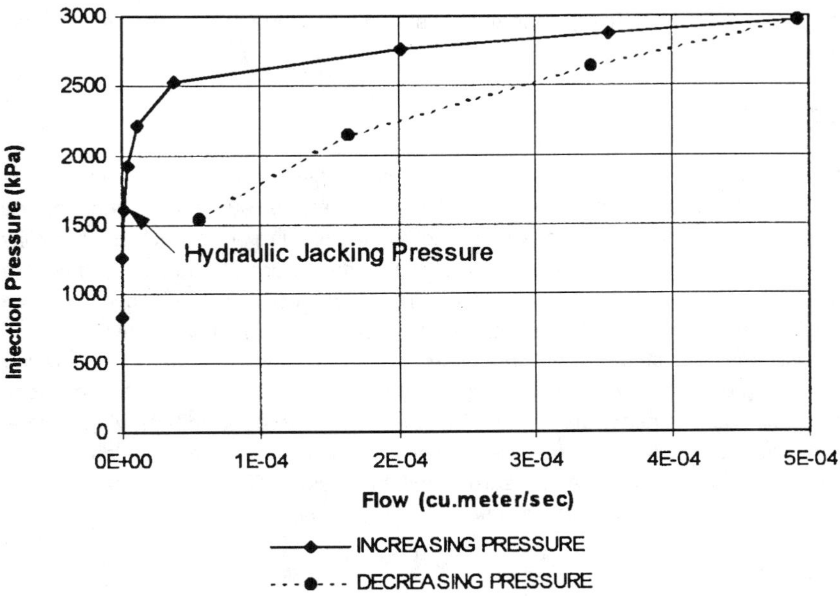

Figure 3. Typical Hydraulic Jacking Test Diagram

The "hydraulic jacking pressure" of critical concern for design is the minimum water pressure which causes opening of joints in the rock mass. These pressures are investigated by hydraulic jacking testing and the results compared to the static hydraulic pressures in the tunnel and shafts corresponding to the HGL. Hydraulic jacking tests are conducted by injecting water at high pressures between straddle packers in exploratory borings along the tunnel. Numerous hydraulic jacking tests were performed at several depth intervals in approximately two dozen deep borings along the alignment of the MWWST facility alignment to investigate the effects of various orientations of joints. This work was carried out with the assistance of a specialty geotechnical consultant, Dr. Gregg Korbin. These tests have become state-of-practice and are necessary for reliable design of pressure tunnels.

An illustration of an actual test conducted in a boring, drilled for MWWST, is shown in Figure 3, where the dramatic decrease of curve slope at pressures beyond the point of hydraulic jacking is directly related to the increase of rock mass permeability due to the opening of joints in the rock mass. Hydraulic jacking testing is described in detail in Hamilton, et. al., 1987, and Johannesson, 1988.

Topographical Characteristics

Localized variations of in situ stress due to topographic phenomena, e.g., steeper slopes, incision-like geological features such as gorges, or varying depth of overburden types of

critical features and their effects on the vertical alignment, is the presence of a "buried valley," or depression of the top-of-rock elevation, located east of Shaft L along the MWWST facility (see Figure 2). This geological feature resulted in locating the deepest point of the tunnel in this area.

In Situ Permeability

The prevailing natural, or in situ, hydraulic conductivity (permeability) of the rock mass is an essential design consideration for any tunnel in rock. Groundwater inflows encountered during construction must be assessed, and is also of particular importance for the feasibility of unlined and concrete lined pressure tunnels. Quantitative data for this assessment is obtained by packer pressure testing during preconstruction design, in conjunction with the hydraulic jacking tests described above. Important descriptive information about permeability characteristics is also obtained by observing drilling fluid behavior during exploratory drilling, and during construction by observing groundwater occurrence in the tunnel. Rock mass permeability assessments along the MWWST alignment, data from previous tunnel construction, and satisfactory performance of existing Boston pressure tunnels indicated that a systematically concrete lined pressure tunnel is feasible. This is discussed in more detail under *Pressure Tunnel Design Principles* later in this paper.

Groundwater Table

The location of the natural groundwater table, relative to the depth of the tunnel, may be of importance for the performance of a concrete or unlined pressure tunnel. An extreme and potentially adverse design situation in regards to leakage from the tunnel would exist where the applicable HGL is high above the ground level, and where permeability of the rock mass is high and the in situ groundwater level low relative to the invert level of a proposed tunnel. For the MWWST facility the groundwater level is near the ground surface and favorable for design of a concrete lined pressure tunnel.

Modulus of Deformation

The magnitude of the modulus of deformation (deformability or stiffness) of the host rock mass may be of significance for design of pressure tunnel lining systems in rock. Deformability of the rock affects any load sharing required for design of impermeable steel linings and performance of reinforced concrete linings designed to control leakage from the tunnel. However, design of steel linings is often governed by external loading due to groundwater pressure on an empty conduit, rather than hoop stress due to internal pressure and any potential load-sharing criteria. Lining design considerations for the MWWST facility are discussed in greater detail under *Pressure Tunnel Design Principals*, later in this paper.

Constructability Considerations

Design of tunnels generally requires thorough consideration of construction methods, costs, logistics, scheduling, sequence, support and lining requirements, boreability, muck disposal, groundwater issues, power requirements, heading length and numerous issues specific to a particular facility due to its environmental or geographical setting or specific serviceability requirements. The following paragraphs will only discuss the most significant constructability issues related to design of urban water supply tunnels in rock and the MWWST facility.

Shaft Sites

Construction of water supply tunnels in areas of generally flat terrain is performed through shafts, which inherently tend to increase in depth with increasing hydraulic head and application of a permeable concrete tunnel lining system. The main construction or work shaft is the focal point of most above ground construction activities in support of underground construction, such as transportation of materials, manpower access, administration, utilities and other services. Identification of suitable shaft sites along the alignment of a long tunnel in an urban/suburban area may present challenges, particularly in residential neighborhoods, due to regulatory and environmental impact considerations.

For the MWWST facility, suitable sites for construction shafts that were already owned or controlled by the Authority were found available along the general alignment of the proposed tunnel, at two locations (Shaft E and Shaft 5A) as shown on Figure 1. The third site, Shaft L, was privately owned, requiring a temporary construction type easement and lease arrangements. The site was selected to allow for construction of two headings, approximately five miles long, from a single construction shaft, following thorough engineering evaluations, including a Value Engineering effort.

Invert Gradient (Slope)

A gentle "uphill" invert slope is always desirable in tunnel excavation and lining to allow for gravity drainage of groundwater inflows away from the excavation heading or area of concrete placement. This also allows for "downhill" transport of tunnel muck. A maximum invert gradient of 0.01 is desirable from an economical and safety aspects of rail haulage during construction. From water tunnel design considerations, including consideration of future unwatering and manned access for inspection and maintenance, a minimum invert gradient of 0.001 is required for gravity drainage in the completed facility. This invert slope should be toward a shaft where suitable disposal of drainage water with minimum preparation and treatment requirements could be arranged at occasions of future entry into the tunnels. These considerations must be balanced with considerations of construction costs by avoiding unnecessary depths of the tunnels and shafts. For the MWWST facility, overall design considerations allowed invert gradients to range between 0.002 and 0.007.

Lining Construction

Where reinforcing of concrete linings is required, it is undesirable to require multiple layers, close spacing, and large bar sizes of reinforcing steel. A minimum concrete lining thickness of 0.3 m (12 in.) is required to achieve the necessary high quality concrete placement, while allowing for all anticipated types of primary support, including steel ribs embedded in the concrete and installation of localized steel lining where required. Changes of lining systems, e.g., from unreinforced concrete to steel linings, is also undesirable, as high incremental costs are associated with such changes. However, special lining treatments and changes in lining type are often necessary as a result of localized ground conditions. It is also important to protect the fresh concrete from groundwater inflows during placement of the concrete lining to reduce the frequency of circumferential shrinkage cracking.

Excavated Tunnel Diameter

The excavated tunnel diameter must accommodate all anticipated types of primary support, while allowing sufficient clearance for installation of any lining system that is anticipated along the tunnel alignment. This dimension should be kept to a minimum, but is most efficiently left to the actual contractor. This is because the contractor has numerous construction considerations that are related the excavated diameter, such as excavation methods, ventilation, transportation and costs of the final lining.

Construction Methods

Feasible tunnel and shaft construction methods are assessed during design to identify any effects on structural components and overall construction costs. Excavation by tunnel boring machine (TBM) and drilling and blasting, and feasible primary support systems are evaluated at a minimum. Ground characteristics which may require consideration in structural design, or development of special lining treatment components (e.g. grouting) are also identified.

Excavation by TBM, and primary support by installation of rock dowels, are anticipated for most of the length of the MWWST. Construction shafts are anticipated to be excavated by conventional shaft sinking methods while the riser shafts will be excavated by raise boring or a combination of raise boring and slashing to minimize construction activities at ground surface.

Pressure Tunnel Design Principles

The specific principles applied to the design of a systematically concrete lined pressure tunnel such as the MWWST facility is discussed briefly in the following paragraphs.

Vertical Alignment

A preliminary vertical alignment of a proposed tunnel is required early during design to provide depths for exploratory borings. To establish this depth for a pressure tunnel several important preliminary design criteria, need to have been already established:

- Design HGL
- Tunnel Horizontal Alignment
- Ground Surface Profile

- Overburden Depth (Range)
- Presence of any "Buried Valleys"
- Locations of Construction Shafts

The minimum depth envelope along a proposed concrete lined or unlined pressure tunnel is established to avoid hydraulic jacking in the rock mass at any location along the tunnel. In flat or gently sloping terrain, and assuming that vertical stress is critical, this is accomplished by application of the "0.5 Cover Criterion" as follows:

$$CR = \frac{TOR - INV}{HGL - INV}$$

where:

CR	=	cover ratio (0.5)
TOR	=	top of rock elevation (modified to account for overburden depth if needed)
INV	=	elevation of the tunnel invert
HGL	=	elevation corresponding to the Static Hydraulic Gradeline

This expression may also be summarized in the following statement: "The internal pressure head must be balanced in the vertical direction by half this amount of rock cover." Application of this criterion results in an apparent factor of safety against the occurrence of hydraulic jacking of approximately 1.3, assuming prevailing densities for most igneous and metamorphic rock materials. As the results of geotechnical characterization become known, e.g., detailed knowledge about hydraulic jacking behavior in the rock mass, and variations of depth of overburden soils, the required minimum depth of a proposed tunnel may be somewhat reduced (or increased) locally.

Following establishment of minimum pressure tunnel depths at topographically critical locations, constructability considerations such as construction shaft location, invert gradients, tunnel heading lengths, and efforts to minimize shaft depth, are all incorporated to complete the preliminary vertical alignment.

Final Lining Design

The design of a systematically concrete lined pressure tunnel requires several important considerations during preconstruction design development, and continuous verification throughout the shafts and tunnels during excavation. Identification of any areas requiring augmentation of the systematic lining by installation of appropriate localized, special lining treatment is required as excavation proceeds. The final step of design may be considered to be verification of performance by pressure testing on completion of construction of the facility. The most significant considerations involved in design of the systematic lining, implementation of special lining treatment components, and performance testing are summarized in the following paragraphs.

Tunnel and shaft final linings may be broadly characterized as impermeable, semi-per-

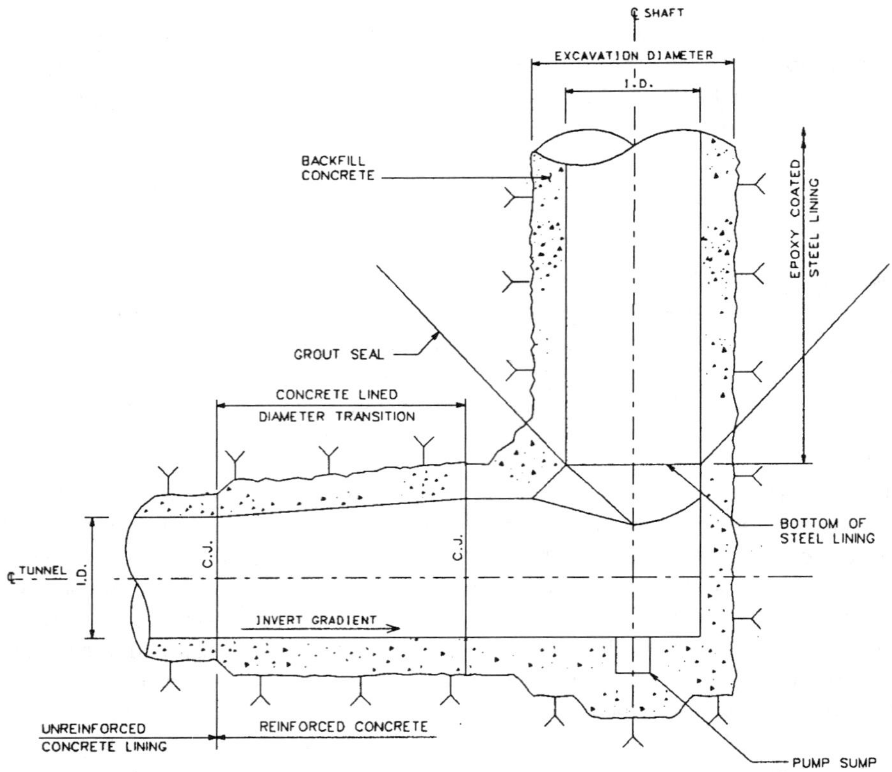

Figure 4. Typical Shaft.Tunnel Intersection

meable or permeable. Impermeable type pressure tunnel linings include jointed steel cylinder pipe linings and concrete cylinder pipe (PCCP and RCCP). These types of lining components are manufactured off -site in segments (sections or cans) and joined in the tunnel or shaft by welded or gasketed joints. Leakage from these types of linings is minimal and concentrated to the field joints. Extensively reinforced, cast-in-place concrete linings may be considered semi-permeable. Occasionally, nominally reinforced linings are installed to reduce infiltration in low pressure sewer tunnels, e.g., CSO tunnels. The distribution of circumferential (transverse) cracks is ameliorated by the introduction of longitudinal reinforcing steel. Nominal reinforcing is installed in pressure tunnels only to control and distribute cracking in areas of low modulus of deformation or in areas of stress concentration, e.g., at conduit intersections, to preserve structural integrity. An unreinforced or nominally reinforced concrete lining may be classified as a permeable lining. The latter types of linings are installed mainly to provide a smooth conduit surface, to protect and support the rock and to facilitate future inspection and maintenance work.

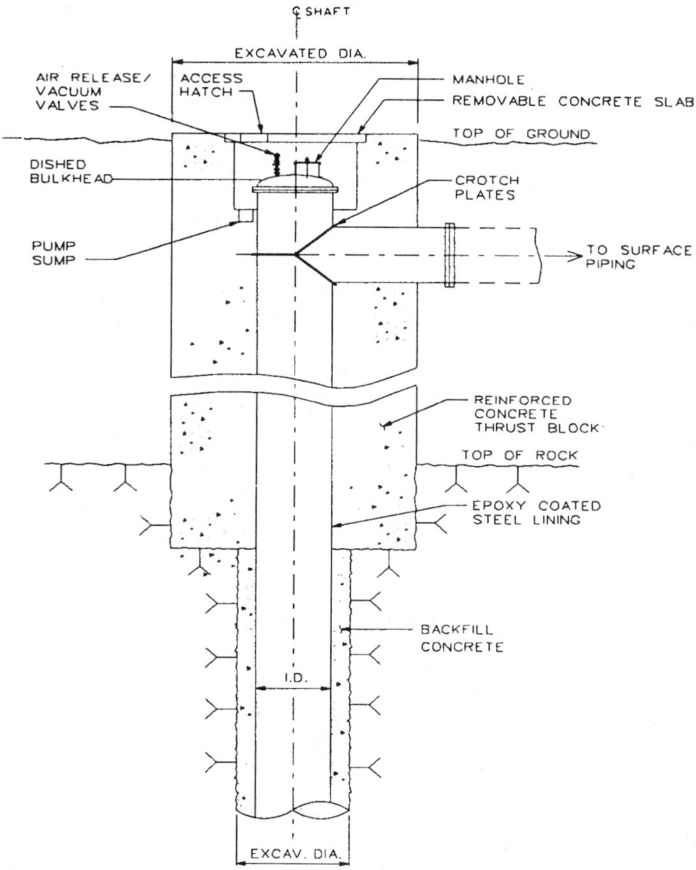

Figure 5. Typical Top of Shaft Structure

Because of the ongoing engineering evaluations during construction, the construction contract must contain mechanisms to deal fairly and efficiently with the required flexibility to install the various linings which are dictated by the actual ground conditions. This requirement is not unique to requirements for final linings in pressure tunnels, but is commonly used to handle tunnel primary support requirements. Accordingly, appropriate alternate final lining types are developed during preconstruction design with estimated quantities provided for bidding. Contractual provisions are employed to delay procurement of tunnel lining system components (e.g. reinforcing steel and steel linings), until the extent of the various lining types has been determined.

SHAFT STRUCTURES: Shaft linings must be of the impermeable type to at least the minimum depth required by the cover criterion. The extent of impermeable lining in a specific shaft also depends on other considerations, including in situ permeability as observed by testing and/or mapping information obtained during excavation. Impermeable linings in shafts are almost exclusively welded steel linings, with internal epoxy coating for corrosion protection. Reliance on ground/lining interaction is commonly not done in design of shaft steel linings. Shaft linings at lower elevations may consist of unreinforced or reinforced concrete as appropriate. Shaft/tunnel intersections may be lined with nominally reinforced concrete, with large diameter steel pipe bends, or constructed as a tee, as required for a specific shaft. All three types of intersections exist in various shafts along the MWWST facility. An example of a water tunnel/shaft intersection is shown in Figure 4.

The top-of-shaft structure must provide thrustblock functions to stabilize the structure vertically due to uplift, and horizontally due to the change of conduit direction on pressurization of the system. Accordingly, the top-of-shaft access manhole is structurally a "blockout" in an otherwise massive concrete thrust block, which is commonly reinforced in the vertical direction for continuity and to provide lateral bending strength. A large diameter, steel pipe tee is commonly embedded in this concrete at the top of the shaft with a horizontal branch for connection to surface level piping, a vertical branch facing upward and provided with a removable bulkhead for future access, and a second vertical branch facing downward and connected with the shaft conduit. A schematic illustration of a top-of-shaft structure is shown on Figure 5.

STEEL LINING: The thickness of steel linings for tunnels and shafts must satisfy three basic design criteria and would be the largest required by:

- Analysis based on internal pressure
- Analysis based on external pressure (buckling)
- Minimum thickness required for handling during construction

Earlier designs commonly added an allowance in thickness for corrosion considerations. Currently, corrosion protection systems are included where high risks for stray currents are present, as recommended in the AWWA M11 standards for steel pipes. Steel linings in tunnels are commonly coated internally with a field applied cement mortar lining in the order of 12.7 mm (1/2-in.) to 19.0 mm (3/4-in.) thick. The alkaline environment from the internal cement mortar lining and external backfill concrete provide the primary corrosion protection system for the steel lining.

The steel lining sections are installed and backfilled externally with concrete of strengths ranging from 10.3 MPa (1,500 psi) to 17.3 MPa (2,500 psi). This backfill concrete is required to have very low shrinkage characteristics and is installed to stringent construction quality requirements.

Analyses based on internal pressure and the common thin cylinder equation, is applied where load sharing (ground interaction) is not considered:

$$t_s = \frac{p_i d}{2f_s}$$

where: t_S = thickness of the cylinder wall
p_i = internal pressure
d = internal diameter
f_S = allowable hoop stress

Thickness arrived from internal pressure analyses is required to be verified for both static pressure and total pressure including transient pressure.

Numerous closed form analysis approaches and formula have been presented during the last several decades to model load sharing between the steel lining, backfill concrete and surrounding rock to provide more economical designs of steel pressure tunnel linings. Available space prevents a detailed discussion of these methods of analysis in this paper. The various methods are summarized by Brekke and Ripley, 1987, and by Moore, 1989.

As in the case of most steel lining components designed for the MWWST facility, steel lining thickness requirements are generally governed by external pressures and the potential for buckling during future unwatering of a particular conduit. Two methods of analysis are currently available for plain cylinder steel linings, by Amstutz, 1970, and Jacobsen, 1974. Both methods are based on prevention of single lobe buckling of an embedded steel cylinder and apply a factor of safety of 1.5 against buckling. In this type of analysis a gap is assumed to exist between the outside surface of the steel lining and the surrounding backfill concrete. This assumed gap allows deformation of the steel lining in response to external pressure, forming a lobe whose stability is analyzed. The magnitude of this gap, which is assumed to account for shrinkage of the backfill concrete and thermal deformations in the steel lining itself, must be the minimum realistically achievable. This results in the requirement of a high quality, low shrink backfill concrete to fill the annular space between the steel lining and the excavated rock surfaces. An illustration of the assumed steel lining shape used in buckling analysis is shown in Figure 6.

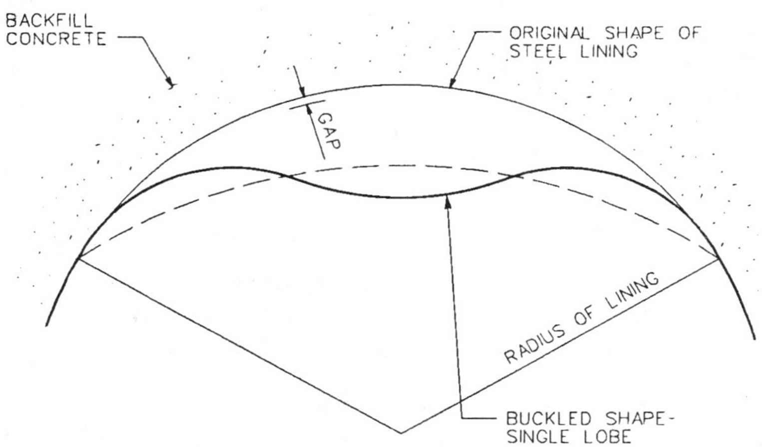

Figure 6. Single Lobe Buckling Principle

Amstutz's method was used almost exclusively during the past several decades, but has been superseded by the more conservative Jacobsen method with the advent of computerized analysis during recent years. The Jacobsen method of analysis was used for design of the relatively limited steel linings in the MWWST facility. Details of analysis for steel linings due to external pressures (buckling) are found in Moore, 1990, and Stutsman, 1993.

The use of external stiffeners to minimize steel lining thicknesses is attractive for economical design of large diameter steel linings subject to very high external pressures. However, the significant disadvantages of external stiffeners, e.g., underground transportation and clearance issues, and issues related to placing high quality backfill concrete, often preclude the use of external stiffeners, except where plain steel lining thickness exceed 38 mm (1.5 in.). Where thicker steel linings are required, space for manned access must be provided between the outside of the steel lining and the rock surface.

Steel linings up to 38 mm (1.5 in.) thick may be installed using external backing bars and internal, single sided, full penetration welding, without application of post weld heat treatment. Both post weld heat treatment and manned access for welding outside the steel lining are expensive requirements which may be avoided using external stiffeners to reduce lining thickness. Pressure relief valves have also been applied in several designs to reduce the required thickness of steel linings for hydropower developments. However, relief valves are not considered reliable in water supply tunnels due to potential deterioration after decades of inactivity. Only plain cylinder pipe steel linings are provided in the MWWST facility.

Steel linings for the small diameter community riser shafts along the MWWST alignment were analyzed as free standing cylinders exposed to external liquid grout pressure during installation. The excavated shaft openings for these structures are too small in diameter to allow manned entry. Installation and welding of these linings are made entirely from the ground surface.

A minimum steel lining wall thickness must be provided to allow practical handling of lining sections during manufacturing, transportation and installation. Two commonly applied rules to assess minimum plate thickness are:

$$t = \frac{D}{288} \quad \text{(PG\&E Rule, D} \leq 1.5\text{ m.)} \qquad t = \frac{D + 0.5}{400} \quad \text{(Bu.Rec. Rule, D} \geq 1.5\text{ m.)}$$

where: D = internal diameter (m)
t = minimum wall thickness (m)

Appropriate stulling and blocking must be applied during handling and installation of the steel linings to prevent damage. Design for minimum lining thickness was not applicable for the MWWST facility components, as external pressure governed the design.

Anchor rings to ensure positive anchorage of the steel linings to the concrete in the top-of-shaft structures, and seepage rings at the bottom of steel linings, were included in all shafts of the MWWST facility. General guidelines for design of these components are provided in Stutsman, 1993.

Economical, commonly available, fine grained, carbon steels and adequate material toughness are required for pressure tunnel steel linings. For the MWWST facility two alternate steels are allowed; A516 Grade 60 and A537 Class I. Either of these steels may be installed as long as the minimum thickness, shown on the drawings, is provided. Fabrication, installation and quality control specifications are required to be in accordance with applicable AWWA, AWS and ASME Boiler and Pressure Vessel Code requirements.

UNREINFORCED CONCRETE LINING: Unreinforced concrete linings are commonly designed for external water pressures, any applicable rock loads, and compressive strength requirements for an unwatered tunnel conduit. A minimum thickness of 0.3 m (12 in.) of 27.6 MPa (4,000 psi) concrete is required for the MWWST facility. Appropriate specification requirements govern mix shrinkage characteristics, protection of fresh concrete from groundwater during placement, and other construction quality issues. Contact grouting of all potential void areas between concrete and rock is also required.

In areas where permeability of the host rock, is of concern for the performance of the tunnel, as observed by mapping of the tunnel and monitoring of groundwater infiltration during excavation, the tunnel lining will be provided with reinforcing steel to control leakage. The amount of reinforcing steel will be determined by the assessed modulus of deformation and applicable internal pressure at the specific location. Discussions of design analysis applying ground/lining interaction for reinforced concrete linings are provided in Moore, 1989.

Locally, the completed, reinforced concrete lining may be further augmented with selective consolidation grouting, as determined to be required during the final lining determination process. At locations in the tunnel where performance of the final lining is determined to be of particular concern based on observations of ground conditions and applicable internal pressure, the concrete lining may be substituted with a limited length of steel lining.

On completion of the MWWST facility concrete lining, some infiltration of water through the lining is anticipated. This infiltration is generally associated with circumferential (transverse) shrinkage cracks and construction joints. The anticipated spacing of circumferential shrinkage cracks in a well constructed cast-in-place unreinforced concrete lining may range from 5 m (15 ft) to 8 m (25 ft). The spacing of construction joints in a lining may range from 46 m (150 ft) to 67 m (220 ft). Pressure grouting at relatively high pressures is prescribed at visible, leaking cracks and joints in the tunnel lining to meet stringent infiltration criteria.

TUNNEL MAPPING AND FINAL LINING DETERMINATION: Design of a systematically unreinforced concrete lined pressure tunnel continues through the excavation phase of construction. It is essential that the entire length of a pressure tunnel is mapped during tunnel excavation by a qualified engineering geologist who records all observations which are of importance for selection of appropriate final lining system components, including:

- Joint Spacings and Orientations
- Joint Aperture/Infillings
- Characteristics of Shears/Shear Zones/Fault Zones
- Rock Durability Information

- Weathering/Alteration Characteristics
- Groundwater Occurrence
- Rock Stability Information/Primary Support Installed
- Rock Type

The mapping information is used by the design engineer to verify design assumptions, and where required, to determine the appropriate lining system augmentation.

Groundwater occurrence information is of particular importance for pressure tunnel lining selection because of its value as an indirect measure of rock mass permeability and hence a measure of potential leakage from the tunnel. Groundwater occurrence requires observation and record keeping throughout the length of a tunnel and at specific locations, as it appears immediately following excavation and periodically as it dissipates with time following excavation.

For the MWWST facility a specific procedure has been established to select the final lining required at any location in the tunnel. The construction manager will map the tunnel. Following review of the tunnel mapping information by the design engineer, areas exhibiting characteristics of concern will be assessed in more detail. It is intended to complete determination of final lining requirements at selected points of progress of excavation in each heading. (e.g., at each one third point). After this initial information is reviewed, the need for in situ testing will be assessed. On completion of these assessments, appropriate final lining components/treatment will be selected for the portion of the heading under examination.

Performance Testing

The MWWST facility will consist of three separate, complete pressure tunnel loops which are inter-connected at the ground surface, where these are also connected to the parallel Hultman Aqueduct transmission system. On completion of the tunnels and shafts, these loops will be filled with water and pressure tested separately. Limited and controlled rates of pressure increase/decrease during filling, testing and unwatering of each complete tunnel loop is intended to limit the overall rates of change of pressure gradients across the final lining components, and thus promote distribution of cracking in concrete linings, and dissipation of external pressures behind steel linings.

The first phase of pressure testing is the filling of the tunnel and shaft conduits. This phase will be followed by "over-pressurization" of the system, i.e., pressure increase beyond complete filling of the loop conduits. For the MWWST facility, the rate of steady pressure change has been limited to pressure changes corresponding to 6 m (20 ft.) of head per hour during all stages of testing. A constant pressure holding period of 14 hours will be introduced at the end of each 60 m (200 ft.) increment of change of head. Also, a also limiting total head change of 60 m (200 ft.) in a 24 hour period is required.

At full test pressure, a holding period of 14 days will be maintained. The steady flow rate into the loop at the end of the 14 day test period may be considered close to the steady leakage rate out of the tunnel loop, although leakage from pressure tunnels has occasionally increased with time due to dissolution or erosion of joint infillings, or because of hydraulic jacking. If the observed rate of leakage is found to be unacceptable, additional final lining work, e.g., consolidation grouting, may be required within the specific tunnel loop. On completion of pressure testing, all conduits will be disinfected prior to commissioning and operation of the MWWST facility.

ACKNOWLEDGMENTS

The MWWST facility planning and design has been produced by a team of designers and scientists led by Sverdrup Civil, Inc., as prime consultant. Jacobs Associates performed the pressure tunnel design under the leadership of Mr. A.M. Petrofsky, Principal (Ret.), and the senior author of this paper. Professor Tor L. Brekke, University of California at Berkeley, provided valuable expertise and guidance in the approaches to design of pressure tunnel facilities.

BIBLIOGRAPHY

American Water Works Association (AWWA), 1989, "Steel Pipe - A Guide for Design and Installation," M11, Manual of Water Supply Practices, AWWA, Denver, 175 pp.

Amstutz, E., 1970, "Buckling of Pressure Shaft and Tunnel Linings," Water and Power Dam Construction, Nov., pp. 391-399.

Brekke, T.L., Ripley, B.D., 1987, "Design Guidelines for Pressure Tunnels and Shafts," EPRI Report AP-5273, June, .

Caspe, H.P., Kim, A.Y., Bergen, L.J., Araujo, J.R., 1994, "Planning for the MWRA MetroWest WaterSupply Tunnel," North American Tunneling '94 Conference, April, Boston.

Caspe, H.P., Kim, A.Y., Bergen, L.J., Araujo, J.R., 1994, "Environmental Planning for the MWRA MetroWest Water Supply Tunnel," Tunnelling '94 Conference, July, London, England.

Hamilton, P., Korbin, G.E., 1987, "Importance of Multiple Stress Measurements in Design of Balsam Meadow Hydroelectric Project," RETC Conference, pp. 850-866.

Jacobsen, S., 1974, "Buckling of Circular Rings and Cylindrical Tubes Under External Pressure," Water Power, Dec., pp. 400-407.

Johannesson. P., Schliess, A., Korbin, G.E., 1988, "Calaveras Project High Head Pressure Tunnel," Journal of Water Power & Dam Construction, Nov., pp. 43-48.

Moore, E.T., Editor, 1989, Volume 2, "Waterways," Chapter 3, "Tunnels and Shafts," Civil Engineering Guidelines for Planning and Designing Hydroelectric Developments, ASCE, New York.

Moore, E.T., 1990, "Designing Steel Tunnel Liners for Hydro Plants," Hydro Review, pp. 72-90.

Stutsman, R.D., Editor, 1993, "Steel Penstocks," ASCE Manuals and Reports on Engineering Practice No. 79, Oct., ASCE, New York, 432 pp.

Zoback, M.L. and Zoback, M., 1980, "State of stress in the conterminous United States," Journal of Geophysical Research, Vol. 85, no. B11, pp. 6113-6156.

Chapter 3

IMPERVIOUS TUNNEL LININGS IN A LAND OF SHAKES

Drupad B. Desai,[1] Chien-Tsun "C.T." Chu,[2] and Ron E. Redd[3]
[1]Vice President and Technical Director of Underground Structures, Daniel, Mann, Johnson, & Mendenhall; [2]Senior Structural Engineer, Daniel, Mann, Johnson, & Mendenhall; [3]Project Manager for Riverside Badlands Tunnel, Metropolitan Water District of Southern California

ABSTRACT

The 12,710-m (41,720-ft) Riverside Badlands Tunnel is located between the most active faults—San Andreas and San Jacinto—in Southern California. To accomplish the impervious final lining design and to accommodate the effects of earthquake ground motions, a new approach was developed and is prescribed in this paper. Two final lining options, welded steel pipe and prestressed concrete cylinder pipe, are considered for the tunnel, which will be excavated under groundwater and ground conditions from soft rock to hard rock at depths exceeding 244 m (800 ft). The buckling strength, serviceability, and structural integrity under seismic effects and other loads were considered in the design.

INTRODUCTION

The Riverside Badlands Tunnel, part of the $1.1 billion Inland Feeder System (DMJM/WCC, 1996) located in one of the most active seismic regions of Southern California, will be excavated under groundwater and ground conditions of alluvium, weak San Timoteo formation, and harder fractured metamorphic rocks, and at depths exceeding 244 m (800 ft). The tunnel, with 366 cm (12 ft) finished diameter, is located between the San Andreas Fault and the San Jacinto Fault. These are among the most active faults in Southern California, with maximum design earthquake magnitudes (Mw) of 8 and 7.5, respectively, expected to occur over the life of the structure. The tunnel is subject to an external groundwater pressure of up to 101 m (330 ft) water head and an internal hydrostatic pressure of about 76 m (250 ft).

Because of the ground characterizations and extremely intensive seismic zone, a new design approach was developed on this project to accommodate the effects of earthquake ground motions and to accomplish the impervious lining design. Two final lining options considered for this facility are (1) welded steel pipe (WSP) and (2) prestressed concrete cylinder pipe (PCCP). In both designs, low-density cellular concrete is used as backfilled material between the final lining and the initial tunnel support. Since most of the WSP installation is under the groundwater table, the buckling strength becomes an important consideration in the pipe design. The critical buckling strength is controlled by the annular gap that forms at the steel-concrete interface and depends on the degree to which the gap may be affected by seismic excitations. In the design of PCCP, the tunnel excavation method is different from the traditional trench excavation as described in American Water Works Association (AWWA, 1992).

SITE CONDITIONS AND GEOLOGIC HAZARDS FROM LAND SHAKING

The profile and geological section of the Riverside Badlands Tunnel is shown in Figure 1. Approximately 80 percent of the tunnel is below the groundwater table. Based on geologic explorations, the tunnel alignment has been divided into 10 reaches consisting of three distinct types of ground: alluvium, weak sedimentary rocks of the San Timoteo Formation, and stronger but sheared and fractured metamorphic rocks as summarized in Table 1. In general, anticipated ground conditions include raveling, running, and flowing conditions in the alluvium; raveling, yielding, and flowing conditions in the San Timoteo Formation; and moderately jointed, blocky and seamy, and crushed conditions in the metamorphic rocks. Several fault zones are expected, consisting of intensely fractured and sheared/brecciated rock and gouge. There is a potential for squeezing conditions in the fault zones, as well as in reaches of high ground cover above weak siltstone and claystone beds in the San Timoteo Formation.

Table 1. Summary of Anticipated Ground Types by Reach*

Reach	Estimated Station Limits		Ground Types
	From	To	
1	1090 + 00	1115 + 00	Alluvium
2	1115 + 00	1142 + 00	San Timoteo Formation (Upper Member)
3	1142 + 00	1213 + 00	Metamorphic Rock
4	1213 + 00	1258 + 00	San Timoteo Formation (Upper Member)
5	1258 + 00	1263 + 00	Alluvium
6	1263 + 00	1329 + 00	San Timoteo Formation (Upper and Middle Members)
7	1329 + 00	1340 + 00	Alluvium
8	1340 + 00	1438 + 00	San Timoteo Formation (Middle and Lower Members)
9	1438 + 00	1469 + 00	Metamorphic Rock
10	1469 + 00	1507 + 00	San Timoteo Formation (Lower and Middle Members)

*Station limits may vary by 100 feet (±)

The tunnel is located between the San Andreas Fault and the San Jacinto Fault, as shown in Figure 2. Considering the seismic setting of the project, and the relative proximity and activity of the San Andreas and San Jacinto faults, it should be expected that the tunnel will be subjected to strong seismic shaking during and following construction. Besides the two major faults, four other minor faults were also considered for deterministic evaluation of ground motion parameters. These are Reservoir Canyon Fault, Banning Fault, Chicken Hill Fault, and Blind Thrust Fault.

SEISMIC CONSIDERATIONS

The effects of earthquakes on tunnels due to the above-mentioned faults may be

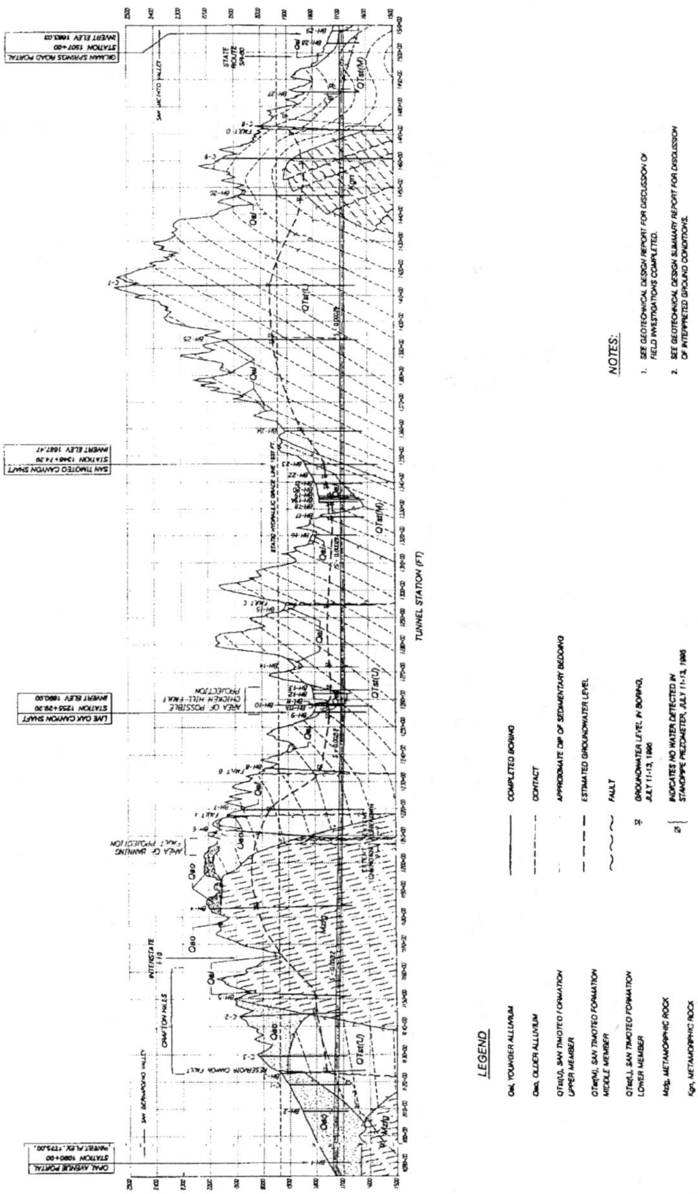

Figure 1. Tunnel Profile

grouped into two categories: ground failure and ground shaking. Ground failure broadly includes various types of ground instability— including faulting, landslides, and liquefaction, which are not discussed in this paper. Ground shaking is the vibration of the ground produced by seismic waves propagating through the earth's crusts. Various types of seismic waves develop as a result of an earthquake, including traveling shear waves and transverse shear waves. Peak horizontal ground velocity and acceleration are seismic shaking parameters for tunnel linings subjected to seismic waves. These parameters were estimated using deterministic methods and maximum design earthquakes. Proposed maximum design earthquake magnitudes (Mw), based on studies completed for Metropolitan's Eastside Reservoir Projects, are 8, 7.5, 6.5, 7, 6.5, and 7 for San Andreas Fault, San Jacinto Fault, Reservoir Canyon Fault, Banning Fault, Chicken Hill Fault, and Blind Thrust Fault, respectively.

The mean values of peak velocities and peak accelerations at the ground surface were estimated for six sites along the tunnel alignment, and attenuated for tunnel depth. Table 2 (DMJM/WCC, 1996) summarizes the locations of the six sites (Site 1 through Site 6), general ground conditions at each site, tunnel depth, and peak accelerations and velocities at ground surface and at tunnel depth. The effects of seismic excitations on the tunnel lining were evaluated by the seismic strains due to peak velocities and accelerations (Merritt, et al, 1990).

Table 2. Estimated Peak Horizontal Ground Motions

Site		General Ground Conditions[1]	Depth[1] m(ft)	Peak Values At Ground Surface		Peak Values At Tunnel Depth	
No.	Name			a_{max} (g's)	v_{max} m/sec (fps)	a_{max} (g's)	v_{max} m/sec (fps)
1	Opal Avenue Portal	Alluvium	9 (30)	0.52	1.40 (4.6)	0.34	1.13 (3.7)
2	Near Banning Fault	San Timoteo Formation and Metamorphic Rock	152 (500)	0.52	0.55 (1.8)	0.36	0.49 (1.6)
3	Live Oak Canyon - Shaft	Alluvium and San Timoteo Formation	34 (110)	0.42	1.19 (3.9)	0.27	0.94 (3.1)
4	San Timoteo Canyon - Shaft	Alluvium and San Timoteo Formation	43 (140)	0.56	1.10 (3.6)	0.36	0.88 (2.9)
5	Maximum Cover	San Timoteo Formation	259 (850)	0.68	0.67 (2.2)	0.48	0.61 (2.0)
6	Gilman Springs Road Portal	San Timoteo Formation	15 (50)	0.72	0.76 (2.5)	0.50	0.70 (2.3)

1. From surface down to tunnel invert

The longitudinal and circumferential strains due to seismic waves are described in the following.

Seismic Waves Propagating Along Tunnel Axis

When seismic waves propagate either along the axis of the tunnel or at a given angle,

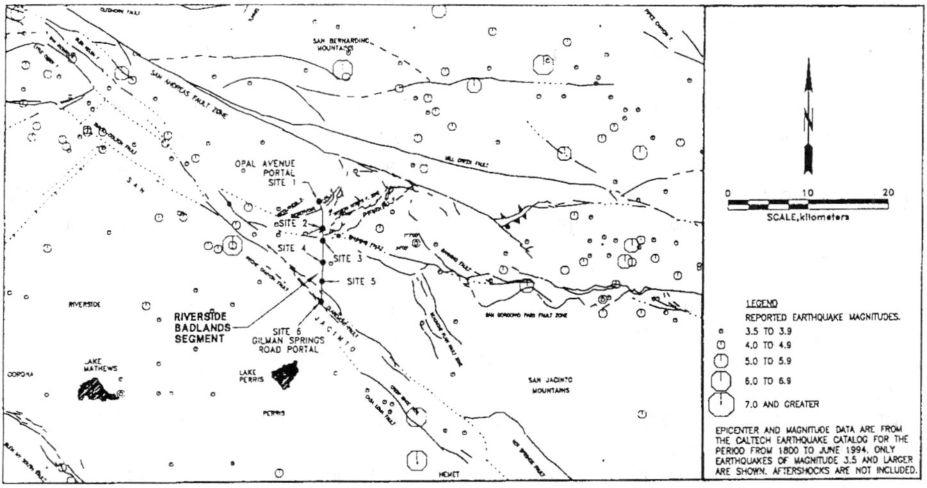

Figure 2. Regional Fault and Epicenter Map

the axial and curvature deformations can be conservatively evaluated by assuming a shear wave at 45 degrees to the tunnel axis. For this case, the maximum longitudinal strain (ε_l) in the lining is (Merritt et al., 1990):

$$\varepsilon_l = \pm \left[\frac{v_{max}}{2c_{se}} + 0.7R \frac{a_{max} g}{c_{se}^2} \right]$$

where:

v_{max}	=	peak particle velocity produced by earthquake
a_{max}	=	peak acceleration, in gravity units
g	=	acceleration of gravity
c_{se}	=	effective shear wave velocity for the value of v_{max}
R	=	mean radius of lining

Seismic Waves Propagating Perpendicular to Tunnel Axis

When waves propagate normal or nearly normal to the tunnel axis, racking of a tunnel may develop. Usually the vertically propagating shear waves produce the most critical racking deformation of the lining. The strains due to racking (ε_{rack}) induced by an impinging shear wave are approximately (Merritt et al., 1990):

$$\varepsilon_{rack,c} = \frac{v_{max}}{c_{se}} \left[2\frac{t}{R} + \frac{3 E_m R}{16 E_l t} \right] \quad \text{[In Compression]}$$

and

$$\varepsilon_{rack,t} = 2 \frac{v_{max}\, t}{c_{se} R} \quad \text{[In Tension]}$$

where:

t	=	thickness of lining
E_m	=	modulus of elasticity of medium
E_l	=	modulus of elasticity of lining

LINING DESIGN CONSIDERATIONS

Tunnel Excavation Methods and Initial Supports

For the site conditions and seismic hazards along the Riverside Badlands Tunnel, a variety of tunneling methods and equipment are considered applicable to the project. These include hand mining methods, soft ground shields and tunneling machines, drill-and-blast methods, roadheaders, and rock tunnel boring machines (TBMs). A combination of methods could be used, due to the range of geologic conditions and the distribution of materials along the proposed alignment.

The tunnel excavation will need to be sized to accommodate the initial support requirements, the specified final linings (including minimum annular space for their installation), and the line and grade tolerances. The anticipated tunnel diameter is about 488 to 518 cm (16 to 17 ft) where a TBM or shield is utilized, and about 488 to 549 cm (16 to 18 ft) where a horseshoe tunnel shape is excavated using drill-and-blast or roadheader methods. Initial support requirements will depend on the selected excavation methods and ground conditions. Options for initial support include steel ribs with lagging or shotcrete, precast concrete segments, and steel liner plates.

Loadings and Required Strength of Final Linings

The design loading parameters and performance criteria for tunnel linings were based on ground conditions. Metropolitan will require a watertight final lining system for the tunnel, consisting of either WSP or PCCP. The two final lining alternatives are designed to take into consideration hydraulic and operational criteria, external soil/rock loads, external and internal hydrostatic pressures, construction loads, and the estimated characteristics of the design basis earthquake.

To evaluate stresses and strains due to fluctuation of groundwater levels, maximum and minimum external hydrostatic pressures were considered in load combinations. Due to geologic conditions, WSP and PCCP final linings were designed for the full internal water pressure without considering any load sharing with the surrounding soil/rock. Fifty percent of ground load and 80% of minimum groundwater pressure were used to estimate maximum lining tensile stress and strain (full internal hydrostatic pressure). Full ground load and maximum groundwater pressure were used to estimate maximum lining compressive stress and strain (internally empty).

WSP was designed in accordance with AISC Allowable Stress Design Method (1989) and AWWA M11 Design Guide (1989). PCCP final lining was designed in accordance with ANSI/AWWA C304 (1992). These design codes were used to determine corresponding allowable stresses and strains, forces, and moments for static and transient load conditions. In addition, the minimum steel thickness (relative to handling stresses) was checked by the criteria developed by Parmakian (1982). For PCCP design, an additional load was considered for the inner concrete core to carry full external hydrostatic pressure acting on the outer surface of the steel cylinder, since the groundwater might infiltrate into the outer concrete core.

For seismic conditions, the allowable stresses will be the corresponding static allowables

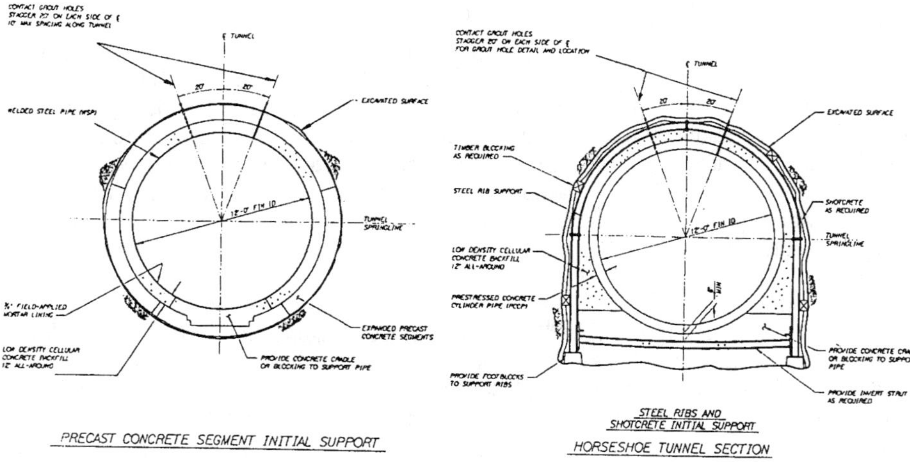

Figure 3. WSP Lining Sections

increased by 67 percent, but not to exceed yield strength for WSP design. The strength limit design criteria stated in ANSI/AWWA C304 (1992) were applied to PCCP seismic conditions, except that the allowable compressive flexure strain was considered to be 0.004. This was based on the maximum flexure compressive strain of 0.004 for a concrete lining (Merritt, 1985).

STEEL TUNNEL LININGS

WSP was developed as a final lining based on the design considerations described above. The construction considerations influencing the final design include tunnel excavation and the installation of initial support. WSP may be installed into the tunnel and set in place with a pipe-placing machine, as shown in Figure 3. A 30-cm (12-in.) nominal annulus between the pipe and the initial supports would be backfilled with low-density cellular concrete. For structural reasons, the backfill must be at least 20 cm (8 in.). The use of low-density cellular concrete for backfilling the annular space is considered advantageous, because (1) it reduces the risk of pipe flotation, which could damage the pipe; (2) it facilitates backfilling of long lengths of pipe in a single and cost- effective operation; and (3) it provides additional corrosion protection. The minimum compressive strength recommended for cellular concrete is 2.76 MPa (400 psi) and the corresponding wet density is 1121 kg/m^3 (70 pcf) with 0.00181 shrinkage ratio.

Design evaluations to determine the minimum plate thickness for WSP final linings indicate that 11.1-mm (7/16-in.) thick steel plate is required for internal pressures, based on an allowable stress of 50 percent of the yield stress for 248 MPa (36 ksi) steel. The plate thickness of 11.1 mm (7/16 in.) is also adequate for internal pressures (including transient effects), based on allowable stress of 75 percent of the yield stress for 248 MPa (36 ksi) steel. A minimum plate thickness of 12.7 mm (1/2 in.) is required for handling conditions based on the criteria proposed by Parmakian (1982) and assuming an acceptable handling stress of 186 MPa (27 ksi).

Buckling Strength

An important consideration in determining the minimum required plate thickness is adequate buckling resistance to withstand external hydrostatic pressures that will act on the pipe when the tunnel is unwatered, i.e., not under internal pressure. Maximum groundwater levels range from about 20 feet to more than 300 feet above the tunnel. There are several methods for estimating the critical buckling pressure of steel pipes subjected to external groundwater pressures. Two of the most widely used methods were developed by Amstutz and Jacobsen (ASCE, 1993). Jacobsen's method is preferred, because the method developed by Amstutz has been shown to be unrealistically conservative in certain situations (ASCE, 1993).

The critical buckling strength of cylindrical pipes is mainly controlled by the following three conditions: (1) the lining slenderness ratio of pipe radius to thickness; (2) the yield stress of steel pipe, which is 248 MPa (36 ksi); (3) the annular gap, which forms at the steel-concrete interface. The annular gap is a combination of four factors:

(1) Shrinkage of the cellular concrete. Given a concrete shrinkage ratio of 0.00181, the gap ratio for cellular concrete is 0.00020.

(2) Thermal contraction. Given a 35°F temperature difference between the ambient temperature of the cellular concrete-rock system at installation and the lowest operating temperature anticipated for the unwatered tunnel, the gap ratio is 0.00023.

(3) Expansion due to internal pressure. Assuming that 50% of the deformation that the concrete-rock (San Timoteo Formation) system would undergo due to 76 m (250 ft) of internal hydrostatic pressure is nonrecoverable, the gap ratio is 0.00007. For the above three factors, the gap ratio is 0.00050.

Seismic Effect

During an earthquake, the cellular concrete will be deformed, and the gap between the steel pipe and cellular concrete may be increased by the possible nonrecoverable deformation in cellular concrete. The maximum radial strain of steel pipe and cellular concrete under seismic ground motions was derived from Metro Rail Transit Consultants (1984) as

$$\varepsilon_{max} = \frac{\Delta D}{D} = \pm \frac{4}{3}\frac{V_{max}}{C_{se}}$$

where: D = diameter of pipe
ΔD = deformation of pipe diameter

If 100 percent of maximum radial deformation of cellular concrete were included in the gap ratio, the total material cost of steel pipe would be increased about 1.28%. Therefore, 100% of the maximum radial deformation is conservatively considered nonrecoverable. The gap ratio due to seismic is 0.00055 for reaches in alluvium and 0.00011 for reaches in both San Timoteo Formation and metamorphic rock. The total gap ratios for the four factors are 0.00105 in alluvium reaches, and 0.00061 in San Timoteo Formation and metamorphic rock reaches.

Using the gap ratios in Jacobsen's method for WSP in alluvium reaches, a 366-cm (12-ft) diameter pipe of 12.7-mm (1/2-in.) plate can withstand an allowable external hydrostatic pressure of about 214 KPa (31 psi), which corresponds to an external head of about 21 m (70 ft) vs. a maximum anticipated external hydrostatic head in alluvium reaches of about 18 m (60 ft). Using a similar approach for San Timoteo Formation and metamorphic rock reaches, a 366-cm (12-ft) diameter pipe of 12.7-, 15.9-, 19-, 22.2-, or 25.4-mm (1/2-, 5/8-, 3/4-, 7/8-, or 1-in.) plate can withstand an allowable external hydrostatic pressure of about 255, 407, 579, 779, or 993 KPa (37, 59, 84, 113, or 144 psi), which correspond to external heads of about 26, 42, 59, 79, or 101 m (86, 137, 194, 260, or 332 ft), respectively.

For the internal and external hydrostatic pressure considered in this project, the design of WSP is indirectly affected by seismic excitations, which may decrease the buckling strength by increasing the gap between the WSP and the concrete.

PCCP TUNNEL LININGS

PCCP selected as a final lining would be carried into the tunnel and set in place with pipe-placing machines following tunnel excavation and the installation of initial support. The annulus between the pipe and the initial support system will be backfilled with low-density cellular concrete, as shown in Figure 4. PCCP is a precast concrete pipe consisting of a high-strength concrete core, an embedded welded steel cylinder, a high-tensile- strength prestressing steel wire wrapped around the concrete core, and a dense mortar coating that encases and protects the wire-wrapped prestressed core from physical damage and external corrosion. Reinforcement for shrinkage is included in the core concrete inside the steel cylinder. The embedded steel cylinder serves as a watertight membrane as well as a structural component of the composite section. The steel bell and spigot joint rings are welded to the steel cylinder ends. Together with the elastomeric gasket, they form a watertight and self-centering joint between sections of pipe.

The PCCP final lining design was evaluated using standard design criteria provided by ANSI/AWWA C304 (1992) for core and coating crack control, core compression control, wire and cylinder stress control, burst and maximum pressure control, for PCCP laid in trenches and subjected to the combined effects of internal pressures, external loads, and pipe and fluid weights. External loads were modified as applicable to the tunnel in lieu of the trench condition, and the load combinations were substantiated by adding seismic provisions for the underground structures.

Design evaluations of the pipe wall composite section indicate that a uniform concrete core thickness of 203 mm (8 in.), mortar coating of 32 mm (1.25 in.), steel cylinder of 1.5 mm (16 gage) for external water head below 70 m (230 feet) and 3.42 mm (10 gage) for over 70 m (230 ft), and corresponding prestressing wire of 4.8-mm (0.192-in.) diameter spaced at 12.4 mm (0.489 in.) and 11 mm (0.434 in.), respectively, are required to satisfy design criteria. A PCCP total wall thickness of 235 mm (9.25 in.) is considered to be satisfactory and is lower than the allowables for static and seismic conditions. The PCCP design in alluvium

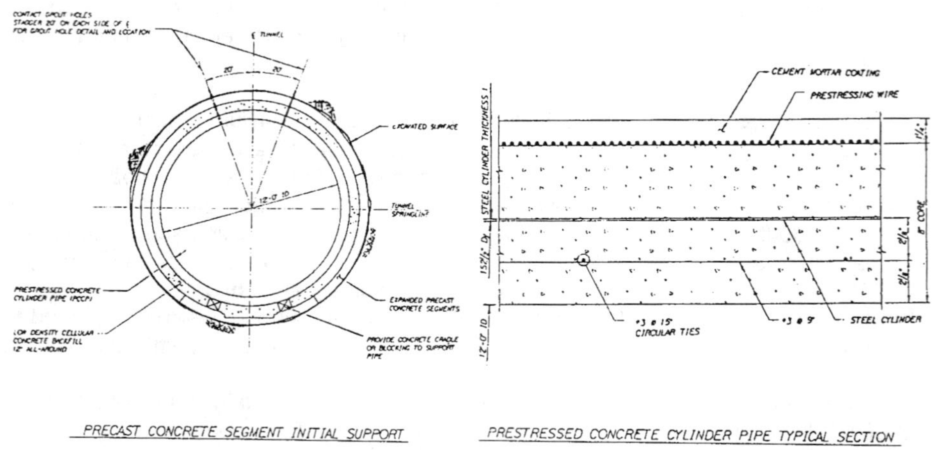

Figure 4. PCCP Lining Sections

reaches was controlled by seismic excitations, in which the tensile strain of mortar coating is critical. The thickness of core concrete is mainly decided by the buckling strength of inner-core concrete and steel cylinder subjected to external hydrostatic pressures.

CONCLUSIONS

The impervious final linings (WSP and PCCP) were designed by the procedures described above for seismic shaking, internal and external hydrostatic pressures, and ground and construction loads. In this approach, the effect of seismic shaking is included not only on the lining strains, but also on the buckling strength of WSP surrounded with low-density cellular concrete. The evaluations indicate that the gap between pipe and concrete may be increased about 100% by seismic excitations for WSP in alluvium reaches, and the gap is a controlling factor of buckling strength. The thicknesses of WSP are mainly decided by the buckling strength under external hydrostatic pressures.

For PCCP in alluvium reaches, the design of mortar coating is controlled by seismic effects. The external hydrostatic pressure is considered to act on the surface of steel cylinder when the underground water infiltrates the outer concrete core. This will increase the thickness of inner-core concrete due to compressive allowable and buckling strength. The approach used in this project will provide the required serviceability and structural integrity under seismic effects and other loads.

ACKNOWLEDGMENTS

The writers greatly acknowledge the contribution of Metropolitan Water District of Southern California: J. Narvaiz, J. Gallanes, and R. Lyons. Thanks also to the following consultants for their continued support and guidance during the final design phase: Dr. T. Brekke, Dr. G. Korbin, and Dr. J. Merritt. The authors also acknowledge assistance by DMJM/Woodward-Clyde workers.

REFERENCES

American Institute of Steel Construction (AISC), 1989, Manual of Steel Construction, Allowable StressDesign, Ninth Edition.

American Society of Civil Engineers (ASCE), 1993, Steel Penstocks, ASCE Manuals and Reports on Engineering Practice No. 79, ASCE, New York, NY.

American Water Works Association (AWWA), 1989, Steel Pipe—A Guide for Design and Installation,AWWA Manual M11, Third Edition.

American Water Works Association, 1992, AWWA Standard for Design of Prestressed Concrete Cylinder Pipe, ANSI/AWWA C304-92.

Daniel, Mann, Johnson, & Mendenhall and Woodward-Clyde Consultants, 1996, "Final Design Report for

Riverside Badlands Tunnel, Inland Feeder Project," June 14, prepared for Metropolitan Water District of Southern California.

Daniel, Mann, Johnson, & Mendenhall and Woodward-Clyde Consultants, 1996, "Geotechnical Design

Summary Report for Riverside Badlands Tunnel, Inland Feeder Project," June 14, prepared for Metropolitan Water District of Southern California.

Merritt, J.L., Monsees, J.E., Desai, D.B., 1990, "Shaking Design in Los Angeles: Earthquake Considerations in Design of Two Major Projects in LA Basin, METRO and NORS," International Symposium on Unique Underground Structures, Proceedings, Vol. 1, Denver, Colorado, pp. 18-1 to 18-27.

Merritt, J.L., Monsees, J.E., and Hendron, A.J., Jr., 1985, "Seismic Design of Underground Structures,"RETC Proceedings, Vol. 1, American Institute of Mining, Metallurgical, and Petroleum Engineers, New York, NY, pp. 104-131.

Metro Rail Transit Consultants, 1984, "Supplemental Criteria for Seismic Design of Underground Structures," Southern California Rapid Transit District, June.

Parmakian, J., 1982, "Minimum Thickness for Handling Steel Pipes," Water Power & Dam Construction, June, pp. 18-19.

2

Ground Support

Chairman
D.B. Desai
DMJM
Los Angeles, CA

Chapter 4

COMPARISON OF PERFORMANCE OF SOIL MIX WALL AND CONVENTIONAL BRACING SYSTEM FOR STATION EXCAVATIONS IN ALLUVIUM

Timothy P. Smirnoff,[1] Anthony Stirbys,[2] and Amanda Elioff[3]

[1]Senior Supervising Engineer, Parsons Brinckerhoff, Los Angeles, CA; [2]Senior Supervising Geologist, De Leuw Cather, Los Angeles, CA; [3]Senior Tunnel Engineer, Parsons Brinckerhoff, Los Angeles, CA

ABSTRACT

The construction of the continuing expansion of the Los Angeles Metro System has required the construction of eight new station structures. One contractor proposed, as a construction alternative for the support of excavation of the Hollywood/Vine Station, the use of soil-mix walls instead of the conventional soldier piles and wood lagging. Soldier piles and shotcrete or wood lagging support systems had been generally used on the other subway station excavations constructed to date throughout the system. Immediately east of that station and in almost exactly the same geology, the Hollywood/Western Station was being excavated at the same time using a conventional support system. This paper compares the performance of these two support systems in the alluvial soils of the Los Angeles Basin.

INTRODUCTION

The continuation of the Los Angeles Metro System Operating Segment 2, which will add over 8 miles of heavy rail line to the existing system, required the construction of eight new stations, [Figure 1]. Past experience has shown that station box excavations could be economically supported using conventional soldier pile and wood lagging and structural steel cross bracing, in conjunction with dewatering where appropriate. The location of the stations in the pubic right-of-way required decking of downtown and major arterial streets. The Hollywood/Vine and Hollywood/Western Stations are located about one quarter mile apart on world-famous Hollywood Boulevard. Given the proximity of the two stations, the site geology was similar. Constructed beneath and within the public right-of way, these two stations have large off-street entrance structures and multiple (12-16) appendages for emergency exits, air ventilation, and blast relief structures. The construction contract documents provide for both conventional excavation bracing and decking schemes and for contractor-proposed design and alternatives for excavation support.

The successful low bidder for the Hollywood/Western Station, the joint venture of Kiewit-Shea, selected the conventional (soldier piles and wood lagging) excavation support approach. This contractor had already successfully used this method on another station. The Kajima-Ray Wilson joint venture, low bidder for the Hollywood/Vine Station, proposed the use of deep soil-mix within the system. After review by the Owner, the Metropolitan Transit Authority (MTA) and its engineers, the latter proposal was accepted. This was the first

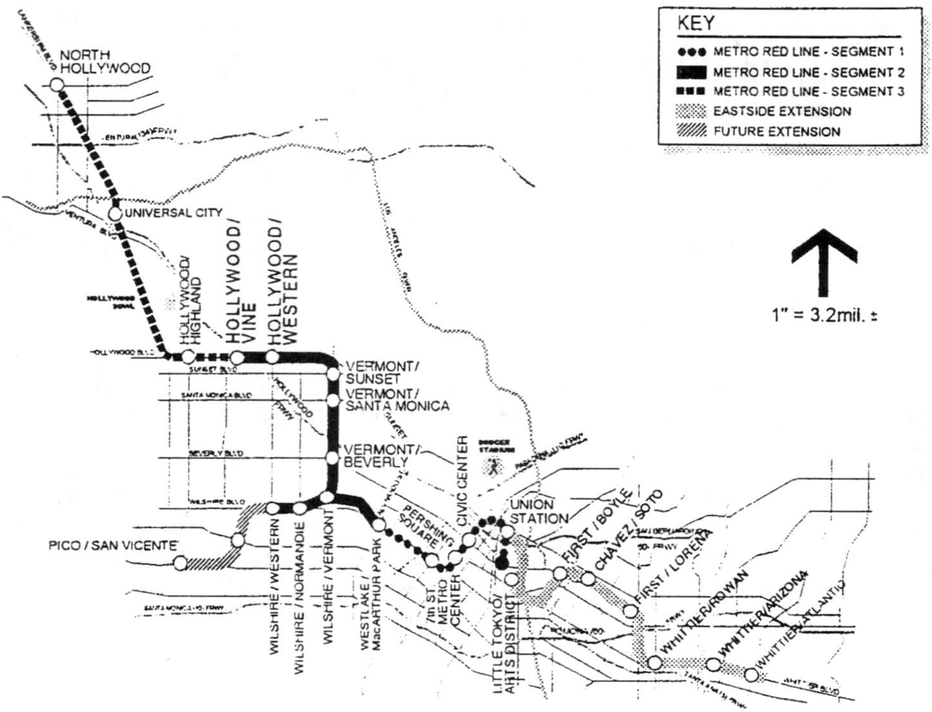

Figure 1. System Map, Los Angeles Metro Red Line

application of such a method for station construction in the Los Angeles area. It was presented as an alternative design; thus no cost or time advantage for its construction was offered as a Value Engineering Cost Proposal. It allowed, in the contractor's opinion, a less encumbered bulk excavation, thereby possibly saving time. This paper compares the performance of the two excavations, the internal bracing force, and the induced movements of the walls and adjacent structures and utilities.

STATION DESCRIPTIONS AND LOCATIONS

Hollywood/Western Station: The cut-and-cover construction required the excavation of a 60- to 70-foot-deep box. Including ancillary facilities at both ends, the station is about 600 feet long and 52 feet in interior width. Located underneath Hollywood Boulevard between the eastern curb of Serrano Avenue to about 78 feet east of Western Avenue, the station has an overall excavated dimension of approximately 62 feet in width. The station is located in a developed commercial and residential area. Along the station alignment, the ground surface slopes mildly from about Elevation 401 at the eastern end to about Elevation 392 at the western end. The bottom slab elevation is about Elevation 338, with a planned platform level at about Elevation 336. The station has an east-to-west gradient of about 0.3 percent.

Hollywood/Vine Station: The Hollywood/Vine Station and integral crossover structure are located beneath Hollywood Boulevard extending from approximately 150 feet east of El Centro Avenue to about 130 feet east of Vine Street. The combined excavation of station and crossover is approximately 890 feet long, with an inside width of approximately 52 feet. The ground surface elevation ranges from about Elevation 379 at the eastern end to about Elevation 375 at the western end of the excavations. The invert slab bottom elevation varies from Elevation 308 to 311, requiring a depth of excavation of from 64 to 71 feet.

The twin running tunnels for the subway line segments were driven through both station reaches prior to construction of the station excavations, and supported initially with precast concrete expanded segments. The tunnel supports were to be removed as part of the station construction.

REGIONAL GEOLOGY

The Metro Rail alignment is located within the Los Angeles Basin, which is composed of several tectonic or structural blocks. The Hollywood/Western and Hollywood/Vine Stations are located in the Central Block and near the boundary of the Northwestern Block. The Central Block is bounded on the north by the Santa-Monica-Hollywood-Raymond Hill fault zones. The Central Block of the Los Angeles Basin is the area adjacent to the Northwestern block, and is underlain by a deep structural depression filled with the following geologic units, in order of deposition:

1. Puente Formation: The Upper Miocene bedrock underlying the area consists pre dominantly of stratified and weakly interbedded claystone, siltstone and sandstone. The materials in the Puente Formation are generally low-strength (weak) rock, with a local presence of hard sandstone beds ranging from fractions of an inch to several feet or more in thickness. The upper portion of Puente is completely weathered and exhibits almost soil-like characteristics, with little or no cementation and without distinguishable bedding planes. The weathered zone is underlain by an oxidized and fresh Puente.

2. Old Alluvium: These Pleistocene sediments consist of granular Alluvium deposited in relatively "swift" water environments and fine-grained Alluvium deposited in relatively "quiet" water environments. The granular Old Alluvium consists of primarily medium-dense to very dense clean sand, silty sands and sandy gravel. The fine-grained Old Alluvium consists of medium-stiff to very hard clay, silty clay, sandy clay, silt and clayey silt.

3. Young Alluvium: These Holocene sediments consist of granular alluvium deposited in relatively "swift" water environments and fine grained Alluvium deposited in a "quiet" water environment. The granular Young Alluvium consists of loose to dense clean sand, silty sand, gravely sand, and sandy gravel, with a potential of local cobbles and boulders. The fine-grained Young Alluvium consists of firm to hard clay, silty clay, silt, clayey silt and a local presence of traces of gravel.

4. Other Structural Geology: The Hollywood Fault zone is located along the base of the Santa Monica Mountains as a scarp-like feature within Old and Young Alluvium.

The fault tends eastward along the south face of the mountain front approximately one quarter mile north and parallel to the Metro Rail alignment along Hollywood Boulevard.

SITE STRATIGRAPHY

Hollywood/Western Station: In general, the stratigraphy of the site area below the street surface consists of Young Alluvium overlying Old Alluvium. In the station area the Young Alluvium varies in thickness from about 23 to 40 feet, and consists of a medium-dense to dense silty sand, sandy silt, and clayey sand interspersed with fine-grained Young Alluvium consisting of medium-stiff to hard silty clay and clayey sand. The Old Alluvium underlying the Young Alluvium is extremely heterogeneous and non-uniform. The Old Alluvium consists predominantly of medium-dense silty sand and clayey sand with trace of gravel, interspersed with fine-grained Old Alluvium consisting predominately of very stiff to hard silty clay and sandy clay.

Hollywood/Vine: The stratigraphy at this station site closely parallels those at Hollywood/Western and consists of predominantly granular alluvium interspersed with layers and pockets of gravel and small amounts of fine-grained alluvium consisting of clayey sand, sandy clay and clayey silt. Within the station area the Young Alluvium is about 20 feet thick at the eastern end and increases to about 80 feet at the western end. Within the immediately adjacent pocket track area, the thickness of Young Alluvium varies between 55 and 80 feet.

The Puente Formation bedrock was not found to the depth of investigation at either station site. Piezometer monitoring of groundwater levels at each of the sites indicates that the water table elevation is below the cut excavations for both stations.

EXCAVATION SUPPORT AND DESIGN CRITERIA

Station construction at both sites was in close proximity to adjacent existing buildings, some considered “sensitive” due to their construction or historic status. Because of the depth of the station excavations, the foundations were within the zone of influence of the station excavations, requiring that the excavations be shored and braced.

The lateral pressure on the excavation support system is a function of the shoring system, construction procedure, and subsurface conditions. The contract specifications recognized the possible use of both flexible and rigid support systems and provided a lateral earth pressure diagram and specific design criteria for each support system. For both flexible and rigid wall support systems such as soldier pile and lagging walls with tiebacks or internal bracing, a design equivalent earth pressure of 30H was specified.

CONVENTIONAL SOLDIER PILE AND LAGGING EXCAVATION SUPPORT

For more flexible soldier pile and lagging support systems, the contractor had to comply with a number of design criteria and specification requirements that were similar to those used throughout this project and on many of the urban subway projects constructed in the past decade:

1. Soldier piles could not be driven but had to be placed in predrilled bored holes with

encasement of the soldier pile with 2500 psi concrete up to the lowest point of the excavation adjacent to the pile, with the remainder of the hole filled with lean concrete.

2. The support of excavation is required to be analyzed for all conditions that might occur during the various stages of construction, including installation, relocation and removal of bracing, soil removal and dewatering of excavation, and concreting of bottom slab.

3. The elements supporting vertical loads as well as lateral pressures are analyzed as members subjected to combined axial and bending loads.

4. The maximum slenderness ratio of struts shall not exceed 120 and the maximum axial stress to which struts may be subjected is limited to 12 ksi.

6. Vertical members of both flexible and rigid systems are designed as hinged at braced levels and at the bottom of the excavation.

7. The support systems were to be free of intermediate columns and, when wales are used, tight bearing is required with star packing for load transfer.

8. Preload internal bracing members including struts to 25 percent of the design compression load occurring during the excavation phase.

The conventional soldier pile and lagging braced excavation was supported by 24-inch-wide flange sections at 8- to 12-foot centers placed in predrilled holes and backfilled with lean concrete. Crosslot bracing using 24- to 36-inch diameter pipe struts was located at 8- to 16-foot centers. Wood decking was supported on deck beams placed just below the existing street grade and tightly wedged in place prior to bulk excavation.

SOIL-MIX SUPPORTED EXCAVATION

Deep soil mix wall construction uses large-diameter, hollow flight augers to penetrate soils and to allow mixing with cement and bentonite materials introduced during penetration and extraction. The result is a soil-cement which is designed to have consistent and defined engineering properties. At Hollywood/Vine, the soil-cement strength was specified to be 100 psi. The equipment includes a soil mixing machine as the drilling/mixing unit and a batch/mixing plant as the flow control and slurry preparation unit. The soil-mixing unit carries augers guided by a vertical steel lead on a track-mounted base. These units typically consist of three auger shafts and auger heads. A multi-axis gearbox distributes the drilling power to the auger shafts. Each auger is equipped with discontinuous auger flights and mixing paddles which provide the required degree of mixing and ensure homogeneity, but prevent vertical transport of soil to the surface. The mixing plant is generally located off-street and supplies the batch-mixed cement-bentonite grout and additives through an automated batching system. Positive displacement pumps supply the grout to each of the injecting augers. Photos 1, 2 and 3 show (respectively) the soil mix machine, augers, and in-place wall.

Photo 1. Deep Soil-mixing Machine

The soil-mix walls were accepted by the owner with several requirements:

1. The design of the struts and wales as well as design lateral pressure shall be in accordance with the contract-specified criteria. The walls were specified as rigid and required a design lateral pressure of 30H.
2. Steel structural sections would be embedded in each lance location or about one every 4 feet.

Photo 2. Soil-mixing Augers

Photo 3. Soil-mix Wall with Embedded Steel Pile Sections

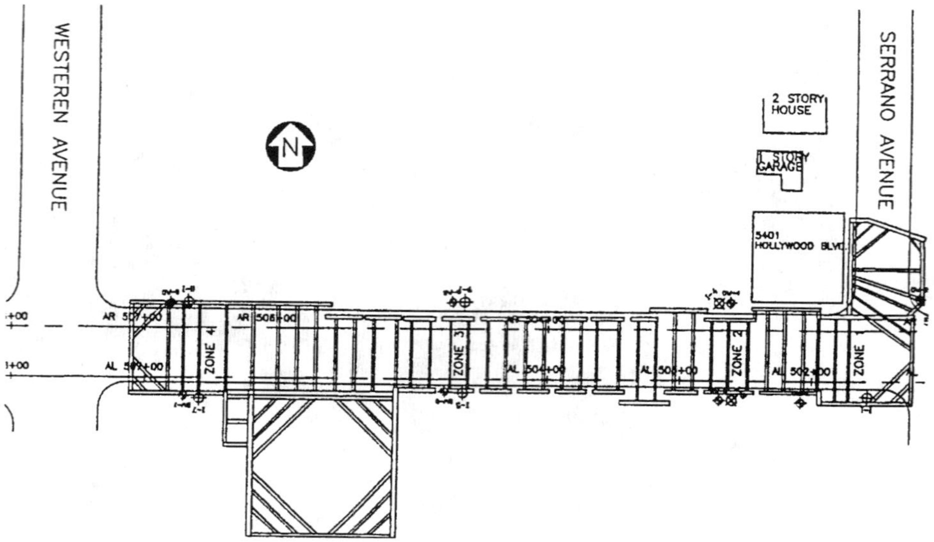

Figure 2. Instrumentation Plan Hollywood/Western Station

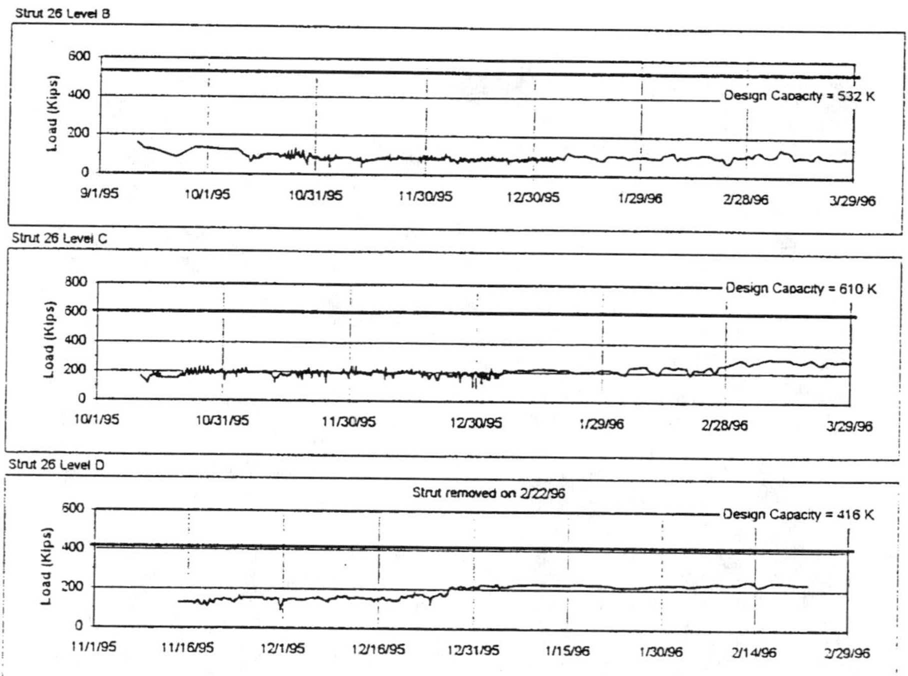

Figure 3. Strut Loads for Instrumentation Zone 3, Hollywood/Western Station

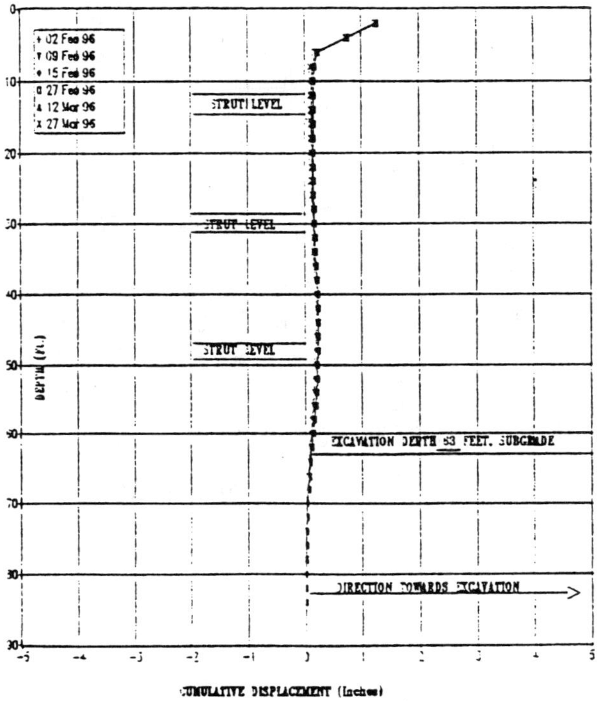

Figure 4. Inclinometer Measurements, Zone 3, Hollywood/Western Station

3. The concrete-soil mix should attain a minimum compressive strength of 100 psi and will not be considered in wall action other than as lagging.

EXCAVATION PERFORMANCE

Because of the urban setting and the close proximity of adjacent structures and utilities, excavation performance and the movements surrounding the station excavations were extensively monitored. Each station had at least three instrumented zones consisting of three sets of struts. Figure 1 shows the Instrumentation layout for the Hollywood/Western Station. Each strut load was monitored by strain gages installed on the struts and read on a periodic basis. Figure 3 shows a portion of the record of strut loads for a typical loading zone. In association with each instrumented zone, two inclinometers, one on each side of the excavation, were installed to monitor horizontal soil movements adjacent to the excavations, as shown in Figure 4. Horizontal wall movements were also taken at periodic intervals to ensure wall stability. Conventional survey points were installed on sidewalks and adjacent buildings to monitor movements. Figures 5 and 6 summarize the results of the instrumentation readings for two such instrumented zones for each of the two stations. The figures indicate equivalent average pressures calculated as the average of the common strut level readings recorded for each of the instrumented zones. The figures show the movements adjacent to each wall induced at the end of the excavation process, and the simplified soil profile.

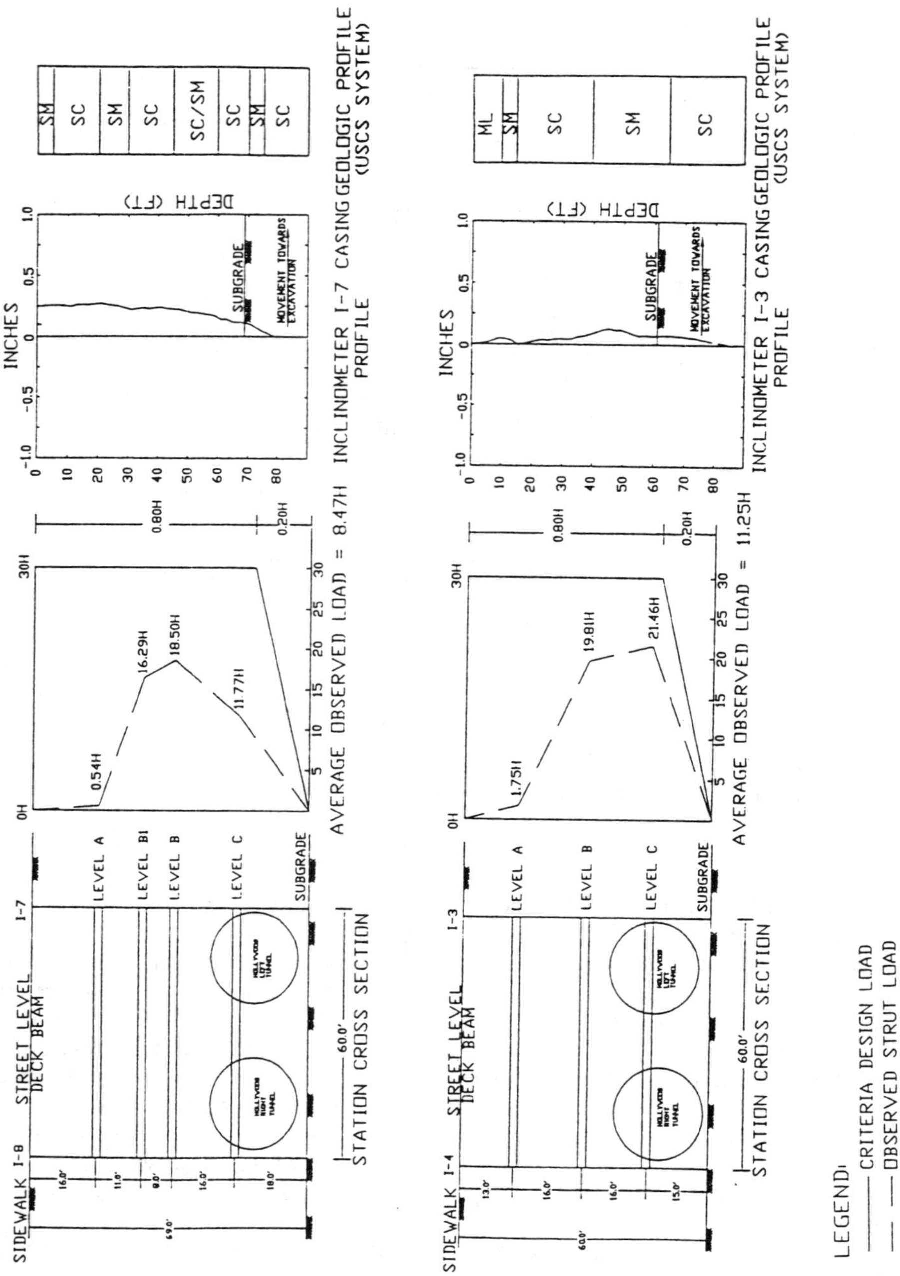

Figure 5. Summary of Observed Instrmentation Data, Instrumentation Zones 2 (above) and 3, Hollywood/Vine Station

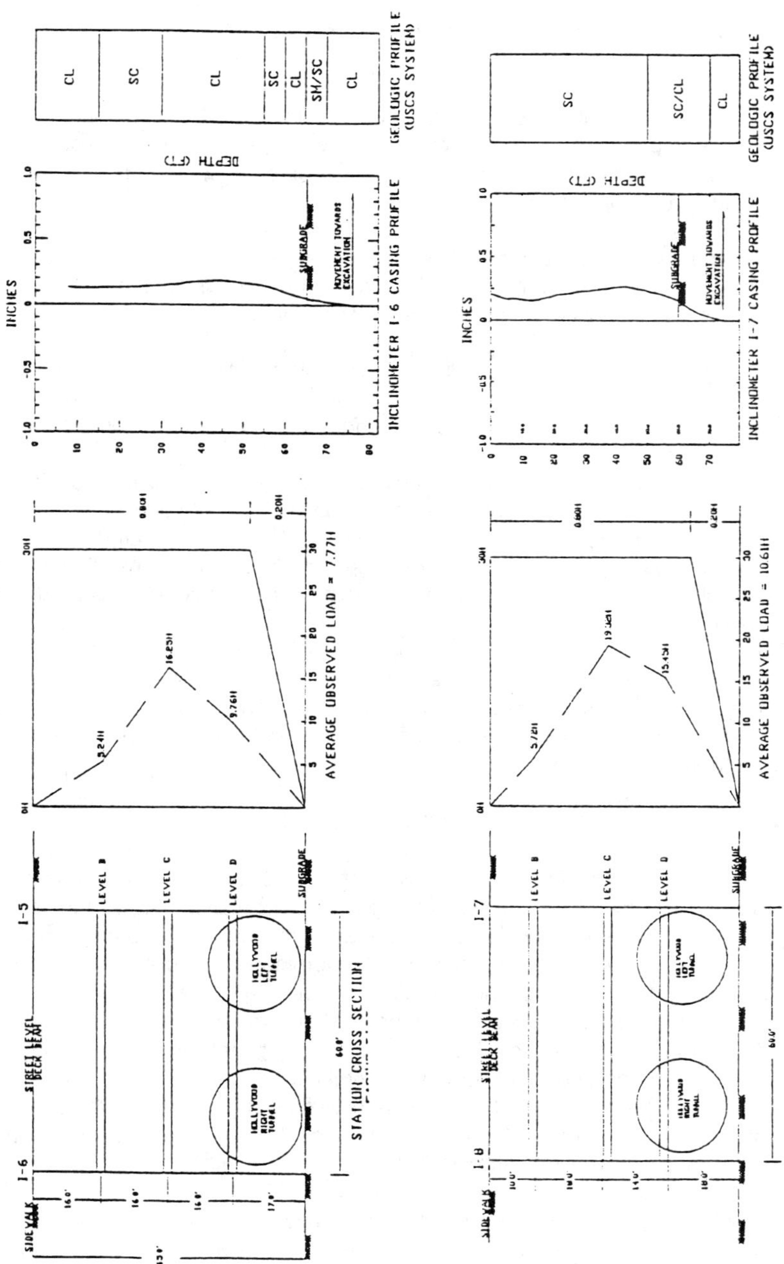

Figure 6. Summary of Observed Instrumentation Data, Instrumentation Zones 2 (above) and 3, Hollywood/Western Station

OBSERVATIONS

Review of the data and construction performance indicates the following observations:

1. The design lateral pressures were conservative in both cases. Such large values of design pressure served the owner in ensuring a lateral excavation support system with sufficient rigidity to minimize the effects on the surrounding building foundations and adjacent utilities.

2. Lateral soil pressures observed in both cases were considerably less, about half that normally used for the flexible conventional wall systems, and in no case was the presumed maximum design pressure exceeded.

3. Wall movements were generally small for both the flexible conventional and the rigid soil-mix wall support systems. Experience from station excavations on this project indicates generally consistent performance and small-magnitude movements.

4. Experience in the field indicated that once the soil-mix walls were installed, excavation could be carried out at a greater rate than by using a conventional wall system. The soil- mix wall provided a stable and clean face for excavation and allowed the use of larger equipment capable of more rapid excavation. Installation of walers and struts was not slowed by the lagging process and could be carried on directly behind the excavation.

5. Critical to the economy of the soil-mix wall system was the early, rapid, and coordinated relocation and support of utilities. Without coordination and utility relocation, wall excavation could not begin and was often delayed, necessitating a number of out-of-sequence moves.

CONCLUSIONS

The use of soil-mix walls to build a subway station box in this urban area was shown to be successful. The soil-mix wall allowed for rapid bulk excavation at rates never witnessed on the project with conventionally performed a station excavation. The major obstacle that slowed the contractor was the inability to properly coordinate and arrange in a timely fashion the location, support, and/or removal of utilities that were affected by the wall construction. Once the soil-mix walls were in place and the decking was installed, the excavation proceeded at a consistent pace without the customary waiting for bracing level installation and without losses of ground during downward excavation. Movements of walls and adjacent structures were generally less then that of the nearby conventionally excavated station. The stiffness of both the support systems resulted in generally small wall movements and correspondingly low apparent soil pressures.

It is unclear that this method of excavation will become popular as an alternative means of support of excavation, because of its reliance on very specialized equipment and because of a local preference for the conventional support technique. The contractor using this soil-mix wall has since bid additional subway station work in Los Angeles but has declined to use this alternative.

ACKNOWLEDGMENTS

The Authors wish to thank the Los Angeles Metropolitan Transportation Authority for granting permission to share this experience with the profession. To the personnel of the contractors' teams, the construction manager, Parsons-Dillingham - especially the Geotechnical Group that monitored the excavation and produced this useful and accurate data go our special thanks.

Chapter 5

HORIZONTAL CHEMICAL GROUTING BENEATH THE HOLLYWOOD FREEWAY, LOS ANGELES METRORAIL C-311

Gary E. Taylor,[1] Joseph K. Cavey,[1] and Michael F. Roach[2]
[1]Hayward Baker, Inc.; [2]Traylor Brothers, Inc.

ABSTRACT

A rigorous horizontal drilling and chemical grouting program was specified beneath the Hollywood Freeway (U.S. Route 101) section of Los Angeles Metrorail Contract C-311. Twin 6.67-m (22-ft.) diameter starter tunnels were to be conventionally mined through a mixed alluvial and bedrock face consisting of silty sand, gravel, and weathered sandstone and siltstone of the Topanga Group. The contact between the alluvium and bedrock dipped toward the Crossover Excavation where the horizontal drilling and grouting operations were performed. The intent of this grouting program was to stabilize the mixed face to minimize ground loss and surface settlement, and to reduce groundwater inflow into the tunnel headings. Given the highly variable mixed face conditions, horizontal drilling up to 86 m (284 ft.) resulted in upward trending trajectories which likely followed the bedrock contact.

Chemical grouting was performed in a primary, secondary and tertiary, three dimensional sequence. Evaluation of the primary and secondary grouting parameters indicated where tertiary grouting was required. Grouting pressures induced a small rock failure at the tunnel heading as well as increased stresses on the tiebacks of the excavation.

INTRODUCTION

The Los Angeles Metrorail Red Line is presently being constructed between North Hollywood and Hollywood Boulevard. The tunnel alignment, having an overall length of approximately 13 km (8 miles), will traverse southward on Lankershim Boulevard to the Universal City Station and continue below the Hollywood Hills, connecting to the Hollywood Boulevard tunnels at the La Brea shaft (B-251 Contract). Metrorail Contract C-311 includes the excavation of a 109 m (360 ft.) long Crossover Shaft, the excavation of the North Access Shaft, excavation of the "Starter Tunnels" between the two shafts, and the construction of TBM-driven tunnels below the Hollywood Hills. As shown in Figure 1, the Starter Section was mined below the Hollywood Freeway, one of the most traveled freeways in Los Angeles, connecting the San Fernando Valley to the Los Angeles basin.

Soil borings indicated that a mixed face condition would be encountered below the Hollywood Freeway, causing concern over settlement of the freeway above. Therefore, conventional tunnel construction methods, consisting of small backhoes (and similar equipment) excavating 1.2 m (4 ft.) lengths followed by shotcreting and steel rib installation, were used to minimize surface settlement. Further, chemical grouting was specified to increase

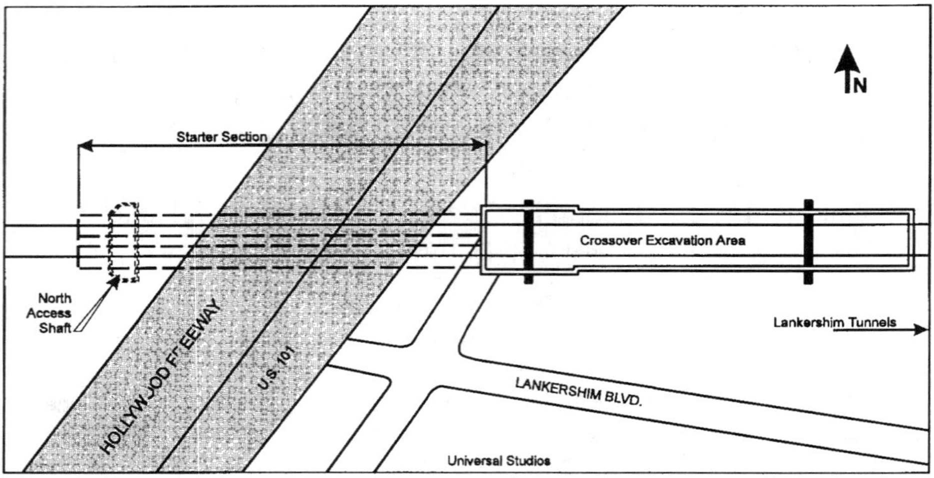

Figure 1. Starter Tunnel Alignment

the stand up time of the alluvium above the tunnel cut line and reduce the inflow of groundwater along the northerly dipping alluvium/bedrock contact.

After the completion of the twin, 91-m (300 ft.) long Starter Tunnels, two Tunnel Boring Machines, each measuring over 82 m (270 ft.) in length, would be pushed through the Starter Tunnels to the North Access Shaft to begin tunneling in the southward direction. Therefore, the timely excavation of the Starter Tunnels was critical to the overall project schedule.

SUBSURFACE CONDITIONS

The Starter Tunnel alignment traversed through and directly below dense silty sand, silt and gravel that overlied the Miocene Topanga Group. The highly weathered Topanga Group dips north/northwest such that the tunnel alignment was nearly perpendicular to the siltstone and sandstone bedding. During the excavation of the Crossover Shaft several large boulders were also encountered directly above the Contact. Although most of the Starter Tunnel excavation would be through the underlying Topanga, granular materials at the crown of the tunnel on the north side (closest to the Crossover shaft) caused consider-

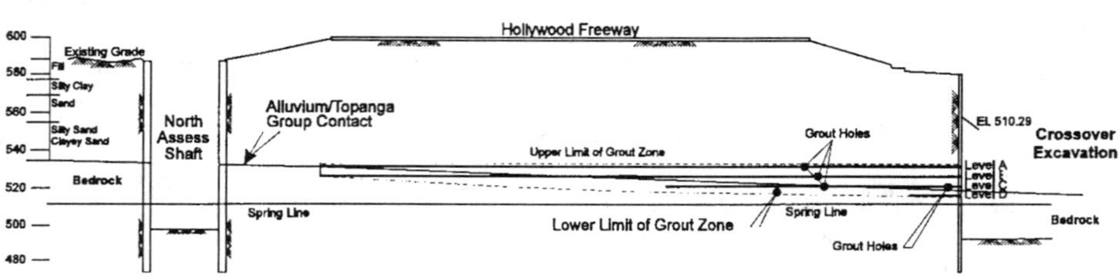

Figure 2. Profile of Subsurface Conditions

able concern. Groundwater, located approximately 6.1m (20 ft.) above the crown of the tunnel, was also expected to cause a problem as running conditions would lead to an unstable face and possible surface settlement of the Lankershim Underpass and adjacent Hollywood Freeway.

Chemical grouting was specified to increase the cohesion of the granular materials directly above the Topanga rock and to create an impermeable umbrella above the tunnels. Successfully implemented on previous Los Angeles Metrorail projects (Gularte, et al, 1991), sodium silicate-based chemical grout injected into the permeable soils above the crown of the tunnel was planned to create a continous hardened mass that would resemble sandstone. The grouted mass, having a specified unconfined compressive strength of 1.38 Mpa (200 psi), would be created by installing grout pipes horizontally above and slightly below the crown of the tunnel, and injecting specially formulated Geloc-4, an N grade sodium silicate chemical grout.

Grout Pipe Locations and Grout Volume Design

The grout zone above the tunnel was not constant due to the northerly dipping Topanga perpendicular to the tunnel alignment. As the profile in Figure 2 shows, the crown of the tunnel in the north 45.5 m (168 ft.) of the Starter Section resided in the alluvium, and would require substantially more soil stabilization than the south section of the alignment. Therefore, 4 levels of pipes were installed to lengths of 86.1 m (274 ft.), 42.4 m (140 ft.), and 12.1 m (40 ft.) to provide a treatment zone that varied in thickness from 3 to 6 m (10 to 20 ft.). The pipes were arranged on a 1.5-m (5 ft.) spacing to cover a 6.67-m (22 ft.) radius arch (centered at the springline) above the crown of the tunnel as shown in Figure 3a.

Based on the locations of the grout pipes, the volume of grout to be injected into each pipe was determined from the estimated porosity of the soil. Using a porosity of 0.30 throughout the grout zone, a unit volume of 307.6 liters of chemical grout per cubic meter (2.3 gallons per cubic ft.) would be necessary to permeate the void space of the soil in the grout zone. After drilling, each grout pipe was surveyed using a horizontal gyro to provide horizontal and vertical pipe deviations from the planned trajectory. Using the survey results, cross sections were generated along every 7.3 m (24 ft.) of tunnel alignment and grout "bulbs" or spheres were drawn to represent theoretical grout travel into the soil adjacent to the point of injection (Baker, 1982). At each section, circles were drawn such that approximately 0.3 m (1 ft.) of overlap was achieved. The overlapping circles represent interlocking grout bulbs designed to form a continuous grout mass that would prevent groundwater flow into the tunnels. Grout volumes in each pipe were then determined by projecting the circle for a distance 7.3 m (24 ft.) as a cylinder along the tunnel alignment, and calculating the required volume of grout to fill the void space of the cylinder of soil.

It should be noted that this grouting program utilized a combination of horizontal drilling techniques combined with horizontal gyro downhole survey methods to produce an "as-built" method of determining design parameters. Horizontal directional drilling and surveying differs in that the drill string is maneuvered into the intended trajectory (typically around curves in the tunnel alignments). Although this as-built method usually requires some redrilling (to cover areas in which the pipe trajectory diverged from the intended location) and the introduction of approximately ten percent more chemical grout than the idealized pattern in order to compensate for the unpredictable trajectories, the cost and production time were

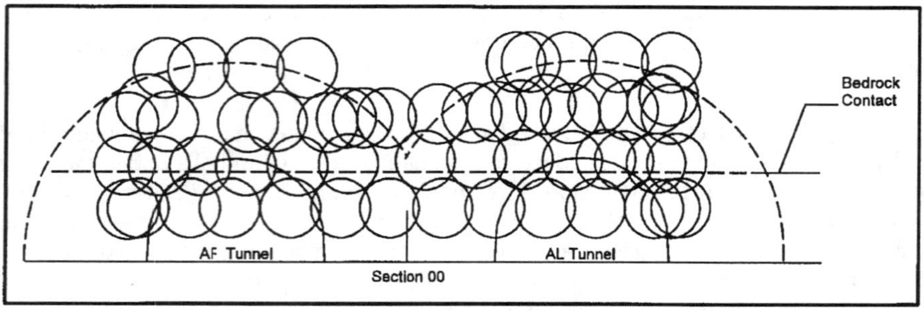

Figure 3a. Grout Pipe Section at the North Wall of Crossover Excavation

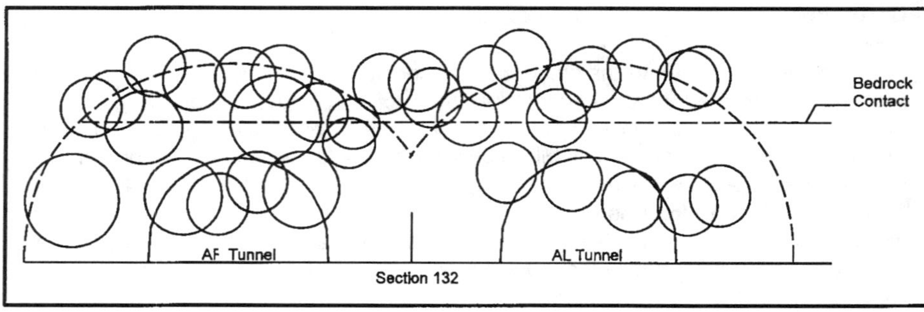

Figure 3b. Grout Pipe Section at 132 ft. from the North Wall of Crossover Excavation

substantially less when compared to utilizing horizontal directional drilling techniques for this application.

A primary and secondary grouting sequence was planned to drive water uniformly away from the tunnels and to ensure a continuous grouted mass. First, grouting was performed in the pillar and progressed outward on both the Alignment Right (AR) and Alignment Left (AL) sides. Therefore, water would be driven down and away from the tunnel crown. Along each pipe a primary/secondary sequence was followed to provide continuous grout interlock. Grout was injected every 1.8 m (6 ft.), or at every fourth grout port with a spacing of 0.45 m (1.5 ft.) on the primary pass of grouting in each pipe. On the secondary pass of grouting, injections were performed in the intermediate locations, and again on a 1.8 m (6 ft.) spacing. The overall grout volume was divided such that 60% was injected into primary locations and 40% was injected into secondary port locations.

Grout Pipe Installation

All drilling and grouting activities were performed from the crossover shaft located north of the freeway. Sleeve port grout pipes (SPGP's), having a 50.8 mm (2-inch) diameter, were installed using rotary duplex drilling methods with water as an internal flush. The 127 mm (5-inch) OD casing was drilled up to lengths of 86 m (284 ft.), and the inner string was withdrawn. Next, PVC SPGP was pushed into the cased hole and capped. A weak cement-bentonite grout was then pumped through the drill head and into the annulus area between the SPGP and the 127-mm (5-inch) casing. The casing was extracted, leaving the grout pipe intimately sealed to the adjacent ground.

After several pipes were installed, gyroscopic surveys were conducted to verify the locations of the grout pipes. The surveys showed that all of the Level A bores had traveled upward, even though some were purposely angled downward to compensate for the upward trending trajectories. In fact, many of the pipes that were angled downward appeared to rise even further above those that aligned level. This rise was attributed to the drill string being deflected off the rock layer and subsequently bowing upward. Additional alignment problems were recognized when the drill string encountered large boulders that either caused severe deviations or precluded advancement. After redrilling several locations, it was recognized that it would be impossible to install pipes such that a perfect array was maintained with less than 1.8 m (6 ft.) of spacing between each pipe. Grout volumes were increased to provide a larger radius of treatment in areas where pipe spacing exceeded 1.8 m (6 ft.). As shown in Figure 3b, design volumes were planned to provide a nearly continuous curtain above the tunnel. A total design volume of 2,740,000 liters (723,910 gallons) was anticipated to be necessary based on cross sections plotted every 7.3 m (24 ft.)

Chemical Grouting

The total actual volume of chemical grout pumped into the grout zone was 2,650,260 liters (700,200 gallons). As discussed above, a primary/secondary grouting sequence was implemented at the start of grouting. Inflatable 50.8-mm (2-inch) packers were pushed to the back of each grout pipe, and injections began at the back port in each pipe. After the primary design volume was injected in each port, the packer was extracted 1.8 m (6 ft.) to the next grout port, and the next design volume was injected. This process was repeated until the packer was withdrawn to excavation wall. The packer was then re-inserted into the grout pipe, and secondary volumes were injected in intersticial locations along the pipe.

After the primary and secondary grouting sequence was completed in several A and B row pipes, grout flows and pressures were compared to evaluate the effectiveness of the chemical grouting program. Primary and secondary grout flows and pressures were compared between 51 and 86 m (168 and 284 ft.) from the Crossover Excavation Wall. The amount of void filling or ground "tightening" was assessed by comparing the primary grout flow rate to the secondary flow rate at similar pressures. Because grouting pressures varied from 4 to 10 bar (60 to 150 psi) in each port, Flow vs. Pressure graphs (FP graphs) were developed to compare the flow and pressure relationships in between stages. A sample Flow vs. Pressure graph (Figure 4) shows that secondary points should be below and to the right of the primary "curve." The curve or range provides a means of evaluating secondary flow rates that were pumped at different pressures. It should be noted that a true comparison of grouting flow rates and pressures must be achieved by FP graphs or comparing pressures at nearly the same flow rates. Because headloss is proportional to the square of the grout flow rate, a linear relationship cannot be used to predict flow rates at varying pressures.

Although the FP graphs were very helpful in determining grouting effectiveness, it was more practical to list primary and secondary flow rates and pressures in a table, and check

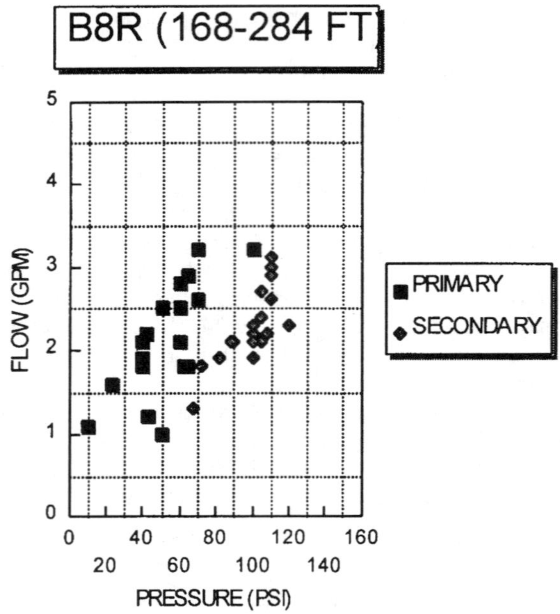

Figure 4. Typical FP Graph

for anomalies. In areas where high flows and/or low pressures were realized in a secondary pass, additional grout was injected in the particular port, and the two adjacent grout ports. Further, repeated refusals, indicating that no grout flow is possible without high pressurescapable of fracturing the ground, can be verified to prevent large areas of soil to be left ungrouted.

Based on the flow and pressure comparisons in the A and B pipes from 51 to 86 m (168 to 274 ft.), it was decided that a primary/secondary/tertiary grouting sequence would be implemented from 0 to 51 m (0 to 168 ft.). This alteration in the grouting program was made to ensure that silty sand and gravel was completely permeated throughout the the thicker grout zone on the north section of the alignment. It was critical that cohesionless material, expected to be near the springline and above, in the first half of the Starter Tunnel alignment be stabilized. This section of the alignment was directly below the Freeway. The tertiary grouting sequence was added by dividing the secondary grouting volume, 40% of the total design volume, in half.

At the completion of grouting, primary, secondary and tertiary flows and pressures were compared. Considerable increase in flow restriction was recognized in between primary and secondary stages, and those ports that showed little or no flow restriction between primary and secondary stages, frequently showed reduced flow in tertiary grouting. Further, several grouting refusals were realized at the completion of grouting. The refusals, defined by pressures exceeding 0.46 bar per meter of overburden (2.0 psi per ft. of overburden) with flow less than 3.79 liters per minute (1 GPM), are indications that the grout zone around a particular pipe is saturated, and grout injection is only possible by fracturing the ground. This phenomenon is suspected to have caused sign)ficant ground movement.

Ground Saturation and Fracturing

As the specified chemical grouting progressed, ground movement was observed by the following: a rock blowout at the heading of the AR tunnel on the North Access Shafl, inueased load cell readings on the South wall of the Crossover Shaft, and considerable ground movement measured in inclinometers at the South wall of the Crossover Shaft.

During an evening shift, leaks were observed while pumping a secondary pipe location near the wall of the North Access Shaft. The grout pipe in which injections were expected to be causing leakage was fumed off, and the leakage stopped. Pumping was restarted in the port, which was approximately 4.8 m (16 ft.) from the wall of the North Access Shaft, aRer the grout was allowed to set for 2 hours. One half hour aRer grouting was restarted, a grout "blow out~ was observed in the heading of the North Access Shaft. The heading had not been advanced more than 1 m (3.3 h.) inside the excavation wall, and had been supported with approximately 102 mm (4 inches) of shotuete. The blow out area was approximately 1.5 m (5 ft.) in diameter, and the rock and soil showed evidence of chemical grout lenses that had fommed in the soil and in between rock fractures.

Review of grouting pressures showed that the adjacent primary port was pumped at a flow rate of 9.5 LPM (2.5 GPM) at 6.9 bar (100 psi). Prior to the blow out, the secondary port was pumped at a flow rate of 3.79 LPM (1 GPM) at the same pressure. Clearly primary grouting permeated the rock/soil interface, and secondary grout flow was restricted by the less permeable soil and rock. The sphere of grout treatment had grown so large, that the path of least resistance was longitudinal (along the pipe) downward into the rock and into the tunnel heading. This experience indicates that grout pressures should be considered when grouting near an open tunnel face. Further, the blowout makes a strong case for grouting in stages: the grout bulb grew and grew until its path of least resistance caused ground fracture. By grouting in a primary/secondary sequence, grout travel was restricted to the most uitical areas, adjacent to the pipe.

When grouting was nearly 60% complete, considerable load increases were recognked on load cells on the south wall of the Crossover Station Shaft. These load increases corresponded to increased lateral wall movements as well. The load cells, installed and locked off for the anchor-braced excavation, read between 667 kN and 823 kN (150 and 185 kips) prior to grouting. Figure 5 shows the load cell locations and inclinometer readings for instrumentation placed on the south wall of the Crossover Excavation. Load Cells at depths 12.7 m (42 ft.), 17.6 m (58 ft.) and 21.2 m (70 ft.) showed 20-25% increases in loading from the beginning to end of grouting. Inciinometers placed inside the grout zone showed 50.8 to 63.5 mm (2 to 2.5 inches) of movement during grouting. Further, movement was recognized by the squeezing of concrete in between channel piles in the area in which the pipes were installed.

STARTER TUNNEL EXCAVATION

Since the Starter Tunnel excavation was on the project's critical path, the contraactor elected to begin this work prior to completion of the grouting program. Once the active grouting zone of the top two rows (A and B) had retreated north 12.1 m (40 ft.), excavation

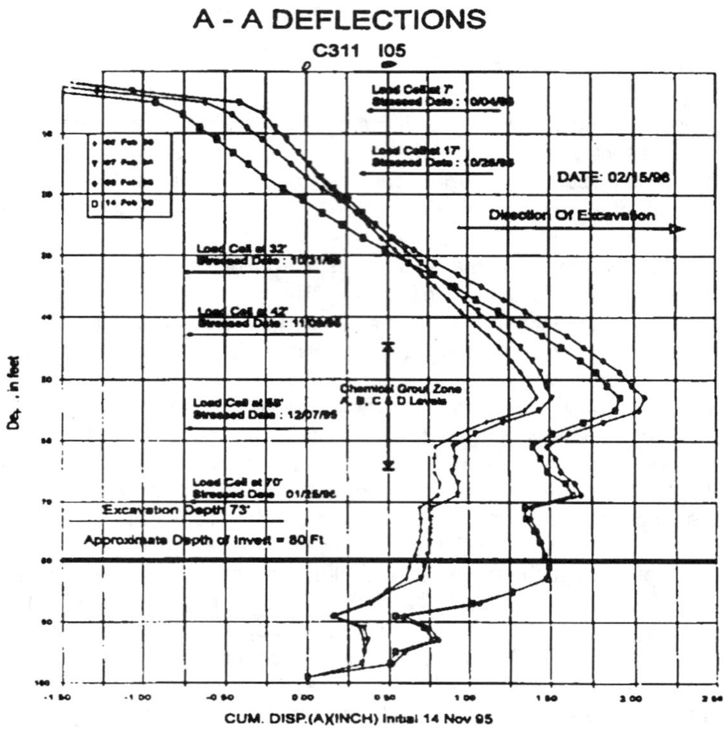

Figure 5. Load Cell Locations and Inclinometer Readings (Courtesy of Jacobs Engineering Group, Mott MacDonald Hatch and ACG Environments, A.J.V.)

began from the North Access Shaft. These tunnels, at 6.67 m (22 ft.) in diameter x 90.9 m (300 ft.) long, were successfully excavated using the Top Heading and Bench method. Initial support in the top heading consisted of a W8X35 arch rib and 101.6 mm (4 inches) of fiber-reinforced shotcrete. Bench support was provided by W8X40 posts and a similar application of shotcrete. As a precaution in the top heading, #9 x 12 ft. Spiling were driven ahead of each 1.2 m (4 ft.) advance.

This initial 6.1 m (20 ft.) was excavated using a hydraulic excavator, with muck removal by a side dump track loader. Since this reach was outside of the grouted zone, additional spiling was installed to prevent ground loss. Once the heading was advanced into the grouted region, a roadheader was required to excavate the stabilized soil and Upper Topanga shales. Following completion of the grouting program, the crews were moved to the larger and more accessible Crossover Shaft to complete the excavation.

The design of the initial support system, coupled with a major transportation artery overhead, dictated that the top heading be excavated in 1.2 m (4 ft.) advances. The bench was subsequently excavated in 2.4 m (8 ft.) advances, resulting in an overall production rate of greater than 1.1 m (3.5 ft.) per day.

CONCLUSION

Chemical grouting provided a dry and stable soil in which very little soil loss occurred during conventional mining. By accurately locating pipes with a gyroscopic survey tool, a precise grouting program was designed and implemented. The primary/secondary/tertiary grouting sequence implemented below the Hollywood Freeway assured a continuous grouted mass. Pressure and flow comparisons between stages of grouting in a pipe were essential to verifing the amount of void filling during grouting. As the soil reached saturation, chemical grout injection caused substantial wall movement, and experience has shown that grout pressure must be monitored when grouting near an unbraced tunnel heading.

REFERENCES

Baker, W.H., (1982), “Planning and Performing Structural Chemical Grouting,” Proceedings of Grouting in Geotechnical Engineering, New Orleans, LA, pp. 515-539.

Gularte, F.B., et al., (1991), “Tunneling Performance of Chemically Grouted Alluvium and Fill: Los Angeles MetroRail Contract A-130,” Proceedings of Rapid Excavation and Tunneling Conference, Seatle, WA, pp. 291-307.

Chapter 6

A REVIEW OF THE GROUTING ON LINE 2 OF THE CAIRO METRO

David W. Campo,[1] Donald P. Richards,[1] and Michel Coudry[2]
[1]Parsons Brinckerhoff International, Cairo; [2]Soletanche Bachy Egypt

ABSTRACT

This paper presents a review of the grouting works of Line 2 of the Greater Cairo Metro System. The Nile River alluvial deposits underlying this ancient city provided for a challenging and sometimes difficult grouting project. Permeation grouting using state-of-the-art computerized grouting techniques, as well as extensive ground treatment by jet grout, was performed for soil consolidation and/or groundwater control. During the project, over 425,000 m^3 of ground was treated, requiring 1.4 million linear meters of drilling and PVC injection pipe, 54,000 m^3 of bentonite cement and 230,000 m^3 of chemical and mineral grout. In addition, over 70,000 m^3 of ground was treated using the jet grout method. Relative to major international transportation projects, the grouting works for Line 2 of the Cairo Metro represent one of the largest worldwide grouting endeavors to date.

INTRODUCTION

Within the shadows of the ancient pyramids of the Giza Plateau is the rapidly expanding city of Cairo, the largest city in Africa and the Middle East, with a population of over 18 million. Cairo Metro's master plan was developed in the early 1970's in order to fulfill the government's commitment to reduce the city's traffic congestion and associated environmental pollution, which have been steadily rising due to a rapidly growing population. The plan consists of a regional line and two urban lines (Fig. 1). The Regional Line 1, completed in 1989, is the first metro in Africa. Constructed using cut-and-cover techniques and extensive grouting for groundwater control, this line is 42.5 km long, 4.5 km of which runs underground through the center city area. The two urban lines, Line 2 and Line 3, will provide service to the densely populated districts of Shubra, Giza and Embaba.

Line 2, consists of three Phases. Phase 1, begins at the Shubra El-Kheima Workshop/ Maintenance Facility and extends 11 km to existing Sadat Station, at the heart of the city center. Interchanges with the Regional Line 1 are provided at Mubarak and Sadat Stations. Phase 2A (5 km) begins at Sadat Station, crosses under the Nile River, and terminates at the new Cairo University Station. Phase 2B, will extend the line an additional 3 km to the south terminus at Giza Suburban Station.

Line 2 includes 6 km of above ground and viaduct section, along with 13 km of underground construction, for a total length of 19 km. The underground section of the project consists of the following:

consists of the following:

- Cut-and-cover tunnel; 1.8 km
- Bored tunnel; 9.3 km (9.4m excavated diameter)
- New Cut-and-cover stations; 10
- Enlargement of existing interchange stations; 2
- Tunnel annex connections; 12

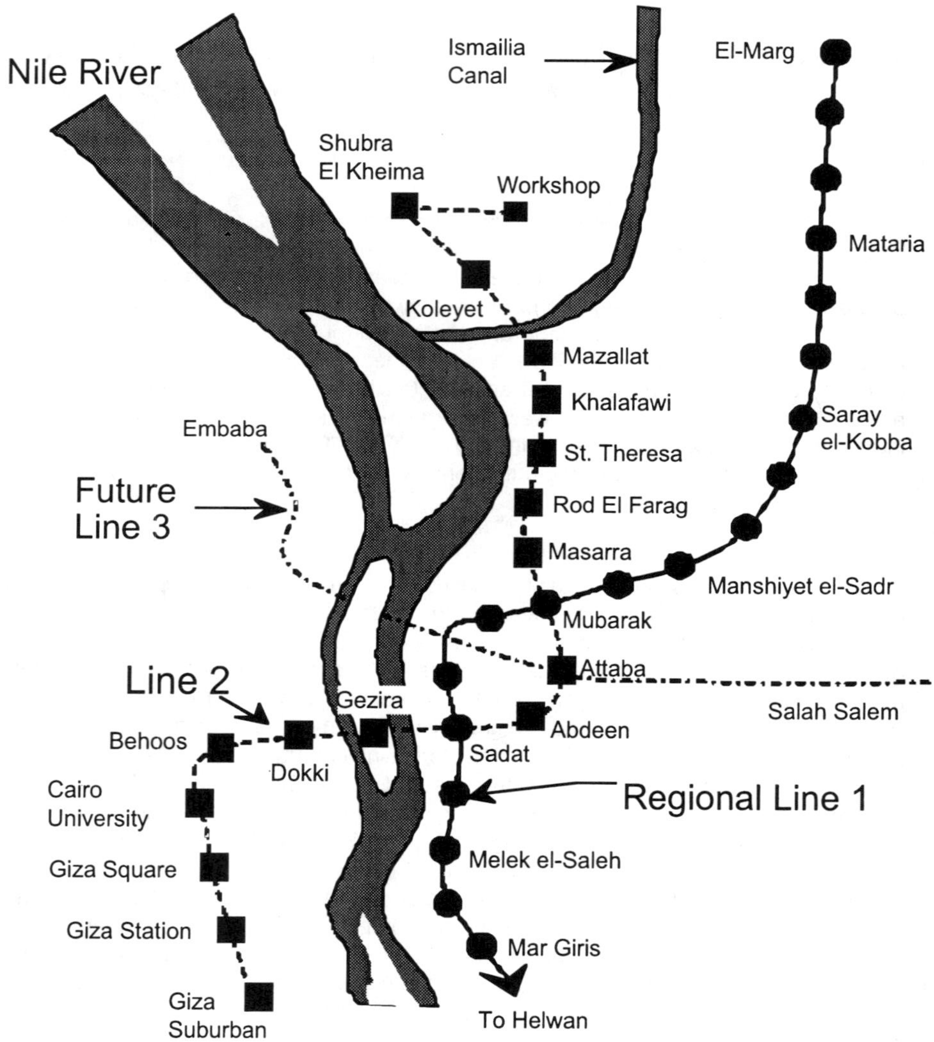

Figure 1. Map of Greater Cairo Metro System

The planning, design and construction of the system are the responsibility of the National Authority for Tunnels (NAT), a division of the Ministry of Transport. For Line 2, NAT is being assisted by the Greater Cairo Metro Consultants (GCMC), a joint-venture of Parsons Brinckerhoff International, Electrowatt Engineering Services and Sabbour Associates. INTERINFRA, a consortium of international contractors, was awarded the Phase 1 work in February 1993. Subsequently, the Phase 2A contract was awarded in October 1995, also to INTERINFRA. Included in the consortium were two French grouting specialty contractors, Soletanche and Bachy, whose joint venture, Soletanche Bachy Egypt (SBE) were responsible for the design, implementation, quality control and testing for all of the grouting works for the project.

In this ancient city of rich traditions, state-of-the-art grouting methods and materials were used in the construction of Line 2 of the Greater Cairo Metro System. Grouting was required for ground consolidation and groundwater control for construction of underground stations, cut-and-cover tunnels, galleries for special connections to the bored tunnel, starting and ending points for the tunnel boring machine, and for the protection of major utilities and adjacent structures.

Grouting for Line 2 was carried out over a 3.5 year period beginning in the Fall of 1993. During this time, approximately 425,000 m^3 of ground was treated, requiring 1.4 million linear meters of drilling and PVC injection pipe, 54,000 m^3 of bentonite/cement grout, and 230,000 m^3 of chemical and mineral grout. In addition, over 70,000 m^3 of ground was treated using the jet grout method. The grouting works for all of Line 2 represent one of the largest worldwide grouting endeavors to date.

PROJECT GROUND CONDITIONS

The city of Cairo lies along the Nile River at the apex of the Nile Delta. Until the completion of the Aswan High Dam in late 1960's, seasonal flooding of the river created thick alluvial flood plain deposits along the Nile valley. In the Cairo area, the thickness of these deposits is approximately 60 to 90m. Both the meandering Nile River and man's occupation of the area for the last 5000 years have had a great effect on the variations in these soil deposits. Within the limits of the project, the soil profile (from the surface downward) consists of:

- Man-made Fill
- Silty Clay and Clayey Silt
- Interbedded Clay, Silt and Sand (transition soils)
- Alluvial Sands (coarse to fine)

The surficial fill layer is relatively thin, generally 2m to 4m along the Metro alignment. In the city center between Mubarak and Sadat Stations, the fill is thicker, ranging between 4m and 10m in thickness. A natural deposit of stiff, overconsolidated, silty clay to clayey silt underlies the fill. The clay deposit includes occasional sand and silt partings and ranges in thickness from 4m to 10m. The undrained shear strength of the clay varied from 75 to 100 kPa.

Beneath the clay is a thick alluvial sand deposit that extends down to bedrock, beneath the limits of the metro construction. The upper few meters of this stratum are part of a transition layer, a highly interbedded stratum of clay, silt and fine sand.

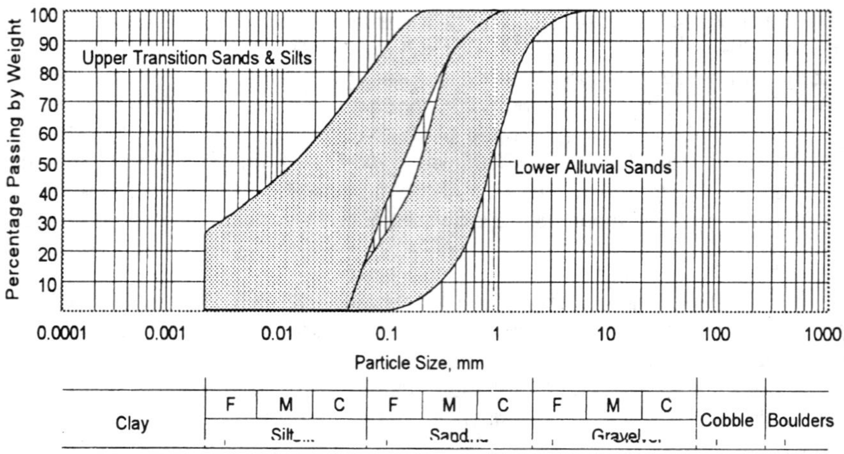

Figure 2. Grain Size Distribution

Below the transition layer the alluvial sand becomes a more uniform coarse to fine sand. It contains frequent lenses of silts and clays, layers of course sands, and increasing amounts of gravel are found with depth. Layers of cobbles and boulders are occasionally encountered at depths greater than 30m. Most of the grouting for the project was within this lower sand alluvium. The limits of grain size distribution for the upper transitional sands and silts, and the lower alluvial sand layers, are shown in Figure 2.

Other soil parameters for the sands within the grouting zones are indicated below:

- Soil Porosity, n; 0.35 to 0.45
- Permeability, k; 10^{-3} m/s to 10^{-5} m/s
- D_{10} Hazen Diameter; Average 0.1mm

Two groundwater regimes exist along the Metro alignment; a perched aquifer above the clay layer charged by leaking water supply pipes, leaking sewers, etc., and a lower aquifer confined within the sand deposit beneath the clay layer. The perched groundwater is shallow, generally 1 to 2m below ground surface. Piezometric levels in the confined aquifer are about 3 or 4m from ground surface and generally respond to fluctuations in the Nile water level.

CONTRACT REQUIREMENTS

The section of the project specifications concerning the grouting works was developed by NAT based on a significant amount of grouting experience gained during the design and construction of the Regional Line 1 (A. Salam, M.E., 1984). Accordingly, the contract documents specified the grouting materials to be used for the various grouting applications on the project. The contract documents also specified the general aspects of the method of injection, requirements for the grouting plant, and control and recording of the grouting parameters. In addition, performance requirements relative to the final permeability and consolidated strength of the treated ground were specified.

Provisions for the use of bentonite/cement, and chemical based sodium silica soft gel (watertightness) and hard gel (consolidation) were included in the tender documents. However, during tender negotiations with NAT, INTERINFRA made known a problem that occurs when the TBM bentonite slurry encounters ground treated with silica gels. A chemical reaction between the slurry and the silicate causes "foaming" of the slurry, resulting in a loss of control over the TBM slurry density; jeopardizing the stability of the tunnel face. Flint, G.R., et. al. (1993) reported this same problem during slurry shield tunnel construction for the Greater Cairo Wastewater Project, a multinational sewerage project that has been on-going in Cairo since the late 1970's. In addition, ground treatment was anticipated in the upper clay and fine transition soils. Effective permeation of these soils is normally not possible using chemical gels. Accordingly, NAT accepted the addition of jet grout and a newly developed mineral grout to the contract documents, thereby providing alternatives to the silica gels where required.

The specifications required that all design and construction documents be submitted to the owner for review and approval. Before using any grout material and/or ground treatment method for the project, a Grouting Method Statement was required. These statements provided details of all aspects of the grouting, including work procedures, drilling and injection methods, grout mix proportions, equipment, methods of testing, and quality control procedures. Design Calculation Notes were also submitted providing detailed design calculations for the specific grouting application. Design drawings were then submitted showing the positions of the grouted zones, grouthole spacing and depth, type of grout required, and details of testing for the treatment efficiency. In addition, a Site Specific Grouting Note was provided which indicated specific grouting parameters for the site, such as the sequence of the grouting, maximum injection volumes and pressures, flowrates, and any special grouting consideration for that particular site.

Once the work began, any modifications to procedures also required approval. Technical justification was required for all proposed modifications. At completion of grouting at any site, a comprehensive grouting report was submitted to the owner for review, to confirm compliance with approved method statements and site notes, and for a technical evaluation of the grouting.

GROUND TREATMENT METHODS

Permeation and jet grouting were the two methods of ground treatment used on the project. Because of the large volumes of grout to be injected over the relatively short construction schedule, high production drill rigs were developed specifically for the project and state-of-the-art computerized grouting systems were used to control the injection process.

Permeation Grouting

Permeation grouting was the preferred injection method for the coarse to fine alluvial sands within the deeper grouted zones. This was accomplished using PVC injection pipes, commonly called "Tube-a-Manchettes" or TAMs, placed in pre-drilled groutholes. The portion of the injection pipe within the grouted zones was equipped with grout "sleeves" at 50cm intervals. The injection pipe grid spacing varied between 1.0m and 1.4m, depending on the depth of drilling, soil permeability and the specific grout type used. By carefully regulating the injection volumes and pressures through each sleeve, "claquage," or fracturing, of the soil was controlled, thereby maintaining uniform permeation and retaining the grout within the intended treatment zone. Welsh, J.P. (1991) provides a detailed description of permeation grouting using TAM's.

The TAMs proved to be the most appropriate injection method for the stratified alluvial deposits. They allowed for greater control over the placement of the grout material, i.e., variable injection quantities, pressures and flowrates in individual sleeves as required by variable soil conditions, multiple phases of injection with different materials, and re-injection when the initial grouting treatment was incomplete or unsuccessful.

Injection Pipes. The inside diameter of the PVC injection pipe was a constant 40mm for it's entire length, however, the pipe wall thickness varied from 4mm to 7.5mm. The heavier walled pipe was used from the tip of the pipe to 5m above the grouted zone, where increased strength was required to limit pipe deformation caused by the high injection pressures within the grouted zones. Above this level, the thinner walled injection pipe extended upward to the ground surface.

At the start of the project, only one type of TAM was envision for use for all grouting applications. However, these PVC injection pipes were ductile and not easily broken and while driving the TBM through the first treated ground perforated with these injection pipes (i.e., the TBM break-in's at Khalafawi and Rod El Farag Stations) long twisted pieces of PVC clogged the cutterhead and the TBM slurry pipe system, causing some minor delays to the tunnelling.

As a result of this problem, special injection pipes made from a more brittle ABS material were used for all subsequent grouting applications where the TBM was required to pass. These pipes had sleeves spaced at 33cm and were assembled in 36cm long pieces using a screw-type connection, making them more conducive to breaking into small pieces during the TBM passing. To further minimize this problem the in-situ pipes were cut into various size pieces using a special cutting tool lowered down the injection pipe after completion of the grouting. Finally, the pipes were filled with a neat cement mix after injection was completed, making them more brittle and easier to break. These improved procedures were successful.

Injection Pipe Drilling. Approximately 35,000 PVC injection pipes were required for all the grouting works for Line 2. At an average depth of 40m, over 1.4 million linear meters of drilling was performed using track mounted DCH and Puntel rotary drilling rigs. These rigs were specially designed for the project to handle extra long drilling rods (6 to 9m) in order to maintain the high production rates required to achieve the construction schedule. Average daily drilling rates of 1500 m/day/machine were achieved. In addition, the rigs were equipped with heavy duty drillrods (114 mm diameter) to limit horizontal deflection of the grouthole during the drilling. Tricone or fish tail type drill bits were used, providing a drillhole diameter of 114 mm or 146 mm, respectively.

The drilling was advanced using continuous low pressure circulation of a bentonite drilling mud, for removal of the soil cuttings and to prevent the collapse of the drillhole. The bentonite was prepared using high turbulence mixers to ensure proper dispersion. The fresh slurry was maintained at a Marsh Viscosity between 40s and 50s and the density between 1.02 and 1.05 kg/cm^3. During the drilling process, the fluid was re-cycled through de-sanding units with vibrating screens. The re-cycled slurry required a Marsh Viscosity less than 50s and a density less than 1.3 kg/cm^3.

At completion of drilling, the non-hardening bentonite drilling fluid was replaced by a bentonite/cement "sleeve grout". The sleeve grout mixture included 250 kg/m^3 of cement, and 35 kg/m^3 of bentonite for stability. The injection pipe was then lowered into the drillhole and the sleeve grout was left to harden for a minimum of 4 days before the injection process could begin. Beginning injection before the sleeve grout was sufficiently hardened resulted in excessive grout resurgence at ground surface, coming upwards through the soft sleeve grout.

The maximum depth of drilling for Line 2 was about 50m. Considering this great drilling depth, there was concern that horizontal deviation of the drillrods might occur due to the presence of hard or gravely soil layers, obstructions, or concrete overbreak adjacent to the diaphragm walls. This would lead to excessive separation of the TAM's, resulting in "windows" of untreated soil within the grouted zones. Despite the heavy drillrods and electronic inclinometers positioned on the drill masts for precise control over the verticality, visual observations in excavations and verticality measurements confirmed that deviations of the groutholes did occur. Accordingly, systematic measurement of the verticality using a small diameter inclinometer was initiated and the results were plotted to determine the actual position of the injection pipe within the treated zone. Normally, the measured deviations were less than 1% of the depth of drilling, which was considered acceptable for the thicker base plugs of the cut-and-cover stations. If the deviation was judged to be excessive, supplemental injection pipes were installed or supplemental grouting in neighboring groutholes was performed.

Computerized Grouting. Proprietary computerized grouting systems from both joint venture grouting specialists were used to control the entire process of injection; SINNUS by Soletanche and EPICEA by Bachy (Figure 3). At the core of these systems is a data acquisition unit controlled by a multi-task processor that monitors in-line flow meters and pressure gauges, and executes pre-programmed grouting instructions (maximum pressure and or volume, maximum and minimum flowrates, etc.). Linked with automated batching plants, these systems are also able to control the mixing of the grout. Each system controlled up to 12 independent and simultaneously operating grout pumps. The entire operation was overseen by a single trained operator from a computer console where all the various grouting parameters for each operating pump were graphically displayed in real-time.

At the end of each shift the data collected during the grouting was transferred to a floppy disk and fed into a data base for each site, as well as an entire project data base. The data base was then used to perform a wide variety of statistical analyses, each of which was tailored to emphasize a particular parameter or group of parameters, such as distribution of grout take or final injection pressures over the entire grouted mass. An additional feature was the ability to produce graphical two-dimensional colored representations of grout parameters within any horizontal and/or vertical slice of the grouted mass. The statistical analyses and graphical representations were used in the evaluation of the grouting work, and for preparation of the final grouting reports for each site. Detailed descriptions of these computerized grouting systems are provided by Guillaud, M., et.al. (1994) and Harris, R.R.W, et.al. (1994).

Without the use of these computerized systems and a sufficient amount of grouting equipment, this massive grouting project would not have been able to be completed within schedule.

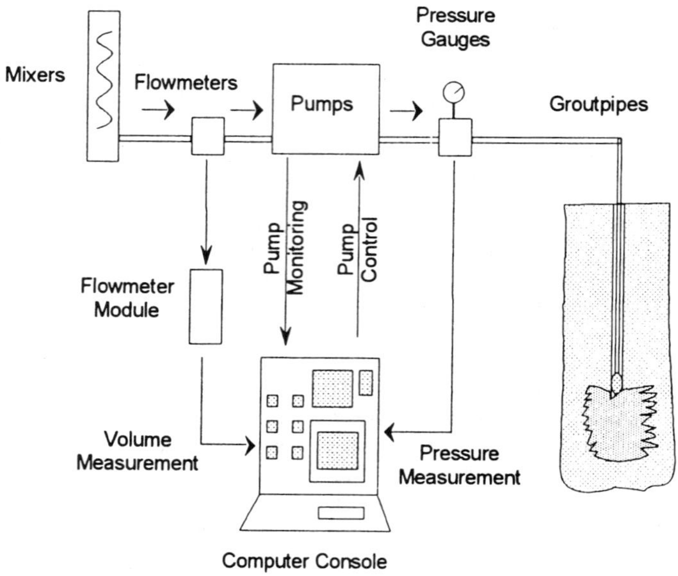

Figure 3. Computerized Grouting Schematic Layout

SBE provided 8 separate computer systems, 68 grout pumps, and 8 mixing units, all housed in movable containers for easy transport from site to site along the alignment. Approximately 720,000 injection points (manchette sleeves) were grouted during the course of the work, each point normally receiving multiple phases of injection. It is estimated that the computers regulated injection at least 2.3 million times over the duration of the project. The average daily injection rate was 230 m^3/day (two, 12-hours shifts).

Jet Grouting

Whereas permeation grouting leaves the soil structure "undisturbed," jet grouting is both a soil replacement and in-situ mixing process, which completely destroys the natural structure of the soil. It consists of injecting cement into the ground at high pressure, eroding the soil and forming a fluid soil/cement suspension creating a nearly homogeneous mass. The process results in the formation of approximately cylindrical columns of grouted soil with the diameter of each column dependent on the soil type, it's resistance to erosion, and the amount of jetting energy applied. Xanthakos, P.P., et. al. (1994), provides a detailed description of the jet grouting method.

During the construction of Line 2, over 70,000 m^3 of ground was treated by the jet grout method, making this project one of the largest single jet grouting operation to date. In addition, the depth of treatment by jet grouting was generally between 25 and 40 m, which was considered to be near the limits of the current technology, particularly for it's intended purpose both of consolidation and watertightness.

The addition of jet grouting to the contract documents provided a method of treatment for the clay and fine soils (interbedded silts and clays) of the transition layer beneath the clay. Effective permeation of these soils is normally not possible using chemical or mineral grouts. Accordingly, jet grout was normally used for the shallower ground treatment applications, such as TBM exit and/or entry points from the underground stations, repairs of major defects in diaphragm wall joints, at connections between diaphragm walls and existing structures, and at openings left in the diaphragm walls for utilities which could not be diverted. In addition, jet grout was used where high strength treatment was required, for example, at underground galleries for special connections to the bored tunnel, and for support underpinning of buildings overlying the tunnel.

All three methods of jet grouting (single, double and triple fluid systems) were used on the project, for installation of over 3900 jet columns. The double fluid system was most often used, producing column diameters between 0.8 and 1.8m. The use of the single fluid system was limited to one site where column inclinations exceeded about 36 degrees from vertical. The range of injection parameters used for the double and triple jet method are given in Table 1.

Table 1. Jet Grout Parameters

Jet Parameter	Double Jet	Triple Jet
Column Diameter, m	1.3-1.8	1.5-1.8
Water Flowrate, liters/min	---	100-110
Air Pressure, bars	12	12
Uplift Speed, cm/s	8-45	7-7.5
Drillrod Rotation, rpm	14	5
Grout Flowrate, liters/min	135-230	70-72
Water Pressure, bars	---	470-500
Grout Density, kg/m^3	1.4-1.52	1.55-1.65
Grout Pressure, bars	400-480	30

Prior to beginning work at each site, a Site Specific Jet Grout Note was submitted, providing all the pertinent jetting parameters for that particular site. The parameters were varied according to soil type, design column diameter, depth of treatment, soil density and other factors. At the start of the work, a series of shallow trial columns were grouted within the upper clay and subsequently excavated to validate the jet parameters used in relation to the measured actual column diameters.

Design strength for the jet treatment was 3 MPa (30 bars). However, test cores performed early in the work found that the in-situ strength of the treatment was generally much higher, averaging about 3 to 4 times this value. The strength results were dependent on the cement content of the grout mix and also the soil type being treated. Higher in-situ strengths were obtained when jetting through the sand as opposed to the clay soils. Accordingly, the cement content of the grout was adjusted between 500 and 750 kg/m^3, depending on the ground treatment required and the type of ground being treated. About 15 kg/m^3 of bentonite, was added to the mix to improve it's stability. The W/C ratio of the grout mix was maintained between 0.75 and 1.

The drilling was performed using the same types of rigs as for the injection pipes, modified to accept the jetting equipment. High pressure Triplex and Obermann pumps capable of delivering grout or water at maximum pressures of 50 to 60 MPa (500 to 600 bars), and maximum flowrates of about 300 to 350 liter/min. A computerized batching plant prepared the grout mixture at a rate of 15 m^3/hr. During the formation of the jet column, the step distance, step time, and rotation speed of the jet monitor were regulated by a pre-programmed computerized control panel located on the rig. Gauges to measure grout and/or water pressures were supplied both at the pump location and the jet rig to ensure that head losses through the high pressure lines were within acceptable limits.

Jet grouting was often performed directly beneath existing utilities and existing buildings, which presented concern regarding rapid and potentially damaging ground heave often associated with the jet process. The most important factor in minimizing ground heave is to ensure the free flow of spoil to the surface during the formation of the jet column. Any restrictions/blockages of the drillhole during the formation of the jet column results in excessive build-up of pressure in the soil, causing the ground to heave.

In order to minimize heave potential, any reductions in air pressures and spoil flow rates were carefully monitored at the rig. When either was observed, the jetting was immediately stopped and the blockage cleared by raising and lowering the jet monitor while jetting at low pressures. To further minimize potential ground heave, “prejetting” of the upper clay zones was made a routine part of the jetting operation. The prejetting was accomplished using a high pressure water jet (200 - 300 bars). This widened the drillhole diameter through the clay, ensuring the free flow of spoil, and thus minimizing the heave potential. Due to the above efforts, heave during jetting was not generally a problem, although one specific site did measure almost 0.5m of heave. This required re-construction of the concrete working platform, but did not cause any other damage. Pre-jetting also helped to ensure that the required diameter of the column was maintained in the stiff clay.

Due to the heat of hydration of the jet grout and negligible sources of heat loss or dissipation, the temperature of the jet grouted mass remained quite high for a period of about 2 months following completion of the work. During injection into groutpipes installed through a recently completed jet grout, mass, these high temperatures softened the PVC injection pipe and caused a high percentage of groutpipe deformation. Water temperatures inside the injection pipes were measured as high as 50°C. The number of deformations were highest at the Mubarak Station TBM break-out (north), where the triple jet system was used to form the jet mass. At this site the groutpipes were installed through the jet mass in order to form the underlying hard gel plug. Considering that the cement content of the treated ground is highest using the triple jet system it was reasonable that the ground temperatures were also highest at this location.

INJECTION MATERIALS

As previously stated, the contract documents specified the injection materials to be used for the various grouting applications on the project. Specifically, provisions were incorporated into the technical documents for the use of an initial injection of bentonite/cement, followed by a sodium silicate soft and/or hard gel. A newly developed mineral grout was added to the technical documents during contract negotiations between NAT and INTERINFRA with the intent to provide the capability of a hard gel equivalent consolidation grout, but with a viscosity near that of water for permeation into relatively fine grained soils.

Bentonite/Cement

Bentonite/cement (B/C), a particulate grout, was specified as a first phase of injection to permeate the larger soils voids of the gravels and coarse sands and restrict the spread of the more expensive second phase sodium silicate chemical gels. The B/C also provided a void-size reducing pre-treatment to minimize syneresis of the sodium silicate gels. Syneresis is the shrinkage of the gel within the soil voids due to the expulsion of water. The initial phase of B/C also improves both the groutpipe sleeve cementation and the contact between the diaphragm or slurry walls and the treated ground. B/C was also used for re-grouting selected sleeves having low pressures (less than 5 bars) during injection with the chemical gels.

The B/C grout was composed of 250 kg of cement, 35 kg of bentonite and 1.5 liters of a fluidizer for every 1m^3 of grout mix. The density was maintained around 1.2kg/liter with a Marsh Viscosity between 35 to 40 seconds. Injection flow rates varied between about 450 and 800 liters/hr. At the start of the project, the target injection quantity was based on an average of 12-13% of the volume of the treated soil, allowing a local maximum quantity of up to 20% in case of open gravel layers. The B/C was injected into each grout sleeve up to the maximum quantity or until a maximum pressure of 8 bars was achieved. If the final pressure was below 8 bars, an additional volume of B/C was injected until the pressure reached 13 bars.

This procedure was found to impede injection of the second phase chemical gel because of excessive amounts of hardened B/C surrounding the sleeve grout. In some cases this prevented the sleeve from opening during gel injection, particularly when injecting into finer soils. Accordingly, the injection procedure was modified to limit the amount of B/C injected by eliminating the B/C re-injection if the final pressure was less than 8 bars. In addition, the maximum injection quantity was reduced from 20% down to about 12%. These modifications improved the absorption of the subsequent chemical gels.

The most significant problem associated with the B/C was the deformation of injection pipes during injection. When this occurred, the packer became stuck inside the pipe, or it was prevented from being lowered past the point of deformation. The frequency of the deformations increased significantly when injecting the B/C into the finer grained soils. The deformation of the pipe occurred during the first few seconds of injection, when the B/C injection pressures were essentially unlimited to allow "cracking" of the sleeve grout. When this became a problem at a particular site, maximum injection pressures were lowered, additional injection pipes were installed (where required) or neighboring injection pipes were over-injected to compensate for the lack of grout volume in the deformed and ungrouted injection pipes.

Sodium Silicate Gels

The smaller soil voids present after B/C grouting were filled by a second phase of injection using a sodium silicate based soft or hard gel, depending on the specific application requirements. The soft gel was specified for watertightness applications only, reducing the permeability to a target value of 1×10^{-6} m/s, and in some instances as low as 1×10^{-7} m/s. When the specific application required soil consolidation/strengthening as well as watertightness, hard gel was specified.

The sodium silicate, SiO_2Na_2O, was prepared in France maintaining a silica/sodium ratio of around 3.2, which forms a basic (alkali) solution particularly suitable for grouting. The material was shipped to Egypt in solid form, requiring in-country treatment with water under high pressure and temperature to return the material to a groutable liquid. When neutralized with certain acid solutions, the hydrated sodium silicate sets to form a gel.

The setting reagent for the soft gel was a mineral based soda bicarbonate. An organic hardening reagent, Rhone Poulene Type 600, was used for the hard gel. The mix proportions of the sodium silicate and hardener were varied to provide, Type A, B, and C hard gels, resulting in design strengths of standard Fountainbleau sand mortar samples of 1.4, 1.0 and 0.3 MPA, respectively. The type of hard gel selected for a particular application was dependent on the required design strength of the treated soil.

The range of the various mix proportions for a 1 m^3 batch of soft and hard gel are shown in Table 2.

Table 2. Mix Proportions for Chemical Grouts

Gel Type	Soda Silica (liters/m^3)	Hardener 600 (liters/m^3)	Bicarbonate Solution (liters/m^3)	Water (liters/m^3)
Soft	140	---	370	490
Hard (A)	450-510	80-100	---	390-470
Hard (B)	425-385	65-85	---	490-550
Hard (C)	315-335	55-65	---	600-630

The rheological and mechanical properties of the gels were controlled by variations in the mix proportions, as follows:

- Alpha ratio = Bicarbonate or hardening reagent /sodium silicate
- Beta ratio = Water / sodium silicate

Increases in the alpha ratio provided shorter setting times and higher strength while increases in the beta ratio resulted in longer setting times and reduced gel strength.

The thickness of the hard gel plugs was designed using a hydraulic gradient across the plug of 5. When determining the stability of the consolidated plug against uplift, lateral friction between peripheral walls and the plug was considered, thus reducing the required plug depth. Unit lateral friction forces were limited to 1/10 of the grout design compressive strength, with a maximum value of 0.05 Mpa.

The Marsh Viscosity of the mix was maintained between 32 and 40 seconds, and the setting times between 30 and 70 minutes. Considering the porosity of the Nile alluvium is between 35 and 45%, the target injection quantity for the gels was based on an average of 38% of the volume of the treated soil, allowing a local maximum quantity of up to 45%. Thus total volume, gel plus B/C, could vary between 49 and 58% of the treated volume. This volume accounts for spillage and other above ground losses, as well as migration beyond the limits of the "design volume" of treated ground. Injection flowrates varied from 250 to 600 liters/hr, and the maximum injection pressure was set between 15 and 20 bars.

The major problem with the use of silica gels in the Cairo environment was controlling their setting times due to the wide variations in temperatures throughout the day. As the temperature warmed, the silicate solution heated to temperatures of up to 40°C inside the storage tanks. As a result, the setting time of the gel mix was significantly reduced requiring frequent adjustments in the mix proportions to maintain the desired setting time. Setting time samples of nearly every grout batch were obtained by the plant operator to determine when adjustments to the mix proportions were required.

This was particularly a problem during the hot summer months when the already heated mixture was pumped through sun-baked delivery lines between the grout pump and the work platform. This caused a rapid build-up of set grout inside the line, requiring them to be cleaned at least once per shift. Although cleaning the lines was quick and easy for the soft gel, it was a difficult and time consuming task for the hard gel.

As a result of this build-up, head losses through the delivery lines had to be determined before injection of each injection pipe in order not to obtain falsely high actual pressure readings during injection. This was accomplished by injecting a small amount of grout through the delivery line with the packer laying open on the working platform. The difference between the pressure reading at the pump and at the end of the delivery line was the head loss through the line. This value was input into the computer that automatically deducted this value from the injection pressure readings. Head losses were measured as high as 10 bars; however, delivery lines were generally cleaned if the head loss reached 3 or 4 bars.

Mineral Grout

To provide an alternative to the hard gel treatment for applications requiring both water-tightness and soil consolidation, the Contractor proposed and NAT accepted the addition of mineral grout to the contract documents. Mineral grout (a proprietary grout, trade named "Silacsol") was developed by Soletanche in the early 1980's in an attempt to address a heightened environmental awareness during the late 1970's, which resulted in the banning of the many extensively used toxic chemical grouts. It is a non-polluting, environmentally friendly, suspension grout material that is able to effectively penetrate the more fine-grained soils which previously required the use of various types of chemical grouts. Mineral grout can permeate soils having a permeability as low as 1×10^{-5} m/s, slightly more penetrable than the silicate-based hard gels, having a practical limit of 5×10^{-5} m/s. Mineral grout viscosity is about two centipoise, only slightly higher than water.

Mineral grout is composed of extremely fine silica and calcium powders mixed with water. A solution of silica and soda is added to advance the setting time, along with a small amount of a dispersing agent to minimize flocculation of the fine powders. Additional minerals are also present in small amounts. The silica and calcium have components of aluminum and magnesium, respectively. When these materials are mixed, the reaction is similar to the hydration of cement.

The components and mix proportions used on the Cairo Metro for $1 m^3$ of mix are as shown in Table 3:

Table 3. Mix Proportions for Mineral Grout

Solution	Composition	Trade Name
Solution of Silicious Components	350 liters	RA
Solution of Calcium Components	595 liters	RB
Solution of silica in soda	49 liters	LS
Dispersing agent	6 liters	AS

Both the silica (RA) and calcium (RB) components are processed such that their particle size distribution is slightly finer than the finest ultra fine cement. On-site, a centrifuge is used to remove silica particles greater than 15 microns (*um*). The grain size distribution of the final mix is checked regularly using a laser diffraction particle size distribution analyzer, a high technology instrument able to measure particle size distribution within the range of 0.1 *um* through 200 *um*. A summary of the various physical properties for the mineral grout mix used on the project is shown in Table 4.

Table 4. Rheological Properties of Mineral Grout

Description	Physical Property
Viscosity	2 cP (water is 1 cP, typical hard gel 4-5 cP)
Setting Time	about 45 to 100 minutes
Density	about 1.13 +/- 0.03
pH	13
Grain Size	d_{95} less than 12 *um*, d_{100} less than 15 *um* (1 *um* = 0.001 mm)

The strength characteristics of the mineral grout are on the same order as bentonite-cement grouts. However, mineral grouts exhibit higher strength than B/C grouts for the same concentration of total dry material. By varying the ratio of RA/RB, a wide range of strengths can be obtained. Test columns of Fountainbleau sand injected with mineral grout gave crushing strengths ranging from 1.5 to 2.0 MPa. However, the in-situ strength of the treated soil as measured by test cores was generally considerably higher, on the order of 5 times the laboratory test samples.

Mineral grouts do not experience syneresis, as opposed to the chemical gels for which substantial expulsion of water results in contraction of the gel and reduction in treatment efficiency. Unlike the silicate-based gels, mineral grout does not exhibit significant creep (sustained deformation under a constant load), and does not "break down" over time; and is therefore considered a permanent form of treatment.

The major disadvantages of mineral grout are similar to those of all suspension type grouts, i.e., maintaining the proper stability, and the potential for excessive filtration and bleeding during injection. Mineral grout also requires a substantial amount of quality control and specialized equipment compared to the use of the standard gels, and for the Cairo soil conditions, was found to be a more expensive treatment material than using the traditional combination of B/C and silica gels, for a given set of boundary conditions and service requirements.

PROJECT GROUTING APPLICATIONS

Grouting was required on Line 2 for ground and ground water control for construction of underground stations and cut-and-cover sections, TBM break-in/out's, annex connections to the bored tunnel, protection of major utilities, support of buildings undermined by the tunnel, and for the crossing of a major irrigation canal. In addition, grouting was used for miscellaneous applications such as repair of major defects in diaphragm wall panels and joints, and to provide ground support at connections between diaphragm walls and existing structures or between two different excavation support systems.

Construction of Underground Stations and Cut-and-Cover Tunnel

Construction of the underground stations, the enlargement the existing interchange stations and cut-and-cover sections, required low permeability plugs to reduce ground water infiltration quantities. These were formed at the base of 1.2m thick perimeter diaphragm walls extending up to 50m in depth (Fig. 4). The more shallow station access entrances and air intake/exhaust shafts were constructed similarly. The specifications required that the final permeability of all grouted plugs be no more than $1x10^{-6}$ m/s. Without the cut-off plug, dewatering quantities would have been excessive and would have allowed exterior ground water lowering, causing potentially damaging building settlements.

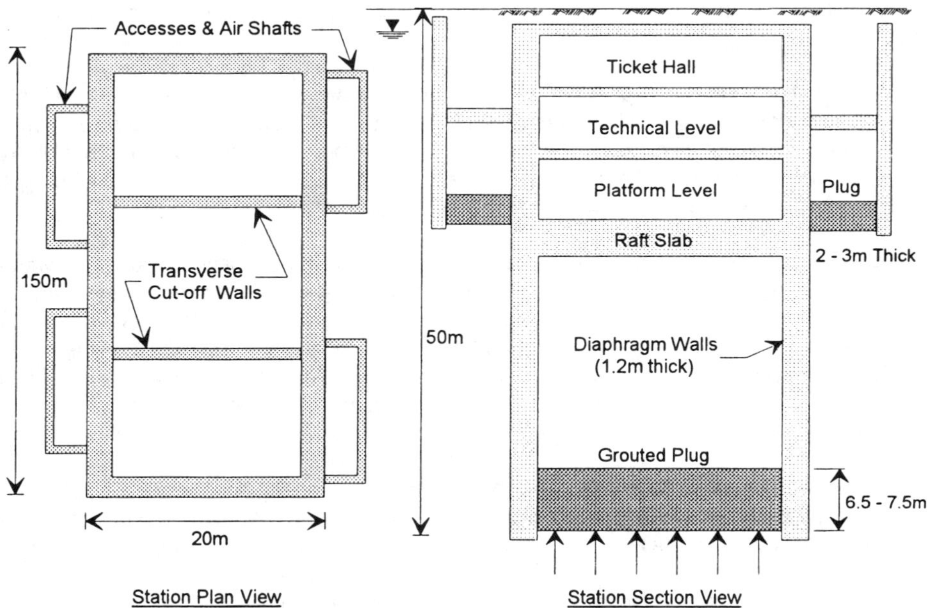

Figure 4. Typical Station Construction

The grouted plugs were designed to reduce the in-situ permeability and were positioned such that the uplift pressure at the base of the plug was balanced by the weight of soil between the bottom level of the plug and the final excavation level, using a factor of safety of about 1. As a result, a soft gel application with a pre-treatment of B/C was appropriate as soil consolidation was not a consideration. The thickness of the soft gel plug was designed using a hydraulic gradient of 3. The combined design criteria resulted in plug thickness ranging from 2 to 7.5m, with bottom of plug depths ranging between 13 and 49m. A minimum plug thickness of 2m was adopted for the shallow station accesses and air shafts.

Within the 150m long stations, two transverse slurry walls were constructed to the toe of the diaphragm walls, dividing the stations into three separate boxes and limiting the size of each individual grouted plug. Along the cut-and-cover section, transverse slurry walls were constructed at about 45 m intervals. Grouting was performed separately for each individual box, followed by a pumping test to confirm that the target permeability was achieved.

The performance of the soft gel plugs was good for Phase 1, and is proceeding similarly as well for Phase 2A. Of the 145 separate boxes formed to date (January 1997) for construction of the Phase 1 and Phase 2A stations boxes (including accesses, and air shafts) and cut-and-cover tunnel sections, 135 boxes, or 93%, achieved the required permeability of $1x10^{-6}$ m/s, or better. The permeability of the "non-conforming" boxes ranged between $1.1x10^{-6}$ and $4x10^{-6}$ m/s. The boxes for which the grouted plugs did not achieve the specified permeability were successfully excavated with only a slightly increased pumping rate, with a few exceptions noted below. For Phase 1, the average B/C and soft gel treatment percentages for the grouted plugs of the large station boxes and the cut-and-cover sections were 13% and 39%, respectively. Phase 2A works are in progress but are expected to reach similar values.

After confirmation of the plug permeability and before top-down construction of the stations and cut-and-cover sections began, all injection pipes were required to be cut 50cm below the final excavation level. This was required to minimize the potential "holes" in the plug resulting from pulling the injection pipes out of the ground when grabbed by the backhoe bucket or by crane-mounted grabs.

The injection pipes were cut using a high pressure water jet device that was lowered down each injection pipe to the required depth. Field testing was performed to determine the required time of cutting, which was generally between 2 and 4 minutes. However, regular field checks of the cutting time were required due to the water jet erosion effect on the nozzle diameter causing a gradual increase in the nozzle diameter which decreased the cutting action of the water jet. Subsequent to cutting, the injection pipes were filled with B/C, making them slightly brittle and more likely to "break-off" during excavation.

During excavation for Line 2, injection pipes were occasionally observed to be pulled-out of the ground. Normally the break point was well above the plug level, and not below the transition joint between the two different thickness injection pipes. However, during excavation at both Mazallat and Sadat Stations, plug leakage was observed after reports of extended lengths of injection pipes being pulled out at excavation level. Fortunately, the leakage was controlled within hours by emergency injection using a water reactive polyurethane grout.

The most serious problem for the Phase 1 project occurred during excavation of the west box of Abdeen Station. Within a few meters of final excavation level, serious leakage was observed, requiring emergency backfilling for bottom stabilization. Over the next five months various investigations and remedial treatments were performed to locate and close the leak. In order to temporarily control the leakage, additional dewatering wells were installed inside the box and piezometers were installed to monitor groundwater levels both inside and outside of the station. A pipe was placed into the main leakage area and a valve was installed for controlling the rate of discharge. The settlements of adjacent buildings were carefully monitored, as well as sand content of the discharged groundwater.

The leakage was finally sealed by pumping concrete into the leakage pipe, however, the event delayed the TBM break-in from Abdeen by nearly 5 months and caused unanticipated additional settlement of several of the adjacent buildings.

TBM Break-in/outs

Line 2 included 22 locations where the TBM entered the ground (break-in) and/or entered into a station (break-out), all requiring soil consolidation and groundwater control. Considering that the treated zones had to extend into the upper clay and transition soils, the ground treatment of these critical zones required jet grouting, as treatment by permeation grout could not be done in the fine grained soils. The design of the break-in/out zones evolved from using a basic jet grout treatment, to a more conservative combination of jet grout, and slurry wall box with a hard gel plug, Fig. 5.

The evolution was a result of problems encountered with the jet grout treatment during exposure of the jet treatment for the TBM break-out (TBM station entry location) at Masarra Station. During the demolition of the diaphragm wall at the Masarra break-out a small discontinuity of untreated soil was encountered at the interface between the treated ground and the diaphragm wall, which extended to the base of the treatment. Because the clean uniform alluvial deposits have no natural cohesion, they are particularly susceptible to erosion when subjected to a hydraulic gradient. As a result, a progressive failure occurred, forcing soil and water under pressure through the discontinuity and into the station.

In order to stabilize the situation, emergency flooding of the station was authorized by NAT. This balanced the water pressures across the break-out securing the situation until remedial grouting was completed, afterwhich the station was dewatered and the TBM break-out completed. A similar failure at Rod El Farag break-out occurred but was contained by emergency grouting from the surface, avoiding the need to flood the station.

Because of these incidents, a more conservative design using a slurry wall box with a

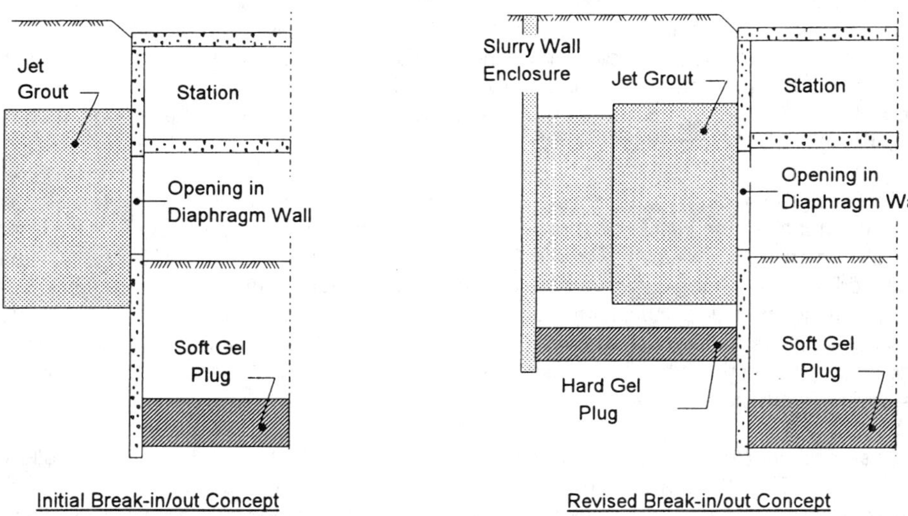

Figure 5. TBM Break-in/out

hard gel lower plug was developed. Test drains were also drilled through the diaphragm wall around the TBM openings to detect treatment discontinuities and/or leakage, allowing selective regrouting to seal the leakage prior to demolition of the opening. Pumping tests were then performed to confirm the permeability of the slurry walled box demonstrating its' watertightness.

Extensive review of the jetting parameters and field procedures was performed to determine the cause of the observed deficiencies in the jet treatment. The most important observation made was that the frequency of the discontinuities increased as a function of depth, and the first evidence of the discontinuities generally occurred at depths greater than about 15m. This led to speculation that vertical deviation of the jet columns was occurring, leaving small "windows" of untreated soil between columns, or possibly the jetting energy was not sufficient to account for the denser soils with depth.

Minor modifications were then made to the jet parameters, including increasing the jetting energy at greater depths to account for the denser soil conditions. In addition, quality control efforts were significantly increased and a more conservative design approach was adopted for jet grout applications. It was decided to consider the jet method appropriate for watertightness applications only for shallow ground treatment less than about 15m. Deeper jet grout treatments were subsequently designed with an envelope of supplementary watertightness grouting, usually using a hard gel plug beneath and a slurry wall perimeter cut-off box.

Tunnel Annex Structures

Tunnel ventilation and dewatering, and train traction power supply, was provided through a series of 13 tunnel annex structures constructed within 5m from the tunnel. These structures, constructed using deep diaphragm walls, required a connection be made between the annex shaft and the completed tunnel at depths ranging from 18 to 28m. Consolidation as well as watertightness grouting was required for the connection gallery to the tunnel as well as a watertight plug for excavation of the shaft. As with the TBM break-in/out's the ground treatment for these structures evolved during the work, from a simple jet grout treatment, to a more conservative combination of jet grout, and slurry wall perimeter box with a hard gel grouted plug that allowed dewatering of the box during hand-mining of the connection gallery, increasing the factor of safety, as noted in Fig. 6.

The design evolution was driven by similar problems encountered with the depth of the jet grout treatment for the initial TBM break-outs. Prior to demolition of the diaphragm walls for the jet treated gallery openings of the first three annex structures 12A, 12B and 16, test drains drilled through the diaphragm wall along the perimeter of the openings brought water and soil at high pressure, indicating a lack of watertightness, which was essential for the safe completion of the gallery connections. As a result of the poor results of the test drains, a blanket of hard gel was provided beneath the jet treated mass at these annexes to seal any discontinuities in the treatment.

At several annex connections mineral grout was selected for the treatment of the gallery connection. Even though the treatment was observed to be non-uniform, with zones of untreated soil present, its' mechanical performance was acceptable and allowed the gallery connections to be made without instability problems. However, the watertightness perfor-

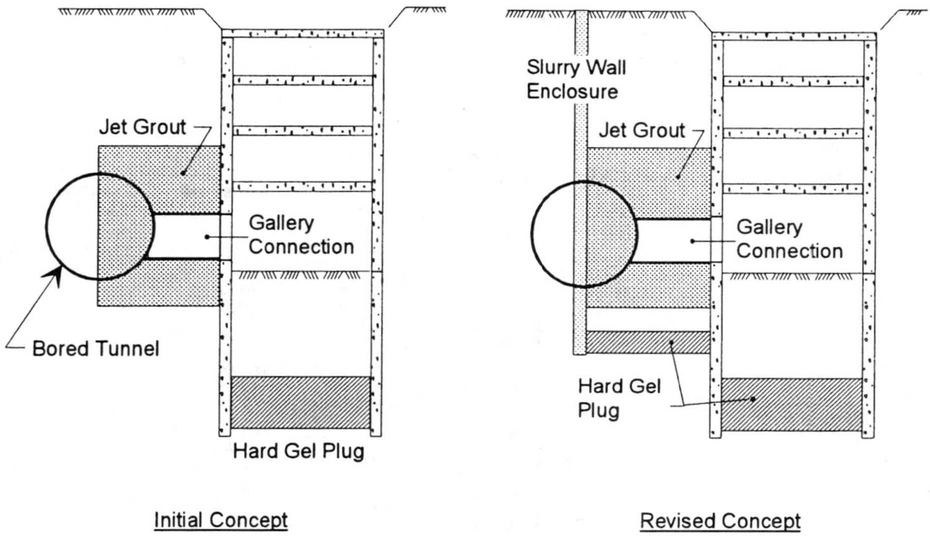

Figure 6. Tunnel Annex Connections

mance of the mineral grout plugs for the shaft and the closed box connection plugs was poor. Of the 9 mineral grouted plugs at the annex structures, only one provided the specified permeability.

Since these findings were not in line with those obtained on previous mineral grouting projects, like the Paris Metro where watertightness performance was proven to be acceptable, extensive field and laboratory investigations were performed by SBE to determine the cause of the lack of proper grout penetration. It appeared that excessive filtering of the grout was occurring during injection, probably caused by the finer particles within the sand layers, where these plugs were located. Subsequently, the mineral grout was finally abandoned for watertightness applications.

It should be noted that slight "foaming" of the TBM bentonite slurry did occur when the TBM passed through ground treated with mineral grout at the annex connections. This was attributed to the adverse chemical reaction between the small percentage (about 5%) of the solution of silica in soda (LS component) added to the mix and the bentonite slurry. Although slight adjustments to the slurry pressure were required while passing through the treated ground, it was not considered to be a problem.

Hand-mined Tunnel at Mazallat Station

North of Mazallat Station, two multi-story residential buildings encroached the limits of the cut-and-cover tunnel section preventing execution of 17m of diaphragm walls, Fig. 7. In order to complete the tunnel, a 3400m^3 block of jet grout was placed beneath the corners of the buildings as temporary underpinning support while multidrift hand-mining techniques were used to construct the tunnel. Core borings inclined through the completed jet grouted

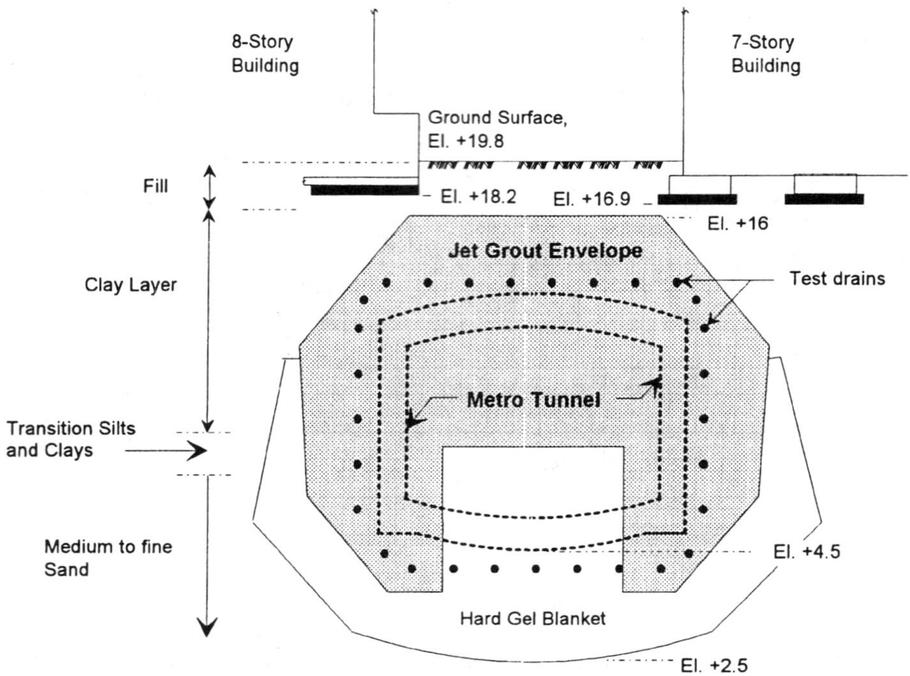

Figure 7. Mazallat Hand-mined Tunnel

mass indicated good results, with recovery values generally over 90%. However, because of the previous difficulties with the jet grout for watertightness applications and the critical nature of the work, a 2m thick hard gel "blanket" was injected, surrounding the jet grout mass to ensure its' watertightness.

Compression test results on selected core samples drilled into the jet treatment were variable, indicating in-situ strengths ranging 3 MPa to 20MPa (average 12 MPa). This was much higher than the 3 MPa required by the design, and almost three times the average crushing strength of spoil samples obtained during the work.

Due to the close proximity to the buildings and the potential for rapid and damaging ground heave, a continuous electronic laser monitoring system was set on each building. When uplift of more than 2mm was recorded, an alarm was triggered which notified the operator to immediately stop the jetting process, alleviating the heave potential. Special care was also taken to monitor the free flow of spoil to the surface, thereby minimizing the potential for large ground heave and building damage. The ground treatment, hand-mining and permanent tunnel were completed without incident, limiting building settlements to under 15mm.

CWO Tunnel Crossing

North of Masarra Station, the Metro tunnel passed 6m beneath the 3.2m diameter Boulac Spine Tunnel, part of the Greater Cairo Wastewater Project, Fig. 8. In order to limit the settlement of the spine tunnel to below the estimated damage threshold, some form of ground consolidation was required. The spine tunnel, bored by TBM methods, is comprised of a 17.5 cm thick concrete segmented lining, inside of which was placed a 53.5cm thick reinforced concrete and brick inner lining. In anticipation of the future crossing, the Cairo Wastewater Organization (CWO) designed the spine tunnel to account for some additional ground movement during the Metro tunnel crossing. At the crossing, a 40m length steel sleeve was cast into the reinforced concrete lining, providing additional resistance to the anticipated settlement.

Before the crossing, extensive finite element modeling was performed to estimate the settlement of the spine tunnel, using surface settlement data accumulated during the Metro tunneling and soil strength parameters determined during the subsurface exploration program. Based on the modeling, it was estimated that without ground consolidation, the spine tunnel would experience significantly more than the 10mm allowable settlement determined from a detailed stress analysis of the composite tunnel section. With ground consolidation/strengthening, the estimated settlement of the spine tunnel was 7mm, less than the allowable. Considering that watertightness was not required for this application, and the known "foaming" problem with the hard gel, mineral grout was selected for the ground treatment.

Approximately 1500m^3 of mineral grout (50% of the treated soil volume) was injected beneath the Spine tunnel to consolidate the supporting soils. As with other sites using the mineral grout, the final injection pressures were low, generally 4 bars or less. For this site, the strength properties of the mineral grout were modified by increasing the RA/RB ratio, resulting in an average compressive strength of the Fountainbleau sand columns of 2.5 MPa.

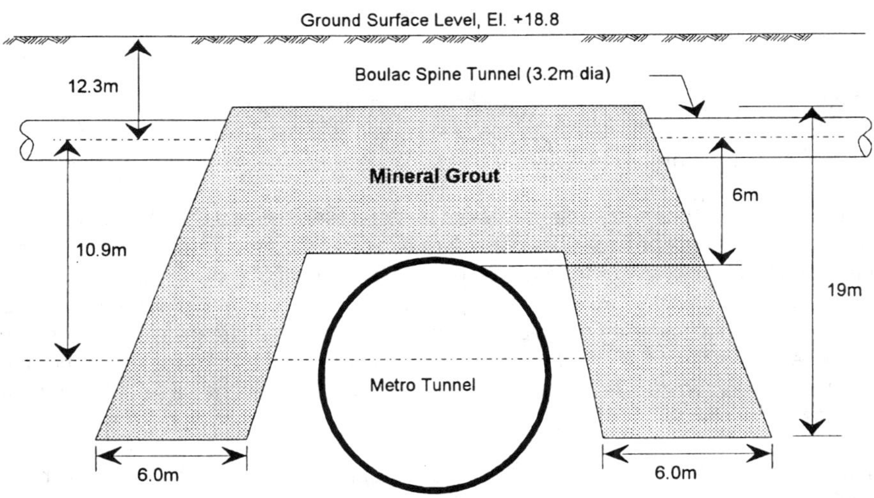

Figure 8. CWO Tunnel Crossing

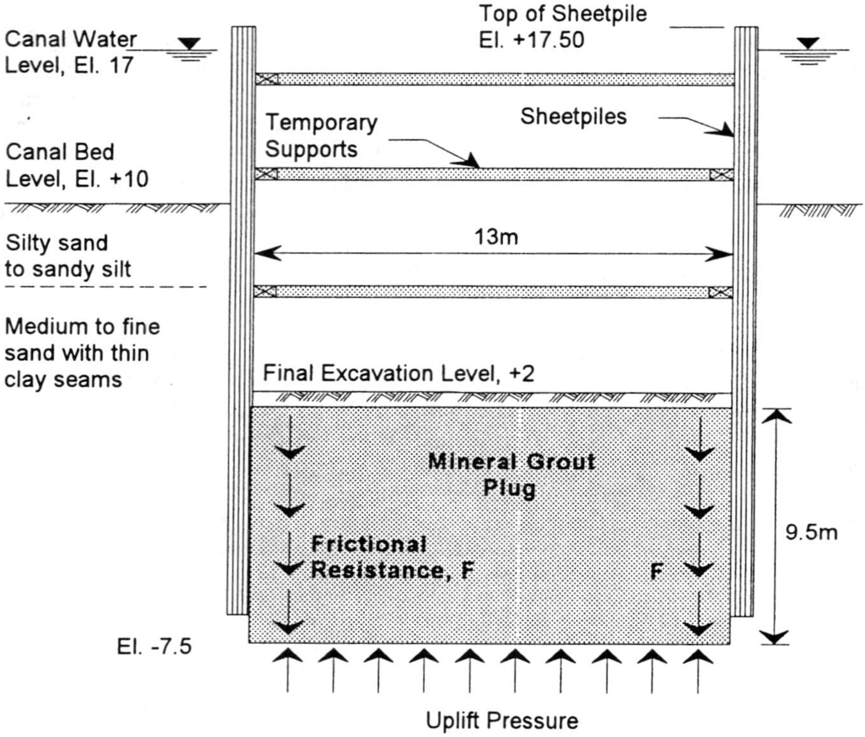

Figure 9. Ismailia Canal Crossing

Ground consolidation using the mineral grout was successful, allowing only 9mm of settlement of the CWO spine tunnel during the crossing of the Metro tunnel.

Ismailia Canal Crossing

Where the Metro tunnel alignment crossed the 90m wide Ismailia Canal north of Mazallat Station, a two-phased cut-and-cover cofferdam construction method was required to maintain waterflow for this major man-made irrigation canal, Fig. 9. This was the first site on the project to use the mineral grout, which was selected because of it's watertightness and strengthening capabilities, and non-creep characteristics. The crossing was constructed using three steel sheetpile cofferdams that were backfilled to provide working platforms for the grouting operation. Mineral grout plugs were then formed at the base of the 24m long sheetpiles.

The plugs were designed for full hydrostatic pressure, balanced by the weight of the plug as well as a component of friction between the plug and the steel sheetpiles (A factor of safety of 1.05 was used). The design strength of the treated ground was required to be 0.5 MPa, and a frictional resistance of 0.05 MPa/m^2 of contact area was used, resulting in a plug thickness of 9.5m.

The first phase to be constructed was the north cofferdam, 44m long and 13m wide. Grouthole spacing was 1.2m by 1.6m and the design called for 40% mineral grout treatment, which was compatible with the estimated soil porosity beneath the canal. B/C grout was injected in groutpipes along the sheetpiles to provide good contact between the sheetpiles and the treated soil, and in the top two grout sleeves of all injection pipes to prevent the mineral grout from migrating upward and outside the limits of the treated zone. A final treatment percentage of 40% of the treated soil volume using the mineral grout was achieved, using a primary/secondary injection sequence for comparison of final injection pressures. Higher pressures in the secondary injection pipes being a general indication of proper void filling.

At completion of injection, permeability tests performed in test borings drilled into the plug and Cone Penetrometer Tests (CPT) tests throughout the upper part of the part of the plug indicated variable ground consolidation, but provided a reasonable indication of plug strength and permeability. Based on this field testing, the excavation started and proceeded until a depth of about 11m when severe boiling was observed at several locations, requiring emergency flooding of the cofferdam for stabilization. Ultimately, the excavation was completed by lowering the hydrostatic pressure at the base of the plug with a series of 20 dewatering wells around the outside perimeter of the cofferdam, pumping a combined rate of 1200 m^3/hr. At the final excavation level, it was evident that sufficient grout overlap between groutpipes had not always been achieved, an indication of poor grout penetration leading to "windows" of untreated soil.

Based on the experience in the north cofferdam, the grouthole spacing was reduced to 1.2m by 1.2m in the southern and central cofferdams, and the minimum design injection percentage was increased to 55% of the treated soil volume (actual treatment percentages were 64% and 67%, respectively). In addition, two additional grout sleeves were placed below the design level of the plug and were left ungrouted until full scale pumping tests were completed inside each cofferdam, allowing injection with hard gel if the pumping tests indicated unacceptable results. The pumping tests in each box indicated permeability's of $2.5x10^{-6}$ m/s (south) and $6x10^{-6}$ m/s (central), and the lowest sleeves were not grouted. Although these permeability's were above the specified permeability, it was thought this was due to a normal amount of leakage through the sheetpiles above the canal bed level, which could be controlled during the excavation.

As a result of the revised grouting, the southern cofferdam was successfully excavated with no additional watertightness measures. However, the excavation in the central cofferdam encountered leakage problems, requiring a similar perimeter dewatering system as the North box in order to complete the excavation.

Miscellaneous Grouting Applications

Grouting was successfully used for miscellaneous applications such as repair of major defects in diaphragm wall panels and joints, temporary soil support where gaps in the continuous diaphragm walls were left to accommodate utilities that could not be diverted. In addition, grouting was used to provide ground support at connections between diaphragm walls and existing structures, or between two different excavation support systems, Fig. 10.

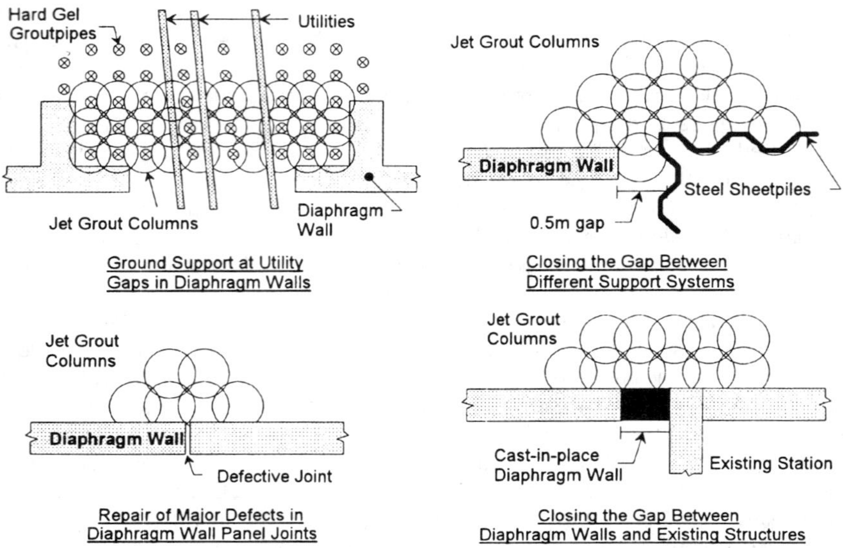

Figure 10. Miscellaneous Grouting Applications

CONCLUSIONS AND RECOMMENDATIONS

The advent of computerized grouting not only makes it possible to control many injection points simultaneously allowing high productivity on fast track jobs, but also allows quick analysis of injection parameters to determine what if any re-grouting may be necessary.

Jet grouting provides a technical solution to ground consolidation in almost any soil type and in a wide variety of geometric and structural applications, but may not be reliable for watertightness below depths of about 15m, without special design for Quality Control considerations.

Mineral grout, while providing an environmentally friendly grouting solution, must be tested in-situ on any new sites to verify the penetration characteristics so that a dependable injection program can be developed.

In many situations where both consolidation and watertightness are required, a combined solution of grout (gel and jet) with slurry walls provides a less risky but still economical ground improvement solution.

ACKNOWLEDGEMENTS

The authors acknowledge the National Authority of Tunnels, for permission to publish this paper, especially Engineer. Sh. El Bedaiwy, Head of NAT Studies Sector, for her continued technical support and cooperation. In addition, they are grateful to Parsons Brinckerhoff International, Soletanche Bachy Egypt, and the joint venture team of INTERINFRA for their support in preparing the paper.

REFERENCES

A. Salam, M.E., 1984, "Grouting of Alluvial Soil in Cairo Underground Metro," Proceedings, Conference, on Tunneling in Soft and Water Bearing Ground, pp. 135-142.

Cambefort, H., 1987, "Grouts and Grouting," Ground Engineering Reference Book, Butterworth, London, pp. 32/1-32/30.

Coe, R.H., et.al., 1994, "A Review of Grouting on the Greater Cairo Wastewater Project," Grouting in the Ground, Thomas Telford, London, pp. 163-182.

Flint, G.R. and Foreman, W., 1991, "Bentonite Tunneling for the Greater Cairo Wastewater Project," Proceedings, International Symposium on Tunneling in Congested Cities, Cairo, Egypt, pp. 231-252.

Guillaud, M., Hamelin, J.P., 1994, "Computerized Grouting Design and Control," Grouting in the Ground, Thomas Telford, London, pp. 215-225.

Harris, R.R.W. and Cottet, P., 1994, "Evaluation of Computer Control and Analysis of Grouting Parameters During Grouting Works," Grouting in the Ground, Thomas Telford, London, pp. 237-245.

Richards, D.P., and Burchell, A.J., 1990, "Engineering Geology Considerations in Planning, Design and Construction of the Cairo Metro," Proceedings, 7th IAEG Conference, Lisbon, pp. 4223-4231.

Welsh, J.P., 1991, "Grouting," Underground Structures - Design and Construction, Sinha, R.S., ed., Elsevier, New York, pp. 328-350.

Xanthakos, P.P., Abramson, L.W., Bruce, D.P., 1994, "Jet Grouting," Ground Control and Improvement, John Wiley & Sons, New York, pp. 580-683.

Chapter 7

CLEARANCE IMPROVEMENTS FOR THE WHITE HAVEN AND ROCKPORT TUNNELS

G.M. Elliott,[1] J.J. Brady,[2] and J.C. May[3]
[1]Consultant; [2]B-U Corporation; [3]Consolidated Rail Corporation

ABSTRACT

Increased clearance was required in two railroad tunnels that were built about the time of the Civil War. Clearance infringements, whether they be excavation tights or caused by ice accumulation, reduce tunnel serviceability. Achieving increased clearance often means disturbing the ground and old ground support structures. Access and transportation needs dictate how and when the work can be carried out, and strongly influence the design approach and the methods used by the contractor to execute the work. This paper explores the pertinent considerations of the owner, the designer, the field inspection team, and the contractor in bringing this project to a successful conclusion.

INTRODUCTION

As those who have undertaken tunnel rehabilitation projects will attest, rehabilitating an existing tunnel can be as challenging, if not more challenging than driving a new tunnel, due to a variety of technical, logistical, and management constraints. The luxury of being able to walk through the existing tunnel to inspect existing conditions is no guarantee that there will be any fewer surprises during construction than driving new tunnels. Frequently, there are no as built drawings, no geologic records, and the possible causes for any observed deterioration of the lining are numerous. The only certainty is that a lining, if present, was constructed for some reason, and that would have related to sub-optimal ground conditions. Each project can be quite unique, and there must be a team effort and an equitable sharing of responsibility if the project is to be completed successfully. Within the project boundaries, the owner, the designer, the field inspection team and contractor must have a clear understanding of the position of each member of the team and the overall project goals. Outside of the project boundaries, each member of the team has also different management objectives to satisfy within their respective organizations that can themselves put pressures on team interactions. A recently completed project will serve as a case history to illustrate some of these interactions and issues, and the importance of close teamwork in achieving the optimum product.

In 1993 the Consolidated Rail Corporation (Conrail) prepared to increase clearances through bridges, and 12 tunnels in Pennsylvania in order to be able to ship containers double stacked on rail cars. With market forces and contractual arrangements requiring all work at 140 locations to be completed in 3 years, an aggressive schedule was set in which to complete the work. Two of the tunnels that were enlarged were the White Haven and Rockport

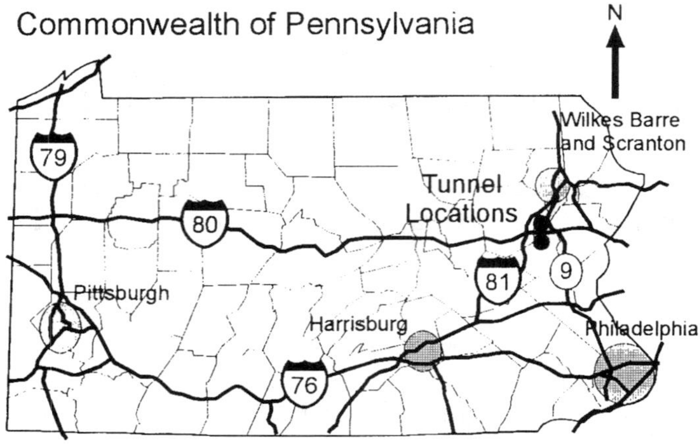

Figure 1. Site Location

Tunnels, located on Consolidated Rail Corporations (Conrail's) Lehigh Line in the Poconos region of Pennsylvania (Figure 1). The two tunnels were located approximately 18 miles apart on a single track section of railroad with a midpoint siding that could be used for basing a rail mounted operation. Access by road was possible to the north portal at White Haven Tunnel, but Rockport was approximately 7 miles from the nearest road access point and deep within the Lehigh River Gorge.

The 527m White Haven Tunnel (Figure 2) was believed to have been constructed in the early 1860s, and is in a horizontally bedded sandstone and siltstone unit, having an unconfined compressive strength in the range from 17 MPa to 117 MPa. Typical dimensions were approximately 5.8m to 6.4m height above top of rail, and 7.6m wide at springline. The tunnel was unsupported except for the 43m of stone masonry lining at the north portal. There was no portal structure at the south portal, and the portal face was blocky and rugged. Small rock spalls from the unlined portions of the tunnel were experienced throughout, but this was no indication of an impending collapse. The tunnel included two unlined and unsupported shafts that had been blocked at the surface, but let significant amounts of water into the tunnel. Ice accumulation during the winter was a significant maintenance problem. The gradient of the tunnel was approximately 0.5% from north to south, and surface water from a small catchment area in the vicinity of the north portal passed through the tunnel. Ice columns that formed against the walls would collapse during warmer days, and form ice dams that blocked the flow of surface and ground water through the tunnel. In severe winters, the level of ice and water in the ditch would rise up as much as 1m and threaten the track. Large icicles also formed from the roof above the track and posed a hazard to train crews. Ice was also considered to be the cause of serious bulging of the sidewalls of the masonry lining just inside the north portal, and this had contributed to severe deterioration of the portal arch.

The 365m Rockport Tunnel (Figure 3) was completed in 1884, and was constructed after part of the original line, which had been located at lower elevations in the gorge, had been washed away in a flood. It was mined through folded sandstone, having a compressive

Figure 2. North Portal, White Haven Tunnel

strength in the range from 69 MPa to 117 MPa, and was laid out on a 5° curve. This tunnel was originally excavated for two tracks, and thus was slightly wider than White Haven Tunnel. Typical dimensions were approximately 6.4m to 8m height above top of rail, and 10m wide at springline. It was also unsupported except for four stone masonry lined sections totaling approximately 73m, including both portals. The condition of the Rockport Tunnel was similar to White Haven, and experienced similar rock spalls and ice accumulation problems. In Rockport Tunnel, the condition of one section of masonry lining showed signs of deterioration where part of the masonry near the footing on one side of the tunnel appeared to have been excavated out by ditch clearing equipment, and thus weakened the arch. Both portals, but particularly the short north portal, showed significant deterioration by cyclic ice accumulation.

The primary project objectives were to increase clearance above the top of rail for a 6.5m high by 3.05m wide clearance envelope. To the extent that ice accumulation within the clearance envelope constitutes an obstruction, and knowing the history of severe ice accumulation in these tunnels, a solution to mitigating ice accumulation and improving drainage was also sought. However, before contracting for the design and construction of the work, a number of project management issues needed to be addressed.

PLANNING STAGE

Having identified the need for the project and set overall project goals, Conrail chose to address and make decisions on several important issues that would strongly influence the

Figure 3. North Portal, Rockport Tunnel

project. Clearly, these decisions were made early in the project when the degree of uncertainty about cost, schedule and technical feasibility was greatest, but they served to define the constraints under which the railroad wished the contract to be completed and construction to be planned and executed. Initial budgets are assessed by Conrail engineers on the basis of a short visual survey of the tunnels. However, the actual cost the railroad would ultimately pay was greatly effected by Conrail's willingness to make key decisions early in the project and share risk.

Geometrically, clearance in a tunnel can be achieved either by raising the roof, lowering the floor, or a combination of both. A decision on whether to go up, or down, or whether to wait and allow to the consultant to make recommendations after a preliminary investigation, was largely dependent upon operational considerations. If the track could not be taken out of service, by rerouting trains onto other lines, it tended to make economic and logistical sense to raise the crown. Also constraints on track profile had also to be evaluated.

Another important issue was drainage. The tunnels already experienced standing water in the ditches, and the height of rail above ditch water levels was minimal. Due to inherent drainage and subgrade problems for track in tunnels, a reduced ballast section is undesirable and the removal of floor rock in situations where the track must be maintained in service is very difficult at best.

There was then the issue of who should take the risk for interruptions to the work while trains pass through the tunnel if construction is to be carried out under "live track". If the railroad did nothing to control its own traffic, then the risk of work interruptions would be taken solely by the contractor and this would be reflected by much higher unit rates. To keep prices down, particularly on this low-volume line, it was preferable for Conrail to adjust train schedules and provide the contractor scheduled work windows of 8-hours in each 24-hour period. However, it was also necessary to write contract language that motivated the contractor to clear the track and leave it safe at the end of each window when the trains need to pass through.

The presence of track structure in a tunnel can be both an aid and a hindrance. If rail equipment is used, access is simplified if a switch and siding is located nearby, in which to clear the track for train operations at the end of each work window. Each days logistics must be carefully planned but benefit afforded by track operating equipment to use larger and higher construction equipment can prove advantageous. For conventional construction equipment, the presence of track rails and limited level floor width create maneuverability problems. Crawler mounted equipment can damage the rails and any equipment movement could cause a potentially dangerous misalignment of the rails. One acceptable solution for this situation is filling in the tunnel floor temporarily to the top of the rail with dunnage or ballast to provide a working platform, which is removed when all other work has been completed. However, there are often only limited sources for such materials and these can be located a considerable distance from the site. Recognizing the potential of contractors to be more cost-effective one way or another, Conrail agreed to permit either method for contractor access.

In the case of the White Haven and Rockport Tunnels, surveys established that up to 1.0m of additional clearance was required, with the White Haven Tunnel requiring the greater enlargement. In both tunnels, there was little to no ballast beneath the rail ties, so sufficient clearance could not be achieved simply by passing an undercutter through the tunnel. To lower the track would require blasting or slow hoe-ramming, which would be a considerably more costly operation. Also drainage was on a very shallow grade, and at White Haven, a culvert north of the north portal presented a constraint on designing the vertical profile. The Lehigh Lline experienced between 6 and 10 freight trains a day that could not be rerouted but could be rescheduled to allow the contractor a regular 8-hour work window each day. So it became readily apparent that the most sensible approach to increasing the clearance would be to raise the roof of the existing tunnel. In order to motivate the contractor to clear the track at the end of each window, and also to share the risk that the railroad would be forthcoming in providing the promised work window each day, three important clauses were written into the specifications and specific pay quantities were bid for contractor track time reduction.

- If the contractor failed to clear the track and leave the tunnel in a safe condition at the end of each work window, then the railroad stipulated a liquidated damages amount that would be deducted off the contractors invoice for every half-hour or part thereof.

- For the occasions when the railroad delayed the start of each work window by more that 30 minutes, the contractor was asked to include with his bid, a price for standby time. This item was paid in increments of 15 minutes of delay starting at the scheduled start time.

- The contractor was also asked to include with his bid a price for downtime should the work window be foreshortened. This too was paid in 15 minute increments of time that the 8-hour word window.

The next set of issues revolved around how to contract the work, and the amount of design interaction that was desired. First there was the issue of how to structure the contracts (e.g., one large design-build contract, separate design and construction contracts, or separate preliminary design, final design, and construction contracts.), and then there is the issue of how to pay (e.g., lump sum; cost plus fixed-fee; or cost plus fixed fee with a maximum cost limit). Conrail had also to decide whether to use its own resources to carry out construction management and inspection, contract the design firm either to undertake the construction management or provide on-site technical assistance to support its own project engineer, or contract an independent construction management firm.

Due to the variability of ground conditions, track time, and construction methods traditionally experienced in railroad tunnel projects, and the lack of information about existing structures, Conrail prefers to separate the design and construction components for tunnel work because it enables the scope of work for both parts to be defined better, and allows them to spread the risk fairly among all members of the project team. Conrail has used a wide variety of contracting methods, including several variations of the design/build concept, and it has been their experience that unless the total scope is well defined and all the technical and quality control standards are well understood, placing all the responsibility on one party for projects as uncertain as tunneling serves only to drive the price up. They see benefits to their contracting procedures in formally specifying contractor interaction as part of the design scope (to address constructability issues), and improving the way they handle value-engineering proposals from contractors.

In the case of the Pennsylvania Clearance Project, Conrail issued two design contracts and four construction contracts for the seven tunnels. Conrail stipulated that the design be bid lump sum, and that the designer also include with his bid, rates to be applied for supplying on-site technical assistance during construction. For construction, Conrail elected to contract variable quantity items on a unit price basis and to use lump sum pricing for those items which were either functional systems or easily quantified. Conrail wanted its own engineering staff to have regular input to the design details, for whilst liability and indemnification clauses can be written into a contract to grant the railroad compensation for design errors, they do not keep trains running if a design failure were to occur. Close interaction with the design details not only helps address operational and maintenance considerations and preferences in a timely fashion, but also helps keep the design focused on what the railroad wishes to be accomplished. For example, although an unlined tunnel ideally would be free of all rock spall, the existing White Haven and Rockport Tunnels historically had had continued non-structural spalling and the design criteria established that the situation could not be cost effectively remedied during enlargement, so rockbolting and shotcreting criteria were established to provide a finished roof with similar stability. The resolution of this issue early in the project had a significant impact on the anticipated means and methods for construction, and consequently on the anticipated scope of design work. If this issue had not been addressed early in the design through close designer-owner interaction, a lot of design effort could have been spent preparing a design that the owner neither wanted nor could afford. Conrail also wanted its own personnel to do the construction management, but envisaged the need for on-site technical help for certain inspection activities.

The last set of issues revolved around the schedule for the work. Should, the project schedule drive itself, or should schedule milestones be imposed from the start. In the first instance, Conrail could have set a milestone by which initial site investigations and conceptual designs of alternative strategies should be completed, and allowed later milestones to be developed as the project proceeded on a step by step fashion. In the case of the Pennsylvania Clearance project, Conrail was forced by the contractual arrangements with the other funding participants and various economic and commercial considerations to complete all 140 locations within three years. Additional constraints such as the number of work sites and their inherent delay to railroad operations within a given route segment similarly dictated specific aggressive scheduling.

The project milestones that were established for the White Haven and Rockport project were as follows. Funding for the work was finally approved in March 1993, and it was desirable to have the work at the White Haven and Rockport Tunnels completed before the depths of the upcoming winter, just eight months later. The consultant was given approximately 2-weeks in April after the start of the three day pre-bid site visit to bid on the design of seven of the Pennsylvania Clearance Tunnels. Following the award of the contract, just under 3-months was allowed to complete and sign contracts, undertake site investigations and preliminary design activities, finalize designs and prepare construction plans and specifications, and an engineer's estimate ready for distribution to contractors. At the design stage, the contractor was then given 3 weeks in the latter part of August, to attend a pre-bid site visit and prepare his bid. The schedule for the construction work was scheduled to take 130 days starting in mid-September in order that the temperature sensitive aspects of the work be finished before the cold Poconos winter set in.

DESIGN STAGE

BID PREPARATION

The tight schedule was felt by all project participants from the outset. From the designers point of view, bid preparation is the make or break point in a project, and it can be disastrous to give too little thought to how the project will be accomplished at the time the bid is prepared, particularly if the bid lump sum. Consequently a condensed bid period puts considerable pressure on the project manager. Numerous technical issues have to be addressed, multiple tasks have to be undertaken, and the risks of each decision have to be weighed. At the same time, it is necessary to stay extremely competitive for there are often numerous other firms keen to get this kind of work. On a fast track project such as the White Haven and Rockport tunnels, early decisions on key project constraints actually helped the engineer since they served to focus the bid preparation effort. The following few paragraphs describe some of the technical issues on the project and the way they were resolved during the design.

The approach to preparing the bid was to review the information collected during the site visit (photographs and notes) and formulate a preliminary assessment of the means and methods of construction which might be anticipated to achieve the project goals. From this point, development of the bid worked backwards to identify alternative design solutions, design calculations, design information needs, and data collection strategies that might be undertaken. Having outlined the tasks that might be undertaken to complete the design, the necessary manpower and equipment resources can be defined and investigated, and manpower and duration estimates for each task can be applied to determine costs. Then followed a detailed scheduling of activities to ensure that the various project milestones for each tunnel could be satisfied, an activity that was crucial on a fast track project such as this. Finally, there was the significant task of preparing the written proposal, which, as was the case for the White Haven and Rockport tunnels, included a description of the preliminary assessments of the means and methods of construction, and the proposed investigation strategies. This is an important area where trust and respect among team members can develop early. The ideas and discussion presented in this section enable the engineer to demonstrate the degree of understanding of the project, and help the client formulate an opinion of the engineer's capabilities and the technical feasibility of the approaches.

INVESTIGATION STRATEGIES

The development of data collection strategies for railroad tunnel projects must account for tight space constraints, and the highly uncertain availability of track time in the tunnel. Generally, no guaranteed track "windows" are provided for site investigation and yet the time available to carry out the investigations is very short. In addition, the floor geometry is seldom flat because of the existence of rail and ditches, and railroads do not like non-railroad equipment being driven over unprotected rail. So it was important to develop flexible and sufficiently comprehensive site investigation strategies that minimized equipment mobilization/demobilization time and maximized the rate of data collection. Whilst it is generally accepted that an extensive and complete pre-design investigation is desirable to minimize the possibilities of encountering unexpected ground conditions, railroad engineers seldom get this luxury, and the designer has to face the possibility that some details may not be resolved prior to construction. Some further investigation and finalization of the design details may need to be included with the construction package.

In the case of the White Haven and Rockport Tunnels, the primary information needs comprised;

- The amount of rock and liner removal that was needed to achieve the design clearances;
- The strength of the rock material and the structural features that controlled stability;
- The nature of the materials behind the masonry in the lined sections;
- Delineation of features that eliminated the use of certain types of ground support elements;
- The rate of groundwater flow into and surface water flow through each tunnel; and
- Locations of groundwater inflows (i.e., locations where ice would most likely form if

the tunnel was left uninsulated).

The means and methods used to address these needs consisted primarily of survey, geologic mapping, and probe hole drilling. Surveys were made of the rail profiles, ditch profiles, and profiles in the crown 1.5m to either side of track centerline. In addition cross-section were surveyed an approximately 15 m intervals and at locations of abrupt geometry changes, such as the ends of lined sections. Geologic mapping of rock outcrops and exposures inside and outside of the tunnel were used to record the structural characteristics of the rock mass, and to record the locations of water inflows that would form ice in winter months. Probe hole drilling consisted of drilling holes, between 35mm to 39mm in diameter, a distance of up to 2.5m using a pneumatic, jackleg, percussion drill, and then carrying a visual inspection of the hole using a fiber-optic borescope. This method was selected over conventional borehole drilling because it minimized the drilling distance, the equipment could be operated off the top of a scissor lift mounted on a high-rail vehicle (the latter enabling simple and rapid deployment of equipment in and out of the tunnel), and several holes could be drilled within a short space of time (thus enabling work to be carried out even during short track windows. Important engineering features, such as discontinuities, contact breaks, rock texture, and points of water ingress intersecting the probe hole, can usually be seen with the borescope, although it is not possible to get an accurate geologic description of the rock mass or samples for laboratory testing. The advantage of the borescope is that it provides an inexpensive continuous visual record of the hole that can be shown to and reviewed by others later. Also, the hole diameter is similar to that typicallly drilled for rockbolts, and so it was also possible to carry out rockbolt tests in several hole locations to support rockbolt design. Rock samples needed for laboratory strength testing were also taken using a portable concrete coring drill.

EXCAVATION AND GROUND SUPPORT IN UNLINED SECTIONS

Given the range of unconfined compressive strength determined for the rock units through laboratory testing, it was anticipated that the clearance improvements in unlined portions of both tunnels would be achieved primarily by the use of controlled trim blasting in the center crown, although the option to use a road-header or hoe ram was available to the contractor. The issue of whether to blast the entire tunnel width in order to restore a regular arch roof for optimum roof stability or whether to limit excavation only to that part of the crown where clearance was required, was also evaluated technically and economically. Technically it made little difference to the redistribution of stresses in the rock around the opening if the excavation was carried out either way. The square shape of the existing opening suggested that weakening of bedding planes in the haunches might have occurred in the past and precipitated subsequent slabbing of rock. Consequently, if the roof was “notched out”, steps needed to be taken to prevent this mechanism reoccurring in the future to the extent practicable. The need for a thin shotcrete layer to inhibit weathering and prevent small rock spalls was evaluated but considered unnecessary on the basis that small rock spalls were acceptable as discussed above, and major rock slabs would be held in place by the rockbolts that would be installed for arch support.

The need for any kind of ground support was evaluated carefully since all indications were that ground support had not been deemed necessary for arch stability in the past. However, calculations indicated that the use of rockbolts would be prudent, though not absolutely necessary, so rock support in the unlined sections of tunnel was provided by installa-

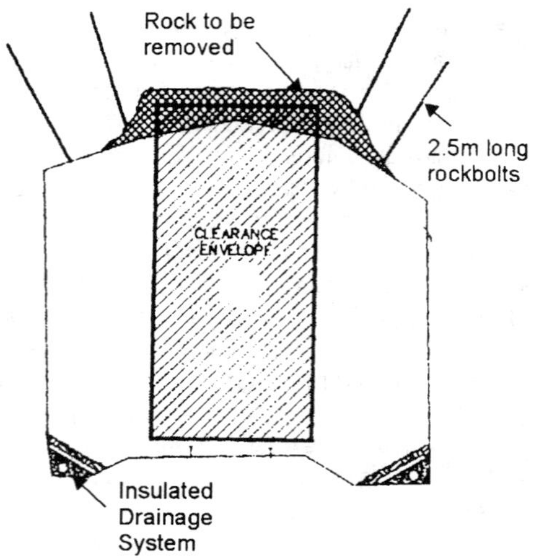

Figure 4. Typical Ground Support in Unlined Sections

tion of 2.5m and 3.7m long, resin-grouted rockbolts to support the less stable curved haunches, as illustrated in Figure 4. Longer rockbolts, spot bolting, and shotcrete were used where additional support was required.

EXCAVATION AND SUPPORT OF PORTALS AND LINED SECTIONS

With the exception of the South Portal at Rockport tunnel, relatively little material needed to be removed from portal or lined sections. However, some repairs were necessary, and the installation of support at the South Portal of White Haven Tunnel was justified.

Support of the south portal at White Haven was provided by 3.7m long rockbolts installed horizontally into the face and a 0.1m thick layer of fiber-reinforced shotcrete. Vertical pattern bolting for approximately 10m inside the portal was also used.

At the north portal, only a small amount of clearance improvement was required and this was restricted to the corners of the clearance envelope only. The amount of section reduction did not lead to an unacceptable reduction in the stability of the lining, and the quantity of material removed did not justify complete replacement with a new arch construction. Since there was also some value to the clearance project for keeping the existing portal structure designs were prepared in which the required clearance would be accomplished by excavating notches into the masonry lining. In order to prevent weathering of the freshly exposed stone, or loosening of remaining stones during the construction of the notches, a thin layer of shotcrete was applied to the surface of the notch. In those sections where the notches penetrated the thickness of the masonry arch by more than a few inches, an 0.1m thick layer of fiber-reinforced shotcrete was applied to the inside of the arch from springline to springline, and rockbolts were installed to help harness the latent strength of the weathered rock material behind.

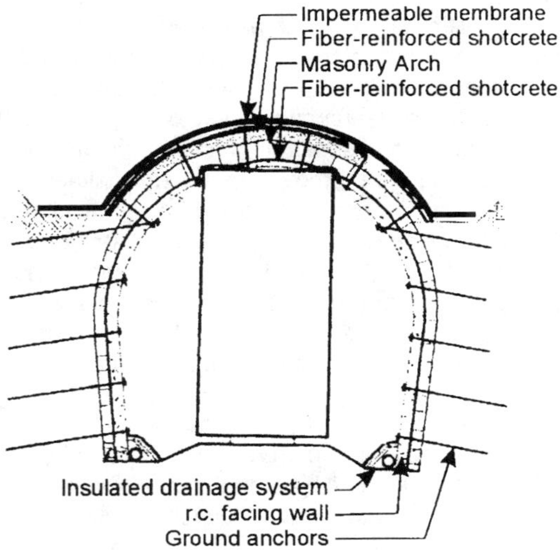

Figure 5. Repair to Bulged Section of Portal Lining

At the very north end of the lining, where the cover was too shallow to anchor bolts and consisted primarily of backfill soil, the walls of the liner over a 5m section had bulged into the tunnel significantly and thfpe arch had suffered severe deterioration. A review of the profile geometry and observations of groundwater in test pits hand dug in the steep slope above the portal concluded that this deterioration of the liner was the result of many years of ice jacking The water falling on the slope above the tunnel appeared to run down the slope at the soil/ rockcontact, and then infiltrate where the ground flattened out, just above the deteriorated section, and drain into the tunnel. Two alternative designs were prepared for bidding. Both alternatives included facing the existing masonry walls from ditch to springline with a 0.3m thick reinforced concrete wall anchored back using passive ground anchors. Above spring line, one alternative was a complete replacement of the arch with with a lattice girder and shotcrete arch constructed on top of the reinforced concrete walls. The second alternative, illustrated in Figure 5, involved applying shotcrete treatment the inside of the arch, and on the outside, the soil was stripped back to expose the outside of the arch to an elevation approximately 2.3m above springline, a 0.3m thick fiber-reinforced shotcrete blanket was placed, and the inner layer of shotcrete was connected through to this outer blanket using short bolts fitted with plates and nuts on both ends. To inhibit further water infiltration at this location, a PVC blanket was incorporated in the backfill at a specified elevation, to route the majority of infiltrating water into the surface water channels to either side of the portal structure. The second alternative was thought to be more difficult to construct, but presented considerably less surface excavation and backfill.

Both portals at Rockport Tunnel required clearance improvement, although at the north portal the required clearance improvement was small, and simple notching seemed appropriate. However, the north portal had experienced deterioration from ice accumulation. A couple of stones in the arch and footing had broken and pieces had fallen out. Cracking and bulging of the portal face was also evident. The north portal was quite short, and was likely built to prevent kinematic block failures, which today would be supported using rockbolts and shotcrete. Designs were prepared for both replacement and repair so that the railroad could chose the least costly alternative. The repair alternative involved applying a layer of fiber-reinforced shotcrete to the inside of the existing masonry work, and installing rockbolts to harness the latent rock mass strength. The required clearance could then be notched or ground out of the perimeter where clearance needed to be improved. The replacement involved constructing a lattice girder and shotcrete arch bearing upon existing masonry walls that would be left in place during arch removal. In the end neither plan was implemented since it was determined that the masonry arch could be removed safely in short sections allowing the exposed rock to be rockbolted directly using 3.0m long rockbolts and subsequently covered with a 0.1m thick layer of fiber-reinforced shotcrete. The portal face was also supported using 3.0m and 3.7m rockbolts drilled horizontally into the rock mass. For clearance reasons at the south portal, there was no alternative but to replace the existing portal and masonry lining with a new nominally 0.3m thick lattice girder and shotcrete arch and reinforced concrete portal face.

The undercut wall inside the Rockport Tunnel was repaired using a 0.1m thick fiber-reinforced shotcrete facing over a third of the tunnel perimeter extending up from the ditch, and tied back into the rock behind using pattern rockbolts.

Figure 6. Completed Tunnel Insulation

INSULATION

The mitigation of the major sources of ice accumulation in the tunnels was accomplished using a method first pioneered by the Swedish railroads and later used in North America, by British Columbia Rail and Canadian Pacific Railroad. More recently, Guildford Transportation has used the method in tunnels in New York and Massachusetts. The method makes use of a closed-cell polyurethane material manufactured by Dow Chemical, marketed under the name Ethafoam Plank. Single-layer applications of the product have an R-value of 26.5 (per W/m^2-ºC) which is sufficient to prevent water flowing down the back of the lining from freezing under the minimum average monthly temperatures in the Poconos. The Ethafoam plank insulation system provides several desirable elements:

- It is flexible so that can be deformed to follow the profile of both regular tunnels and irregular tunnel profiles;
- It is comparatively inexpensive (the contract price for ditch to ditch insulation was of $662.50 per foot run of tunnel. This cost is sensitive to the ability of the contractor to develop a working system to do the work, and as such can vary considerably between contractors); and
- It is easily patched and repaired in the event leaks develop or vandalism occurs, or the rock deteriorates behind the lining.

Insulation was designed to be installed only in those sections of the tunnel which exhibited the most significant inflows of ground-water (the project started too late in the winter for the locations of ice accumulation to be mapped prior to design), and was attached to the surface of the rock or masonry lining by means of short, 12.5mm diameter, galvanized steel pins set into rock on a 1m square grid that easily pierced the Ethafoam upon applied pressure. Experience communicated by previous users of this installation method stressed the importance of not expanding the grid of pins much greater than 1m. A curved plate and nut were used to hold the material in place.

As applied in the White Haven and Rockport Tunnels, (Figure 6) the 50mm thick Ethafoam insulation was fabricated into 2.5m by 9.75m panels. Joints between the panels were sealed using 6.5mm thick strips of the same product, which were heat-welded in place using an industrial hot air gun, thereby forming a continuously sealed insulation layer. Ends of the insulated sections of the tunnel were caulked using insulation and secured against the air blast of passing trains using galvanized mine strapping.

Concern was raised during the design about the possibility of the insulation material being fuel for a tunnel fire, ignited either by sparks coming out of the engine exhaust or by the heat generated by an engine that might come to a halt within the tunnel. Temperature data measured at different locations above the stack suggested that the air temperature at the crown of the tunnel above a stationary engine would be close to or a little above the melting point of the insulating material. Data was available to suggest that although the Ethafoam would primarily melt if temperatures rose too high, it was possible for the material to burn. Therefore, although the precipitating mechanisms were unlikely to occur, there was a desire to prevent sparks coming into contact with the Ethafoam and to diffuse the heat over a larger area so that cooling by air passing through the tunnel would prevent melting. To

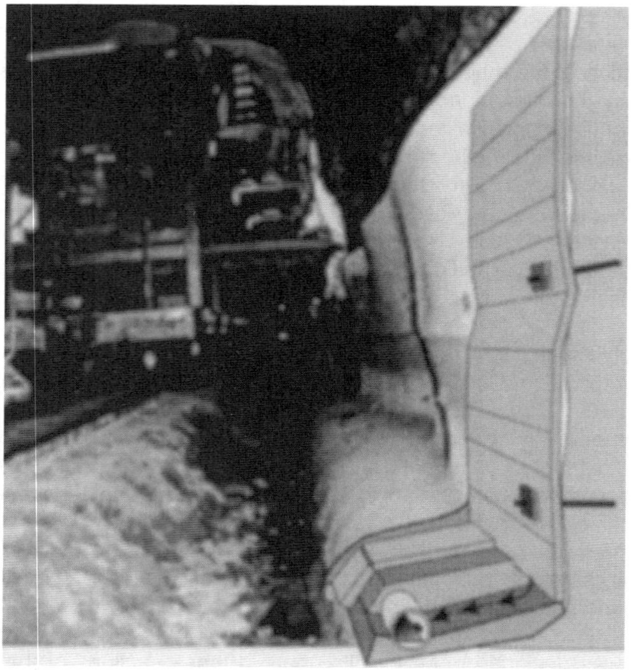

Figure 7. Insulated Drainage System Schematic

address this issue, a spark deflector sheet, consisting of 5-mildead forged soft aluminum laminated to 12.5mm thick polyethylene foam, was specified to be attached to the insulation layer 3m to either side of the center of track.

DRAINAGE

Although previous users of this insulating method did not find a need to insulate the water flowing in the ditches, there was uncertainty in view of the shallow grades and low flows in the ditches in both tunnels, about whether water in the ditches would not remain unfrozen before reaching the portals where the accumulation of ice was no longer a concern. The concern was that if water froze in the ditches before reaching the portal, water discharging at the bottom of the insulated sections would back-up, freeze, and possibly jack the insulation layer away from the rock wall. Consequently, an enclosed insulated drainage system was designed for each tunnel that was integrated with the insulated lining system.

The insulated drainage system designed for White Haven Tunnel (Figure 7) consisted of a perforated 0.3m. diameter PVC pipe resting on a lean concrete bedding with the perforations towards the wall of the tunnel. Water running down the back of the insulation lining would discharge onto the lean concrete and run straight into the drainage pipe. The pipe was covered with coarse aggregate for protection, a layer of Ethafoam for insulation, and a 0.1m thick layer of unreinforced shotcrete to weigh down and protect the insulation. Pipe cleanouts were located at approximately 122m on center. Surface flows entering the White Haven tunnel at the north portal were also diverted into the insulated drainage system.

At Rockport Tunnel, the ditch grade is barely 0.5%, and there is no grade over a 150 ft section at the north end. Concern was expressed that that the insulated ditch drainage design used at White Haven would be inadequate for use at Rockport, and an alternative flume system was designed whereby water discharging the insulation lining would first flow on a 1% grade concrete channel that would discharge down vertical inlets into a more elaborately insulated pipe embedded within the flume itself. During construction new information came to light that downplayed the importance of the issues upon which this design was based, and in view of the practical difficulties of its construction raised by the contractor, the design was abandoned in favor of the White Haven tunnel design.

Evidence for the success of this drainage system to prevent excessive ice accumulation in the ditches was provided during the first winter after construction. The ditches down-gradient of the portals remained open when in previous winters they had been covered with a mantel of ice and snow.

TEAMWORK

The team approach to the project was integrated with the design through design meetings at the 30% and 60% stages, and through the use of weekly status faxes sent by the engineer to the client. The objective of the 30% meeting was to report on the results of the site investigations and to present and discuss design concepts for addressing the various technical aspects of the project. This is typically the point of the project where product samples are on display and where models can be prepared to demonstrate practical considerations of novel design components or different construction methods. The objective of the 60% design meeting was to review drawing layouts and preliminary cost estimates and to discuss

Figure 8. Open Ditches in Winter!

construction contracting issues. Status faxes were also used to report evaluations of issues raised at meetings or during telephone conversations and to solicit decisions on the design details.

The foundations for integrating the prospective contractor into the project team were laid early through considerations given to the way information was organized and communicated in the plans and specifications. Thought was given to emphasizing the type of information that contractors would need to find and assimilate early in the bidding process, and to simplify the task of finding detailed information by people working with the documents on a daily basis during construction. For instance:

- Separate drawing packages, incorporating duplicated detail sheets were prepared for each tunnel so that the drawings package would consist of two independent halves, one for each tunnel in the contract.

- The Control of Work section, was used to present the available geologic conditions (i.e., the materials section), to describe the work, reference hazards and precautions, and to identify sequence and schedule constraints; and to describe clearly for each pay item in the bid activities that are included and not included.

- In developing the list of bid items, careful consideration was given to how work might be measured by those working in the field and how partial payments might be assessed, thus making it easier for payment issues to be addressed during construction. For instance, in the bid, rock excavation was bid by the unit volume. Wording was incorporated into the measurement and payment section of the specifications to explain that the owner and contractor would jointly assess and agree on the excavation quantity at the start of construction, and from that point on, it would become a lump sum item qualifying for partial payments by the proportional length of tunnel excavated.

- Efforts were made to minimize the burden of preparing, reviewing and approving submittals, by reviewing the intent of the submittals written into each specification and including only those that were deemed to give real value to the project. Whilst most engineers and construction managers might complain about contractors' reluctance towards the task of preparing submittals, no contractor appreciates overly onerous or unreasonable submittal requirements that seem to add no value to the project. Also no contractor appreciates an engineer who takes the full review period stipulated in the specifications to return approvals or comments. It usually makes for improved team relations if submittal requirements and review procedures are designed to simplify preparation and review time, and actually add value to the project. Clearly it is beneficial if some method can be devised that encourages and rewards everyone involved in the task of preparing, reviewing, and approving submittals.

- The submittals section was used not only to identify what submittals needed to be made, but grouped these submittals by when they were due, and described in some detail the required content of each submittal. By organizing the submittals section in this way, it was thought that it would (a) help the contractor to assess the manpower required to keep up with submittals, and (b) provide an easy checklist that all submittals are made at the appropriate times, and contained the necessary information.

- No attempt was made to finalize every last detail, especially those details that are better left for detailing when the contractor's means and methods are known, or lend themselves to proprietary solutions. These details can be completed either by allowing the contractor to propose a design through the submittal or value-engineering process, or simply acknowledging that the details will be finalized by the engineer during construction with input from the contractor.

CONSTRUCTION STAGE

BID PREPARATION AND MOBILIZATION

To a contractor, no two projects are exactly alike. This is especially true of projects involving the rehabilitation of railroad tunnels built over 100 years ago. The contractor needs time to plan and mobilize for the project. The more unique the project, the longer the time needed to plan and mobilize, particularly if unusual materials are used, such as the Ethafoam for the insulation system, or if special equipment needs to be fabricated to execute the work efficiently, such as the work platform constructed to install the insulation lining. Bid preparation is the make-or-break time for contractors as well as designers, and contractors also bemoan the pressures of a very condensed bid period. Too little time spent assimilating the information provided in the bid documents and formulating a plan of action can translate into disaster from both a profit and schedule perspective, and contractor might offset the shortage of time to fully develop ideas by increased unit prices for the work. However, the contractor's plight is eased somewhat if important and relevant information is openly disclosed in the bid package and if the site visit is conducted in an open manner. In this case, the information in the package was well laid out, and key issues and precautions for the contractor to heed were clearly stated.

Bid preparation starts with a review of the plans and specifications to assimilate the scope of work and key factors that will ultimately control the way in which the work is executed. Then follows one or more site inspections, and finally back to the plans and specifications for bid preparation. The contractor has to prepare an approach for executing the work and identify key pieces of equipment that might need to be bought, leased, rented, or purpose built. It is also necessary to work out the logistics of integrating all the construction components together so that the project milestones are met. Having formulated the means and methods that will used, it is then necessary to translate and spread the costs and profit among the pay items listed in the bid documents. It is also necessary to work out carefully the lead times need for materials as well as the design and manufacture of purpose built equipment.

One of the key dilemmas that contractors face when translating the expected costs among the pay items is the assumptions that should be used, particularly about pay items that are set up to pay for both scheduled and discretionary activities. Often a contract will include unit price items that are to be applied whether the work is carried out as part of a scheduled production activity or at the engineers direction. However the cost of the work associated with that bid item can vary greatly depending upon whether it is carried out in sequence with planned activities or demanded at an unscheduled time. For instance, rock bolts by the foot is an item included in the contract. However, a 3.5m long, 31mm diameter resin encapsulated rock bolt can be installed for say $36/m, or $138 per bolt, when installed as part of a pattern. However, the cost to install a bolt "at the direction of the Engineer" when the installation crew and equipment have been mobilized elsewhere on the project, can be over a thousand dollars. The same is true for shotcrete. The ratio of the activity cost if it is directed out of the sequence to the cost if it is applied in sequence, can exceed 5 to 1. The amount to which a contractor must "weight" his bid must include an assessment as to whether common sense will be used when directing the use of discretionary quantities. Generally it is preferable that pay items for production items be separated from discretionary items, even if they go by the same description.

In the process of developing a strategy for executing the work being bid, a contractor may also identify an area where an alternative, less expensive or more reliable method might be worth proposing under potential value engineering clauses. The effort required to prepare a submittal on such an idea adds further burden to the contractor at a critical period of the project.

In the case of the White Haven project, it was decided at an early stage to mount a rail based construction operation, but the work required a considerable amount of specialized job built equipment to be designed and constructed. Some of the unique features of particular interest from the construction point of view on this project that affected costs and schedule included:

- repairs and clearance improvement at the north portal of White Haven Tunnel;
- the insulation layer;
- the insulated drainage system; and
- construction of the new portal at the south end of Rockport Tunnel.

These features and the construction methods used in their construction are discussed in the following sections.

ROCK EXCAVATION

Rock can be removed economically by use of a road header or drill and blast. Road headers come in many sizes and shapes. It would have been possible to excavate almost all the rock in the White haven tunnel using a road header, and indeed it was tried. However, the machine that was used was slightly too small and it needed to work off a stable platform. To reach the proper height, the road header was mounted on a railroad flat car. However, the flat car had not been provided with adequate bracing to the walls and floor of the tunnel, and thus it was unable to mine the rock at the desired production rate. Consequently, this method was abandoned, and the project was completed by trim blasting. For this project, a Gardner Denver Universal III Jumbo with two JTT1 booms and PR55 drills was mounted on a flat car (Figure 9). A rolling deck was constructed on the same car in front of the jumbo to facilitate the loading of holes with explosives and wiring of detonators. A non-electric system was specified to be used to initiate the blast. The rolling deck was also used to scale and bar down loose rock after the blast.

Figure 9. Layout of Rock Excavation Car

The blast rock was shot directly into an air-dump car, and then hauled out of the tunnel to be deposited along the railroad. R.O.W. at a location designated by the owner. This worked reasonably well, except not all the shot rock landed in the dump car, and the dump car was destroyed. Replacement of such a piece of equipment is significant and has to be factored into the excavation price. On a previous project the contractor had designed a wood planked gondola car with folding wings. Once the car was positioned under the area to be blasted the wings would be folded out against the rock ribs using built in hydraulic cylinders. This set up caught a greater proportion of the fly rock, but required considerable daily repairs, and in the end it too was destroyed. Another alternative to this method of rock removal is to ballast the ditches and plank the tracks with heavy timbers. Then blast directly onto the timbers, pick up the shot rock using a front end loader, and tram it out of the tunnel to a waste area. This method tends to be costly and requires much greater periods for muck removal.

There are other methods of rock removal such as drilling holes and filling with an expanding agent, such as Bright Star, or drilling holes and using hydraulic expanders. The former method was used to do some rock splitting and removal above the south portal of the Rockport Tunnel. It worked well but was slow and expensive for general rock excavation, as is the experience of using hydraulic expanders.

NORTH PORTAL WHITE HAVEN

In areas where the liner was to be “notched”, the critical construction issue was the development of an efficient method to remove the necessary portion of the masonry arch, but not more than was needed. Too much removed would weaken the arch. Too little and the double stack cars would not clear. This required very exact measurement. A template car was built that provided direct measurement, steel to steel, no springs, from track to top of

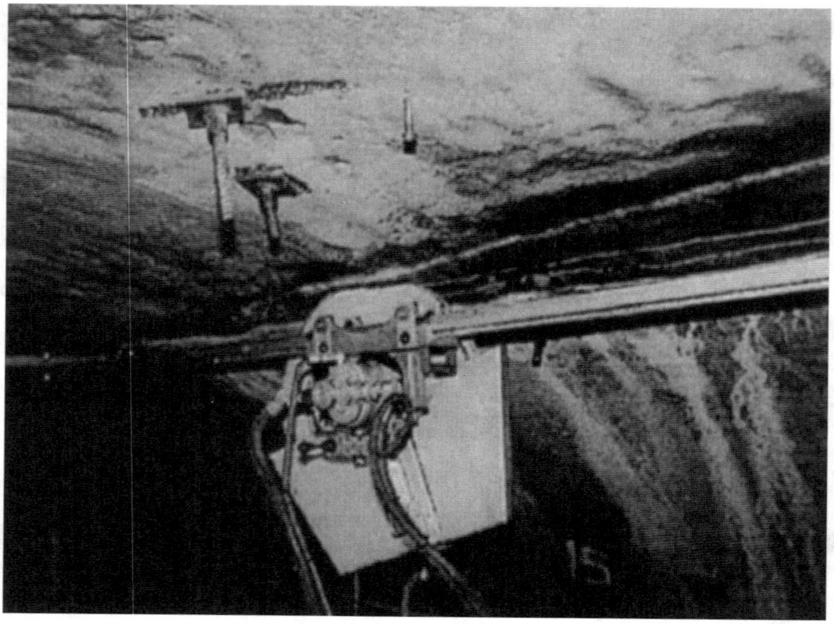

Figure 10. Notch Construction Using Hydraulic Track Saw

template. In this fashion we could determine the exact width and depth of the notch. The selected method involved cutting a series of slots parallel with the axis of the tunnel, and then splitting the masonry between pairs of slots using a pneumatic spader. The slots were cut with a hydraulic track saw moving on a track that was bolted to the stone arch and orientated parallel to the line of the slot to be cut (Figure 10). The saw cuts were spaced approximately 50mm to 75mm apart. were cut to varying depths in order to develop the proper profile for the notch. This method involved minimal vibration of the arch and resulted in quite a clean finished notch profile.

This notching was not performed until the upper portion of the arch had been excavated, blown clean, and a 12-in layer of shotcrete applied. Rockbolts were then drilled through the masonry arch and the 12-in shotcrete layer. Plates and nuts were added to top and bottom to form a sandwich. The excess bolt on top side of arch was trimmed off. An additional layer of shotcrete was applied to embed the bolt and nut. This was done so the bolts would not pierce the geomembrane which was next applied. The whole area was then restored to its original contours and seeded. On inspection the following winter, there was no ice accumulation in the repaired section of tunnel lining.

INSTALLATION OF TUNNEL INSULATION

A review of the plans identified four distinct situations for installing the insulating liner:

1. Full arch to toe drain on a regular masonry wall or rock wall;

2. Full arch to toe drain on an irregular rock surface;

3. Half sections from spring line to toe drain on regular surface; and

4. Half sections from spring line to toe drain on irregular surface.

The key to successfully carrying out this aspect of the construction was the design and construction of a piece of equipment that could efficiently handle all four cases. The regular masonry and rock surfaces were expected to be straight forward. The real problem was the irregular surfaces, and areas where roof elevations were high as a result of the rock breaking back to weaker planes and discontinuities. The wall and arch could vary from the center line of track anywhere from 3.6m to 6.1m. The same was true of the arch. The car also had to fit through the existing portals, and when parked on a contractors rail siding, had to allow for the passage of freight trains (Figure 11). For this project, a multi-deck work platform, consisting of a series of frames containing sliding work decks, was constructed and mounted

Figure 11. Construction Train Parked in Siding

on a 16m rail flat car. The work decks could expand from 3m wide to 10m wide once in place. Additional height above the top deck was handled with ladders and scaffold. In this way work could be carried out on a full 16m of tunnel at one time, and at four different levels. Level one was from ditch to first sliding deck, approximately 2.1m; level 2 was the next 2.1m from deck 1 to deck 2; level 3 was the 2.1m above deck 2; and level 4 was on planking laid across scaffolding towers erected on the top deck. The goal was to install a complete 16m long arch segment in a shift, including all fixing pins, insulation from ditch to ditch, cross-panel stripping, and spark deflector. This would be almost 279m^2 per shift. This target was never accomplished, although production got close in the regularly shaped masonry sections (Figure 12). The main underestimation was the time required to drill and install the 12.5mm galvanized hanging pins. The foam had a fair degree of flexibility but could not be made to follow the contours exactly (Figure 13). This became even more difficult as the spark deflector also was to fit on the same pins and was much narrower. But despite the difficulty in installation, the purpose built work platform worked and was very cost effective compared with alternative solutions. Lessons were learned, and if done again improvements on installation in the irregular sections could be made.

FIELD INSPECTION

The field inspection team is responsible to insure that the plans and specifications are followed. However, it is equally as important, seeing no plans are 100% perfect, that the field inspection team handle the inevitable changes with a fair and even hand. To this end, having on-site technical assistance provided by members of the design team during construction facilitates the procedure of seeing that the design intent is achieved even though minor details might be waived. Also under the value engineering clause, it is the field inspection team's responsibility to expedite the review of the contractor's proposed changes, by working closely with the design team and owner to ensure a quick, fair, and open-minded review.

Often, a contractor sees an opportunity to propose a value engineering (VE) alternative. Such a proposal either better lends itself to the means and methods that contractor has available, or reflects experience of having constructed different designs to address the same design issue and evaluated which works better. Invariably the VE proposal is expected to result in a net savings to the owner. A VE proposal presented to the combined entity is viewed as a challenge to their design (offends their pride of authorship). Much effort is expended to show why the original design was superior to the contractors proposal. It takes a very "big" firm to review these value engineering proposals fairly without letting pride of authorship cloud good engineering. But when "fairly" reviewed, the entire project team, including contractor and designer, can more quickly understand the full ramifications of the proposal. Do they satisfy the design intent? Do they meet the owner's long term requirements? Do they reduce the uncertainty, and hence "risk" in the expected performance. The Engineer has some flexibility to implement the VE clauses which can also be exercised. For instance some VE proposals may yield only minor cost savings, but provide a disproportionate improvement in quality or reliability. In such instances, the contractor might be granted all the cost savings, rather than just the proportion specified in the contract documents. Such an action would be another means of improving team relations and the spirit of cooperation.

On the White Haven and Rockport tunnels project, there was generally good communication between all parties during construction, and this significantly contributed to dealing with and resolving several complicated issues in the field. Representatives of each member of the project team met weekly to discuss progress, upcoming activities and construction related issues. At these meetings, the contractor would often raise a design issue that he felt might be better addressed by an alternative method. The designer shared with him the design intent and rationale for the design presented in the plans, and the owner shared the operational and maintenance concerns. With this information in hand, the contractor could give the idea more thought, and perhaps present a more detailed proposal at some later meeting. This would then be reviewed against the design intent and other key performance goals, and would be either accepted or rejected with explanation or request for modification. If a contractor's propsal was accepted, a change order would be quickly forthcoming so that the contractor could immediately incorporate the accepted modification into his construction schedule. As a result of implementing this open system of communication, the job proceeded smoothly and without any major conflicts.

CONCLUSIONS

This paper has reported on preparation, design, and construction of a clearance improvement project in two railroad tunnels. The project not only required rock removal and ground support, but incorporated solutions to mitigating ice accumulation, liner repairs, portal reconstruction, and drainage improvement. This project was successfully completed in 1994. The authors were respectively the design teams project manager, the contractor, and the owner's project manager, and each has contributed his own perspective on the execution of the project, and presented and discussed some of the key issues that characterized the design and construction. The owner received the facility he required at a reasonable cost. This could only have been accomplished on this unique job through a true team effort.

This job ran smoothly because all parties sat down on a regular basis, reviewed schedule and progress, and kept design intent in the foreground at all times. All were open to suggestions and alternate proposals from other team members, whether it was alternate designs suggested by the contractor, or alternate construction procedures suggested by the designer. They weren't all accepted. But they were received openly, without prejudice, and were fairly reviewed. Those thought to be valuable were handled very quickly and made part of the contract.

3

TBM Case Histories I

Chairman
D. Rogstad
Frontier-Kemper
Evansville, IN

Chapter 8

ANALYSIS OF TBM PERFORMANCE AT THE RECORD SETTING RIVER MOUNTAINS TUNNEL #2

Bill McCormick,[1] Eric Gilmore,[2] Ron Tudor,[3] Brad Bush,[3] and Gregory E. Korbin[4]

[1]Carollo, Black & Veatch, currently with Kumar and Associates; [2]Carollo, Black & Veatch; [3]Kajima Engineering and Construction; [4]Tunnel Consultant

INTRODUCTION

Las Vegas has recently been termed the "Construction Capital of the Country". It has also been acknowledged as the fastest growing city in the country for several years running. The rapid growth improves the economic picture of the region but it also imparts unusual demands on the area's infrastructure. In 1994 the Southern Nevada Water Authority (SNWA) was working on a master plan to address the rapid growth. A long range analysis revealed the need for additional water treatment capacity and delivery systems. Short term projections showed that the current system would reach capacity in the summer of 1997 with the possibility of demand exceeding the supply.

The River Mountains Tunnel #2 is part of a $150M scheme to quickly upgrade the system to meet the short term demand, improve the overall system operation and allow for a smooth transition into the larger additions being planned for the future. The SNWA selected Carollo/Black & Veatch (CBV) to provide detailed design and construction management for the upgrades. The River Mountains Contracts consisted of five separate contracts broken out separately to stream line a very aggressive (just 34 months) design and construction schedule. Necessary work included four miles of excavated and lined tunnel, connections from existing pipelines to and from the new tunnel including large diameter (to 3.66m (144 in)) valves, a 18 927cu m (5 million gal) regulating reservoir, a surge tank and a 170 343 cu m (45 million gal) reservoir to add additional storage to the system.

Notice to Proceed was given to Kajima Engineering and Construction (Kajima) on May 22, 1995 for the construction of the River Mountains Tunnel #2. The design called for a tunnel to be driven 6070m (19 915ft) in rock and lined to a finished diameter of 3.71m (12.17ft). Excavation began on December 14, 1995 and holed through on April 30, 1996. Lining began on June 25, 1996 and a notice of Substantial Completion was issued on December 13, 1996. This paper describes the measures taken to complete the tunnel in 18 months from Notice to Proceed to Substantial Completion.

THE RIVER MOUNTAINS SITE

The River Mountains are a low range of rugged mountains separating the Las Vegas Valley from its main supply of water, the Colorado River at Lake Mead. They are volcanic in origin and the harsh desert environment has left them as expanses of exposed bedrock rising 305m (1000ft) above the valley with little or no vegetation. An existing 3.05m (10ft) diameter tunnel through the range is part of the current system that treats water from Lake Mead and delivers it to Las Vegas. The River Mountains Tunnel #1 was constructed from 1968 to 1970 by Utah Construction and Mining Company for the United States Bureau of Reclamation. The River Mountains Tunnel #2 was constructed 45.7m (150ft) north and parallel to the existing tunnel. The new tunnel slopes from an elevation of 581.6m (1908ft) at the east portal to 575.5m (1888ft) at the west portal, matching the slope of the existing tunnel. Lake Mead is 5km (3mi) east of the east portal and the Las Vegas Valley is 5km (3mi) west of the west portal.

Similar methods were used for excavation and lining and records of those activities aided both the design and construction of the new tunnel. A comparison of the construction of the two tunnels will be included later in this paper.

TUNNEL GEOLOGY

The River Mountains consist of faulted, NE and NW trending, Tertiary volcanic rocks. They are flanked on the north, west and east by Tertiary sedimentary rocks and Quaternary alluvium. A wealth of information pertaining to the site geology was available from the design and construction of the first tunnel. This included logs from 14 borings drilled along the existing alignment, surface geologic mapping and a geologic profile of the existing tunnel. Records of the previous tunnel excavation including location and type of support, rock stability and excavation rates were also available. Existing data as well as data from seven new borings and geophysical studies were used in siting the new alignment and preparing the Geotechnical Design Summary Report (GDSR). Twenty six geologic units were described in the original geologic mapping. From those, rocks with similar engineering characteristics were grouped together and resulted in four engineering geologic units being presented in the GDSR. The units in order of first occurrence in the excavation are: Conglomerate, Agglomerate, Rhyolite, and Rhyodacite. Intrusive dikes were also mapped over a small percentage of the excavation. Table 1 shows engineering data collected for the different engineering geologic units along with predicted and as excavated lengths of the different rock types (Intrusive dikes account for less that 2% of the excavation and are not included in Table 1).

The tunnel alignments are above the groundwater table and no water was encountered during the first excavation. However, stories of leakage from Tunnel #1 during its first year of operation worried both designers and the construction team. Although open joints up to 0.3m (1ft) wide were encountered during the excavation for Tunnel #2, no water was encountered.

TBM SPECIFICATION

As part of the contract documents, CBV listed minimum requirements for the tunnel boring machine (TBM) to be used on the project, as summarized in Table 2. In addition to those minimum requirements, the TBM was also required to have hydraulically controlled roof supports for cutterhead stabilization and to control over break, a finger shield that allowed for rock bolting immediately behind the head, a gripper footprint pressure low enough to prevent damage to the tunnel walls and a dust scrubber system including multiple cutterhead dust seals. By contract, Kajima was also required to maintain access to a list of critical spare parts including a spare main bearing, bearing seals, bull gear and drive motor with gearbox.

Table 1. Engineering Properties of Engineering Geologic Units. (Intrusive Dikes Were Anticipated in 3.5% of the Excavation and Encountered in 1.8%)

PARAMETER	AVERAGE VALUES FOR ENGINEERING PROPERTIES			
	Rhyodacite	Agglomerate	Rhyolite	Conglomerate
Bulk Density (g/cm³) (pcf)	2.18 (136)	1.99 (124)	2.10 (131)	2.23 (139)
Moisture Content (%)	2.9	3.8	1.5	1.5
Young's Modulus (GPa) (x 10⁶ psi)	8.62 (1.25)	5.38 (0.78)	20.27 (2.94)	9.03 (1.31)
Unconfined Compressive Strength (MPa) (x 10³ psi)	43.32 (6.28)	20.10 (2.92)	32.59 (4.73)	24.40 (3.54)
RQD	73	53	66	47
Fracture Spacing (m) (ft)	0.20 (0.67)	0.13 (0.42)	0.20 (0.67)	0.1 (0.33)
RMR Classification	Good to Fair	Fair	Good to Fair	Fair to Poor
Q-System Classification	Good	Fair to Poor	Good to Fair	Fair to Poor
Anticipated Footage (m) (ft)	2034 (6672)	1821 (5974)	1821 (5974)	182 (598)
Actual Footage (m) (ft)	1875 (6150)	1908 (6260)	1977 (6487)	198 (650)

Table 2. Minimum Specified and Supplied TBM Charactertistics

	SPECIFIED MINIMUM	SUPPLIED
CUTTERHEAD POWER (kW) (hp)	671 (900)	932 (1250)
CUTTERHEAD TORQUE (Nm) (ft-lb)	677 910 (500 000)	772 817 (570 000)
CUTTERHEAD THRUST(kN) (lbf)	178/cutter (40 000 /cutter)	178/cutter (40 000 /cutter)
TBM MUCKING CAPACITY (m/hr) (ft/h)	7.63 (25)	10.67 (35)
MUCK HANDLING BEHIND TBM (m/h) (ft/h)	7.63 (peak) (25)	9.14 (peak) (30)
(m/h) (ft/h)	6.10 (sustained) (20)	9.14 (sustained) (30)

The CBV design team realized that the Owners required schedule would likely preclude the use of a new TBM. A survey of available used TBM's indicated that machines were available that could be suitably rebuilt to meet the requirements of the job.

Figure 1. Lowering TBM Head onto Cutterhead Support and Side Supports

CONTRACTORS TBM

While preparing their bid, Kajima indeed found that time did not allow for consideration of a new TBM. Therefore, during preparation of their work plan, Kajima planned to use rebuilt Jarva Mark 12 for the excavation. At that time the machine was boring a 1219m (4000ft) long, 4.37m (14.33ft) diameter tunnel in Quincy, MA. That machine was expected to complete that drive in less than two months and be ready to begin mining the River Mountains Tunnel #2 by mid-October 1995.

Kajima submitted their work plan based on the use of the Jarva machine but continued to study the issue. Based on the geotechnical information and the data from the previous excavation it was obvious that the rock would cut well. Advances in cutter and cutterhead design could provide for advance rates significantly higher than the 32.92m/d (108 ft/d) averages achieved during the first excavation. As part of their continued study, Kajima compared the Jarva Mark 12 to a Robbins 129-182 that was also available. The key issue of the comparison was the muck handling capacity. Analysis of head geometry, buckets and main beams of the two machines was done. It was demonstrated that the Robbins could be modified to handle a larger volume of muck and therefore provide potentially higher advance rates than the Jarva. In mid-June Kajima informed CBV that they were proposing to modify their work plan to include use of the rebuilt Robbins model 129-182 in place of the Jarva Mark 12. Kajima provided CBV with information showing that the rebuilt Robbins TBM would meet specification requirements and the change was approved. Kajima had been working with BORETEC of Solon, Ohio on the TBM comparison and employed them to perform the rebuild. To insure that the TBM was rebuilt to their satisfaction, equipment superintendents from Kajima worked in the BORETEC shop during the rebuild.

The rebuild started by increasing the diameter of the cutterhead from 3.66m (12ft) to 4.34m (14.25ft). Cutter head design and balance was checked by an outside consultant. Kajima had decided to go with an open face design and the increase in diameter also allowed for an increase in the muck bucket size. Muck bucket design was optimized through the use of a wooden model. The main beam hopper and conveyor were enlarged and strengthened to handle the increased volume of muck to be delivered by the larger buckets.

Figure 2. TBM Prepared to Begin Mining at the West Portal

Additional space was obtained in the main beam by removing the existing conveyor and building a new one to use all available space. Deep throated top rollers were installed on mounts built into holes cut through the walls of the main bean and accessible from outside it. Traditional return rollers were not used. Instead, guide rollers and wear plates were installed where necessary. A 660mm (26in) belt was fit into the main beam along with skirting to provide maximum filling capacity and to prevent spillage. A moveable head pulley and hydraulically controlled tail pulley section provided the necessary capacity for adjustment.

Other improvements to increase production capacity included replacing all 29 standard 394mm (15 1/2 in) cutters with 432mm (17in) discs, replacing 149kW (200hp) drive motors with 186kW (250hp) units and adding a fifth 186kW (250hp) motor to increase torque. Cutterhead rotation speed was set to 12 rpm. To keep TBM advance as continuous as possible various schemes for rapid regrip were studied. Included were options for double grippers and single grippers with fast reset. A less elaborate system which including single grippers controlled by high capacity flow valves was used. This system was designed for a quick reset of the grippers and allowed for regripping to be done in less than 45 seconds including steering corrections.

For rock bolt installation, Gardner Denver model 83 drills were mounted on each side of the TBM behind the cutter head. Hydraulically controlled slides allowed the drills to be moved independent of the TBM. This allowed rock bolts to be drilled while the TBM was advancing, even at rates as high as 12m/h (40 ft/h). Drills were adjustable and able to drill the top 120° of the tunnel.

CONVEYORS

Kajima chose to use a 610mm (24in) wide Long-Airdox continuous conveyor system for muck removal behind the TBM. A magazine type storage system with capacity for storing 457m (1500ft) of belt was used. Additional belt was added to the system in 305m (1000ft) lengths. The extra 152m (500ft) of system storage allowed Kajima flexibility in terms of interruption of mining for adding belt. Mechanical splices were used for additions of belt, an operation that typically took less than 2 hours to perform. To reduce wear, mechanical splices were replaced with vulcanized splices on weekends. Three in-line booster drives were installed at appropriate locations as the tunnel advanced. All drives were built to fit onto the tunnel rib at springline without additional excavation and were installed on weekend shifts. Some delays were seen for start up and testing of the booster drives. The main and booster drives were interlocked by electrical control cable (dupe line) and set to start up in succession. Once outside, the muck was transferred to an inclined stacker belt which moved the muck to the spoil area. The system was rated to handle 363tonne (400ton) of muck per hour but operation showed it was capable of handling additional amounts.

VENTILATION

Ventilation was provided by Spendrup Fan Co. of Clifton, CO and consisted of four 149 kW (200hp) axilvane turbine fans mounted within the tunnel at the west portal. The fan system blew air to the heading through a 1.22m (48in) flexible bagline system. A bagline canister was mounted on the last section of trailing gear and fed out 183m (600ft) of ventilation line automatically as the TBM advanced. TBM advance was stopped to change an empty canister with a second one that was filled and waiting outside. Changing bagline canisters delayed mining approximately 30 minutes.

A wet scrubber system was used to draw dust laden air from the cutterhead. Clean air was discharged from the scrubber down station of the main ventilation system outlet for proper circulation. With TBM advance rates over 9.14m/h (30ft/h) the wet scrubber was often overwhelmed and had to be cleaned frequently.

Kajima originally thought that booster fans would be needed within the tunnel to provide adequate ventilation for the full excavation. The ventilation was pushed to the limit and at the end of the excavation was still providing 14.2m^3/s (30 000cfm) of air flow in the tunnel without booster fans.

THE EXCAVATION

Erection of the TBM began on November 21, 1995 after delivery of the cutterhead. A launching trench had been excavated while awaiting delivery of the TBM that allowed the west portal. machine to be erected within two feet of the portal face (sta 0+00). Crews from BORETEC and Kajima worked two shifts per day assembling the TBM and backup equipment from November 21 until December 14, 1995 when it began the excavation. Mining progressed using three - 8 hour shifts with shift changes at the heading. Kajima initially used a schedule of five mining days and two maintenance shifts on Saturday.

The first three weeks of the mining were "stop and go" as the TBM and trailing gear advanced into the tunnel to allow the final pieces of the continuous conveyor and the ventilation system to be installed. As the bugs were worked out of the system the potential of the TBM became apparent. By sta 4+57 (15+00) the excavation had encountered all types of rock described in the GDSR and TBM performance remained essentially constant in terms of the thrust and torque required to cut the rock. In the third week of the excavation daily advance rates near 61m/d (200 ft/d) were achieved. By the fifth week daily rates near 91.4m/d (300ft/d) were common and by week 8 crews were disappointed with anything less than 30.5m (100ft) per shift. During this time additional problems were encountered and corrected. The TBM was now advancing at rates of 7.6m/h (25ft/h) and above which brought out shortcomings in the conveyor systems. First to appear was an unbalanced advancing tail piece, the first car of the trailing gear. The drag created when the TBM advanced faster than the conveyor could pay out belt caused the advancing tail piece to rotate counterclockwise and push the rail up the opposite rib of the tunnel. Hydraulic rams used to adjust the position of the advancing tailpiece were over stressed and broke. Two days were spent redesigning the advancing tailpiece, adding counterweights to overcome the rotation and additional hydraulic rams. Two days were also lost while the mainbeam conveyor head pulley was redesigned and rebuilt. The delivered system used short hydraulic rams and a system of cables and pulleys to adjust the head pulley. Instantaneous penetration rates near 40 ft/hr caused spillage, even in the enlarged hopper, and lead to binding of the belt and broken mechanical splices. The pulleys and cables proved inadequate for mainbeam cleanup and belt adjustments. A more reliable system of longer hydraulic rams was fabricated and installed. This made routine cleaning and maintenance of the TBM conveyor more efficient.

During week 9, after mining the 1610m (5280ft), crews established their first world record by mining 148.4m (487ft) over three shifts, exceeding the old record of 143.9m (472ft) by 4.5m (15ft). The following weeks were slowed by installation of the first booster drive for

Figure 3. Project Manager Waiting for Workers to Emerge from the Excavation After Holing-thru

the continuous conveyor, but days with production over 122m (400ft) became common. During the week of February 26th crews broke their own daily record by advancing 150m (492ft) and added the weekly record by mining 636.7m (2089ft) between February 26 and March 1, 1996.

Starting the beginning of week 16, mining changed from three eight hour shifts to two ten hour shifts with four hours each morning dedicated to maintenance. This improved availability of the TBM by approximately one hour per day.

The final full week of the excavation saw crews setting still another world record for TBM advance when they excavated 83.2m (273ft) in one ten hour shift. After a tense final 30 hours of the excavation which saw multiple mechanical splice failures on the continuous conveyor, the crews holed-thru six weeks early and were greeted by more than one hundred spectators including SNWA managers and guests.

GROUND SUPPORT

Less than 5 percent of Tunnel #1 was supported and that was done entirely with rock bolts. Tunnel #2 was designed 0.6m (2ft) larger in diameter which CBV felt could lead to greater support requirements. CBV therefore specified a minimum pattern of rock bolts and described several other support systems that might be required. A conservative approach toward evaluation of support requirements was used to insure prospective contractors were prepared for the worst. For the most part, all rock types proved to be self supporting and rock bolts were installed more in anticipation of holding the slickline during concreting than required for support.

The excavation for the River Mountains Tunnel #1 encountered the worst ground in the second half of the excavation and this was true of Tunnel #2 as well. In the second half of the drive, crews encountered three short zones of blocky/slabby ground. Over the final six weeks of the excavation these zones slowed the excavation by a total of three shifts as crews installed light support consisting of split set rock bolts, wire mesh and mine straps. Some additional support work was required well after the TBM had mined past these areas as stress relief in parts of the excavated rock caused the crown and ribs to loosen. In all, the excavation required less than half the support described in the GDSR.

TBM PERFORMANCE

CBV provided full time inspection throughout the excavation. When the TBM was excavating, the inspectors collected performance data including elapsed time for each push, average drive motor amperage per push and average propel pressure per push. This information was shared with Kajima and helped assess rock conditions and the performance of their TBM. This information quickly revealed that the TBM was well suited, if not over designed, for the sedimentary and volcanic rock at the site. It was apparent that the full thrust and horsepower of the TBM would not be used and that production was limited by the capacity of the muck handling system.

For analysis the TBM performance data was input into a spreadsheet program with database capabilities. The push length and elapsed time were used to calculate the instantaneous penetration rate (IPR), the average amperage was converted to TBM horsepower and the average propel pressure was used to calculate the thrust per cutter (with correction for drag).

Table 3 contains a summary of that data for the different rock types excavated. The data is slightly skewed in that the conglomerate was excavated while crews were in the first part of the learning curve. They were not pushing the TBM hard enough to get a true picture of the optimum performance.

Table 3 reveals slight differences in thrust and power requirements related to the strength of the rock, however, the overall penetration rate achieved was essentially the same for each rock type since the controlling factor was the capacity of the conveyors to remove muck. The data show that the conveyor system could handle no more than 9.75m/h (32ft/h) of continuous advance despite only utilizing 65% of installed TBM power and less than 50 % of its thrust capacity, on average.

Table 3. Excavation Database Summary

ROCK TYPE	LENGTH (m) (ft)	(%)	AMPS/ MOTOR	POWER avg kW (avg hp)	PROPEL MPa (psi)	THRUST/ CUTTER kN (kips)	IPR m/h (ft/h)
CONGLOMERATE	198 (650)	3.3	117.9	276.7 (371.0)	9.37 (1359.1)	27.6 (6.2)	5.88 (19.3)
AGGLOMERATE	1908 (6260)	31.4	174.6	550.3 (737.9)	16.6 (2401.2)	70.3 (15.8)	9.66 (31.7)
RHYOLITE	1977 (6487)	32.6	180.0	563 (755.0)	18.0 (2612.9)	78.7 (17.7)	9.85 (32.3)
RHYODACITE	1875 (6150)	30.9	187.9	612 (820.7)	19.6 (2841.8)	88.1 (19.8)	9.24 (30.3)
INTRUSIVE	112 (368)	1.8	139.5	403.3 (540.8)	13.0 (1879.9)	48.9 (11.0)	9.51 (31.2)

AVAILABILITY

The TBM was available 88.1% of the time. The major factor affecting availability was the TBM (main beam) conveyor. Even after the rebuild of the head pulley the system remained a high maintenance item. At IPR's in excess of 10.7m/h (35 ft/h) the belt and hopper were usually running full and spillage of fines into the main beam resulted. If the main beam was not cleaned regularly, cuts in the belt and torn mechanical splices resulted.

Cutter wear was minimal during the excavation. None of the 29 cutters were changed until 1864m (6114ft) into the drive and then it was only the 8 gauge cutters. Through the course of the excavation a total of 30 cutters were changed. All gauge cutters were changed once with 4 being changed twice, face cutters 5 through 21 were changed once and all four quad cutters lasted the entire excavation!

UTILIZATION

The main reason for not mining when the TBM was available was the continuous conveyor. Most problems with the conveyor were electrical caused by "ghost current" in the control cable (dupe line). This would cause operator's control lights to be on when the conveyor was off, off when it was on and at one point the only way to start or stop the conveyor was at the main drive outside the west portal. The problem was solved by simply adding signal boosters to the long runs of cable, but finding the problem was time consuming. Installation of booster drives also caused down time. Each booster had its own controls and transformer which had to be connected to and synchronized with the main drive and other boosters. By the installation of the third booster, crews were able to perform the work on Saturday with only minor disruption for start up on Monday morning.

CONCRETE LINING

CBV gave three options for lining the tunnel, steel lining or Reinforced Concrete Pipe (RCP) both with cellular concrete backfill and Cast-In-Place (CIP) concrete. Kajima chose

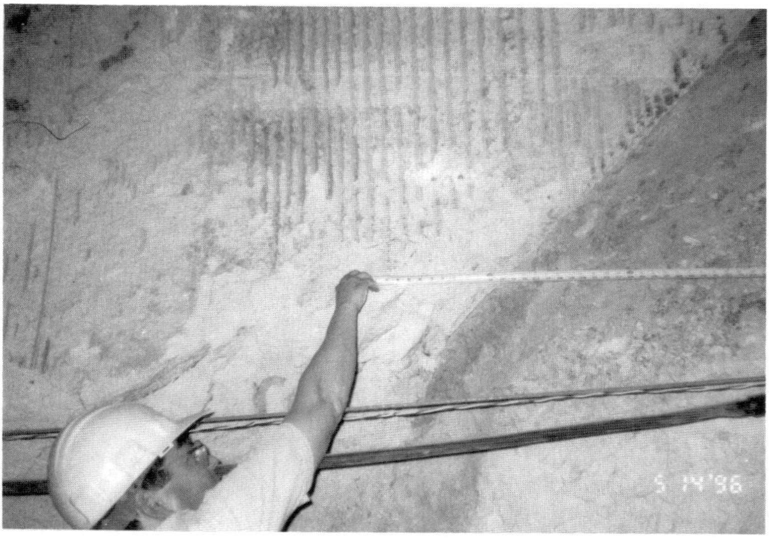

Figure 4. Indentations Left by Outside Gauge Cutter Showing Penetration Rate of 25.4mm/rev (1in/rev)

the CIP option which proved to be far more economical.

Concrete in Tunnel #1 had regular circumferencial cracking and significant leakage when first watered up. Records showed that the tunnel had sealed and become essentially water tight after less than one year of service. This information lead CBV and the owner to

Figure 5. Concrete Pumping Equipment and Crew Placing Final 91m (300ft) of the Tunnel Lining

Figure 6. Concrete Form Carrier Moving Forms into the Tunnel for the First Pour

accept some cracking and therefore initial leakage and allow continuous placement of concrete by specification. Also allowed were sloping joints which had been used on the first tunnel with satisfactory results. Both allowances were felt to maintain the quality of the lining and at the same time give prospective contractors the ability to meet the required schedule.

Kajima purchased 128m (420ft) of invert forms and 119m (390ft) of arch forms to form the 3.71m (12.17ft) lining and a Schwing 5000 pump to place the concrete. The pump and form carrier were diesel driven. A local concrete supply company, CSR, was employed to build and operate an on-site batch plant capable of suppling 115cu m (150cy) of concrete per hour for the lining. Two months were spent assembling forms, modifying the concrete lining equipment, installing a California switch in the tunnel and building the batch plant.

On June 20, 1996 Kajima attempted their first placement. The full 119m (390ft) of forms were erected in the east end of the tunnel. The concrete pump was set with 183m (600ft) of 152mm (6in) slickline that extended to the steel transition piece to be cast into the first pour. A concrete delivery system consisted of the batch plant capable of batching 4.6cu m (6cy) of concrete at a time and 3 concrete trains each consisting of one 13.6tonne (15ton) Plymouth locomotive with three 4.6cu m (6cy) moran cars. The initial haul time from batch plant to the pump, a distance of 5852m (19 200ft) was 45 minutes. The first placement included 61m (200ft) of tightly spaced rebar that gave the largely inexperienced crew problems. After four pours and 73m (240ft) of plugged slickline, the first 119m (390ft) of lining was finally placed.

After this troublesome start, Kajima crews worked their way up the learning curve. A two crew system was used with concrete placement during day shift and tunnel cleanup and form movement on night shift. Both crews typically worked 10 to 12 hours per shift to accomplish the work. By week 4 crews were moving and filling 82m (270ft) of forms. By week seven consistent cycles of 91m (300ft) days were seen. By the end of the lining activities crews were moving 110m (360ft) of forms and placing over 420 cu m (550cy) of concrete per day. Crews finished the lining in 17 weeks averaging 75m (246ft) per day. Average rate of concrete delivery was 36cu m/h (47 cy/h) with a maximum of 44cu m/h (58cy/h).

A TALE OF TWO TUNNELS

The River Mountains Tunnel #1 was constructed by Utah Construction and Mining Company for The Bureau of Reclamation between 1968 and 1970. They excavated the tunnel using the then relatively new technology of a TBM. The lining was also cast-in-place. Table 4 provides a direct comparison of the many similar features of the two projects.

Not shown in Table 4 are the support requirements of the two tunnels. In all, Tunnel # 2 required less support and zones with fall out were less severe than reported in the first excavation. It is felt that stroke length and gripper position play the greatest part in this difference. The Jarva Mark 11 had two sets of grippers in a cruciform configuration. The top grippers loosened the weaker rock in the arch lead and to greater fallout. The 0.6m (2ft) stroke did not allow the grippers to span fallout areas. The rebuilt Robbins TBM used a single set of grippers set at springline. The 1.22m (4ft) stroke of the machine allowed the grippers to span the short stretches of weakened rock.

CONCLUSIONS

Rarely does a contractor have the opportunity to drive a tunnel in near perfect conditions. Given the dry, essentially self supporting, easily excavated, non-abrasive ground in a rural environment, Kajima was able to optimize tunnel excavation and lining to achieve world record rates.

Significant Achievements

- World record advance rates: Best Shift - 83.2m (273ft)
Best Day - 150m (492ft)
Best 5-day week - 636.7m (2089ft)
- No lost time accidents.
- Project completed on time and under budget.

Table 4. Comparison of River Mountains Tunnels #1 and #2

	RIVER MOUNTAINS TUNNEL #1	RIVER MOUNTAINS TUNNEL #2
DATE	1968-1970	1995-1997
LENGTH (m) (ft)	6087 (19 970)	6070 (19 915)
DIAMETER -Excavated/Lined(m) (ft)	3.66/3.08 (12/10.1)	4.34/3.71 (14.25/12.17)
BID PRICE NUMBER OF BIDS NEXT LOW BID HIGH BID	\$3,946,619 20 \$5,299,445 \$9,215,094	\$20,233,465 6 \$20,561,900 \$26,834,000
MINING		
TBM	JARVA Mk11-1200 (Rebuilt)	ROBBINS 129-182 (Rebuilt)
WEIGHT(tonne) (ton)	59 (65)	109 (120)
POWER (kW) (hp)	298 (400)	932 (1250)
THRUST	2758kN @ 11.0MPa (620 000 lbs @ 1600 psi)	5160kN @ 29MPa (1 160 000 lbs @ 4200psi)
TORQUE	230 500Nm (170 000 ft-lbs)	772 817Nm (570 000 ft-lbs)
STROKE (m) (ft)	0.61 (2)	1.22 (4)
CUTTERHEAD SPEED	9.3 rpm	12 rpm
MUCK REMOVAL	Muck Train	Continuous Conveyor
BEST SHIFT(m) (ft)	31.7 (104)	83.2 (273)
BEST DAY(m) (ft)	89.3 (293)	150 (492)
BEST 5-DAY WEEK(m) (ft)	339.6 (1114)	636.7 (2089)
BEST MONTH(m) (ft)	1012.3 (3321)	1639.5 (5379)
DAILY AVERAGE(m/d) (ft/d)	32.9 (108)	62.5 (205)
PENETRATION RATE RANGES (m/h) (ft/h)	0.61-8.2 (2-27)	3.05-10.4 (10-34)

	RIVER MOUNTAINS TUNNEL #1	RIVER MOUNTAINS TUNNEL #2
AVAILABILITY	80%	88%
UTILIZATION	55%	36%
CONCRETE		
LENGTH OF FORMS(m) (ft)	73.2 (240)	119 (390)
CONCRETE PUMP	Pressweld Gun	Schwing 5000
SLICKLINE	107m long, 152mm dia 350 ft long , 6in dia	183m long, 152mm dia 600 ft long, 6in dia
CONCRETE DELIVERY	Moran cars	Moran cars
DAILY AVERAGE(m/d) (ft/d)	52 (171)	75 (246)

Lessons Learned

- TBM technology is available to produce rapid rates of excavation. For greater advance rates, muck removal technology will need to be advanced as well.
- Long concrete placements with sloping joints can be used to increase production without sacrificing the quality of the work.
- The RMR and Q-Rating rock classification systems provide conversative estimates of ground support in tunnels driven with modern TBM's.

Chapter 9

PUNK ROCK IN PORTLAND

Joseph P. Gildner,[1] Dan Nowak,[2] Don Z. Painter,[3]
Gordon F. Revey,[4] Pat Wilmoth,[5] and Shane Yanagisawa[3]
[1]Tri-Met Transportation District of Oregon; [2]Construction and Tunneling Services;
[3]Frontier-Kemper/Traylor Brothers; [4]GEOTEK andf Associates, Inc.;
[5]Parsons Brinckerhoff Construction Services

ABSTRACT

Frontier/Traylor is presently completing construction of twin tunnels for the Tri-County Metropolitan Transit District of Oregon's (Tri-Met) westside lightrail system. The tunnels are each 4,550 m (14,935 ft) in length and 6.48 m (21.29 ft) in diameter. The 1/3 of each tunnel was driven from the west portal using drill and blast methods and the remaining 2/3 was driven from the east portal using a TBM. The paper describes:

The difficulties encountered by the TBM mining the weathered Columbia River Basalts.

The execution of drill and blast work despite challenging ground conditions, stringent noise and vibrations specifications, and public relations problems.

The excavation and lining of the Washington Park Station 79 m (260 ft) below the park zoo.

INTRODUCTION

Three separate papers were originally written to describe very different aspects of the underground works on the Tri-Met Westside LIght Rail Project in Portland, Oregon. At RETC's request the papers were consolidated. By necessity the three sections on TBM excavation, drill & blast excavation, and station cavern excavation are brief but they do show the challenging nature of the project and how the challenges were met.

The project is part of the 29 km (18 mile) long westside extension being constructed by the Tri-County Metropolitan Transportation District of Oregon. The work is being performed by Frontier/Traylor, a joint venture of Frontier Kemper Constructors and Traylor Brothers. Lead designer for the project is Parsons Brinckerhoff. Notice to Proceed was issued in July 1993 and the project is expected to be completed by September 1997. By October 1996 all the excavation had been completed. As of January 1997, the concrete lining and trackwork are well underway.

As shown on Figure 1, the tunnels penetrate through the West Hills of Portland in a generally east-west alignment. Each tunnel is about 4,550 m (14,935 ft) long and 6.48 m (21 ft.) in diameter. For contract purposes, the tunnels is divided into three sections called Reaches A, B, & C. Each reach is about 1.5 km (5,000 ft) long. The contract requires that

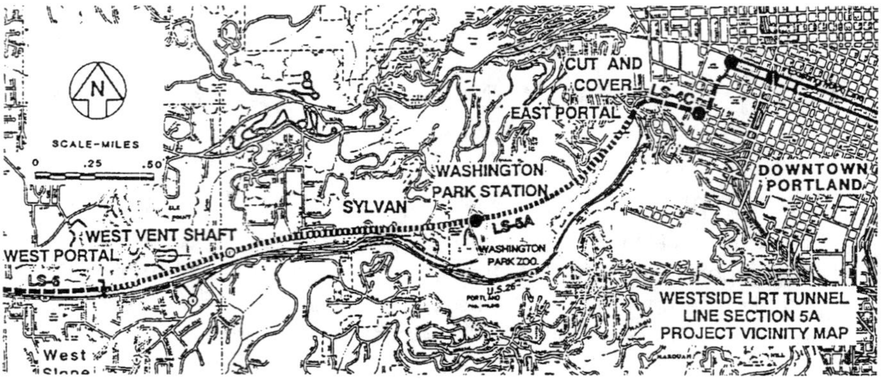

Figure 1. Tunnel Alignment

the first 1,584 m (5,200 ft.) from the west portal (Reach A) be excavated by drill and shoot methods. The first 1,533 m (5,035 ft.) from the east portal (Reach C) was required to be excavated by a new TBM using a conveyor belt for muck removal. Reach C is a continous 5% grade uphill from the east portal. The center 1,432 m (4,700 ft) of Reach B could be excavated by either drill & blast or TBM at the contractor's option however the contract schedule left no choice but to continue use of the TBM from Reach C.

In addition to the tunnels, the underground works include excavation and concrete lining of the Washington Park Station which is at the interface between Reaches B and C. The subway station is about 79 m (260 ft) underground and is the deepest in North America. Two large shafts provide access from the surface via elevators. Other structures include a ventilation shaft over the Reach A tunnels. The shafts were sunk from the surface using drill & blast although raise boring and slashing were originally intended.

PROJECT GEOLOGY

The different excavation methods required for each reach were a result of both geological and environmental conditions. The geological conditions are related to three major formations along the alignment, the Columbia River Basalts (CRB), Boring Lava, and Portland Hills Silt (PHS). Figure 2 shows a simplified geological profile and tunnel alignment.

Reach A is the western third of the alignment and is 60% Boring Lava, 25% CRB, and 15% PHS. The Boring Lava is good quality material consisting of medium high to high strength, slightly weathered to fresh basalt. Joints are randomly oriented cooling type fractures with spacing from close to wide. The PHS varies from low strength siltstone to stiff-to-hard silts and clayey silts with a generally massive structure. The CRB has very low to high strength, is slightly weathered to completely weathered, and has very close to wide spaced joints. The alignment passes through 3 fault zones. Due to the variable nature of these formations, the reach was deemed unsuitable for TBM excavation and the contract required excavation by conventional methods.

Reach B is the middle third of the tunnel alignment, and could be excavated by either

Figure 2

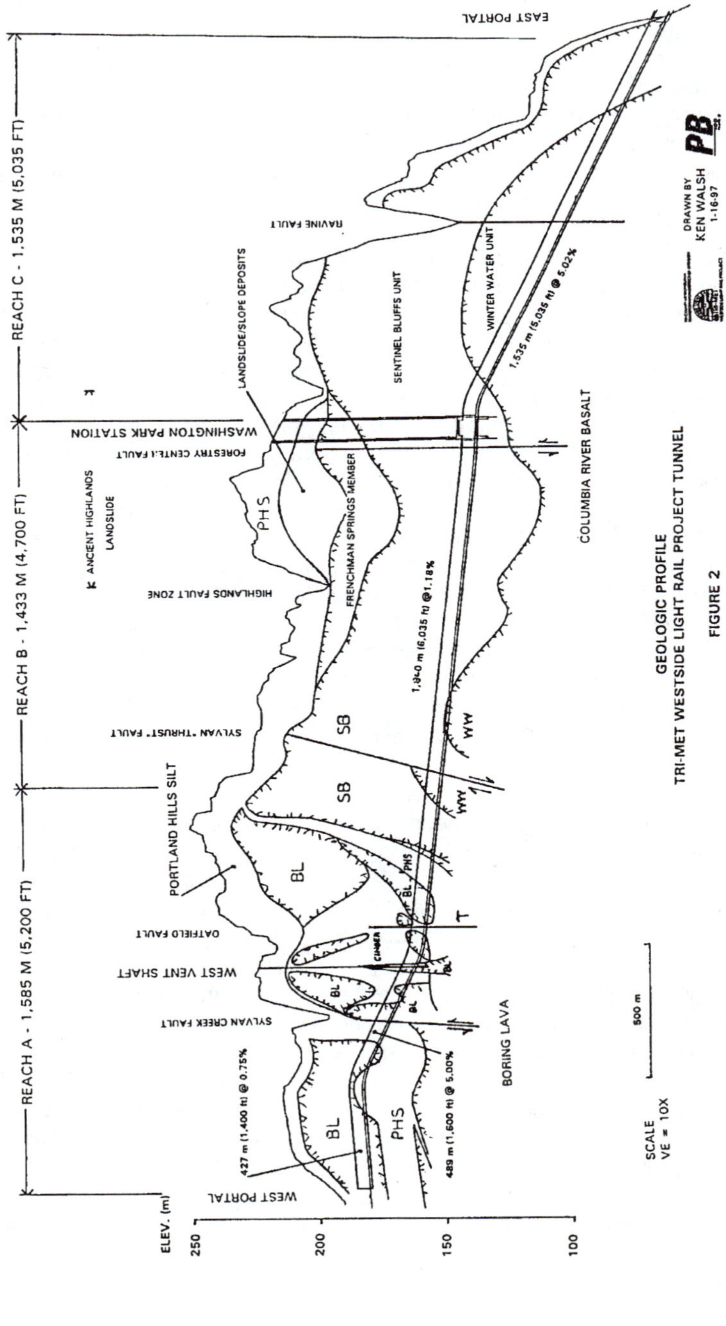

GEOLOGIC PROFILE
TRI-MET WESTSIDE LIGHT RAIL PROJECT TUNNEL

FIGURE 2

drill & blast or TBM. The entire 1,432 m (4,700 ft.) length of Reach B is within the Sentinel Bluffs. The eastern 2/3 of Reach B passes through a gentle syncline structure, while the western 1/3 is within an anticline.

The Reach B section of the tunnels is entirely within the Sentinel Bluffs unit which has similar rock mass characteristics to the west end of Reach C. The central section of Reach B was predicted to have more lower quality rock due to peripheral effects of a major thrust fault zone parallel to the alignment. A flow breccia zone 1.5 to 5 m (5 to 17 ft) thick was expected within the tunnel horizon for a length of 457 m (1,500 ft). Tunneling in Reach B ranged from poor in a wet and faulted section to good in the reaches of more competent rock. A blocky basalt flow base overlying the flow breccia increased cutter maintenance on the TBM.

Reach C is the eastern third of the tunnels and was required to be excavated by TBM. Reach C is entirely within two units of the Columbia River Basalt except for a short section at the east end of the starter tunnels which is Portland Hills Silt. The alignment passes through an anticline which is a result of the tectonic activity which formed the West Hills (Tualatin Mountains) of Portland. This anticline is the dominant geologic structure throughout most of Reach C. The tunnels start at the east end in the Sentinel Bluffs unit of the CRB. At 335 m (1,100 ft) into the drive the tunnels pass into the Winter Water unit of the CRB. After passing through 1,036 m (3,400 ft) of Winter Water, the tunnels go back into the Sentinel Bluffs, which is the CRB unit present throughout the remainding 152 m (499 ft) of Reach C.

The eastern 133 m (435 ft) of the tunnels was designated as the starter tunnels which would be mined conventionally before the TBM was launched for the drive. The first 10 m (33 ft) of the starter tunnels was full face Portland Hills Silt with mixed face silt and CRB for the next 43 m (141 ft). The final 80 m (263 ft.) of the starter tunnels is full face CRB. The CRB Sentinel Bluffs in this area was expected to be slightly to moderately weathered and low to medium high strength, with very close to close jointing. Many of the joints in the eastern section of the Sentinel Bluffs were expected to have silt and clay infilling to thicknesses of less than 0.5 cm (0.20 in). A 3 m (10 ft) thick flow breccia was also expected in this area.

The remaining 202 m (663 ft) portion of the eastern Sentinel Bluffs was thought to be similar to that in the starter tunnels but with increasingly better quality as depth of overburden increased. The entire 335 m (1,100 ft) section of the eastern Sentinel Bluffs proved to be difficult ground for tunneling because the joint infilling was saturated and provided very little friction between the basalt blocks causing the ìstandup timeî to be very short.

The Winter Water unit was expected to be medium high strength, slightly to very slightly weathered, with very closely to closely spaced, slightly rough joints. This ground is better suited for mechanical excavation and, except for a short faulted section in the middle of the 1,036 m (3,400 ft) stretch, tunneling proceeded well through the Winter Water. The faulted section of ground was very similar to the eastern Sentinel Bluffs in that joint infilling was detrimental to ìstandup time.î The first notable quantity of groundwater was encountered near the fault.

The west end of Reach C is in the Sentinel Bluffs unit again, but the quality of this section is much better than the eastern portion of this unit. The rock in this area is described medium high to high strength, slightly weathered to fresh, with joints spaced very close to close. The joints in this area are described as being locally mineral healed. Tunneling conditions were good throughout this section.

TBM EXCAVATION - BORE-REGARD VS. PORTLAND'S WEST HILLS

A new TBM was purchased from Construction & Tunneling Services. At a party to celebrate the machine's unveiling, the TBM was christened Bore-Regard. Little did anyone at the party know that Bore-Regard and his mining crews would soon become locked in a two bout heavy-weight fight with the West Hills of Portland.

In This Corner - West Hills of Portland

There were two tunnels to mine, each 2,966 m (9,735 ft.) long and about 6.48 m (21 ft.) in diameter. At the end of each tunnel the TBM would have to be disassembled inside the tunnel and backed out via the east portal. The tunnel route was expected to run through two layers of slightly weathered to fresh Columbia River Basalt. The fight program (Geotechnical Design Summary Review) described the basalt as having a low RQD but the fractures were thought to be cooling type joints, limited in extent, wavy, and exhibiting a high degree of interlocking. Standup time was considered adequate for hardrock TBM operation.

The contract specified three different levels of ground support which ranged from rockbolts only all the way up to W6x25 circular ribs spaced at 4 ft. o.c. supporting wire mat lagging. The contract assumed that ribs and wire mat lagging would be used for only 16% of the tunnel. In the end, 99.2% of the tunnel used ribs and lagging.

Contender Specifications

Diameter	21'-3 1/2"	6.49 m
Cutterhead Drive		
Type	variable speed hydraulic	
Power	2,100 hp	1,567 kw
Speed	0 to 6.4 rpm	
Torque Max. Speed	1,240 ft-kips	1,680 kN-m
Max. Torque 0-3.7 rpm	2,480 ft-kips	3,360 kN-m
Intemittent torque 0-2.5 rpm	3,470 ft-kips	4,700 kN-m
Number Hydraulic Pumps	6	
Number Hydraulic Drives	10	
Cutters - Backloading style		
Quantity and Size	43 @ 17 inch	43 @ 432 mm
Capacity	55,000 lbs	245 kN
Propel System		
Grippers	4 oversize shoes, right/left & top/bottom	
Max thrust in 2 shoe mode	2,375 kips	10,564 kN
Max thrust in 4 shoe mode	4,750 kips	21,128 kN
No. thrust cylinders	2 per shoe, total of 8	

Stroke	24 inches	610 mm
Fighting Weight	500 tons	450 tons

As can be seen on Figure 3 there was no main beam on Bore-Regard. Instead, four oversize gripper shoes formed a shield which supported, steered, and propeled a fully shielded cutterhead. The grippers provided nearly complete 360 degree ground support. Ground suppport was installed from shields attached to the top and side grippers. Therefore the grippers never had to step over the ring beams and the grippers could be stroked without regard to ring beam location.

The forward end of the gripper structure telescoped into the forward shield via four pairs of propel cylinders. Each pair of cylinders projected in a V configuration from a gripper shoe into the cutterhead. Because the cylinders were mounted at an angle to the machine axis, they also provided torque reaction Four gripper cylinders provided gripping forces in the horizontal direction and four more gripped in the vertical direction. Each pair of grippers could be operated independently.

Direction and roll control of the forward shield were provided by the primary thrust circuit. When boring straight ahead, an equal amount of oil was delivered to each cylinder. With this arrangement the pressure in each cylinder automatically adjusted itself to compensate for uneven drag on the forward shield and kept the cutterhead moving along squarely. To steer, an increased volume of oil was added to the appropriate cylinders. To roll the forward shield, oil was added to every other cylinder. The operator had three basic controls: steer left/right, steer up/down, and roll clockwise/counterclockwise.

The gripper/propel system had two operating modes. The normal mode was to continuously advance. Propel and regrip were alternated between the horizontal and vertical grippers so the TBM did not stop mining waiting for a regrip. As there was no need to wait for regrip, the short stroke of the cylinders was not a problem. The short stroke was a help in limiting the amount of unsupported ground between the shield and grippers. In the continuous mode gripper ground pressure was about 1,729 kPa (250 psi.)

The second operating mode (cyclic mode) was used in bad ground where high thrust forces were required to overcome high ground pressures on the shield or where low gripper pressures were desireable. All eight gripper and propel cylinders were used to advance the cutterhead. Gripper ground pressure was cut by about half but the TBM had to stop for regripping.

The TBM forward shield assembly consisted of the cutterhead, bearing/drive structure, and the shield. Commonly called a "flex shield", it was hinged along one side and attached to three hydraulic cylinders on the opposite side. The front shield was called a flex shield because it consists of a 5 cm (2 in) thick steel shell that could be hydraulically actuated to expand or contract. In the expanded mode the shield provided active ground support immediately behind the cutterhead. For starting up in squeezing ground the shield and gripper shoes could be contracted to release the machine from a trapped position.

The cutterhead had a flat face dressed with 43 cutters. A notable feature was the minimal projection of the cutterhead in front of the flexshield, just 914 mm (36 in) and later improved to 610 mm (24 in).

Figure 3

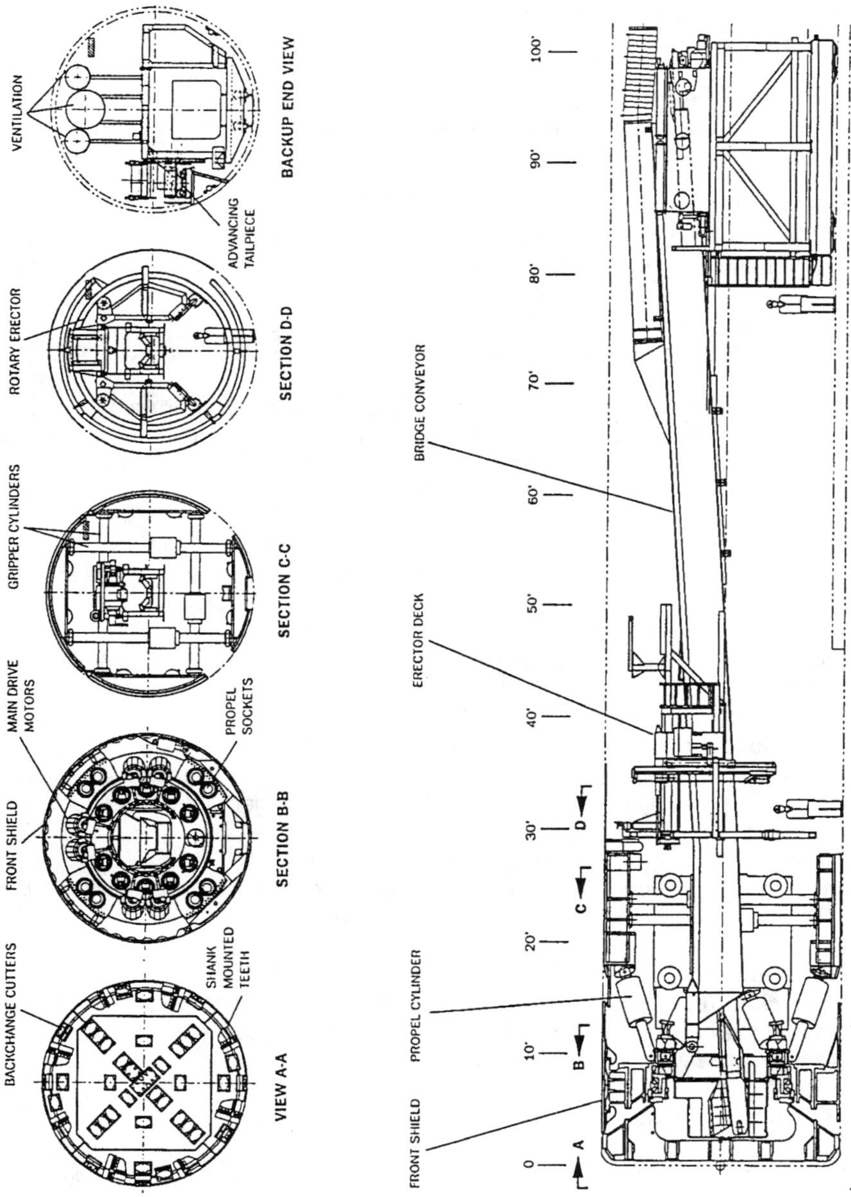

Figure 3 - Bore-Regard Revealed

FIGHT CHRONOLOGY

First Bout - Round 1 - West Hills Win - Stunning Upset!!!

Bore-Regard starts mining August 4, 1994 and stops mining September 22, 1994 after only 47 m (155 ft.) of tunnel are driven. Can't lay a glove on the West Hills. Bookies lay 10 to 1 odds against Bore-Regard.

Timeout - September 22 to December 12, 1994

Bore-Regard catches breath and gets modified grippers, additional torque, and poor man's slurry system.

First Bout - Round 2 - A Draw

Bore-Regard struggles 275 m (1,250 ft) to good ground. Modifications improve performance but cutterhead has severe wear, bucket teeth knocked out - looks bad for contender.

Timeout - April 12 to April 27, 1995

Cutterhead buckets are rebuilt and protective wear plating is put on cutter saddles

First Bout - Round 3 - Bore-Regard Wins - Cutterhead face badly bruised

Tunnel holed through on December 29, 1995 after enduring unmapped fault zones, long sections of raveling ground, massive groundwater, cutter failures, split bucket lips. Bookies win on point spread.

Bore-Regard accepts challenge for rematch from West Hills!!! - spends 3 months rebuilding December 30, 1995 through April 4, 1996 comes with added face protection.

Second Bout - Round 1 - West Hills Lose

A little help from Bore-Regard's corner, 330 m (1,100 ft) of the worst ground is mined using a backhoe, ribs and spiling, while Bore-Regard is still in the other tunnel.

Second Bout - Round 2 - Bore-Regard Wins

Starts mining April 4, 1996 and holes through at the now excavated zoo subway station 1,176 m (3,919 ft) away on May 21, 1996. Cutterhead faceplates severely worn.

Time out - May 21 through June 7, 1996

Bore-Regard cutterhead replated around perimeter, additional steel wear blocks welded to face. TBM dollied 120 m (400 ft.) through station cavern.

Second Bout - Round 3 - Bore-Regard Delivers Knockout Blow!!!

Holes through on August 14 after mining 1,265 m (4,218 ft.). Bore-Regard retires from fight scene and grows daisies out in Gresham, Oregon. Announces he's ready to fight for tunneling crown with any and all contenders.

Round by Round Fight Analysis

As can be seen on Table 1 and Figure 4, Bore-Regard learned from his first fight with the West Hills and came back strong in the second bout.

First Bout - Round 1 - West Hills Win - Stunning Upset!

Table 1. Fight Statistics

FRONTIER/TRAYLOR J.V.

TRI-MET WESTSIDE LIGHT RAIL PROJECT

PORTLAND, OREGON

BORE-REGARD VS. WEST HILLS OF PORTLAND

	1st BOUT WB TUNNEL	2nd BOUT EB TUNNEL
EXCAVATION TIME	61 WKS	20 WKS
TOTAL FOOTAGE	2,947 M 9,675 FT	2,478 M 8,135 FT
WEEKLY AVERAGE	48 M 159 FT	124 M 407 FT
DAILY AVERAGE	8 M 26 FT	21 M 68 FT
MINING TIME	1,719 HRS	1,144 HRS
AVG. PENETRATION RATE	5.63 FT/HR	7.11 FT/HR
CUTTER CHANGES	290 EA	51 EA

WORKING 6 DAYS/WEEK, 3 SHIFTS/DAY

WB TUNNEL STATISTICS DO NOT INCLUDE 3 MONTH SHUTDOWN

A starter tunnel was excavated conventionally and Bore-Regard was brought to the face. The TBM began mining the Westbound (WB) tunnel on August 4, 1994 and immediately encountered problems. By September 22, 1994 when the TBM was shut down, only 47 m (155 ft) had been mined. The TBM was expected to mine at an average daily rate of 36 m/day (120 ft./day). Instead, an average daily rate of only 1.2 m/day (4 ft/day) was achieved with the best days advance being 4.5 m (15 ft). The TBM was shut down after a review of the cores and new core data which showed that the same kind of ground could be expected for another 270 m (900 ft.). Although the cores were correctly logged, the ground's behaviour in reaction to the TBM cutterhead was unanticipated.

The basic problem appeared to be the short stand-up time of the rock which prevented the TBM from mining a full face. The ground being mined was a weathered basalt with numerous randomly oriented cooling fractures. The fractures were very tight but there was no cementation between the fractures. Many of the fractures had a thin coating of mud. When the TBM cutterhead was brought in contact with the face, the material was not cut but instead ravelled from the face in an uncontrolled manner. Large sections of the face sloughed off onto the cutterhead, sometimes locking up the cutterhead. The raveling proceeded ahead

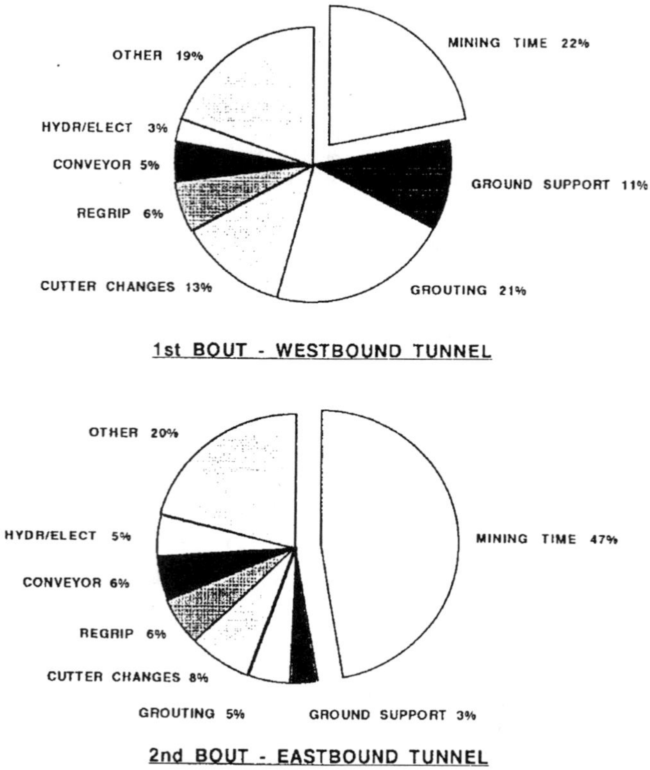

Figure 4. Bore-Regard vs. West Hills Performance

of the TBM and the TBM ingested more material per lineal foot of tunnel advanced than the neat volume of the tunnel. Rapidly, a sloping face formed in front of the machine. The sloping face caused the arch to ravel to a diameter larger than the TBM from about 9 to 3 o'clock. Voids as high as 6 m (20 ft) were created above the TBM.

The raveling characteristic was exacerbated by the 5% uphill tunnel grade. When a TBM is mining uphill, the bottom of the TBM must be ahead of the top thus undercutting the face. With a 6.49 m (21.3 ft.) diameter TBM on a 5% grade, the TBM could be undercutting the face by as much as 432 mm (17 in). The overall effect was similar to mining into a 6.5 m (21 ft.) high wall of unmortared bricks, taking bricks out of the bottom of the wall until the wall collapses. This hypothesis was confirmed by the ground behaviour in the parallel eastbound tunnel where the first 330 m (1,100 ft.) was later mined with a backhoe to sound rock. Even though the EB tunnel was mined heading and bench, a vertical face could never be maintained and a marignally stable angle of repose for the face was about 11 degrees from the vertical.

Timeout for Strategic Modifications- September 22 to December 12, 1994

Reversing Drive - Additional hydraulic valving, piping, and controls were added to allow cutterhead reverse rotation.

Three Shoe Steering - The TBM had two pairs of grippers: a top & bottom pair and a right and left pair. The TBM was originally configured to operate in either an alternating two shoe mode in good ground or a four shoe mode in bad ground. The raveling face extended up above the top of the TBM so there was nothing for the top gripper to push against. The grippers were modified so that the side grippers thrust out first and then hydraulic rams from the side grippers thrust down on the bottom shoe to lock it in place. The top shoe just floated above the bottom one. The TBM steering was improved since the head could now be steered upwards more easily in bad ground.

Slurry - Perhaps one of the most radical changes was to make a poor man's EPB machine. The original TBM configuration lent itself to replacing the dust seals around the main bearing support with water tight plating. Thus the TBM was modified to hold a liquid head up to 4 m (13 ft) above the invert. The TBM was started with water only in the head. The trapped water quickly churned into a slurry which densified to about 1,600 kg/m^3 (100 pcf). The slurry built up because the cutterhead was turned slowly. Inside the cutterhead, paddles guided the muck to a muck chute. As the each paddle rose out of the slurry, the water and fines ran off leaving only the larger rocks which slid off the cutterhead paddle and into the muck hopper. Theoretically the unsupported face was reduced from 6.3 to 2.4 m (21 to 8 ft) in height.

Additional Drive - The TBM was built with an inspection port that could be replaced with an 11th hydraulic drive. The spare drive was installed boosting torque 10%.

The TBM resumed mining on December 12, 1994. Although progress was slow, the TBM's exhibited improved steering and and improved ground penetration. The major surprise was that with the slurry, the TBM operated best in reverse! The slurry sloshing in and out of the cutterhead was actually transporting the muck through the cutterhead. The slurry cut down on the number of times the head was locked up by face wedge failures. When a wedge of rock did fall against the face and lock it up, the TBM only needed to be backed up a few inches instead of the 1.2 to 2.4 m (4 to 8 ft.) previously required. Unfortunately, the slurry material also made a mess both in the tunnel, at the muck pile, and at the landfill. A better solution needed to be found.

First Bout - Round 2 - A Draw

Voids were still being created in front and above the TBM. Past experience showed that the TBM could continue operations with voids as high as 1.2 m (4 ft.) above the TBM. However when the voids grew to more than 1.2 m (4 ft) high, advancing the TBM would increase the void above the TBM rapidly. When this occurred, the solution was to fill the void with a foamed cement grout acclerated with sodium silicate. The shielded construction of the TBM allowed the cutterhead and shield to be sealed off within a shift and grout was pumped through the cutterhead and into the void. The grout set up quickly enough that the TBM could resume mining as soon as the grout pipes were disconnected. The TBM would mine through the grout at the face while the walls were left standing. Slurry use was discontinued after the TBM made it through the first 330 m (1,100 ft of tunnel). The foamed grout was used repeatedly in the rest of the tunnel and by the end of the westbound tunnel over

1,235 m^3 (1,544 cy) of grout had been injected at 37 different locations.

Timeout - April 12 to April 27, 1995

After the TBM penetrated into a short stretch of good ground 600 m (2,000 ft) from the portal, the cutterhead was inspected. Heavy wear was noted on the cutter saddles and cutterhead face. The machine was stopped while an extensive program was undertaken to put abrasion resistant wear plate around the cutter saddles.

First Bout - Round Three - Bore-Regard Wins - Cutterhead Face Badly Bruised

The loose blocky ground continued to severly wear the cutters and saddles until the end of the tunnel. Cutter breakage became a serious problem with one cutter being replaced for every 9.3 m (31 ft) of mining. The cutters were too brittle for the blocky ground and suffered extensive cracking and chipping. Cutter failures were also caused by loss of the swaged retainer rings which were beat out of shape by loose ground grinding against the cutter hubs.

Water in the tunnel also complicated mining. Two unmapped faults were crossed which had peak flows of up to 126 l/sec (2,000 gpm). The high water flows washed in extensive amounts of debris and hampered the installation of ring beams. Slow progress was made only because of the steady work be the mining crews.

On December 31, 1995 the TBM holed through into the drill and blast portion of the tunnel. The head was severely worn with up to 100 mm (4 in) of steel having been worn away at the perimeter. The TBM was disassembled in the tunnel and taken outside for an extensive rebuilding and cutterhead faceplating.

Bore-Regard accepts challenge for rematch from West Hills!!!

Bore-Regard was rebuilt from December 30, 1995 through April 4, 1996. Improvements included:

Cutterhead Modifications - Additional faceplating was added on the center section of the head and on the segments reducing the distance the cutters protruded from the head, from the original 114 mm to 76 mm (4.5 to 3in) on the center and down to 25 mm (1 in) in the gauge area. The plating was sacrificial and was intended to resist abrasive contact with the loose muck. The additional plating was not intended to support the face by reducing the clearance to the face since the ground had already failed by the time it touched the face. Extensive hardfacing of the outter third of the cutterhead perimeter was also done. Four additional openings were cut into the middle third of the face. The idea was to reduce the distance the muck had to travel before entering the face and thus reduce the amount of wear caused by muck rubbing against the face.

Cutter Modifications - Changes to the cutters included changing the hubs, cutter rings, and retainers. Hubs were changed to provide a 29% larger contact area. Rings were changed to another grade of material, from a much harder alloy, to resist cracking. Retainers were changed from the swaged ring design to the typical square inset retainer.

Extended Flex Shield - The flex shield was replaced, and the new shield extended out over the cutterhead 305 mm (12 in) further to reduce overcut in the crown of the tunnel (the rear of the cutterhead was reduced in diameter to accomodate). The original cutterhead projected out from the shield 914 mm (36 in). The modifications reduced the projection to 610 mm (24 in). This was expected to reduce torque demand, increasing performance.

Second Bout - Round 1 - West Hills Lose

The longest stretch of bad ground was the first 330 m (1,100 ft.). This ground was mined using a small backhoe while Bore-Regard was still in the other tunnel. Most of the ground was soft enough that the spiling could be driven in with a small hoe-ram. The bench was excavated with a Cat 988 front end loader.

Second Bout - Round 2 - Bore-Regard Wins

After being assembled at the portal Bore-Regard was moved up to the face and commenced mining on April 4, 1996. The TBM holed through at the now excavated zoo subway station 1,193 m (3,919 ft) away on May 21, 1996.

TBM modifications were responsible for part of the improvement but the big reduction in groundwater in the second tunnel also played a significant part. The first tunnel was very effective in dewatering the ground to be mined. The ground still had significant raveling areas which weighed heavily on the steel sets, but Bore-Regard was able to handle the ground with relative ease and no void filling grout was required.

Cutter performance was much improved over the first tunnel. The added plating reduced the occurence of stalls by reducing the "bite" of the cutters, when they contacted large blocks. The combination of a larger seating area and a better retainer design kept the rings from walking off the hubs, and the softer ring material nearly eliminated ring breakage. Eastbound cutter changes averaged 1 cutter per 48 m (159 ft) of tunnel, vs 1 cutter per 10 m (33 ft) in the westbound tunnel.

Time Out - May 21 through June 7, 1996

Heavy wear was noted on the head at the outter third of the face and the perimeter. This was probably caused by the reduced cutterhead to ground clearances which restricted muck flow to the buckets. Large grooves were worn into the outer areas of the head, and eight cutter housings immediately inboard of this area had to be changed. Special wear resistant steel blocks were welded in a radial pattern on the face. Welded onto the heads perimeter were precurve strips of wear resistant steel covered with hard facing. After work was completed Bore-Regard was moved on dollies 122 m (400 ft.) through the zoo subway station cavern and relaunched.

Second Bout - Round 3 - Bore-Regard delivers knockout blow!!!

Bore-Regard holed through on August 14, 1996 after mining 1,285 m (4,218 ft.). Along the way there were 15 sections of ground which received 352 m^3 (440 cy) of grout. The TBM showed the advantage of the four shoe design when moving past cross passages which had already been mined from the first tunnel.

In the worst material mining advance rates were reduced due to longer regrip times. The raveled ground was not consolidated enough at the crown to allow the flex shield to anchor the head in the bore. As a consequence the TBM could mine in the 4 shoe mode but the shoes had to be regripped in pairs rather than simultaneously to prevent the head from pulling back.

In the loose, ravelling ground a the large amount of spillage between the gripper shoes and the flex shield, and the shoes themselves, lead to a lot of downtime for hand mucking. Several schemes were investigated for cleaning up this spillage, but most involved dumping the muck into the cutterhead. No viable alternative to hand mucking was ever developed.

Bore-Regard Retires - Final Comments

In general, the four-shoe gripper system of this machine was probably the only design that would have been able to complete these tunnels. The ability to grip 360 degrees was essential in the geology encountered, as a tail thrust system is not possible when using ring steel. Another advantage was that in better conditions, the machine could be operated by alternating between horizontal and vertical grippers, resulting in nearly continuous operation. Bore-Regard steered quite well in all ground once the 3 shoe boost trick was developed. Variable cutterhead speed with full torque available in a stalled condition was essential in the geology encountered on this project. The variable speed lessened the impact on gear reducers and bull gear.

Bore - Regard had a best day of 61 m (201 ft.) and a best 8 hr shift of 29 m (96 ft.) while setting ring steel and wire mat lagging. These high production rates were due to the arrangement of the shield and shoes which allowed the rings to be erected at the tail of the TBM without regard to gripper shoe position.

Bore-Regard has now retired to a storage yard in Gresham, Oregon where he spends the days watching daisies grow. Who knows what will happen next? Perhaps he will come out of retirement if the right tunneling challenge comes along!

DRILL AND BLAST EXCAVATION

Conventional drill-blast methods were used to excavate 3,170 meters (10,400 ft) of tunnel, nineteen cross passages, three shafts, and an underground rail station. All of this blasting work occurred in or under densely populated areas of Portland and its western suburbs. The urban proximity of this blasting work created many public relations and blast design challenges. This section focuses on the blast design measures and physical efforts used to control blast vibration and over-pressure levels measured at the many public and private buildings near the Tri-Met drill blast excavations.

West Portal U-Wall Excavation

Tri-Metís tunnel plan included a U shaped sinking cut to establish a steeply sloped face for the west portal entries. After all the topsoil and dirt were removed, the depth of rock to track grade in the U-Wall cut gradually increased from 1.5 to 9.75 m (5 to 32 ft) over a length of approximately 260.6 m (855 ft). The top soil was removed to provide firm footing for drilling and other work in the cut that occurred in the winter months — Oregonís rainy season. As it later turned out, it was very wet in the cut during the drilling and loading operations. We debated leaving some of the dirt over the rock to provide a natural blasting mat, but if we had done this we would have lost many blastholes due to collar squeezing in the water saturated soil. Moreover, hole loading would have been extremely difficult because as it was, some hole collars were under a foot of standing water.

Our rationale for planning the shots in the U-Wall was based on determining all of the blasting goals, prioritizing them, and then designing the blasts based on priority. In most complex blast applications, it is usually impossible to fully satisfy all of the design objectives. Therefore, some sort of trade-off analysis is needed to balance the design sacrifices based on priority. In all blast design work there are usually two general types of goals; safety goals and operational goals. For instance, we may desire minus 30.5 cm (12 in) fragmentation as an operational objective and no flyrock as a safety goal. Since the safety goals must have the highest priority, we might increase stemming and sacrifice the fragmentation goal. Our design goals for the U-Wall surface blasts, ranked by priority were:

1. Control Flyrock — no rock farther than 7.6 m (25 ft) from the blast perimeter.
2. Keep maximum vibration level at structures below 12.7 mm/s (0.5 in/s).
3. Keep Airblast level below 105 dB C-scale at 7.6 m (25 ft) from structures.
4. Plan blasts that could be loaded in one day.
5. Stop traffic on U.S. Highway 26 for no more than 15 minutes during blasting.
7. Produce material that can be excavated to grade by a 1.53 m^3 (2 yd^3) backhoe.
8. Produce smooth final walls.
9. Design for 0.47 kg/m^3 (0.8 lb/yd^3) powder factor.

Controlling flyrock was our greatest concern because the Oregon Department of Transportation (ODOT) required that our blasts occur between the hours of 11:00 PM and 2:00 AM. So we are blasting late at night, with poor visibility, 30.5 m (100 ft) from Highway 26, and within 76.2 m (250 ft) of occupied homes with sound asleep residents. The homeowners were also recently pre-conditioned by a recent earthquake and the winter Olympics stories about blasting damage to homes above the underground hockey rink in Lillehammer, Norway. We were also limited to blasting the U-Wall and two other knobs of basalt in a total of ten blasts. However, the knobs were removed mechanically with a D-10 impact ripper and a hoe-ram.

U-Wall Blast Design

The U-wall excavation was divided into five separate blasts. The first shot was relatively small; 12.2 by 12.2 meters (40 by 40 ft) square, with about 1.5 m (5 ft) of rock from surface to track grade. This first ìtestî shot was purposely small so most of it could be covered with blasting mats to evaluate the efforts to limit rock movement. It was accepted that the later shots would be too large to cover with blasting mats, hence loads that would limit rock movement without surface cover needed to be developed. To control surface movement, a minimum of 2.1 m (7 ft) of clean crushed stone stemming was used in all of the 63.5 mm (2.5 in) diameter blastholes. All other design elements were adjusted to meet this firm stemming requirement. In the shallow 1.5 m (5 ft) cuts, 1.2 m (4 ft) of subdrill was used to gain enough hole depth to meet the 2.1 m (7 ft) stemming requirement. This conservative subdrilling also helped satisfy the requirement to completely break all the shots to grade and eliminate any secondary blasting. Considering the hole loads at the different cut depths, drill patterns were adjusted to achieve an average powder factor of 0.47 kg/m³ (0.8lb/yd³). The second shot had 583 holes in it, but the holes were relatively shallow — from 2.7 to 5.2 m (9 to 17 ft) deep. The number of holes was decreased in each of the next three shots as the cut deepened and loading became more difficult. The vibration control criteria necessitated the need to delay deck load the blastholes in the last two shots. The number of blastholes in the decked shots was reduced to facilitate single shift loading.

Noise and Vibration Control

For the U-Wall blasts, the vibration limit was 12.7 mm/s (0.5 in/s) measured at 61 m (200 ft) or at the nearest structure. The potential for damaging a structure at 12.7 mm/s (0.5 in/s) is extremely low, however complaints are based on perception not fact. To provide a comfortable safety factor and reduce actual complaints, a design limit of 6.35 mm/s (0.25 in/s) was used to establish actual load limits. Using typical ground response constants developed by the US Bureau of Mines, it was determined that a scaled distance of 57 (imperial units) would keep ground motion velocity levels below 6.35 mm/s (0.25 in/s).

U-Wall Blast Loads

Pre-splitting was considered as a controlled blasting method for the final 6:1 sloped walls. However effective pre-splitting would have required an extra blast that made it difficult to meet the five total blast limit. Instead a ìpost splittingî or trim blasting technique was used, whereby lightly loaded holes drilled along the final walls could be integrated into the adjacent production shots. To minimize premature ground movement or column disruptions in the later firing trim charges, we placed a row of lightly loaded buffer charges between the final trim holes and the nearest blastholes. The general loads for the blasthole, buffer, and trim holes are shown in the drawing below. We believed we could control the lighter trim and buffer charges with less than 2.13 m (7 ft) of stemming. To control rock movement, 1.2 m (4 ft) of crushed stone stemming was used in the buffer holes and 1 m (3 ft) was used in the trim holes.

All of the explosives chosen had to have very high shock resistance to prevent precompression failures and propagation. For the blasthole loads we chose a packaged ammonium nitrate based blasting agent, packaged in water tight plastic cartridges, and fueled with nitro-ethane. These stiff in 50 mm (2 in) diameter cartridges weighed 2.25 kg (5 lb) each, and they loaded extremely well through the muddy and wet hole collars. The buffer holes were loaded with 42.5 g/m (200 grain/ft) cord with 150 g (.33 lb) PENTEX boosters taped on at 1.2 m (4 ft) centers. Trim holes were lightly loaded with 42.5 g/m (200 grain/ft) cord. We were concerned about the potential for cut offs in the buffer and trim holes because of their proximity to the outside blastholes. To ensure their complete detonation, we double primed them — top and bottom.

U-Wall Drill Pattern

Variable drill patterns were used to maintain an average 0.47 km/m³ (0.8 lb/yd³) powder factor in the U-Wall blasts. A 1.5 by 1.8 m (5 by 6 ft) pattern was used in the 1.5 m (5 ft) cuts, and it gradually increased to 1.8 by 2.4 m (6 by 8 ft) as the cut gained depth — up to 9.75 m (32 ft). All blastholes were drilled vertically, while the buffer and trim holes were

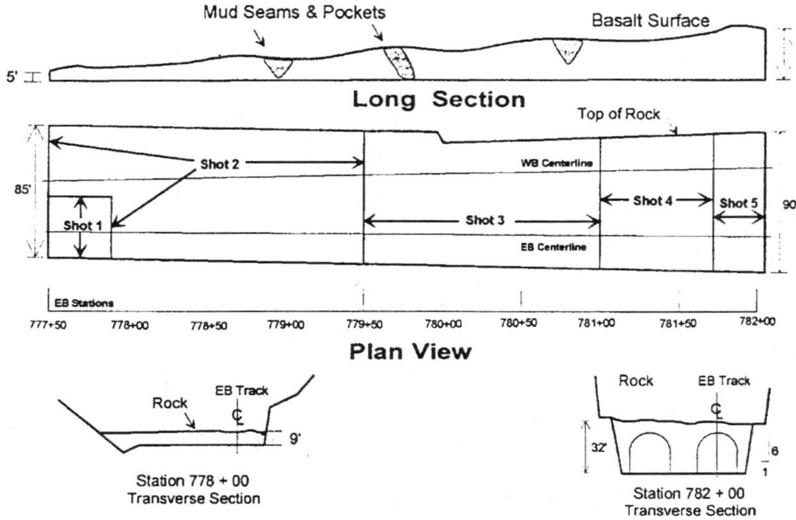

Figure 5. West Portal U-Wall Blasting Sequence

angled to match the six to one final wall slope. The trim holes were drilled on 51 cm (2 ft) centers in the 1.5 to 4 m (5 to 13 ft) section of the cut. This spacing was increased to 76 cm (3 ft) in the last three shots — all deeper than 4 m (13 ft). The burden between the trim and buffer holes was 76 cm (3 ft) in all five shots. Buffer hole spacing equaled the adjacent blasthole spacing that varied with cut depth. Since the blastholes were vertical and the buffer holes were angled 6:1, their burden varied with depth. We tried to maintain a minimum burden of 51 cm (2 ft) at the toe, and no more than 2.1 m (7 ft) at the collar. The shot partitioning and drilling layout for shot number 5, the final face-up shot, is shown in the Figures 5 and 6

U-Wall Blast Initiation

One of our design requirements was to design shots that could be loaded, tied in, and checked — all in one day. We also expected very wet and sloppy loading conditions and we knew we would have shots with over 500 holes; so we desired a system that was easy to tie in and easy to inspect. Our vibration control requirements also mandated that we use a system that would provide one blasthole per delay timing in shots with up to 40 rows of 13 holes. To meet these requirements, we used a nonelectric initiation system with 500 ms in-hole delays integrated to a 25 ms surface delay. All of the blasts were delayed to fire up the center, in a flat chevron, from the shallow to the deep end. From the middle of each row, the 25 ms surface delays were sequentially connected from hole to hole, or deck to deck, towards the perimeter. The trim holes, loaded with regular 500 ms units, were tied, in bunches of two or three, to the buffer hole at the end of each row. A single 42 ms — surface delay — was used between the center hole and the first hole on one side to create an 8 ms timing offset between the two sides. Parallel combinations of 200 and 42 ms surface delays were used to establish row to row timing. Timing was varied to ensure that no more than one blasthole, or blasthole deck fired within any 8 ms timing window. Our weight limits allowed

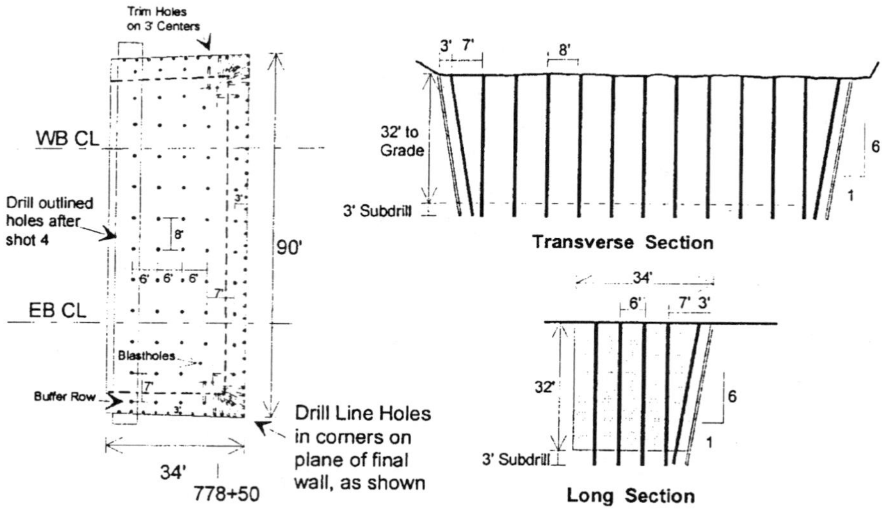

Figure 6. West Portal U-Wall Blast #5 Layout

three of the lightly loaded trim or buffer holes to fire with any of the blasthole charges. Primary initiation was provided by A nonelectric lead-in line, spliced to a surface delay detonator at the first hole.

U-Wall Blast Results

Rock movement was very minimal on four of the five shots. Some mud blew out of the third shot and landed about 15.2 m (50 ft) away on highway 26. No damage was done and the mud was quickly cleared by hand and traffic resumed within fifteen minutes after blasting. An inspection of the shot revealed that we had a soft mud pocket within the hard basalt where the blowouts occurred. The last two shots had hard rock throughout the collar zone so we continued to use seven feet of crushed stone stemming in the blastholes. As expected, the fragmentation in the unloaded seven foot collar region was blocky and tight. However, all five U-Wall shots did swell up and dug well at the floor grade once the blocky collar region was peeled off with the 1.2 kg/m^3 (2 yd^3) back-hoe. In hindsight, it was agreed that digging conditions would have improved if the drill patterns would have been tightened to achieve a powder factor around 0.6 kg/m^3 (1.0 lb/yd^3). All of the blasts broke to the final design perimeter with little or no overbreak. Half cast holes were visible on all the final walls.

The actual vibration levels were reasonably close to the levels predicted by the USBM attenuation formula. The following chart gives the actual and predicted vibration levels for the U-Wall blasts. These readings were recorded by Instantel MiniMate Seismographs that were located in various locations around the five U-Wall blasts. The seismographs were placed at sidewalk locations and not immediately next to structures. The log-log graph correlation between the predicted and actual data curves is very good. The slopes are similar and it appears that a K constant higher than the standard 160, would have closely predicted the actual vibration values.

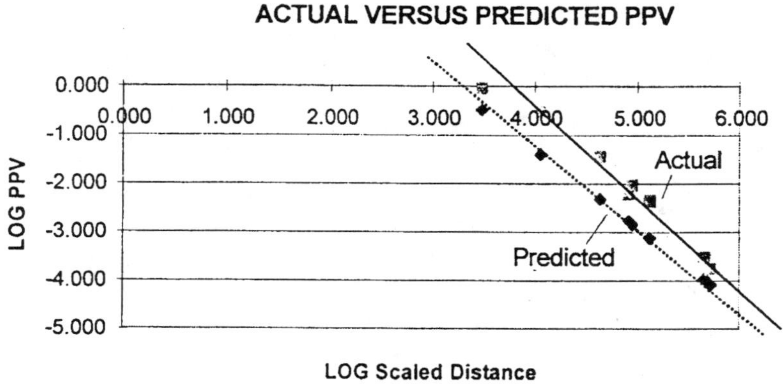

Figure 7. Vibration Level Curves

Tunnel Blasting

Two parallel full face tunnels were driven from the Tri-Met west portals. Round depths in rock and mixed faces varied from four to eight feet. In many sections of the excavated of tunnel, the face, or part of it, was advanced by mechanical digging with a Schaeff loader. The ground varied from very hard basalt to soft clay-like mudstone. The basalt portions of faces were loaded and blasted whereas the mudstone was dug out without blasting. In the mixed faces, the mudstone was often dug out to provide open slash relief for the rock portions of the face. At some locations well beyond the portals there has been as little as 21.3 m (70 ft) of cover above the tunnels as they progressed under a mortuary and public residences.

A typical 94 hole round using 44.5 mm (1.75 in) diameter blastholes is shown in the following figure. Packaged emulsion high explosives in 32 x 400 cm (1.25 x 16 in) cartridges, were used as the primary explosive in these rounds. The Smoothwall trim holes were loaded with semi-gelatin dynamite trim cartridges in 19 x 609 cm (0.75 x 24 in) sticks, primed by a single stick of the high explosive emulsion. All holes were initiated with a series of up to 20 different long period delay detonators, clipped to a 3.8 g/m (18 grain/ft) looped cord trunkline at the face.

Tunnel and Shaft Blasting Noise and Vibration Control

Controlling blasting noise and vibration was very challenging at the Tri-Met project due to the proximity of the excavations to numerous urban structures. The project specifications over-pressure limit, measured in C scale, was 93 dB at night and 105 dB during the day. The linear peak airblast limit, measured with traditional blast recording instruments, was 129 dB. For the tunnel blasts, the C scale airblast values ranged from 4 to 11 dB below the concurrent linear peak values. So for this project, all of the mitigation efforts were aimed at meeting the relatively low C scale limits. Airblast and vibration from the early rounds, fired during the day and next to the busiest highway in Oregon, caused few complaints. For night blasting, the lower noise limit of 93 dB—C scale was harder to meet. To reduce airblast and vibration levels, up to 20 different delays were used to lighten the mass of explosives firing at any one

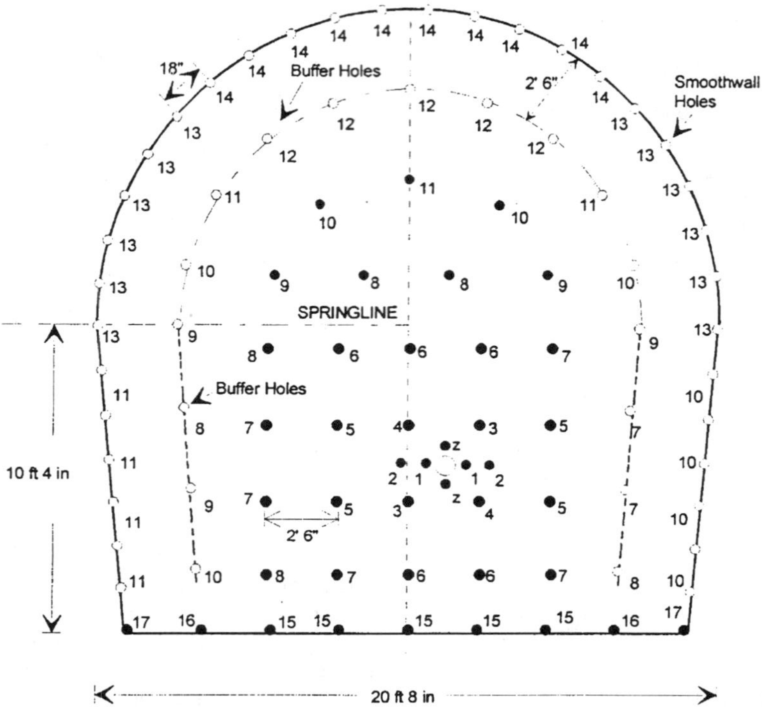

Figure 8. Typical Tunnel Round

time throughout the blast duration. This measure helped, but it alone was not enough to meet the 93 dB—C scale limit for the tunnel blasts near the west portals. To further mitigate noise levels, several drive-through rubber flap curtains were installed in the tunnels. However, the low frequency blast energy traveled through them with very little reduction. The predominant frequency of the Tri-Met over pressure waves has been 12 Hertz. To further mitigate the noise, two sets of center opening hinged steel doors were installed inside each portal. Moreover, an extra cross passage was driven between the two tunnels to allow the air over pressure pulses to vent from one tunnel to the other during blasts. Installation of the doors and cross passage, along with increasing distance, dropped the average C scale values from 98 dB to 93 dB.

The blasting in the three shafts also presented unique vibration and airblast control challenges. The west vent shaft collar was excavated within 30 m (100 ft) of an occupied apartment building and the two large station access shafts were excavated in the Portland zoo parking lot next to the forestry center building. Charges for these blasts were limited by using a vibration prediction equation that was derived from a regression analysis of all the previous tunnel blast data. To meet these small charge weight per delay limits, the shaft rounds were split into quadrants that were separately delayed by 42 ms surface delays connected between individual cord trunklines. By using this hook-up technique, the effective

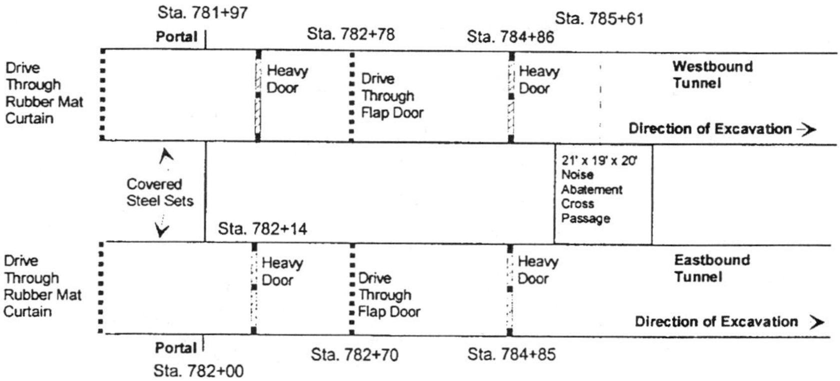

Figure 9. Blast Noise Mitigation Measures

charge weight per 8 ms delay was reduced by a factor of four. Surface trunklines were also covered with sand and special shaft covers fabricated with wood and rubber matting were placed over the shaft collars. These measures effectively reduced noise and contained any flying rock. Moreover, measured ground vibration never exceeded the self imposed limit of 6.4 mm/s (0.25 in/s) at nearby structures and blasting never interfered with normal operations at the zoo or forestry center. A typical controlled blast shaft round hook-up is shown in Figure 10.

WASHINGTON PARK STATION EXCAVATION

The Washington Park Station is the only underground station for Tri-Met's light rail system and it is the deepest mass transit station in North America to date. At an average depth of 80 m (260 feet) the station is located approximately 1,524 m (5,000 feet) and 3,048 m (10,000 feet) from the east and west portals respectively. The station and its supporting facilities have been designed with three distinct functional features.

The first functional elements start at the track and platform level which was designed as a twin tube binocular configuration with inbound and outbound platforms separated by a center rock pillar about 18.2 m (60 ft) wide (Figures 11 & 12). The two platforms are connected by diagonal cross passages that lead to the elevator lobbies at the east and west ends. The diagonal passageways provide a direct pathway and a clear view of the elevator lobby from both ends of the platform. The station platforms are approximately 61 m (200 feet) long to accommodate Tri-Met's two-car light rail vehicles. Each platform spans 9.8 m (32 ft) and the height of the platform varies from approximately 6.7 to 11.9 m (22 to 39 ft).

On both sides of each platform waiting area the station design included adjoining "tail" tunnel sections that extend 29 m (95 ft). At approximately the mid-point of the tail tunnels, the designers have positioned the emergency ventilation plenums. These plenums are perpendicular to the running bores and they connect with backside of the shafts. The unfinished dimensions of the plenums include a width of 7.0 m (23 feet) a length of 28.3 m (93 feet) and a height of 5.9 m (19.5 feet). The combination of excavations for the tail tunnels, plenum and shaft creates two perimeter rock pillars on either side of each lobby. The dimen-

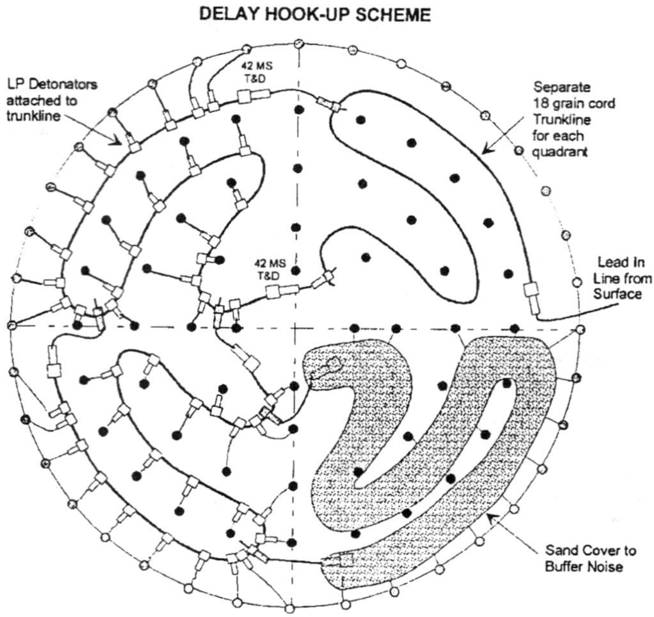

Figure 10. Controlled Shaft Blast Hook-up

sions of each of these rock pillars is roughly 8.2 m (27 feet) by 12.2 m (40 feet).

The second functional element consists of two shafts that are spaced 85.3 m (280 feet) apart which is critical to the layout of the headhouse structures at the surface. Both shafts have an inside diameter of 9.45 m (31 feet) and each shaft houses two elevator hoistways, exit stairs, electrical and mechanical chases, station air supply or return ducts, and the emergency tunnel ventilation fan ducts. At the bottom of each shaft is an electrical room descending another 4.6 m (15 ft) below the floor slab of the lobby.

The headhouse structure comprises the third group of functional spaces. The design layout and arrangement of the structure responds to the desires of the owner and the community to create a pedestrian/transit link across an expansive parking lot while optimizing the positioning of critical power supply and ventilation equipment in proximity to the shafts. The layout of the headhouse includes two underground levels to house the large ventilation equipment, and surface level facilities house the majority of the electrical and mechanical equipment and public spaces.

Original Construction Considerations

The station is located in Washington Park in proximity to a number of institutions including the metropolitan zoo, which is Oregon's #1 paid tourist attraction. Because of the amount of pedestrian/vehicular traffic at the surface, Tri-Met established a project goal to limit the amount of muck that must be hauled from the surface, particularly the hard-rock

Figure 11

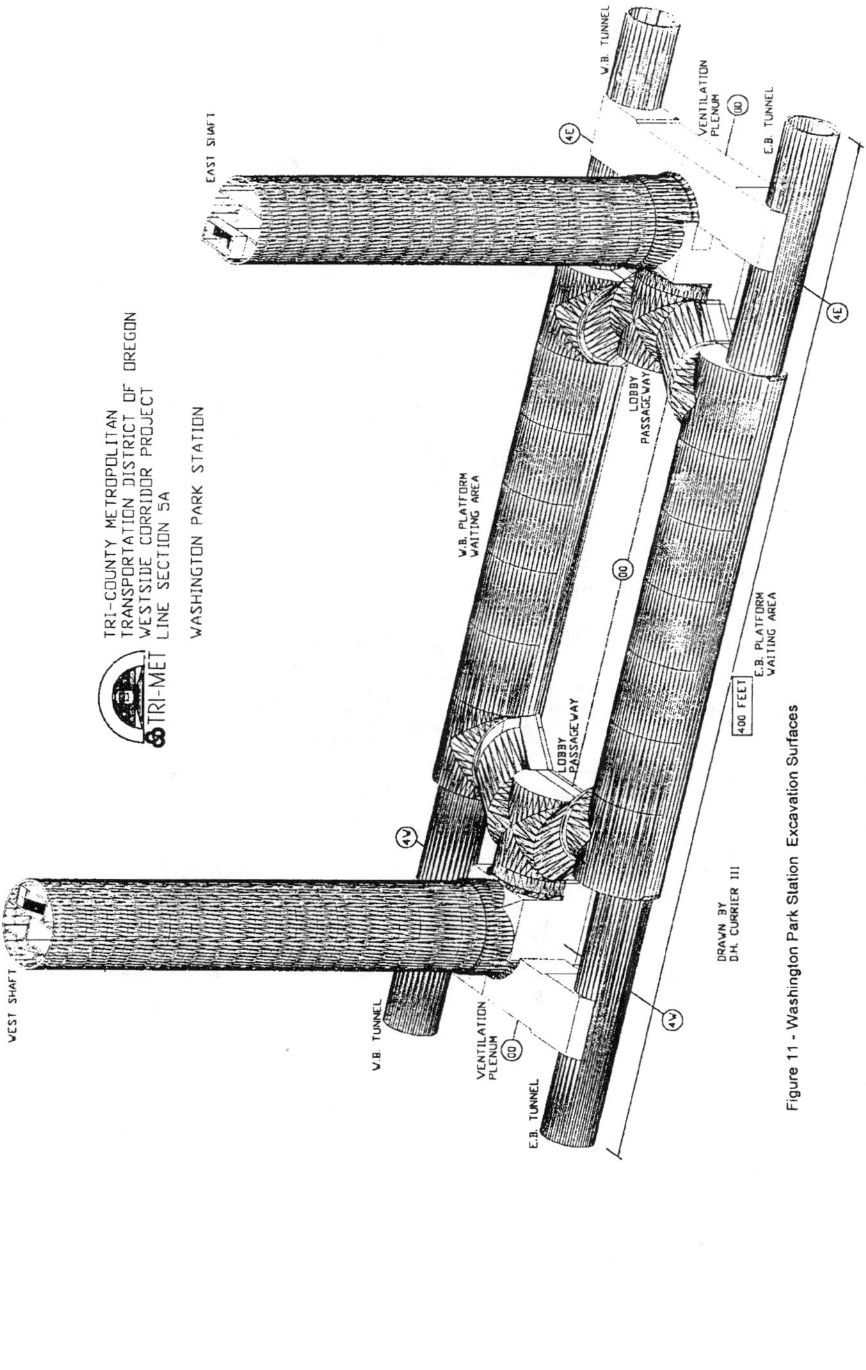

Figure 11 - Washington Park Station Excavation Surfaces

Figure 12

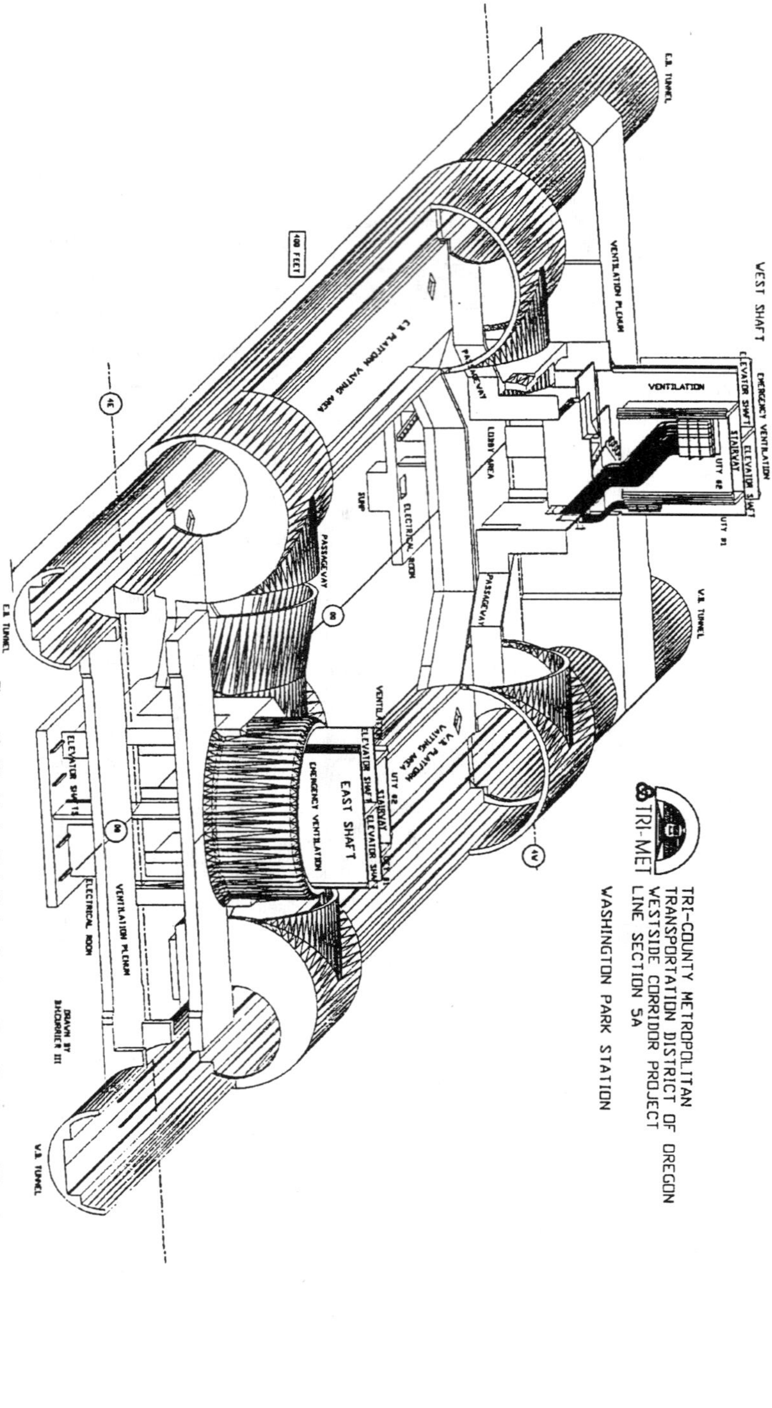

Figure 12 - Washington Park Station Cut-Away View of Underground Structure

portions of the station platform rooms and passageways and the shafts. Achieving these goals required the station platform level would be mined from the first TBM heading to reach the station from the east portal and the muck would be transported to the portal for disposal. Likewise, for the shafts, the specified excavation sequence included sinking the each shaft at a full diameter for a distance of 12 to 18.3 m (40 to 60 ft) below the collar to a depth where the rock is competent, Below this point, it was assumed that all shaft muck (about 107,000 kN (12,000 tons) per shaft) would drop to the running tunnels some 55 m (180 ft) below through a single raise excavated within the boundary of each shaft.

The station is also located in an ancient landslide mass which extends approximately 793 m (2,600 ft) from its toe to the headscarp and the width varies from 366 to 1,005 m (1,200 to 3300 ft) and covers more than 506,000 sq m (125 acres). Design studies were conducted for static and seismic loading conditions and showed no clear evidence of any landslide related movement at the headhouse structure site. To verify these conclusions, a series of inclinometers where installed both north and south of the excavation in vicinity to the station to monitor lateral ground movements.

For the headhouse structure, Tri-Met established limits in the contract work whereby Frontier/Traylor and their subcontractors would only excavate and install the temporary shoring. The follow-on contractor was assigned the task to build the actual structure within the excavation. Tri-Met established these limits in the work to allow more time to implement a more economical re-design package for the headhouse structure

Actual Construction Activities

Impact of Tunnel Delays: During construction of the first tunnel, significant delays developed due to ravelling ground conditions that stalled the progress of the tunnel boring machine. Because of the delays, the entire sequence of station construction was modified to build the station concurrent with the tunnel work. Both shafts were sunk the full depth using conventional mining techniques and the platform rooms and passageways were excavated via the shafts. These changes dramatically increased the amount of blasting within city limits and near the zoo, which required a complete overhaul in the blasting permit addressing the impacts on both humans and animal species.

Headhouse Construction: Frontier/Traylor and the subcontractors started the excavation in December 1995, using a temporary soil nail shoring system to support the upper soils (i.e. Portland Hills Silt, landslide deposit, Pliocene sedimentary unit and the decomposed basalt). The limits of the excavation are approximately 110 m (360 ft) long and 27 m (90 ft) wide and extends to a depth of about 11 m (36 ft). Shortly after the excavation commenced, deep-seated movements at the ancient landslide shear zone were detected in the inclinometers installed around the perimeter of the headhouse. The translational movements detected by the inclinometers pinpointed the location of the shear zone. Along the east side the bottom of the excavation approached to within 4.6 to 6.1 m (15 to 20 ft) of the slip plane. Similarly, along the west side the bottom of the excavation approached to within 12.2 to 13.7 m (40 to 45 ft) of the slip-plane.

The inclinometer movements indicated that the headhouse excavation triggered the development of a local reactivation within the central body of the ancient landslide. The excavation created a slot in the slide mass which, in turn, created a location for the landslide to "toe-out" into the base of the excavation. A complete review of the instruments indicated that the slide had been reactivated up to 122 m (400 ft) upslope (north) of the excavation, and from 122 to 213 m (400 to 700 ft) to the northwest. Displacements measured in the inclinometers ranged from 1.3 to 5 cm (0.5 to 2 inches), increasing towards the excavation.

Because of this unanticipated movements, the excavation and construction sequence was altered and remedial measures were installed to stabilize the local reactivation and prevent additional movements. Remedial measures included the installation of 14 horizontal drains extending north and northwest of the excavation. Eleven of the drains have been installed from inside the excavation at the west end, approximately 9 m (30 ft) below the ground surface. These drains will be permanent facilities cast into the lower level of the headhouse. The length of these drains vary from 91 to 274 m (300 to 900 ft) and two of the drains produced initial flow rates of 4.4 to 7.6 liters/sec (70 to 120 gal/min). Total flow rates dropped within several weeks to approximately 1.6 liters/sec (25 gal/min).

Piezometers northwest of the headhouse showed a dramatic drop in the groundwater levels of as much as 6 m (20 ft) within weeks of the drain installation. The drains installed to the north and north east of the headhouse reduced groundwater levels typically 1.5 to 3 m (5 to 10 ft). The tangible benefits of the drains allowed the excavation of the west side to continue as planned without concern for local failure. Conversely, at the east side of the excavation the risk of unacceptable ground movements if the east side progressed further. Without additional remedial measures, the creep movements could slowly increase to 7.6 cm (3 inches) with local cracking and settling of the ground surface.

In February 1995, the headhouse excavation at the east end was suspended in favor of starting the shaft excavation. During the suspension a number of geotechnical consultants reviewed the potential problem and agreed that additional support was required to arrest the creep movement as headhouse excavations continued for another 7.6 m (25 ft). Alternative support methods were suggested (reticulated pin piles, tangent pile walls) but it was agreed that a row of shear piles along the upslope wall was the most economical solution. The wall included 19 shear piles consisting of heavily reinforced auger cast piles socketed into basalt bedrock below the shear zone. The cantilever piles were designed to resist shear movement at the toe of the local reactivation and allow the excavation to proceed to final grade. Moreover, the foundation of the building was redesigned to incorporate the shear piles into the foundation thus serving a dual purpose and the net change was only one pile more than the original foundation design.

The excavation of the east side resumed in December 1995, and the work continued through the wet winter season until April 1996. Four of the shear piles were installed with inclinometers to monitor the deflection of the piles as they were loaded by slide movements. Total deflections at the top of the piles varied from 1.3 to 3.3 cm (0.5 to 1.3 inches) during the final excavation. Upon reviewing the instrumentation data Tri-Met's consultants determined that the movements were sensitive to each particular bench cut within the excavation and/or changes in the elevation of the groundwater elevations above the slip-plane in the slide area above the excavation.

After reviewing the instrumentation data, Tri-Met authorized additional changes in the contract for Frontier/Traylor and their subcontractors. Work continued on the headhouse and the foundation and exterior walls were constructed and struts were installed in the east and west cells. At the east cell this required concurrent activity with shaft and cavern excavations.

Shaft Mining Construction: As mentioned previously, construction of the east shaft commenced before the headhouse excavation was complete. The shaft work started when the headhouse excavation was less than 40% complete or approximately 3 m (10 ft) below the ground surface. Construction of the west shaft started after the headhouse excavation reached a depth of 9 m (30 ft) below the ground surface. For both shafts, conventional soft-ground excavation methods using a small backhoe was suitable to excavate the upper soils and decomposed basalt to a depth below the slip. The outside diameter of each shaft was 10.4 m (34 ft) and the soft ground was supported by 7 gage liner plate 61 cm (24 inches) high and W10 x 54 ring beams spaced on 1.2 m (4 ft) centers. In each shaft the slip-plane was encountered in proximity to the top of rock. The thickness of the plane was relatively uniform in both shafts and approximately 1 m (3 ft) thick.

At approximately 19.5 m (64 ft) below the ground surface both shafts encountered the top of the basalt flows identified as Columbia River Basalt. The basalt profile varied from very low to medium high strength, slightly to moderately severe weathering. Jointing in the rock varied as did the degree of vesicularity. The layers also included flow breccia zones of hard soil-like material in between the major basalt flows.

A revocable permit to use explosives was established by the City of Portland for station shafts and platforms. The permit established noise and vibration criteria to protect the adjacent facilities that were within 30 m (100 ft) and the nearest residences that were within 305 m (1000 ft) of the shafts. Table 2 summarizes the noise criteria.

In consideration of the human response factors to blasting, the City prohibited blasting in the shafts between midnight and 6 AM. Because the platform was so deep, Tri-Met and Frontier/Traylor convinced the City to allow blasting around the clock. The parties established a target value of 0.6 cm/sec (0.25 in/sec) between the hours of midnight and 6 AM. The goal was that no single occupied residential structure within 61 m (200 ft) experience more than ten (10) exceedances of the target value. In addition, no single event could exceed 1.3 cm/sec (0.5 in/sec).

In consideration of the animals inside the nearby zoo, a physiological and behavior study was conducted to assess the impact of noise and vibration. Six species were selected (Asian elephant, black rhino, spotted owl, penguin, langur, and mangabey) for study before, during and after blasting. A program was established to assess cortisol (indicator of acute stress) and progesterone (reproductive function in females). The program called for routine blood, urine and fecal assays together with close circuit television monitoring systems for the elephants and rhinos. Thankfully, this was not required for human response factors.

To control noise, Frontier/Traylor constructed a two shaft lids of wood and steel. The lids were constructed with W6 x 25 steel beams and timbers (7.6 cm x 30.5 cm (3 in x 12 in)) tied together with W10 x 54 ring steel. Each lid weighed approximately 244 kg/m^2 (50 psf) and the lids were secured atop the shaft collar prior to each blast.

Based on the variable geologic conditions, Frontier/Traylor generally detonated full-diameter blast rounds limiting the depth advance to 1.2 m (4 ft). An overdrilled pyramid cut

Table 2. Noise Criteria

Time of Day	dBC (max)[1]	Max. P.P.V. (cm/sec)[2]
7 AM - 7 PM	105	1.3 (0.5 in/sec)
7 PM - 10 PM	98	1.3
10 PM - Midnight	93	1.3
Midnight - 6 AM Shaft	No Blasting	No Blasting
Midnight - 6 AM Platform	93	1.3/0.6 (0.5/0.25 in/sec)
6 AM - 7 AM	98	1.3

[1] dBC: (Noise Criteria) Decibel on the C-scale as measured on the C-Weighted network (Slow Response) of a sound level meter. To approximate the overall sound pressure level (SPL) the C-weighted scale (dBC) is used. In comparison to A-scale (dBA) this scale is less sensitive to lower and higher frequencies. Therefore, it is a better gauge of the overall SPL. Also, Oregon's Department of Environmental Quality (DEQ) uses the C-weighted scale to restrict the level of "impulse sounds" like blasting. DEQ's limitations were adopted for the noise criteria. Noise measurements between the hours of 7PM and 7AM were taken at the nearest occupied residential structure from the point of blast. Noise measurements between the hours of 7AM and 7PM were taken at the nearest occupied structure (commercial or residential) from the point of blast.

[2] P.P.V. (Vibration Criteria) The contract documents specified a limit of 1.3 cm/sec (0.5 in/sec) at a distance of 61 m (200 ft), or the nearest occupied structure, whichever is less. Frontier/Traylor established a goal to limit vibrations to 0.6 cm/sec (0.25 in/sec). The City agreed to this goal. These limits were established recognizing that the U.S Bureau of Mines recommends higher limits for frequencies that are generally 40 to 70 Hz. The specified limit was 25% of the USBM recommendation and the goal was 12%.

was used to provide center relief and create a sump for dewatering. The typical pattern included 4.1 cm (1 5/8 in) diameter blastholes and the layout included smoothwall perimeter holes drilled on 61 cm (24 in) centers with 76 cm (30 in) burden to the adjacent ring of blast holes. A sensitive emulsion explosive product was used for the primary holes and a semi-gelatin dynamite was used for the perimeter holes. The typical initiation system included eighteen (18) trunkline attached to long delay detonators combined with short delay trunkline detonators to lessen the number of holes per delay. Overall, the typical maximum charge weight per delay was 2.5 kg (5.5 lbs) to keep the specified peak particle velocity of 1.3 cm/sec (0.5 in/sec) at the nearest structure.

Following each round, the sidewalls of the shaft were immediately supported with shotcrete and dowels. In both shafts, the majority of the exposed rock was moderate to medium strength and the initial support consisted of eighteen (18), 2.4 m (8 ft) swellex bolts placed in a 1.8 m by 1.8 m (6 ft by 6 ft) pattern with 5.1 cm (2 in) of fiber reinforced shotcrete. For the remainder of the shafts, where the rock was considered low to moderate strength the initial support included twenty-one (21), 3 m (10 ft) cemented-grouted dowels (No. 8 bars) placed on 1.5 m by 1.5 m (5 ft by 5 ft) pattern. In between the dowels the rock was supported with 7.6 cm (3 in) (minimum) of fiber reinforced concrete.

In each shaft and throughout the platform passageways Frontier/Traylor used a fiber reinforced shotcrete. The specifications called for shotcrete with a compressive strength of 3.45 MPa (5,000 psi). Frontier provided shotcrete with 41 kg (90 lbs) of fiber reinforcement for every 0.76 m^3 (1 yd^3). Actual measured strengths from cores taken from the shotcrete indicated that the shotcrete typically reached strengths of 5.5 MPa (8,000 psi).

Platform Mining Construction: Once both shafts reached the cavern level excavation crews proceeded immediately to mine the lobby passageways and the ventilation plenum. At the east shaft, the geologic conditions included a flow breccia zone within 3 m (10 ft) of the roof of the adjacent passageway and there was a sustained groundwater flow into the excavation. Prior to the start of passageway blasting a number of drain holes were pre-drilled about 4.5 m (15 ft) to reduce the flow and ensure both complete encapsulation of the grout around the dowels and avoid weakening of the grout by groundwater dilution.

In all the passageways and openings in the platform the crews used top heading and bench techniques to safely advance the mining operations. In some of the flow breccia and low strength rock, crews used a hoe-ram to chip out the rock. In the harder rock, crews used jack-leg drills and jumbo drills to cut 3.8 and 4.5 cm (1-1/2 and 1-3/4 in) diameter holes to various depths in the basalt. In the moderate to medium strength rock, the round length was typically 1.8 m (6 ft) and the perimeter holes were typically drilled on 0.5 to 0.6 m (18 to 24 in) centers with a minimum of 0.8 m (30 in) of burden to the adjacent row of production holes. In the low to moderate strength basalt, the round lengths were generally limited to 1.2 m (4 ft) and, in some cases the perimeter hole spacing was limited to 0.3 m (12 in).

To minimize overbreak, the explosive distribution in the higher strength rock was typically 0.34 kg (0.75 lbs) explosives in each 1.8 m (6 ft) hole. At a spacing of 0.5 m (18 in) this provided a distribution of about 38 g (0.083 lbs) of explosive for each 0.09 m^2 (1 ft^2) of perimeter area. At a spacing of 0.6 m (24 in) the distribution was 28 g (0.063 lbs) per 0.9 m^2 (1 ft^2). In the production holes, the typical loading was 2.3 g (5 lbs) in a hole. Using a 1 m by 1 m (3 ft by 3 ft) pattern, this provided a local powder factor of about 1.4 kg/m^3 (2.4 lbs/yd^3).

In the plenum, elevator lobby and diagonal platform passageways, the crews installed spiling in the arch out ahead of the blasting round. The spiling generally consisted of #11 (Grade 60) bars each 4.6 m (15 ft) long and positioned at the discretion of the crew foreman. In order to minimize the amount of overbreak to the spiling, the crews had to limit both the charge weight of the bottom charges and the number of simultaneous detonations.

The flat roof in each plenum was initially supported by a combination of rock dowels and shotcrete. The roof was initially supported by combinations of friction dowels, mine straps and rock bolts. In the later case, the typical pattern included six (6) #9 (Grade 60) rock dowels each 3.7 m (12 ft) long with a nominal spacing of 1.2 m by 1.2 m (4 ft by 4 ft). Each sidewall was supported with four (4) #9 (Grade 60) rock dowels each generally 2.4 m (8 ft) long with the same nominal spacing as in the roof. Between the dowels, the crews applied 5.1 cm (2 in) of shotcrete between mid-height of the sidewalls and invert and 10.2 cm (4 in) of shotcrete in the upper roof and upper half of the sidewalls. A number of rock dowels were tested at the outset of the mining and each dowel successfully passed the pull test of 89 kN (20 kips).

In each elevator lobby, the initial supports included: a) two rows of 3.2 cm (1-1/4 in) diameter (Grade 150) bars, 15.2 m (50 ft) long with a nominal spacing of 2.4 m by 2.4 m (8 ft by 8 ft), and, b) #11 (Grade 60) dowels, 6.1 m (20 ft) long with a nominal spacing of 1.2 m by 1.2 m (4 ft by 4 ft). The sidewalls were initially supported by #9 (Grade 60) dowels, 3.7 m (12 t) long with a nominal spacing of 1.2 m by 1.2 m (4 ft by 4 ft). The initial support in the crown included a minimum of 15 cm (6 in) of fiber reinforced shotcrete; in the sidewalls the shotcrete thickness was decreased to minimum of 7.6 cm (3 in).

For the inbound platform cavern, the crews used a three-stage advance of the heading in the form of a top center cut, staggered top side slashes, and a lower bench. Because of the variable geologic conditions in the arch, the crews generally limited the center top heading to a width of about 4 m (13 ft) (approximately one-third platform width) which allowed for a round length of 2.4 m (8 ft) (i.e 0.6 x width of the cut (as specified)). The crews did not install spiling during the excavation of the top heading. In the cavern above the springline, the crews installed twelve (12) #9 (Grade 60) dowels 3.7 m (12 ft) long with a nominal spacing of 1.5 m (5 ft) circumferential by 1.2 m (4 ft) longitudinal. In compliance with the specifications the top heading slash cuts were advanced at least 15 m (50 ft) beyond the face of the bench cut with at least 7.6 m (25 ft) between the face positions in each slash cut.

As crews began mining the platform for the station concurrent work in the outbound tunnel bore passing along the north side of the station. The tunnel crews were approximately 61 m (200 ft) west of the station platform when shaft crews started on the brow of the top heading of the lobby and the crown of the plenum from the east shaft. Tunneling in the outbound bore would continue for another 5-1/2 months with the TBM as the crews advanced another 1250 m (4100 ft) westward to "hole-through". By the time the outbound tunnel had "holed-through" the station crews had successfully mined and initially supported roughly 46% of the platform level and completed 100% of the shaft work. Work completed in the platform level included the inbound cavern and the connecting passageways and elevator lobbies, and both plenums.

Mining of the outbound cavern was completed after the TBM and back-up systems were removed from the bored tunnel. Enlarging the pre-excavated tunnel bore was done with slash rounds to enlarge the roof to the design dimensions of the cavern. The crews used lightly loaded charges equivalent to the perimeter hole charge weights to minimize the amount of overbreak in the crown. The platform waiting area was excavated using slash and bench techniques similar to the inbound platform cavern

In total, approximately eighteen (18) months were required to substantially complete the excavation for the station. During this period, Frontier/Traylor's crew successfully and safely detonated over 667 kN (150,000 lbs) of explosives initiating over 250 rounds in the process. Each shot was monitored for noise and vibration and the target value of 0.6 cm/sec (0.25 in/sec) within 61 m (200 ft) was achieved each time. Overall, 19,000 m^3 (25,000 yd^3) of soil and rock were transported to the surface through the shafts the hauled offsite. The crews installed over 13,000 m (43,000 ft) of rock dowels and spiles and applied 4,600 m^3 (6,000 yd^3) of shotcrete support.

Approximately one month after the station mining was substantially complete the TBM "holed-through" into the inbound platform cavern arriving from the east portal. Frontier/Traylor took advantage of this opening to complete some repairs and adjustments on the cutterhead before pulling the machine through the cavern to the west end of the station where it was launched to complete the remaining 1,250 m (4,100 ft) of tunnel. The TBM completed its run approximately three months after leaving the station and it took another month to clear the TBM and its backup system from the platform cavern.

Final Concrete Lining (Station Platform): Once the platform excavation work was substantially complete, Frontier/Traylor and the subcontractors set to work on lining the entire platform. Because of delays in the project, Tri-Met established a critical path in the sequence of work in the station where the west shaft is turned-over first followed by the east shaft, the outbound cavern and adjoining passageways and lobby areas, and finally the inbound platform cavern. This sequence of work in the station was done concurrently with the final concrete work (i.e. invert, arch and track work) in the running tunnel. Since the tunnels were inaccessible to the station all the access for the labor, equipment and materials was limited to the two shafts.

To initiate this work Frontier/Traylor and the subcontractor installed a PVC membrane lining in all of the platform openings (including the electrical rooms) and up into a portion of each shaft extending 24 m (80 ft) up form the bottom of the shafts. The platform level is located approximately 4.6 m (15 ft) below the regional groundwater table in the formations of the basalts. Typical groundwater flows into the platform area were measured at roughly 6.3 liters/sec (100 gal/min). The membrane was placed on top of a felt cushion anchored to the shotcrete lining and the entire waterproofing system is tied into the subfloor drainage system which carries the water by gravity to the east portal.

The installation of the membrane lining throughout the platform prohibited the use of rock anchors as supports for the concrete forms. Therefore, the formwork had to be internally braced and self supporting. A number of steel and wood forming systems were designed and assembled by Frontier/Traylor and their subcontractors. EFCO steel form systems were used to form the walls for the electrical room, plenum, elevator lobby and passageways. These forms were also used to cast the perimeter walls in the shaft-station transition which extended up approximately 6.1 m (20 ft). In the shaft, the crews used a combination of EFCO and wood forms for the interior partition walls for the elevator cabs, exit stairs, electrical/mechanical chases, and ventilation ducts. For the upper walls and arch in the diagonal passageways and elevator lobby, Frontier/Traylor built wood forms. Finally, in the platform cavern, the crews used a Cifa form.

One major challenge in the passageways was the installation and tying together of the steel reinforcement at the intersection of the passageways and the elevator lobby. Frontier/Traylor staff had to build a computer model to identify the multiple intersecting "groin" lines in the arches. Each arch line required a simultaneous solution of two independent quadratic equations in order to establish the proper dimensions. The crews used as many as three rotating laser survey systems simultaneously to identify the groin line locations in the opening. The crews used threaded anchor penetrations through the membrane to secure template bars to position the steel reinforcement.

To deliver concrete to the platform level, Frontier/Traylor used access from both the shafts and set of dropholes drilled immediately east of the station limits. Depending upon the pour and the access, the concrete was delivered in either buckets or pumped from the surface. In some cases the concrete had to be pumped down 80 m (over 260 ft), pushed as much as 100 m (over 325 ft) and raise as much as 10 m (over 30 ft). Frontier/Traylor used a pumpable mix that included Type I/II cememt, Type F flyash, and 3/4-inch rounded rock and natural sand. The batch plant also used admixtures which included a mid-range water reducer and a superplasticizer to maintain a pumpable mix.

Final Concrete Lining (Station Shaft): Once the concrete lining reached the shaft-station transition elevation crews intitated a slip operation to cast the final concrete lining. Starting at this elevation, Tri-Met modified the design of the stairs from cast-in-place to precast. The units were broken down into individual risers (2.4 m by 5.8 m (8 ft by 19 ft) and each unit weighed 169 kN (38 kips). The crew slip-lined the shafts using a three-level plywood platform that contained a 1.2 m (4 ft) form for concrete on its second level. The stair units were hoisted in place one flight at a time and the crews attached three stair units in place before raising the slip-line platform. The crews used eighteen (18) hydraulic jacks to move the form 15 cm (6 in) per hour and 3.7 m (12 ft) every 24 hours. In addition to the stairs the crews had to simultaneously install five (5) signal/communications boxes, each weighing 6.7 kN (1.5 kips) and containing 22 separate conduits. The crew also had to simultaneously install five (5) traction electrification boxes 4.5 kN (1 kip) and containing 9 separate conduits. Over 2500 m (8000 ft) of conduits had to be install and each individual piece was 3 m (10 ft) long.

REFERENCES

Minor, M., 1994, "Noise Monitoring Report," Frontier/Traylor J.V. Document.

Nicholls, H.R. and Duvall W.I., 1971,"Blasting Vibrations and Their Effects on Structures", US Bureau of Mines Bulletin 656 .

Revey, G.F. and Painter, 1995, "Close in Blasting at the Tri-Met Light Rail tunnels in Portland, Oregon", International Society of Explosives Engineers Proceedings, Nashville, TN.

Schmitz, J.,1994, September, "Tunnel project a meeting of the mines", Pacific Builder & Engineer.

Tri-County Metropolitan Transportation District of Oregon, 1993, "Westside Corridor Project Line Section 5A Westside LRT Tunnel Contract Drawings and Specifications", Portland, Oregon.

Tri-County Metropolitan Transportation District of Oregon, 1993, "Geotechnical Design Summary Report", Portland, Oregon.

Chapter 10

TARP — THE DES PLAINES TUNNEL SYSTEM (NORTH LEG) A CASE HISTORY

Robert Rautenberg and Paul McDermott
Kenny Construction Company

ABSTRACT

This paper will discuss the construction of 15.9 km (9.88 mi) of tunnels on a project in Chicago, Illinois. The tunnels were driven utilizing two rock and four soft ground TBMs ranging in sizes from 7.39 m (24.25 ft) in rock down to 1.95 m (77 in) in soft ground. In addition to the mining, there are 63 upper level structures and 16 dropshafts averaging 61.5 m (202 ft) deep included on this project. Many diverse mining, excavation, support and lining techniques were employed on this project.

TUNNEL AND RESERVOIR PLAN

Construction began on the Metropolitan Water Reclamation District of Greater Chicago's (MWRDGC) Tunnel and Reservoir Plan (TARP) in the early 1970's as a solution to the problems of pollution and localized flooding caused by combined sewer overflows and development along the existing flood plains. Upon completion, the TARP System will serve an area of 971 sq. km (375 sq. miles) consisting of Metropolitan Chicago and 51 neighboring suburbs.

The MWRD has planned the TARP System in two phases, the first being aimed at pollution control and the second at reducing flooding. Phase I which is currently under construction employs a network of some 175.4 km (109 miles) of tunnels ranging in size from 2.67 m (8.75 ft) to 10.06 m (33 ft) in finished diameter. Besides the tunnels, there are three pumping stations, 252 dropshafts, various control and diversion structures, and numerous upper level sewers. Phase II of the system will include an additional 34 km (21 miles) of tunnels and three major surface reservoirs with a total capacity of 153 million cu m (5.4 billion cubic feet).

The concept of Phase I is to divert the combined sewer overflows through the upper level diversion and control structures and into the dropshafts along the route of the tunnel. The dropshafts then convey the overflow into the various "deep tunnels" which flow to the pumping stations. At the pumping stations, as the capacity becomes available, the MWRD pumps the overflows up to local treatment facilities and then discharge the treated water into the local waterways.

TARP is divided into four separate subsystems which follow natural drainage basins:

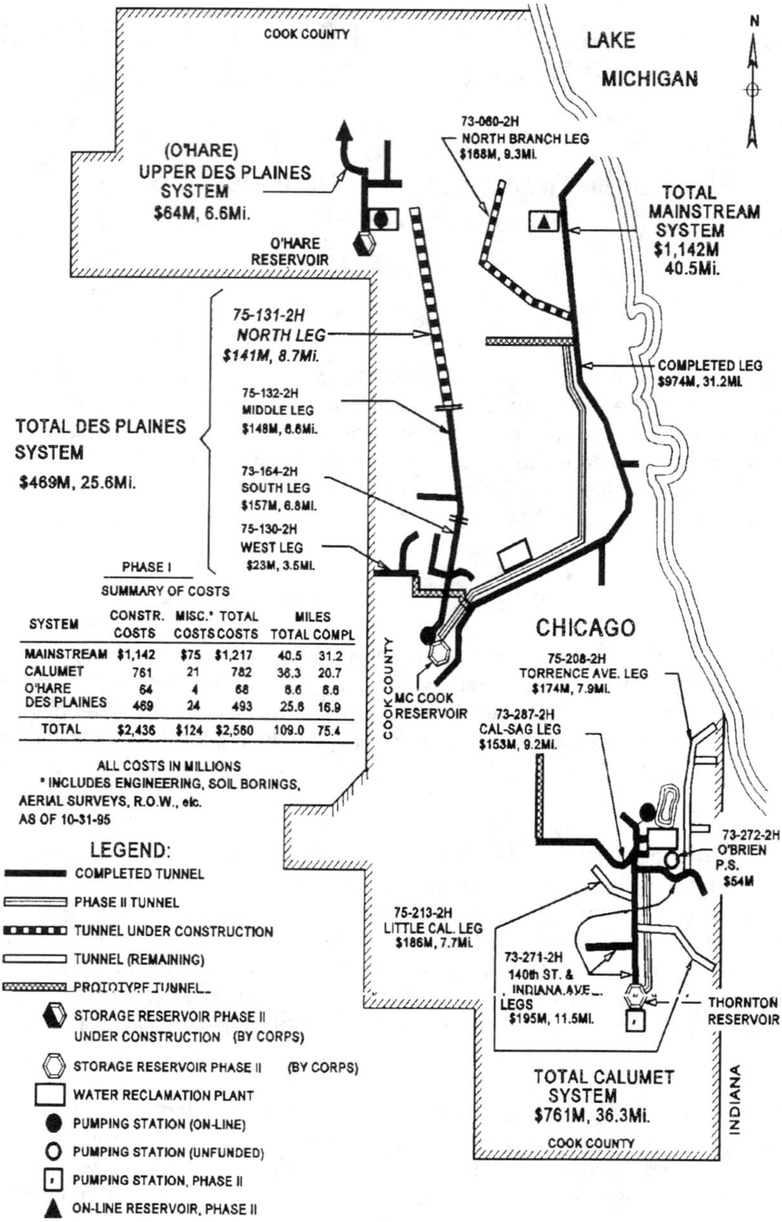

Figure 1. MWRDGC — TARP Status as of July 1996

the Mainstream, the Calumet, the O'Hare and the Des Plaines (See Figure 1). To date, 150.3 km (93.4 mi) of tunnels or 86% of Phase I is either completed or under construction representing a total construction cost to date of approximately $2.36 billion (US).

NORTH LEG PROJECT DESCRIPTION

The North Leg Tunnel is the final leg of the 41.2 km (25.6 miles) Des Plaines System and will provide relief for portions of Chicago and five neighboring communities. Upon its completion which is set for January 1, 1999, it will connect with the two previously completed sections that have been in operation since 1993. The $141,120,000(US) [See reference section for actual bid results.] contract was awarded to the Joint Venture of Kenny/Kiewit/Shea (KKS) on December 21, 1994 and crews mobilized immediately to begin site preparation.

The project as designed is in some ways two completely independent projects. The northern portion of the project is located completely in the overburden which is generally made up of stiff cohesive soils. Four soft ground TBMs were used and the method of tunnel supports varied from 4.72 m (15.5 ft) in diameter ribs & lagging to jacked pipe with inside diameters of 1.52 m, 2.13 m and 2.44 m (60, 84 and 96 in). In addition to the mining, numerous shafts were excavated and supported by using soldier pile and lagging, circular rings and boards, concrete caissons or steel beams and wood sheeting. The southern portion of the project in addition to having the same types of soft ground excavations, also required extensive rock excavation and utilized two hard rock TBM's, blind hole shaft drilling and raise bore shaft drilling, along with conventional drilling and blasting.

The mainline 6.71 m (22 ft) finished diameter tunnel is being mined out of an existing 9.14 m (30 ft) in diameter construction shaft and will extend outward some 10.78 km (6.7 miles) where it will be removed from a 9.14 m (30 ft) in diameter dropshaft that is to be constructed under this contract. The tunnel is to be driven at an average depth of 66.1 m (217 ft) below the surface within the Silurian System of Dolomite Limestone. Original exploratory borings show the unconfined compressive strength of the rock to range between 95 MPa (13,800 psi) and 196 MPa (28,400 psi) with the majority of the rock being about 145 MPa (21,100 psi).

The first 10.0 km (32,800 ft) of mining in rock is expected to be typical of previous tunneling operations within the Chicago area, generally sound rock with some thinly bedded joints that may at times carry some water. The final 808 m (2,650 ft) of mining is expected to fall within an area known as the "Des Plaines Disturbance Zone" which is believed to be the result of a meteorite impact that disturbed the flat lying strata of Silurian System. Subsurface investigations have indicated that a formation of shale will be encountered in this area originating in the invert and gradually becoming full face. Unconfined compressive strengths of the shale are anticipated to be around 76 MPa (11,000 psi). Contract documents require the installation of steel ribs, wire mesh and shotcrete in this area as it is mined. Along the route, the mainline tunnel is intersected by 15 dropshafts which are interconnected to diversion and regulating structures and local upper level sewers. Just before entering the "Des Plaines Disturbance Zone" the tunnel will intersect the second rock tunnel which is a 1.83 m (6 ft) finished diameter tunnel 410 m (1,344 ft) in length.

Because of the poor rock quality found within the disturbance zone on the northern portion of the project, the MWRD decided to raise the final 3,243.4 m (10,641 ft) up into the overburden to an average depth of 21.0 m (69 ft) within a stratum of generally stiff cohesive clay. The 3.66 m (12 ft) finished diameter tunnel is intersected at four locations by drop structures which will carry overflow from ten existing sewers and collecting structures. This northern portion of the project is connected to the much deeper southern portion of the project at the C-2 location by means of a double gated regulating structure and a 9.14 m (30 ft) diameter dropshaft.

Shaft Excavation

The horizontal alignment of the tunnel follows the general path of the Des Plaines River. As a result, the dropshafts and drop structures located along the route lie in close proximity to and in some instances within the designated wetland area of the river. Relying on experience in the Chicago land area, KKS realized that one of two problems could at some point lead to a major disaster. The first factor considered is the fact that the work area is prone to flooding from the Des Plaines River due to heavy rains or snow melt. If at any time the river was to rise above a shaft that was open to the tunnel, a major disaster would be at hand. Secondly, if liner plates and rings were to be used as we had done in the past, our ability to grout off any infiltration from the river might be cost prohibitive.

As a solution, a concrete caisson was considered as a support system for the overburden soils down to the rock interface for all of the dropshafts. Two criteria had to be met in order to make this process economically feasible. First, the caisson would have to be designed in such a way that it could be used as the final lining of the overburden portion of the shaft. Secondly, the caisson would have to be designed in such a way that would allow the inlet connection of the shaft to be hand mined from the inside of the shaft over to the regulating structures at a later date.

After several designs were submitted and reviewed, a final system was developed allowing for the caisson to act as a final liner. In order to meet contractual requirements, a 3.05 m (10 ft) and a 3.66 m (12 ft) system were developed. At a later date when it came time to excavate the inlet portion of the shaft, removing portions of the reinforced concrete caisson would be quite labor intensive. In order to construct the inlet to the required dimensions, an opening spanning about 178° of the caisson would have to be removed to a height of either 7.32 or 9.14 m (24 or 30 ft). To simplify this excavation, the opening was bulkheaded and smooth back steel liner plates were assembled to cover the opening. When the inlet excavation begins, crews will simply cut away the steel liner plates and proceed with the excavation (See Photo No. 1).

Shafts were set up with a launch pad and guide beam system and the concrete was placed in twenty foot lifts. Incorporated into each initial concrete pour was a 12.7 mm (1/2 in) steel cutting edge that would be left in place at the rock interface. As the inner and outer forms were stripped, the material was clammed out of the center and the caisson was sunk to the interface. Upon landing the caisson, a concrete bottom was poured and we then grouted the shaft off through 38 mm (1 1/2 in) grout pipes that had been installed in the concrete lining (See Photo No. 2).

At the time of this writing, all caissons have been successfully completed within budget with the exception of two shafts at which a large number of boulders were encountered and required some additional work.

Photo No. 1 - Caisson with inside form in place and reinforcing steel completed. Reinforcing has been omitted on the left side whichwill be the opening for the inlet excavation. In the bottom left corner smooth back liner plates are being built.

Photo No. 2 - Excavation operation of the caisson with material being clammed from the center. Caisson is sinking under it's own weight while being held plumb with four guide beams.

Upon completion of the caisson, crews then excavated the rock portion of the dropshafts. Contract documents prohibited blasting in the vertical portion of the dropshafts that are in the finished diameter range of 2.13 m to 2.74 m (7-9 ft). Contract documents also state that emergency egress must be available within 1,219 m (4,000 ft) of the heading. On the rock portion of the project, there is a total of 15 dropshafts. Of the 15 dropshafts, nine are required for emergency egress and the balance could be completed at anytime. As had been done on previous TARP projects, those shafts required for emergency egress were excavated using blindhole or downhole drills. Case Foundation Co. drilled nine total shafts on the project ranging from 2.84 m to 3.45 m (9.33 to 11.33 ft).

As Case Foundation Co. finished drilling the barrel or vertical portion of the shaft, KKS drill & blast crews followed closely behind to excavate the boot portion and exit conduit to the centerline of the tunnel. The last 3.66 m (12 ft) of exit conduit was generally excavated at 1.83 m (6 ft) high rather than the planned 3.66 m (12 ft) to allow for the TBM to pass with enough competent rock to grip. Currently, excavation of all boots and exit conduits that were downhole drilled is complete. Before the main line tunnel concrete begins, the balance of the exit conduits and stairs will be excavated from within the tunnel (See Figure 2).

The excavation of the remaining dropshafts was evaluated and it was decided to excavate the exit conduit and boot portions of these shafts from the tunnel. At each shaft location, a bridge was constructed across the tunnel at spring line allowing the drill & blast operations to proceed as the mainline mining was being done. All muck from the drill & blast work was hauled out to the main shaft and hoisted. Once the drill & blast is completed, American Mine Services will mobilize and raisebore the remaining shafts at either 2.84 m (9.33 ft) or 3.45 m (11.33 ft) in diameter.

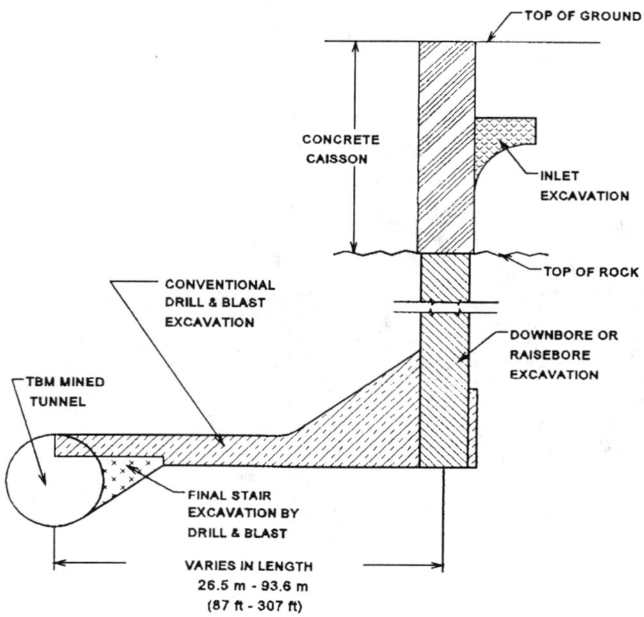

Figure 2. Shaft and Boot Excavation Sequence

By far the most challenging of all shafting work on the project involved the C-2 Construction Shaft. It is at this location that a 9.14 m (30 ft) finished diameter drop shaft is to be constructed and will act as the receiving shaft for the 7.39 m (24.25 ft) rock machine. In addition, a regulating structure containing two 2.74 m x 2.74 m (9 ft x 9 ft) electric sluice gates and a 5.49 m (18 ft) radius inlet spillway into the dropshaft is to be built here. All mining operations for the 4.72 m (15.5 ft) soft ground tunnel would also have to be set up here.

To meet all of the above criteria, an elliptical shaft 20.12 m x 30.18 m (66 ft x 99 ft) was excavated that would serve as the mining shaft for the upper level tunnel upon completion of Stage I excavation and a retrieval shaft for the lower level tunnel under Stage II. Stage I excavation required the entire shaft to be excavated to a depth of 27.1 m (89 ft) or the invert of the 4.72 m (15.5 ft) tunnel. To support the excavation, forty three-total 610 mm (24 in) soldiers beams were placed and grouted in 914 mm (36 in) diameter drilled holes. As the excavation proceeded, 102 mm (4 in) nominal hardwood lagging was installed between the piles. Of the forty three-total soldier beams, twenty-nine were 30.48 m (100 ft) in length and the balance, which were located in the area of the future 9.14 m (30 ft) dropshaft, were 35.0 m (115 ft) in length and rested on the rock interface. Five levels of bracing were installed as the excavation proceeded, each level consisting of a pair of 356 mm (14 in) elliptical walers or ring beams that were made up of 12 segments. Upon completion, nearly 266 tonnes (293 tons) of soldier beams and walers were installed during the Stage one excavation (See Photo No. 3).

Upon completion of the mining and concreting of the 3.66 m (12 ft) tunnel, Stage II excavation was begun on the southern portion of the shaft. Stage II excavation entailed excavating the southern portion of the shaft another 7.32 m (24 ft) to the rock interface using steel rings and wood sheeting. The first 3.66 m (12 ft) was excavated at 16.15 m (53 ft)

Photo No. 3 - Aerial view of 20.12 m x 30.18 m (66 ft x 99 ft) elliptical shaft during the 4.72 m (15.5 ft) mining operation.

Photo No .4 - Looking into the elliptical shaft during stage two excavation. The 16.15 m (53 ft) and 10.06 m (33 ft) diameter shafts are seen at the top of thephoto. In the foreground is the concrete operation for the regulating structure.

diameter to allow for the construction of the inlet spillway. The final 3.66 m (12 ft) was then stepped in to 10.06 m (33 ft) in diameter. As noted previously, the rock interface was shale and would require wire mesh, rock bolts and shotcrete for support. A shaft of 10.06 m (33 ft) in diameter was drilled and blasted another 15.85 m (52 ft) down to the top of the boot and then concrete lined at 9.14 m (30 ft) to the bottom of the inlet spillway. At the time of this writing, crews are drilling and blasting the boot while using shotcrete, wire mesh and rock bolts for support. As the Stage II excavation is being carried out, concrete work on the adjacent regulating structure within the shaft is also being completed (See Photo No. 4).

Once crews completed the Stage I excavation of the C-2 Construction Shaft, they were moved over to dropshaft 3R to excavate another elliptical shaft which was slightly smaller at 13.72 m x 24.99 m (45 ft x 82 ft) and a total depth of 18.59 m (61 ft). Support of this excavation was similar to that previous noted except that the soldier beams were 22.55 m (74 ft) long 356 mm (14 in) H-piles. Excavation of this shaft also had to be carried out in

Photo No. 5 - Setting up the 1.96 m (77 in) TBM within the elliptical shaft at DS-3R for a 243.8 m (800 ft) run of 1.52 m (60 in) jacked pipe.

stages, this time however it was three. Initial excavation was to a depth of 7.32 m (24 ft) feet at which a jacked pipe operation was set up for approximately 243.8 m (800 ft) of 1.52 m (60 in) reinforced concrete pipe (See Photo No. 5). Stage II excavation proceeded to a total depth of 15.24 m (50 ft) at which we hand mined and jacked approximately 41.15 m (135 ft) of 1.22 m (48 in) reinforced concrete pipe. The third and final stage was to a total depth of 18.29 m (60 ft) and it was at this elevation that the jacked pipe operation was set up for 474.8 m (1558 ft) of 2.44 m (96 in) pipe that intersected the 4.72 m (15.5 ft) mined tunnel. After all of the mining was completed, two connecting structures and one regulating structure were then constructed within this elliptical shaft. When all of the concrete and backfill operations were completed, all of the soldier beams were extracted for reuse by means of a jacking system that was designed and built on site.

In the construction of the 63 upper level structures, a total of 53 different shafts are to be excavated of which, 25 will be completed using soldier beams and lagging. After subcontracting the drilling for the soldier beams on the two elliptical shafts, it was obvious that in order to maintain and have control over our schedule, we would have to complete the drilling operation with our own forces. To accomplish this task, a Watson 2000 mobile drill rig was purchased having the capability of drilling to a depth of 24.38 m (80 ft). With this drill rig, 610 mm (24 in) diameter holes are drilled and 356 mm (14 in) beams are set and grouted in place. All of the shafts are excavated at a width of 7.32 m (24 ft) and lengths of either 9.75 m (32 ft) or 15.85 m (52 ft) to allow for the use of standardized steel bracing sets. As prior noted, all soldier beams will be extracted for reuse upon completion of work at each site.

The balance of the upper level shafts are excavated as either circular or rectangular and excavated conventionally by clamming with wood sheeting and steel rings or beams for support. In larger diameter shafts, a backhoe is being used along with a clam bucket for excavation (See Photo No. 6).

Photo No. 6 - Excavation of a large circular shaft using a backhoe and clam bucket. Supports are steel rings and wood sheeting.

Tunnel Excavation

All mining operations for the 7.39 m (24.25 ft) tunnel was to be done out of the existing 9.14 m (30 ft) diameter construction shaft. Under a prior contract, a 9.75 m (32 ft) diameter tunnel was driven beyond the center of the shaft approximately 57.9 m (190 ft). The tunnel was then concrete lined to 9.14 m (30 ft) to a point just upstream of the shaft where a reinforced concrete bulkhead was installed to eliminate the infiltration of ground water from the unlined portion of the tunnel.

Contract documents required that a new 2.13 m (7 ft) diameter vent shaft be constructed over the existing tunnel approximately 137.1 m (450 ft) downstream of the existing construction shaft before the start of any work in the tunnel. Upon completion of the vent shaft, a new reinforced concrete bulkhead was to be built between the vent shaft and the existing construction shaft. The bulkhead was necessary to separate the current project from the existing tunnel which was in full operation. Once the bulkhead was installed, the original bulkhead could be removed and construction of the starter tunnel could begin.

Due the anticipated delivery of the TBM, it was decided to excavate an additional shaft over the existing tunnel that would serve as a muck shaft. Excavation of this shaft just upstream of the bulkhead would allow the vertical belt system and all of the conveyor system to be installed, tested and ready for operation prior to the TBM arriving on site (See Photo No. 7).

Excavation of the overburden portion of the vent and muck shafts was the proving ground for our proposed caisson shafting operation. Identical 3.05 m (10 ft) inside diameter caissons with 305 mm (12 in) thick walls were sunk approximately 21 m (69 ft) to the rock interface. Once the shafts were completed and grouted, American Mine Services was mobilized and pulled two 2.84 m (9.33 ft) diameter raises.

Photo No. 7 - View looking toward concrete bulkhead in tail tunnel showing belt storage unit assembled prior to the start of mining.

The muck handling system for the 7.39 m (24.25 ft) tunnel consisted of a Long-Airdox horizontal conveyor system and a Lakeshore vertical belt which had been used on a previous TARP project. Due to the smaller diameter and depth of the tunnel, the power train of the Lakeshore system was revised and a new Flexowall belt was purchased. With all shafting and bulkhead work completed, the installation of the vertical conveyor belt, transfer tower, splice platform, main drive and 548.6 m (1800 ft) belt storage unit was then installed and tested by February 1996.

The TBM of choice for the mining was an angle cylinder, open gripper machine that utilized the inner cutter head, drive cartridge and units, gripper cylinders and miscellaneous electrical and hydraulic components from a Robbins 9.85 m (32.33 ft) main beam machine used on two previous TARP projects (See Photo No. 8). Construction and Tunneling Services (CTS) handled the design of the machine and the rebuild was carried out by KKS forces in Harvey, Illinois.

The design of the TBM eliminated the main beam structure of the original machine and replaced it with a 21.34 m (70 ft) long bridge structure. The bridge was pinned on the front end to the cutterhead and rested on the upper deck of trailing deck number one. The bridge structure weighed approximately 39 tonne (43 ton) and was used to support not only the operator's console and the hydraulic platform but also supported the 86 tonne (95 ton) gripper carriage. The gripper carriage traveled along the outside edge of the bridge structure on four 457 mm (18 in) diameter wheels as the TBM cycled forward to a maximum stroke of 2.28 m (7.5 ft). In addition to the elimination of the main beam structure, the eight propel cylinders from the previous TBM were replaced with four 559 mm (22 in) diameter thrust / propel cylinders that provided the diagonal thrust for the TBM. Conversion to this configuration facilitated the redesign of trailing deck number one as it was used with the machine on

Photo No. 8 - 7.39 m (24.25 ft) TBM during rebuild operations. Diagonal thrust cylinders can be seen between the drive canister and the gripper pad.

its previous TARP project. In addition to supporting the advancing tail piece for the horizontal conveyor, the deck now had to be capable of supporting the load of the bridge structure on its top deck which was approximately 87 tonne (96 ton). Boretec handled the design and layout of trailing deck number one which was fabricated while the TBM was being rebuilt.

Alignment control of the TBM was the final design concern. Since the head of the TBM pivoted about an 203 mm (8 in) pin on the bridge structure and the bridge structure rested on trailing deck number one, it was not possible to fix two targets for laser guidance. A ZED modified 261 laser guidance system was thus selected. Once the machine was assembled and moved to the face, a survey of the cutterhead was conducted and all information regarding roll, attitude and elevation of the cutterhead was entered into the ZED system. A Leica motorized theodolite equipped with a laser beam was the instrument of choice and once all the bugs were worked out of the system, the laser was generally moved ahead in 122-152 m (400-500 ft) increments.

The basic specifications of the redesigned TBM are as follows:

Cutterhead			
	Speed	7 rpm	
	Power	2,461 kW	(3,300 hp)
	Torque	3,050.6 kN.m	(2,250,000 lbf-ft)
	Number of Cutters	Center (4) 394 mm	(15.5 in)
		Face/gage (44) 432 mm	(17 in)
	Loading	24,948 kg	(55,000 lbs)
Thrust System			
	Maximum Thrust	1,097,693 kg	(2,420,000 lb)
	Nominal Thrust	1,415,207 kg	(3,120,000 lb)

System Pressure	17.2 MPa	(2,500 psi)
Max. Advance Rate	23.88 mm/rev	(0.94 in/rev)

Electrical

Total Power	2,759 kW	(3,700 hp)
Primary Voltage	13,200 volts	
Secondary Voltage	575/460 volts	

Shipments of the completed machine began to arrive on site in March of 1996 approximately one year from the start of the redesign. Prior to the arrival of the TBM, all 54.86 m (180 ft) of trailing gear was assembled underground and was installed in the tail tunnel along side of the belt storage system (See Photo No. 9). Assembly of the machine began on March 7, 1996 and mining began just 33 days later on April 9, 1996. To date, several jobsite records have been posted namely:

Best Shift	35.66 m	(117 ft)
Best Day	99.06 m	(325 ft)
Best Week	376.12 m	(1,234 ft)
Best Month	1,499.00 m	(4,918 ft)

A reliable communication interface between the TBM and the conveyor system has played an essential role to the continued progress of the mining operation. In all seven conveyors make up the muck handling system including the TBM conveyor, a transfer conveyor, the mainline horizontal belt, the vertical belt, a shuttle belt, a 243.8 m (800 ft) overland belt and a radial stacker. The TBM and the entire conveyor system are controlled by state of the art logic controllers that maintain constant communications between each other. In addition to the main drive unit, four 295 kW (400 hp) inline booster units were installed as well as one 147.5 kW (200 hp) return booster. The system is designed to handle 726 tonne (800 ton) per hour.

Mining on the 3.66 m (12 ft) finished diameter tunnel was done with a Lovat 4.72 m

Photo No. 9 - View also looking into the tail tunnel showing the fully assembled trailing deck alongside the belt storage unit prior to TBM assembly.

Photo No. 10 - 4.72 m (15.5 ft) TBM holing through after mining 3,243.4 m (10,641 ft) of tunnel.

(15.5 ft) TBM. This machine was used previously on the Interisland Tunnel Project in Boston by the joint venture of Kiewit/Atkinson/Kenny. The machine was shipped to the jobsite where minor hydraulic and electrical reconditioning was conducted prior to its use. The machine was installed on July 21, 1995 and holed through after mining 3,243.4 m (10,641 ft) on January 24, 1996 (See Photo No. 10). Supports for the tunnel included 102 mm (4 in) ring beams on 1.22 m (4 ft) centers with 76 mm (3 in) nominal hardwood lagging between the rings. Muck handling was done using a conventional rail haulage system with 13.76 cu m (18 cu yd) muck boxes that were hoisted to the surface and dumped. Average production rate on the two shift operation was 18.29 m (60 ft) per shift.

Once the mining was completed, the tunnel was concrete lined to a finished diameter of 3.66 m (12 ft) using a telescopic form manufactured by EFCO. A total of 54.86 m (180 ft) of form was successfully cycled on a daily basis with the concrete being injected into the form through guillotine valves. All of the concrete was pumped from the surface using a Schwing 5000 HPR concrete pump and 152mm (6 in) diameter placing line that reached a maximum length of 914.4 m (3,000 ft).

As was the case with the rock tunnel, the 1,219.2 m (4,000 ft) emergency egress requirement was also in effect on this tunnel. A total of three drop structures were designed with drifts intersecting the tunnel at 75 angles, one of which was directly under the Des Plaines River with approximately 7.93 m (26 ft) of cover to riverbed. Two of the drifts were handmined, one at 4.72 m (15.5 ft) diameter using ribs and lagging and one by jacking 2.44 m (96 in) reinforced concrete pipe. Both hand mined drifts were mined from the drop structures to the tunnel with the final 0.91 m (3 ft) being mined from the mainline tunnel. The third drift, while not required for emergency egress, was specified to be mined from the tunnel and then concrete lined. KKS proposed an alternate to the MWRD of jacking 2.44 m (96 in) reinforced concrete pipe the entire length from the drop structure. The mining operation was sequenced in such a way that once the mainline mining was sufficiently past the intersection, it was shut down to set up for the retrieval of the 2.97 m (117 in) TBM. When the TBM

Photo No. 11 - 2.97 m (117 in) TBM holing through into the 4.72 m (15.5 ft) tunnel and being jacked out onto a transporter for removal.

reached the mainline tunnel it was jacked out into the tunnel onto a transporter, rotated and taken out to the main shaft for removal and mainline mining resumed after a six shift shutdown (See Photo No. 11).

As previously noted, the project entailed a total of four soft ground and two hard rock TBMs. The second rock TBM was a Jarva Mark 6 that had been used on three previous jobs in the Chicago area. The machine was resized by Kenny Construction to a diameter of 2.47 m (8.12 ft) and delivered to the site in November of 1996. Mining was done on a two-shift basis using conventional rail haulage for the muck system. The TBM mined a total of 470.6 m (1544 ft) to the intercept point of the mainline tunnel and was then backed up and removed from the same shaft. Once the mainline tunnel passes, crews will begin placing 1.83 m (72 in) reinforced concrete pipe in the tunnel and then grouting annulus.

In addition to the 4.72 m (15.5 ft) mainline machine and the 2.97 m (117 in) machine used to jack the 2.44 m (96 in) pipe, a 1.96 m (77 in) Lovat and a 2.59 m (102 in) Decker machine were also used. The 1.96 m (77 in) Lovat machine was used to jack 1.52 m (60 in) pipe for three different runs of 244, 235 and 70 m (800, 772 and 230 ft). The 2.59 m (102 in) Decker TBM was used to jack 2.13 m (84 in) pipe on runs of 166 m and 46 m (543 ft and 151 ft). The 4.72 m (15.5 ft) TBM was used on one other much shorter run of 114 m (375 ft) run for a connecting sewer at one of the drop shafts. Original plans called for the installation of a 3.35 m (11 ft) connecting sewer which we originally planned to construct by jacking 3.35 m (11 ft) reinforced concrete pipe with a 4.06 m (13.33 ft) mining machine. Contract plans showed the placement of a 134k volt oil filled transmission conduit directly above the tunnel with a separation of about 4.27 m (14 ft). In addition, there was an out cropping of rock in the area which put the top of rock only 7.62 m (25 ft) ft below the ground surface and the invert of the tunnel about 0.3 m (1 ft) above the rock. The situation posed two concerns. First, if we encountered any cap rock, blasting in close proximity to the electrical line would be very difficult. Second was jacking the pipe over the shot rock. Because of these two concerns, it was decided to give the MWRD and oversized connecting sewer at 3.66 m (12 ft) and mine the tunnel with the 4.72 m (15.5 ft) TBM which we felt could mine through any cap rock that

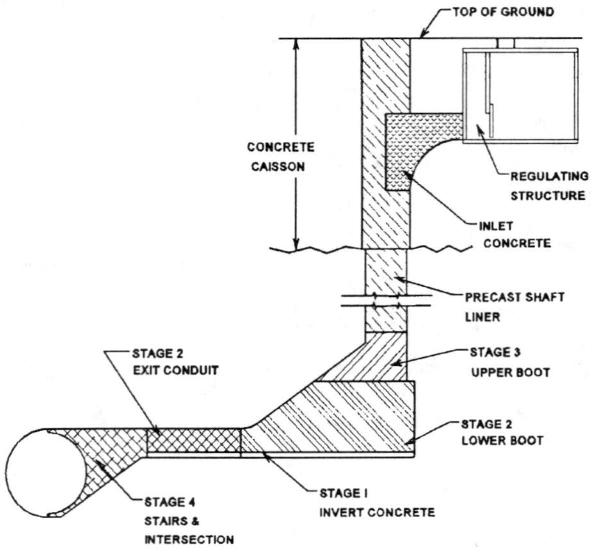

Figure 3. Boot and Shaft Concrete Sequence

might be encountered. The tunnel was supported with ribs and lagging just as it had been on the mainline soft ground tunnel. The mining of the tunnel was scheduled so that the completion of the mining coincided with the completion of the 3.66 m (12 ft) tunnel lining operation. Crews finished lining the mainline tunnel and moved in immediately to line the short connecting sewer.

CONCRETE WORK

Concrete work on the connecting and regulating structures was done in most cases immediately after their respective shafts were excavated. In general, two crews would move between locations and structures to tie the reinforcing steel while as many as four crews would form and pour the structures. As the structures were completed, all bracing and support systems were removed and salvaged for reuse on other shafts.

Boot shaft concrete has been divided up into four different stages. The first stage of concrete includes tieing the reinforcing steel and pouring the invert of the boot and exit conduit. Before moving to a new location, the crews also tied the reinforcing steel for the exit conduit and the lower portion of the boot. Stage two of the operation poured the exit conduit with a steel box culvert form and then formed and poured a portion of the boot. The form system for the boots is an adjustable steel form that allows us to make one mass pour accounting for about 75% of the total concrete in the boot. To date all of the smaller size boots have been completed up to stage two and the mass concrete poured averages about 183.5 cu m (240 cu yd). Stage three operation uses aluminum hand set forms to tie the top of the boot into the circular shaft. The fourth and final stage is to pour the stairs of the exit conduit along with the connection to the tunnel (See Figure 3). While the first three stages are accomplished with crews being serviced entirely from the surface, stage four will be completed entirely from the tunnel.

As stage three of the concrete work in the boots is completed, crews will then begin the task of setting and grouting precast shaft liners in all of the vertical portions of the shafts. Liners for the shaft are made up of reinforced concrete pipe in diameters of 2.13 m and 2.74 m (7 and 9 ft) and are set in place from the top of the boot to the bottom of the inlet structure and grouted. As crews reach the bottom of the inlet structure, the pipe will be decked off and excavation of the inlet structure over to the regulating structures will then begin. Upon completion of this excavation, crews will begin forming and pouring the inlet structures to complete the shafts.

As noted earlier, concrete lining of the 3.66 m (12 ft) diameter soft ground tunnel has been completed. Plans are currently under way for the lining of the 6.71 m (22 ft) rock tunnel. At the present time, Ewing Records is building 73.15 m (240 ft) of 6.71 m (22 ft) full round telescoping forms which are scheduled for a March 1, 1997 delivery. Success on a previous TARP project along with the availability of equipment has dictated the use of the identical concrete placement system for this operation. A 137.6 cu m (180 cu yd) per hour central mix batch plant will be set up at the main construction shaft and drop concrete down a 254 mm (10 in) drop pipe into a 27.5 cu m (36 cu yd) batch hopper. The hopper will discharge the concrete into one of four trains, each being made up of a 22.7 tonne (25 ton) locomotive and three 9.17 cu m (12 cu yd) Maxon agitor cars. As the concrete operation nears the main shaft and the haul distance decreases, one of the four trains will be eliminated. The concrete will travel through the tunnel to a California switch or pour deck that will be set up approximately 122 m (400 ft) in front of the concrete form. At this point, the trains will be shuttled to discharge into one of two 7.65 cu m (10 cu yd) Maxon remix cars. At the discharge end of the remix cars will be a Schwing 5000 concrete pump that will pump the concrete to a Putzmeister placer for injection through guillotine doors into the form.

PROJECT STATUS

To date, approximately 1,707 m (5,600 ft) of mainline rock tunnel remains to be mined, 808 m (2650 ft) of which is anticipated to be in shale requiring steel ribs and shotcrete for support. Anticipated completion of mining is March 1997. As far as the soft ground tunneling is concerned, while several short hand mined tunnels remain, one final 2.13 (84 in) jacked pipe run of 46 m (150 ft) remains and is currently scheduled for the summer of 1997. Once the TBM is removed from the C-2 drop shaft, the concrete operation for the 6.71 m (22 ft) tunnel will begin. Concrete work on the structures is proceeding right along with about 40 of the 63 total structures completed. When all of the structures are completed, subcontractors will mobilize to begin the installation of the electrical and mechanical control equipment.

REFERENCES

Budd, T., and Rautenberg, R., 1995, "Calumet Tunnel Project, A Case History", Proceedings, Rapid Excavation and Tunneling Conference, San Francisco, California, pp. 247-267

Knoerle, Bender, Stone & Associates, Inc., (1976), "Des Plaines River System Tunnels and Shafts of the Tunnel and Reservoir Plan, Part 1", Prepared for The Metropolitan Sanitary District of Greater Chicago.

De Leuw, Cather & Company, (1976), "Geotechnical Design Report, Volume One", Prepared for The Metropolitan Sanitary District of Greater Chicago.

BID RESULTS

Kenny/Kiewit/Shea	$ 141,520,000.00 (US)
Healy/Dillingham/PCL	$ 169,132,000.00 (US)
Perini/Tutor-Saliba	$ 171,369,000.00 (US)
Traylor Brothers/Frontier-Kemper	$ 176,677,000.00 (US)
Kajima/McHugh/Granite	$ 184,487,000.00 (US)

Chapter 11

HIGH PERFORMANCE TUNNEL BORING MACHINE FOR QUEENS WATER TUNNEL, NO. 3: A DESIGN AND CASE HISTORY

Brian B. Khalighi and Joseph J. Diehl
Atlas Copco Robbins Inc. and Grow Tunneling Corp.

ABSTRACT

This paper discusses the Queens Water Tunnel, No. 3, Stage 2, being driven by a new Robbins High Performance (HP) Tunnel Boring Machine (TBM).The paper is divided into four sections. In the INTRODUCTION section, a short overview of the scope of the project is presented. In the GEOLOGY section, the geological condition along the tunnel drive is briefly introduced. The TBM DESIGN section discusses the various design criteria and the resulting machine specifications. The paper concludes, in the TBM OPERATION section, with a case history describing the tunneling operation from start-up to the time of the writing of this paper.

INTRODUCTION

Water Tunnel No 3, a project of the New York City Department of Environmental Protection (DEP), will improve fresh water distribution throughout the city. After completion, it will allow for the much needed maintenance of two existing tunnels that have been operating continuously since 1917 and 1936. When finished, the tunnels supply about 90% of the water consumed by the city and its residents, currently at 5.7 million m((1.5 billion gal) a day. Water Tunnel No. 3 ranks as the largest and most costly construction project in history of New York City. Total excavation involves 93 km (58 mile) of tunnel with construction divided into four stages. Stage 1, a 21 km (13 mile) long drill and blast tunnel is complete. Stage 2, now underway, is divided into three sections. It comprises a total of 24 km (15 mile) of interconnected tunnels, extending from the end of stage 1 tunnel, to serve the boroughs of Brooklyn, Queens, and Manhattan. Stages 3 and 4 are presently being planned and designed. In March of 1995, the Joint Venture of Grow-Perini-Skanska was low bidder at $172,467,000 for the construction of Water Tunnel No. 3, Stage 2, underneath the borough of Queens in New York City.

GEOLOGY

The project consists of boring of more than 7.6 km (25,000 ft) of tunnel at 7.06 m (23 ft 2 in) diameter through hard and jointed formations of varying rock types including biotite-hornblende gneiss intermixed with granitic gneiss, amphibolitic, pegmatite, and biotite schist. one section of rhyolitic schist has been confirmed by test to have an unconfined compressive strength of 275 MPa (39,000 psi). Cherchar Abrasivity Index for some typical samples is in the 3.3 to 4.3 range indicating a fairly high abrasivity. After completion boring,

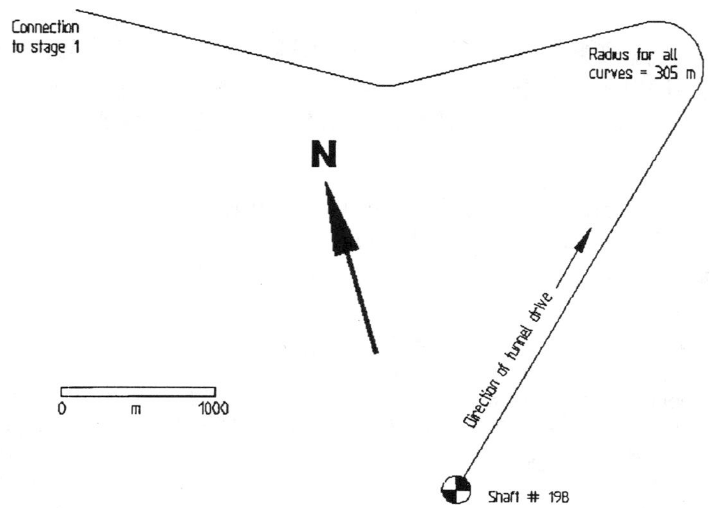

Figure 1. Queens Tunnel Route

the tunnel will be concrete lined to a finished diameter of 6.1 m (20 ft 0 in).

The tunnel begins at shaft 19B, which is 9.1 m (30 ft) diameter and 204 m (670 ft) deep, and advances NE at an average slope down of 4.2%. The shaft forms the only point of access and egress during excavation.

As indicated in Fig. 1, the tunnel negotiates a 116(curve to the left after approximately 3 km (1.9 mile) and then continues in a westerly direction, with two more curves, before it connect to the Stage 1 tunnel.

TBM DESIGN

TBM Selection

Based on the geological conditions expected to exist along the tunnel , it was clear that the machine chosen would be of the High Performance (HP) open hard rock type. These types of machines are well suited for competent ground conditions where support will be occasional and consists mainly of rock bolts, ring beams, or wire mesh. Due to the open design, access to the walls and crown is relatively easy, allowing for the placement of supports as required.

The Robbins HP open hard rock TBM, as shown in Figure 2., weighs 610 metric tons (1,345,000 lb) approximately and has a diameter of 7.06 m (23 ft 0 in). The machine can be modified to any diameter between 6.50 m (21 ft 4 in) and 8.50 m (27 ft 11 in), making it suitable for future hard rock tunneling projects. The TBM consists of three major components, the Cutterhead/Cutterhead support, Gripper, and Main Beam assemblies.

The cutterhead support houses the main bearing, main drive gear, seals, and the cutterhead drive train. These consist of an electric motor, gear reducer, and hydraulic clutch. The cutterhead support assembly also includes the front, side, and roof supports. These supports firmly stabilize the cutterhead during boring and ensure precise cutter tracking.

The main beam assembly consists of the main beam and rear section which is connected to the main beam by a bolted flange. The forward end of the main beam assembly is bolted to the cutterhead support. The aft end of the main beam is supported by the gripper assembly during the boring cycle. The gripper assembly consists of the gripper cylinder, gripper carrier, and gripper shoes. This assembly provides reaction to the forward thrust of the machine. The main beam can be moved up, down, or sideways relative to the tunnel axis throughout the full working stroke. This is how the TBM is steered. The function of the rear section is to support the weight of the machine during the regrip cycle, when the gripper carrier is being moved forward to begin the next boring cycle. Additionally, the back-up towing system is connected to the rear section legs.

Table 1. Specifications for 235-282 HP TBM

TBM Diameter	7.06 m (23 ft 2 in)
Diameter Range	6.50 m (21 ft 4 in) to 8.50 m (27 ft 11 in)
Cutter Size	482.6 mm (19 in)
Cutter load capacity	311 kN (70,000 lb)
Number of Cutters	50
Cutterhead Thrust	15,550 kN (3,500,000 lb)
Cutterhead Power	10 x 315 = 3150 kW (4220 hp)
Cutterhead Torque	3624 kNm (2,669,000 lb-ft)
CutterheadSpeed	8.3 rpm

Cutters

The 50 cutters installed on this TBM are of Wedge Lock type and have a disc diameter of 482.6 mm (19 in). Compared to the traditional V Block mounted cutters, the wedge lock cutters are simpler to install, more efficient, less expensive, and require less time for cutter change. The wedge lock not only provides a stable mounting of the cutter shaft with proper line contact on the seat, rather than the point contact seen in the V block mounting, it also takes the load off the mounting bolts. It also takes less time to mount a wedge lock cutter than a V block mounted one as there are less number of parts to handle.

Cutterhead

The cutterhead has a diameter of 7.06 m (23 ft 2 in) with new cutters. The construction of the cutterhead consists of a cylindrical center section and two outer cutterhead segments. 50 front loading cutter housings are strategically located and welded to the face of inner and outer cutterhead segments to provide the optimum cutter spacing. The outer cutterhead segments are bolted onto the inner cutterhead and then field welded at the tunnel site.

Main Bearing and Drive System

The main bearing, as shown in Figure 3, is a three element roller bearing. This is

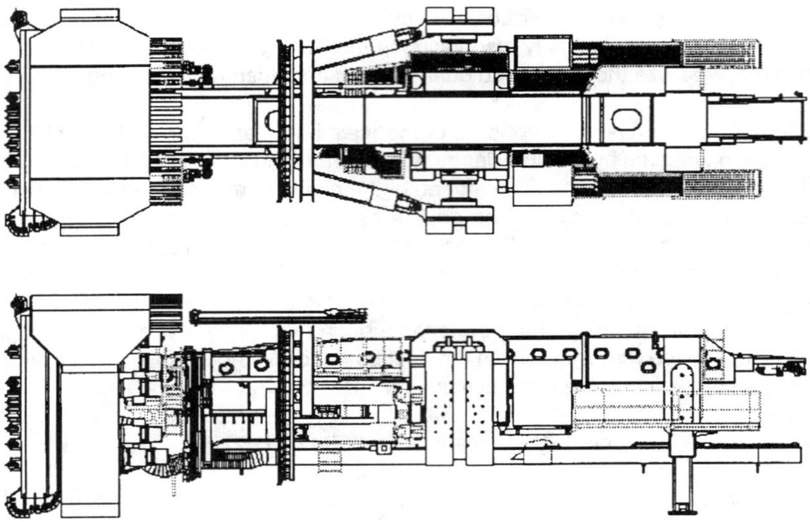

Figure 2. Plan and Side View of HP TBM 235-282

similar in design to the bearings used on the nine Robbins HP TBMs built to date. The bearing is capable of sustaining the high thrust loads realized under HP operation.

The three element bearing design is very well suited for sustaining the eccentric loads which can occur during steering of the machine or in blocky face conditions.

Each drive train, as shown in Figure 4, consists of a single speed electric motor, gear reducer, and a hydraulic actuated friction clutch mounted to the rear of the train for easy access. The drive trains all mesh on a common main drive gear.

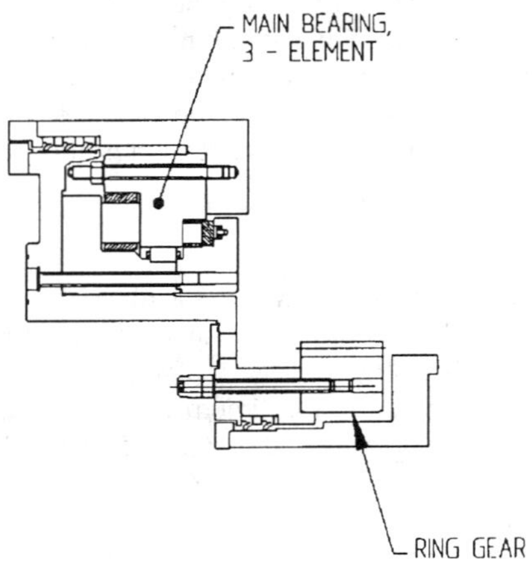

Figure 3. Main Bearing, Ring Gear, Seals

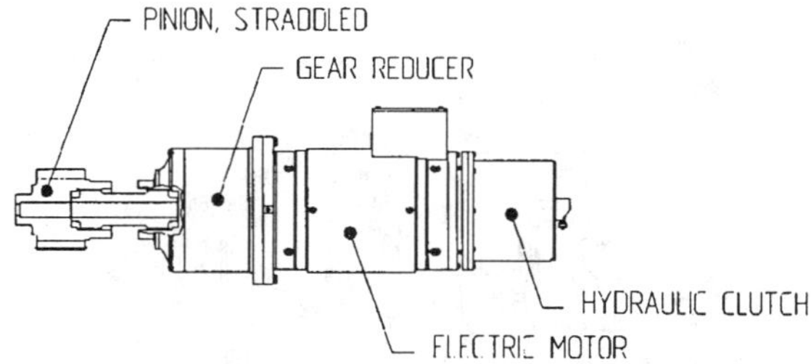

Figure 4. Drive Train

This provides a rugged system, capable of delivering very high breakout torque as well as sustaining a high running torque.

The anticipated rock conditions are very well suited to this system, since the tunnel is expected to be in fairly uniform and competent ground.

Should the cutterhead get stuck in blocky ground conditions, the high breakout torque can be very useful. By momentarily applying approximately two times the running torque, the head can in many cases be broken free. This is done by bringing the motors up to full speed and then taking them off line. The clutches are engaged and the rotational inertia of the motors is transferred to the cutterhead.

The machine currently has 10 drive trains, each rated at 315 kW (422 hp). Total installed cutterhead power is therefore 3150 kW (4220 hp).

The speed of the cutterhead is 8.3 rpm. This provides a cutterhead peripheral velocity of 184 m/min (604 fpm).

Gripper and Thrust System

The thrust system consists of four propel cylinders that provide the required cutterhead thrust at a pressure which is lower than the hydraulic system design pressure. This further increases the safety factor already designed into the hydraulic system. Hence, the system wear factors are decreased.

The gripper system, as shown in Figure 5, consists of a large bore, double rod gripper cylinder and two gripper shoe assemblies. The gripper cylinder is capable of providing a gripping force which is many times greater than the thrust force of the machine. This ensures a rigid anchoring system during forward propulsion of the TBM with virtually no slippage of the gripper shoes on the tunnel wall. The contact area of gripper shoes are large enough to maintain the ground pressure at low enough levels. The gripper shoes are slotted to accommodate the systematic rock bolt pattern along the length of the tunnel.

Auxiliary Systems

The TBM is equipped with a hydraulic service hoist assembly with a capacity of 4 tons.

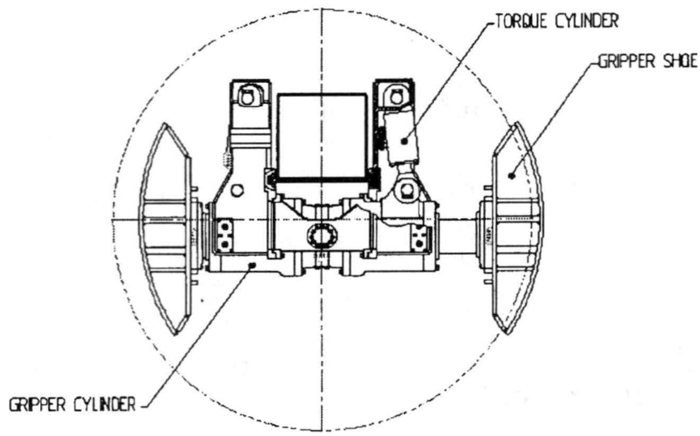

Figure 5. Gripper Cylinder and Gripper Shoes

The hoist is used for transport of cutters and supplies from the front end of the back-up system to the area behind the cutterhead support.

Ground Support System

The TBM is also equipped with roof and probe drilling systems. The roof bolting system, as shown in Figure 6, consists of two Atlas Copco COP1238 rock drills mounted to the TBM in such a way that rock bolts can be installed in the upper half of the tunnel cross section as the TBM moves forward.

The probe drilling system, as shown in Figure 7, consists of an Atlas Copco COP1238

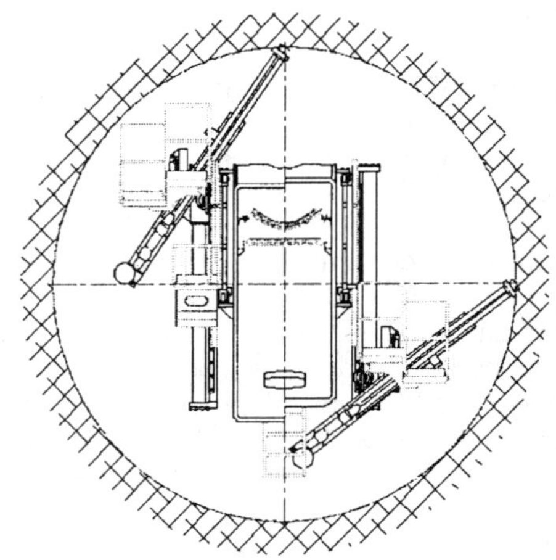

Figure 6. Roof Drills

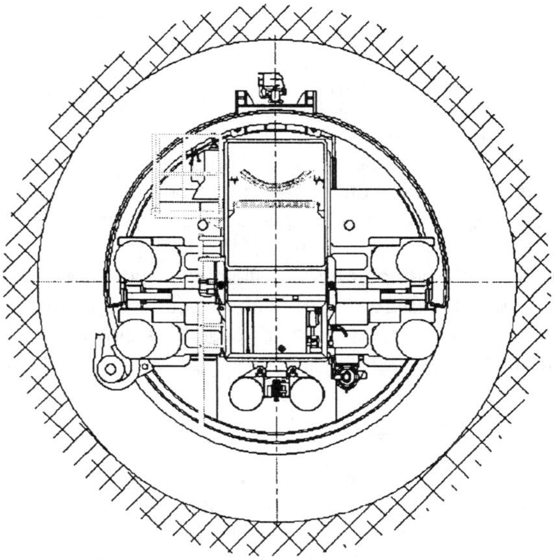

Figure 7. Probe Drill Circular Track Arrangement

rock drill, mounted to the main beam via a circular track, that is used to probe ahead of the TBM either through the face or above the cutterhead at a slight angle to detect any water or differing rock conditions.

Back-up System

Behind the TBM is the 55 m (180 ft) back-up system manufactured by Boretec Inc. This system, which is continuously towed by the TBM, consists of a bridge between the rear section of the TBM, a transfer conveyor, 6 rolling gantry cars with upper and lower decks, and one ramp section.

The lower deck of gantry car number six contains a single rail track, coupled to a turn out assembly on car five, thus allowing a double track setup on cars two through four. Supplies such as rails, continuous conveyor components, fan lines, etc. can be stored on one side, leaving the other side free for trains. Attached to deck one is the continuos conveyor advancing tail piece

The upper deck of car two carries the TBM transformer/electrical cabinet assemblies. The upper deck of the remaining cars contain an air scrubber and fans, fresh water tank and pumps, advancing tail piece power pack, emergency generator for dewatering pumps, high voltage cable storage, and ventilation fan line erector.

A hopper on the back-up system transfers the muck from the bridge conveyor, located in between the TBM and the back-up system, to the continuous conveyor that transport the muck out of the tunnel.

TBM OPERATION

TBM Manufacture

The TBM was designed and manufactured in approximately 10 months. Throughout the designing and manufacturing of the machine several meetings were held between Atlas Copco Robbins Inc., Grow-Perini-Skanska, and outside consultants to review the progress of the operation in order to produce the best possible product. Manufacturing of major structural components and assembly of the TBM was subcontracted to Markham and Company, Ltd. of Chesterfield, England. Under the terms of the contract, Markham built and assembled the TBM according to the drawings and specifications that were provided by Atlas Copco Robbins Inc., and under the supervision of Atlas Copco Robbins and Grow-Perini-Skanska personnel. The machine was shipped to Port Elizabeth, NJ, in June of 1996 and transported to the job site by both truck and barge.

Site Assembly

All components of the TBM were lowered to the bottom of a 204 m (670 ft) deep, and 9.1 m (30 ft) diameter shaft. Prior to the TBM assembly, an erection chamber was drilled and blasted at the base of the shaft. A 107 m (350 ft) long starter tunnel and a 55 m (180 ft) long tail tunnel were also constructed. A mud slab was poured in the entire invert with two each C15x40 channels embedded in the slab and centered and spaced at 2.3 m (7 ft 7 in). Gripper walls were poured in the last 20 m (66 ft) of the starter tunnel from the invert to 6 m (19 ft 6 in) above the mud slab. Walls were built to provide support for the gripper shoes during the initial tunnel boring.

The TBM cutterhead was manufactured in 3 pieces which were welded together at the job site. A framed tent was erected over the cutterhead to protect the welding operation from the elements and a local welding company was contracted to perform the welding which required 63 m (206 ft) of fillet weld. A detailed sequence of assembly of major TBM components was developed prior to lowering of the major components down the shaft for the final assembly. The five major heavy components of the TBM are, main beam at 85.2 tons (188,000 lb), cutterhead support and main bearing and seals assembly at 110 tons (242,000 lb), gripper carrier at 67 tons (148,000 lb), rear support at 34 tons (75,000 lb), and cutterhead at 79 tons (175,000 lb). These components were lowered by using an American 480B winch mounted on a specially designed frame. Two Manitowac M250 cranes, with rated capacity of 300 tons each and with 46 m (150 ft) long booms, were used to pick the winch and frame assembly, hold them over the load so that the load could be connected to the winch, and pick the entire winch, frame, and load assembly, and set the frame over the shaft. After placement of the frame over the shaft, the winch was used to lower the load to the bottom of the shaft.

Once at the bottom of the shaft, the major components were set on dollies with Hillman rollers attached to the bottom of each dolly. The dollies rode in the channels which were embedded in the mud slab. Air tuggers were installed in the starter and tail tunnels and were used to move the TBM components in either direction to connect the pieces.

Once the five major heavy components were lowered and connected to each other, one Manitowac M250 with three parts of line, with total capacity of 41 tons (90,000 lb), was used to lower and assemble the remaining components.

Upon delivery of the back-up system components to the job site, they were set up on a 55 m (180 ft) long section of track and as many back-up components as possible were installed. Deck one was lowered first, by the Manitowac M250, and was set on the track at the bottom of the shaft and pulled into the tail tunnel. The bridge between the TBM and deck one was lowered and attached to the TBM rear section and deck one. The continuous conveyor advancing tail piece and the bridge conveyor were installed next. Decks two through six were installed in the same manner and the back-up system components were installed on the decks while cars were at the bottom of the shaft. After completion of assembly of each car, it was advanced toward the heading by the air tuggers in the starter tunnel, and connected to the rest of the system.

Muck Transport System

In the Queens tunnel, a continuous conveyor system, supplied by Long Airdox and with a belt width of 610 mm (24 in) is used to transport the muck from the TBM back to the shaft 19B at a rate of 550 tons (1,213,000 lb) per hour with a belt speed of 4.5 m/s (880 fpm). The tunneling crew will install conveyor mounting brackets at regular intervals on the left hand side wall of the tunnel, in the area between the TBM and the back-up system. The advancing tail piece of the conveyor which is installed on the first deck of the back-up, is continuously pulled forward by the TBM. As the tunnel is driven, the conveyor is extended with 450 m (1500 ft) long belt sections, allowing the TBM to advance 225 m (750 ft) at a time. Each new section is quickly spliced in at the combined storage/take up unit located in the starter tunnel.

At the bottom of the shaft, the continuous conveyor deposits the muck, via a hopper, onto a Vertical Pocket Lift Conveyor. This conveyor is designed and supplied by the Lakeshore Mining Company. The steel cord belt, supplied by Trellex of Germany, is capable of elevating the muck to the top of the shaft at a rate of 570 tons (1,257,000 lb) per hour with a belt speed of 2.5 m/s (492 fpm).

The muck coming out of the shaft is dumped onto a 125 m (410 ft) long transfer conveyor and is deposited 20 m (66 ft) off the ground in a 2,250 m^3 (2943 yd^3) muck pile alongside the Long Island Railroad (LIR) siding located on the site.

TBM Performance

At the time of writing of this paper the TBM had just started the boring operation and no meaningful performance data were available.

SUMMARY

The Queens Water Tunnel, No. 3, Stage 2, is a phase in the largest and most costly construction project in the history of New York City. This paper illustrates the planning and the multitude of equipment that are involved in this enormous tunneling project.

4

Conventional Excavation

Chairman
L.M. Lenahan
McNally Tunneling Corp.
Westlake, OH

Chapter 12

BLAST PARAMETERS FOR THE NAGAGAMI HYDRO-ELECTRIC PROJECT TAILRACE PLUG BLAST, UTILIZING LONG ROUND JUMBO TECHNOLOGY

Joseph Keenan
Technical Representative, ICI Superior Explosives, Thunder Bay, Ontario, Canada

ABSTRACT

In March of 1994, Dynatec Mining Limited entered into a contract with PCL Civil Constructors to carry out the underground excavations for the Algonquin Power Corporation's 15 Megawatt Nagagami hydro-electric power station located near Hearst, Ontario, Canada. Dynatec completed the work for this project in early October 1995.

Dynatec Mining Limited's work included the excavation of a 175 meter long penstock driven at a grade of -25% to a 4000 m^3 powerhouse located 50 meters below the intake elevation. A 1200 meter long tailrace was driven from the Nagagami River to connect with the powerhouse. Two 50 meter raises were driven from the powerhouse to surface to provide access for equipment and personnel. The excavation of the tailrace tunnel (5.8m x 5.9m) was completed using Dynatec's Long Round Jumbo Technology.

INTRODUCTION

On November 30, 1995 the Algonquin Power Corporation brought the $40 million Nagagami hydro-electric project on line. Throughout the development of this project several unique issues concerning design, size and habitat protection needed to be addressed. The tailrace plug blast was one of these unusual situations. This blast was challenging because the rock plug to be removed had two faces, one on the tailrace side and one on the river side. The river face was very high and shear and was located on a bend in the Nagagami River. The river face characteristics made it impossible to utilize a coffer dam. The use of a coffer dam in the river meant that workers and equipment would be positioned beneath a high shear wall with trees and rocks overhead. The damming of the river would have also created environmental issues surrounding the well being of fish and other wildlife. It was for these reasons that a new and innovative plug blast design needed to be developed. The final plug blast design was based on modifying the drift round used in everyday tunneling at the Nagagami Project. This drift round design reduced many of the environmental concerns surrounding the damming of the river and all of the concerns about working under the high river side face. The drift round design's only limitation was that once the blast was fired the tailrace tunnel could not be re-entered. This limitation meant that the blasting of the tailrace plug was a "one shot deal".

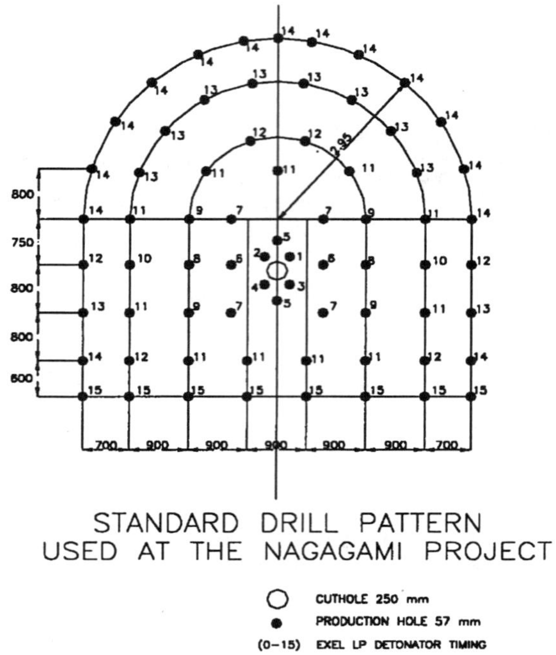

Figure 1

DYNATEC MINING LIMITED'S LONG ROUND JUMBO

In order to drill off this new and unique plug blast design, Dynatec Mining Limited utilized their Long Round Jumbo Technology. The manufacture of the Dynatec Long Round Jumbo was completed in September of 1991 by Continuous Mining Systems. This Jumbo unit was designed to tackle large drifts up to 6 meters by 6 meters at an extended advance of up to 9 meters per round. The drill units located on each of the two booms are capable of drilling blastholes ranging from 45 mm to 57 mm while the affixed frontal ITH drill is capable of drilling a 250 mm reamer hole. The Long Round Jumbo drill was used in everyday drifting at the Nagagami hydro-electric project where it averaged 5.5 meters of advance per round over the duration of the project.

DRILL PATTERN DESIGN PARAMETERS

The drill design parameters chosen for the plug blast were based on Dynatec's drill pattern used in day to day drifting at the Nagagami project. (Fig. 1) This drill pattern was altered to incorporate two extra rows of lifter holes to achieve the desired final void. The production holes for this blast were drilled to a 57 mm diameter while the reamer holes were drilled to a 250 mm diameter. The production holes for this blast were designed to be drilled on a 0.9 m burden and a 0.8 m spacing. The perimeter holes for this blast were drilled on a reduced burden and spacing which equated to a 0.7 m burden and a 0.6 m spacing.

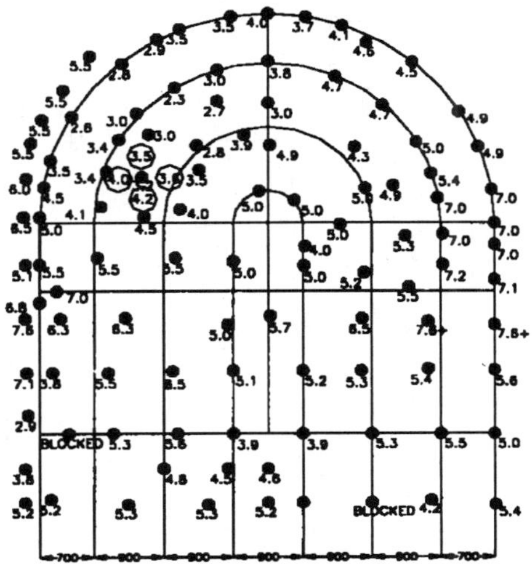

TAILRACE PLUGBLAST
PRODUCTION HOLE LENGTHS

○ CUTHOLE 250 mm
● PRODUCTION HOLE 57 mm
5.3 HOLE LENGTH IN METRES

Figure 2

DETERMINING ROCK PLUG THICKNESS

Prior to drilling the production holes for this blast it was necessary to determine the thickness of the rock plug to be blasted. To determine the thickness of the rock plug, the jumbo drilled several test holes into the rock face. When the jumbo drill steel broke through into the river, the thickness of the rock in that area could be determined by measuring the amount of drill steel in the hole at the time of break through. (Fig. 2) After testing, the test holes were plugged with either a grout plug or grouting cement. This measuring system was adopted because of the difficulty in accurately surveying the river side of the plug wall. The test hole data which was gathered provided the jumbo drill operator with sufficient information to drill the production and reamer holes without breaking into the river. The production holes, for this blast, were drilled approximately 0.5 meter to 1 meter shorter than test holes. The hole depth data also provided information which was required for cut hole placement. The four large reamer holes were drilled into the plug face at the shallowest point. This procedure reduced the amount of rock to be removed by the cut area of the blast.

VIBRATION CONCERNS

The vibration parameters developed for this blast were derived from the PCL Civil Constructors Technical Specification Manual. The manual stated that the peak particle velocity of any rock plugs blasted must be limited to prevent damage to existing structures or installed equipment. This limitation would require the use of reduced explosive charges and increased numbers of delays. Due to the great distance between the tailrace plug and the

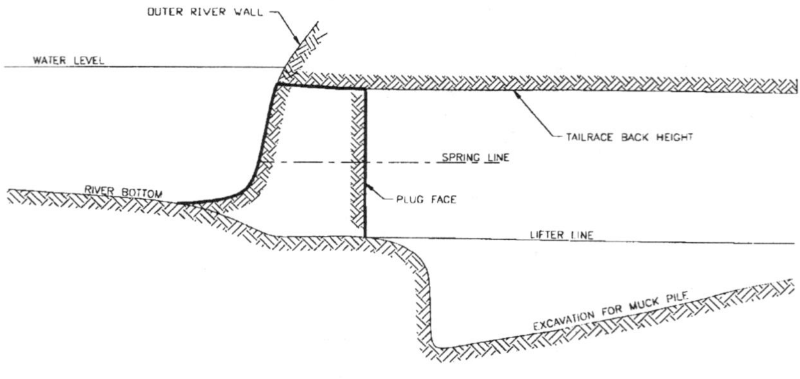

CROSS-SECTION OF TAILRACE PLUG

Figure 3

turbine room (approx. 1.2 Km) the threat of vibration damage to the concrete installations in the turbine room was limited. However, a vibration concern did still existed because of possible blast vibration effects on the high river wall. Blast vibration could possibly deteriorate the high river wall during the blast and put rocks, trees and silt into the river. These objects in the river could cause obstructions which would cause environmental concerns. For this reason the kilograms per delay were kept to a minimum.

BLAST CONCUSSION CONCERNS

As with any large tunnel blast the possibility exists that the concussion from the blast may cause damage to existing structures. Originally, the Nagagami blast was designed to be detonated with the tailrace tunnel full of water. This design was developed in response to concerns on muck pile control. This design was later dropped because of the water hammer effect due to hydrostatic pressure on the turbine room. The water hammer effect is created by the pressures released from the blast and water's inability to be compressed. This physical characteristic can create a destructive force which would be capable of damaging the tailrace tunnel and turbine room after the detonation.

MUCK PILE CONTROL

The rock plug to be removed from the tailrace tunnel had a volume of 223.6 m^3. In order to avoid building a coffer dam in the river to muck out the broken rock, the plug blast design had to be capable of expelling the broken rock from the tunnel mouth against the pressures of on rushing water. This design created questions concerning the removal of the rock from the river floor after the blast and the environmental effects which may be created by plugging a section of the river. This design was determined not to be feasible due to the increasing number of variables that appeared after every calculation. In order to solve the muck pile displacement problem an equivalent rock volume to that of the plug was removed from the floor of the tunnel. This excavation was achieved through benching back the floor of the tailrace tunnel. This approach allowed the broken rock of the plug blast to settle into the void leaving the mouth of the tailrace tunnel free and clear.(Fig 3.)

EXPLOSIVE PRODUCT SUMMARY

In considering the explosives to be used in this blast it was important to be aware that there was only one chance to blast this plug. In order to achieve an extra level of insurance, specific explosives suitable to the task were selected. They were as follows:

ICI POWERditch 1000 Explosives, 50 x 400. The POWERditch 1000 made up the majority of the explosives used in this blast. This product was chosen because it is an easy to load, hard shell packaged, Ammonia Gelatin Dynamite which can be used in underwater blasting. This explosive ensures superior breaking power under severe conditions, it is resistant to dynamic precompression, has a high detonation pressure and has excellent performance under high static pressure. This product also contains a phlagent which reduces the chances of hole to hole propagation.

ICI PRIMAFLEX, 85 grams/m Detonating Cord. The PRIMAFLEX detonating cord was used to trace the full length of the perimeter holes in the blast. This product offers a continuous length of decoupled charge which is easy to use and is rugged and abrasion resistant to side penetration of oil and water.

ICI PENTEX 8, Cast Pentolite Boosters. PENTEX 8 boosters were used to double prime all of the holes except the perimeter holes which will be single primed. This booster offers a detonation pressure which is almost twice that of POWERditch 1000. The PENTEX 8 booster is also very reliable and has excellent water resistance.

ICI TIMEMASTER Electric Detonator. The TIMEMASTER electric detonator offers a 4 meter copper legwire unit with 15 delay numbers using progressive interval delay timing. The use of the TIMEMASTER detonator provides long intervals between delays with the highest delay being 8000 ms.(Fig. 4) The use of electric detonators also provides the additional benefit of being able to constantly monitor the resistance and continuity of each series.

POWDER FACTOR CALCULATIONS

PLUG BLAST DIMENSIONS

Tunnel Width	Ht. to Arch	Arch Radius	Total Height	Average Hole Depth
5.9 m	4.0 m	2.95 m	6.95 m	6.0 m

TUNNEL ARCH VOLUME
($2.95\ m^2$ X 3.1416 X 6.0)/2 = $82.02\ m^3$

BOTTOM OF TUNNEL VOLUME
5.9 m X 4.0 m X 6.0 m = 141.6 m^3

TOTAL TUNNEL VOLUME
82.02 m^3 + 141.6 m^3 = 223.6 m^3

TOTAL QUANTITY OF EXPLOSIVE REQUIRED
906.3 kg

TOTAL POWDER FACTOR
906.3 kg/ 223.6 m^3 = 4.05 kg/m^3

CUT AREA POWDER FACTOR

The cut area of the plug blast was loaded using 30 cm dowels to separate each cartridge of POWERditch 1000. This procedure decreased the cut area powder factor by 50%.

CUT VOLUME
1.2 m X 1.2 m X 3.6 m = 5.2 m^3

TOTAL QUANTITY OF EXPLOSIVE USED IN THE CUT AREA
25.7 kg

CUT POWDER FACTOR
25.7 kg/ 5.2 m^3 = 4.94 kg/m^3

TOTAL KILOGRAM OF EXPLOSIVE PER DELAY

TIMEMASTER Electric Detonator Delay Number	Time In Milliseconds (ms)	Kilograms Of Explosive Per Delay (kg)
1.00	500.00	2.20
2.00	1,000.00	4.40
3.00	1,500.00	2.80
4.00	2,000.00	3.30
5.00	2,500.00	6.0
6.00	3,000.00	7.0
7.00	3,500.00	65.40
8.00	4,000.00	81.70
9.00	4,500.00	85.70
10.00	5,000.00	125.40
11.00	5,500.00	123.30
12.00	6,000.00	178.80
13.00	6,500.00	82.40
14.00	7,000.00	86.20
15.00	8,000.00	52.40

LOADING PROCEDURES

Cut Production Holes: The cut holes refer to the holes which are located directly around the 4 reamer holes. As with any drift round it is imperative that the cut void be maintained in order for the blast to be successful. Due to the length of the reamer holes and the close proximity of the cut holes, the possibility of "freezing the cut" was an obstacle. The expression "freezing the cut" refers to the general concern that an excessive powder factor around the reamer holes may limit the ability of the rock to be expelled out and away from the plug face. A frozen cut would eliminate the void which is necessary for the rest of the blast to break into.

In order to avoid a frozen cut, the cut production holes were loaded using 30 cm wooden dowels to space every cartridge of POWERditch 1000. This action effectively reduced the quantity of explosive in the cut holes by 50%. To ensure the complete detonation of the POWERditch 1000, which was separated by dowel spacers, a strand of ICI CORDTEX 50 detonating cord was used to trace the entire length of the explosive column. Each hole was also plugged at the collar using a plastic birdie plug which was used to confine the explosive in each hole.

Production Holes: The production holes for this blast were loaded using a full column of POWERditch 1000. These holes were loaded to a 2 foot collar and plugged using a plastic birdie plug.

Perimeter Holes: For this plug blast the final wall conditions were very important because of the inability to access the tailrace after the blast. For this reason extra precautions were made to ensure that the final perimeter walls were stable. For this blast, both the burden and spacing of the perimeter holes were reduced leaving each hole with less breaking work to do. The explosive chosen for perimeter control was PRIMAFLEX 85 grams/meter detonating cord. The perimeter holes were loaded using either a single or double strand of PRIMAFLEX detonating cord leaving a 2 foot collar. The holes drilled with larger burdens had two strands of PRIMAFLEX placed in them while the tighter burden holes had only one strand. Each perimeter hole was toe primed using a single cartridge of POWERditch 1000. These holes were also plugged using a plastic birdie plug to confine the PRIMAFLEX at the collar of the hole.

Detonator Placement: The detonator placement in this blast proved to be very complicated due to the fact that the blast design called for the blast to break both inwards into the tailrace tunnel and outwards into the river. This design parameter meant that the blast had to break in two opposing directions simultaneously. Because of the opposing directions of the blast, the detonator placement for each hole was very important. If a detonator was placed only in the toe of each hole, the detonator would stand the chance of being ripped out by the violent actions of adjoining holes and detonate unattached to its explosive column in the river. This scenario also held true if each hole was only collar primed. The solution to this challenge was to place the detonators as close to the center of each hole as possible. This location would provide the detonators with some protection during the detonation.

ELECTRIC DETONATOR HOOK UP PROCEDURE

The electric TIMEMASTER detonators which double primed each blast hole, as seen

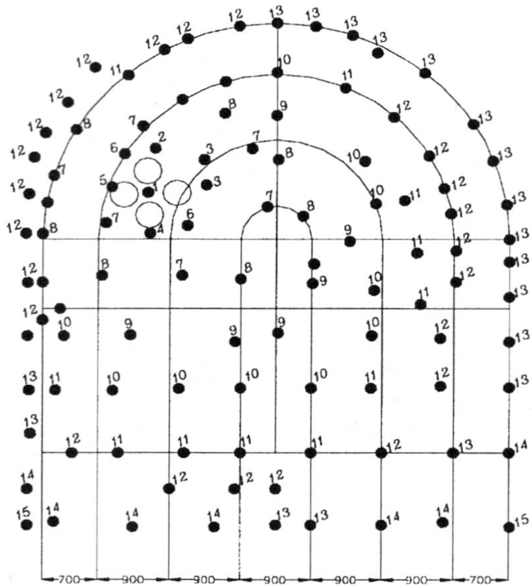

TAILRACE PLUGBLAST
IN THE HOLE DELAYS USING
ICI TIMEMASTER ELECTRIC DETONATORS

○ CUTHOLE 250 mm
● PRODUCTION HOLE 57 mm
14 IN THE HOLE DELAY

Figure 4

in Figure 4 were broken into 6 equal series. From these series, 5 series contained 34 detonators (17 holes) while 1 series contained 26 detonators (13 holes) plus an additional 8 dummy detonators to balance the series. The balancing of the series brought the total number of detonators to 204 units. The blast was connected using 100 meters of #23 gauge buswire wire which in turn was connected to 500 meters of #14 gauge blasting line. The end of the blasting line was connected to an ICI MK II 1000 Shot Blasting Machine. This machine has an operating voltage of 400 volts and a rated energy of 88 Joules.

ELECTRIC DETONATOR CALCULATIONS

Buswire Resistance = Rb

Rb = 6.7 ohms

Firing Line Resistance = Rf

Rf = 4.15 ohms

Total Number of Series

S = 6 series of 34 detonators

Detonator Circuit Resistance = Rd

Rd = 34 detonators per series x 1.8 ohms per detonator/ 6 series
Rd = 10.2 ohms

Total Circuit Resistance = Rt

Rt = 6.7 ohms + 4.15 ohms + 10.2 ohms
Rt = 21.05 ohms

ICI MK II 1000 Shot Blasting Machine

According to the MK II 1000 Shot Blasting Machine firing recommendations, no more than 50 detonators can be placed in a single series and when utilizing 6 series the total resistance cannot exceed 42 ohms. Therefore, because the plug blast has only 34 detonators per series and the total resistance of the 6 series is only 21.05 ohms, the MK II blasting machine can successfully detonate this blast.

CONCLUSION

The plug blast was successfully fired in the early morning hours on October 16, 1995. The 1.2 km tailrace tunnel took approximately 20 minutes to fill with water. The force of the blast did not fill the river with churned up sediments or debris from trees or rocks which were perched on the steep rock face. The plug blast concluded the work to be done on the tailrace tunnel and put the Nagagami-Algonquin Power Corporation in a position to start producing electricity. The Nagagami project is now producing power to be used in the main grid in Northern Ontario. There have not been any problems with the tailrace output.

ACKNOWLEDGMENTS

The Author would like to congratulate and thank all parties involved in the blasting of the tailrace plug at the Nagagami Hydro-Electric Project. The author would also like to thank all the people who assisted in the writing of this paper.

Nagagami-Algonquin Power Corporation

Chris Jarratt - Project Manager

PCL Civil Constructors (Canada) Inc.

Ed Neuman - Project Manager
George Whaley - Project Superintendent
Roger Moore - Project Engineer

Dynatec Mining Limited

Bill Shaver	-	President
Syd DeVries	-	Area Manager
Al Gorman	-	Project Manager
Tom Maule	-	Project Captain

McCaw Drilling And Blasting

Irwin McCaw	-	President
Barry Beck	-	Blaster

ICI Explosives

Yves Newman	-	Construction Manager
Chris Kelly	-	Technical Supervisor
Mike Lefrancois	-	Initiations Specialist

BIBLIOGRAPHY

Atlas Powder Company, 1987, "Initiation Systems," Explosives And Rock Blasting, Atlas Powder Company, Dallas,Texas, pp. 79-138.

Dynatec Mining Limited, 1995, "Nagagami Hydroelectric Power Project," Dynatec International Semi-Annual News, p. 3.

Kerr, D., 1996, "Nagagami Hydro: In Partnership With The First Nations," Hydro Review, Feb., pp. 22-28.

Pavel, G., 1991, "Report On Testing Of The Dynatec Long Round Jumbo At Noranda Minerals Inc., Mines Gaspe Division," Nov., Dynatec Mining Limited, Murdochville, Quebec.

Chapter 13

THE SM 3 POWER PROJECT DESIGN CHARACTERISTICS AND EXCAVATION METHODS

Claude Boisjoly,[1] Don Nguyen,[1] and Pierre de Courval[2]
[1]Hydro-Québec; [2]SNC-Lavalin Inc.

ABSTRACT

The Sainte-Marguerite-3 (SM 3) hydro project, presently under construction, requires the excavation of an underground powerhouse and large water conveyance tunnels. This paper describes the planning of the project, main features of underground structures, excavation techniques and ground support.

INTRODUCTION

The Sainte-Marguerite-3 hydroelectric project is located on the Sainte-Marguerite river in Québec, some 80 km north of Sept-Îles and 800 km northeast of Montréal (Figure 1). Hydro-Québec is the owner and manager and SNC-Lavalin, the designer.

The project consists of a 170 m high rockfill dam, a diversion tunnel, a spillway located in a secondary valley and an 8.2 km long, 16.5 m high power tunnel leading to a 3 unit underground powerhouse with an installed capacity of 1395 MW.

The underground structures, including all tunnels and chambers, represent a total length of 10 km with a volume in excess of 2,000,000 m^3 while open cut excavations for the portals, canals, spillway and dam foundations represent a further 2,300,000 m^3. In addition, a considerable amount of fill and excavation was required for the 100 km long penetration and access road from Sept-Îles.

Construction started in 1994 and the first two units will be put into service towards the end of the year 2001. Civil works for the third unit will also be carried out, but the unit itself will only be built and installed in the future, when needed. Access roads, diversion tunnel and main cofferdam works began in 1994 while access tunnels to the powerhouse and headrace tunnel began in 1996. All excavations are done by drill and blast; a TBM excavated alternative was considered for the headrace tunnel.

The paper describes the design characteristics of the underground civil structures, and the methods and overall planning that were put into place to ensure the success of the project.

GEOLOGY AND INVESTIGATIONS

The crystalline, igneous and metamorphic rocks which underlie the SM 3 area of Québec

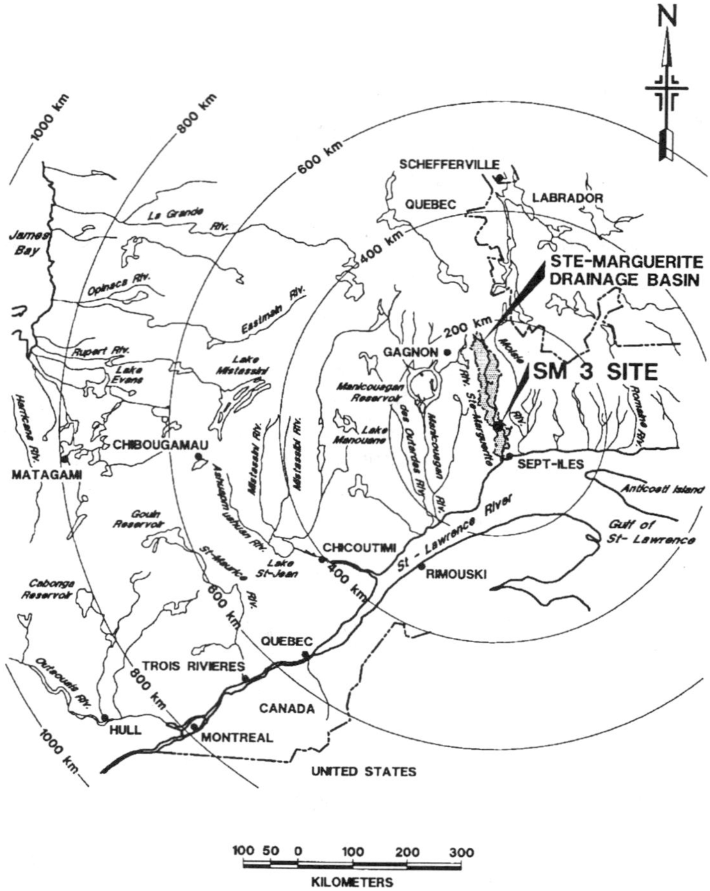

Figure 1. Project Location

belong to the Grenville geological province and form part of the Canadian Shield. The bedrock generally consists of a series of migmatized and, more rarely, banded gneisses (Ref. 1). Locally, these series have been intruded by small to very large masses of anorthosites, gabbros and granites. With the exception of the spillway and intake, located in migmatic gneisses, most of the SM 3 civil structures, including the dam, headrace, powerhouse and tailrace lie within a large anorthosite batholith whose dimensions are in the order of 8 km in the east-west direction by 12 km in the north-south direction. Locally within the batholith, and especially along its margins, large masses of gabbros are found in association with the anorthosites; granitic and pegmatitic dykes intersect the rock mass. Steeply dipping to sub vertical northwest and northeast joint systems are dominant; other steep dipping joint systems have north-south, east-west and north-east/south-west orientations. Large exfoliation joints related to glacial rebound are also found close to surface and are mostly parallel to the topography; these are often open with sand or clay infillings.

The anorthosites and gabbros at the site are characterized by low permeabilities, in the range of a few Lugeon units, and high RQD values, usually 95 to 100 %. Results of over 50 uniaxial tests indicate compressive strength averages of 161 MPa for saturated samples and 205 MPa for unsaturated samples, with a peak at 282 MPa. Cerchar abrasivity tests indicate very abrasive to extremely abrasive rocks with index values ranging from 2.9 to 4.3. The in situ stresses at powerhouse level were determined by hydro-fracturing tests done in a 260 m deep borehole; the results are as follows:

Minimum horizontal stress: σ_h = 11,0 (± 1,4) MPa acting in the N34°W direction (± 17°)

Maximum horizontal stress: σ_H = 17,7 (± 3,1) MPa acting in the N56°E direction (± 17°)

The northeastern direction of the maximum horizontal stress, as well as the compressive nature of the measured stresses are in the range of most measurements in eastern North America. Vertical stress is assumed equal to the gravitational stress.

As with any large batholith, the rock mass is intersected by faults and shear zones, but these are relatively few. Only one shear zone was encountered during excavation of the 978 m long diversion tunnel and, based on air photo-interpretation and drilling results, only one significant fault and less than a dozen small shear zones are anticipated along the entire 8286 m horizontal length of the power tunnel.

Excluding the intake and powerhouse areas, drilling along the headrace corridor has been limited to only two holes, both at the sites of strong lineaments identified by photo-interpretation. This relatively small quantity of exploration drilling, representing less than 0.1 m of core per meter of tunnel, results from the numerous rock exposures and generally thin overburden in the highlands which enabled efficient photo-interpretation and geologic mapping

Conditions encountered to date along the completed diversion tunnel and the on-going access tunnels confirm the anticipated good quality of the rock.

SCHEDULING OF THE PROJECT

In order to commission the first two units at the end of the year 2001, the main dam has to reach level 375 m by July 1998, so that reservoir filling can start. Main dam construction thus had to begin in the spring of 1996. As a consequence, the cofferdam and diversion tunnel, as well as the main camp had to be constructed prior to completion of the main access road. All equipment and supplies were hauled up the existing narrow logging road where steep grades exceeding 15 % and severe winter conditions were dealt with by means of convoys of tractor trailers equipped with chains on the driving wheels assisted by wheeled loaders.

The underground excavations were subdivided into six main contracts:

- diversion tunnel;
- upstream section of the headrace tunnel including the intake;
- mid-section of the headrace tunnel including adit n° 2;

- downstream section of the headrace tunnel including the three penstocks, the drainage gallery and the 329 m high surge shaft and chamber;
- access tunnels to the powerhouse, adit n° 1 to the headrace tunnel, machine hall top heading, gate chamber and pilot shaft of n° 3 bus bar shaft;
- machine hall benching, three bus bar shafts, draft tubes, collector and tailrace tunnel.

MAIN DAM AND OPEN CUT STRUCTURES

The first permanent civil structure awarded was the 6,500,000 m³ main dam; it has a maximum height of 170 m from crest at elevation 410 to rock foundations at lowest point of 240, and is the highest rockfill dam in Québec. The crest length is only 380 m which is indicative of the deep and narrow Sainte-Marguerite valley. The core of the dam is made of impervious till resting on bedrock; a thickness of over 15 m of fine sands and 20 to 25 m of coarser alluvium were excavated under the core and filters. The filters consist of natural and fabricated granular materials; rockfill is obtained from quarries and from the spillway excavations. Other contracts for open cut excavation which will be awarded include the switchyard and the tailrace channel.

UNDERGROUND STRUCTURES

Diversion Tunnel

The diversion tunnel has a length of 978 m, a width of 9 m and a height of 15.1 m; the gated upstream portal is concreted while the downstream portal is not. The heading and benching excavations were completed in 30 weeks. Underground rock conditions were excellent with generally homogenous rock and an absence of heavily fractured zones; only one sheared and brecciated zone, less than 10 m wide, was intercepted. Total seepage along the full length of the tunnel was in the order of 1 liter/second. Construction difficulties were encountered during winter excavation of the upstream portal, mainly related to the stabilization of a large talus slope lying in the portal area. The talus, originating from overlying unstable cliffs, is up to 10 meters in thickness and contains many large elements of a hundred to a few hundred cubic meters. Although most of the talus is located above and to the sides of the portal, its base had to be excavated to accommodate the portal. Stabilization was obtained by covering the talus with over 10,000 m² of anchored chain link mesh put in place by helicopter to limit on-ground activity. In addition, a series of 5 boulder retaining walls 4 to 5 m high and consisting of anchored steel beams, wood lagging and a sand cushion were put in place above the portal. Problems were encountered at the downstream portal following blasting of the 4200 m³ rock plug, just prior to river diversion. Following this blast, a major rock slide occurred which completely blocked the outlet channel. The slide, in the order of 18,000 m³, occurred on a large planar gouge-filled joint dipping at 37 degrees towards the excavation and daylighting towards the base of the lowest bench; remedial works including excavation and additional reinforcement of the portal took nearly two months but did not affect overall progress.

Headrace

A distinguishing feature of the SM 3 project is the 8286 m long unlined pressurized

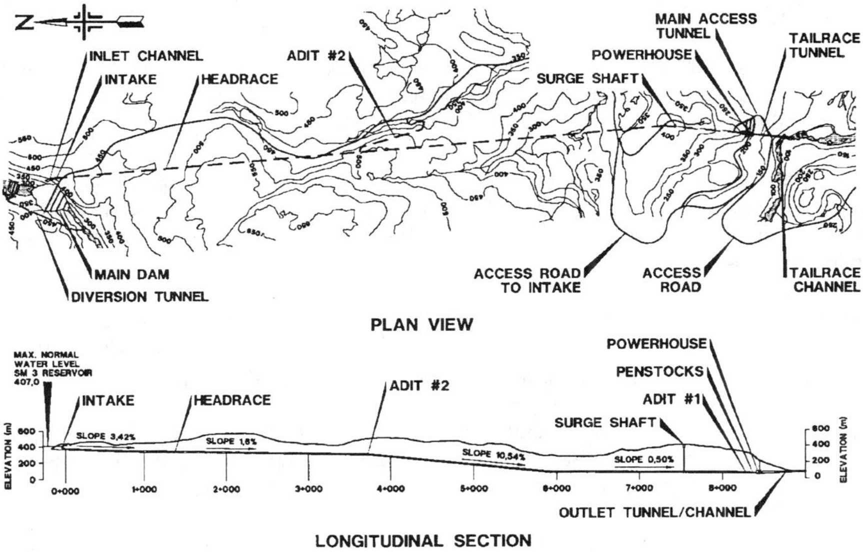

Figure 2. Project Description

power tunnel (Figure 2); it connects the intake, close to the main dam, to the powerhouse almost in a straight line. A small 10 degree kink is observed in the tunnel alignment at the surge shaft. The reason for this layout is to locate the surge basin at the highest possible elevation, 418 m, to avoid a high, above-ground structure, and as close as possible to the powerhouse.

The power tunnel profile is divided in three sections:

- between the intake and adit n° 2, the slope is low, 3.42 % and 1.60 %, in order that the tunnel be kept as high as possible to minimize the length of adit n° 2;
- downstream of the adit n° 2 junction with the power tunnel, between PK 3+825 and 5+925, the slope increases to 10.54 % in order to obtain sufficient cover under a topographic low at PK 6+250;
- downstream of PK 5+925, the slope is kept to the minimum of 0.50 % required for gravity drainage.

The power tunnel is of inverted U-shape with a height of 16.5 m and width of 11.5 m; its arch-height to tunnel-width ratio is 0.25 and its sectional area is 179.8 m^2. Excavation of the tunnel will be by full face heading of 8 m height and benching of 8.5 m height. It is expected that the Manning coefficient will be in the order of 0.030.

The power tunnel and associated structures are to be awarded in three contracts to accelerate excavation.

The contract for the upstream 963 m of the tunnel will also include the excavation and concreting of the inlet channel and intake structure.

The 5225 m long central section of the tunnel also includes the 834 m long adit n° 2.

The downstream section, 2098 m long, will also include the excavation of the manifold, the three penstocks, the drainage gallery, the 279 m high, 7.5 m diameter surge shaft and the 50 m high 18 m diameter surge chamber.

The use of a tunnel boring machine (TBM) to excavate the power tunnel was studied. Assuming a Manning coefficient of 0.015 for the TBM excavated tunnel, a diameter of 10.5 m for a sectional area of 86.6 m^2 would have been required. The method was not retained for the following reasons:

- In view of the combination of large tunnel diameter and very hard and abrasive rocks, the TBM excavated alternative did not offer an economic advantage, but was felt to represent a greater risk.
- Contractors in Québec do not have much experience with large diameter TBM's but have extensive experience with the drill and blast method in hard rock and the prices were expected to be more competitive.
- The drill and blast method offers more flexibility in scheduling and contracting and requires only minimal ground support due to anticipated excellent rock conditions.

Downstream Access Tunnels and Powerhouse

This contract includes the permanent access tunnel to the powerhouse, the temporary access tunnel (adit n° 1) to the downstream section of the power tunnel, the access tunnel to

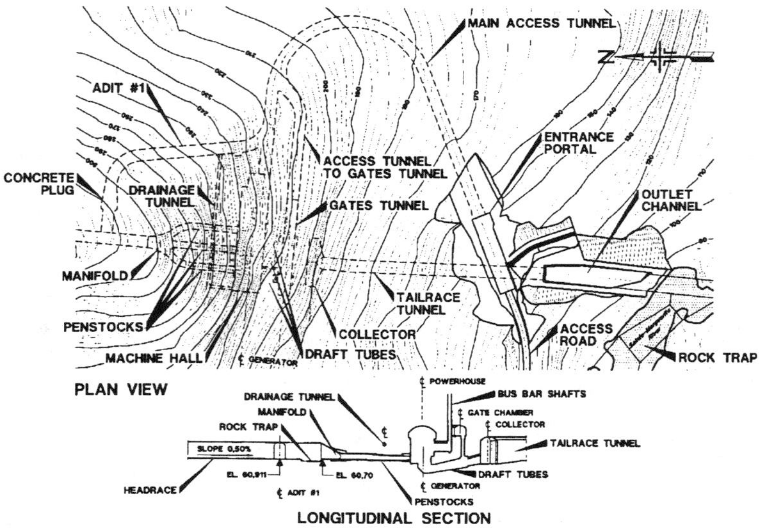

Figure 3. Powerhouse Area

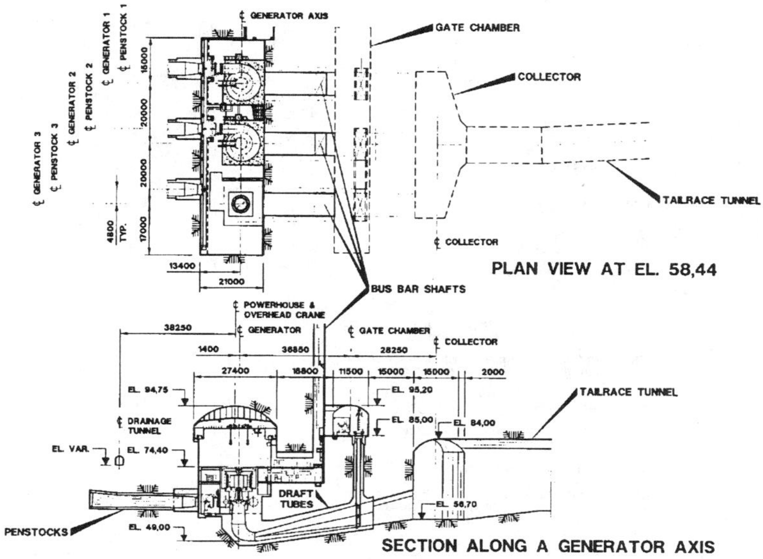

Figure 4. Powerhouse

the gate chamber as well as the gate chamber and the top heading of the powerhouse (Figure 3). All access tunnels are 11.5 m wide and 8 m high; the top heading of the machine hall is 27.4 m wide by 10 m high.

The top heading of the powerhouse will be excavated in three stages, a full face heading of 11.5 m width and a slash on each side.

Power Contract

This contract will include the benching excavation of the machine hall, the excavation of the three draft tubes, the collector, the tailrace tunnel and portal and the excavation of the three bus bars shafts (Figure 4). Special care will be given to detailed excavation of the rock ledges of the 600 ton capacity overhead crane (Figure 5); these ledges serve as a seat for the lightly reinforced concrete beam of the overhead crane.

The powerhouse is 106 m long and, below the overhead crane rock ledges, 24 m wide; a total height of 37 m is to be excavated in this contract. The total height of the powerhouse, from crown to the invert of the draft tubes, is 47 m.

The total length of the collector and free flow tailrace tunnel is 241 m, the tailrace tunnel is 12 m wide by 24 m high.

EXCAVATION CRITERIA

For underground as well as open cut excavations, Hydro-Québec specifies the following general blasting criteria:

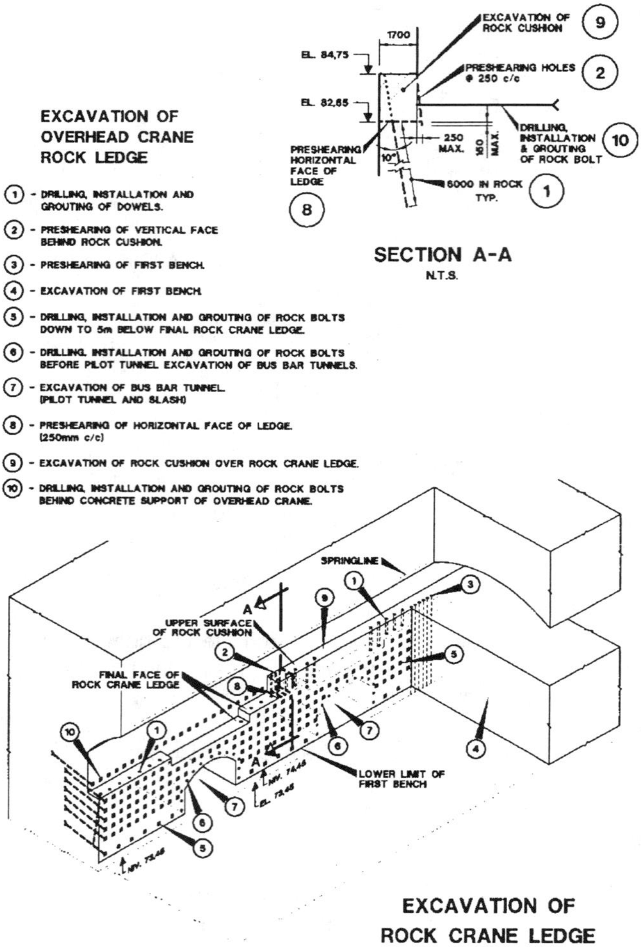

Figure 5

Open Cut

Maximum bench height — 10 m

Maximum hole diameters:

- perimeter and buffer holes — 65 mm max.
- primary blasting holes — 100 mm max.

Spacing of perimeter holes — 600 mm c/c

Loading of perimeter holes — max. 0.65kg per m² of presplit surface plus 1.25 kg max. bottom charge

Two rows of buffer holes are specified, the first 900 mm distant from the perimeter holes and the second at 1800 mm from the first row of buffer holes; hole to hole spacing in both cases is 1500 mm.

The use of bulk explosives is not permitted in perimeter holes nor in the first row of buffer holes.

Underground, Top Heading

Maximum length of a round	4.0 m
Maximum hole diameters for all holes (perimeter, buffer and primary blasting):	51 mm max.
Perimeter holes:	
. maximum flare (for a 4.0 m round)	300 mm
. spacing	600 mm c/c
. loading	max. 0.65kg per m^2 of presplit surface plus 0.5 kg max. bottom charge

One row of buffer holes is specified 900 mm from the perimeter holes row with a hole spacing of 1500 mm. Bulk explosives are not permitted in buffer nor perimeter holes.

Underground, Benching

Maximum height of a bench	10 m
Maximum length of a bench	10 m
Maximum hole diameters for all holes (perimeter, buffer and primary blasting):	65 mm max.
Perimeter holes:	
. spacing	600 mm. max.
. loading	max. 0.65kg per m^2 of presplit surface plus 0.5 kg max. bottom charge

Two rows of buffer holes are specified, the first at 900 mm from the perimeter holes row and the second at 1500 mm from the first row of buffer holes; spacing is 1500 mm hole to hole. The use of bulk explosives is not permitted in perimeter holes nor the first row of buffer holes.

All these criteria were successfully used in the James Bay project Phase 2; results obtained since the beginning of the SM 3 project are also very good.

CONSOLIDATION

Three types of consolidation are specified on the SM 3 project: wire mesh, rock bolts and shotcrete.

Wire Mesh

In all rock excavations, chain link wire mesh is used to ensure the protection of workers during construction and operation. In open cut excavations, wire mesh is systematically applied at portal faces and in areas where structural concrete is required. Elsewhere in open cut excavations, wire mesh installation is determined by Hydro-Québec's geologists, depending on geological conditions. The mesh is removed prior to concreting; similarly, in hydraulic channels as well as in the surge pool, the mesh lying below maximum water level is removed at the end of construction, mainly due to problems of ice formation.

In underground excavations, wire mesh is installed in the crown and upper 1.5 m of sidewalls of all tunnels and adits and all large chambers, including the machine hall. In the machine hall, wire mesh also covers the walls above side openings, such as bus bar galleries, penstocks and draft tubes. Wire mesh is also installed in the surge shaft and chamber. All wire mesh is removed at the end of construction in the last 754 m of the power tunnel from PK 7+520 to 8+274, as well as in the surge shaft and chamber.

Rock Bolts

Permanent rock bolts are of the hollow core type with expansion shell anchoring device; all permanent rock bolts are grouted with cement. Resin type rock bolts are not permitted.

Areas of pattern bolting and spot bolting are specified based on anticipated geological conditions, the geometry of the openings and the experience in that type of rock.

Generally, permanent rock bolts are 4 to 6 m in length; for pattern bolting, they are generally positioned along a 2.0 x 2.0 or 2.3 x 2.3 m square grid. Bolts of 4 m length are most frequently used, while 6 m long bolts are used in areas of larger spans such as in the machine hall crown and tunnel intersections.

In open cut excavations, pattern rock bolting is specified on portal walls as well as on the sidewalls to a distance of 30 to 50 m from the portals.

In underground openings, pattern rock bolting is specified near portals, at tunnel intersections, at the collars of the surge and bus bar shafts and in the crown of all tunnels in the vicinity of the powerhouse, such as the main access tunnel, the gate chamber, the draft tubes, the penstocks and the collector.

The proportion of pattern bolting sections represents less than 2 % of the total length of the power tunnel, diversion and tailrace tunnels and access adits. Elsewhere, spot bolting is specified. The density of bolts put in place in the crown in spot bolted areas corresponds to 30 % of the density of bolts in pattern bolted areas. In small tunnels and shafts, spot bolting is generally specified with both lengths of 2 and 3 m.

Due to the size of the powerhouse, 27.4 m wide at the top, pattern bolting is specified; rock bolts are 6 m long with a pattern of 2 x 2 m. For the overhead crane rock ledges, rows of 6 m long rock bolts are specified as shown on figure 5. Also excavation sequences to create the rock ledges are detailed in this figure. This excavation method was developed for the James Bay project and gave good results.

Shotcrete

Shotcrete is only specified in the final 52 m of the power tunnel, in the manifold and in the penstocks. The shotcrete is put in layers of either 50 mm thickness or 100 mm thickness, in which case welded wire mesh is also specified.

CONCLUSION

Excavations of the permanent underground structures have begun and initial results are excellent. The specifications appear well suited to the type of work, and at this time, no problems are anticipated regarding excavation and reinforcement.

ACKNOWLEDGEMENT

The authors wish to thank Hydro-Québec and SNC-Lavalin for permission to publish this paper and their colleagues for their support.

REFERENCES

1. Franconi A., Sharma K. N. M. And Laurin A. F., 1975, "Betsiamites and Moisie Rivers area (Grenville 1968-1969)", Geological Report-162, Ministère des Richesses Naturelles, Québec, Québec.
2. SNC-Shawinigan inc., 1995, "Exploration Géologique et Géotechnique 1994", Aménagement Hydroélectrique Sainte-Marguerite-3/Site SM 3, Vol. 1., Montréal, Québec.
3. Haimson B. C., Lee M. Y., Feknous N. and de Courval P.,1996, "Stress Measurements at the Site of the SM 3 Hydroelectric Scheme near Sept-Îles, Québec", Int. J. Rock Mech. Min. Sci. & Geomech. Abstr., Vol. 33, N° 5, pp. 487-497.

5

New Technology

Chairman
L. Ozdemir
Colorado School of Mines
Golden, CO

Chapter 14

ISLAIS CREEK CONTRACT E: TUNNELING THROUGH JET GROUTED BAY MUD

D.N. Adams,[1] M.J. Robison,[2] and M.M. Wong[3]
[1]Senior Engineer, Jacobs Associates, San Francisco, CA;
[2]Senior Project Manager, Jordon, Jones & Goulding, Inc., Charleston, SC;
[3]Department of Public Works, City and County of San Francisco, CA

ABSTRACT

Islais Creek Contract E, a transport/storage tunnel in San Francisco, was visited on a field trip during construction by delegates to the 1995 RETC. This paper is a follow-up to that field trip. The project's two tunnels are 3.7 m and 3.2 m finished diameter, 155 m and 77 m long, respectively. The contractor used 1,000 overlapping jet grout columns to pre-treat the weak Bay Mud along the tunnel alignments, then mined them with an open faced tunnel shield, eliminating the need for compressed air tunneling. This paper documents the jet grouting program and the tunneling operation used on the project.

INTRODUCTION

Like many cities in the United States, San Francisco has a combined sewer system which collects urban runoff and sewerage in the same pipe. This system can be a major source of pollution during wet weather. During even the slightest rainfall, untreated wastewater may overflow into San Francisco Bay or the Pacific Ocean, causing officials to close beaches and ban fishing. In the early 1970's, the City and County of San Francisco embarked on a major program to control wet weather sewage overflows. The program, known as the Clean Water Program, consists of a series of facilities, including underground transport/storage structures, pumping stations and treatment plants.

One portion of the Clean Water Program is the Islais Creek Facilities, which consists of a series of underground box structures and tunnels. The Islais Creek Transport/Storage System will collect and store the excess flows for later treatment at the City's Southeast Treatment Plant, prior to discharging into the Bay.

The challenges that faced the designers and constructors of Islais Creek Contract E were numerous. First, Contract E had to cross beneath the existing Caltrain/Amtrak Railroad embankment and the viaducts supporting Caltrans' Interstate Highway I-280. Second, the system had to be built within the weak Bay Mud that exists in the Islais Creek area. The Bay Mud in this area is recognized as some of the weakest in the entire San Francisco Bay Area. In fact, the excavation support system for the nearby Davidson Box Sewer, which was built in the late 1970s, failed rather dramatically during construction (Clough and Reed, 1984). Third, the system had to include a control structure to divert flow from the existing Three Compartment Alemeny Sewer (Alemeny Sewer) into the new facilities. As a result of these

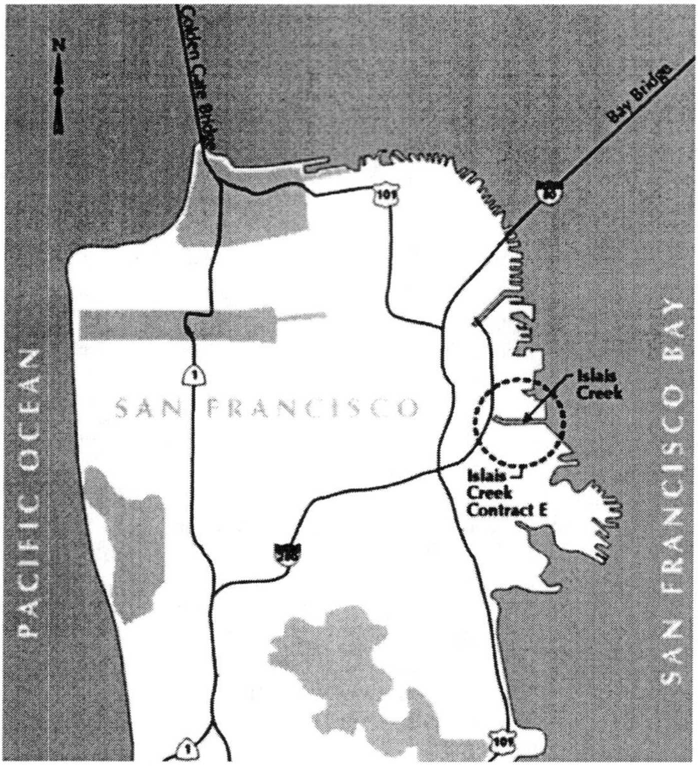

Figure 2. Islais Creek Contract E Site Location Plan

challenges, the design of Contract E included compressed air tunneling for the undercrossing of the railroad and highway viaducts; deep concrete box structures built within and founded on thick steel and concrete diaphragm walls, laterally supported by jet grouted Bay Mud (Adams and Robison, 1997); and top-down construction to facilitate connection between the Alemeny Sewer and the new facility.

Following award of the contract, the general contractor and its specialty geotechnical subcontractor proposed that jet grouting be used to pre-treat the soft Bay Mud along the tunnel alignments and that the tunnels be excavated using an open-face tunnel shield equipped with a roadheader. Following a demonstration test section, the proposal was accepted, the Bay Mud along the tunnel alignments was jet grouted, and the tunnels were excavated with the open-face tunnel shield without the use of compressed air. This paper describes the construction and performance of the two soft ground tunnels which were constructed through jet grout-treated, weak Bay Mud.

PROJECT DESCRIPTION

Islais Creek Contract E is a 340 m long component of the Islais Creek Facilities, which was constructed between January 1994 and April 1996. The project is located in the southeast corner of San Francisco, adjacent to the west and south sides of Islais Creek. A site location plan is shown in Figure 1, Islais Creek Contract E - Site Location Plan.

Contract E is located at the downstream end of the Islais Creek System, which feeds combined sewer overflows to the Southeast Treatment Plant. The project, approximately 340 m in length, consists of the following main features: the 155 m long, 4.6 m excavated diameter, 3.7 m finished diameter Davidson Tunnel; the 77 m long, 4.1 m excavated diameter, 3.2 m finished diameter Undercrossing Tunnel; and 110 m of Transport /Storage Box Structures constructed as open cut.

A layout showing the various features of Islais Creek Contract E is shown in Figure 2, Layout Plan of Islais Creek Contract E. In this figure, the previously existing structures are shown with dashed lines. The structures built under Contract E are shown in solid lines. This figure also shows the existing Caltrain/Amtrak Railroad and Interstate I-280, which bisect the project.

The Davidson Tunnel is located adjacent to and under many structures, including an active 1.83 m diameter sewer, the 7.5 m wide by 10.4 m deep Davidson Box Sewer, water, gas, telephone and high voltage electric lines, and several surface structures.

The Undercrossing Tunnel crosses under the City's 1.68 m diameter soil-supported Force Main, three viaducts belonging to Interstate I-280, the Caltrain/Amtrak Railroad, and several electrical and fiber optic utilities.

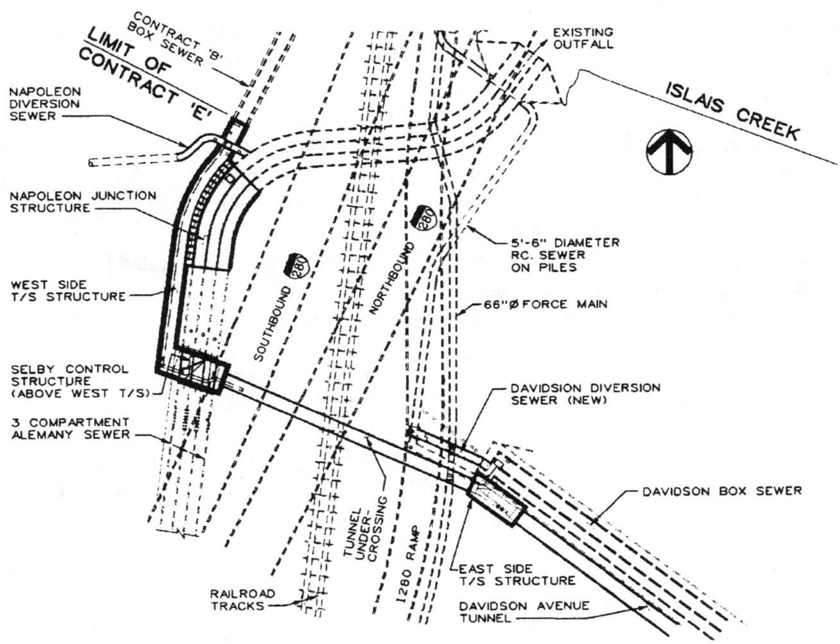

Figure 2. Layout Plan of Islais Creek Contract E

The 91.5 m West Transport/Storage Box crosses under and connects at two locations with the City's existing Alemeny Sewer. The 18.5 m East Transport/Storage Box connects to the Davison Box Sewer and was sized to serve as the tunnel access shaft.

Finished invert elevations across the project range from 12.8 m to 14.3 m below ground surface, with maximum excavation depths of 15.2 m.

GEOLOGICAL SETTING AND GEOTECHNICAL FEATURES

Prior to about 1890, the site area contained an embayment surrounded by tidal marshland. Islais Creek was a wide, meandering waterway surrounded by hills rising 76 to 92 m above the basin. A land reclamation operation took place in the 1930s, and most of the area was filled with 3 to 6 m of sand, rock fill and miscellaneous waste. The soil formations relevant to the design and construction of the project are Fill and Bay Mud.

FILL: The characteristics of this layer are highly variable. This surface layer consists mainly of brown, medium dense, poor to well graded sand and gravel, clayey sand, clayey silt and silty sand; boulders and cobbles are frequently encountered. Fill was encountered to depths of 2 to 3 m below ground surface, within the excavation areas.

BAY MUD: This is the most critical formation for the design and construction and underlies the project for its entire length. It falls into the MH or CH category of the Unified Soil Classification System and can therefore be designated as a plastic silty clay or clayey silt with trace of organics, shells and sand. The consistency ranges from soft to medium stiff in the upper portions of the deposit, to medium stiff to stiff in the lower portions. The upper portion (6 m) has a higher water content, liquid limit and plasticity index and a lower unit weight than the mud below. The undrained shear strength varies considerably along the project alignment as summarized in Table 1.

Consolidation test data indicates that the Bay Mud ranges from normally consolidated to slightly overconsolidated, with overconsolidation ratios ranging from 1.05 to 1.62. Consolidation of the Bay Mud in the Islais Creek area is an ongoing process that has caused considerable settlements (more than 0.6 m in some areas). Consolidation of the zone below the railroad embankment reflects the surcharge load of the embankment itself. The embankment continues to accelerate consolidation of the underlying Bay Mud, resulting in settlements of the ground surface and increased shear strength.

Groundwater along the project alignment was encountered at depths ranging from 1.5 to 2.1 m below ground surface and may have been affected by tides.

Table 1. Undrained shear strength (S_u) of native Bay Mud and unconfined compressive strength (U_c) after jet grouting treatment in kPa.

	S_u (kPa)	U_c (kPa)
Undercrossing Tunnel		
west side	29	2140
east side	23	2430
under embankment	48	5070
Davidson Tunnel	18	4520

TUNNEL DESIGN

Original Design

The original project design called for tunneling to be performed under compressed air. The designer of the project, Jacobs Associates of San Francisco, indicated that the Tunnel Face Stability Factor (($_v$/S_u) ranged from 5 to 9 for both tunnel alignments. A value of 6 or more typically indicates squeezing ground conditions. They also reported that the tunnels would be driven through numerous subsurface obstructions, including timber and concrete piles. To account for the squeezing ground and the anticipated obstructions, they specified use of compressed air and a compartmentalized breast boarding shield for tunneling. The specified primary support was gasketed steel liner plates.

Jet Grouting Proposal

Shortly after award of the $21.7 million contract, the General Contractor, Kajima Engineering and Construction, Pasadena, CA, and their Geotechnical Subcontractor, Nicholson-Rodio Joint Venture, Bridgeville, PA, submitted a Construction Incentive Change Proposal (CICP) to use jet grouting to pre-treat the soft Bay Mud along both tunnel alignments, excavate the jet grouted soils with a road header, and install steel ribs and wood lagging as the primary tunnel lining.

Jet grouting is a soil improvement technique which utilizes fluids jetted at very high pressure to fracture, partially replace and simultaneously mix soil in situ with a stabilizing agent. The proposal, based on 2-D and 3-D F.E.M. analysis, called for full face treatment, extending 1.2 m around the excavation, as shown in Figure 3, Cross Section of Tunnel and Jet Grout Columns. This mass would have a minimum unconfined compressive strength of 830 kPa, approximately 12 times the strength of the Bay Mud. At that strength, it would

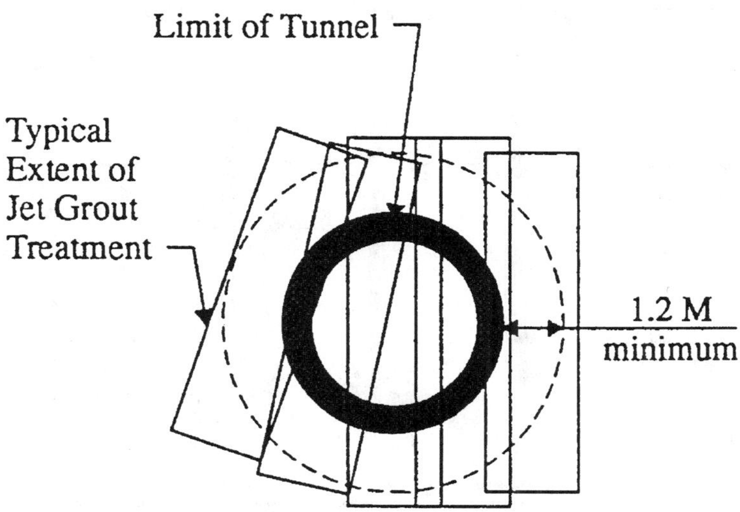

Figure 3. Cross Section of Tunnel and Jet Grout Columns

provide the necessary face support, and upon excavation, create a ring of support sufficient for the placement of the temporary lining. The F.E.M. analysis predicted settlements at the ground surface on the order of 10 mm, after tunneling. The proposal also included construction of a Jet Grout Test Program, to confirm the proposed jet grouted soil design parameters.

The City, recognizing the reduced risk of loss of ground, health problems for miners, and fire, accepted the CICP for a gross savings of $190,000. The City's acceptance was contingent on the results of a Jet Grouting Test Program (JGTP). The City's designer amended the contractor's proposed temporary lining of steel ribs and lagging to include 9 m of bolted steel liner plate to be installed at the beginning and end of each tunnel.

JET GROUTING PROPOSAL

Purpose

The purpose of the JGTP was to show that jet grouted columns could be installed to create a continuous mass of treated soil in the weak clay formation; that the design strength of 830 kPa for the grouted soil could be achieved; and that heave and settlement of the ground and adjacent structures could be controlled.

The test program consisted of 12 jet grouted columns installed using the Rodinjet technique and equipment. Six of the test columns were single-fluid (RJ1) and six were double-fluid (RJ2), using grout and air. The two-fluid method of jet grouting is illustrated in Figure 4, Jet Grouting Procedures Using Two-Fluid Method.

Test Description

The test columns were installed below the fill level to depths ranging from 4.6 to 8.5 m in two days. The exposure, examination, and sampling of the test columns required the installation of a shoring system composed of sheet piles and bracing.

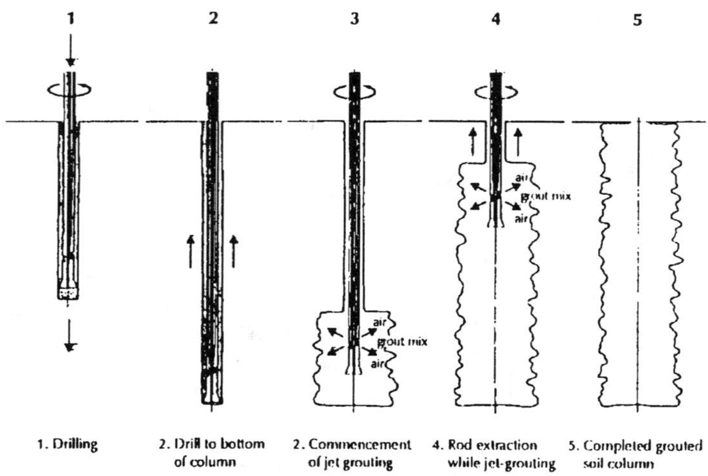

Figure 4. Jet Grouting Procedures Using Two-Fluid Method

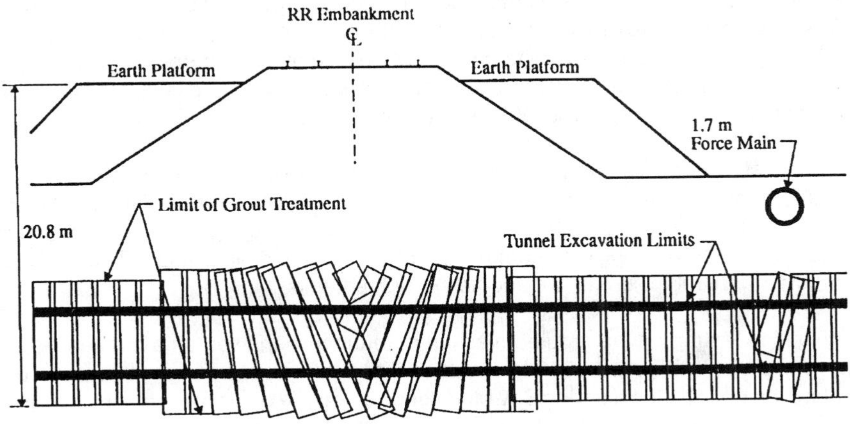

Figure 5. Profile of Undercrossing Tunnel

An automatic batching plant was calibrated to obtain a constant water/cement ratio for the grout of 0.83 by weight, at a production rate up to 20 m^3 per hour. Type I/II low alkali cement was employed to prepare the grout mix, resulting in a viscosity of 32 Marsh seconds and a unit weight of 1570 gr/l. The Uc strength of the grout averaged 18 MPa at 7 days and 33 MPa at 28 days. A variety of grouting parameters were tested, resulting in cylinders of uniformly grouted clay.

At completion of jet grouting, a 9 m x 9 m sheet-pile supported test pit was excavated to expose the top 1.2 m of each column. Diameters ranged from 0.8 to 1 m for the single-fluid (RJ1) system and from 2.4 to 2.9 m for the double-fluid system (RJ2). Laboratory testing (ASTM C30) on the grouted soil showed an average unconfined compressive strength of 3400 kPa for RJ1and 1400 kPa for RJ2 samples.

Based on the test results, a grouting pressure of 400 bar was selected, together with an air pressure of 9 bar. A conservative grout column spacing of 1.8 m was chosen to account for the numerous subsurface obstructions. This layout resulted in 7500 meters of RJ2 columns, with a replacement factor (volume of grout injected per volume of soil treated) of 38 percent, and up to 400 kg of cement per cubic meter of grouted soil.

PRODUCTION JET GROUTING

Production jet grouting started on the Davidson Tunnel. The 155 m stretch was grouted with 676 columns. The existing 1.83 m sewer on piles overlying the excavation alignment was underpinned by jet grouting to avoid settlements during the tunneling operation.

The jet grouting operation then moved to the Undercrossing Tunnel, where treatment required installation of columns inclined up to 30 degrees from vertical, as shown on the profile in Figure 5, Profile of Undercrossing Tunnel. The 1.68 m force main was heaved 140 mm during jet grouting of the first 6 or 7 jet grout columns beneath it. After a revised installation procedure was proposed and successfully tested, jet grouting resumed beneath the force main. The revised installation procedures included precutting and pretreatment of the fill. Precutting and pretreatment are used to control ground deformations. Precutting uses increased flow and pressure of fluids during drilling to maintain an open hole while jetting. Pretreatment is the systematic stabilization of the soils above the design jet grout zone to allow for drilling and jetting with continuous flow of spoils.

The installation of jet grouted columns beneath the railroad embankment required the use of a temporary earth platform on both sides of the embankment, as illustrated in Figure 5. The inclined columns immediately underneath the tracks were grouted through precased holes to prevent the washing of the embankment soils by the drilling fluids. The 77 m stretch was grouted with 306 columns, with no heave of the tracks.

Routine testing on the grouted mass in both tunnels showed an average unconfined compressive strength above 3500 kPa. Unconfined compressive strength test results on core samples taken in the differing Bay Mud zones along the tunnel alignments are shown in Table 1. Rock Quality Designation (RQD) for the cores averaged greater than 60 percent.

TUNNELING EQUIPMENT

Both tunnels were excavated with a Dosco/Terra Form, open-faced, road header tunneling machine. The machine was made of two components - the shield and the boom cutter. The shield assembly consisted of a front section and a tail section. The body of the front section was equipped with a 120 mm hood, a boom cutter frame, twelve 1350 kN x 1.7 m stroke thrust jacks, two hydraulic roll stabilizers, and eight 1100 kN x 0.8 m stroke breasting jacks. The tail section was equipped with a rib erector and expander, which had an expansion jacking capacity of 260 kN. A 17 m long conveyor belt brought muck from the face out over the top of a trailing frame into 6 m^3 cars, which were transported by locomotive to the tunnel shaft.

The boom cutter was ring mounted on a rail carriage and was powered by a 190 horse power Poclain MS50 direct drive hydraulic motor. It had a 118 mm cutter head with 20 drag bits located radially around it, and was capable of extending 2.3 m in front of the shield. Head rotation was 60 rpm at 32 MPa during normal operations. The torque at normal operations was 2237 kg-m, with a force delivered at each bit of 58 kN. The overall length of the 4.6 m diameter machine was 5.2 m, and it weighed approximately 624 kN.

Two different outer shields, 4.6 m and 4.1 m in diameter, were used on the project. Both shields were equipped with the same boom cutter.

TUNNEL EXCAVATION AND MACHINE PERFORMANCE

The Davidson Tunnel excavation commenced on February 14, 1995. During the first 30 working days, the tunnel advanced an average of 0.8 m per day. By April 29, the full trailing gear was installed. The remaining 129 m of tunnel excavation had an advance rate of 5 m per 10 hour shift. An average of 6.1 m per shift occurred during the last 15 shifts leading to hole-thru on May 8, 1995.

The tunnel machine was removed from the Davidson Tunnel and the boom cutter was re-assembled within the second 4.1 m diameter tunnel shield. Tunneling for the Undercrossing

Tunnel commenced on June 22, 1995. During the first 13 work days, the average advance rate was 1.5 m per 10 hour shift. After assembly of the trailing gear, the average advance rate was about 3.7 m per shift, with hole-thru on August 3, 1995.

Typically, the road header boom was used to cut approximately 1.2 m to 2.4 m in front of the shield, as measured from the bottom of the face to the top, respectively. At the beginning of a typical excavation cycle, muck was collected on the bottom of the shield. After excavating the bottom half of the face, mucking and excavating were simultaneously performed by the road header feeding the conveyor with the muck cars in place. The excavated space stood unsupported for the entire excavation and mucking period. After the 1.2 to 2.4 m long cylindrical space was clear of muck - often an operation which required miners to work in the "bald zone", the machine was pushed into the excavation with the thrust jacks, which reacted against the primary lining. During excavation, primary lining, typically steel ribs and lagging, were stood within the tail section of the machine. As the machine was shoved forward, the primary lining was shoved out against the previous set, which was then expanded. During the shove, machine steering was performed by varying the pressures on specific thrust jacks.

Throughout excavation of both tunnels, the jet grouted soils in the face and crown were stable. Throughout much of the Davidson Tunnel, the jet grouted soil was completely continuous. Often, the temporary lining was not fully engaged. Throughout the Undercrossing Tunnel, discontinuities in jet grout treatment occurred around timber piles. During one excavation sequence, a third of the tunnel face was "untreated"; however, the clay in these "untreated" areas was very stiff and uncharacteristically dry. In no cases did instability of the face occur.

Tunnel muck consisted of chips of the jet grouted Bay Mud ranging in size from 3 to 70 mm, which behaved like siltstone or mudstone chips. Generally, muck was lifted to the ground surface and temporarily stockpiled near the shaft. During wet weather, the muck tended to become more soft in consistency. It was loaded into trucks directly from the stockpile with a standard bucket loader for off-site disposal. Most of the tunnel muck was re-used by the landscaping industry. By comparison, native Bay Mud excavated during construction of the many cut and cover structures was difficult to handle and costly to dispose.

Groundwater seepage was approximately 0.32 l/sec at the Davidson Tunnel, and 0.12 l/sec at the Undercrossing Tunnel.

A total of 41 piles were encountered during both tunnel excavations. Precast concrete piles and timber piles were broken up and removed from the face with no impact on the excavation operations. However, the wood in the tunnel muck did make disposal more difficult. In addition, at several locations, the remaining timber "stub" at the limits of the excavation complicated the steerage of the machine. At two different locations steel H-piles were encountered. These two piles did stop excavation, but stability of the jet grouted soil mass at the face made removal by hand relatively quick, resulting in very little impact on production.

During the excavation, a strong ammonia odor was typically released into the air inside the tunnel. Based on analysis of air monitoring and sampling, it was determined that during excavation, airborne ammonia concentrations ranged from 7 to 20 mg/m^3. After ex-

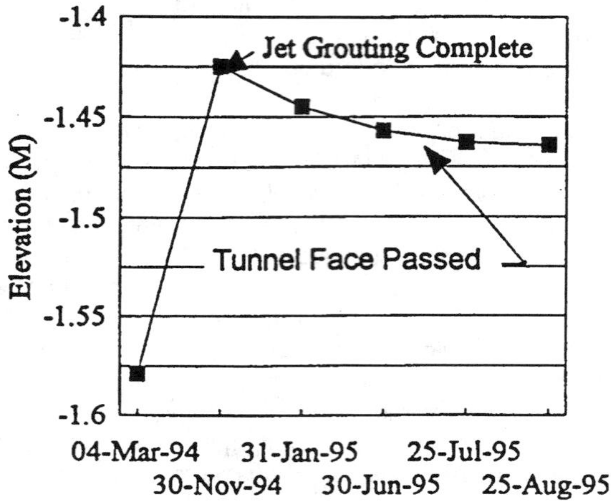

Figure 6. Summary of Settlement Data

cavation stopped, airborne concentrations of ammonia were reduced immediately to a range of 0.7 to 2.6 mg/m^3. Organic cartridge respirators were required to be worn during excavation, and full circulation ventilation was run continuously to the tunnel face throughout the tunnel operations.

INSTRUMENTATION DATA

Instrumentation included surface settlement points, lateral deformation points, and inclinometers located across the entire project site. An instrumentation program was developed and implemented by Haley & Aldrich Inc. of San Francisco, who served as the project construction managers.

Perhaps the most representative settlement data was obtained on the City's 1.68 m force main, where it crossed over the tunnel centerline. Figure 6, Summary of Settlement Data, presents a plot of the elevation readings taken on the force main over the project duration. The settlement which occurred after jet grouting and before tunneling is most likely due to consolidation settlement of the soft Bay Mud below the temporary earth platforms and the railroad embankment. Also shown on the plot are significant project milestones. As indicated, the overall settlement of the force main after tunneling passed under it is of the order of 6 mm. This data corresponds well with other data obtained on the site. Generally along the Davidson Tunnel centerline, settlements ranged from 5 to 15 mm at the five undisturbed settlement points.

Settlement data for the Undercrossing Tunnel is more difficult to generalize due to the use of the temporary earth platforms, the continuous settlement of the railroad embankment (documented prior to the start of construction), and the repeated track re-ballasting effort which destroyed settlement points. Points on the tracks, which were monitored before tunneling and after completion of the tunnel lining, indicate overall settlement ranging from 12.7 to 19 mm. Other settlement points above the Undercrossing Tunnel indicate similar results.

Three inclinometers were installed within 4.6 m of the tunnel centerline after jet grout-

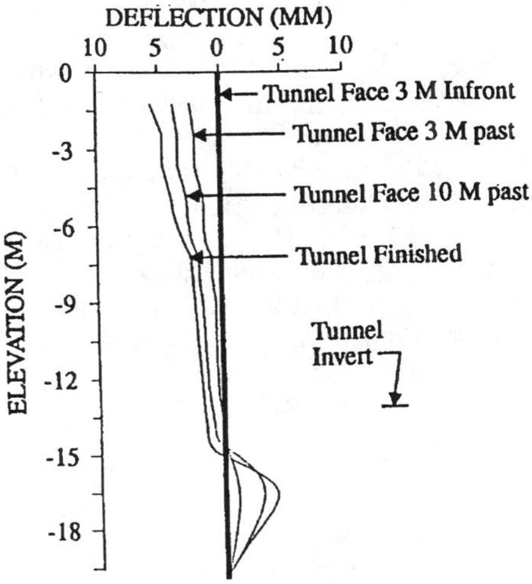

Figure 7. Plot of Inclinometer I-9A

ing was completed. The inclinometer data was obtained daily, immediately before and after the machine passed, and then weekly, when the machine was within 30 m of the instrument. The data, in conjunction with observations made during tunnelling, indicate that the amount of ground movement into the tunnel was negligible. A plot of Inclinometer I-9A, which was located 4 m from the center of the Undercrossing Tunnel is presented in Figure 7, Plot of Inclinometer I-9A. It is representative of inclinometer data obtained from the other inclinometers as well.

SUMMARY AND CONCLUSIONS

Jet grouting was successfully used to improve native soft clay to 12 times its in situ strength and eliminate squeezing ground conditions for construction of 232 m of 3.7 and 3.2 m finished diameter tunnels. One thousand, 8 m long, double-fluid jet grout columns were installed from the ground surface to below tunnel invert in advance of tunneling operations. The tunnels were excavated with an open face shield using a boom-mounted road header to cut and remove the grouted soils.

This method eliminated the use of compressed air for face stabilization, along with its associated risks, including loss of ground, health problems for miners, and fire. The jet grouted mass provided an initial ring of support which allowed for installation of the primary tunnel lining. It also facilitated excavation around and removal of numerous obstructions without impacting tunnel production.

As the jet grouting operation progressed, the ground heave which occurred initially was controlled. Overall movements at the ground surface during tunnelling were minimal.

This project was the first full-face application of jet grout pre-treatment of weak clay in the industry, and resulted in lower costs than originally estimated.

REFERENCES

Adams, D.N., Robison, M.J., 1997, "Construction and Performance of Jet Grout Supported SPTC Walls in Weak Clay ", Journel of the Boston Society of Civil Engineers, Civil Engineering Practice, Volume 12, Number 1.

Clough, G.W., Reed, M.W., January 1984, "Measured Behavior of Braced Wall in Very Soft Clay", ASCE Journal of Geotechnical Engineering, Vol 110, No. 1.

Pellegrino, G., Adams, D.N., 1996, "The Use of Jet Grouting to Improve Soft Clays for Open Face Tunnelling", Proceedings, International Society of Soil Mechanics and Foundation Engineering, Internation Symposium of Technical Committee TC28, Geotechnical Aspects of Underground Construction in Soft Ground, pp. 355-360.

Chapter 15

AN EXPERIMENTAL STUDY ON DEVELOPMENT OF OVAL SECTION TBM

Sohta Takatsu,[1] Taira Emoto,[2] Hikozou Imaoka,[3]
Takeshi Suzuki,[4] and Yoshihito Minami[5]
[1]Hazama Corporation; [2]Advanced Construction Technology Center; [3]Sato Kogyo Co., Ltd.; [4]Maeda Corporation; and [5]Komatsu Ltd.

An Oval Section TBM (tunnel boring machine) enables excavation of tunnel with elliptical cross section by inclining the cutterhead. It is developed to utilize the proven conventional TBM technology with minimum modifications. Various types of tunnels which are preferred to have non-circular cross section can be economically excavated, reducing both overbreak and the amount of muck produced.

On the other hand, an inclined cutterhead may create a number of new problems, such as in excavation efficiency and steering, that are unknown in excavation works using a circular TBM. This paper reports on verification tests carried out to examine the extent of such problems using experimental models of the machine in an attempt to confirm the feasibility of the Oval Section TBM.

INTRODUCTION

Recently, the use of TBMs has increasingly been popular in excavation works for tunnels in mountainous regions of Japan. Traditionally used for aqueduct tunnel works, it has not hitherto been common to employ TBMs in excavation works for development of highways or railways. It is because highways and railways need large cross sections, and moreover, the circular cross section of TBM results in unnecessarily large excavation volume that incurs a loss of rationality in a variety of aspects, such as low economical efficiency.

In shield tunneling technology, they have developed a variety of non-circular tunneling machines, such as multi-face, elliptical and box shielding machines. It is proposed that TBM tunneling can be better utilized in hard rocks if a method for elliptical excavation is developed.

The development of Oval Section TBM, which enables economical excavation of oval section, will certainly helps to rationalize non-circular tunneling works in mountainous regions.

In this research, "Oval Section TBM Method" was developed based on the proven TBM technology without greatly modifying the conventional machine, which was made possible by inclining the cutterhead against the tunnel axis. The purpose of this study is, therefore, to develop a method for elliptical excavation and attempt to put it into practice.

OUTLINE OF THE METHOD AND ITS PROBLEM

Outline of the Method

The difference between an Oval Section TBM and a conventional circular TBM is simple; the former has an inclined cutterhead against the tunnel axis. Characteristics of the Oval Section TBM are as follows:

Enabling Economical Excavation. Application of the Oval Section TBM Method to various types of tunnels that need elliptical cross sections results in economical excavation, reducing both overbreak and the amount of muck produced. For example, if an Oval Section TBM with an inclination angle of 0.70 rad (40°) is applied in an excavation work for a double-lane highway as shown in Figure 2, excavation volume is reduced by 20% compared with that in a circular excavation.

Stabilizing Effects Produced by the Elliptical Cross Section and Inclined Cutting Face. Tunnels with elliptical cross sections are supposed to ensure better stability than the conventional horseshoe-shaped tunnels because the transmission of stresses becomes smoother. In addition, the inclined cutterhead has a stabilizing effect on the cutting face to the same extent as the ring cut method (Figure 3).

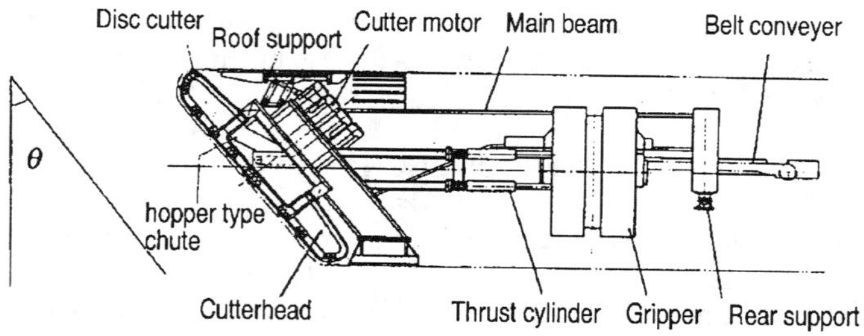

Figure 1. Mechanism of Oval Section TBM

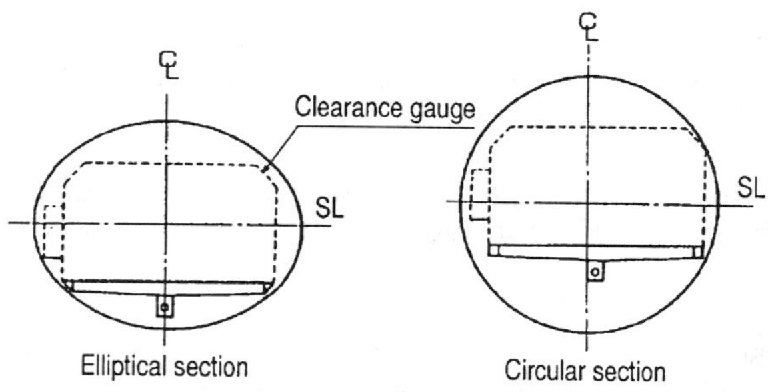

Figure 2. Comparison of Inner Section of Double-Lane Highway

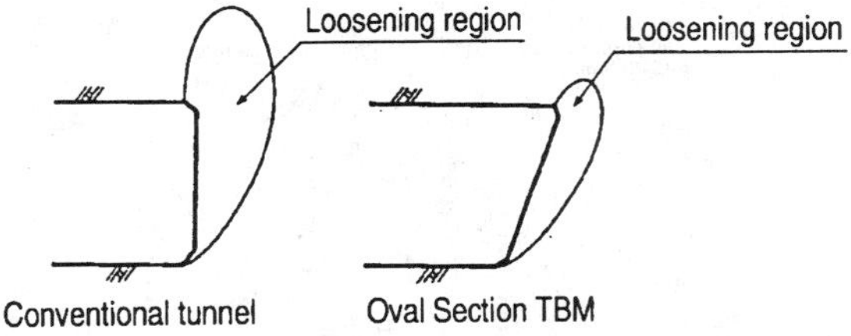

Figure 3. Stabilization Effects of Cutting Face

Examination of Practical Application and Utilization

Economical Application. If the Oval Section TBM Method is applied in the excavation of tunnels for a variety of uses, overbreak is kept minimum; the excavation thus becomes more economical than by the conventional method. For example, excavation of a double-lane highway tunnel is supposed to benefit from the application of this method to a great extent. Therefore, an appropriate dimension of a cross section for this type of tunnel was studied, followed by a study on the most economical inner section in the Oval Section TBM Method.

Other Application. Excavating a pilot tunnel with the Oval Section TBM improves workability because a wide base course is created as a result. If, for example, an 8.0 m-diameter Oval Section TBM with an inclination angle of 0.70 rad (40°), mucking operation inside the drift by a dump truck of 20 ton would become an easy operation.

Problems of the Method and the Objectives in This Paper

When a cutterhead is angled and a thrust is applied parallel with the tunnel axis as in the conventional method, the thrust transmitted to the cutter can be divided into two forces: a true thrust working at a right angle to the cutting face and a sliding force working parallel

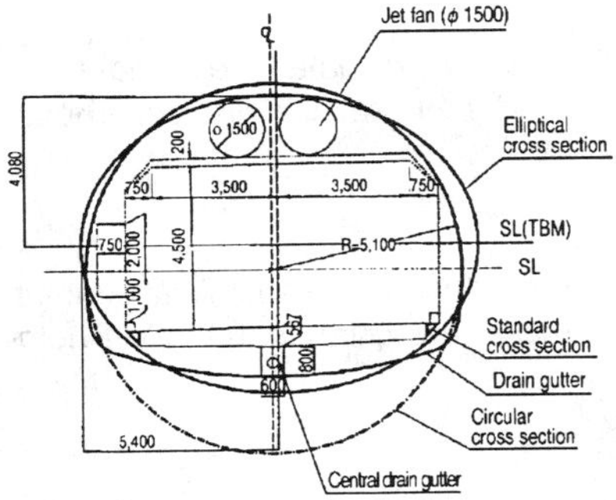

Dimensions of Oval Section TBM

Inclination angle of cutting face		0.72rad (41°)
Diameter	Major axis	5.4 m
	Minor axis	4.1 m
Inner section area		69.1 m²

Comparison of inner section areas

	Sectional area	Area ratio to the standard section
Elliptical section	69.1 m²	1.02
Circular section (r=5.1 m)	81.7 m²	1.21
Standard section	67.6 m²	1.00

Figure 4. Comparison of Inner Sections

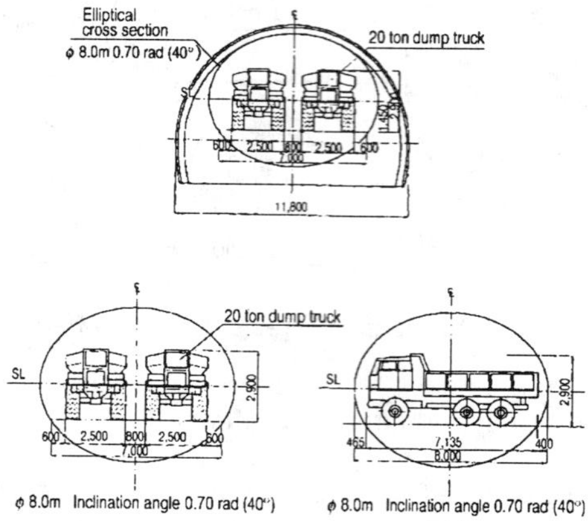

Figure 5. Location of Dump Truck

with the cutting face. Therefore, it is important to study differences between the conventional circular TBM method and the Oval Section TBM Method in terms of excavation efficiency.

Accordingly, a number of tests were carried out to clarify and compare the excavation efficiency of the conventional circular TBM method and the Oval Section TBM Method, studying the following items:

(1) Relation between excavation thrust and excavation speed;
(2) Relation between excavation torque and excavation speed;
(3) Relation between lifting force and steering;
(4) Comparison of cutter loads;
(5) Stability of cutting face; and
(6) Muck removing mechanism

EXPERIMENTAL EXCAVATION

Objective of Tests

The objective of the experimental excavation is to obtain basic data on elliptical excavation and confirm the feasibility of the Oval Section TBM Method with inclined cutterhead.

Specifications for Test Apparatus

Photograph 1 shows the Oval Section TBM, and Table 1 gives the specification for it. The specifications for the test specimen (artificial bed rocks) are given in Table 2. Photograph 2 shows the outward form of the specimen.

Verification Items

The experimental excavation can be divided into three sub-experiments, that is, excavation with an Oval Section TBM with an inclination angle of 0.53 rad (30°) and 0.70 rad

Photograph 1. Oval Section TBM (Inclination Angle 0.70 rad (4°))

Table 1. Specifications for Oval Section TBM

Name of Items	Major Specifications		
	Vertical type	Inclination angle (0.53 rad (30°))	Inclination angle (0.70 rad (40°))
(1) Excavation diameter (major axis x minor axis)	2,400 mm (circular)	2,400 mm x 2,150 mm	2,400 mm x 1,970 mm
(2) Weight	16,000 kg	13,720 kg	15,093 kg
(3) Cutterhead torque	117.6 kNm	117.6 kNm	117.6 kNm
(4) Revolution speed	0.24 rad/s		
(5) Disc cutter	12" (305 mm) x 22 pieces		
(6) Hydraulic motor	1248 kg x 3		
(7) Thrust jack	60 ton - 1050 mm - st x 4		

Table 2. Specifications for Artificial Concrete Bed Rocks

Name of bed rocks	Main dimensions
Artificial concrete bed rocks	
(1) Type of cement	Ordinary cement
(2) Size of coarse aggregates	Maximum size: 40 mm
(3) Slump	12 cm - 15 cm
(4) Concrete strength	Uncontained compressive strength (Eq.) qu = 58.8 MPa
With Boulder inserted bed rocks	
(1) Size of boulder	Maximum size: 300 mm
(2) Boulder Strength	Uncontained compressive strength (Eq.) qu = 117.7 MPa
(3) Mortar strength	Uncontained compressive strength (Eq.) qu = 117.7 MPa

(40°), and excavation with a circular TBM; all of these results were compared to each other on excavation performance.

Driving Resistance at the Cutterhead. Driving resistance at the cutterhead was measured to obtain the torque actually used in the excavation. Here, the driving resistance is defined to be the torque obtained when the cutterhead is rotated without contacting to the face.

Photograph 2. Test Specimen

Measurement of Thrust Resistance. To calculate the thrust used in an actual excavation, it is necessary to measure resistance forces of excavation incurred by the friction between the concrete walls and supporting parts such as a supporting jack and a vertical shoe. Therefore, resistance forces were measured and the friction coefficient between the shoe and the concrete walls was calculated.

Test Results

No-Load Test

MEASUREMENT OF DRIVING RESISTANCE AT THE CUTTERHEAD: Table 3 gives the measurement results.

MEASUREMENT RESULTS OF RESISTANCE FORCES OF EXCAVATION: Table 4 gives the measurement results.

Results of Tests on Artificial Concrete Bed Rocks

RELATION BETWEEN EXCAVATION SPEED AND EXCAVATION TORQUE: Figure 6 shows the relation between excavation speed and effective excavation torque (= excavation torque - driving resistance at the cutterhead) of the excavations carried out with TBMs with inclination angle of 0.53 rad (30°), 0.70 rad (40°), and 0 rad (0°). The solid line shown in the figure represents the calculation values derived from the empirical formula for the circular excavation of the artificial concrete bed rocks (qu: uncontained compressive strength = 58.8 MPa).

Though variable to a certain extent, with the variable strength of the artificial concrete bed rocks being considered, it has been shown that the effective excavation torques of inclination angles 0.53 rad (30°) and 0.70 rad (40°) show reasonable correlations. It was found that all the measured values tend to be a little smaller than the calculated values.

Table 3. Driving Resistance of Cutterhead

	Driving resistance of cutterhead
Inclination angle 0.53 rad (30°)	31.4 kNm
Inclination angle 0.70 rad (40°)	30.4 kNm
Vertical type (Circular TBM)	31.4 kNm

Table 4. Friction Coefficient

	Friction coefficient
Inclination angle 0.53 rad (30°)	0.560
Inclination angle 0.70 rad (40°)	0.624
Vertical type (Circular TBM)	0.560

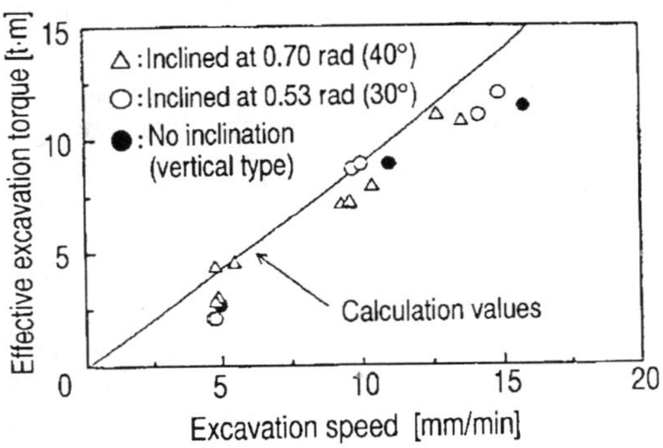

Figure 6. Relation between Excavation Speed and Effective Excavation Torque

It can therefore be assumed that the calculation values for the conventional circular TBM can sufficiently be regarded as the effective excavation torque of elliptical excavation without incurring problems in the actual operation.

RELATION BETWEEN THE EXCAVATION SPEED AND THE THRUST: Figure 7 gives the relation between excavation speed and effective thrust (= excavation thrust - resisting forces of excavation) of the excavations carried out with TBMs with inclination angle of 0.53 rad (30°), 0.70 rad (40°), and 0 rad (0°). The solid line shown in the figure represents the calculation values derived from the empirical formula which corresponds to the circular excavation of the artificial concrete bed rocks (qu: uncontained compressive strength = 58.8 MPa). It has been shown that the effective thrust of inclination angles 0.53 rad (30°) and 0.70 rad (40°) agree with the calculation values.

It can therefore be concluded that the calculation values for the conventional circular TBM can sufficiently be regarded as the effective thrust of elliptical excavation without incurring problems in the actual operation.

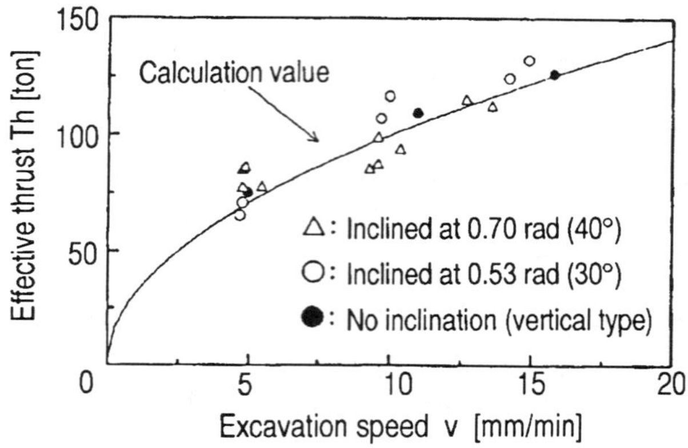

Figure 7. Relation between Excavation Speed and Effective Thrust

LIFTING FORCE: In the excavation of elliptical cross section, an lifting force occurs on the front of the cutterhead which pushes the test apparatus upwards owing to the inclined cutterhead. Therefore, a roof supporting jack should work to keep balance during the excavation. Figure 8 shows the forces working during the elliptical excavation.

The magnitude of a force working in the direction of the tunnel axis $F_c(x)$ was obtained from the calculation values of the effective thrust mentioned earlier (i.e. the calculation values derived from the empirical formula for to the circular excavation). The magnitude of an lifting force $F_c(y)$ was derived from the relation between the force in the direction of tunnel axis and lifting force obtained by applying each disc cutter to the relational expression shown in Figure 9.

Where

F_r :	Designed roof supporting force
F_s :	Side supporting force
F_{RS} :	Rear supporting force
Th_1 :	Thrust jack force
W_g :	Weight of test apparatus
$F_c(x)$:	Force working on cutterhead in the direction of tunnel axis
$F_c(y)$:	Lifting force on cutterhead
X_c :	Distance to the cutterhead center
X_w :	Distance to the barycenter
X_{RS} :	Distance to rear support
X_r :	Distance to roof support

Figure. 10 gives the comparison between the calculated roof supporting force ($F_{r'}$) and the measured roof supporting force (F_r). The measured value agree with the design roof supporting force, i.e. ,the calculation value. Thus, it can be assumed that the roof supporting force can be designed according to the expression below.

$$F_{r'} = (F_c(y) \cdot X_c - W_g \cdot X_w - F_{RS} \cdot X_{es}) \ / \ X_r \quad (1)$$

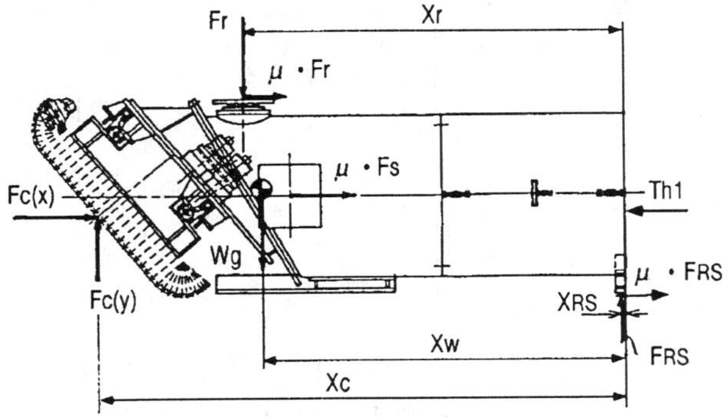

Figure 8. Schematic Diagram of Forces Working During Elliptical Excavation

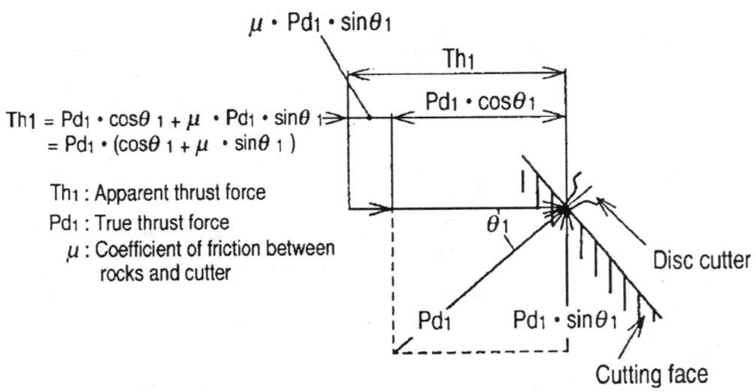

Figure 9. Schematic Digram of Forces Working on Disc Cutter

MEASUREMENT RESULTS OF DIRECTIVITY: As described above, an lifting force occurs on the cutterhead. To counter this force, the roof supporting jack suppresses the cutterhead. Figure 11 is the records of pitching angles to the excavation distance and vertical displacements of the target. The pitching angles were within $\pm$ 0.5° and the vertical displacements were within $\pm$ 2 mm, showing that the test excavation was well controlled.

Changes in pitching angles occurred at the beginning of the excavation when the speed was accelerated with a thrust, and the pitching angle did not change during the excavation. Therefore, it can safely be said that the excavation alignment was leveled throughout the excavation. The excavation alignment was adjusted downwards when applying a larger roof supporting force while it was adjusted upwards when applying a smaller roof supporting force. It was thus confirmed that the pitching can be controlled by changing the roof supporting force.

Figure 12 gives the records of rolling angles to the excavation distance and horizontal displacements of the target. It is seen from the figure that a horizontal displacement of 5 mm occurred on the right hand side before the left hand side had a horizontal displacement. This happened because the machine was not steered at the beginning but suddenly directed towards left to see if it was possible to steer horizontally.

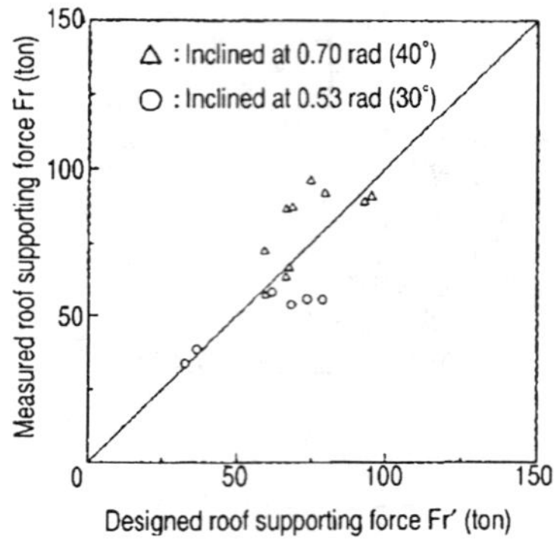

Figure 10. Comparison between Measured and Calculated Values of Roof Suppopting Force

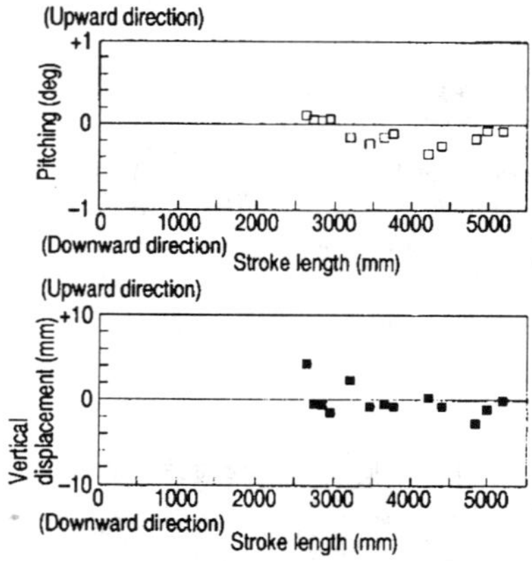

Figure 11. Measurement Results of Pitching and Vertical Displacment

A force occurs in the yawing direction due to the inclined cutterhead in the elliptical excavation, but it is relatively small compared with the side supporting force. Therefore, it can be concluded that the horizontal steering method employed in the conventional circular TBM can be used for the elliptical excavation without incurring a problem.

CONSIDERATIONS AND EXPERIMENTS FOR THE STABILITY OF CUTTING FACE

Advantages of Angled Cutting Face (Flat Elipse)

When excavating a flat elliptical cross section by the Oval Section TBM Method, it is assumed that the stability of cutting face is improved to the same extent as in the ring cut method because the resultant cutting face is angled. The danger of cutting face being fallen is assumed to be reduced. To confirm this, the stabilizing effect of angled cutting face was

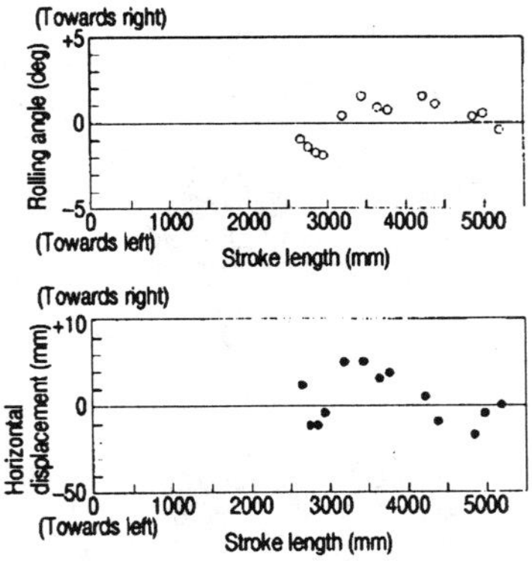

Figure 12. Measurement Results of Rolling Angle and Horizontal Displacement

examined using Murayama's formula.

(1) Analytical method: Calculation of sliding materials using the expression proposed by Murayama.

(2) Rock mass conditions:

Overburden	H = 2D (D = excavation diameter)
Weight per volume unit	$\gamma = 9.8$ kN/m^3
Cohesion	C = 98 kPa
Internal friction angle	$\phi = 0.53$ rad (30°)

*N.B. The rock mass conditions were assumed to be poor, which may appear occasionally during excavation.

(3) Comparison factor: Sizes of sliding materials

(4) Compared items: Vertical cutting face, cutting face angled at 0.35 rad (20°) and cutting face angled at 0.53 rad (30°)

(5) Results

The analytical results are in Figure 13.

(6) Consideration

Under these rock mass conditions, the more angled cutting face is, the smaller the amount of sliding materials becomes. When the angle becomes larger than 0.53 rad (30°), loosening region of the cutting face almost disappears.

Tests for the Stability of Cutting Face

Test Objective. Cutting face might fall when encountering a fault fracture zone during excavation. As there might be a difference between the stability of angled cutting face and vertical one under such rock mass conditions, a laboratory test was carried out with a scale model for the following two purposes:

(i) To understand the difference between the fall mechanisms of angled cutting face and vertical one visually.
(ii) To understand the displacement in the rock mass near the cutting face quantitatively.

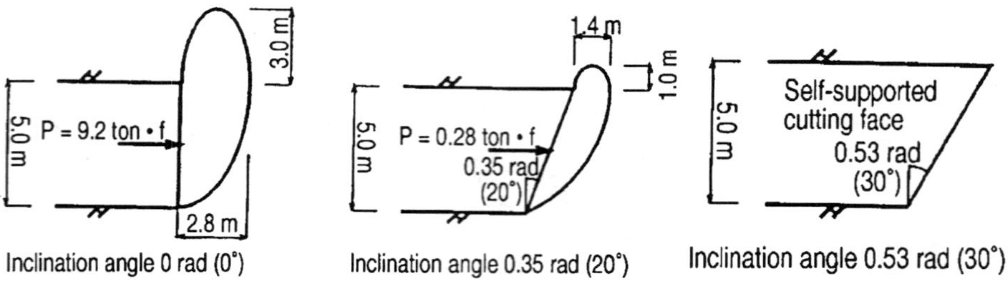

Figure 13. Loosening Region

Test Conditions

SCALE MODEL: A scale model was designed with regard to efficiency and an required overburden of 5D (50 cm). Figure 14 is the drawing of the scale model.

TEST APPARATUS: The tank containing the rock mass was made of steel, 80 cm high, 100 cm wide and 10 cm deep. The front side was a transparent acrylic board with a grid pattern of 2 cm square. There was a void space of 10 cm x 10 cm which represents a tunnel. Excavation was represented by making a free face with a sliding door installed in this void space. Photograph 3 shows the outward form of the test apparatus.

TEST METHOD: The test was carried out with the following procedure:

1. Sands were dropped from above to make a sand layer of 5 cm thick.

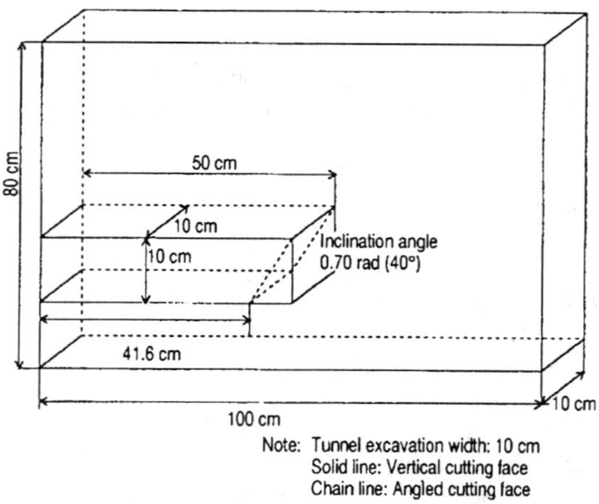

Figure 14. Scale Model for the Test

Photograph 3. Test Apparatus (Vertical Cutting Face)

2. Impacts were applied on the tank to compact the sand layer.
3. The above procedure was repeated, setting gauge plugs when necessary until an overburden of 5D was obtained.
4. The sliding door representing a cutting face (both vertical and angled) is slid at a speed of 1 mm/ min.
5. When the sliding door reached the required displacement, the conditions of the sand layer near the cutting face was checked visually and recorded with a 8 mm video recorder.
6. The images taken by the video recorder were input in a personal computer and displacements at gauge plugs were analyzed.

Test Results. The following insights into the stability of cutting face were gained through the comparison between the images obtained from this test (photographs 4, 5, 6, and 7):

1. Angled cutting face makes it capable of withstanding under rock mass conditions that cause falling in the case of vertical cutting face. Therefore, angling the cutting face greatly reduces the danger of failure even when encountering a weak layer of a fault fracture zone in the driving of TBM.

2. It was confirmed that under rock mass conditions in which even an angled cutting face cannot escape falling, the area of failure is extremely small comparing with that of a vertical cutting face. Thus, it can be assumed that should a failure happens, rehabilitation procedure will be more simplified than in the case of the vertical cutting face.

MUCK REMOVING TEST

Test Objective

Because the cutterhead is inclined in the Oval Section TBM Method, it is questionable if muck in a bucket can sufficiently be discharged onto the belt conveyor when using the same muck removing mechanism as that in the conventional method.

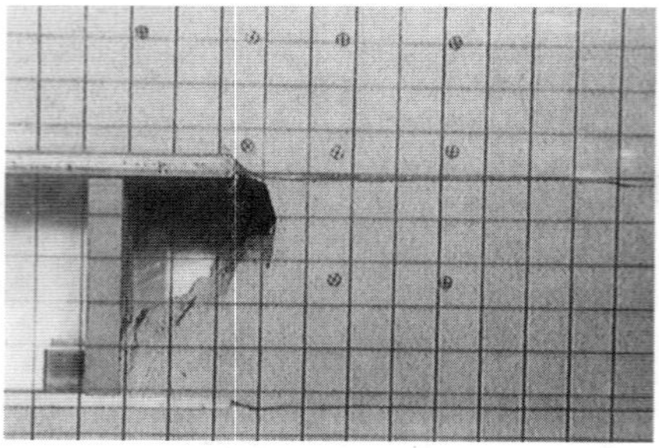

Photograph 4. Vertical Face: Water Content 0.7%

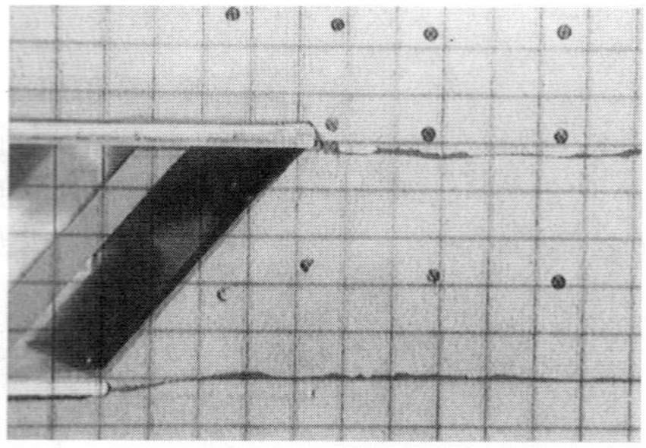

Photograph 5. Angled Cutting Face: Water Content 0.7%

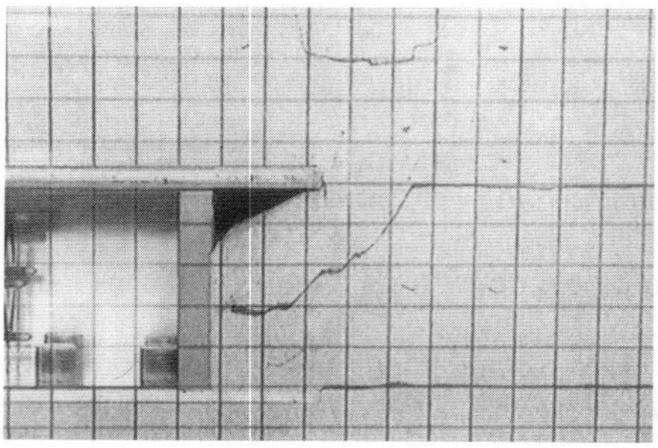

Photograph 6. Vertical Cutting Face: Water Content 0.0%

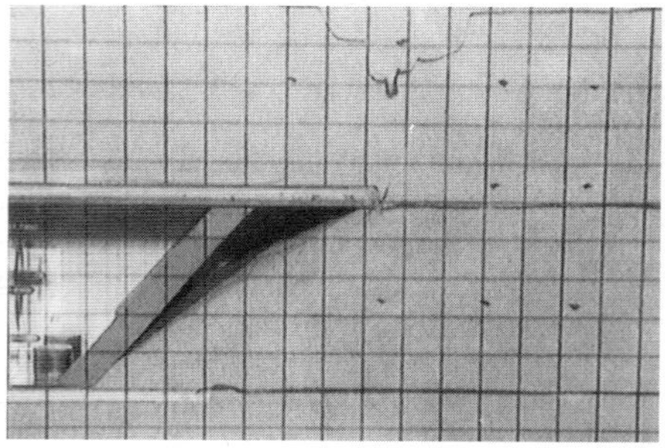

Photograph 7. Angled Cutting Face: Water Content 0.0%

To overcome this problem, a muck removing mechanism suitable for the Oval Section TBM Method was developed, and a number of tests were conducted to verify its intake and removal performances.

The test was carried out with a scale model of 6 m diameter cutterhead with the inclination angle of 0.70 rad (40°) to obtain the following items:

1) Muck removing ratio
2) Amount of muck left out
3) Maximum revolution speed
4) Cutterhead torque

The following cutterhead configurations were proposed and tested.

1) Proposal A: The maximum angle of the slider is 0.53 rad (30°)[*1] (The angle of the perimeter was 0 rad (0°) (level))
2) Proposal B_1: The maximum angle of the slider is 0.70 rad (40°). Muck collecting plates installed.
3) Proposal B_2: The maximum angle of the slider is 0.70 rad (40°). Muck collecting plates installed (around the perimeter only). Baffle plate installed (around the perimeter only, inclination angle of 0.26 rad (15°)).

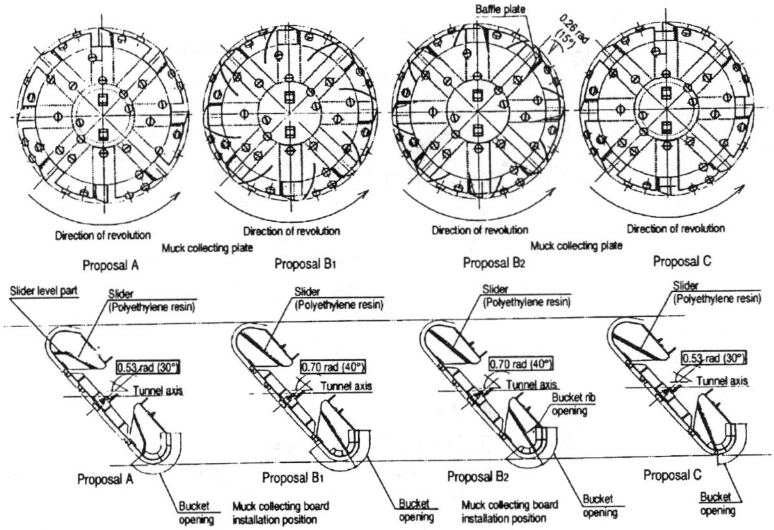

Figure 15. Proposed Configurations of Cutterhead

4) Proposal C: The maximum angle of the slider is 0.53 rad (30°)

[*1] Angle to the horizontal plane

CONCLUSION

1) From the measurement results of muck removing ratio, it was found that the cutterhead configurations B_1, B_2, C and A had high muck removal performance in order of mention.

2) The cutting head configuration of the proposal C had a smaller amount of muck left under the front side of the cutting face when the revolution speed of the cutterhead of the ϕ 6 m machine was smaller than 0.40 to 0.46 rad/s. However, at a revolution speed higher than that range, cutterhead configurations of the proposals B_1 and B_2 had a smaller amount of muck left. The reason for this seems to be that the angle of the slider in the proposal C (0.53 rad (30°)) was smaller than that in the proposals B_1 and B_2 (0.70 rad (40°)), because when the revolution speed is high, muck inside a bucket cannot be easily dropped onto the belt conveyor.

3) Muck seemed to be left around the collecting board in front of the cutting face only in the proposal B_1 as inferred from the peeling condition of paint on the cutterhead. The visual observation also indicated that the configuration of collecting board should take the one employed in the proposal B_1.

4) It was found that the cutterhead torques of the proposals B_1 and B_2 were a little larger than that of the proposal C. This seemed to be owing to an increase in the revolutional resistance of the muck collecting board. However, it was shown that the percentage of muck intake torque to the nominal torque was 13% at maximum

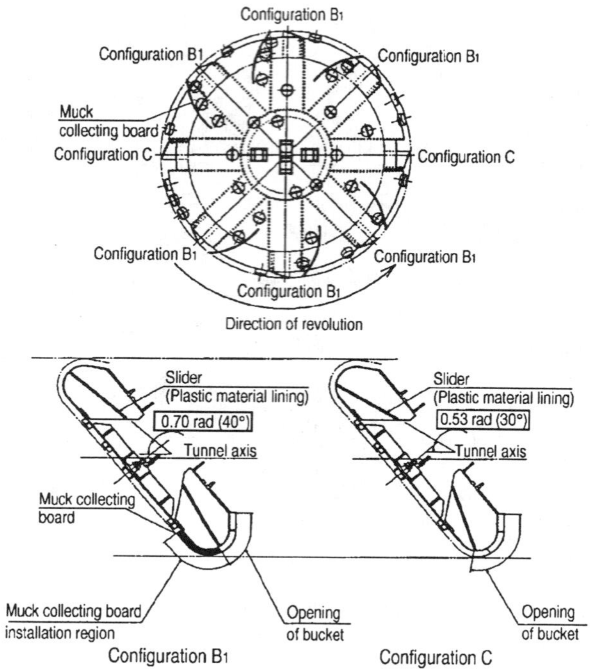

Figure 16. Proposed Cutterhead Based on Test Results

when conversed into the ϕ 6 m machine. Because the design percentage is 20%, it can safely be concluded that this was not particularly problematic.

5) The maximum revolution speed at which all the muck taken inside a bucket can be dropped onto the belt conveyor was 0.31 to 0.42 rad/s in the proposal C while it was more than 0.65 rad/s in the proposals B_1 and B_2. It is concluded therefore that the proposal C has a potential danger of detention and clogging of muck in the cutterhead of the ϕ 6 m machine whose revolution speed is 0.65 rad/s.

Considering the results described above, it was confirmed that if the cutting head configuration in the proposal B_1 is used as the model configuration, it is possible to remove muck when using the ϕ 6 m machine whose cutting face has an inclination angle of 0.70 rad (40°).

Also, it is proposed that the following measures can reduce both the amount of left-out muck in front of the cutterhead that concentrates around the face of the collecting board and the abrasion of the disc cutter caused by a secondary breakage of muck:

1) Adaptation of the proposal C, which has a high performance in muck intake in a low range of revolution speed, for about a quarter of the whole number of buckets to the basic configuration of the proposal B_1.

2) Reduction of abrasion of the disc cutter due to a secondary breakage of muck by installing the disc cutter at the back of the muck collecting board.

CONCLUDING REMARKS

Experimental excavations were carried out on artificial concrete with or without boulder inserted bed rocks using Oval Section TBMs with a cutterhead diameter of 2.4 m and inclination angles of 0.70 rad (40°) and 0.53 rad (30°). The results of this test can be summarized as follows:

1) The breaking pattern shown in the photograph of the artificial boulder block (Photograph 8) indicates that the machine carried out successive breaking as in the ordinary breaking mechanism, regardless of the inclination angle of cutterhead.

2) When inclining the cutterhead forward at an angle of θ, the effective thrust applied on the cutterhead was the thrust x cos θ, which showed that the measured value and the theoretical value agreed with each other. It was confirmed through the measuring of cutter loads that the circular TBM and the Oval Section TBM have the same excavation speed for the same excavation volume in the tunnel axis direction if a same thrust is applied. Therefore, there occurred no losses in the excavation speed when using the Oval Section TBM.

3) It was confirmed through the measuring of hydraulic pressure of the cutter motor that there were no losses in the excavation torque, and the following theoretical formula for the circular TBM was found to be applicable:

$$T = Pd \times N \times rm \quad (2)$$

where

T: Excavation torque
Pd: Loads on the disc cutters
N: Number of the disc cutters
rm: Average size of the disc cutters-mounting area

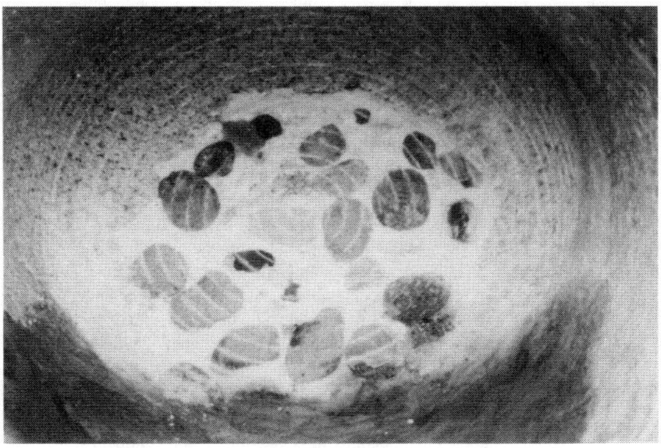

Photograph 8. Breaking Pattern of Boulder Rock

4) As the effective thrust was represented by a cos*q* parameter, the lifting force was also represented by a sinθ parameter. The roof supporting was confirmed to be steered with a force smaller than sinθ x thrust.

5) The load working on the cutter bearing measured a value proportional to the thrust in both the circular and elliptical excavations. The same principle applied for pitching and rolling.

6) For the use of Oval Section TBM cutterhead, adding muck collecting plates on the circular TBM cutterhead enables enhanced muck removal performance.

With all the expected problems concerning the elliptical excavation being considered as described above, it has been confirmed that the Oval Section TBM follows the same theories as the conventional circular TBM without incurring a problem in excavation efficiency.

Therefore, it is concluded that a new TBM method has successfully been developed which enables both a suitable configuration for the designed cross section with a minimum overbreak and a superior stabilization performance.

REFERENCES

Muraoka, Sakuro, 1996, "Tunneling Earth Pressure" (in Japanese), Proceedings of the Third Tunnel Engineering Symposium, Japan Society of Civil Engineers

Chapter 16

THE TECHNOLOGY OF AUTOMATED GUIDANCE FOR TBM'S AND ROAD HEADERS

Alwin Poltinger[1] and W.L. Hollinshead[2]
[1]DYWIDAG; [2]Precision Centerline, Inc.

GENERAL

Tunneling by hand digging is as old as the history of humans. Mechanized tunneling did not begin to replace manual methods until around the beginning of the 19th century (Brunel patent, 1818). It brought with it a higher degree of safety for the workers and a much higher rate of boring. But for any kind of tunneling, an accurate and feasible method of survey must be accomplished during mining. That means, that for each tunnel project the most efficient survey method must be devised. The method to be defined depends on the spatial curve, diameter, speed of tunneling method, purpose of tunnel and contractual accuracy.

The final accuracy of a finished tunnel is heavily dependent upon three critical elements:

1. Quality of the above ground survey
2. Tunnel traverse and directional transfer from above ground
3. Guidance for the mining job.

Combining The Survey And The Guidance System

Since the guidance system (GS) is one of the critical factors in the determination of the tunnel final accuracy, it is necessary to analyze how the directional control and tunnel traverse measurements can be effectively integrated into a GS. Any guidance system, whether manual or high technology, must continually detect and compensate for movement of its theodolite or laser station standpoint. Movement beyond a preset allowable amount must generate a warning or error message to the user. Additionally, optical refraction can interfere with the results.

Refraction will have an effect on any kind of optical GS. Therefore, careful selection of target and standpoint locations can minimize, but not completely eliminate, the influence of refraction. Some GS technologies, such as Gyro-based systems, are immune to the effects of refraction, but have other, methodical errors which will be described later in this paper.

GUIDANCE

Overview

Why is a guidance system necessary? A GS, whether manual or automatic, compares the design direction and slope of a tunnel with the current position of the mining

maparameters can be derived:

1. 3-D projective curves
2. stroke of push rams or profiling information
3. adjustment of cutterhead or cutting arm
4. waterproof tapered lining.

These parameters are subsequently used to adjust thrust balance, machine articulation, etc., to steer the machine to, or maintain, the proper heading. This steering may be accomplished by providing the necessary parameters to a human machine operator or by automatic on-line techniques and used in a closed-loop autopilot system. This shows that a GS can detect an imperfect drive in real time and show this immediately, but it cannot influence the drive directly.

Criteria Of Influence

From the beginning, the owner of a tunnel should develop the technique of guidance based on the general requirements of the anticipated mining job. After the overall 3-D-accuracy of the tunnel is defined, the following questions should be asked:

1. Is the guidance system self checking
2. what redundant measurements are required
3. is the accuracy of each sensor or measuring device commensurate with the overall requirements
4. does the GS show absolute values
5. in which time frame are results shown
6. how is the information shown to the machine driver
7. is a database available and how can one maximize the utility of information available
8. is the GS error tolerance well within the overall error budget?

Automated Guidance Advantages

The advent of high technology to the automated guidance of boring machines brings many advantages to owners and tunnel contractors. These advantages, coupled with daily operational cost savings, usually more than offset the higher cost of an automatic system over a manual system. The prominent advantages of automatic measuring and guidance systems are as follows:

- Maintaining highest mined accuracy to the design centerline and slope at optimum drive speeds. This includes minimization of wandering and porpoising
- Reduction of designed overcut because of minimized fluctuations in the mined centerline.
- providing a pre-calculated and displayed optimized mining path to return to centerline when unforeseen events create machine positions and orientations outside the norms.
- Real-time, on-line measurement of critical tunneling data and creation and maintenance of a comprehensive database for this data.

- Provides the means for comprehensive current or post analysis of all phases of tunneling status
- Provides the capability of forwarding measured parameters to autopiloting systems.
- Provides the means to optimize the selection and orientation of tapered tunnel liner segments on an as-mined basis

Common Guidance System Characteristics

All types of automatic guidance systems have several features in common:

Stored Alignment: The design alignment for the entire job is usually stored in memory, with current measured position and attitude continually compared to this alignment.

Measurement Sensors: A complete range of sensors measures the machine location and orientation. These sensors are usually interfaced directly to the system computer.

Driver's Display: An electronic display, either digital or graphic, is provided so that the driver can readily interpret the current machine position and orientation and initiate proper guidance procedures.

Types Of Automatic Guidance Systems

Laser Based: The well known ZED-system uses a laser which is detected by an active, two-plane target mounted on the TBM. The active target detects the entry angle and position of the laser beam, from which machine location and orientation is calculated. The laser beam is generated from either: (1) a fixed generator on a wall-mounted laser station, or; (2) through the eye-piece and lens of a wall-mounted servo theodolite, which can follow the TBM-mounted target. The TBM manufacturer Herrenknecht, from Germany, now also offers such a system.

Gyro Based: To eliminate the influence of refraction, some GS manufacturers incorporate gyroscopes mounted on the TBM as the primary sensor of machine position and orientation. Guidance systems from Tokimec and BGT use this concept.

North seeking gyros intrinsically measure the direction of its fixed axis against NORTH. There is no intrinsic tie-in to the overall survey or the tunnel traverse. Since the calibration of gyros drifts from day to day, and because the real as-mined chainage is not directly measured, a backup survey must be performed after each shift to provide re-calibration information for the gyro so that the system results stay within an acceptable accuracy.

Theodolite Based With Automatic Target Seeking: State of the art systems are now using servo theodolites which incorporate on-board self-seeking target acquisition capabilities (Figure 1). The targets are normal EDM prisms which are inexpensive and very robust. This paper will focus on the configuration, function and application of a *theodolite based system with automatic target seeking* (TATS). The manufacturer in this example is Dyckerhoff & Widmann (DYWIDAG).

CONFIGURATION: The primary components of a TATS system are: (1) a Leica

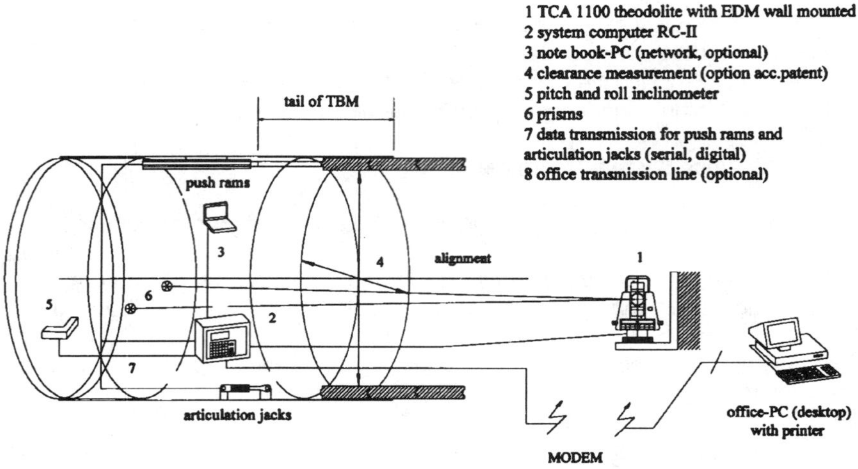

Figure 1. TBM Guidance System (TATS)

TCA1100 motorized theodolite, with Automatic Target Recognition (ATR), mounted on the wall of the most recently mined tunnel; (2) a dual axis inclinometer to directly measure TBM pitch and roll; (3) two or more TBM-mounted EDM prisms which function as primary targets; (4) additional EDM prisms mounted at known locations in the just-finished drive to function as reference points, and; (5) a system computer with proprietary software to orchestrate the measurements, calculate and display the results and maintain a comprehensive historical database. All components are robust and designed for operation in harsh environments.

THE SURVEY WINDOW: The survey window on a TBM mining machine is the sight path from the theodolite station to the TBM targets. Obviously, this path must be unobstructed for the theodolite to successfully measure to the targets. Since this path must run reasonably parallel with the machine axis, the many physical demands of the TBM and its backup gear constantly try to "steal" space from the survey window. But conventional EDM prisms require a minimum radial spacing to ensure that both prisms are not simultaneously "seen" by the theodolite detection system, a condition which could create erroneous results. The spacing demands of the machine-mounted EDM prisms in a TATS system are in direct opposition to the requirements to minimize survey window size.

THE SHUTTERED PRISM: To accommodate these conflicting demands, a "shuttered" prism system has been developed by DYWIDAG to allow a survey window of the smallest possible dimensions. A shuttered prism utilizes an electrically controlled mechanical actuator to cover and uncover (shutter) each prism upon command of the system computer. Since two prisms are used, the prisms are driven from "fail safe" circuitry which ensures that, at any time, only the prism to which the theodolite is pointed will be uncovered. This mutually exclusive setup ensures correct operation at initial startup and even after a power failure.

OPERATIONAL CONCEPT: The TCA1100 theodolite seeks and measures the primary target prisms on a frequent basis, at a rate set by the system user. A two-axis precision inclinometer, mounted on the TBM measures the machine pitch and roll. The theodolite and inclinometer measurements are combined by the system computer to yield a direct measurement of tunnel station at any time, in addition to providing a very accurate corresponding picture of the position and orientation of the TBM relative to the design alignment. This information is presented in a quick-view graphical form, as well as in a concise full information format, to the machine operator for guidance purposes.

OPTIMIZED PROJECTED PATH: An optimized projected path curve is also presented on the graphical display. This curve considers job-specific parameters, such as minimum mining radius capability, and shows the operator the optimum path back to the alignment or proper orientation. This capability minimizes the likelihood of the occurrence of serious problems, such as impaction of the machine, due to aggressive oversteer.

SYSTEM ACCURACY: When measuring to two primary targets, the system accuracy is within 3 mm/100m, or, 0.04 in./100 ft.

SYSTEM DISPLAY: The system display is a computer monitor presenting a comprehensive, yet concise, graphical display combined with traditional digital information. The display utilizes a central "bullseye" to provide "at-a-glance" guidance capability, and yet provides simultaneous plan and profile views of the alignment, together with location and orientation of the machine relative to that alignment. Additionally, a green/yellow/red changing color scheme is used to automatically provide early warning to the operator if the TBM exceeds job-specific, preset limits for excessive: (1) horizontal or vertical lead, or; (2) offset from the alignment.

SYSTEM SELF-CHECKS: The stability of the theodolite station is continually self-monitored by the system. The motorized TCA1100 theodolite periodically seeks and measures to backsight prisms located at known points on the just-completed tunnel wall. If the readings to these stable backsights indicate that movement of the station point has occurred, an error message is immediately displayed on the operator's screen. Other system self-checks are continuously performed in the background, and error messages presented if necessary.

RING CALCULATION FUNCTION: The integrity and stress point management of segmented liner joints is maximized through the use of ring calculation software integrated into the TATS system. This software calculates an optimized allowable predictive path, considering: (1) overall current TBM position and orientation information; (2) measured clearance information, and; (3) geometry of the tapered rings

MISCELLANEOUS: The TATS guidance system also provides:

- Comprehensive Database — An historical database of each measured point is maintained in the system computer. This database may be readily accessed to provide a printed graphic of any section or a display of the history of any section. This database may also be exported for transfer to a remote computer.

- Above-ground On-line Display — The DYWIDAG TATS system has the capability to provide an above-ground, on-line monitor which provides the same display as seen by the TBM operator. Additionally, the full database and parameters menus may be accessed from this above-ground computer.

APPLICATIONS

Fully Automated Machine Guidance

Since the TATS system is fully automatic, requiring no personnel attention for operation, it only becomes natural to consider combining the TATS system with TBM's which may be controlled remotely, providing a TBM which operates in a fully "closed loop", or manless, mode. Technology may be further applied to provide a control loop which is *adaptive* in response to certain variables such as soil conditions, TBM handling and tunnel curvature.

Examples of such closed loop systems are the HyperPilot system from HyperCube (Seymour, Harlan R.) and the CAP system from France. These systems contain all the elements to interface to various TBM control points to enable full machine control. Additionally, HyperPilot contains an "expert" system which is adaptive in nature, monitoring the TBM historical response to various commands and drive conditions to develop, or "fine tune", the optimum operating envelope for the machine and its TATS/HyperPilot control system. Because sensors are read every 5-10 seconds, response of the TBM to ram thrust commands are instantly known, and the full envelope is developed, and monitored in an iterative manner. This process yields a system which can automatically respond to changing situations much faster than an "open loop" system using human interaction as the loop-closing element.

The adaptive control loop integrates two different software methodologies, namely algorithmic for the analytical system, and heuristic for the expert system. The analytical system uses statistics and control theory to actively control the machine. The expert system monitors the progress of the dig, massaging the output of the analytical system, if necessary. This is reactive control. The two systems are merged in order to compute the best and safest steering commands.

This autopiloting system has provided significant cost savings to its customers. The TBM operator is relieved of his non-stop control task. This allows work to progress faster, since the operator can do other tasks in the tunnel, or leave the tunnel altogether, which means that there is one less human at risk if there is an accident.

The software has proven to be very accurate. During straight line or gently curved sections it typically keeps the TBM within a couple of mm of the planned path. At the beginning of tight curves (such as those cut by articulated TBM's) an initial error in the 20 mm range may accumulate, but the error typically subsides to less then 5 mm during the last 80 % of the curve.

This software completely and reliably automates the steering of a TBM by reading sensor inputs from the tunnel and writing steering commands back to the TBM. This type of control system using the HyperPilot has been used extensively in over forty (40) tunnels in Japan, resulting in minimization of manpower and high accuracy of bore.

The TATS system and the HyperPilot system are frequently checking themselves against

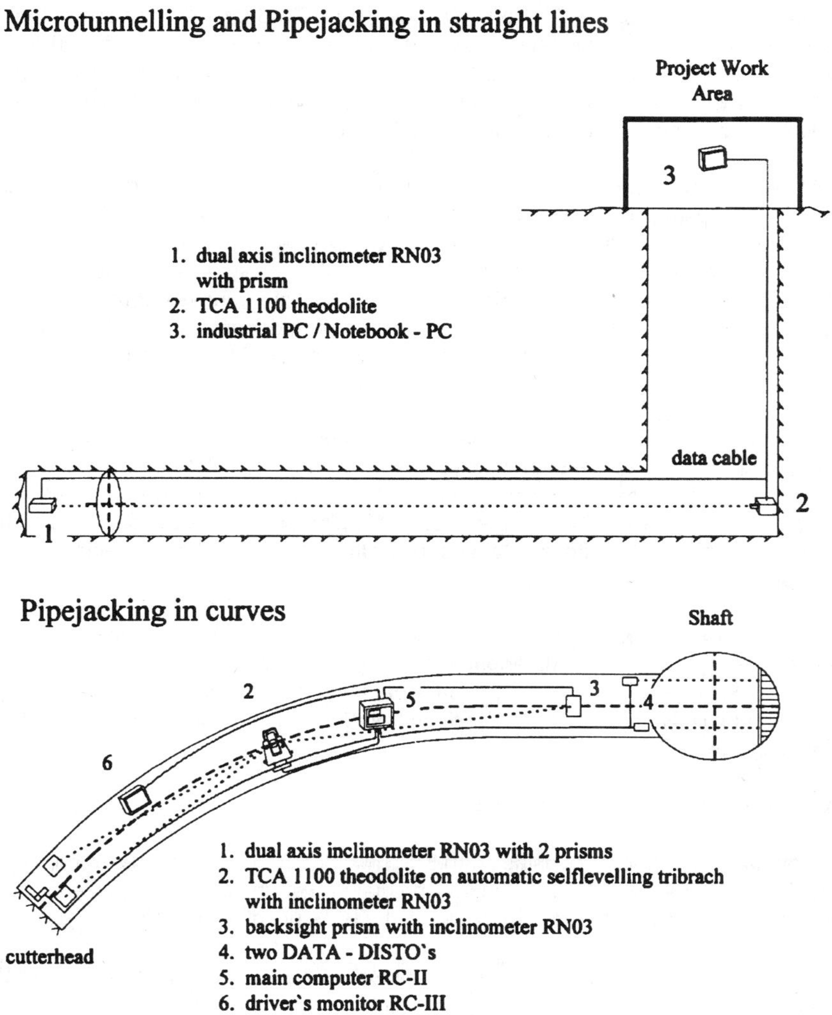

Figure 2. Leica-DYWIDAG-Guidance System

each other. This results in

1. reduction in operating personal costs
2. extensive guidance and educational assistance to machine drivers
3. reduction in anticipated overcut because of minimized wandering.

Pipejacking Along A Curved Alignment

From the TATS system has evolved a unique system for the automatic guidance of pipejacking machines along a 3-D curved alignment (TATS:3-D), or for straight alignments longer than about 400m. Uniquely, all the system components are located and mounted within the driven pipe and move along the design alignment as the installation progresses (Figure 2). The system uses the knowledge of the driven path like an "expert" system, applying historical performance to build the future operation.

At the beginning of the bore/shove, the theodolite station is located at the launching shaft where the system 'learns' where the pipe is pushed during the startup phase. In that early stage it is used exactly as a conventional TATS system. After about 50 m, the theodolite is installed in the most-recently shoved pipe on a self-leveling base together with a 2-axis inclinometer. The theodolite, now an integral part of the pipe, moves continuously as the new piping is jacked. This "sliding" theodolite now repeatedly measures the angle to the TBM-prisms from its moving station, now using its original, fixed station standpoint as a backsight.

Since the theodolite directly measures the distance from the theodolite to the backsight prism in the shaft, the exact 3-D-theodolite position can be calculated from the previously measured and stored positions of the TBM during the bore. Since the pipes are always following the bore cut by the TBM, the position of the pipe and its "passenger" theodolite are very accurately known at all times. Consequently, the 3-D position of the measured prisms on the TBM may always be calculated.

After mining another 50 m or so, the backsight-prism is now also mounted to a position inside the latest pipe, together with another inclinometer to continuously monitor the on-line roll at the backsight station. Now, the entire measuring apparatus moves with the pipe, measuring and calculating new TBM positions from formerly 'learned' ones. Installed systems have shown an accuracy of 2 cm after 600 m.

Guidance of Microtunneling In Hardrock

Pilot research tunnels are using MiniTBM's (MTBM) for geological research. The size of these TBM's are between 1 - 2 m and may be manned or unmanned. A guidance system has been developed by DYWIDAG for such small TBM's used in rock (Figure 3).

Large Bores: In diameters of up to two meters, a system based on the TATS3-D GS can be used. However, instead of having the equipment fixed in a pipe, the theodolite and backsight target are now mounted on the MTBM trailing gear, whose axis position within the bore may vary as the bore progresses. Consequently, the current position of the theodolite and backsight target must be repeatedly determined for the GS calculations to be meaningful. This is accomplished by measuring two cross-section profiles of the bore, one at a position close to the station of the theodolite and the other close to the backsight. This gives the position of the theodolite axis, and backsight target, inside the well-known location of the bored tunnel.

Each bore profile is defined with an array of small EDM's which measure to surfaces without the need of prisms. Three or four EDM's are utilized at each plane. The measurements of the EDM's radially to the rock define the radial location of the trailing gear axis at any time, relative to the axis of the bored tunnel. Because the position of each EDM is known and fixed relative to the axis of the trailing gear, the absolute radial position of the

Curved Alignment Microtunneling With Hardrock MTBM

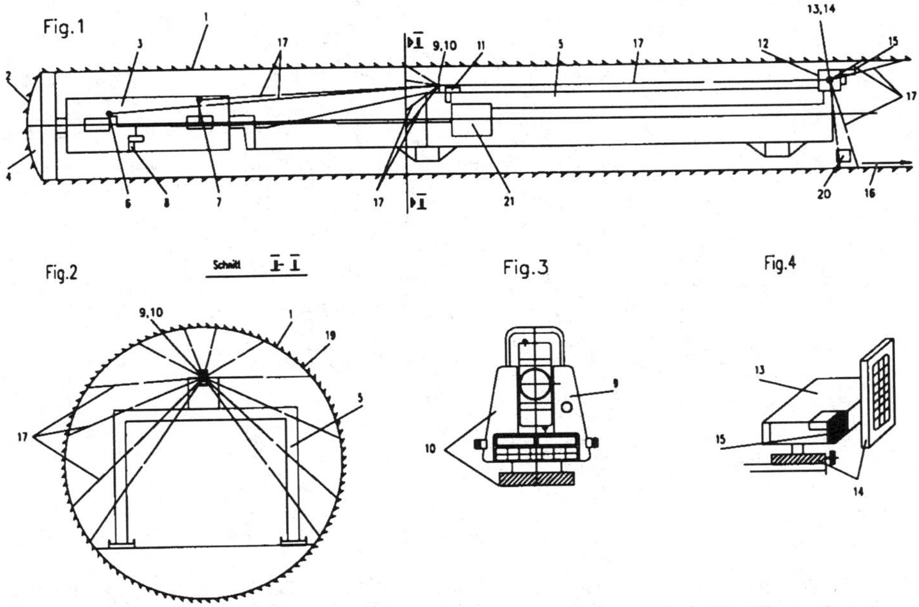

3=TBM main beam; 6/7=TBM Prisms; 8/11/15=Inclinometers; 9/10=Theodolite and Profiling equipment; 12/13/14=Backsight Prism and profiling equipment; 16=Chainage measurement;

Figure 3

theodolite and backsights may now be easily calculated using standard trigonometric procedures. Consequently, the location of the cutterhead, calculated from measurements to the targets on the MTBM, is fully known in absolute terms.

Small Bores: For smaller bores, when the height of the theodolite, and its self-leveling base, is too large to be accommodated, the system is modified to obtain cutterhead position by the measurement of profiles only. Assuming that the main beam of the MTBM is rigid, two measured cross section profiles of the main beam will define the location of its axis within the bored tunnel. Since the relationship of the cutterhead and the main beam axes are known and fixed, the location of the cutterhead axis can be readily calculated by 3-D extrapolation. The profiles are measured with EDM's as previously discussed (Figures 4 & 5).

Closing The Loop: In both cases, the as-mined axis is appended to the database. It is a logical step to apply these data to the "expert" system to provide closed loop control of the machine. Research trials are currently underway for testing of this fully automatic control.

Unmanned Microtunneling

Straight Drives: Micromachines from 200 mm diam. up to 1600 mm are normally driven in straight lines for distances of up to 300 m. from a launching shaft. For that purpose

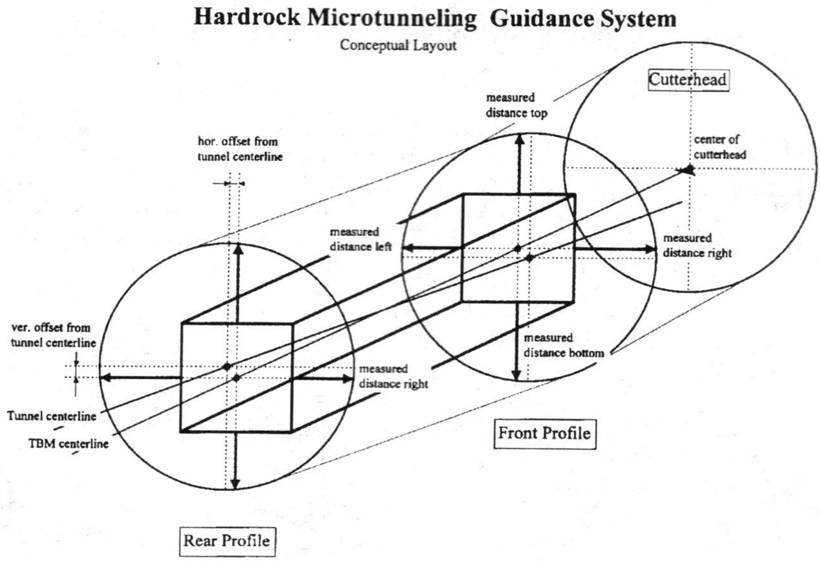

Figure 4

a TATS system can be used easily. If a cutterhead prism is mounted near the axis inside the machine, no exit cables (video or data cable) of any kind are necessary because an inclinometer is not required. The automatic target recognition feature of the TCA1100 theodolite provides all the information which is necessary for microtunneling. The horizontal and vertical lead is calculated from the axis direction of the most recently driven 30 to 50 cm of tunnel, which is continually updated with the newest values. It is simple and easy to use.

Curved Drives: The newest application of technology is to enable guiding micromachines through curved alignments. In this scenario, a video based automatic 3-D-traverse measurement is used which incorporates fixed CCD arrays inside the pipe (Figure 5). This system uses the same "learning method" as the TATS:3-D system described previously. A research system is currently undergoing installation with testing to begin immediately. A curve of virtually any radius can be driven if there is room enough for the sensors inside the pipe. This Video-Traverse-System (VTS) technology is used as a quality analysis system in other industries and now becoming available to microtunneling.

Guidance Systems For Road Headers

Guidance of road headers (RH) is a natural extension of the philosophy involved in the guidance of TBM's and MTBM's. The fundamental technology used is that incorporated by the TATS system.

Utilization begins with initialization of the system. i.e., obtaining the current position, direction, and pitch and roll of the road header chassis. This is accomplished in the general nature of a TATS system as used for a TBM GS. However, because of the capability for faster movements and lateral motion of the machine, a radio link for data transfer is installed

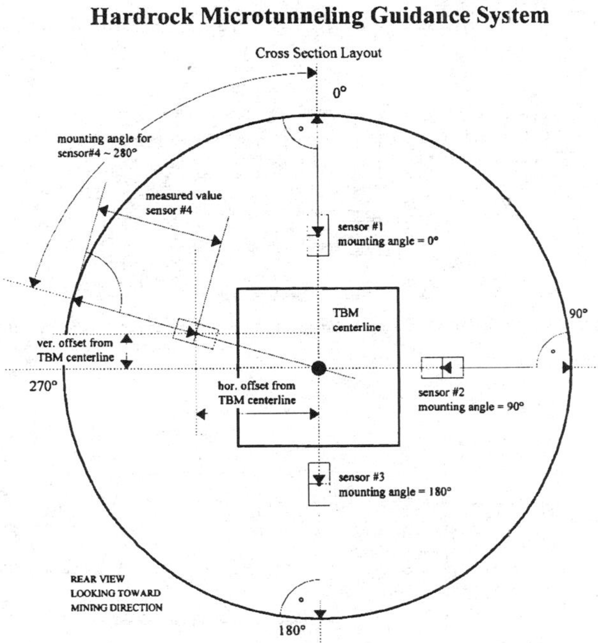

Figure 5

between the system computer, which is mounted on the machine, and the theodolite station, thus eliminating the unwieldy interconnecting cable (Figure 6).

The designed tunnel alignment, together with the design profiles, have been previously loaded into the onboard system computer, which is hardwired to a data display console situated in front of the RH operator. An image of the desired tunnel profile at the current chainage point is displayed for the operator. Electronically superimposed upon this display, with the proper scale, is an icon which represents the real-time two-dimensional position of the RH cutterhead. This icon moves immediately (with less than 1/2-second lag time) to follow the movements of the cutterhead, as commanded by the RH operator. Consequently, the RH operator "sees" at all times the position of his cutterhead relative to the outline of the desired tunnel profile, thus ensuring that an accurate profile will be cut at all stations of the tunnel.

The accuracy of the profile as mined by the RH operator is further enhanced by the use of changeable colors on the display (green, yellow and red) to "fine tune" the location of the cutterhead as it approaches the boundary of the design profile. Green, seen for the majority of the mining within the profile, means that the cutterhead is inside the profile and not closer than 10 cm (4") to the design boundary of the profile. When the cutterhead intrudes into this 10 cm internal warning zone, the color switches to yellow, and the calculated offset of the tangent line to the profile is also shown on the operator's screen. If the cutterhead moves outside the profile design boundary, the color switches to red and the calculated offset is shown there as well.

Guidance System for Road Headers

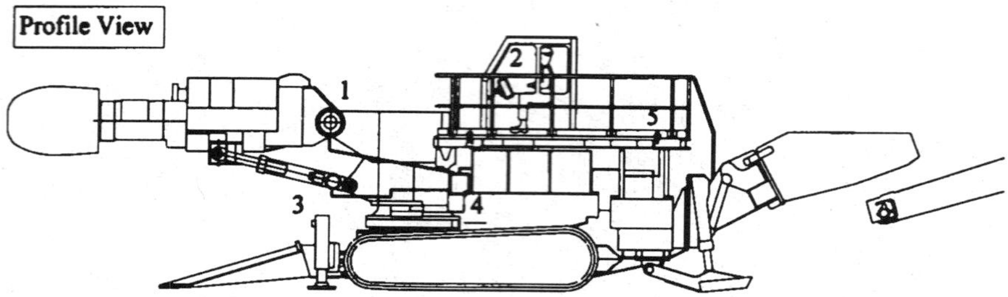

1:incremental sensor, 2:screen, 3:elongation sensor, 4:databox, 5:prisms, 6:inclinomter, 7:incremental sensor, 8:radio link

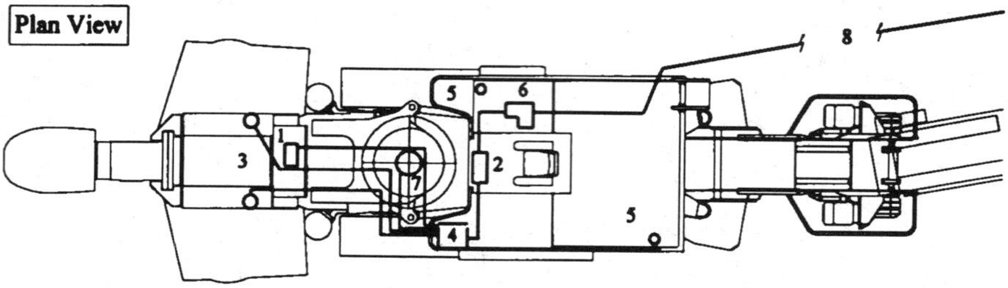

1:incremental sensor, 2:screen, 3:elongation sensor, 4:databox, 5:prisms, 6:inclinomter, 7:incremental sensor, 8:radio link

Figure 6

The combination of (1) overlaying of the cutterhead dynamic position onto the pertinent design profile, and; (2) using the color operator alert system, results in the following advantages for the mining company:

- Elimination of post profiling
- Increased accuracy and uniformity of as-cut profiles
- Minimization of amount of shotcrete required
- Higher speed during cutting

Further productivity gains may be achieved if an SPS is inserted in the cutterhead control loop, and the color change points are used to semi-automatically control the movements of the cutterhead. Cutting speed is reduced if the color switches to yellow and will be disabled, unless overridden, if in a red area.

This type of road header GS can also be established on drilling jumbos or excavators in tunnels.

CONCLUSION

It is shown that technology permits the development of modular guidance system components for all types of slow moving mining applications, including microtunneling straight or curved lines, pipe jacking in 3-D alignments, TBM's of all sizes, EPB and slurry shield machines and road headers and drilling jumbos. These modules automate and simplify the current manual application of procedures from the fields of surveying, civil engineering, and mechanical engineering in the mining of tunnels. Rapid evolution in the fields of electronics and computers has allowed development of a truly "universal" guidance system.

The future of guidance system technology is here now. The application of this technology:

1. Allows machine operators to increase their focus on shared duties
2. Provides on-line feedback to project management as to overall job accuracy andprogress
3. Improves accuracy and consistency of cut
4. Provides cost savings in areas of personnel, procedures and material required

BIBLIOGRAPHY

Seymour, Harlan R., Technical Document, "An Automated TBM Steering Program", Hypercube, Inc., Los Angeles, California, USA

Viadric, Tehnical Documentation, Wente & Thiedig

6

TBM Case Histories II

Chairman
M. Gay
Kiewit Construction Co.
Omaha, NE

Chapter 17

TARP - COMPARATIVE STUDY OF THE MINING AND CONCRETE LINING OPERATIONS FOR TUNNELS WITH DIAMETERS OVER 9 M (30 FT)

Luminita Calin and Joseph Fernandez
Parsons Brinckerhoff Quade and Douglas, Inc., Chicago, Illinois

INTRODUCTION

The Tunnel and Reservoir Plan (TARP), commonly known as the Deep Tunnel Project, is the daring and unique concept developed by the Metropolitan Water Reclamation District of Greater Chicago (MWRDGC) as a two-phase solution to the problem of waterway pollution and basement flooding due to sewer system backups created by the combined sewer overflows (CSOs).

Phase I, primarily for pollution control, includes the tunnels, shafts, connecting sewers and structures. These Phase I facilities intercept, during wet weather events, the CSOs within the MWRDGC TARP service area. The tunnels and shafts provide temporary storage of the CSOs for subsequent treatment at Water Reclamation Plants (WRP) before discharging into waterways. Ultimately, upon completion of Phase II, the tunnels, located from 45 m (150 ft) to 100 m (330 ft) below ground, will convey the CSOs to reservoirs for storage until WRP capacity is available.

Phase II, which includes the reservoirs, pumping facilities, additional tunnels and shafts, has been planned to serve a dual purpose. The reservoirs provide storage of excess flows during wet weather events, those beyond the WRP capacity. Following the wet weather events, the stored flows can be conveyed, at controlled rates, to the WRPs for full treatment. These excess flows previously would have resulted in both overflows into the waterways and eventually backups in the local sewer systems. The storage capacity of the reservoirs also will promote the full utilization of the conveyance capacity of the tunnels and shafts, which in turn results in optimizing the conveyance capacity of the existing sewer systems. Therefore pollution control is further enhanced by allowing for capture of a greater volume of CSOs and greatly minimizing backups in the local sewer systems.

The TARP project consists of four systems: O'Hare, Mainstream, Des Plaines, and Calumet, covering a total area of 971 km^2 (375 mi^2) and serving the City of Chicago and 51 neighboring communities as shown in Figure 1.

With only two contracts left before completion of the Phase I tunnels, the 12.7 km (7.9 mi) Torrence Avenue leg, to be awarded in 1997, and the 12.4 km (7.7 mi) Little Calumet leg unfunded as of yet, both part of the Calumet System, the pollution control phase of the project is now 86 % complete (Table 1).

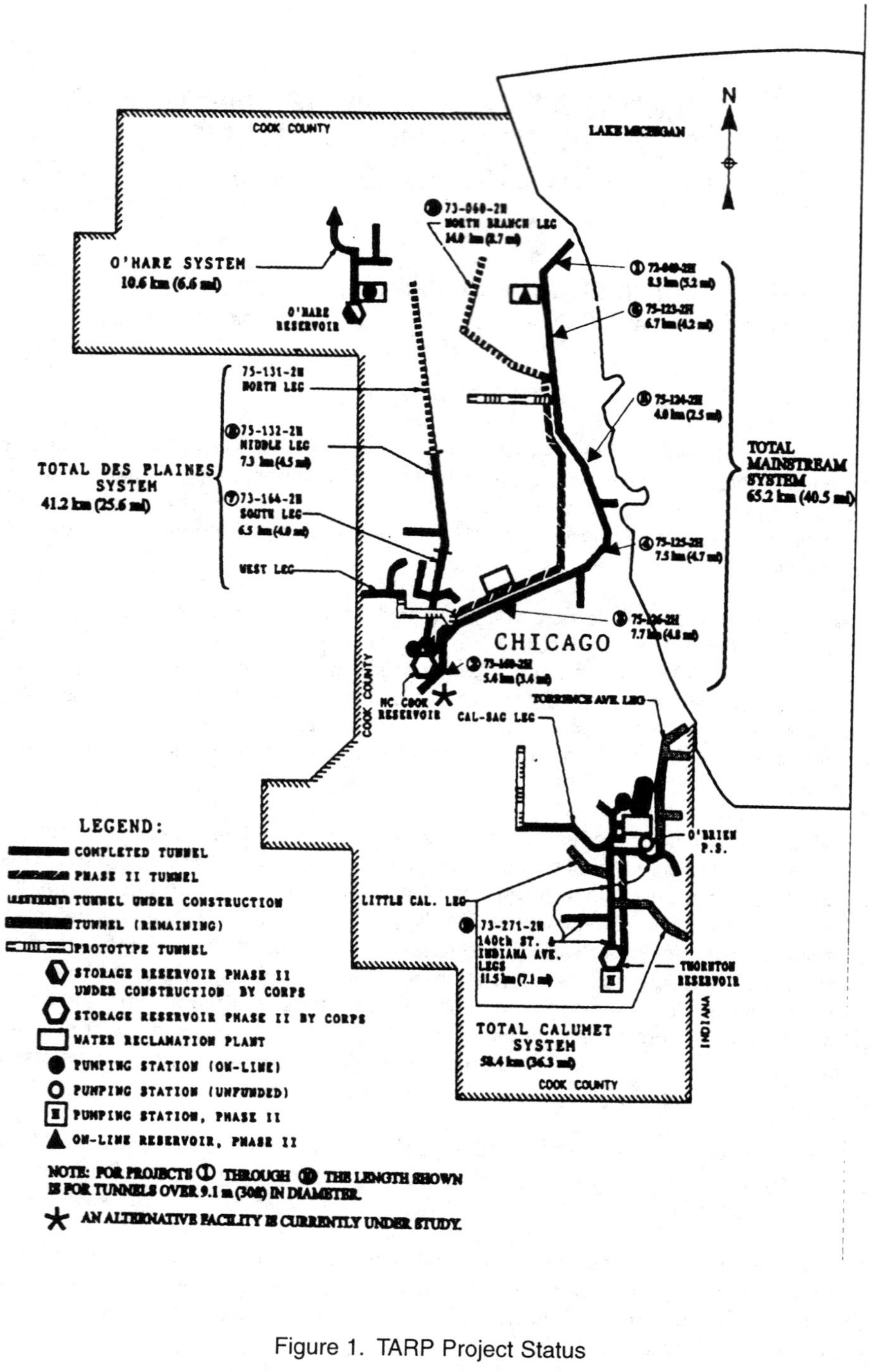

Figure 1. TARP Project Status

Table 1. TARP Phase 1 Status

System	Length km (mi)	Completed km (mi)	Operational km (mi)
Mainstream	65.2 (40.5)	65.2* (40.5)	50.2 (31.2)
Calumet	58.4 (36.3)	33.3 (20.7)	33.3 (20.7)
O'Hare	10.6 (6.6)	10.6 (6.6)	10.6 (6.6)
Des Plaines	41.2 (25.6)	41.2* (25.6)	27.2 (16.9)
TOTAL	175.4 (109.0)	150.3 (93.4)	121.3 (75.4)

Note: * The North Branch of the Mainstream System and The North Leg of the Des Plaines System are currently under construction, but mining is completed.

This paper will describe the TARP large tunnels over 9 m (30 ft) in diameter, which total 82 km (51 mi) out of the 150 km (93 mi) completed to date. These are as follows:

Mainstream System - Addison St. to Wilmette — awarded in 1975
① Contract 72-049-2H

Mainstream System - 59th St. to Addison St. — awarded during 1977-1978
② Contract 73-160-2H
③ Contract 75-126-2H

④ Contract 75-125-2H
⑤ Contract 75-124-2H
⑥ Contract 75-123-2H

Des Plaines System - 59th St. to Fullerton Ave. — awarded in 1988
⑦ Contract 73-164-2H
⑧ Contract 75-132-2H

Calumet System - 140th St. and Indiana Ave. legs — awarded in 1990
⑨ Contract 73-271-2H

Mainstream System - North Branch — awarded in 1992
⑩ Contract 73-060-2H

GEOLOGY OF THE CHICAGO TARP AREA

The host rock for the TARP is the Silurian dolomite which constitutes the bedrock throughout the area. The generalized stratigraphy of the Chicago TARP area is presented in Figure 2. The tunnels are approximately 45 m (150 ft) to 100 m (330 ft) below existing grade elevation. The top of rock projects about 3 m (10 ft) to 30 m (100 ft) below ground surface. Glacial drift and lake bed deposits overlay the bedrock. The strata dip at a low angle in an easterly direction. Very few exceptions occur within localized gentle folds superimposed on the regional dip. Glacial erosion has truncated the top of rock with the result that the total thickness of the bedrock diminishes from east to west.

SYSTEM	SERIES OR GROUP	FORMATION	MEMBER	THICKNESS METER (FEET)		DESCRIPTION
				MAX.	MIN.	
QUATERNARY	Pleistocene			40 (130)	3 (10)	Overburden: glacial drift
						MAJOR UNCONFORMITY
SILURIAN	Niagaran	Racine		110 (360)	0	Dolomite, light gray. Reef facies: crystalline, massive and porous. Interreef facies: argillaceous, thin bedded, with scattered porous chert nodules and thin chert beds.
		Sugar Run		6 (20)	0	Dolomite, light gray, argillaceous, dense, occasional chert nodules.
		Joliet	Romeo	7 (22)	5 (17)	Dolomite, light gray, fine grained, crystalline, dense.
			Markgaf	10 (32)	6 (20)	Dolomite, light gray to bluish gray, argillaceous, fine grained, some chert, some thin greenish gray shale partings.
			Brandon Bridge	7 (22)	0	Dolomite, pinkish gray, fine grained, red and green shale partings.
	Alexandrian	Kankakee		24 (79)	6 (20)	Dolomite, gray to pink, medium grained, scattered chert nodules, prominent green shale partings.
		Edgewood		33 (107)	6 (20)	Dolomite, cherty, light gray to gray, fine to medium grained. Becomes darker in color and more silty and shaly with depth, especially where thick.
						MAJOR UNCONFORMITY
ORDOVICIAN	Maquoketa	Neda		5 (15)	0	Shale, red, hematitic, oolitic.
		Brainard		40 (132)	0	Shale, greenish gray, medium soft, thin interbeds of light gray dolomite. Slakes.

Figure 2. Generalized Stratigraphy of the Chicago TARP Area

The Silurian dolomite has very good tunneling properties. The dolomite is slightly argillaceous, homogeneous and very strong. The unconfined compressive strength varies between 124 MPa (18,000 psi) and 234 MPa (34,000 psi). Bedding planes are generally not very pronounced. Two orthogonal joint systems, a northeast and a northwest set, prevail but the joints are widely spaced. In the exploratory borings, which were spaced about 600 m (2,000 ft) apart along the tunnel alignment, the mean RQD values varied between 91 and 98, showing excellent rock quality. The underground openings are generally self-supporting. Groundwater inflow is mainly through joints and bedding planes. Inflow rate is quite variable. Occasional shear zones contributed inflows in the order of 60 l/sec (1,000 gpm) over 150 m (500 ft) of tunnel, but the average inflow is 0.3 to 0.6 l/sec (5 to 10 gpm) per 150 m (500 ft) of tunnel.

Underlying the Silurian dolomite is the Ordovician shale which was to be avoided to the maximum extent possible. However, almost 3 km (1.9 mi) of the Mainstream tunnels were mined through shale. On exposure to air, the shale commonly deteriorates into disk-like fragments and irregularly shaped fragments because of air slaking and stress relief. When wetted, the shale becomes soft and plastic. The unconfined compressive strength varies between 1 MPa (145 psi) and 28 MPa (4,000 psi).

PROJECT GEOLOGY AND MINING OPERATIONS

A summary of the Tunnel Boring Machine's (TBM) performance on all ten projects is presented in Table 2. The average progress rate data are based on eight-hour-shifts, three-shift-days and five-day-weeks. Details on the TBMs' specifications and the mucking system employed by each Contractor are presented in Table 3.

Following is a short narrative on the geology, and highlights on the mining for each

Table 2. Summary of TBMS' Performance

Contract	Contractor	Award Year	Tunnel Length km (mi)	Tunnel Diameter m (ft-in)	Shift Avg. m (ft)	Shift Best m (ft)	Day Avg. m (ft)	Day Best m (ft)	Week Avg. m (ft)	Week Best m (ft)	Month Avg. m (ft)	Month Best m (ft)
72-049-2H	Kenny-Paschen-S&M	1975	8.3 (5.2)	9.17 (30-1)	5.7 (18.7)	18.3 (60.0)	16.8 (55.2)	51 (167)	82 (270)	194 (637)	333 (1,091)	750 (2,461)
73-160-2H	M/K-Kenny-Paschen-S&M	1977	5.4 (3.4)	10.77 (35-4)	6.6 (21.6)	14.9 (49.0)	18.6 (61.0)	36 (118)	89 (291)	144 (472)	361 (1,183)	551 (1,809)
75-126-2H	Heally-Ball-Horn	1977	7.7 (4.8)	10.74 (35-3)	5.1 (16.8)	13.5 (44.3)	15.4 (50.6)	37 (120)	77 (253)	161 (529)	309 (1,014)	545 (1,789)
75-125-2H	Paschen-M/K-Kenny	1977	7.5 (4.7)	9.83 (32-3)	5.1 (16.8)	17.8 (58.5)	14.6 (47.9)	45 (146)	84 (274)	186 (611)	327 (1,074)	686 (2,252)
75-124-2H	Shea-Kiewit-Shea	1978	4.0 (2.5)	9.86 (32-4)	6.1 (20.1)	19.2 (63.0)	17.8 (58.3)	44 (143)	85 (278)	165 (542)	332 (1,088)	573 (1,881)
75-123-2H	Ball-Healy-Horn	1978	6.7 (4.2)	9.86 (32-4)	7.3 (24.1)	15.8 (52.0)	19.8 (65.1)	41 (136)	108 (354)	183 (600)	418 (1,370)	617 (2,024)
73-164-2H	M/K-Paschen	1988	6.5 (4.0)	10.77 (35-4)	6.4 (20.9)	26.8 (88.0)	17.7 (58.0)	49 (160)	91 (298)	193 (634)	380 (1,246)	686 (2,251)
75-132-2H	Impregilo-Ebasco-Losinger	1988	7.3 (4.6)	9.83 (32-3)	8.6 (28.1)	26.5 (87.0)	23.9 (78.3)	63 (207)	128 (421)	276 (905)	488 (1,602)	753 (2,472)
73-271-2H	Kenny-Kiewit-Shea	1990	11.5 (7.1)	9.86 (32-4)	11.6 (38.1)	29.3 (96.0)	34.7 (113.7)	72 (236)	182 (596)	324 (1,064)	715 (2,346)	1,093 (3,586)
73-060-2H	Perini-ICA-O&G	1992	14.0 (8.7)	9.83 (32-3)	7.6 (24.9)	24.1 (79.0)	21.7 (71.3)	59 (193)	110 (360)	278 (911)	465 (1,524)	807 (2,648)

project.

- Mainstream System
  ① Contract 72-049-2H

The Addison to Howard segment, part of the Addison to Wilmette contract comprises the northernmost segment of the Mainstream System. Awarded in 1975, this was the first TARP large diameter tunnel. It was excavated entirely through Silurian dolomite and the formations penetrated include the Racine, Joliet and Kankakee formations.

The tunnel intercepted a total of fifteen crevices filled with varved clay. Some of these interceptions were full-face exposures and the average exposed width in the tunnel was 8 m (26 ft). At one time, when intercepting a full-face exposure, the clay filling squeezed around the TBM cutterhead and the Contractor had to hand-excavate in front of the machine in order to free the cutterhead. Considerable temporary support was needed to stabilize the tunnel arch. The support consisted of steel ribs, wood lagging and application of shotcrete. The presence of the clay-filled crevices provided a major source of delays.

The Contractor completed the excavation of this 8.3 km (5.2 mi) tunnel in two years, from August of 1977 to August of 1979.

- Mainstream System
  ② Contract 73-160-2H

The geology of this contract was noteworthy because of the wide spacing between joints and the minimal groundwater infiltration which was estimated at about 9 l/sec (150 gpm) before grouting. Stability problems were minimal, with very few overbreaks. The tunnel penetrated the Joliet, Kankakee and Edgewood formations.

The Contractor completed the excavation of this 5.7 km (3.5 mi) tunnel in one year,

Table 3. Comparison of Significant TBM Specifications Data and Mucking System

Project	Award Year	TBM Manufacturer	Diameter m (ft-in)	Power KW (HP)	Thrust KN (lbs)	Speed rpm	Mucking System
72-049-2H	1975	Jarva	9.17 (30-1)	1,790 (2,400)	13,344 (3,000,000)	4	Conventional rail muck cars and inclined conveyor
73-160-2H	1977	Robbins	10.77 (35-4)	1,790 (2,400)	12,454 (2,800,000)	3.6	Conventional rail muck cars and hoisting
75-126-2H	1977	Robbins	10.74 (35-3)	1,790 (2,400)	12,454 (2,800,000)	3.6	Conventional rail muck cars and hoisting
75-125-2H	1977	Jarva	9.83 (32-3)	1,790 (2,400)	13,344 (3,00,000)	4.5	Conventional rail muck cars and vertical conveyor
75-124-2H	1978	Robbins	9.86 (32-4)	1,790 (2,400)	11,565 (2,600,000)	3.9	Conventional rail muck cars and hoisting
75-123-2H	1978	Robbins	9.86 (32-4)	1,790 (2,400)	11,565 (2,600,000)	3.9	Conventional rail muck cars and hoisting
73-164-2H	1988	Robbins	10.77 (35-4)	2,237 (3,000)	13,010 (2,925,000)	4.5	Continuous (total) conveyor
75-132-2H	1988	Robbins	9.83 (32-3)	2,908 (3,900)	13,455 (3,025,000)	5.4	Conventional rail muck cars and hoisting
73-271-2H	1990	Robbins	9.86 (32-4)	2,685 (3,600)	12,899 (2,900,000)	5.4	Continuous (total) conveyor
73-060-2H	1992	Robbins	9.83 (32-3)	NA	NA	NA	Continuous (total) conveyor

from July of 1979 to July of 1980, and attained progress rates comparable to a same diameter TBM on a TARP contract awarded in 1988 (73-164-2H).

- Mainstream System
  ③ Contract 75-126-2H

The tunnel penetrated the strata from the Romeo member of the Joliet formation to the rising Brainard formation. The average joint spacing of 42 m (137 ft) was the widest spacing of the Mainstream System. However, a total groundwater inflow of about 32 l/sec (500 gpm) occurred in the southwest reach of the tunnel where the tunnel was excavated through the Romeo and Markgraf members of the Joliet formation.

The significant feature of this contract was the occurrence of the Brainard shale at the north end of the project for approximately 2.8 km (1.7 mi). The shale attained a maximum thickness of 8.5 m (28 ft). Because of its susceptibility to slaking when in contact with air and water, the Contractor was required to apply a 9 cm (3.5 in) layer of reinforced shotcrete below the springline when the shale was exposed less than 2.7 m (9 ft) in the invert. When the shale rose more than 2.7 m (9 ft) above invert, full-circle, reinforced-shotcrete was required. This resulted in a much lower excavation rate, as the mining had to be stopped daily after 15 to 20 m (50 to 70 ft) of advance, for shotcreting operations.

It took just 14 months to excavate 5.0 km (3.1 mi) through dolomite and almost one year for the remaining 2.8 km (1.7 mi) through shale.

- Mainstream System
  ④ Contract 75-125-2H

The tunnel penetrated strata from the Racine formation to the Brainard shale. The average joint spacing was 21 m (70 ft). A total groundwater inflow of about 22 l/sec (350 gpm) was estimated before grouting.

The Brainard shale, encountered southwest of this contract, continued to be exposed in the tunnel for about 200 m (650 ft). The maximum height of 5.6 m (18.5 ft) above invert was noted and because of the anticipated susceptibility to slaking when in contact with air and water, the contract required application of a layer of 9 cm (3.5 in) of shotcrete.

- Mainstream System
  ⑤ Contract 75-124-2H

The tunnel penetrated the Racine formation and the Romeo and Markgraf members of the Joliet formation. The average joint spacing was 34 m (113 ft). Total groundwater inflow before grouting was approximately 38 l/sec (600 gpm). A fault zone, 122 m (400 ft) wide and two prominent shear zones containing breccia and gouge material contributed heavy inflows.

Conservative rock support was installed in these areas. Steel ribs at 1.5-m (5-ft) centers were installed with additional rock bolts between the steel ribs. In addition, steel straps and wire mesh were also installed locally, in areas of clay seams or clay pockets. The support was supplemented by wood lagging in highly fractured areas. Minor overbreaks occurred, especially in the crown, at intersecting joints and bedding planes.

- Mainstream System
  ⑥ Contract 75-123-2H

The tunnel was advanced through the Racine formation and the Romeo, Markgraf and Brandon Bridge members of the Joliet formation. The average joint spacing was 27 m (87 ft). Groundwater inflow was significant, particularly at the south end of the project. A total inflow of approximately 34 l/sec (540 gpm) entered the tunnel through joints, bedding planes and some vugs.

- Des Plaines System
  ⑦ Contract 73-164-2H

The tunnel was mined partly through the Markgraf and Brandon Bridge members of the Joliet formation and almost entirely through the Kankakee formation. The total groundwater inflow was very low, approximately 16 l/sec (250 gpm).

Claystone "pockets" up to 9 m (30 ft) in width were encountered at various locations. In contact with the humid air and water in the tunnel the claystone started to spall a few months after exposure. This damaged the conveyor belt and the Contractor decided to clean the loose material from these pockets and apply shotcrete.

A major rockfall, amounting to a total of 92 m^3 (120 yd^3) occurred in an area of very complicated geology. The area required heavy reinforcement and the overbreak was filled with concrete behind the steel ribs installed to stabilize the tunnel crown.

Mining was delayed from June 4, 1990, to August 16, 1990, (2 1/2 months) because the main bearing of the TBM failed and had to be replaced.

This was the first TARP project to use a total (continuous) conveyor system for muck removal. The mucking system was designed to transport 1,089 tonnes/hr (1,200 tons/hr) of rock from the cutterhead to the surface rockpile. The Contractor excavated a 2.4-m (8-ft) - diameter shaft dedicated exclusively for the vertical conveyor. The total conveyor system worked very well, with very few shut-downs.

- Des Plaines System

  ⑧ Contract 75-132-2H

Mining was delayed on three occasions for a total of approximately two months due to problems with the hoisting system. Among the four contracts awarded after 1988, this is the only contract that utilized the conventional rail cars and hoisting mucking system.

The tunnel projected within Racine, Joliet and Kankakee formations. Significant groundwater inflow was encountered at several locations, with an estimated total inflow of 50 l/sec (800 gpm) before grouting. In most instances the groundwater inflow was adequately controlled with cement grouting. However, several highly fractured areas up to 150 m (500 ft) in width required extensive joint treatment. In an area of approximately 300 m (1,000 ft) the Contractor could not control the inflow by grouting and steel sheet panning had to be placed against the rock before placing the reinforced concrete lining. The Contractor spent almost ten months (March to December of 1992) on the grouting/panning operations.

Clay or claystone occurred as fill in joints or in "pockets". Some of the clay pockets started to deteriorate over time and the Contractor was required to apply shotcrete as an interim operation.

With all the above-mentioned geological difficulties encountered, the new Robbins TBM performed very well. Driven by thirteen motors of 224 kW (300 HP) each, with the pressure on the thrust jack increased to 13,455 kN (3,025,000 lbs) and the cutterhead speed boosted from 4.5 to 5.4 rpm, this was the most powerful machine used on a TARP project to this date.

- Calumet System

  ⑨ Contract 73-271-2H

The 11.5-km (7.1-mi) -tunnel was mined entirely through the Silurian dolomite. Even though the average joint spacing was only 12 m (40 ft), most joints were tight and dry and only approximately 20% were waterbearing. The total inflow before grouting was 25 l/sec (400 gpm) with an average inflow rate of 0.3 l/sec over 150 m (5 gpm per 500 ft) of tunnel.

The rock mass condition throughout the project was good to excellent with very little need for rock reinforcement.

The significant feature of this contract was the project layout itself. From the southernmost point, a 10-m (34-ft) -excavated-diameter construction shaft, the 6.2-km (3.9-mi) -long Indiana Ave. leg proceeds north to another 10-m (35-ft) -excavated-diameter work shaft, which was the connecting point to an existing tunnel. From the work shaft, the 140th St. leg,

5.3 km (3.3 mi) long, proceeds east.

Due in part to only minimal site restrictions, the Contractor was able to optimize the locations where muck was brought to the surface. The Contractor could perform work under this contract applying very beneficial logistics for both mining and concrete lining operations.

Excavation proceeded from the construction shaft to the work shaft, then the vertical conveyor was removed from the construction shaft and installed in an additional 2.7-m (9-ft) -diameter shaft, exclusively used for mucking and downdrilled by the Contractor at his own expense about 100 m (330 ft) east of the work shaft. The mining continued in an easterly direction until completion, when the TBM was backed-up to the work shaft and lifted out.

This very well engineered scheme had the following benefits: the maximum length of horizontal conveyor was halved by using the mid-line shaft, thus greatly reducing the probability for shut-downs; the area around the work shaft was open for the drilling and blasting operations for the connection to the existing tunnel; and, most importantly, it allowed the mining of the 140th St. leg and the concrete lining of the Indiana Ave. leg to proceed concurrently.

This sequence greatly expedited the construction and this project was completed over one year ahead of schedule.

The TBM utilized on this contract was previously used on Contract 75-124-2H in 1978. The Contractor improved the design of the main bearing, made design changes on the cutterhead and redesigned the complete back-up system, merging a total conveyor muck handling system from the cutterhead to the surface storage pile.

The TBM performed better than one could have thought achievable and all previous records of mining for a tunnel diameter of this size were broken. The mining progress was highlighted by:

Best 8 hour-shift	29 m (96 ft)
Best 3 shift-day	72 m (236 ft)
Best 5 day-week	324 m (1,064 ft)
Best month	1,093 m (3,586 ft)

- Mainstream System
  ⑩ Contract 73-060-2H

The TBM was previously used on project 75-132-2H in 1988. The Contractor rebuilt the TBM but the new specifications of the TBM are not readily available.

The geology in the project area did not present a problem for mining. The tunnel was mined completely through Silurian dolomite, from the Racine formation to the top of the Kankakee formation. Groundwater inflow was minimal, as was the requirement for additional rock reinforcement.

Mining started in October of 1993 and was completed in April of 1996, with progress rates less than the ones achieved in 1988 on contract 75-132-2H utilizing the same TBM.

Mining was delayed from August 1994 to September 1994, and from March 1995 to May 1995 (a total of approximately 3 months) because the main bearing of the TBM failed and had to be replaced.

CONCRETE LINING

A summary of some significant concrete lining data, such as method used, form length, duration of the lining operation and monthly progress is presented in Table 4.

Following is a short narrative on the concrete lining operations for each contract.

- Mainstream System
  ① Contract 72-049-2H

Awarded in 1975, this contract did not require the tunnel be concrete lined. The first geotechnical investigations determined that the tunnel would be mined through sound dolomite, with widely spaced discontinuities, without stability problems or problems caused by possible erosion by flowing water. However, due to the unexpected geology encountered during mining, the decision was made to install a concrete lining invert under another contract.

After 1975 it was decided that all tunnels over 9 m (30 ft) in diameter be completely lined with concrete to a thickness of 25 to 30.5 cm (10 to 12 in). The concrete lining was to provide a hydraulically smooth surface and to assure long-term stability to substantially eliminate both the incidence and volume of future rockfalls, therefore, minimizing maintenance costs.

For the five contracts of the Mainstream System awarded in 1977-1978, the contract documents required the tunnels be concrete lined and the minimum thickness specified was 30.5 cm (12 in)

- Mainstream System
  ② Contract 73-160-2H

The batch plant was located on the surface at the construction shaft. The delivery to the concrete form was by three trains made up of a locomotive and three agitator cars, to a Schwing pump which delivered the concrete through a slickline to the forms.

The Contractor used the full-circle method of placing the concrete in a continuous pour, five days a week, resulting in one sloping joint at the end of each week. The continuous pour resulted in a higher rate of progress and lesser number of construction joints but it required twice the length of concrete forms as the vertical bulkhead method in order to sustain a continuous operation

The Contractor bought 100 m (330 ft) of form manufactured by Blaw-Knox in eleven 9.1 m (30 ft) sections.

The 5,731 m (18,804 ft) of 10-m (33-ft) -inside-diameter tunnel was lined in approximately seven months. The average daily placement rate was 480 m^3 (630 yd^3) of concrete.

Table 4. Comparison of Significant Concrete Lining Data

Project	Award Year	Main Contractor	Tunnel length m (ft)	Method used	Form length m (ft)	Duration of lining operation (months)*	Average monthly progress m (ft)
72-049-2H	1975	Kenny	8,312 (27,269)	UNLINED	NA	NA	NA
73-160-2H	1977	Morrison Knudsen	5,408 (17,744)	Full-circle Continuous pour	100 (330)	7	773 (2,535)
75-126-2H	1977	Heally	7,725 (25,345)	Arch-first (during mining) Vertical bulkhead	27.4 (90)	A 17 I 8	454 (1,490) 966 (3,168)
75-125-2H	1977	Paschen	7,526 (24,692)	Full-circle Vertical bulkhead	45.7 (150)	14	538 (1,764)
75-124-2H	1978	Shea	3,978 (13,052)	Full-circle Vertical bulkhead	45.7 (150)	8	497 (1,631)
75-123-2H	1978	Ball	6,682 (21,922)	Arch first Continuous pour	91 (300)	A 7 I 3	955 (3,132) 2,227 (7,307)
73-164-2H	1988	Morrison Knudsen	6,454 (21,174)	Full-circle Vertical bulkhead	30 (100)	NA	NA
75-132-2H	1988	Impregilo	7,323 (24,024)	Full-circle Vertical bulkhead	30 (100)	15	488 (1,602)
73-271-2H	1990	Kenny	11,497 (37,719)	Injection	58.5 (192)	11	1,045 (3,429)
73-060-2H	1992	Perini	13,991 (45,901)	Invert first	55 (180)	NA	NA

NOTE: * Does not include mobilization and form assembly, extended periods without concrete pours (strikes, clean-up) or atypical pours (tail/starter tunnels). Time shown is from first pour to substantial completion of lining. A = arch I = invert

- Mainstream System
  ③ Contract 75-126-2H

The Contractor chose to use the "arch first" method of lining in which a 290° portion of the top arch was lined first, leaving a 70° arch at the invert unlined. This method allowed the mining and lining to proceed concurrently.

The arch forms were manufactured by Blaw-Knox in three 9.1 m (30 ft) sections with a vertical bulkhead.

The Contractor chose to use a movable batch plant manufactured by Ross Uniplant and located in the tunnel. The plant was moved at approximately 140-m (460-ft) -intervals, one week's normal lining production, so that the length of the slickline did not become excessive. The plant delivered the concrete to a Schwing pump and the concrete was then pumped through a slickline and to the forms.

The 7,921 m (25,897 ft) of 10-m (33-ft) -inside-diameter tunnel was lined in approximately three years (July 1980 to June 1983), the arch section in two years and advancing from south to north and the invert section in one year and advancing from north to south. The average daily placement rate was 200 m³ (260 yd³).

- Mainstream System
  ④ Contract 75-125-2H

A batch plant located on the surface at one of the dropshafts delivered the concrete to the forms by three trains made up of a locomotive and three agitator cars. The train delivered the concrete to a Schwing pump which delivered it through a slickline to the forms.

The Contractor used the full-circle method of placing concrete. The total length of the form was 45.7 m (150 ft), manufactured by Blaw-Knox Corp., and made up of five sections of 9.1 m (30 ft) each.

This 7,678 m (25,189 ft) of 9.1-m (30-ft) -inside-diameter tunnel was lined in approximately two years (May 1981 to June 1983) and the average daily placement rate was 290 m³ (380 yd³).

- Mainstream System
  ⑤ Contract 75-124-2H

The batch plant for this contract was located at the surface of the construction shaft site. The concrete was dropped to a hopper at the bottom of the shaft used to fill agitator cars which were trammed through the tunnel to the Schwing pump in front of the forms.

The Contractor used full-circle forms manufactured by Blaw-Knox, of a total length of 45.5 m (150 ft) and made up of five sections of 9.1 m (30 ft) each.

This 4,184 m (13,726 ft) of 9.1-m (30-ft) -inside-diameter tunnel was lined from June of 1981 to August of 1983 with an average daily placement rate of 290 m³ (380 yd³).

- Mainstream System
  ⑥ Contract 75-123-2H

The Contractor chose to use the "arch first" method of lining placement. The 290° arch was placed using the continuous pour, five days a week, and began after all excavation was completed. The invert was placed after the arch section was completed. The 290° arch forms were furnished by Blaw-Knox, in ten sections of 9.1 m (30 ft) each with vertical bulkheads. The vertical bulkhead was used only at the end of each week.

Concrete for the tunnel lining was supplied from two sources. A batch plant used exclusively for this contract and which had to be supplemented by trucking concrete from a second plant. The concrete was transported through the tunnel in agitator cars powered by a locomotive. A four-car train carried the concrete to a Schwing pump which pumped it through the slickline to the forms.

This 6,891 m (22,607 ft) of 9.1-m (30-ft) -inside-diameter tunnel was completely lined, both arch and invert sections in less than one year. The average daily placement rate was 535 m³ (700 yd³).

- Des Plaines System
  ⑦ Contract 73-164-2H and ⑧ Contract 75-132-2H

The design of the South and Middle legs of the Des Plaines system differs from earlier TARP projects. The municipalities along the tunnel alignment utilized well water as the sole source of water supply. Although all of the wells were located in the lower aquifer and the intermediate Maquoketa shale acts as a barrier between the upper and lower aquifers, there were concerns regarding possible exfiltration under a surcharged condition. In order to address these concerns, the required concrete lining was designed to be reinforced with a layer of steel for crack control.

Both contractors opted for the conventional slickline method, full-circle concrete lining. The length of the steel forms was 30.5 m (100 ft). The only difference was the concrete conveyance to the forms.

One contractor dropped the concrete from the batch plant at the main construction shaft, through a drop pipe installed in an adjacent temporary shaft previously used as a muck shaft for the vertical conveyor. At the bottom of the shaft agitator cars were used to transport the concrete through the tunnel and to the concrete pump at the forms.

The other contractor hauled the concrete to various shafts along the tunnel alignment and pumped the concrete through the tunnel to a second concrete pump at the form.

- Calumet System

  ⑨ Contract 73-271-2H

The contract documents specified unreinforced concrete with a minimum thickness of 25 cm (10 in). The concrete was placed in full-circle, telescoping the forms with vertical full height bulkheads, using the multiple port injection method. This was the first time this method was used on a TARP project. The average daily placement rate was 580 m^3 (760 yd^3).

The concrete was hauled through the tunnel in agitator cars. At the pour deck in front of the form the agitators were discharged into remixers prior to being discharged into a Schwing pump. The pump discharged the concrete via a placing line into a Putzmeister placer. The concrete was then injected by the placer through designated injection ports in the form. The form was manufactured by Burke and its length was 58.5 m (192 ft) with an additional 9.75 m (32 ft) of invert.

The technique and equipment used, together with the sequence chosen, led the Contractor to a world record for concrete placement. Between January 24, 1994, and March 24, 1994, for 44 consecutive days, the Contractor placed 58.5 m (192 ft) of 33-cm (13-in) -thick concrete liner in the 9.1-m (30 ft-2 in) -finished diameter tunnel. This achievement is unprecedented for a tunnel diameter of this size.

- Mainstream System

  ⑩ Contract 73-060-2H

The Contractor chose to place the 30.5-cm (12-in) -minimum-thickness concrete liner using the "invert first" method of placement. This was the first time this method was used on a TARP project. The intent was to place the 70° invert and the 290° arch while mining was in progress.

The initial plan was as follows: the invert was to be placed with the use of a simple steel invert screed and poured several weeks in advance of the arch concrete operations in order to allow sufficient setting time necessary to support the arch forms and the form carriers. The concrete was to be pumped through the nearest drop shaft and via a slickline pipe behind the 70° invert curb form. Protection of the concrete in the invert was to be provided under the horizontal conveyor from possible damage during placement operations. Access to the ongoing mining and lining operations was to be diverted through various shafts in order to eliminate any traffic over the newly poured concrete.

For the arch placement the Contractor is using five form sections of 11 m (36 ft) each, manufactured by Concrete Form Consultants, Ltd., each with its own carrier, being able to pour a maximum of 55 m (180 ft) of arch in one day. The concrete pump is set at the nearest drop shaft location ahead of the concrete operations and the concrete is discharged through a slickline behind the form.

The Contractor started the invert pour in July, 1995. The mining was approximately 4 km (2.5 mi) away from completion and the Contractor experienced great difficulty in placing concrete according to plan. Concrete tests revealed unacceptable voids and segregation. A special paving machine with power screed was used to finish the invert concrete. For the next seven weeks the Contractor poured 1.75 km (1.1 mi) of the 70° invert with a very low progress rate of 250 m (820 ft) per week. In September of 1995 the invert pour work was suspended and the Contractor started assembly of the arch forms. The first arch pour was on October 24, 1995 when mining was approximately 2 km (1.25 mi) away from completion. For the next 26 weeks, until completion of the mining on April 28, 1996, the Contractor poured 1.4 km (0.9 mi) of arch at an average weekly rate of less than 55 m (180 ft). From this point on, after disassembling and removing the TBM from the tunnel, the Contractor has not experienced the difficulties imposed by the simultaneous mining operations (such as conveyor system, utilities, etc.).

The Contractor is currently pouring both invert and arch sections and has attained maximum weekly rates as high as 1,100 m (3,610 ft) for the invert pour and 275 m (900 ft) for the arch pour.

CONCLUSION

Between 1975 and 1979, six contracts on the Mainstream System were awarded and included the use of the largest TBMs ever manufactured. Construction of the main tunnels was interrupted between 1979 and 1988 due in part to reduced funding levels. Beginning in 1988, five major contracts have been awarded and two are currently under construction. One of these projects, the North leg of the Des Plaines System, is using a 7.3-m (24-ft)-diameter TBM and is not included in our study.

TARP has demonstrated that large diameter hard rock TBMs are cost effective. In late 70's, machines up to 10.8 m (35 ft-4 in) in diameter set mining records and achieved utilization rates of over 50%. In late 80's and early 90's, new machines of the same diameter and also rebuilt TBMs broke the previous mining records and utilization rose to over 60%. This increase can be related to the improvements of the back-up systems as well as refinements in the machines themselves. One important innovation has been the use of a continuous conveyor system instead of the conventional rail-mounted muck cars to remove the muck from the tunnel to the surface rock pile.

Great credit has been and will continue to be given to the visionary Metropolitan Water Reclamation District of Greater Chicago for its contribution to the advancement of the tunneling technology and experience. Over the past two decades, nationally recognized tunnel contractors converged in Chicago and brought in not only hard work and dedication but also innovation, improvement of the TBM performance, with the manufacturers' help, and advancement of construction management techniques.

REFERENCES

Braam, P.H.M., 1997, Private Communication.

Budd, T.H., Rautenberg, R.R., June, 1995, "Calumet Tunnel Project, a case history," Proceedings. SME, pp. 247-267.

Calin, L.P., Sakamoto, L.S., April, 1996, "Calumet Tunnel System," Proceedings, North American Tunneling and International Tunneling Association, Vol. 1, pp. 51-59.

D'Appolonia, 1981, "Construction Report, Engineering Geology, Rock Reinforcement and Grouting - Tunnels and Shafts, Addison Street to Wilmette," September, MSDGC.

Envirodyne Engineers, Inc., 1993, "Geotechnical Report, TARP, Des Plaines Tunnel System, Contract 73-164-2H, South Leg - 59th Street to Cermak Road," August, MWRDGC.

Envirodyne Engineers, Inc., 1994, "Geotechnical Report, TARP, Des Plaines Tunnel System, Contract 75-132-2H, Middle Leg - Cermak Road to Fullerton Avenue," April, MWRDGC.

Harza Engineering Company, 1984, "TARP, Mainstream System, 59th Street to Addison Street, Construction Report, Volume I, Project Administration and Construction Methods," May, MSDGC.

Harza Engineering Company, 1984, "TARP, Mainstream System, 59th Street to Addison Street, Construction Report, Volume II, Geology and Hydrogeology", May, MSDGC.

Parsons Brinckerhoff, Inc., 1995, "As-built Geotechnical Report, TARP, Calumet Tunnel System, Contract 73-271-2H, 140th Street and Indiana Ave. Legs," March, MWRDGC.

MSDGC, 1983, "Tunnel Construction Reports," Vol. 2, Jan.

MWRDGC, 1993 to 1997, "Resident Engineer's weekly reports for Contract 73-060-2H," No.1 to No.227.

MWRDGC, 1996, "Summary of Rock Tunneling Machine Performances," June.

Robbins, R.J., 1993, "Tunnel Boring Systems for the Nineties," Underground Space '93, pp. 6-12.

Westfall, D.E., Sakamoto, L.S., Goyal, B., June, 1993, "TARP's Contribution to Hard Rock Mining," Proceedings, Second international symposium on mine mechanization and automation, Lulea, Sweden.

Willman, H.B., 1971, "Summary of the Geology of the Chicago Area," Illinois State Geological Survey, Circular 460, Urbana, Illinois.

Willman, H.B., 1973, "Rock Stratigraphy of the Silurian System in Northeastern and Northwestern Illinois," Illinois State Geological Survey Circular 479, Urbana, Illinois.

Chapter 18

BENBROOK CONNECTION TUNNEL FORT WORTH, TEXAS

J.M. Kennedy[1] and C.A. Wiggins[2]
[1]7K Construction Co.; [2]Jenny Engineering Corp.

ABSTRACT

The Benbrook Connection tunnel forms approximately 6,666 m (19,560 ft) of 17,678 m (58,000 ft) of raw water pipeline connecting Lake Benbrook with the Rolling Hills Water Treatment Plant. This connection is designed to minimize costs to the owners, the Tarrant Regional Water District, for providing raw water to its' various clients. The geological formations through which the tunnel passes and the design requirements resulted in somewhat unique applications of initial ground support systems which were modified by the contractor to optimize the mining operations. The pipe sections, fabricated in 12 m (40 ft) sections, are concrete lined and coated steel pipe. Construction of the work shaft and tail tunnel was also modified to suit the contractors methods. A Lovat universal TBM was chosen to excavate through the various rock types. The pipe sections will be transported using beam carriers.

INTRODUCTION

The Tarrant Regional Water District (TRWD), prior to October 1996 known as the Tarrant County Water Control and Improvement District No. 1 (TCWCID#1) built and manages four mapr lakes in the region surrounding the Dallas/Fort Worth metroplex. To the northwest there is Lake Bridgeport in Jack and Wise counties and the Eagle Mountain Lake in Wise and Tarrant counties. To the southeast there is the Cedar Creek Reservoir in Henderson and Kaufman counties and the Richland Chambers Reservoir in Navarro and Limestones counties.

The TRWD supplies raw water. Its main customers are three main Water Treatment Plants owned by the City of Fort Worth, smaller plants at Arlington and Mansfield and maintains the water level at Lake Arlington for use by the City of Arlington, the Trinity River Authority and the Texas Utility power generation plant.

Transmission of water from the Cedar Creek and Richland-Chambers reservoirs is facilitated by means of 121 km (75 miles) each of 183 cm (72 inch) and 229 cm (90 inch) pipeline, respectively, operated by the TRWD. Water is pumped over the rise of the Midlothian Hills to the balancing reservoirs at Kennedale just southwest of Fort Worth. From here it is distributed to the Rolling Hills treatment plant and/or to Lake Arlington.

The cost to the TRWD for pumping water is based on a sliding scale of flat rates for a given maximum quantity of water pumped at any time within the year. With increasing water

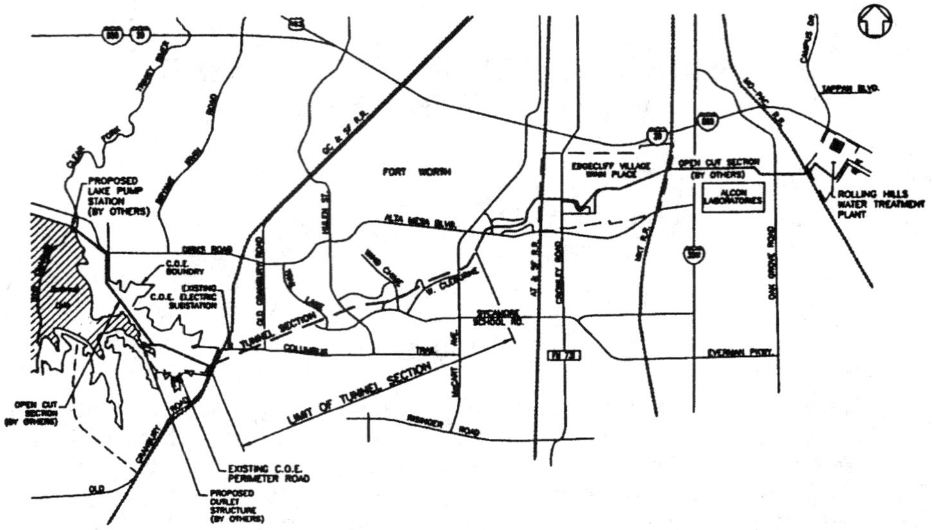

Figure 1. Location Plan (reproduced from Geotechnical Basis of Design and of Construction Specification)

consumption in the service area, and particularly in the summer months, the cost to the TRWD of pumping water from the reservoirs in east Texas was in danger of moving to the next higher utility rate. This increased cost, which would be applied throughout the year would be significant and would be passed on to the consumer. To avoid this occurrence a scheme using the nearby, Army Corps of Engineers owned, Lake Benbrook was developed. Financial studies indicated that costs to consumers with the latter alternative would be significantly less.

The Benbrook Connection project will divert excess raw water from the Rolling Hills water treatment plant for storage in Lake Benbrook in off-peak water usage periods, during winter months. During summer months, when water usage peaks, the water from Lake Benbrook will be used to offset required pumping from the reservoirs in East Texas, thereby maintaining maximum pumping rates below that which would cause an increase in the TRWD utility rates. Additionally water from Lake Benbrook will be used to maintain water levels in the Clear Fork of the Trinity river which provides raw water to the Holly treatment plant in west Fort Worth. The Benbrook connection is a necessary component of the Districts system to be able to provide Fort Worth with water during droughts.

Project Engineers for the entire Benbrook pipeline is Freese and Nichols, Engs. Inc. of Fort Worth, Texas with Jenny Eng. Corp. of Springfield, New Jersey sub-consulting on the design and providing construction services for the connection tunnel contract. 7K Construction Co., a subsidiary of Kenny Construction of Chicago, is the Contractor.

OPEN-CUT VS. TUNNELING

The Rolling Hills Water Treatment plant is separated from Lake Benbrook by a region

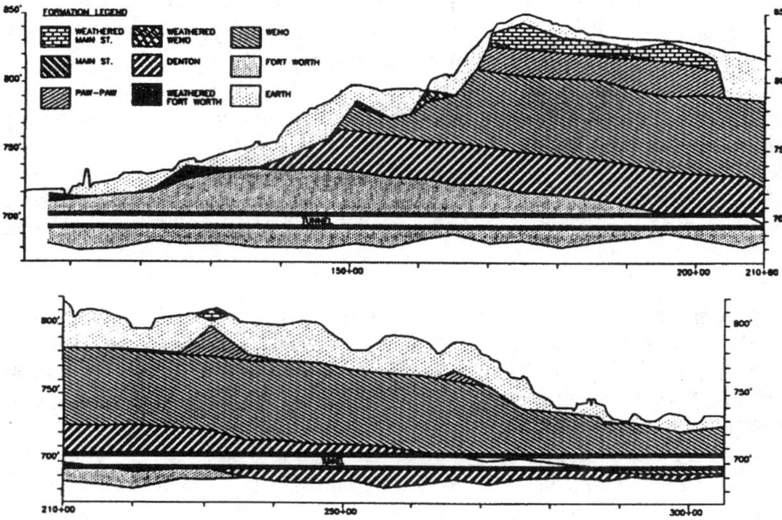

Figure 2. Geologic Profile (reproduced from Geotechnical Basis of Design and of Construction Specification)

of Hills some 43 m (140 ft) approximately above the elevation of the Lake. The installed pipeline will have a total length of 17678 m (58,000 ft) approximately. The first 3749 m (11,000 ft) approximately, from Lake Benbrook to the west tunnel staging area, and the last 9338 m (27,400 ft) approximately, from the east tunnel staging area to the Rolling Hills water treatment plant, of pipeline will be installed by open cut.

The available options to the TRWD to traverse the region of high ground would be to open cut the pipe generally following the topography or to tunnel beneath the region of high ground. While the open cut option would require less initial capital expenditure it would be necessary to pump water both to and from Lake Benbrook. The tunnel option, while initially more expensive, would allow water to flow by gravity to Lake Benbrook and require pumping only from Lake Benbrook to the Rolling Hills water treatment plant. A cost comparison of the two options over the projected life of the project indicated that the tunnel option would be the least expensive. Additionally the choice of tunnel over open cut meant that the complicated logistics of minimking the impact of going through a rather well-to-do residential neighborhood with open trenches would be avoided

GEOLOGY

The selected tunnel alignment connects staging areas to the west on undeveloped property owned by the Corps of Engineer just west of Old Granbury Road and to the east on undeveloped property at South Creek Park, to the south-east of the McCart St. Alta Mesa Blvd. intersection. The mined length of the tunnel will be approximately 6666 m (19,560 ft) and will slope up from west to east at approximately .01%. The tunnel alignment will encounter three rock formations going from west to east, the Fort Worth, the Denton Clay and the Weno.

Both the Fort Worth and the Weno formations are a gray limestone interbedded with shaley limestone and soft dark gray shale layers. The strengths and physical characteristics of each of the various layers in both formations are similar.

The average strength of each formation is therefore dependent of the relative proportion of each of the constituents in each formation. Strength tests of the unweathered limestone ranged from 12.4EE10^6 Pa (1800 psi) to 38.6EE10^6 Pa (5600 psi) approximately with an average strength of 25.5EE10^6 Pa (3700 psi) approximately. The Weno in general contains about 50% greater content of the softer shaley layers compared to the Fort Worth.

The Denton Clay is a more homogeneous formation comprising a limey shale with interbedded limestone and fossil layers. The rock is however significantly softer with strengths ranging from 0.28EE10^6 to 9.7EE10^6 Pa (41 to 1400 psi) approximately with an average strength of 2.8EE10^6 Pa (400 psi) approximately. This formation does exhibit significant susceptibility to weathering if the insitu characteristics are allowed to vary.

Separating the Denton Clay and the Weno exists a Fossiliferous Marker-Bed approximately 0.91 to 1.83 m (3 to 6 ft) thick. This layer contains numerous shells and fossil fragments and was considered a potential source of water infiltration into the tunnel opening.

Within the alignment the rock mass is unweathered and yielded very low levels of permeability, in the range 1x10^{-7} cm/sec. Additionally RQD values were generally in the good to excellent rock quality range.

INITIAL GROUND SUPPORTS

Type A Tunnel

Within the first 3048 m (10,000 ft) approximately through the Fort Worth formation the initial ground support system designed used ribs and timber lagging. The reason for the choice of ribs and lagging was the potential for the shaley layers to weather and the horizontally layered rock to delaminate with the consequent danger from spelling rock. The specification additionally provided for the application of a protective layer of shotcrete should the deterioration of the shaley constituents dictate additional protective measures. The Contractor, 7K Construction Corp., chose to modify the initial support system by including a reinforced concrete 90 degree invert segment. As a result the Contractor realized an increased efficiency in the operation of the rail mounted equipment used in all construction processes. The remaining 270 degrees of the initial rib and lagging support system was installed as specified.

The additional protective measures have not been required. There is some spelling of the rock within the Fort Worth limestone but it is limited and contained by the almost full timber lagging boards provided.

Type B Tunnel

For the remaining approximately 2914 m (9,560 ft) of tunnel the alignment will pass through the 2152 m (7,060 ft), approximately, of the Denton Clay and 762 m (2,500 ft),

Type A Support — 90 deg. Invert Plus Ribs and Lagging

approximately, of the Weno formation. From the Contractors perspective the somewhat controversial choice of grouted full precast concrete segments for initial ground support was made. The primary motive for the use of full concrete segments in the Denton Clay was concern prompted by the previously mentioned susceptibility for instability exhibited by this formation when exposed to alternate drying and wetting cycles. The intent was to maintain the insitu conditions existing to the greatest extent practicable after cutting through the Denton Clay. By installing and grouting behind the precast segments as mining progressed an effective sealing of the exposed material from the effects of weathering both during and after construction would be achieved

Another factor that contributed to the decision to use grouted precast segments was the expressed desire of the owners to make this stretch of pipe line as secure from failure as it could be engineered. This consideration was a result of the topography that located the line as much as 43 m (140 ft) below ground level. Clearly access to this section of the line to effect repairs if there was a failure would not be as relatively simple a task as with pipes laid in open cut trenches. Both the owner and the Engineers were aware that this was an application of a "belt and suspenders" solution but felt that the long term security offered was worth the extra expense.

For the final 762 m (2,500 ft), approximately, in the Weno formation the precast segment initial ground support system was continued because of the existence of the fossiliferous bed between the Denton and the Weno. Additionally the alignment does not move sufficiently far away from the transition with the Denton Clay to eliminate concerns for the relative ease of water transmission with a return to a rib and lagging support system. In any case the Weno as stated previously contained significantly greater amounts of the sensitive shaley layers. For these reasons the decision was made to extend the grouted precast segments to the end of the tunnel.

FINAL LINING

The options available to the owners for a final lining were to use a lined and coated steel pipe, a prestressed cylindrical concrete pipe (PCCP) or a cast-in-place reinforced concrete lining. Given the length of tunnel to be lined and that the pipe would be subjected to pumping pressures the cast-in-place lining was not a competitive option. The owners, as a result of their past poor experiences with PCCP and given the relative inaccessibility of the pipeline for much of the length of the tunnel, decided to use the steel pipe option.

The steel pipe was designed to withstand tensile stresses generated by the pumping cycle to which the pipe would be subjected delivering water from Lake Benbrook to the Rolling Hills water treatment plant. During gravity flow ground support would be provided by both the substantial initial ground support systems installed as well as the concrete backfill between the steel pipe and the initial supports. Additionally the steel pipe was sized based on a design approach used to size steel linings in penstock design. The underlying philosophy in this design is the gap which may be formed between the steel pipe and the backfill concrete. The gap subject to external water pressure can create a local buckling failure of the pipe.

It was also determined that the steel pipe would be both lined and coated with a cement mortar to AWWA standards for long term protection of the pipe. Pipe joints are to be internally lap welded slip joints in the field and finished intemally with a field applied cement mortar. Additionally a cathodic protection system will be applied to the externally exposed steel pipe at locations made accessible by the alignment holes and connected to surface monitoring boxes.

SHAFT EXCAVATION

Original plans called for the Work shaft to be 7.3 m (24 ft) diameter liner plates and shotcreted rock-bolts. To facilitate installation of the TBM and 229 cm (90 in) steel pipe it was decided that a 12.2 m x 7.3 m (40 ft x 24 ft) rectangular shaft would be substituted. A Link

Work Shaft Support Bracing

Tail Tunnel Support

Belt 5800 with a 4536 Kg (10,000 lbs) breaker was used to break the rock and trim the shaft walls. The top 4.6 m (15 ft) of the excavation was in unstable rock, therefore supports were needed. Due to the proximity of the shaft to the creek next to the work site, steel interlocking sheeting was chosen for the supports. The pieces of the sheeting were laced and welded together on top of the ground and placed into the shaft like gang forms onto the preassembled bracing.

The rock in the tail tunnel appeared to be harder than what could be efficiently excavated by conventional mining since blasting was not allowed on the project. The Link Belt 5800 with breaker was used again for the 28.0 m (92 ft) of 3.7 m (12 ft) horseshoe section. After the supports were installed, cement treated sand was backfilled to springline and compacted. A concrete cap was placed to .3 m (1 ft) above the crown. The rest was backfilled with excavated materials.

A 30.5 cm (12 in) mudcoat was poured in the shaft for TBM placement. Two 30.5 cm (12 in) beams were buried in the mudcoat to weld thrust beams when the TBM was launched.

TUNNEL EXCAVATION

The TBM chosen to do the project needed to be able to excavate the Fort Worth limestone and the Denton Clay which was considerably softer. It had to erect the ribs and

Lovat RM 131 RL/SE

lagging supports as well as the precast segmental liner. A muck ring could be installed for the EPB mode if the ground indicated. The TBM is a Lovat RM 131 RUSE with interchangeable disk cutters and rippers. The back loading system was set up initially with the 33 cm (13 in) Palmieri disk cutters to cut the approximately $4.1EE10^6$ Pa (6000 psi) maximum rock. It proved to be the right choice since the same cutters were used until the entire face had changed to the Denton Clay. At that time the rippers were installed to cut the clay in larger pieces.

The TBM was delivered April 12, 1996 and set in the launch pad the same day. It was assembled in the following three days. The TBM was buried April 17 and the trailing gear

TBM Trailing Gear

Switch in Type A Tunnel

Type B — Precast Segments

was added as the machine advanced. The support system chosen in the type "A" section was a 90 degree precast invert section and a 270 degree ribs and lagging section. The expanded sets prevented any spelling of the shale and limestone seams as well as providing thrust support for the TBM. The 1.5 m (5 ft) advance took approximately 10 - 12 minutes and an additional 3 - 4 minutes to erect the supports. Muck was removed via five car trains pulled by eight ton Plymouth locomotives.

Inserts were cast into the inverts to bolt the ribs after erection. Inserts were also placed for the rails to be attached directly to the invert without the use of ties. This aided installation of the rail and also cleanup prior to pipe placement. The segments were kept in alignment by

40 ft. Section of 90" Pipe

fiberglass rods placed into the holes cast into the ends. The rods helped to counter the rolling effect of the TBM by pinning all the inverts together.

Ventilation was provide by four 37.3EE10^3 W (50 hp) fans on the TBM gantry and exhausted via a Schauenburg 61.0 cm (24 in) cassette system. It was attached to a cable suspended from the crown of the tunnel. Alignment holes, 1.2 m (4 ft) in diameter were drilled along the tunnel route at locations which required future cathodic protection wiring. The holes were placed approximately 914.4 m (3000 ft) apart and a portable ventilation system was placed on the top of ground to provide additional air from the shaft to that point.

In the type "B" tunnel section the supports were to be a full ring precast segmental liner. After the contract was awarded an attempt was made to eliminate the requirement for the segmental liner, since in the opinion of 7K it was not necessary. The Owner desired the additional factor of safety of protection that the liner provided and did not accept the proposed change. The segments were manufactured by Gifford-Hill Company of Grand Prairie, Texas. The six piece ring consisted of an invert that matched the previous sections used in the type "A" section, two 72 degree blocks, two 66 degree blocks and a key. The joints were placed in reference which made installation much easier. To date, the most rings placed in a ten hour shift has been 30.

Since, to date, the mined material has been dry and grouting has followed the trailing gear it has not been necessary to have tail seals at the TBM. This contributed to the erection procedure by allowing flexibility during bolting of the gasketed segments. The rings were held in place after passing out of the tail shield by use of a plastic pin, similar to a form spud, which was screwed into the lifting holes of the segments. Grout was injected through the lifting holes after the TBM gantry had passed the segment. The sand/cement mix was made at the shaft and transported to the heading on every third train. The grout car was a 2.7 m^3 (3.5 cy) muck car converted for grouting and attached above a Transcrete America HP 45 concrete pump on a flat car. The grout was circulated in the box via the pump as it traveled through the tunnel thus avoiding separation prior to placement.

The TBM will exit 6666 m (19,560 ft) from the Work Shaft in an exit shaft in a City of Fort Worth park. The remainder of the pipe line to the Rolling Hills Treatment Plant, and to Lake Benbrook, will be open cut under another contract.

PIPE INSTALLATION

The pipe to be installed is a 229 cm (90 in) mortar lined and coated steel pipe as manufactured by North American Pipe of Saginaw, Texas. The joints are 12.2 m (40 ft) long expanded bell and spigot. The pipes will be placed from the work shaft, starting at the exit shaft and working back. The joints will be transported into the tunnel on a beam carrier, dropped on a set of rollers and pushed home where they will be welded. The annulus will be filled with grout after the pipe placement has advanced sufficiently to prevent delays to the grouting operation.

CONCLUSIONS

The Benbrook Connection project is a good example of the potential efficacy of tunnel construction methods over cut and cover for pipe placement. The owners were able to realize economic advantage with the tunnel option as well as minimizing surface disruptions during the construction process.

A well thought out construction process and choice of equipment by the contractor has seen the project progressing reasonably well to date. There has been some slippage behind projected progress, as of the writing of this paper, but it is expected that the project will be brought in on time and within budget.

BIBLIOGRAPHY

June 1995, "Geotechnical Basis of Design and of Construction Specification" by Freese and Nichols, Inc. and Jenny Engineering Corp..

September 1991, "Benbrook Connection Project Preliminary Design Report" by Freese and Nichols. Preliminary Draft.

October 1990, "Regional Water Supply Plan" by Freese and Nichols, Inc. and Alan Plummer and Assoc., Inc.

September 1984, "Report on Capability to meet Potential Water Demands: 1985 -1990" by Freese and Nichols.

June 1982, "Repor on the use of Lake Benbrook for Regulating Storage" by Freese and Nichols.

Chapter 19

CONSTRUCTION OF CITY TUNNEL NO. 3, STAGE 2 FROM BROOKLYN TO QUEENS

Anthony Del Vescovo[1] and Jim W. Rosteck[2]
[1]Schiavone Construction Co., Inc.; [2]J. F. Shea Co., Inc.

ABSTRACT

In December of 1992, the City of New York, Department of Environmental Protection, solicited bids for Contract 543B. The low bidder was a joint venture of Schiavone Construction Co., Inc. and the J. F. Shea Co., Inc. When completed, the project provides a 4.88 meter diameter water tunnel 8,840 meters long. This paper presents a case history of the equipment and construction methods utilized for this, the first portion of the New York City Water Tunnel constructed by use of a tunnel boring machine.

INTRODUCTION

New York City Water Tunnel History

In the early 1800's, New York City discovered it could not rely solely on its wells, springs, and streams to supply water. City officials had to devise a plan to provide water for its rapidly growing population. The plan was to deliver water from areas outside the City. Today New York City's entire water supply relies upon the reservoirs located upstate. The system is a complex one of aqueducts, reservoirs, tunnels, shafts and pipelines. From its farthest point, the water travels 201 km (125 miles) to reach its ultimate destination in New York City. Today the water system carries approximately 65.7 m3/s (1.5 billion gallons of water per day) to its citizens. The force of gravity transports the water to its final destination.

The New York City water supply system consists of three major components; the Croton System, the Catskill System, and the Delaware System. The first part of the Croton System was placed into service in 1842, over 150 years ago. Later, the system was expanded by constructing a second aqueduct. By the year 1898, the Croton System was able to supply 10.5 m^3/s (240 mgd).

The major component of the Catskill System, City Tunnel No. 1, took 23 years to construct. It was started in 1905 and completed in 1928. The overall length is 29 km (18 miles), with a finished diameter of 4.6 meters (15 feet) decreasing to 3.4 meters (11 feet) at its terminus. All tunnel excavation was performed utilizing the drill and blast method. At its deepest point the tunnel is approximately 274 meters (900 feet) deep into rock. With the completion of City Tunnel No. 1, the system was able to provide 20.6 m^3/s (470 mgd) of water to the Bronx, Manhattan, and a small part of Brooklyn.

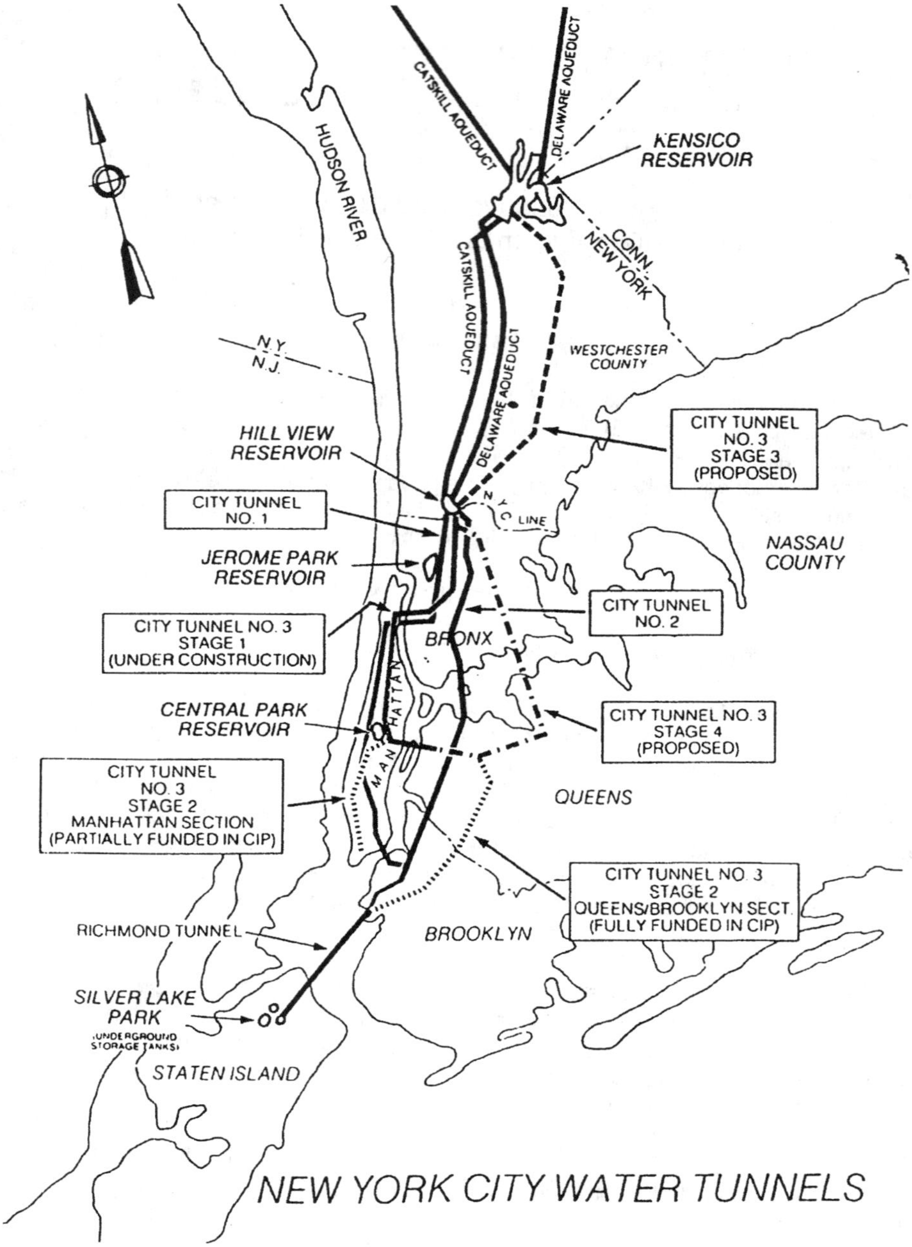
CATSKILL AQUEDUCT
DELAWARE AQUEDUCT
KENSICO RESERVOIR
HUDSON RIVER
CONN.
NEW YORK
N.Y.
N.J.
WESTCHESTER COUNTY
HILL VIEW RESERVOIR
CITY TUNNEL NO. 3 STAGE 3 (PROPOSED)
CITY TUNNEL NO. 1
N.Y.C. LINE
NASSAU COUNTY
JEROME PARK RESERVOIR
CITY TUNNEL NO. 2
CITY TUNNEL NO. 3 STAGE 1 (UNDER CONSTRUCTION)
BRONX
MANHATTAN
CENTRAL PARK RESERVOIR
CITY TUNNEL NO. 3 STAGE 4 (PROPOSED)
CITY TUNNEL NO. 3 STAGE 2 MANHATTAN SECTION (PARTIALLY FUNDED IN CIP)
QUEENS
CITY TUNNEL NO. 3 STAGE 2 QUEENS/BROOKLYN SECT. (FULLY FUNDED IN CIP)
RICHMOND TUNNEL
BROOKLYN
SILVER LAKE PARK
(UNDERGROUND STORAGE TANKS)
STATEN ISLAND
NEW YORK CITY WATER TUNNELS

Figure 1

After completion of City Tunnel No. 1, the City's population was growing even faster. Offcials determined another tunnel would be needed to supply a suffcient amount of water. City Tunnel No. 2, still part of the Catskill System, was designed to bring water to the Bronx, Queens, and Brooklyn. Although the tunnel was a little longer than City Tunnel No. 1, its construction time was much shorter. The tunnel was 32 km (20 miles) long, with the longest segment having a diameter of 5.18 meters (17 feet). Construction began in 1928 and was completed in 1936. This tunnel was also excavated utilizing the drill and blast method. Because the City let out many contracts at one time, the construction period was accelerated. Contractors sank shafts 183 to 274 meters (600 to 900 feet) deep and then drove headings in opposite directions, approximately 2.4 km (1.5 miles) each way. With many contractors excavating and meeting up with each other at the same time, City Tunnel No. 2 was completed in just eight years.

The Delaware System was the third major component of the City's water supply. This system brought water from many reservoirs including the Cannonsville reservoir, 201 km (125 miles) from the City. The Delaware System consisted of about 129 km (80 miles) of tunnels and 145 km (90 miles) of aqueducts to bring additional water to City Tunnels No. 1 and No. 2. The work was completed in 28 years, starting in 1936 and ending in 1964. During this time, construction was delayed as a result of World War II as well as litigation that ended with a Supreme Court decision. This decision restricted the amount of water allowed to flow from the Delaware River to the City, in order to maintain a minimum flow in the Delaware River. The Delaware System supplied an additional 25.4 m^3/s (580 mgd) of water to the City.

By 1954, the Board of Water Supply realized that due to the ever increasing City population, City Tunnels No. 1 and No. 2 were at maximum capacity and that it was impossible to shut any major part of the tunnel system down for repairs, without disrupting service. Oficials decided to construct a third tunnel. City Tunnel No. 3 would satisfy the growing demand for water in the east and south areas of the City, and allow City Tunnel No. 1 or No. 2 to be shut down for inspection, and if necessary, rehabilitation. It would reduce peak flows in City Tunnels No. 1 and No. 2., thus minimizing low water pressure conditions. City Tunnel No. 3 was scheduled to be constructed in several stages.

Stage 1 of City tunnel No. 3 was constructed between 1970 and 1984. Construction was halted for a few years because of New York City's financial crisis and litigation with one of the water tunnel contractors. The completed tunnel is 21 km (13 miles) long, intersects 16 shafts and has three large subsurface valve chambers. The major segment of the tunnel has a finished diameter of 7.3 meters (24 feet). Like Tunnel No. 1 and No. 2, this tunnel was excavated in bedrock about 244 meters (800 ft) deep. Excavation was carried out utilizing the drill and blast method.

Stage 2 of City Tunnel No. 3 is currently under construction. It is composed of two sections, the Brooklyn-Queens section and the Manhattan section. The Brooklyn-Queens section is 18 km (11 miles) long and will connect Stage 1 of City Tunnel No. 3 to Staten Island. The finished diameter on the Queens side is 6.1 meters (20 feet) and the finished diameter on the Brooklyn side is 4.9 meters (16 feet). The Manhattan section will be 3.0 meters (10 feet) in diameter and will be 11 km (7 miles) long. It will connect to Stage 1 at its northernmost point and will terminate in lower Manhattan at its southernmost point.

CHRONOLOGY OF BID AND AWARD

By the late 1980's, work had begun on a few shafts for City Tunnel No. 3, Stage 2. By 1992, five shafts had been completed for the Brooklyn section. At that time, the New York City Department of Environmental Protection (NYCDEP) decided to proceed with construction of the next section of tunnel. In early November of 1992, the NYCDEP released bid documents for City Tunnel No. 3, Stage 2, Brooklyn to Queens, Contract 543B. This tunnel would be 8.9 km (5.5 miles) long and intersect with the five previously built shafts. Many contractors took out bid documents for the project which had a bid date of January 14, 1993. The bidders were:

- A joint venture of Schiavone Construction Inc. of Secaucus, New Jersey and the J. F. Shea Co., Inc. of Walnut, California (Schiavone/Shea)
- A joint venture of Grow Tunneling, Perini, and Skanska
- Kiska Construction

Although the bid documents did not specify how the tunnel was to be excavated, the NYCDEP anticipated that this would be done using a tunnel boring machine. The tunnel permit restricted muck removal and concrete deliveries to the main work site located at Shaft 23B. It was determined that the labor cost and time to do the work from one shaft using drill and blast methods would be prohibitive when compared to the cost of a TBM drive. An additional cost for drill and blast is extra concrete required to fill overbreak caused by blasting. Also, the rock support requirements would be greater for drill and blast due to the loosening of the rock caused by blasting.

This was to be the first New York City water tunnel successfully driven using a TBM. Another first in the NYCDEP's history was the reliance on an OCIP or "wrap-up" insurance program. The City was to purchase the worker's compensation and liability insurance for this project. With worker's compensation rates for tunnel workers in New York approaching 100 percent, this would greatly reduce the contractor's bid.

On the morning of January 14, 1993, bids were submitted by the three contractors. When the bids wer opened, Schiavone/Shea was the apparent low bidder with a bid of $138,768,000, approximately $2,000,000 under the second bidder, the Grow-Perini-Skanska JV.

Schiavone/Shea received the Notice of Award on February 18, 1993. They received the Order to Commence Work on May 3, 1993. The Contract Completion date was February 10, 1998.

GENERAL JOB DESCRIPTION

The project consisted of excavating a tunnel which was 5.8 meters (19 feet) in diameter, 8.9 km (5.5 miles) long and placing a concrete lining for a finished diameter of 4.9 meters (16 feet). The tunnel would start 168 meters (550 feet) below ground surface, at the bottom of existing Shaft 23B in Red Hook, Brooklyn. It would travel northeast on a downward slope of 0.41 percent. This slope was to be maintained throughout the entire tunnel. Problems were expected since two major faults zones were predicted to cross the tunnel only 762 meters (2,500 feet) from the start. These faults had produced 0.21 m^3/s and 0.07 m^3/s (1,100 gpm and 3,400 gpm) of water during construction of City Tunnel No. 2. Provisions would have to be made to grout off or pump large volumes of water.

After 1,534 meters (5,030 feet), the tunnel curved 45 degrees to the left. The curve length was 240 meters (789 feet) and had a radius of 305 meters (1,000 feet). Only 34 meters (112 feet) from the end of the curve, the tunnel passed beneath Shaft 22B. The tunnel was straight 2,616 meters (8,584 feet) to Shaft 21B and then had two small curves within the 2,155 meters (7,071 feet) to Shaft 20B. From Shaft 20B, it was straight 2,043 meters (6,703 feet) to the end of the tunnel. Shaft 19B would be 61 meters (200 feet) from the end of the tunnel.

The Tunnel Permit for the project had several restrictions. All excavated material was to be delivered to the Staten Island Landfill which is under the control of the New York City Department of Sanitation. It was to be used for haul roads and as cover for the garbage.

Another restriction was that all concrete was to be supplied from Shaft 23B, the start of the tunnel. This meant that the concrete would have to be dropped down Shaft 23B, travel 9 km (5.5 miles) to its initial point of placement, and be placed within two hours of batching.

SITE SET-UP

The location of the project site was at 701 Hicks Street in Brooklyn, New York. Brooklyn is a very congested and mostly residential area. The site to be used by the contractor was a small empty lot in the shape of a triangle. The two legs of the triangle were 61 meters by 91 meters (200 feet by 300 feet). This gave the contractor a total of 2,775 square meters (30,000 square feet) to work with, less than an acre of land. Shaft 23B was located at the narrowest part of the iot making access more difficult. Access was constrained by requiring all trucks to enter from one end of the site, travel through the site adjacent to the working shaft and out a gate on the other side of the site. This limited the work site, because a haul road was to be maintained through the site, throughout the project. See site plan drawing. (See Figure 2)

The area in which the site was located in, called the Redhook section, was a high crime area. The previous contractor for Shaft 23B had problems with both break-ins and theft. To attempt to make the site secure, Schiavone/Shea constructed perimeter fencing consisting of soldier piles with pre-cast concrete planks between the flanges and a double row of razor wire on top. The fence was over 3.7 meters (12 feet) high. For further protection, a security guard was on site around the clock.

Some of the items located on the site were a maintenance shop, electrical substation, first aid trailer change houses for workers, compressor bank, water settling tanks, emergency generator, vertical belt, and crane. Once these items were in place, there was little space left for storage of spare parts, equipment, utilities, material, and muck storage. A yard in Newark, New Jersey was used for storage and equipment repair with daily deliveries to the job.

Since the muck storage bin was very small, the maximum amount of muck it could

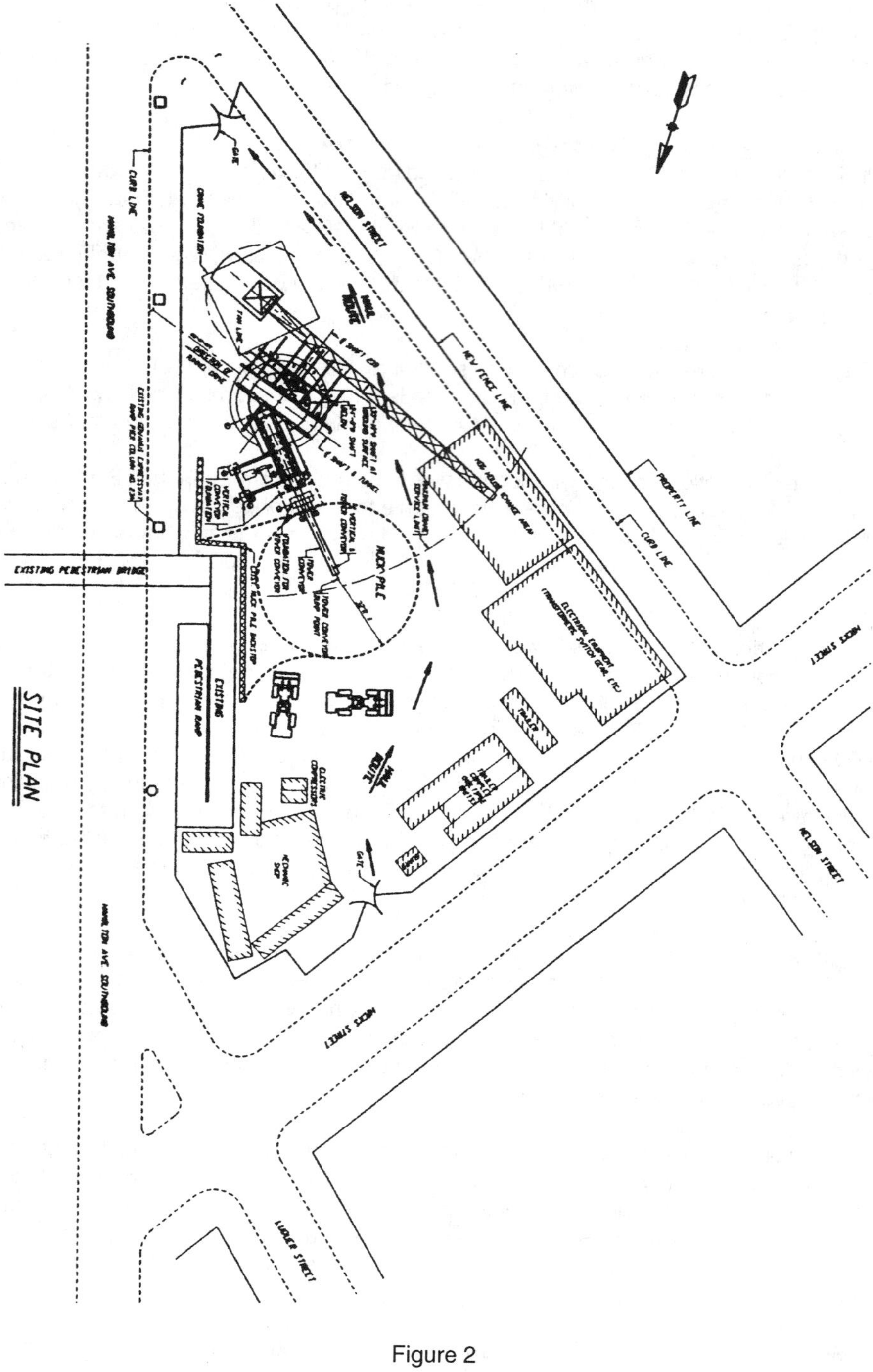

Figure 2

hold was 24.4 meters (80 feet) of tunnel excavation. Therefore, the trucking of muck became very critical. After excavation had started, and muck was stored in the muck bin, it was discovered that the pedestrian walkway adjacent to the bin was settling. The pedestrian walkway was underpinned, jacked, and monitored for the remainder of the project. The settlement also resulted in a smaller muck bin. Now the muck pile was limited in holding only about 20 meters (65 feet) of tunnel excavation.

With the site filled to capacity, no space was left for the City's office complex or the Contractor's office. This forced the Contractor to rent a small lot across the street where both offfices could be located. Another nearby lot was rented and used for employee parking.

The item that took the most time in the set up of the site was the electrical substation which required three independent 33,000 volt Con-Edison service feeders to supply it. Although only one feeder was needed, three feeders were used to have a redundant system to eliminate the possibility of the electrical service being cut off during the mining or concrete operations. In New York, Con-Edison is constantly updating service in different parts of the City. To do this, certain feeders are shut-down. Twice during the tunnel drive, the Contractor was down to one feeder. Three transformers reduced the 33,000 volts to 13,000 volts.

GEOLOGY

As mentioned previously, City Tunnel No. 3 lies approximately 183 meters (600 feet) below the ground surface. This includes an average of 61 meters (200 feet) of overburden and over 122 meters (400 feet) of rock. The rock formation that City Tunnel No. 3 lies in was formerly called "Brooklyn Injection Gneiss" but is now called the "Hartland Formation". Its two basic rock types are felsic gneiss and mafic gneiss. The rock was found to be very hard, strong, and abrasive. Also, there is been a complex intermixing of metamorphic and igneous rock throughout the tunnel. There are high contents of garnets throughout the formation, which indicates a high degree of metamorphism. Other rock types in the tunnel are schists, amphibolites, diabase, pegmatite, gabbro, diorite gneiss, and gneissic granite. The rock's unconfined compressive strengths ranged from 138 kPa (20,000 psi) to 276 kPa (40,000 psi). More importantly, the rock was found to be very abrasive. Testing done during mining had Cerchar Abrasivity Indexes approaching six.

During construction, the rock mass quality was found to be generally good to excellent, with some areas of fractured rock. Prebid documents predicted approximately 18 fault zones, most of them small. Most of the faults proved to be located as predicted. The NYCDEP predictions were made by looking at the records for the construction of City Tunnel No. 2 in the same general area and projecting the faults encountered there into City Tunnel No. 3. Of these predicted faults, most caused little disturbance to the tunneling operation. Extra rock bolts, straps and channel were adequate for mining through these areas. Two of the faults were very blocky with a lot of crushed and ravelly rock. In these areas, the Contractor installed steel ribs and lagging.

Water inflow was of major concern to the Contractor. It was predicted by the NYCDEP that the contractor could expect flow rates of up to 0.19 m^3/sec (3,000 gpm) within shallow dipping fault zones. Most of this was predicted to occur in two fault zones.

DRILL AND BLAST WORK

The contract called for construction of 57 meters (188 feet) of 7.3 meter (24 foot) diameter finished tunnel and a 15 meter (50 foot) long transition going from 7.3 meter (24 foot) diameter to a 4.9 meter (16 foot) finished tunnel diameter. The larger diameter was specified by the Owner in anticipation that the contractor would require more room at the bottom of the shaft for set-up of the equipment.

What was excavated was a 8.5 meter (28 foot) horseshoe section, a 15 meter (50 foot) long transition section and 58 meters (190 feet) of 5.8 meter (19 foot) horseshoe startertunnel. Also constructed was a 9 meter by 3 meter by 3 meter (30 foot by 10 foot by 10 foot) deep sump for the tunnel dewatering pumps as well as a 4 meter by 12 meter by 6 meter (13 foot by 40 foot by 20 foot) deep muck hopper pit with a ramp for an inclined conveyor.

The drill and blast work started by using jack legs to drill out development rounds at the shaft bottom. When enough room was opened up a Joy HRB-11-Ms, two boom air jumbo was used to start driving the top heading.

The 8.5 meter (28 foot) horseshoe was driven by the heading and bench method. The top heading was down to spring line and was excavated blasting ten foot rounds. A Joy HRB-11-Ms (two boom air jumbo) and a Tamrock DHS-105B (hydraulic one boom jumbo) were used side by side for drilling the rounds.

After blasting, a Kamatsu 73D track loader loaded 9 m^3 (12 cubic yard) round steel buckets. The crane, a 227 metric ton capacity Link Belt 2300B hoisted and dumped the buckets. Two buckets were used with one being hoisted while one was loaded.

After completion of the top heading, the bench was excavated. The bench was drilled horizontally. A tail tunnel was also excavated with the tail tunnel drilling being done while the heading was being mucked. The starter tunnel was excavated single heading using the Tamrock jumbo. Lastly, the muck hopper pit and the sump were excavated.

After completion of the drill and blast excavation, a concrete work slab was poured throughout the length of the drill and blast area. Walls were poured for the sump and the hopper area.

For primary support of the rock, fully encapsulated No. 8 Dywidags were installed. Strapping and mesh was also installed as required for additional support.

EQUIPMENT SELECTION

TBM 203-205

At the time of the bid, the sponsor, Schiavone Construction had in its equipment inventory Robbins tunnel boring machine No. 203-205. This machine was built in 1979 for the 63rd Street Subway Tunnel in New York City. Later it was rebuilt and modified for use in Montreal, Canada. It was decided to use the major components of this machine as the basis for a new machine for Contract 543B.

There had been advances and improvements in the design and construction of TBM's since 203-205 was built. The primary difference between this machine and newer machines was the cutters. When 203205 was designed, the 394 mm (15.5 inch) diameter cutters had a rated load capacity of 178 kN (40,000 lbs). The latest 432 mm (17 inch) and 483 mm (19 inch) diameter cutters have rated load capacities of 245 kN (55,000 lbs) on up to 311 kN (70,000 lbs). In hard rock, such as that found in New York, the penetration rate of the machine is dependent upon the amount of load put on the cutters. More load gives more penetration. However, to put more load on the cutters requires more propel thrust. Also, the more the cutter penetrates, the more drag it creates which increases the amount of power required. The cutter selected was the 432 mm (17 inch) cutter.

Table 1 compares the specifications of the original 203-205 to the new 203-205 machine.

Table 1

FEATURE	ORIGINAL 203-205		NEW 203-205	
DIAMETER	6.14 meters	(20 ft 2 in)	5.76 meters	(18 ft 11in)
CUTTERHEAD POWER	895 kW	(1200 hp)	2,088 kW	(2800 hp)
SPEED	5.36 rpm		8.90 rpm	
CUTTER CAPACITY	178 kN	(40,000 lb)	289 kN	(65,000 lb)
THRUST	8,006 kN	(1,800,000 lb)	17,712 kN	4,000,000 (lb)
ELECTRIC TRANSFORMER		2 - 1,000 kVA		2- 2,500 kVA
CUTTER SIZE	394 mm	(15.5 in)	432 mm	(17 in)

The major components of the old machine, the cutterhead support and inner cutterhead, the main beam, and the gripper carrier became the backbone of a new machine.

Construction Tunneling Services designed the new machine. The TBM was not only designed for the Brooklyn to Queens tunnel, but also to be increased 1.9 meters (four feet) in diameter should the joint venture be successful in obtaining the follow-up contract, the Queens Tunnel.

New gripper shoes, bottom shoe, roof and side supports, and thrust lugs were fabricated. Propel cylinders from the 9.8 meter (32 foot) diameter Kenny-Kiewet-Shea machine in Chicago were used. Other major changes included; reinforcing the main beam, a new high capacity bearing, a new bull gear, a new bucket ring and cutter housings, and six propel cylinders versus the original four cylinders

Eight new 261 kW (350 hp) 575 volt motors were installed. New gearboxes were built and installed. New hydraulic, lube oil, and electrical systems were also installed.

The fabrications were jobbed out. Assembly and testing was done at the Bethlehem Corporation located in Easton, Pennsylvania. Schiavone/Shea employees were responsible for procurement, planning, and supervision of the construction of the new TBM.

When assembly and testing was completed, the new high power, high thrust TBM was broken down and shipped to the job.

Trailing Gear

The back-up equipment consisted of 89 meters (292 feet) of single track trailing gear with a conveyor overhead. On the trailing gear was the electrical transformers as well as the lube oil and hydraulic systems. An 89 meter (200 foot) long 762 mm (30 inch) wide conveyor transported the muck over the top of the trailing gear to load muck cars near the back of the trailing gear. This trailing gear had been used on previous projects in Milwaukee, Wisconsin and Chicago, lilinois.

Behind the single track trailing gear, 89 meters (292 feet) of double track trailing gear was located. This trailing gear had been used in Montreal and originally had a superstructure to earry a conveyor and tripper to load muck cars. For 543B, only the bottom section was used and this functioned as a ~California" switch. While one muck train was being loaded, the empty would be waiting on the second track. This trailing gear was built like an open top boat with steel punchings serving as ballast to prevent over-turning due to the unbalancing influence of a loaded muck train.

Track

The main trackage was single track and consisted of 85 pound ASCE rail laid on steel ties. This was installed between the TBM and the trailing gear with the trailing gear being pulled along on the rail.

Muck Transport and Removal

For muck haulage, six Mulhauser side dump muck cars were hauled by 22 metric ton diesel Schoma locomotives. This equipment was the same as that used on the Chunnel. Each car had the capacity to hold 0.30 meters (one foot) of excavated tunnel, thus a six car train corresponded to the 1.8 meter (six foot) stroke of the TBM's propel cylinders. The six car trains weighed 200 metric tons.

The muck trains would travel back to the shaft and were automatically dumped by moving the loaded cars through a Mulhauser car dump. The muck would fall into a 12 meter (40 foot) long hopper and be fed out of the hopper by a Nico apron feeder. The feeder dumped onto a 48 meter (160 foot) long inclined conveyor that carried the muck upward and discharged into a vertical belt conveyor.

The vertical belt conveyor was supplied by Lakeshore Mining and had a capacity of 550 metric tons (500 tons) per hour. The muck was carried vertically 168 meters (550 feet) and discharged onto an outfeed conveyor. (See Figure 3) The muck piled up and was loaded into dump trucks by a Cat 966E loader. The muck was then hauled to the City of New York landfill, Fresh Kills, for disposal.

Personnel Hoisting

For personnel transport in and out of the tunnel, a personnel hoist was purchased. This was an "Alimak" style hoist with a four legged mast holding a gear rack. The hoist had a 13,200 kg (6,000 lb) capacity and could hold 25 workers. Transportation in and out of the heading was by means of a man car and locomotive.

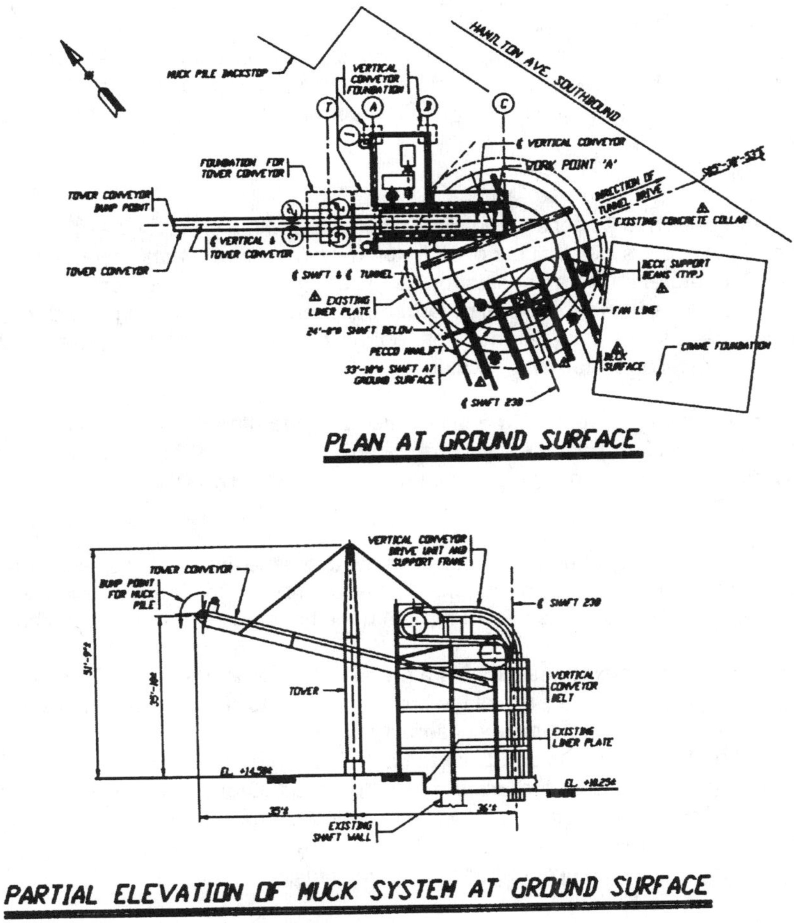

Figure 3

Fire Protection

The need for one item of equipment that was not anticipated during bidding was a fire suppression system for the vertical belt conveyor. During job mobilization, vertical belts in Chicago and Boston caught fire and caused lengthy shutdowns of the projects. As a result, a sprinkler system was installed. The system consisted of two vertical down pipes with open nozzles installed on 3.8 meter (8 foot) centers. A sensor wire was hung in the shaft and if the temperature in the shaft exceeded 49° C (120° F), a solenoid operated valve opened and the belt was flooded with 0.11 m³/sec (1,800 gpm) of water.

Ventilation

The tunnel was ventilated using tandem Joy axi-vane fans mounted on the trailing gear. Air was blown out of the tunnel by the fans and into a 1,524 mm (60 inch) diameter flexible fan line. A cassette held 152 meters (500 feet) of fan line. The cassette was mounted on the rear of the trailing gear and the line would play out behind as the tunnel advanced. The air would go through the fan line, into an elbow and up the nearest shaft. The air being blown out of the tunnel was replaced by fresh air coming from the main work shaft (Shaft 23B).

Water Control

The Brooklyn to Queens tunnel was constructed with a grade of 0.41 percent, a grade change of 4.1 meters per thousand. Unfortunately, the grade change was required to be downhill. Therefore, it was a constant battle to keep the water away from the face. Pumps were always carried with the machine at the heading. Water was pumped through a discharge line back to the sump at the shaft. As mining progressed, intermediate sumps and pumps were installed to cut-off the water.

For dewatering or removing groundwater from the tunnel, three identical Ash pumps set-ups were installed. Each set-up consisted of a 75 kW (100 hp) charging pump and a 150 kW (200 hp) pump to push the water up to the surface. The total pumping capacity was 0.19 m^3/s (3,000 gpm).

On week-ends, a pump watch was required to ensure that should a malfunction occur, the pumps could be repaired so that the TBM would not be flooded. To keep watch on the pumps, remote video cameras were installed at the heading and at the main sump.

The contract required maintaining a probe hole at least 9 meters (30 feet) in front of the machine. The primary purpose of the probe hole was to check for potential groundwater in front of the machine. If a substantial amount of water was encountered, additional holes were drilled and grout was pumped into the holes. At several locations, several shifts were spent grouting off inflows so that mining could proceed. Grouting was done using a Hany piston pump. The grouting equipment was mounted on a flat car and would be brought into the heading when needed. Any water that could be eliminated, allowed more time before flooding of the TBM would occur should the pumps fail.

Rock Support

Rock support was furnished by 2.44 meter (8 foot) long Swellex bolts. Drilling was done immediately behind the roof support of the TBM using two Atlas Copco 1238 roof bolters. Each drill installed two bolts per side. Mounted on a plaftorm behind the grippers, Boart hydraulic drills installed a bolt on each side near spring line. The contract required installation of at least two bolts per foot of tunnel.

Additional support was supplied by C6x8.2 rolled channels installed to bridge between bolts. Tensar fabric mesh was also installed. In four locations, full circle ribs were installed.

MINING DESCRIPTION

Start-Up

After installation of the mining equipment, TBM tunnel excavation began. The TBM started mining 130 meters (428 feet) from the center of Shaft 23B. There was only enough space for a short set-up of the trailing gear since the overall length of the TBM and trailing gear was over 204 meters (690 feet) long. The short set-up used only enough trailing gear to carry the equipment required to run the TBM. Using the short set-up meant that the sections of the trailing gear allowing a six car muck train to be loaded would have to be eliminated along with the trailing "California" switch. Therefore, the mining began with short two car muck trains. The short set-up was used to mine 246 meters (807 feet) of tunnel. Mining progressed well during this time.

Production Mining — See Figure 4 for general arrangement of equipment at shaft bottom.

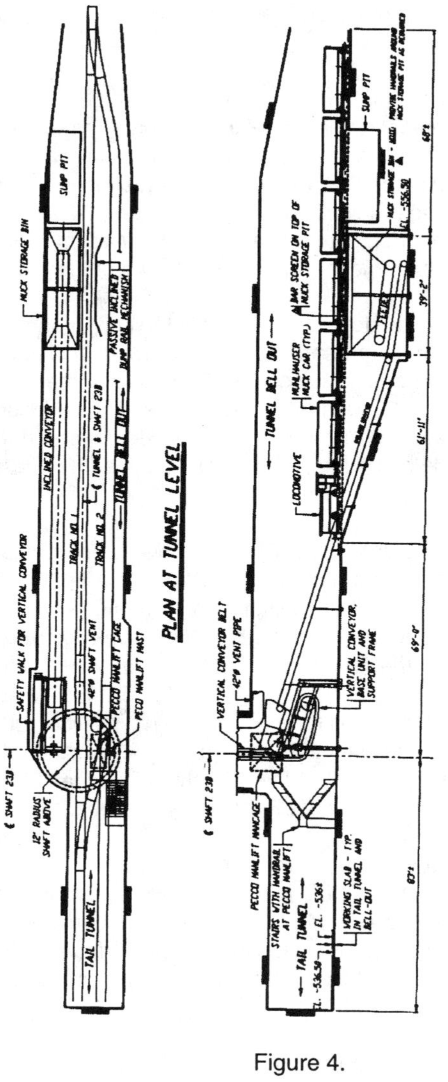

Figure 4

Figure 4.

At this point, the remaining trailing gear was installed and production mining began. Mining began using two six car muck trains. For rock support, contract specifications called for two rock bolts per foot of tunnel. A six bolt pattern every 0.9 meters (three feet) of tunnel was agreed to by the Contractor and the Owner. Extra bolts and five inch steel strapping were installed in areas of blocky or fractured ground. As mining progressed, the tunnel excavation reached the area in which high water inflows were expected. While drilling a probe hole, 1.89 liters/sec (30 gpm) of water was found. The contractor decided to grout the area off. Five additional holes were drilled and all the holes produced 1.58 liters/sec (25 gpm) or more, never cutting down the inflow at the other holes. Packers were placed in the holes and a thin mix of cement and water was used to grout off the area. A Hany grout pump was used for this operation. After 13 days of grouting, mining resumed. The contractor mined through the fault and a very minimal amount of water infiltrated the tunnel. Grout was found within the seams in the rock.

About 1.8 kilometers (1.13 miles) into the tunnel, was the first intersection of a down shaft (Shaft 22B). The shaft intersected the tunnel at the crown and was not designed to be offset to the tunnel. This meant that the survey must be correct in order to intersect it properly. Making things more difficult was that there was a curve just prior to Shaft 22B. The curve had a 305 meter (1,000 foot) radius and was approximately 241 meters (790 feet) long with only 38 meters (125 feet) of straight tunnel after the curve for making corrections. The center of the shaft was within 5 cm (2 inches) of the centerline of the tunnel.

At shaft 22B, mining was halted for minor repairs to the cutterhead. New edges had to be installed on the buckets of the cutterhead. Some minor welding was done on the cutter housings. After mining started up again, the geology of the rock changed and mining became more diffcult. Penetration rates decreased and cutter consumption increased. The rock varied greatly and there was a three to one difference between the highest and the lowest penetration rate. Because of the high cutter wear, it became necessary to stop after every other train and check cutters.

The excavation continued to shaft 21B. Shafl 21B was also intersected within 5 cm (two inches). Minor repairs were performed on the TBM at this location. The mining progressed suffficiently at this point to install a California switch midway between the heading and Shaft 23B. This was done to allow a third train to haul muck with the other two trains, minimizing the chances of having the TBM wait for muck trains while mining.

Mining continued with periodic stops for grouting off water infiltration. A fault was soon reached, which the Contractor thought he might excavate through using rolled steel channel for additional rock support. After mining only a few feet, the rock began to pop. The rock was stressed and 24 sets of ring beams were installed at a spacing of 1.2 meters (4 feet).

There was also one area that had decomposed rock fallout near the left gripper. The decomposed rock was so soft that when the grippers were extended outward, the “rock” was squeezed out leaving a void. This area was formed up with steel channels and wood lagging. Shotcrete was used to fill the void and supply a bearing surface for the gripper. Mining continued until tunnel intersected Shaft 20B. Maintenance work on the buckets was done and mining resumed.

From Shaft 20B to the end of the tunnel was a straight run without any curves. There were many faults within the last 2,042 meters (6,700 feet) of mining. Two unexpected faults caused great cliffficulty in mining. They contained a large amount of decomposed rock or clay. The first area was approximately 24 meters (80 feet) long and the second was approximately 12 meters (40 feet) long. Mining through these areas proved to be very difficult. Steel ring beams had to be installed with steel lagging all the way to spring line. The TBM could only mine one foot at a time. After mining a foot, the miners climbed through the cutterhead and hand mucked the buckets. Another foot was mined and the buckets would be cleaned again. This sequence continued until the TBM was back into competent rock. Because these conditions were not anticipated by the Contractor or bv the NYCDFP, Change Orders were issued to compensate the Contractor for additional costs.

When the mining was within 152 meters (500 feet) from the end of the tunnel, more problems occurred. Water was found in a probe hole and grouting was done. The grouting operation took several days to complete and used over 1,800 bags of cement. The total amount of water infiltration in the tunnel at the end of mining was approximately 60 liters/sec (950 gpm). Mining continued until 107 meters (350 feet) from the end. Fractured and blocky ground was encountered requiring steel ring beams for support. Thirteen sets were used in this area. The grouting and ring steel delayed the holing through about three weeks. Finally, the TBM mined into the tail tunnel for the Queens Tunnel at Shaft 19B.

CONCLUSION

Excavation of City Tunnel No. 3 was successfully completed. The TBM performed very well, considering the extremely strong and abrasive rock that was mined. However, it turned out be not just a tunnel driving job, it was a cutter changing job. The hard, abrasive rocks in the Brooklyn Queens Tunnel wore out more than 4,500 cutters. Within a year 543B should be totally completed which will bring the City one step closer to the completion of City Tunnel No. 3.

7

Difficult Ground

Chairman
R.L. Shufflebotham
S.A. Healy Co.
Plano, TX

Chapter 20

CONSTRUCTION OF THE LEWIS STREET CSO TUNNEL, NASHVILLE, TENNESSEE

Patrick J. Doig
Assistant Construction Manager, W.L.Hailey & Co. Inc.

ABSTRACT

The Lewis Street CSO Tunnel is an important element in the efforts of the Metropolitan Government of Nashville and Davidson County (Metro) to improve the treatment of wastewater in the area.

The tunnel is approximately 1,675m (5,500ft) long with a finished diameter of 2,600mm (102inches). 116m (380ft) of tunnel was constructed in soil with the balance being in limestone rock. Excavation was with a 3,700 mm (12ft) diameter Jarva hydraulic shield in soil and a Jarva Mark 12 TBM in rock. The permanent lining of the tunnel comprised concrete pipe with steel end-rings and a vinyl liner.

This paper describes the construction of the project with particular emphasis on the excavation and lining of the tunnel.

INTRODUCTION

The Metro government is under a mandate from the Tennessee Department of the Environment to improve the quality of discharges into two local waterways, the Cumberland River and Brown's Creek. Metro has contracted with Consoer Townsend & Associates, Inc. to design and supervise the construction of an upgrade to the area's sewage treatment facilities.

One element in the revised system is the Lewis St. Combined Sewer Overflow Tunnel. This tunnel was designed to act both as a conduit to the new detention basin being constructed at the Driftwood site and as a temporary storage facility. The tunnel is 1,675m (5,500ft) in length with a 2,600mm (102inch) finished diameter. Under normal circumstances, flows in the tunnel would be minimal and would require a very much smaller pipe. However, under storm conditions the pipe would fill and would provide some storage capacity. Flow in the pipe was designed to be by gravity and so a vinyl liner was specified to reduce the friction under low flow conditions. Additionally the liner was intended as a protective barrier against hydrog

Also included with the tunnel are three diversion structures that will pick up flows from existing facilities. These are located at Winfrey St. and Trimble St. where they intersect with Lewis St. and at the extremity of the tunnel in Lewis Street itself.

The contract for the construction of the Lewis St. project was bid on November 17th, 1994. There were five bidders and the low bid was submitted by a joint venture of W.L.Hailey & Co. Inc. of Nashville, Tennessee and Frontier-Kemper Constructors, Inc. of Evansville, Indiana. W.L. Hailey were the managing partner. The bid amount was $13,053,700 with the second bidder approximately three percent higher. The contract started on February 21st, 1995.

DESCRIPTION

The Lewis St. tunnel is located about three kilometers to the south and east of downtown Nashville, just south of the Cumberland River. The area is a mix of businesses, mostly service industries and low-income residences. There was good access to the site from adjacent freeways.

The tunnel is located under Lewis St. for most of its length. It also passes under Interstate 40 and U.S. Highways 70N and 70S and the Nashville and Eastern Railroad.

The tunnel is oriented in a north-south direction with the Driftwood Basin, being constructed under a separate contract at the north end and the Lewis Street structure at the south end. The northern terminus of the tunnel is at station 1+86 and the southern terminus at station 56+85 (all stations are in feet). Access to the tunnel during construction was from a main work shaft located at station 6+75. The Winfrey Street structure is at station 31+95 and the Trimble Street structure at station 41+26.

The depth of the tunnel varies from 8m (25ft) at the northern end to 21m (70ft) where it passes under the interstate.

GEOLOGY

The geotechnical report produced by Ogden Environmental and Energy Services of Nashville described the likely geological conditions as follows:

Subsurface conditions through which the proposed sewer will pass include an approximately 113m (370ft) long reach of intermixed, soft to very soft and loose, silty sand, sandy clay and silty clay within the northern, flood plain portion of the alignment. A high ground water table also exists within that area and the proposed sewer invert elevation is below the static water table throughout the alignment. Beginning at approximate Station 5+30, located at about the center of the railroad right-of-way, and progressing southward, there is a change of material from soil to bedrock. The contact between rock and soil is not vertical but slopes downward to the north at roughly a 2H:IV inclination. The transition from soil to rock will occur gradually over about a 9m (30ft) length with a possibility of boulders or deep, soil-filled slots in the bedrock on either side of the transition zone.

After the transition into bedrock two different bedrock types will host the sewer. From the soil/bedrock interface at about Station 5+30 to approximately Station 45+00, thin, interbedded shale and limestone of the Hermitage Formation will be penetrated within the tunnel zone. That rock type has an average composition of 65% calcium carbonate, 23% silica and 10% magnesium carbonate. The average unconfined compressive strength is about 90mpa (13,000 psi). Weathering of the Hermitage Formation is generally minor but isolated zones do exhibit significant weathering such as open or clay-filled seams due to solutioning processes. From about Station 45+00 to the southern terminus of the project at Station 56+85, the tunnel zone will penetrate densely crystalline limestone of the Carters Limestone. The Carters Limestone consists of upper and lower units, both of which will be encountered. The upper, thin-bedded portion of the Carters is separated from the lower, massive Carters by a several inch thick interval of bentonitic clay, The average composition of the Carters Limestone is 81% calcium carbonate, 10% magnesium carbonate and 7% silica. The average unconfined compressive strength is about 117mpa (17,000 psi). Solution weathering can be intense within the Carters Limestone and several moderately sized (less than 600mm (2ft)) cavities were encountered.

The majority of the rock tunnel should encounter relatively sound, unweathered rock. However, small zones and reaches of weathered rock and soil-filled or open cavities will likely occur over isolated areas. More widespread bedrock weathering should be anticipated where rock cover above the tunnel is less than 3m (10ft) such as near the soil/rock transition and in the vicinity of Trimble Street and southward. Also, a thin, horizontal zone of bentonite clay will be encountered along a section of tunnel at the south end of the project between Stations 46+00 to 56+85.

Static ground water levels were observed to be above the proposed sewer elevation throughout the alignment. As such, excavations and tunneling operations at the sewer elevation will encounter wet conditions. Along the bedrock portion of the alignment, the more significant ground water inflows will be restricted to those which occur along existing fractures and openings in the rock mass. Sound, intact rock should be relatively dry. As weaknesses are encountered, localized flows could be on the order of several hundred gallons per minute. In rare instances, larger weaknesses or cavities could produce flows on the order of several thousand gallons per minute.

The diversion structures will require excavations ranging from about 6m (20ft) to 13m (42ft) in depth. Those excavations will generally penetrate a thin interval of soil followed by weathered bedrock and then competent bedrock. The soil cover includes random fill and residual clays ranging from about 1m (3ft) to 4m (14ft) in thickness at the various structures, The weathered bedrock interval appears to extend a minimum of two feet below the general rock surface and in some instances, can extend as much as 3m (10ft) below the general rock surface, Allowable bearing capacities for the bedrock will range from 20 KSF in weathered rock to 100 KSF in sound rock.

INITIAL DEVELOPMENT

Excavation of the main work shaft was completed during the month of March, 1995. This operation was delayed somewhat while final easements were obtained. The shaft was 10m (32ft) in diameter and was supported through the overlying soils with ribs and timber lagging. Support in the rock was with rock bolts. The shaft was approximately 11m (35ft) deep.

Tunnels were then driven by conventional means in either direction from the shaft. To the north, the tunnel was driven to the point where soil was just apparent in the crown at station 5+63. The tunnel was then halted while a soft ground shield was obtained for the remainder of the excavation.

Figure 1. Looking Toward Main Work Shaft from Soft Ground Starter Tunnel

To the south, a starter tunnel for the TBM was driven to station 8+17, a distance of 38m (126ft) from the edge of the shaft. A 4m (13ft) length of this tunnel was then concrete lined to 3,700mm (144in) diameter to accommodate the TBM that was to be used for the drive.

The Winfrey and Trimble Street shafts for the diversion and access structures were also excavated during the early months of the contract so that they would be ready to receive the TBM as it passed through. The Lewis Street shaft was completed some months later.

EXCAVATION OF ROCK TUNNEL

The excavation of the rock tunnel was carried out with a Jarva Mark 12 TBM owned by Frontier Kemper. This machine was delivered to the site during May, 1995 and started mining on May 29th. It was halted when it reached station 10+03 on June 23rd for installation of all back-up gear and replacement of temporary utilities with permanent ones. This initial start up phase of 57m (186ft) of tunnel had taken 20 days.

The machine cut a 3,700mm (144in) bore and had a 73mm (24in) stroke. It was an older "X" style machine which was supported by the grippers against the walls. Later "T" style Jarva TBM's had a leg that extended to the invert to support the machine. Muck was discharged from the conveyor behind the machine into muck cars which were lifted to surface with a crane for dumping. Diesel locomotives were used to pull a train of cars which was sufficient to receive the muck from two strokes of the TBM.

This initial length of tunnel had included a 12m (40ft) long section at station 8+80 which had been difficult to negotiate. It comprised mud seams and solution cavities which contained boulders and detritus. Water inflows from these features measured up to 20l/sec

(100 gpm). The tunnel was supported through this area with ribs and lagging and rock bolts.

Installation work was completed on July 12th and the TBM began advancing again on that date. Thereafter mining was conducted on two ten-hour shifts.

The 664m (2,181ft) run of tunnel from station 10+03 to the Winfrey Street shaft at station 31+84 was completed on September 7th. During this period, 80 shifts were worked for an average advance of 8.3m (27ft) per shift. Best day's production was 41m (135ft) and best week was 158m (517ft). The main impediment to production during this period was the inflow of water. Water had been intercepted at a number of points along the drive and the combined ingress reached 78l/sec (350gpm). Some support was also needed at each of these locations.

When traversing the shaft at Winfrey Street, loose rock at the upstream face of the shaft had caused some damage to the TBM and this required over a week to repair. The machine eventually restarted on September 18th.

The run to the Trimble Street shaft at station 41+40 was 291m (956ft) in length. This section required careful control of the machine as the alignment followed an "S" curve. The TBM holed out at the structure on October 19th after 48 shifts of work. Average production was 6.1m (20ft) per shift which reflected the difficulty of the mining. Best day during this run was 32m (106ft).

Once again, problems were experienced with loose rock at the shaft face and the machine was delayed a further week before mining could restart on October 26th.

The final run of tunnel to the Lewis Street structure at station 56+85 was completed on December 20th, 1995. This length of tunnel measured 471m (1,545ft) and was completed in 76 shifts at an average rate of production of 6.1m (20ft) per shift. Best day's production was 34m (112ft).

As the geotechnical report had anticipated, the Carters limestone was present in much of this run with its upper and lower units separated by a thick band of bentonitic clay. This clay layer proved to be very problematic. A pattern of split set rockbolts was routinely installed in this area but where the layer was 1m (3ft) or less below the roof of the excavation, it caused fall out. This became so extensive in one area that partial rings of liner plates had to be installed to ensure the safety of the tunnel.

In summary, the 1,484m (4,868ft) of tunnel driven with the TBM was mined in 224 shifts at an average production of 6.6m (22ft) per shift. Difficult ground conditions, high water inflows and slow downs at the intermediate shafts had been the main limitations on production.

EXCAVATION OF SOFT GROUND TUNNEL

A Jarva 3,750mm (147in) diameter hydraulic shield was purchased for mining of the soft ground tunnel and was delivered to the site in August, 1995. The shield featured 12 shove rams and was equipped with an overhanging hood and face tables for ground support. Muck removal was by hand mucking on to a conveyor which discharged into a single

Figure 2. Jarva Mark 12 TBM Breaking Through at Lewis Street Shaft

muck car. The shield had a 1,200mm (4ft) stroke and was approximately 4.6m (15ft) long. It was not equipped with any articulation but did have extending fins to compensate for roll.

Prior to starting the mining of the soft ground section of tunnel, ground control measures were instituted from surface. Dewatering wells were installed along most of the tunnel and these were developed so that the water table was lowered sufficiently to allow mining to proceed in relatively dry conditions. In addition, the tunnel passed under a high pressure gas line which had to be supported. This was achieved by driving soldier piles to the rock and then supporting the pipe with beams tied to these piles.

Mining with the shield began on September 11th. For the first 16m (53ft) through the rock/soil interface and beyond, this was an extremely tortuous process. The need to blast in the face while supporting running soils in the roof required considerable care. It was not until January 2nd that excavation could proceed at a more regular pace. Support of the tunnel throughout was with steel ribs at 1,200mm (4ft) centers with timber lagging.

Thereafter, the shield progressed somewhat slowly until reaching the exit shaft at station 1+86 on March 13th. Average production for 115m (377ft) of tunnel over 27 weeks had been only 4.3m (14ft) per week with a best week of 18.3m (60ft).

The tunnel alignment over this stretch was a slow continuous curve. This had caused problems for the control of the shield. Because of its length and its rear mounted shove rams, the shield did not respond quickly to changes in direction. The shield was frequently off-line and corrections were made on a continuous basis.

Figure 3. Jarva Shield After Removal

Additionally, the weight of the shield and particularly the concentration of the weight at the front end caused problems with grade. Once the tunnel was completed, a program of remining had to be instituted to give an acceptable alignment for the pipe. Even so, it was ultimately necessary to manufacture bevel pieces of pipe for this length to be able to string the pipe through the curving tunnel.

PIPE INSTALLATION

With the completion of the rock tunnel in December, 1995, preparations were undertaken for the pipe installation. This involved removing all the service lines, cleaning the tunnel invert, completing any support and then laying the finished rail that would support the pipe. The first pipe was installed on February 26th, 1996.

The 2,600mm (102") diameter Reinforced Concrete Pipe was designed and manufactured by Sherman-Dixie Concrete Industries of Nashville, Tennessee. The pipe featured steel end rings and an Ameron T-Lok liner which had to be welded at the completion of the installation.

The pipe came in 3,650mm (12ft) joints which each weighed 19,000kg (42,000lbs) and was manufactured at Sherman-Dixie's plant in Nashville. The size of pipe presented some logistical challenges. The trucker was required to have special overwidth permits which restricted transportation of the pipe from the plant to the job to daylight hours. With

Figure 4. Pipe Carrier

limited space available at the main shaft and installation planned for two shifts, it was necessary to very carefully coordinate the delivery schedule.

For transporting the pipe in the tunnel, a special pipe carrier was designed and built by the contractor. This ran on twin rail bogies and featured a main needle beam arrangement on which the pipe rested. A curved plate was attached to the top of the beam to support the pipe. At one end of the carrier was located a rubber bull wheel that ran on the interior of the pipe when the beam was being threaded through the pipe. The invert of the pipe was temporarily covered with a sheet of neoprene during transportation in the tunnel to protect the Ameron T-Lock Lining.

To allow precise placement of the pipe, the carrier was equipped with hydraulic cylinders which allowed the carrier to move the pipe into any position. The hydraulics were operated by an on-board Plyrnouth diesel locomotive engine.

Installation of the pipe in the bored tunnel was conducted on a two-shift operation. Station 7+11 at the main work shaft was reached on April 18th. During this period, 1,516m (4,974ft) of pipe was installed in 8 weeks at an average production of 190m (620ft) per week. Best week's production was 362m (1,188ft) and best day's production was 99m (324ft).

Installation of the pipe in the soft ground tunnel did not occur until later in the year and was more difficult due to the tighter tolerance and the curvature of the tunnel.

Once the pipe had been installed, backfilling of the annulus with grout took place. The specification called for lightweight grout fill with a unit weight of 1,060kg/cm (65pcf) and a minimum strength of 1.4 mpa (200psi). A subcontract for this operation was let to Elastizell of St. Louis, Inc. (ESL). It was expected that 4,800 cm (6,400 cy) of grout would be needed between the Lewis Street structure and the main work shaft.

Figure 5. Preparing for Welding of Pipe Joints

ESL started on site in early May and worked discontinuously until early September. The method for placing the grout involved pumping through downholes positioned at approximately 183m (600ft) centers. Daily production on a single shift was sometimes in excess of 150cm (200cy) but the average was somewhat lower. Problems were experienced in displacing the residual water that continued to flow in the annulus.

Finally, welding of the T-Lok liners was completed under a subcontract. The liner had a single lap joint and was welded by application of heat. Additionally, repairs were made to the liner at this time where damage had been done during the installation process.

CONCLUSION

The tunnel was substantially completed in October, 1996. Final completion, including connection of existing facilities to the tunnel could not be achieved due to delays to the Driftwood Basin at the discharge end of the tunnel. The contractor for this facility had defaulted in early 1996 and a new contract had to be let for completion. At the time of writing it was expected that the full system would not be operational until late 1997.

The main feature of the contract, the boring of the tunnel in rock proved to be a difficult process. The prevailing ground conditions were not conducive to a high-speed operation. Frequent encounters with voids carrying large flows of water caused interruptions to the mining that negatively affected the ability to build up any momentum. However, the pipe installation process, which was well planned and engineered gave balance to the overall performance.

BIBLIOGRAPHY

Ogden Environmental & Energy Services: Report of Geotechnical Investigation for Proposed Lewis Street CSO Project, October 27th, 1993.

Chapter 21

CONSTRUCTION OF THE INTER-ISLAND TUNNEL BOSTON HARBOR PROJECT

D.W. Ferguson,[1] R.L. Shufflebotham,[2] T.L. Page,[3] and A.J. Stewart[4]
[1]Project Manager, [2]Project Engineer, [3]Engineer for SA. Healy/Modern Continental JV.; [4]Area Manager, for Stone & Webster Civil and Transportation Services

ABSTRACT:

The Inter-Island Tunnel is a 7.7 kilometer (25,160 foot) long by 3.5 meter (11.5 feet) diameter concrete lined tunnel through the Cambridge formation below Boston Harbor. Constructed as part of the Boston Harbor Project, this tunnel will bring untreated waste water from the MWRA's South System headworks facility at Nut Island in Quincy, MA to the new Deer Island treatment plant. A myriad of problems delayed the construction of the tunnel. These problems included salt water inflows of over 22,000 liters (5,812 gallons) per minute, a major fire in the access shaft, weak rock and soil collapsing on the tunnel boring machine, and hand mining operations.

INTRODUCTION

In 1982 the city of Quincy filed a Law suit in Massachusetts Superior Court, charging that wastewater discharges to the harbor violated state law. This suit began a series of legal actions that culminated in the 1986 federal court order by a U.S. District Court Judge, who found that discharges to Boston Harbor violated the Federal Clean Water Act. The Court ordered the construction of a new primary and secondary treatment facilities and issued a scheduling order outlining a timetable for the cleanup of the harbor. This action created the Massachusetts Water Resources Authority (MWRA) a state agency empowered to carry out the court order.

The 22 communities in the MWRA's southern sewer system sent their wastewater to an antiquated primary treatment plant on Nut Island in Quincy. The MWRA decided to construct a cross harbor tunnel, rather than build a new treatment plant, to convey wastewater from a headwork's at Nut Island to the new Deer Island treatment plant. The following paper describes the construction of the Inter-Island Tunnel.

DESCRIPTION OF PROJECT AND DESIGN CONSIDERATIONS

<u>The tunnel design</u> contract was awarded to the Sverdrup Corporation in February 1989. The design was to be completed by July, 1990. The MWRA's construction schedule anticipated the tunnel contract award by December 1990 and completion of construction in November 1994. This schedule met one of the milestones set by the Federal Court order.

The Inter-Island tunnel is a 7.7 kilometer (25,160 ft) long by 3.5 meter (11.5 ft) diam-

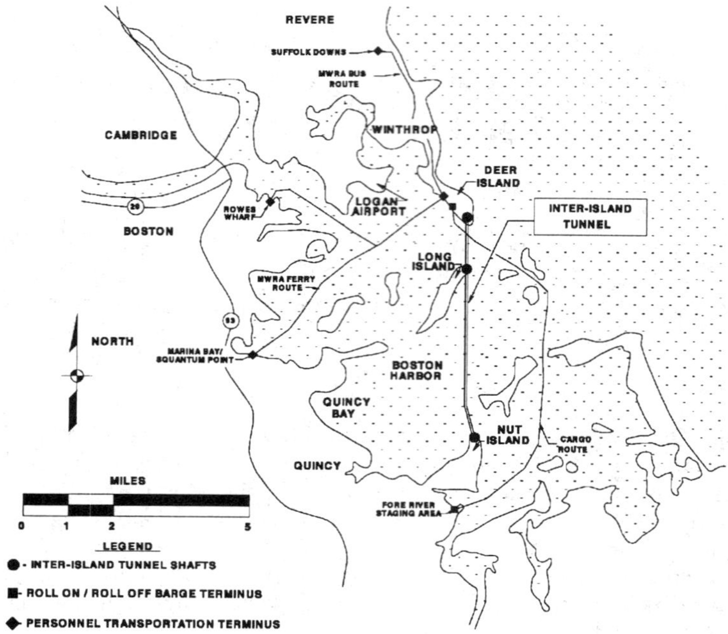

Figure 1. The Inter-Island Tunnel Project Location Plan

eter concrete lined gravity tunnel bored through solid bedrock below Boston Harbor. A 16-foot diameter shaft excavated on Nut Island will direct the screened and de-gritted wastewater from the Nut Island Headworks to the tunnel. The tunnel is capable of conveying up to 1514 million liters per day (400 mgd) of the MWRA's South System's wastewater flows to the shaft located at Deer Island. A third shaft, located at Long Island, accepts untreated wastewater from facilities there. See figure 1 for the location of the project.

SLUDGE RESIDUE from wastewater treatment is barged from the Deer Island plant to the MWRA's sludge processing plant constructed in Quincy MA. Long range plans are to pump the 1.5 million liters per day (400,000 gpd) of sludge through a pipeline running the 13 kilometers (8 miles) from Deer Island to Quincy. Therefore two 35.5 centimeter (14 in) diameter ductile iron pipes, installed and concrete encased in a secondary lining within the tunnel, will transport sludge from Deer Island to Nut Island. A future tunnel will complete the sludge line to the MWRA processing plant located at the Fore River Shipyard.

GEOLOGY

The host rock for the tunnel is the Cambridge Argillite. The Cambridge Argillite is the uppermost formation in the Boston Basin complex, a down faulted body of folded and faulted sedimentary—volcanic rock underlying the Boston harbor region. Made up of silt-size particles of quartz, feldspar, sericite, chlorite and kaolinite, the argillite is typically a fine grained laminar bedded rock similar in composition to shale but harder, more consolidated and not fissile. The argillite is generally light to dark gray but purple and greenish beds occur as well and were encountered in the tunnel. The tunnel intersected a number of diabase dikes, a medium to dark gray or greenish gray hard rock comprised of feldspars and marfic silicates. These diabase dikes cut along and across the argillite in the tunnel and were often broken and blocky. The argillite although generally competent and hard was known at a number of places in the area to demonstrate severe alteration or kaolinization resulting in decomposition of the rock to a soft, whitish clay—like material. This phenomenon proved to have a tremendous impact on the construction of the tunnel.

ADVERTISING AND BIDDING OF THE PROJECT

Special contractual provisions

A Geologic Design Summary Report (GDSR) was included as a concise presentation of geotechnical design and construction issues and was the sole contractual geotechnical baseline report.

A Disputes Review Board (DRB) provision and a change of condition's clause were included in the contract to resolve any disputes efficiently and to avoid delays in the work.

Liquidated Damages of $13,000 per calendar day would be assessed if construction milestones were not met.

A hard rock tunnel boring machine (TBM) capable of attaining excavation rates of 183 meters (600 ft) per week in Argillite while installing rock bolt support.

Site restrictions

ALL ACCESS TO DEER ISLAND is by a MWRA contracted transportation systems. Personnel transportation consists of busses, queued at an off site parking lot, that travel to the Deer Island peninsula through the town of Winthrop; and water ferries with terminus at Quincy, downtown Boston and Deer Island. All equipment and materials required for construction are brought to Deer Island by water using the roll-on/roll-off vessels departing from a staging area at Quincy twice a day.

THE WORK SITE for the tunnel construction consists of an area of only 2832 square meters (.7 acres), due to a limited amount of land and the large amount of construction on Deer Island.

S.A. Healy/Modern Continental 's bid

The S.A. Healy Co. and The Modern Continental Construction Co. joint venture (Healy/Modern) was founded to take advantage of each company's experience, Healy's expertise in tunnel construction and Modern Continental's expertise in the Boston area construction market.

Water inflow and ground support assumptions made by Healy/Modern. In preparing their bid Healy/Modern anticipated a relatively dry tunnel with overall TBM excavation rates of 32 meters per day (126 lf/day) and concrete lining rates of 33 meters per day (130 ft/day).

The GDSR stated the groundwater inflow estimate at 695 liters per kilometer (56 gpm/ 1000 ft). An analysis made by Healy/Modern's geological consultant found that steel ground support assumptions in the bid documents were based upon drill and blast tunnel experience in the area and did not consider ground support advantages achieved with a tunnel boring machine. Healy/Modern only included in their bid the cost to install 31% of the ring beam support called for in the bid documents.

The as bid schedule included duration's of twelve months for mobilization and shaft sinking; thirteen months for tunnel excavation and ten months for the tunnel lining. A nine month duration was included to install and encase the sludge line pipes; the connecting conduit and the shaft lining. The total duration anticipated was 44 months, meeting the contractually specified milestones.

The bids were submitted on February 27, 1991. Five bids were submitted ranging in value from $72.8 million to $103 million. The Engineer's estimate was $93.2 million. The difference between first and second bidders was 3.6%.

Notice To Proceed was issued to Healy/Modern on April 12, 1991.

CONSTRUCTION OF THE INTER — ISLAND TUNNEL

The MWRA had a contract with ICF Kaiser Engineers, Inc. of Oakland, California, for the planning and pre-construction management of the project. In the spring of 1990 the MWRA evaluated proposals from construction management firms to manage, on behalf of the MWRA, the day to day construction activities of Boston Harbor Project. ICF Kaiser Engineers was awarded the contract for construction manager (CM) and became responsible for project management, including supervising and coordinating the efforts of the prime contractors working on the project. All engineering and construction firms working on the project report to the MWRA through the construction manager. Stone & Webster Civil and Transportation Services was a subcontractor for ICF Kaiser on the tunnel work.

Shaft excavation operations

Overburden support consisted of slurry walls, installed at both the Deer Island and Nut Island shafts to support the excavation of the overburden. A subcontract for this work was awarded to ICOS Corp. of America. Both support structures were constructed using unreinforced concrete, socketed 1.5 meter (5 ft) into rock. The slurry walls consisted of interlocking panels that were 2.75 meter (9 ft) long, and 76 centimeter (30 in) wide. The Nut Island shaft consisted of eight panels 31.4 meters (103 ft) deep. This shape allowed for a shaft diameter opening of 6.7 meters (23 ft). The Deer Island shaft consisted of nine panels 41.5 meters (136 ft) deep. This shape allowed for a shaft diameter opening of 7.6 meters (25 ft). The panels were rectangular trenches excavated with a clamshell bucket. As the excavation progressed, the trenches were filled with slurry. A chisel was used to excavate the rock. After completion of the excavation of each panel, 27.5 megapascals (4000 psi) concrete was tremied into the trench displacing the slurry.

The Nut Island slurry wall was completed in 3 months on a single shift basis. The Deer Island slurry wall was completed in 1.25 months using two cranes on a single ten hour shift basis.

THE OVERBURDEN EXCAVATION subcontract for the interior overburden excavation was awarded to Millgard Corporation. They used a 2.4 meter (8 ft) diameter earth auger supplemented with hand excavation to remove the overburden. The Nut Island and Deer Island shafts were excavated to rock ledge in approximately two months each.

A major problem in the Deer Island shaft was a hole at the bottom of one of the panels allowing ground water inflows of 1500 liters per minute (400 gpm) to inundate the excavation. Extensive grouting was employed to reduce the flow to acceptable limits causing a 14 day delay in the work.

Rock excavation of the Nut Island shaft was 5.8 meters (19 ft) diameter and 41.7 meters (137 ft) deep. After an aborted attempt with a vertical boring machine, conventional drill and blast methods were employed to continue the excavation. Using the bench method, 1.8 meter (6 ft) rounds were drilled with sinking hammers, loaded, shot, and then mucked out with a 630 Eimco loader into a skip hoisted by a 100 ton crane. Temporary rock support incorporated 1.8 meter (6 ft) epoxy encapsulated rock bolts and 10 centimeter (4 in) of shotcrete. The conventional drill and blast production for the remaining 35.5 meters (113 feet) of shaft at Nut Island took 2.5 months.

Rock excavation of the Deer Island shaft was a 6.4 meter (21 foot) diameter and 38 meters (125 feet) deep. The shaft was sunk using a combination of full face and bench shaft sinking, drill and blast, techniques. Temporary rock support incorporated 1.8 meter (6 ft) epoxy encapsulated rock bolts, welded wire fabric, and 10 centimeter (4 in) of shotcrete. The excavation took 4 months on a two shift basis.

The Long Island shaft excavation was specified to have a minimum shaft diameter of 51 centimeters (20 in), and a maximum diameter of 1.2 meters (48 in). Healy/Modern chose to excavate the shaft to the maximum diameter facilitating its use as a ventilation shaft during tunnel construction.

The depth of the shaft was a total of 76 meters (249 ft). The shaft excavation was subcontracted to The Millgard Corporation. A casing was installed through the 30 meters (99 ft) of overburden into rock using conventional rotary drill methods. The remaining rock excavation was performed using a reverse circulation drill attachment and a 1.2 meter (48 in) diameter rock drilling head mounted on a crane. The average penetration rate for the shaft was 53 centimeters per hour (21 in/hr).

TUNNEL EXCAVATIONS

A connecting conduit links the Deer Island shaft with the South System Pumping Station. The conduit was excavated to 4.3 meter (14 ft) diameter, with a distance of 41 meters (135 ft) long. The soft ground tunnel was excavated in glacial till overburden using steel liner plate for support. An agreement with the pumping station contractor allowed this work to be done from the pump station end of the conduit, thereby removing the activity from the critical path and cutting 1 month off the schedule.

Rock tunnel excavation specifications required a high speed hard rock TBM. The TBM used was refurbished, and Healy owned. It was a 4.2 meter (13.7 ft) diameter Robbins hard rock main beam machine, model 147-210. The Cutter head turned ten revolutions per minute and was powered by 895 kilowatts (1200 hp) with propel thrust capabilities of 264,000 kilograms (707,000 lb). The machine was equipped with hydraulic rock bolting drills and a 120 degree finger shield for arch support. It had a proven, high speed excavation, track record on more than 32 kilometers (20 miles) of tunnel. In anticipation of blocky ground the cutter head and buckets were breasted.

A PROBE DRILL was added to the TBM's configuration for this project, and with it came a limited access face grouting capability as required by the specification. The TBM backup system was designed for rail haulage and consisted of a single track loading bridge pulling a traveling passing track. Four eight car trains of 4.6 cubic meter (6 cy) capacity muck cars pulled by 16.5 metric ton (15 US ton) locomotives made up the rail haulage fleet.

AT THE DEER ISLAND shaft— trains were dropped and hydraulically indexed through a car dumper. The side dump muck cars discharged into a hopper below the track grade. Muck was metered onto a pocket belt that fed a 453.5 metric ton (500 US ton) per hour vertical belt that extended 100 meters (330 ft) up the shaft. A 60 centimeter (24 in) stacker moved the excavated materials from the top of the shaft to the muck disposal area adjacent to the work site.

Excavation Progress Station 10+00 to station 185+00, the first 4.4 KM of tunnel excavation.

In January of 1993 the TBM was launched with a minimal amount of backup. Utilizing lift off muck boxes hoisted by a 125 ton crane, the excavation proceed to station 20+76. This allowed the final installation of the TBM backup, shaft mucking equipment, and track work. From the onset, TBM excavation was impacted by extremely high ground water inflows, as much as three times the expected amount. The high pressure streams of water created problems with electrical components, washed silts from the cutter head into the invert causing silting of water pumps and interference with rolling stock.

In April of 1993 the shaft conveyor belts and all TBM backup components were operational. By station 42+00, 1 kilometer (3200 ft) of excavation, water inflows had exceeded 5300 liters per minute (1400 gpm); the maximum rate of inflows that Healy/Modern had understood the GDSR to project for the entire 7.7 kilometers (4.8 miles) of tunnel. A change condition notice was forwarded to the MWRA.

By August, with 2.4 kilometers (1.5 miles) excavated, the tunnel intersected the Long Island shaft. Water inflows exceeded 12,100 liters per minute (3200 gpm), surpassing the original designed and specified emergency shaft pumping capacity. The excess water inflows were more than the original sized discharge line could handle. A larger diameter line was ordered. Salt water was 15 centimeters (6 in) over the mainline rail throughout the tunnel. It flowed toward the Deer Island shaft with a velocity of 1.2 meters per second (4 ft/sec). This created havoc with the rolling stock, wheel bearing failures, derails, and slowed muck trains delaying production. All work in the tunnel became more difficult. 9 degree centigrade (48 °F) water temperature had a chilling effect on all operations.

In September 1993 Healy/Modern undertook a rock grouting program in the excavated tunnel to reduce inflows. Since the tunnel ventilation was now through the Long Island shaft, Healy/Modern removed the ventilation duct from the 2.2 kilometer (7200 ft) of tunnel allowing open access to the tunnel for grouting.

A traveling jumbo that supported grout pumps and supplies which allowed rail traffic underneath was utilized. Holes were drilled with feedleg drills. Moyno and piston type pumps with colloidal mixers were used to mix and place the grout. Grouting commenced from the Deer Island shaft towards the TBM, concentrating on zones of high inflows. Initially, random oriented spot drilling of holes was used to intersect individual features of high inflows. Later, circumferential pattern drilling was used over long lengths of fissured argillite.

During October of 1993 to handle the ever increasing water inflows more tunnel sumps were excavated. The discharge line was increased to 30 CM (12 inch) diameter in the first 1.8 kilometers (6000 ft) of tunnel.

In November of 1993, while mobilizing to upgrade the Deer Island shaft pumping capacity to 31,040 liters per minute (8200 gpm), a power outage occurred which caused the shaft to become flooded and the electrical motors inundated. The shaft ultimately flooded to a height of 4.9 meters (16 ft) before flows could be reversed. The November 1993 flood resulted in a 15 day delay. Excavation resumed but was continually hampered by more and more water.

In February, 1994, after much correspondence between the MWRA and Healy/Modern a dispute, regarding water inflow change of conditions, was heard by the Disputes Resolution Board (DRB). The board advised that there was a differing site condition and payment of certain direct cost was due to Healy/Modern. However production impacts could not be ascertained prior to completion of the tunnel excavation.

In March of 1994, with 5.3 kilometers (3.3 miles) of tunnel excavated, water inflows had risen to 20,060 liters per minute (5300 gpm). If uncontrolled, this inflow would flood the shaft to a height of 2 meters (6.5 ft) in 45 minutes. Since it would be impossible to evacuate the heading crews out of the Deer Island shaft in that time period, we had the potential of an extremely hazardous condition.

A STOP EXCAVATION ORDER was issued by the Construction Manager to reduce water inflows to an acceptable level.

A door was added to the ventilation elbow located at the base of the Long Island shaft. A man cage and crane were stationed at the top of the shaft providing an emergency escape route in the event the Deer Island shaft flooded. Again the water discharge line was increased in diameter. The pipe was increased to 41 centimeters (16 in) diameter for the first 1.8 kilometers (6000 ft) of tunnel and to 30 centimeters (12 in) in the next 3 kilometers (2 miles).

Grouting operations took precedence in the tunnel. Several grouting consultants were employed to suggest means and methods to stop some of the inflows.

In April of 1994 the excavation resumed with an emphasis on face grouting. The face grouting bid item quantities were quickly exhausted. The additional work was compensated by time and material payments.

All indications in the contract hinted that little or no face grouting would be required, so the TBM was originally equipped with limited face grouting capabilities. When it was apparent that more face grouting was needed Healy/Modern augmented the existing equipment by mounting a second drill.

The grouting equipment consisted of a colloidal mixer, an agitating holding tank, and a double action piston pump. All were mounted on a flat car, for ease of mobilization to the heading. Bagged cement was also transported in by flat car.

For the 42 working days in April and May of 1994 the average daily excavation production was only 5.5 meters (18 ft) per day. Tunnel excavation was stopped 12 times to face grout. In that time period 90 cubic meters (3200 cf) of grout was place through the face.

During the 8 months spent grouting limited success was achieved. The quantity of hole drilled was 5,802 meters (19,037 ft). Material consumption was 148 metric tons (163 short tons) of Portland cements, 26.5 metric tons (29.2 short tons) of micro-fine cement, 5,560 kg (12,254 lbs) of polyurethane, 8,816 kg (4,000 lbs) of thixotropic stabilizer, and 4,164 liters (1,100 gal) of sodium silicate solution.

The rock grouting program in the excavated tunnel had the effect of moving the inflows around the tunnel diameter, and up or down station in the tunnel. The rock was permeable to water through interconnected thin fractures and bedding planes. Because of this the cement grout take was poor even with microfine cement. The operation was curtailed in June of 1994 when it was determined that little had been achieved. At that time ground water inflows were 17,870 liters per minute (4720 gpm).

Face grouting continued to be utilized when probe hole or feature flows exceeded 190 Liters per minute (50 gpm).

TBM penetration rates in the first 5.3 kilometers (3.3 miles) of tunnel excavation averaged 4 meters per hour (13.4 Ft/Hr) as projected in Healy/Modern's estimate. However the utilization of the TBM was only 21%, yielding an average of 20 meters (68ft) of tunnel excavated per day. This was due largely to the problems associated with the extremely high inflows of salt water. These flows were pervasive. Inflows ranged from .3 to 1.15 liters per meter (.1 to 1 gal/lf) of tunnel excavation for the entire length of tunnel.

Healy/Modern submitted it's claim for delays from the differing site condition for the excavation to station 140+00. A dispute developed between the MWRA and Healy/Modern regarding what the GDSR's prediction of ground water inflow. Unfortunately rather than being precise, the GDSR's estimate of inflow was open to interpretation. In June of 1994 the DRB heard the dispute and determined that the incremental flows stated in the GDSR accumulated to a total inflow of 7427 liters per minute (1962 GPM) for the entire length of the tunnel.

Healy/Modern resubmitted it's existing excess water claims recalculating the impact to excavation operations using the DRB's recommended incremental flows.

Shaft fire

Disaster struck on June 15, 1994 at 10:30 am when a fire started on the vertical conveyor belt approximately 75 feet above the bottom of the Deer Island shaft. In the process of replacing a conveyor bearing located at the top of the shaft a molten nut fell down the shaft into the conveyor shroud igniting the conveyor belt. By 11:00 am the shaft was completely engulfed in an inferno of smoke and flames billowing 15 meters (50 ft) in the air.

At the time of the fire, forty-one individuals were underground. Due to the farsightedness of the emergency escape plan and it's flawless execution, all underground personnel proceeded to the Long Island vent shaft and were evacuated, without injury, using the crane and man cage stationed at the top of that shaft.

The fire destroyed all electrical power drops in the shaft, halting the water pumping operations. At that time, water inflows into the tunnel were at approximately 16,280 Liters per minute (4300 gpm).

Because of the intense heat of the fire, the structural stability of the Deer Island slurry wall concrete became a concern. Engineers were brought in to make an analysis of the structure before re-entry of personnel.

It took 2.5 days before de-watering operations reversed the inflow. The majority of the tunnel was flooded. 75,800,000 liters (20,000,000 gal) of sea water had entered the tunnel. The tunnel was completely submerged to station 146+21 a distance of 4.2 KM (13,621 ft). Because of the slope of the tunnel water was only 30 centimeters (1 ft) deep at the tunnel face.

Repairs to tunnel support systems

The bottom of the shaft was a disaster. A 4.5 meter (16 ft.) high conical pile of debris was all that remained of the vertical belt. In the tunnel, ventilation duct, air and water utilities along with the discharge lines had floated off their pin restraints and were deposited in the invert. Due to the salt content of the water all electrical apparatus that was submerged had to be replaced. Restoration of the tunnel took until September 8, 1994.

Replacement of conveyor system

Rather than replace the vertical conveyor, shaft guides were installed. An 18 cubic meter (24 cy) skip hoisted by a 125 ton crane transported the muck up the shaft. The original muck handling system at the dump pit was intact and after rehabilitation was modified to feed the skip.

STATION 185+00 TO 26+160, THE LAST 3.3 KM OF TUNNEL EXCAVATION

Shortly after restarting the TBM in September of 1995, Healy/Modern began having excavation and support problems with soft rock. In areas of rock with strengths too low to be considered rock, 2.1 megapascals (300 psi) or less, the TBM's grippers were rendered useless and unable to grip causing delays in excavation. By station 196+00 rock quality had deteriorated to the point that support was exclusively rib and Coeur D'Alene lagging. Healy/Modern deemed useless the wire mat lagging as designed in the contract ; due to it's inability to prevent raveling of soil like ground and it's obstructing miners from being able to crib and block ground that had over broke as much as 2.1 meters (7 ft).

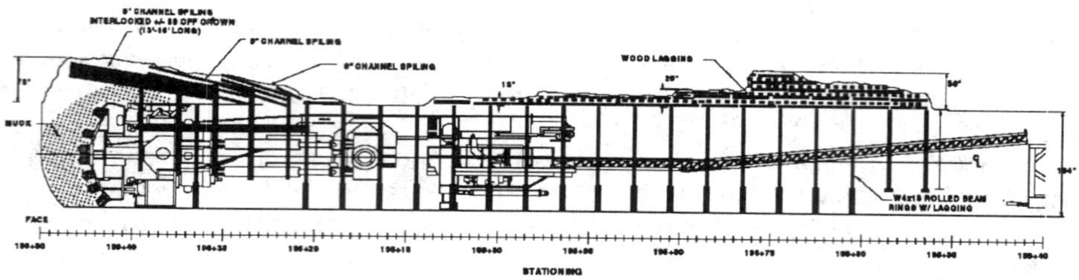

Figure 2. Profile Drawing Showing Enlargement of the Tunnel Over the TBM

Soft ground at station 196+50 to 200 +50

Despite a gallant effort to keep the machine moving forward, On October 9, 1994 the inevitable occurred. During the second shift the tunnel boring machine cutter head became jammed. A large section of the face and arch had slipped down on the cutter head preventing rotation.

After thorough inspections and meetings with the designer, the MWRA concluded that the collapse was a change of condition. In the interim Healy/Modern developed a plan to re-mine and support the arch starting approximately 20 feet behind the face.

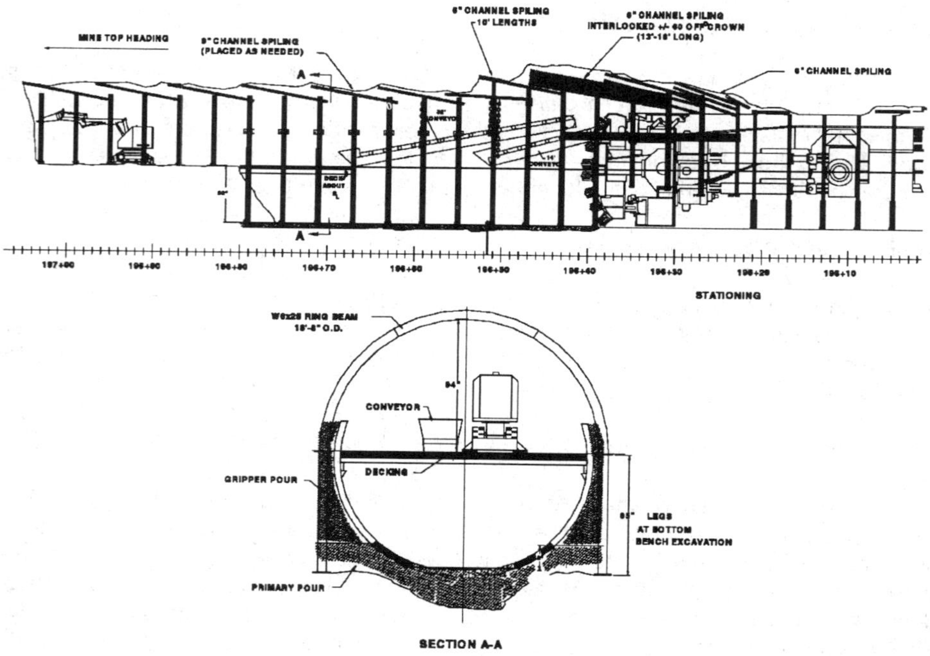

Figure 3. Profile of the Excavation in Front of the TBM

Hand mining around and ahead of Tunnel Boring Machine

Using jump set and crown bar support methods, miners opened up a tunnel arch 6.1 meters (20 ft) in diameter. After increasing the diameter of the tunnel, allowing access to and freeing of the tunnel boring machine cutter head, a top heading was driven forward past the caved section. The distance was roughly 7 meters (23 ft). Excavation was by hand using steel spiling and ring beam methods of support. See figure 2.

A HORIZONTAL BORED HOLE was drilled 92 Meters (302 feet) long and cores taken to determine rock conditions beyond the existing face. A dispute developed as to whether the material found in the cores constituted a differing site condition and secondly, would it have the stand up time required for the tunnel boring machine excavation.

AN EMERGENCY MEETING OF THE DRG was called. Their recommendation was that the risks and impact of another collapse was so great, and the grounds of a massive differing site condition so obvious that a decision to continue the hand mined excavation to station 197+62 on a time and material payment basis was imperative.

HAND MINING CONTINUED on a time and material payment basis. In order to speed up the operation two hydraulic robots, Brokk models 100 and 150 combination breaker and back hoe, were maneuvered around the TBM and used to excavate material and place it on an extendible conveyor that fed into the TBM conveyor on to the back up system. The overall length of the hand mined tunnel section was 40 meters (132 feet). See figure 3.

AFTER THE BOTTOM BENCH WAS EXCAVATED, concrete was poured from the invert to springline using tunnel ribs as the inside form. The TBM was walked forward and re-launched on April 12, 1995

Re-starting tunnel boring machine

Because of the potential of another collapse TBM excavation continued on a 24 hour per day, 7 days per week basis until May 19, station 200+02. At station 200+48 the tunnel finally reached competent rock.

THROUGHOUT THIS REACH of tunnel rock, with strengths to low to support gripper pressures, yielded causing rib deformation and distortion of the tunnel diameter. All excavation support installed to station 200+48 was rib and wood lagging.

The status of excess water inflow claims

The MWRA and Healy/Modern had not been able to resolve the water impacts claim with regard to quantum. In May of 1995 the DRB heard its third dispute involving water inflow change of condition claims. This time, a measured mile needed to be established in order to quantify the delay damages to Healy/Modern. Although none of the tunnel was unimpacted there was a reach, slightly less than a mile of tunnel, that by removing known impacts a measured mile could be developed. The board made it's recommendation regarding the measured mile. By applying the boards recommendations, negotiations between the MWRA and Healy/Modern led to the settlement of the major outstanding claims regarding excess water impacts.

Figure 4. Photograph of TBM in Hand Mined Tunnel

Arch panning

In June of 1995, after a failed grouting effort, a decision was made to develop a method to install the concrete lining with the existing water inflows. Due to the large quantity of panning to install, 24,900 square meters (268,000 square feet) and the fact that the success of the job hinged upon the handling of the water during the lining process, a fire resistant plastic panning with an annulus of .75 cm (.3 inches) was developed. Healy/Modern began installing the arch panning to channel water inflows to the invert. This was anchored to the rock by .5 cm (.188 inch) diameter Rawl drives and grommeted on a 60 cm. by 60 cm. (2 ft. by 2 ft.) pattern.

Hole through

From July of 1995 to the hole through at the Nut Island shaft in November 1995 water inflows ranged from 15,520 to 19,870 liters per minute (4100 gpm) to (5250 gpm). In July of 1995 the MWRA had issued a unilateral change order for delays caused by the excess water inflows while excavating the last 1.6 Km (1 mile) of tunnel. The tunnel excavation was completed November 4, 1995.

During the excavation phase of the work over 20 billion liters (5 billion gallons) of salt water inflows were pumped out of the tunnel and treated before being discharged into the harbor. Ring beam support was installed in only 8% of the tunnel length. However due to a massive differing site condition, the excavation of 2% of the tunnel length impacted the

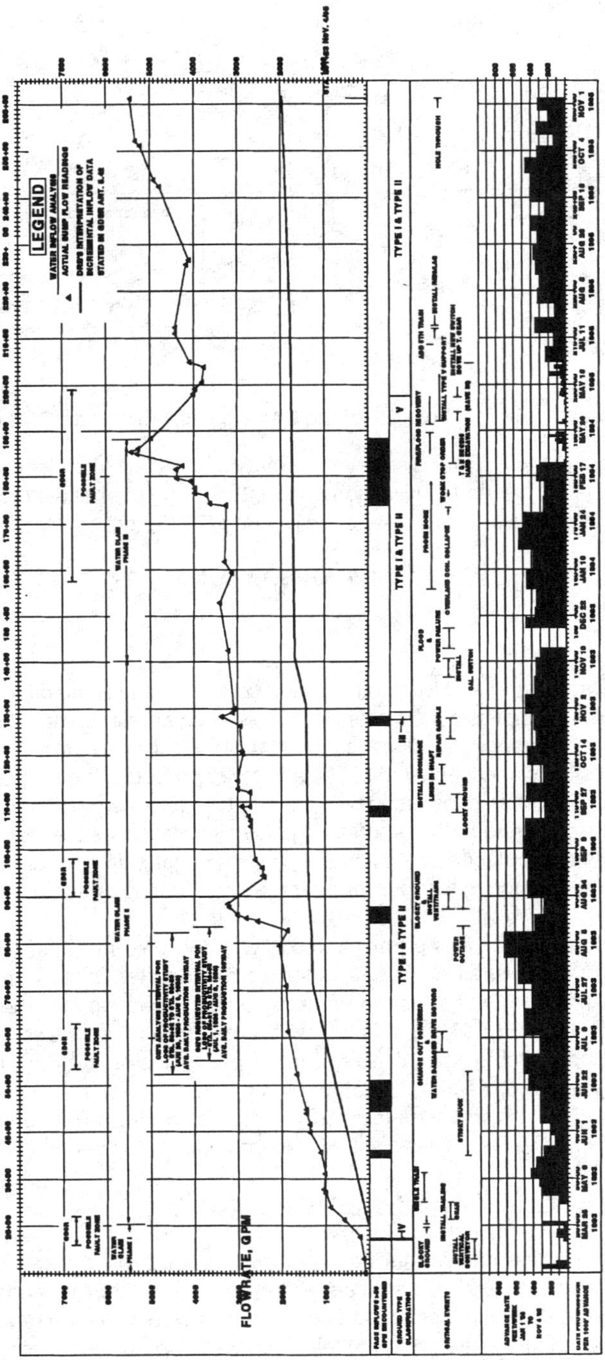

Figure 5. Graph of Water Inflows, Excavation Production and Conditions

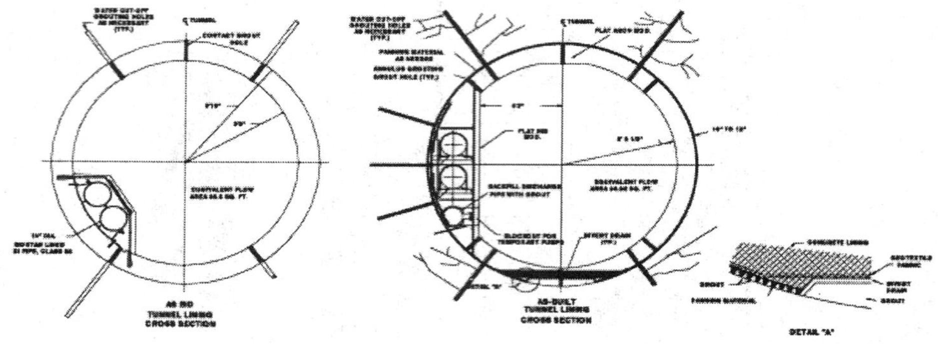

Figure 6. Tunnel Cross Sections Showing as Bid and as Built Lining

schedule by eight months. The tunnel boring machine had an overall penetration rate of 4 meters per hour (13 ft/hr), and a utilization rate of 23%. The overall average for the excavation completed by the TBM was 21.9 meters per day (72 lf/day).

LINING OPERATIONS

Shaft Lining

Both the Deer Island and Nut Island shafts receive a reinforced cast in place concrete lining. The Nut Island shaft is a 4.9 meter (16 ft) finished diameter shaft. There is a transition elbow at the tunnel intersection. Near the top of the shaft, there is a transition to the connecting conduit to the Nut Island Headworks. The shaft was poured using a collapsible steel form in 6 m (20 ft) lifts. The shaft excavation was halted at the lower elevation of shaft concrete in order to place a starter ring for the shaft lining. Successive lifts were poured on top of one another. The lining was placed in 2.75 months, including the time required to excavate the 3.5 meters square opening in the slurry wall for the connecting conduit transition.

The Deer Island shaft has a transition elbow at the bottom of the shaft. The bottom 66 meters (218 ft) of the shaft has a 3.4 meter (11 ft) diameter finish, with a transition to a connecting conduit to the South System Pumping Station. The top 11.9 meters (39 ft) of the shaft has a 4.9 meter (16 ft) diameter finish, with a transition to the surge storage structure. Healy/Modern is proposing to use a reinforced precast pipe as the final lining in the 3.4 meter (11 ft) diameter section. The annulus between the pipe and slurry wall will be backfilled with concrete. The 4.9 meter (16 ft) form from the Nut Island shaft will be utilized at Deer Island also. The Deer Island shaft is schedule for completion in October 1997.

Modification of tunnel lining configuration

In order to mitigate delays that had occurred on the project Healy/Modern, early in 1994, proposed and the MWRA accepted a plan to incorporate the sludge lines into the tunnel liner. This meant that the concreted cross section would change from a circle to a "D" shape. Incorporating this activity into the initial tunnel lining and eliminating the secondary copncrete pour activity trimmed approximately 5 months off the original schedule.

Figure 7. Photograph of Sludge Line Installation

Sludge line installation

A pipe support system was installed prior to completing the excavation, allowing the two 53 centimeter (14 inch) ductile iron sludge lines to be installed in the tunnel during the time period that the TBM was being removed and the tunnel concreting works installed at the Nut Island shaft.

Utilizing a locomotive mounted log loading boom and rail transportation of the pipes, 15.2 kilometers (50,000 lf) of 35.5 centimeter (14 in) ductile iron sludge lines were installed in 42 work days. See figure 7.

Tunnel lining

In July 1995 negotiations began with regard to the impacts of the excess water differing site condition on the concrete work. The concrete lining was to be placed from the upper end of the tunnel, Nut Island, downstream toward Deer Island.

A METHOD HAD BEEN DEVISED to allow the water to drain through the concrete placement area. This consisted of panning as described above in conjunction with an invert drain system. The invert drain was fabricated from welded wire fabric channels measuring 30 centimeters (1ft) wide by 10 centimeters (4 in.) high and covered with a geotextile fabric. The drain was bolted to the invert. A double channel could handle approximately 600 gpm of invert water flows. When inflows in any reach of tunnel were higher than 600 gpm the drain had to be relieved up stream in the lined tunnel. The tunnel dewatering discharge lines were left in place and stubbed out through the lining. A dam was kept just upstream of the tunnel forms. All upstream water is pumped into the concrete imbedded discharge line, past the forms and placing bridge and down tunnel to the next available tunnel sump.

THE FINAL WORK ACTIVITY would be to shut off the water inflows by grout the panning, invert drain, and imbedded discharge lines.

Lining began the end of January, 1996. The negotiations covering the costs of delays due to the excess water inflows and extra work during the concrete lining operations were completed by early February, 1996 and a change order was issued.

Concreting methods and equipment

The concrete equipment consisted of 81 meters (270 ft) of "D" shaped telescopic tunnel forms. A self propelled traveling concrete placing bridge stationed 300 meters (1000 ft.) downstream of the leading end of the tunnel form. Concrete traveled through an overhead 20 centimeter (8 in.) diameter slickline into the form. A drop line and hopper system at the Deer Island shaft supplied the ready mix to three trains consisting of four 4.6 cubic meter (6 cy) capacity Moran agitator cars.

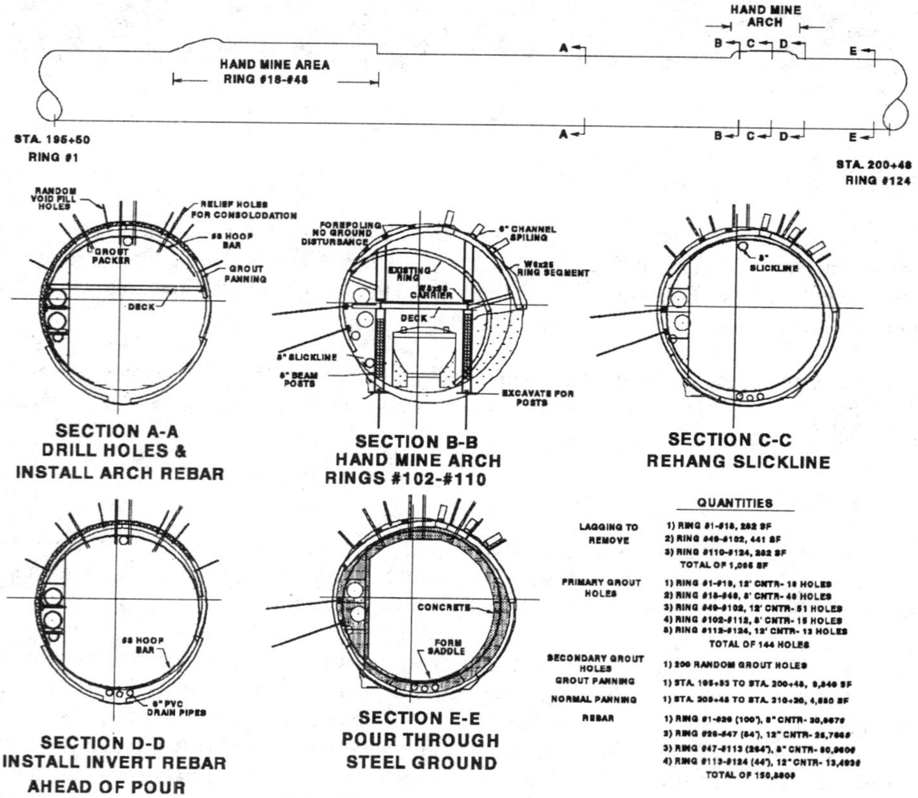

Figure 8. Cross Sections of Tunnel in Soft Ground Concrete Work

Delays to lining operations in the soft ground

The saga of the soft ground did not end with the tunnel excavation. Due to the deformation that occurred while gripping the TBM through and the encroachment of the wood lagging in the web of the ring steel, clearances and thickness of lining acceptable to the designer could not be achieved with normal lining placement.

This fact was brought to light well before tunnel excavation was finished and concrete lining was started. Many options were designed, analyzed and presented to the designers. A 20 centimeter (8 in) thick wall with added concrete reinforcing steel, use of 41.3 megapascals (6000 psi) concrete instead of 27.5 megapascals (4000 psi), excavating the invert, jacking the sets with the 485 metric tons (500 tons) available from the concrete placing bridge gripper system, and lastly installing a 2.7 meter (9 ft) diameter steel lining through the 152 meter (500 ft) zone.

It was not until after the invert steel was removed, prior to invert excavation, and the concrete placing spread pulled through that the designer detailed what was acceptable. In the interim between removal of the invert supports and the final lining being place in this section; the tunnel experienced a small ground slip over eight rings. Prompting the MWRA to go forward with invert excavation to achieve the required design section.

A PLAN WAS DEVISED that would allow the remining of the slipped ground, while excavating the invert to springline for the 152 meters (500 ft) of deformed and tight cross section tunnel. Another scheme was devised to allow crews to pour the 300 meters (1000 ft) of tunnel upstream from the soft ground area while both of the previous tasks were on going; thereby taking a large portion of the remedial work off the critical path.

BECAUSE OF THE TIGHT tolerances and soil like ground conditions, concrete forms could not be set on spuds as usual. Pedestals were designed, installed and set to line and grade to assist with the placement and movement of forms.

CONCERN WAS EXPRESSED by the designer that consolidation of the voids between the wood blocking and broken ground could not be achieved. A three stage grouting program was outlined to achieve the desired results. Because of the great amount of steel installed in the arch. A spring loaded packer design was created to allow an unrestricted passage to the voids in the lagging.

Back-fill Grouting

The designer required the annulus spaces in the tunnel to be backfilled with cement grout. These spaces consisted of the invert drain, voids behind the panning, and the discharge lines. Because of the large volume of material required, and the inefficiencies that would have occurred using bagged material in the tunnel, Healy/Modern decided to pump the grout from the surface at Nut and Long Island. This meant that the grout would have to be pumped 1.2 killometers (10,000 ft).

The Hany Corporation was asked to design a plant to meet our requirements. Their engineers specified the flow characteristics the grout would require in order to be pumped the distance of 1.2 kilometers (10,000 ft). Healy/Modern did an extensive testing program

Figures 9 & 10. Photographs of Excavation in Soft Ground for Lining Thickness

on grout mixes to come up with a mix with an acceptable viscosity that would develop the required strength of 27.5 megpascals (4000 psi). A 0.7 to 1 by weight water/cement ratio was determined to be acceptable, both in strength and pumping requirements.

The equipment designed and built consists of a 81.6 metric ton (90 ton) cement silo, a 800 liter (211 gal) colloidal mixer with an automatic weigh batching system, and a 1200 liter (317 gal) agitating holding tank. A high pressure piston pump, connected to a 1 in. steel grout line delivers the mix to the tunnel. The pressure required to deliver the grout 1.2 kilometers (10,000 ft) is 10.3 megapascals (1500 psi). A unique coupling supplied by ACME of America connect the 2.5 centimeter (1 in) grout pipes. The pipe size was chosen in order to keep the velocity of the grout high to prevent separation. The grout is pumped from the surface to a holding tank in the tunnel that feeds a Chemgrout Moyno placing pump.

To perform the work, a 150 to 300 meter (500 to 1000 ft) length of tunnel is grouted. Relief holes are drilled every 15 meters (50 ft) in the invert, with the invert drain dammed off on the downstream end of the section. Grouting proceeds from down station to up station. Holes are then drilled every 10 ft in the arch for contact grouting, and in required areas for panning backfill. Contact and panning grouting proceeds simultaneously from up station to down station. After invert grout, and contact and panning operations are complete, significant seepage in the section is grouted through spot holes in the area of seepage. An "anti-washout" additive is added to the grout in the invert drain grout to prevent dilution of the grout, and a superplastisizer is added to help the grout flow in the line.

Progress averages approximately 30 meters (100 ft) of tunnel grouted per day. No delivery lines have been plugged, and the goal of pumping 1.2 kilometers (10,000 ft) exceeded. After concrete lining operations pass beyond the Long Island shaft, the grout operations will be stationed there.

SUMMARY & CONCLUSIONS

The Inter-Island tunnel as conceived and bid was a simple project. All tunnel projects are evolved. Unforeseen situations are encountered and innovative and unique solutions are developed and implemented so that The Work can continue.

This project was hampered by two massive differing site conditions.

The contract has doubled in cost and has over run 70% in time. A great amount of effort has been wasted and animosity created debating a vague and inconsistent Geologic Design Summary Report. None of this was to the benefit of the owner or contractor.

The state of the art is such that Geological Engineering cannot predict specifically every time what will be encountered. This does not diminish the obligation of the designer and owner to draw a definitive baseline on which to determine the merits of a differing site condition.

Tunnel projects are successful when all parties do what they are best at, designers design, construction managers manage, and contractors build. Each party has to be held accountable for their part.

Disputes resolutions process

This project's completion by Healy/Modern can be largely attributed to the effective use of the Dispute Resolution Board process. Without its guidance, the wide chasm caused by a flawed Geological Design Summary Report could not have been bridged in the absence of years of litigation. This coupled with a tremendous amount of never say die and inventiveness, is what allowed this project to have a successful completion.

ACKNOWLEDGMENTS

The authors wish to acknowledge the following individuals for there contributions to the successful completion of the Inter—Island tunnel.

Kenneth L. Anderson, Treasurer - Modern Continental Construction Co. Inc.
Jay E. Carlson, Senior Vice President - S.A. Healy Company
Douglas B. Mac Donald- Executive Director, MWRA
Peter M. Douglass, Chairman - Disputes Review Board
Stewart H. Bartholomew - Disputes Review Board
James L. Mann - Disputes Review Board
Ronald E. Heuer - Geotechnical Consultant
Safdar Gill, President - Ground Engineering Inc.
Joseph Griffith, Construction Manager Director, MWRA
Scott Shylanski, Engineer - S.A. Healy/Modern Continental Joint Venture
Peter J. Tarkoy, PHD. Geotechnical and Construction Consultant

Credit for photographs:
Regina Villa Associates Photographs by Karen-Jane Dodge Courtesy of MWRA

Chapter 22

HILLTOP INTERCEPTOR CONTRACT G STORAGE AND CONVEYANCE TUNNEL CLEVELAND, OHIO

Richard Switalski,[1] Thomas Sisley,[2] and Jozef Zurawski[3]
[1]Manager, Northeast Ohio Regional Sewer District; [2]Vice President, Dodson-Stilson, Inc.; [3]President, Dawn Engineering, Inc.

ABSTRACT

Hilltop Interceptor Contract G is one of the segments of the wastewater control system in the greater Cleveland area. The tunnel is a relief interceptor constructed by the Northeast Ohio Regional Sewer District to minimize combined sewer overflows in the Easterly Service Area.

The design study identified the need for a 3.35 m (132 in) diameter, 3.35 km (11,000 ft) in-line storage conveyance tunnel to be constructed at depths of 30 to 70 m (100 to 230 ft). Six structures for access, future flow drop, work shafts and overflow control were required to complete the system. The interceptor reduced the design peak flow from 12.5 m^3/s (285 mgd) to 6.1 m^3/s (140 mgd) by detaining in line 22,800 m^3 (six million gallons) of wastewater.

BACKGROUND

The Northeast Ohio Regional Sewer District (NEORSD) was created by court order in 1972 at which time, it assumed the responsibility for the completion of wastewater management planning for the Easterly Wastewater Treatment Plant Service Area. At that time, the need for an interceptor system to serve the Easterly Separate Sewer Area was identified, as a part of the master sanitary sewer plan to minimize pollutant overflows to Lake Erie and tributary streams. The plan identified a total of 37 km (23 miles) of interceptor system (Heights/Hilltop Interceptors) to help alleviate overflows and basement flooding (See Figure 1).

Hilltop Interceptor Contract G was originally anticipated to consist of 490 m (1,600 ft) of 2 m (78 in) diameter sewer constructed by tunneling and 2,400 m (8,000 ft) of 1.7 m (66 in) diameter sewer constructed by open cut. The interceptor was also required to have 12.5 m^3/s (285 mgd) peak carrying capacity. However, the receiving treatment plant could handle only 6.1 m^3/s (140 mgd) from the Heights/Hilltop Interceptors discharging into Contract G.

As a result, the initial task of the design team (Dodson-Stilson, Inc. and Dawn Engineering, Inc.) was to evaluate the potential design alternatives for the detention of flows. The life cycle cost study and research concluded that the construction of a deep in-line storage tunnel would result in the most economical solution.

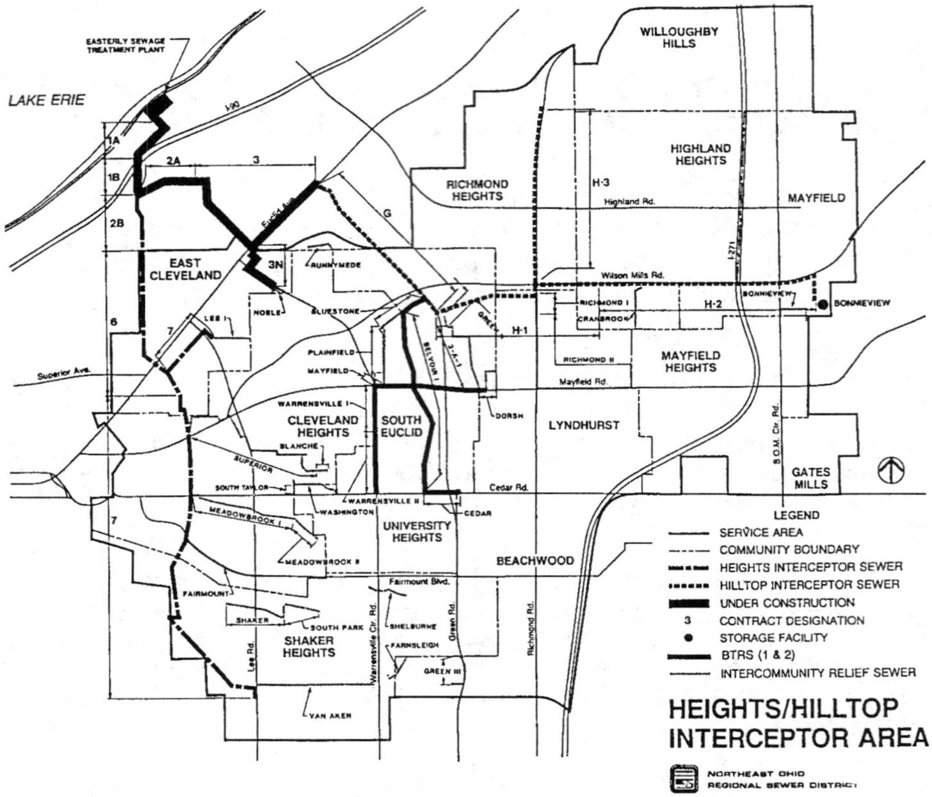

Figure 1

DESIGN AND HYDRAULIC TESTING

The design of the Contract G was based on a five year, one hour storm event. Initial design concepts included the consideration of a near surface storage facility consisting of a concrete lined detention tank. However, the concept of a storage-conveyance tunnel proved to be most economical. Also, due to the depth of the tunnel, it is feasible to construct additional detention tanks next to the interceptor should the need for expansion be identified in future studies.

Once the general design concept was completed, the detail design proceeded in order to meet certain parameters that would minimize maintenance costs of the Contract G, as well as to allow for the wide selection of profiles for major upstream tributary interceptors: three Contracts H and two Bluestone Road Contracts (BTRS). A preliminary subsurface investigation conducted for these two upstream tributary legs revealed that natural oil saturated rock deposits needed to be avoided by selecting tunnel profiles below or above these oil bearing strata.

From the maintenance point of view, it was desirable to have a non-mechanical outlet system to control the discharge from the Contract G to the maximum of 6.1 m^3/s (140 mgd) and still allow for overflow should the actual rainfall event exceeded the design peak flows.

OVERFLOW STRUCTURE

A concept of an overflow structure was born based on ideas taken from typical detention basin's outfall structures used for surface drainage systems:

1. Control outlet discharge rate by an orifice device.
2. Provide an overflow weir to allow for controlled discharge of flows exceeding the design rate.

However, early in the design stage, it was recognized that to design an outlet control device operating under 32 m (105 ft) of a hydrostatic head and discharging into the existing tunnel would require hydraulic modeling. An empirical design approach was considered too risky without guaranteed performance results for the $750,000 structure. The critical issue was to assure stable overflow conditions and still provide acceptable flow characteristics in the shaft and the downstream tunnel.

The design was initiated on a system that would detain incoming flows in the tunnel by controlling the outlet rate through a small diameter pipe (the throttle pipe) at the overflow structure. When the incoming flows exceed the outlet discharge, the water level continues to rise, utilizing the tunnel as a detention tank, until the outlet rate exceeds the inflow rate, at which time the storage in the tunnel is depleted.

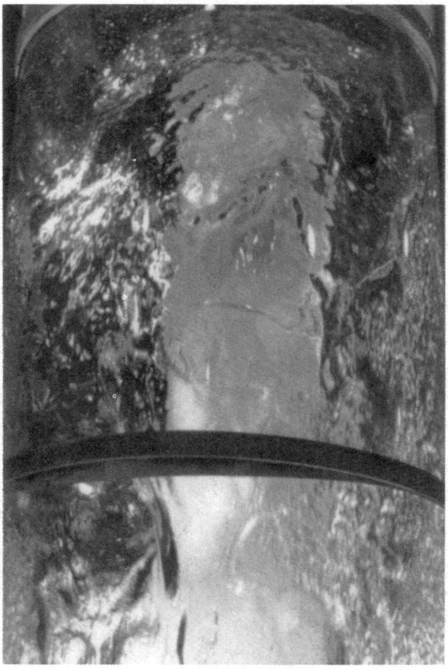

Photograph 1

Photograph 2

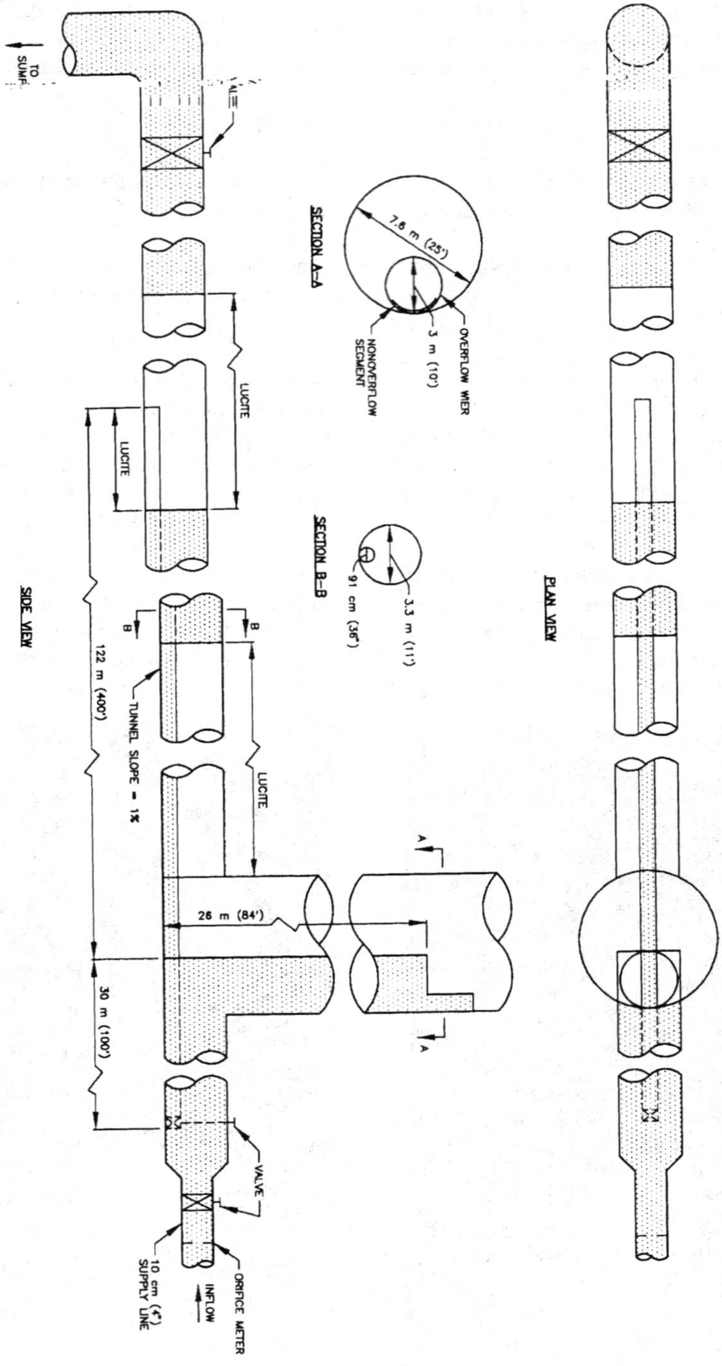
TO SUMP
SECTION A–A
7.6 m (25')
OVERFLOW WIER
3 m (10')
NONOVERFLOW SEGMENT
SECTION B–B
3.3 m (11')
91 cm (36")
PLAN VIEW
SIDE VIEW
LUCITE
LUCITE
LUCITE
TUNNEL SLOPE = 1%
122 m (400')
26 m (84')
30 m (100')
VALVE
ORIFICE METER
INFLOW
10 cm (4") SUPPLY LINE
A
A
B
B

Figure 2

From the flow monitoring data which was incorporated into the SWMM program, a hydrograph of the design storm was developed to calculate the size, depth and profile of the tunnel, length and size of the throttle pipe and the volume of the desired storage capacity. Once the system was sized, the data along with conceptual construction details were sent to the Iowa Institute of Hydraulic Research (IIHR) for modeling. As a result of the IIHR modeling, a smaller diameter throttle pipe was incorporated into the system to decrease its length. A reducer was added to the end of the throttle pipe, in the opposite direction of the flow, with a short larger diameter pipe at the terminus of the throttle pipe to reduce the exiting flow velocities (See Figures 2 and 3).

The IIHR also studied the shape and width of the overflow weir to produce stable flow conditions in the event that the design storm runoff rate is exceeded or the throttle pipe becomes clogged. The modeling concluded that during the worse case conditions, about 1.2 m (4 ft) of flow would over top the weir within the overflow structure shaft (See Photographs 1 & 2). The study also concluded that baffles to control the exiting velocities were not needed. The findings from the IIHR modeling were incorporated to finalize the geometry of the full scale structure.

CONTRACT G SYSTEM

When the design of the interceptor was completed, the Contract G system comprised of approximately 3.58 km (11,750 ft) of 3.35 m (132 in) diameter storage conveyance tunnel at a slope of 1.00% with 53 m (174 ft) of 0.76 m (33 in) diameter throttle piping with reducer and 3 m (10 ft) of 0.9 m (36 in) diameter outlet pipe. The system has a peak discharge rate of 6.35 m^3/s (145 mgd) and a storage capacity of 24,600 m^3 (6.5 million gal.). The discharge from the throttle pipes and the overflow weir is directly into a downstream tunnel completed under previous contracts. Three access shafts, one deaeration chamber, two work shafts, drop pipe and access adit complete the Contract G system.

ACCESS AND WORK SHAFTS

In addition to provisions for future connections, the project took into consideration maintenance requirements and future tunneling. Only initial supports were designed for the installation at Work Shaft No. 2 and the adjacent 23 m (75 ft) of the tunnel excavation. The unlined tunnel section would serve as a tail tunnel for backing muck trains during the excavation of the upstream interceptor. The unlined shaft excavation was also utilized by the upstream designers for installation of a drop structure.

The NEORSD Maintenance Department has a requirement for large diameter interceptors which dictates that the spacing between the access shafts or manholes can not exceed 915 m (3,000 ft). This distance is set by the safety equipment reach capacity, such as winches and cables, as well as by emergency breathing air packs which allow sufficient time for a 460 m (1,500 ft) walk within the tunnel. Also, at this spacing, ventilation of the system can be accomplished in reasonable time prior to the entry of personnel into the tunnel.

This spacing requirement, at times, locates an access shaft in an area where available land area is limited, not available or the surface access shaft covers are in the street and/or roadway with heavy traffic. This was the case on the Contract G requiring the designers to

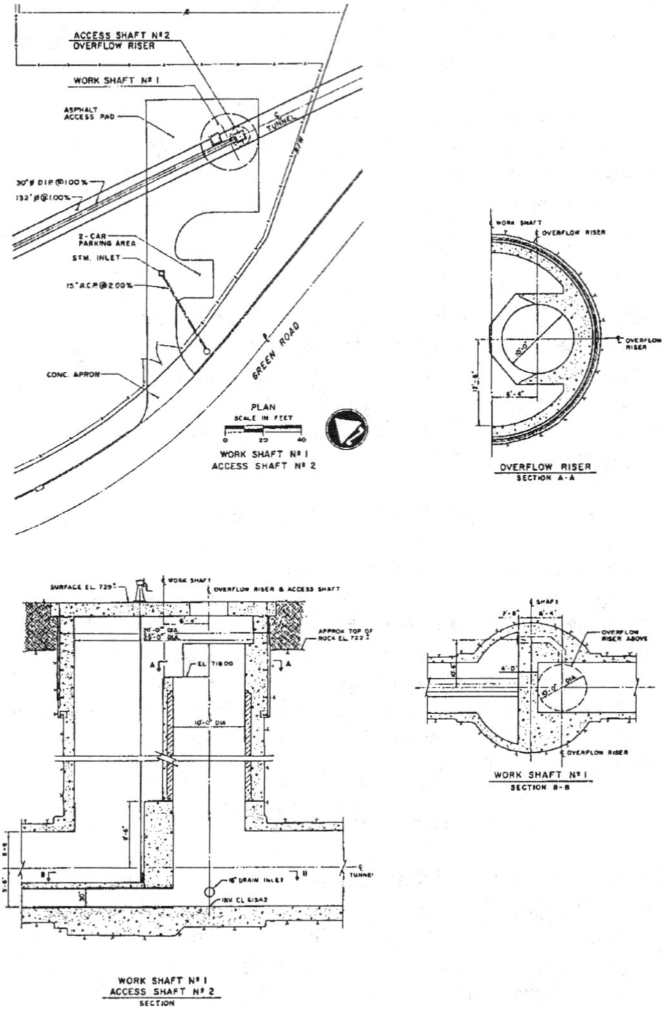

Figure 3

utilize an offset access shaft arrangement (See Figure 6). In addition, the maintenance access to the tunnel is provided by the NEORSD through the use of a Maintenance Trailer operating from the surface. The trailer has a man- cage which allows personnel to be lowered and picked up directly from the shaft inverts (at tunnel level) and allows for air testing equipment to be lowered to the tunnel level prior to the entry of personnel. The cage can also be used to lower air testing monitors any other safety or support equipment (See Figure 7).

GEOTECHNICAL CONDITIONS

The subsurface investigation program was conducted in two phases, during which a total of 12 test borings were performed. The first phase of the investigation aided the de-

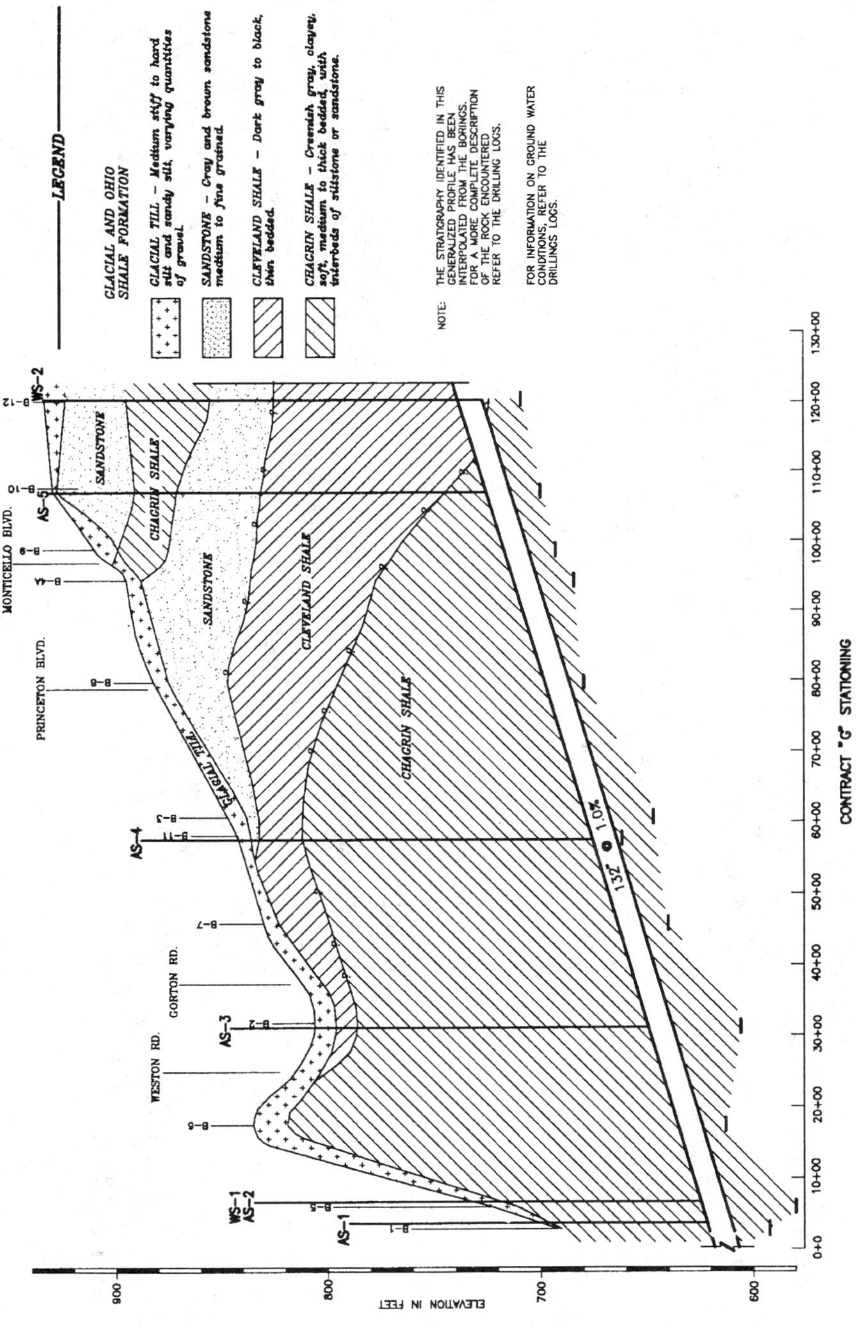
LEGEND
GLACIAL AND OHIO SHALE FORMATION
GLACIAL TILL - Medium stiff to hard silt and sandy silt, varying quantities of gravel.
SANDSTONE - Gray and brown sandstone medium to fine grained.
CLEVELAND SHALE - Dark gray to black, thin bedded.
CHAGRIN SHALE - Greenish gray, clayey, soft, medium to thick bedded, with interbeds of siltstone or sandstone.
NOTE: THE STRATIGRAPHY IDENTIFIED IN THIS GENERALIZED PROFILE HAS BEEN INTERPOLATED FROM THE BORINGS. FOR A MORE COMPLETE DESCRIPTION OF THE ROCK ENCOUNTERED REFER TO THE DRILLING LOGS.
FOR INFORMATION ON GROUND WATER CONDITIONS, REFER TO THE DRILLINGS LOGS.
MONTICELLO BLVD.
PRINCETON BLVD.
GORTON RD.
WESTON RD.
GLACIAL TILL
SANDSTONE
CHAGRIN SHALE
CLEVELAND SHALE
132" @ 1.0%
CONTRACT "G" STATIONING
ELEVATION IN FEET

Figure 4

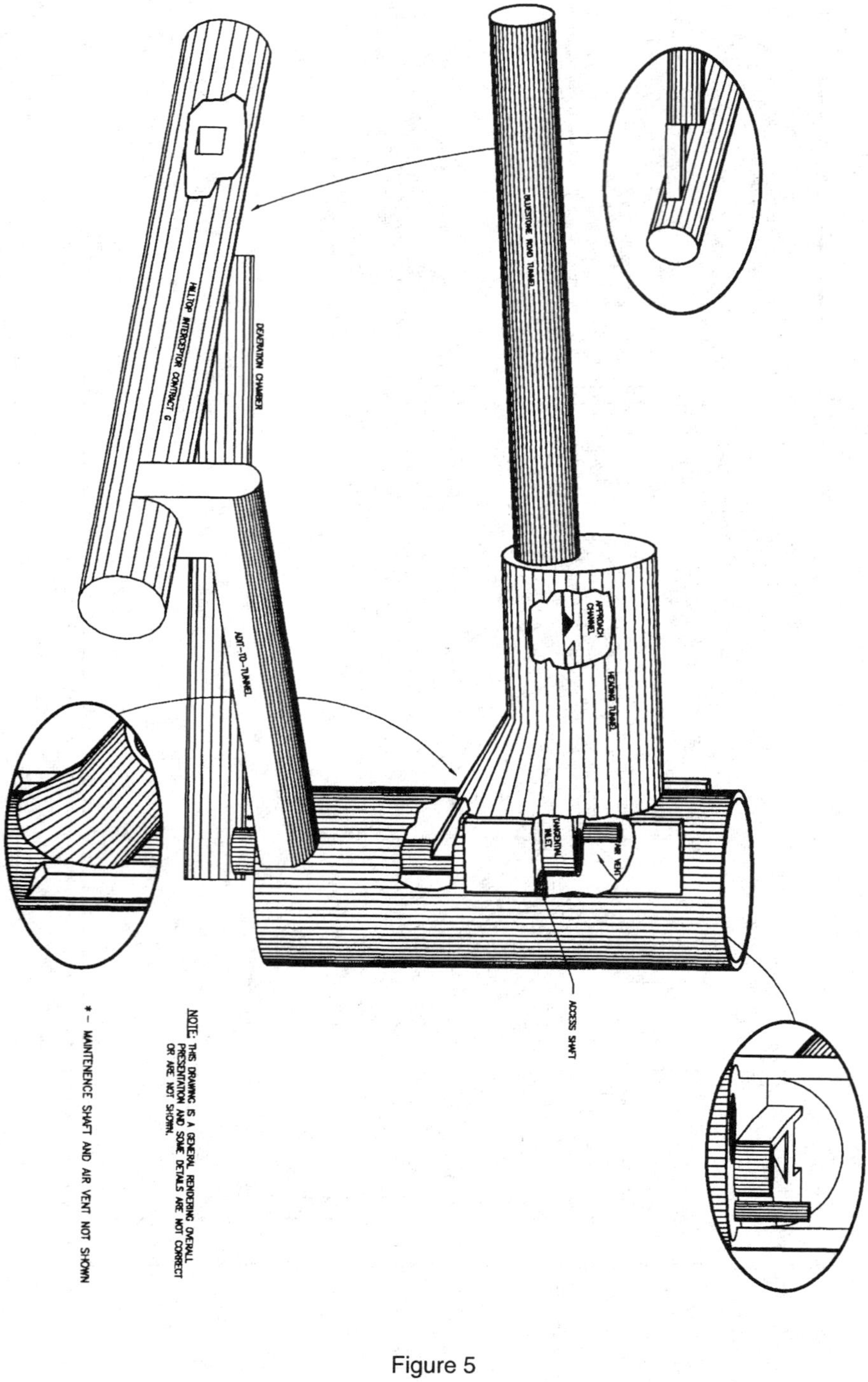
BLUESTONE ROAD TUNNEL
HILLTOP INTERCEPTOR CONTRACT G
DEAERATION CHAMBER
ADIT-TO-TUNNEL
APPROACH CHANNEL
HEADING TUNNEL
TANGENTIAL INLET
AIR VENT
ACCESS SHAFT
NOTE: THIS DRAWING IS A GENERAL RENDERING OVERALL PRESENTATION AND SOME DETAILS ARE NOT CORRECT OR ARE NOT SHOWN.
* - MAINTENENCE SHAFT AND AIR VENT NOT SHOWN

Figure 5

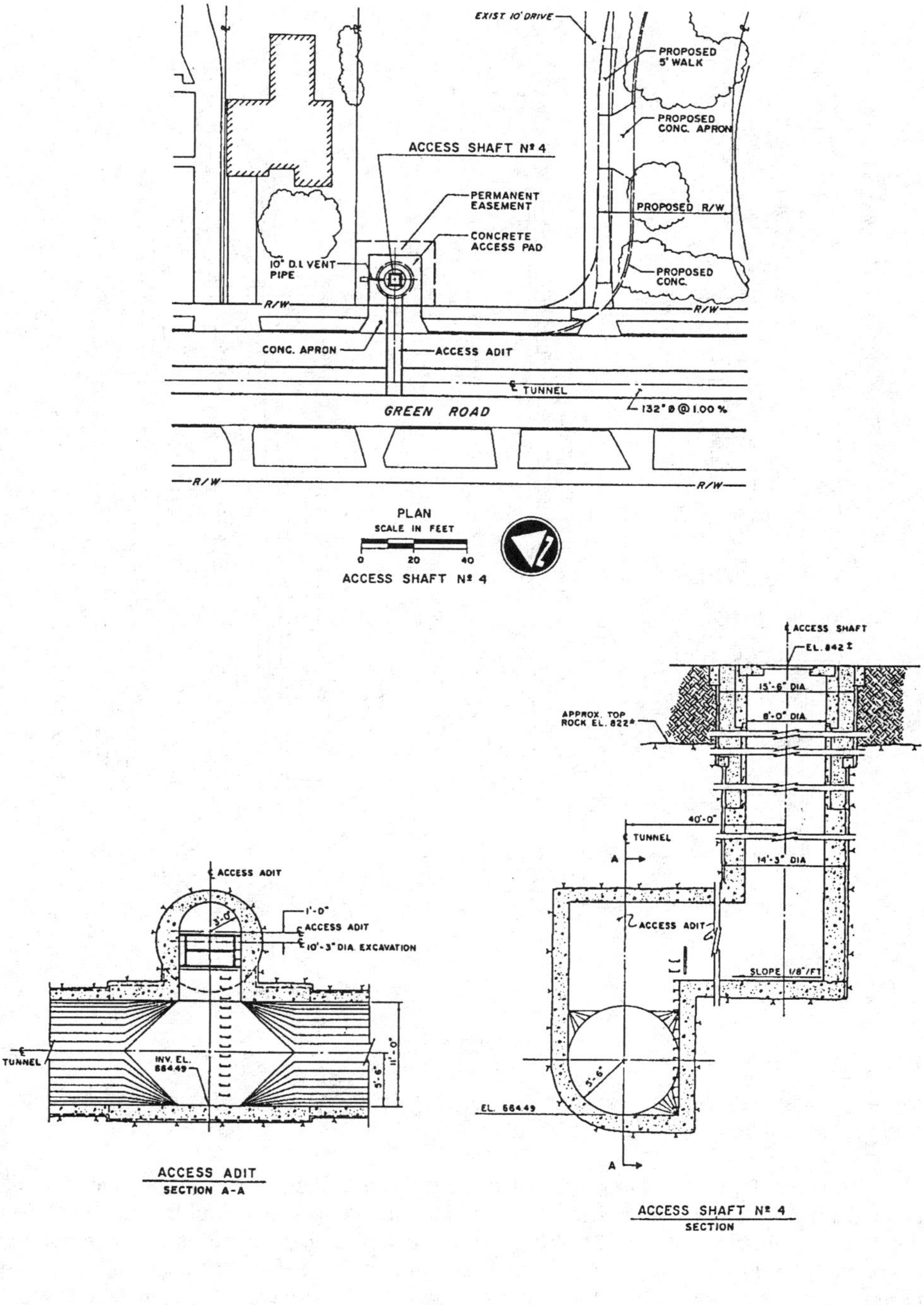
EXIST 10' DRIVE
PROPOSED 5' WALK
PROPOSED CONC. APRON
ACCESS SHAFT Nº 4
PERMANENT EASEMENT
PROPOSED R/W
CONCRETE ACCESS PAD
10" D.I. VENT PIPE
PROPOSED CONC.
R/W
CONC. APRON
ACCESS ADIT
℄ TUNNEL
132" ⌀ @ 1.00 %
GREEN ROAD
R/W
PLAN
SCALE IN FEET
0 20 40
ACCESS SHAFT Nº 4
℄ ACCESS SHAFT
EL. 842 ±
15'-6" DIA.
8'-0" DIA.
APPROX. TOP ROCK EL. 822 ±
40'-0"
℄ TUNNEL
14'-3" DIA.
A
ACCESS ADIT
SLOPE 1/8"/FT
5'-6"
EL. 664.49
A
℄ ACCESS ADIT
3'-0"
1'-0"
℄ ACCESS ADIT
℄ 10'-3" DIA. EXCAVATION
℄ TUNNEL
INV. EL. 664.49
5'-6"
11'-0"
ACCESS ADIT
SECTION A-A
ACCESS SHAFT Nº 4
SECTION

Figure 6

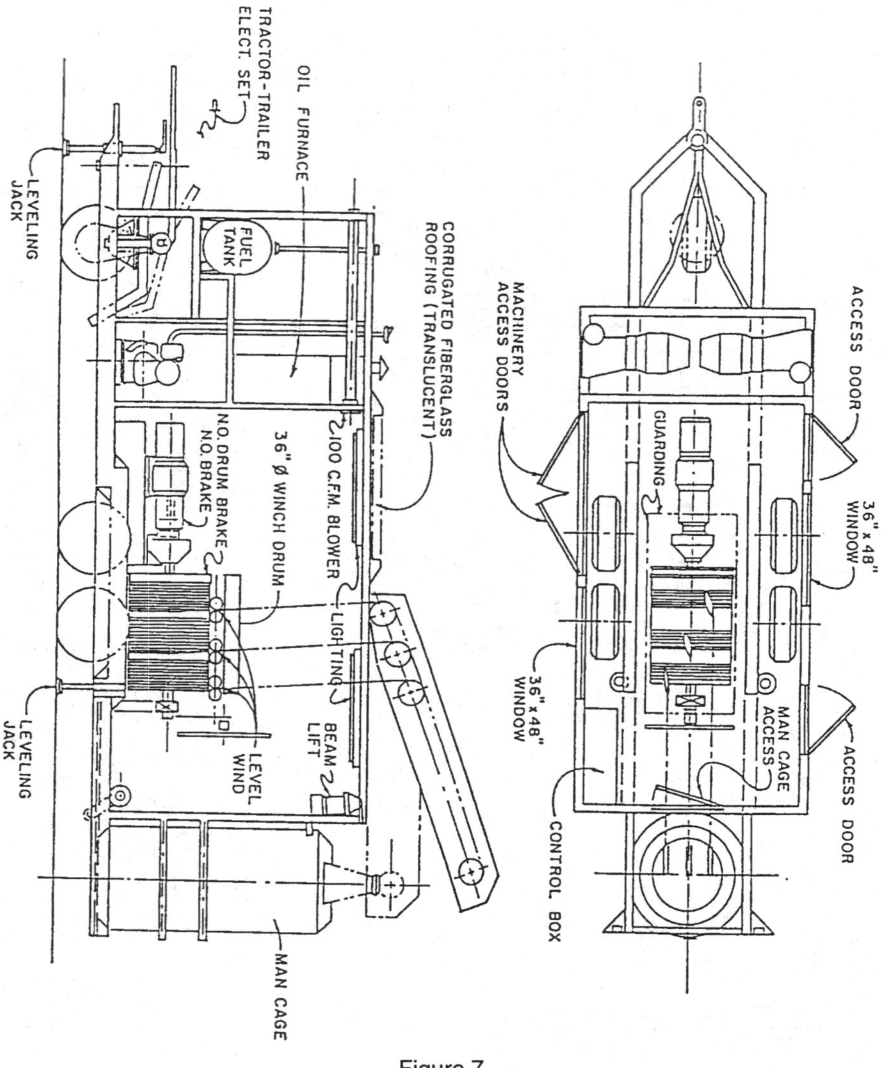

Figure 7

signers in their selection of the Contract G profile which allowed for future flexibility in the siting of two upstream interceptors.

The tunnel profile chosen founded the tunnel in the Chagrin Shale formation for most of its alignment, terminating in Cleveland Shale at the upstream limit of the contract. The depth of the tunnel was from 30 to 70 m (100 to 230 ft) below the surface. Three borings encountered natural gas deposits which dissipated shortly after penetration with the core barrel, and two borings encountered natural oil deposits (oil stains) in the sandstone stratum (See Figure 4).

The rock was tested for unconfined compressive strength, resulting in strength values between 1,110 to 7,560 t/m^2 (1,580 to 10,750 psi), with an average value of 3,230 t/m^2 (4,600 psi). Both the Cleveland and Chagrin Shales are known to be fissile formations. Core samples recovered from the test locations easily parted along horizontal or near horizontal laminae. This phenomenon was profoundly noted in some core samples when they were sprayed with water, an audible "popping noise" could be heard.

Hydrostatic groundwater levels were recorded in the upper glacial till stratum or immediately within the weathered shale rock formation at depths from 2.7 to 7.6 m (9 to 25 ft).

FUTURE CONNECTIONS

Contract G is the downstream leg of the Easterly Service Area interceptors. Two major upstream interceptors were in the planning stage as the Contract G design work continued. These interceptors are two Bluestone Road (BTRS) and three Heights Hilltop tunnel projects. The Heights Hilltop was anticipated to connect near the invert of the upstream limit of the Contract G. As a result, a work shaft supported only by initial supports was designed to be constructed. However, the Bluestone invert connection was uncertain since preliminary investigation for this section of the system indicated presence of oil deposits within the rock at the Contract G profile. Based on this preliminary information, it was decided to construct a work shaft, access adit and deaeration chamber for the future connection of the BTRS contracts. It was obvious that the Bluestone connection would be made higher than the Contract G profile to avoid tunneling through oil saturated rock (See Figure 5).

As a result, the size and configuration of the drop structure for the future BTRS leg was needed to complete the design on Contract G. From studies performed by IIHR for the NEORSD Southwest Interceptor System, a vortex-flow tangential inlet and a vertical circular drop shaft were selected to convey the flows from the anticipated higher elevations of the BTRS interceptor. Once the geometry of the future drop structure was known, the dearation chamber and details for an access adit were completed.

The design of the BTRS connection included provisions to leave the work shaft supported only by initial supports for future tunneling within the upper 58 m (192 ft) of he Contract G profile. The work shaft required for tunneling was large enough to house a future access shaft to Contract G in the same excavation envelope. The final drop structure conveyed 2.7 m^3/s (62 mgd) through a total drop of 43 m (140 ft).

CONSTRUCTION

The final bid package consisted of detailed design for shafts and tunnel (final and initial excavation supports), shaft sites, site restorations, details, geotechnical report and other information necessary to complete the package. The contract was bid on in November of 1990 with KM&M, Joint Venture as the successful contractor. The construction started in January of 1991 with the excavation of Work Shaft No.1/Access Shaft No.2 (Overflow Structure).

The shaft excavation on the project utilized a vertical boring machine (VBM) and drill and blast methods. The VBM was developed by the Kassouf Company of Cleveland, Ohio based on their first designs used in previous NEORSD projects. The machine is capable of

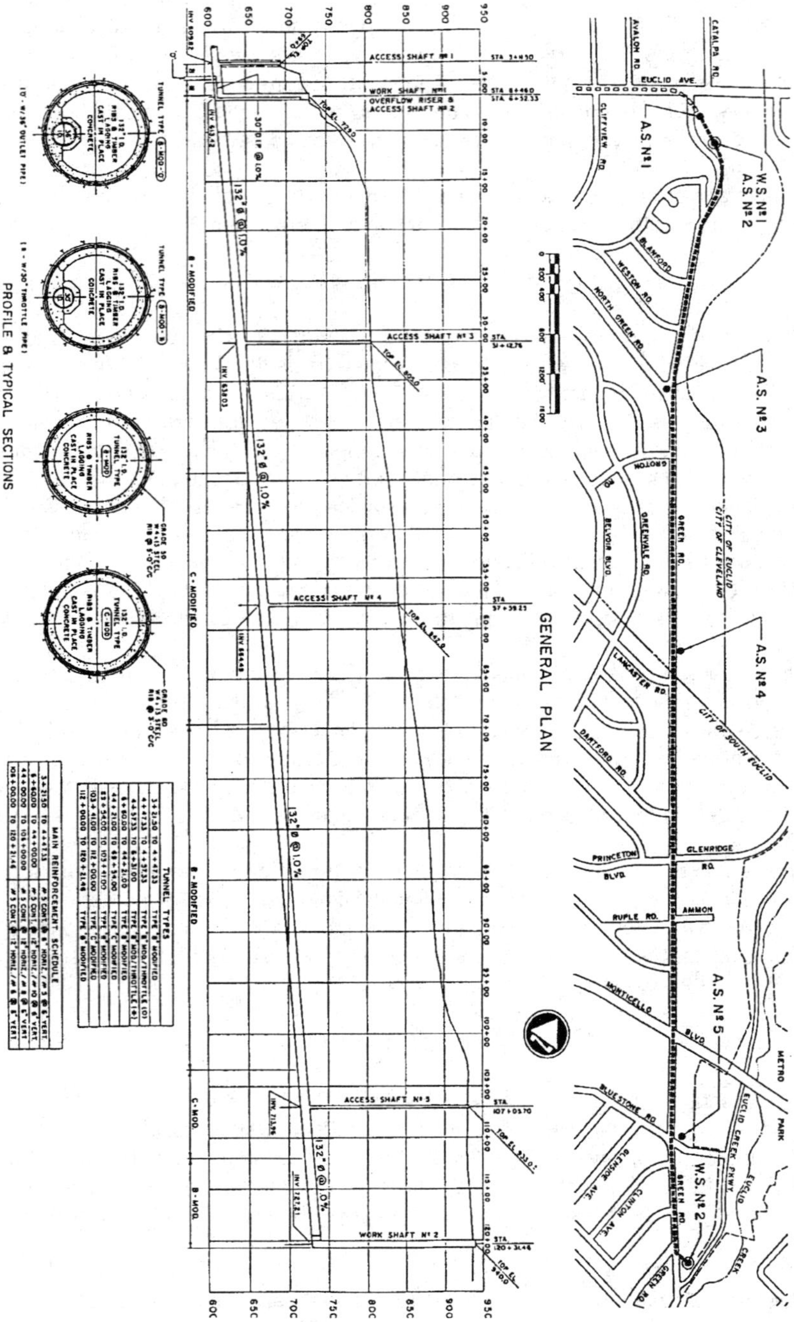
GENERAL PLAN
PROFILE & TYPICAL SECTIONS
A.S. No 1
W.S. No 1
A.S. No 2
A.S. No 3
A.S. No 4
A.S. No 5
W.S. No 2
EUCLID AVE.
AVALON RD
CATALPA RD
CLIFFVIEW RD
BLANFORD
WESTON RD
NORTH GREEN RD
GROTON
BELVOIR BLVD
GREENVALE RD
GREEN RD
CITY OF EUCLID
CITY OF CLEVELAND
LANCASTER RD
CITY OF SOUTH EUCLID
DARTFORD RD
PRINCETON BLVD
GLENRIDGE RD
RUPLE RD
AMMON
MONTICELLO BLVD
METRO PARK
BLUESTONE RD
EUCLID CREEK PKWY
EUCLID CREEK
CLINTON AVE
ACCESS SHAFT No 1
WORK SHAFT No 1
OVERFLOW RISER & ACCESS SHAFT No 2
ACCESS SHAFT No 3
ACCESS SHAFT No 4
ACCESS SHAFT No 5
WORK SHAFT No 2
132" Ø @ 1.0%
B - MODIFIED
C - MODIFIED
C - MOD.
B - MOD.
TUNNEL TYPES
MAIN REINFORCEMENT SCHEDULE
600
650
700
750
800
850
900
950

Figure 8

excavating downward into the shales and other soft rocks. The two stage cutterhead is capable of moving rock cuttings toward the center, where they are lifted by a vertical auger to a skip for hoisting to the surface. As the machine excavates downward, the shaft's initial supports can be installed from the VBM's platform.

The machine proved to be a successful piece of equipment, eliminating complaints associated with drill and blast methods. However, the program of shaft sinking came to a halt when the sandstone stratum was intercepted (Access Shaft No. 5 and Work Shaft No. 2). The method of excavation was switched to drill and blast after the initial 21 m (68 ft) of vertical decent. Once into the sandstone, the pick cutters were unable to penetrate the rock without unacceptable wear. Up to this point, the average excavation progress in rock by the VBM was 1.56 m (5.12 ft) per day. Continuation with the drill and blast methods resulted in a total of 42.7 m (140.2 ft) vertical excavation with an average daily advance of 1.38 m (4.52 ft).

Similarly, at Work Shaft No. 2, the excavation started with the VBM to a depth of 30 m (96 ft) with a daily advance of 1.22 m (4.0 ft). The next 30 m (99 ft) was excavated by drill and blast with an average daily advance of 1.31 m (4.2 ft). Again, the VBM's progress into sandstone was difficult with high wear on the cutters.

The following table depicts excavations progress for other shafts on the projects:

<u>**Access Shaft No. 2/ Work Shaft No. 1**</u>

Excavation in soft ground:	6.7 m (22 ft)
5.5 m (18 ft) diameter VBM excavation downward:	30 m (99 ft)
VBM - Average advance per day:	3.8 m (12.4 ft)
7.3 m (28 ft) diameter VBM overreaming upward:	32 m (104 ft)
VBM - Average upward overreaming advance per day:	8 m (26 ft)

Access Shaft	Soil Depth	Rock Depth	Avg. Advance per day	Dia. in rock
No. 3	6 m (20 ft)	41 m (135 ft)	1.5 m (5 ft) soil 5.9 m (19.4 ft) VBM	4.7 m (15.5 ft)
No. 4	5.8 m (19 ft)	4.45 m (146 ft)	1.4 m (4.75 ft) soil 6.3 m (20.8 ft) VBM	4.7 m (15.5 ft)

Photograph 3 shows the VBM prior to employment into the shaft.

The tunnel initial and final supports were designed taking into consideration geotechnical and tunnel hydraulic operating conditions. The lower section of the tunnel operates under internal pressure resulting from the storage of the inflow almost to depth of 36 m (120 ft). As a result, the final lining of the lower section of the tunnel profile was designed to handle internal hydrostatic pressure. The reinforced concrete lining was designed to resist the cyclic loading that the tunnel is exposed to under each rainfall occurrence.

The final lining in the upper section of the profile was designed as an unreinforced

Photograph 3

concrete lining. An alternate for a final lining utilizing precast pipe was also selected for the different hydrostatic loadings anticipated.

The initial supports consisted of pattern rock dowel installation with ribs designed for areas were rock quality was expected to be less than ideal. The initial supports were designed, detailed and quantities were estimated and listed in the Bid Forms for pricing purposes.

The successful bidder for the project was KM&M, a Joint Venture consisting of three local Cleveland area contractors: Kassouf Co., Mole Constructors, Inc.; and Murry Hill Construction Co. After evaluation of the alternatives, it was the Contractor's decision to utilize the cast-in-place final lining option. The decision was based on costs. The precast pipe lining was found to be more costly when placement and pipe transportation costs were finalized.

The initial lining was changed to steel ribs and lagging for the entire length of the tunnel. This change was initiated by the requirements of providing a thrust reaction for the tunnel boring machine (TBM).

Photograph 4

The TBM selected was by Lovat Tunnel Equipment, Inc. and came equipped with drag cutters and a rear thrust ring requiring ribs and lagging as thrust reaction (Photograph 4). The excavation of the main tunnel started on September 6, 1991 and was completed on February 11, 1992. This resulted in a total of 158 calendar mining days (or 162 mining shifts). The following is the summary of the TBM mining record for the project:

Average feet per shift (beginning 9/6/91)		71.60 (22 m)
Average sets of ribs and lagging installed per shaft		17.30
Total footage mined		11,671.96 (3,558 m)
Total calendar days		158
Total mining shifts		163
Highest footage mined - one shift 1st shift	11/11/91	115.70 (35 m)
Highest footage mined - one work day	12/20/91	223.50 (68 m)
Highest footage mined - one work week	12/02/91 thru 12/06/91	931.60 (284 m)

The adits for the access shaft and the deaeration chamber were excavated by a smaller TBM without any trailing gear to avoid drilling and blasting. After the excavation, reinforcing steel was placed in the adits and dearation chamber, followed by the placement of concrete (Photograph 5).

The final lining in the tunnel was cast utilizing a 64 m (210 ft) long form. The average daily placement was 55 m (180 ft) with 230 m^3 (300 cy) of concrete cast. The placement started at the upper limit of the tunnel profile. Cleaning of the invert, reinforcement placement and rail removal was performed ahead of the concreting operation. Forms were stripped when concrete reached 70 kg/cm^2 (1,000 psi). After the concrete reached its full strength, the lining was contact grouted in the crown area. The tunnel excavation was inspected for rock overbreak prior to the commencement of concrete work to identify high points for placement of the vent pipes.

Photograph 5

For the overflow structure, shaft, and outfall sections of the tunnel (throttle pipe), the concrete mixture for final lining contained microsilica. The microsilica additive produced concrete of a higher density with more resistance to abrasion with an average strength of 530 kg/cm^2 (7,500 psi).

The throttle pipe consisted of DIP piping, anchored down with reinforcing steel to the final lining and was completely encased in concrete (Photograph 6).

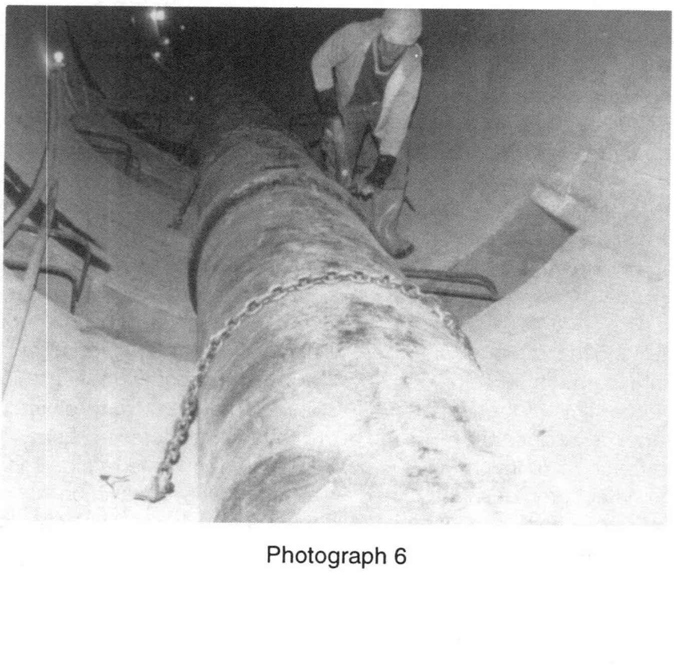

Photograph 6

Photograph 7

CONCLUSIONS

The Contract G proved to be one of the most successful tunnel projects in the Cleveland area. The geotechnical conditions and initial support type requirements were predicted during the design stage. The construction on the project was authorized to proceed on January 9, 1991 with the completion deadline of June 22, 1993. The actual completion was on December 1, 1993. The delay resulted from a change order requiring the Contractor to dispose of the rock muck saturated with natural oil in the same manner as hazardous waste. The muck was tested and the disposal locations were documented.

The revised initial supports system of ribs and lagging proved to be successful and did not promote rock loosening (Photograph 7). The ribs were expanded to provide full contact with the excavated rock surface.

The original bid price was $16,084,517. The final cost of the project was $16,225,741. Total extra work on the contract amounted to $905,039 and the total deductions were $763,814. this resulted in a net increase to the contract bid price of 0.88 percent.

On March 17, 1996, Dawn Engineering, Inc. received a National Honor Award from the Consulting Engineers Council of New Jersey which allowed the project to compete as a finalist for the ACEC's National Award.

ACKNOWLEDGEMENT

The authors would like to acknowledge contributions of Michael Cheetham, the Northeast Ohio Regional Sewer District's Resident Engineer on the project for his assistance in providing the historical data collected during the construction of the tunnel. We would also like to thank Lucy Glodkowski, Lisa Di Franza, Garry Ferrari, Hillary Javora, Charles Polesovsky, Daniel McCormack for their assistance in the preparation of this paper.

8

Geotechnical

Chairman
C. Daugherty
Parsons Brinckerhoff Quade & Douglas
Boston, MA

Chapter 23

BOTTOM STABILITY OF SHAFTS WITH COHESIVE SOIL PLUGS OVERLYING AN ARTESIAN AQUIFER

Steven W. Hunt[1] and Safdar A. Gill[2]
[1]Principal, STS Consultants, Ltd., Milwaukee, Wisconsin;
[2]Principal, Ground Engineering Consultants, Inc., Northbrook, Illinois

ABSTRACT

Design and construction of temporary shafts for underground facilities often require evaluation of bottom stability due to rupture (blow-ins) caused by water pressure acting on a bottom plug. This paper provides an overview of shaft bottom stability evaluation methods for the case of a relatively watertight soil plug which overlies an artesian aquifer. Methods of considering plug perimeter shear, plug buckling and pressure relief are discussed. Minimum factors of safety are suggested. Future use of the shaft subgrade is also considered. Bottom plug failure case histories at several shafts are evaluated. This paper should be useful to project designers, ground support system design engineers and contractors involved with construction of shafts for underground facilities.

INTRODUCTION

Excavation and earth retaining structure system designers should consider stability of the excavation bottom when designing ground support and control of water systems. Considerations should include both the need for the subgrade to support structures or facilities to be constructed as well as the safety of the temporary excavation support system. The soil in the excavation bottom may be weakened or move due to one or more of the following factors:

- Heave from stress relief following excavation.
- Softening/swelling of cohesive soils from adsorption of water.
- Loosening caused by construction equipment.
- Heave of soft cohesive soils from bearing pressures of adjacent soil.
- Heave or boils from quick conditions due to excess upward seepage.
- Heave or bottom buckling of cohesive soils underlain by an artesian aquifer.
- Lost ground from piping of fines with groundwater inflows.
- Voids created from removal of sheeting.
- Densification of loose granular soils during vibratory extraction of sheeting.

A problem or failure occurs when the degree of weakening or movement cause the subgrade to loose its intended support function, when excessive movement causes damage to internal or external facilities or when the ground support system collapses.

In order to properly assess excavation bottom stability, designers need to know the following input factors:

- Pre-construction soil stratigraphy and properties including strength, unit weight, compressibility.
- Subgrade exposure time and duration of excavation phases.
- Aquifer and aquitard conditions including hydraulic conductivity, continuity and water levels.
- Control of water plans including isolation and dewatering components.
- Required post-excavation strength and compressibility of the subgrade to support new structure.
- Movement restraints for the subgrade and adjacent soils as required to protect adjacent property.

The cost of the earth support and control of water system and the risk of system failure depend greatly on the combination of ground conditions and site constraints that exist. Contractors may accept higher risks of performance problems if ground support system costs can be significantly reduced. In order to assess these risks and keep support system costs to a minimum, sufficiently rigorous bottom stability assessments are needed. This paper primarily addresses aspects of bottom stability analyses for the case where the excavation bottom plug includes an aquiclude or aquitard underlain by an aquifer such as a clay layer underlain by sand.

LITERATURE REVIEW

Base Heave From Adjacent Soil Pressure

Deep braced excavations in soft to firm clays are subject to base heave failures from overstressing the base soil in shear which resists adjacent surcharge pressures. In addition to soil strength and excavation depth, the stiffness and depth of the excavation bracing system are the most important factors in determining if this type of base heave is a factor. Overviews of this condition and methods of analysis are presented in Article 14-6 of Bowles (1996) and Articles 37.3 and 54.1 to 54.6 of Terzaghi et.al. (1996). There are many other readily available references on this subject. The factor of safety (FS) against base heave due to adjacent surcharge pressure should generally be at least 1.5. Bowles (1996) recommends a minimum FS of 1.2 to 1.5 for isotropic soils and 1.4 to 1.5 for anisotropic soils. Terzaghi et. al. (1996) suggests a minimum FS of 1.5, but warns that as the factor of safety against basal heave decreases from over 2.0 to less than 1.3, lateral movements and adjacent settlements will tend to increase and the bracing system stiffness will have decreasing ability to reduce movements. Therefore, they recommend a minimum factor of safety of 1.5.

Bottom Softening and Swelling

Clayey soils in an excavation bottom plug have a tendency to soften and swell due to vertical pressure relief. Negative pore water pressures resulting from excavation unloading create a higher attraction to absorb water. If water from groundwater seepage or from precipitation is allowed to pond on the excavation bottom, softening and swelling will occur. This action may reduce the subgrade strength below values considered acceptable for support of the future structure. It may also reduce the bottom plug resistance to lateral and uplift pressures. Generally, bottom softening and swelling can be sufficiently minimized by methods such as:

- Careful excavation technique to minimize soil disturbance and remolding.
- Careful channeling and sumping to prevent ponded water accumulations.
- Isolating the excavation bottom from free water by use of concrete mudmats.
- Surcharging the bottom with a geotextile and stone base.
- Prompt placement of new structure foundations and slabs.

If the excavation base contains expansive clays such as montmorillonite, more careful analysis and measures may be necessary. Article 16.10 of Terzaghi et.al.(1996) discusses heave of expansive clays.

Loosening from Construction Traffic

Initially competent subgrade soils can also become unsatisfactory from construction traffic such as bulldozers and backhoes. Loosened soil may also soften in the presence of water. Construction vibrations in the presence of water may cause the subgrade to "pump" and loose strength. Placement of a mudmat or gravel base course will help to prevent loosening from construction traffic.

Heave and Quick Conditions from Upward Seepage

Upward seepage through the excavation bottom will develop if a water pressure head exists and if the bottom plug soil is granular (an aquifer), mixed granular and cohesive soil (an aquitard) or is cohesive and contains seepage paths through cracks, joints, granular inclusions or other discontinuities. As the upward seepage pressure approaches the downward gravity pressure from the weight of the soil in the flow path, the soil will loose strength and eventually become "quick" and begin boiling. This condition is generally a problem. The excavation subgrade may quickly loose strength and increase in compressibility. This may result in an unacceptable condition for support of the future structure. If the bottom plug is acting as a lower excavation strut, loss of strength could result in large inward movements causing overstressing of walls and bracing. This condition could lead to a collapse of the excavation. Loss of ground associated with piping of fines, heave with boils and increased lateral movements of the support system may also cause excessive damage to adjacent structures.

Seeping water imparts a force on individual soil grains in the flow path through friction. This force may be computed as:

$$F = \gamma_w i V \quad (1)$$

where γw is the unit weight of water, *i* is the hydraulic gradient which is equal to the water pressure head, h, across the flow path divided by the flow path length, L, and V is the flow path volume which is the cross-sectional area, A, of the flow path times the length, L (Cedergren , 1967, Article 3.4).

The seepage pressure from flowing water may be computed as:

$$P_W = h\gamma_W \quad (2)$$

or,

$$P_w = iz\gamma_w \quad (3)$$

where, z is the depth below the bottom plug surface and *i* is the gradient at depth z. When the seepage pressure becomes equal to the effective vertical pressure , γ' z, a quick condition results and,

$$\gamma' z = iz\,\gamma_w = h\,\gamma_w \quad (4)$$

The critical hydraulic gradient when quick conditions develop is defined as:

$$i_c = \frac{\gamma'}{\gamma_w} \quad (5)$$

For a typical granular soil with γ_{sat} = 19 kN/m^3, the effective unit weight would be $\gamma_{sat} - \gamma_w$ = 19.0-9.8 = 9.2 kN/m^3. The critical gradient for this soil is then i_c = 9.2/9.8 = 0.94. For most granular soils, a quick condition and boils will develop when the hydraulic gradient increases to approximately 1 or more.

In order to prevent boils and detrimental loosening of bottom plug soils, the hydraulic gradient through the plug must be reduced so that the factor of safety against instability from a quick condition is adequate. When determining the appropriate factor of safety against quick conditions, a designer should consider: the subgrade condition needed for support of the future structure, the importance of the bottom plug as strut of the ground support system and the need to minimize settlements of adjacent property. The safety factors recommended by various authors in literature that we reviewed are listed in Table 1.

Table 1. Safety Factors Against Quick Conditions

Reference	Recommended Minimum Safety Factor	Comment
NAVFAC DM 7.1, p 267	1.5 - 2.0	Exit gradients less than 0.5 such that: $i_c/i \leq 2.0$
Canadian Foundation Engineering Manual, 1985, p 130	1.5 - 2.0	$i_c/i \leq$ 2 to 30
Febesh and Woolworth, 1984, pp 8-33 - 8-36	1.5 - 2.0	Same as NAVFAC DM 7.1
Cedergren, 1967, p 226	2.0 - 2.5	
Bowles, 1996, p 815-816	1.25	
Dismuke, 1975, p 469	1.5	

Simple charts for estimating the cutoff wall depth that is needed to provide factors of safety against piping (quick conditions) in the range of 1 to 2 are provided in NAVFAC DM 7.1 (Anon., 1982). These charts are commonly referenced by others (Febesh and Woolworth, 1984; Dismuke, 1975) A different chart is presented in Article 29.3 of the Canadian Foundation Engineering Manual (Anon., 1985). The cutoff wall concept presumes relatively watertight walls without discontinuities or "windows." If discontinuities such as non-interlocked or split sheet pile joints or windows in slurry diaphragm walls are suspected, then these preceding factors of safety may not be adequate unless shorter flow path lengths are evaluated.

Instead of using charts, hydraulic gradients or seepage pressures may be estimated from flow nets and then compared to the critical hydraulic gradient for the worst soil condition. The flow net method of analysis is described in Articles 3.2 and 10.6 of Cedergren et. al. (1967), Article 2.6 of Peck et. al (1975), Articles 23 and 24 of Terzaghi et. al. (1996) and many other references.

The preceding methods of analysis generally presume that the excavation bottom plug material is granular with relatively uniform flow. In many soils, seepage tends to concentrate in more permeable channels or zones. Such channels and zones may be oriented in any direction and may vary greatly in size. Of particular interest is the case where a channel(s) of sand and/or gravel penetrates through an otherwise cohesive and low permeability bottom plug. Although the overall bottom plug may act as an aquitard to produce limited inflow, the seepage concentrated in more permeable channels should be checked to determine a factor of safety against boiling or quick conditions. The seepage pressure and critical hydraulic gradient within a channel or "pipe" may be computed using the equations cited above. The cross-sectional area of the flow path ultimately cancels in these equations. This means that the factor of safety against a boil or blow-in may be computed by estimating the channel flow path length, the unit weight of the channel soil and water pressure head or pressure difference which acts on the channel. This method neglects friction losses which occur around the sides of the channel (i.e., the walls of the pipe). Results for a channel flow evaluation will tend to be a little more conservative than for the case of upward seepage throughout the bottom plug. Accordingly, slightly smaller factors of safety against quick conditions may be acceptable for channel flow

Lost Ground by Piping of Fines

Seepage through the bottom plug may also cause subsurface erosion, lost ground and excessive movements if fines are "piped" or transported with the moving water. If muddy and/or silty water is flowing into the excavation through the bottom plug, then detrimental piping is occurring. Piping of fines may also occur at excavation bottom sumps, pressure relief wells, along the face of cutoff walls or through wall joints or discontinuities. Continued inflows with subsurface erosion will cause a gradually increasing volume of lost ground and increased risk of ground support system collapse or serious ground and adjacent facility movements. Once discovered, measures should be promptly taken to stop subsurface erosion. These measures may include the following:

- Installation of a graded filter or filter fabric and surcharge stone over the zone where piping of fines is observed.
- Deactivation of well pumping followed by reduction of well screen opening size and/or filter size.
- Reduction of hydraulic gradients by adding more pressure relief wells or by partial flooding the excavation to reduce the differential head.

Bottom Heave, Blow-In or Buckling of a Cohesive Plug Overlying an Artesian Aquifer

Cohesive Plug Failure Modes. When the bottom plug of an excavation includes a continuous layer of low permeability soil such as silty clay or clayey silt, and is underlain by an artesian aquifer, upward hydrostatic pressures may cause bottom instability. Common types of bottom instability include:

- Heave of entire plug mass (like a plunger in a tube) without significant seepage or boils.
- Heave with tensile cracking leading to blows or boils.
- Blows or boils through joints, channel inclusions, old wells or boreholes or other aquiclude discontinuities.

Analysis of bottom instability from discontinuities was discussed in the previous section. Methods of evaluating the case where seepage does not exist or is very small and therefore hydrostatic pressure is exerted on the base of the aquiclude layer of the plug are discussed below.

Cohesive Plug Heave if No Perimeter Shear. The factor of safety (FS) of a cohesive bottom plug (aquiclude) against heave from hydrostatic uplift pressure is commonly computed by dividing the total weight of the plug by the uplift force on the base of the aquiclude layer:

$$FS = \frac{\gamma_t \, T A_b}{\gamma_w \, H_b \, A_b} \qquad (6)$$

or,

$$FS = \frac{\gamma_t \, T}{\gamma_w \, H_b} \qquad (7)$$

where, γ_t is the total average unit weight of the soil plug above the base of the aquiclude layer, T is the average thickness of the clay plug above the base of the aquiclude layer (including portions extending below the bottom of excavation sheeting or walls), A_b is the cross-sectional area of the plug base, γ_w is the unit weight of water and H_b is the water head corresponding to the water pressure acting on the base of the aquiclude layer. The resistance to heave from perimeter shear is ignored. Some factors of safety recommended in the literature for this method of analysis are listed in Table 2. The minimum factors of safety range from 1.0 to 1.43 when perimeter shear is ignored and if the width of the excavation, D, is not large relative to the bottom plug thickness, i.e., $D/T \leq 1.5$. These factors of safety are lower than generally cited for the quick condition case because perimeter shear is being ignored.

Table 2. Safety Factors of Cohesive Bottom Plug Against Blow-In

Reference	Recommended Minimum Safety Factor	Comment
NAVFAC DM 7.1, p 270	1.0	Perimeter shear ignored
Canadian Foundation Engineering Manual, 1985, p 1	1.43	Perimeter shear ignored
Teng, 1962, p 104	1.3	Perimeter shear ignored
Bowles, 1996, p 815-816	1./25	Perimeter shear ignored
Febesh and Woolworth, 1984, pp 8-33 - 8-36	1.0	Perimeter shear ignored
Febesh and Woolworth, 1984, pp 8-33 - 8-36	Not Stated	Include full perimeter shear D/T ≤ 1.5

Cohesive Plug Heave if Perimeter Shear Included. When circumstances justify a more thorough evaluation of the factor of safety against blow-in, perimeter shear should be considered. The factor of safety against heave of the plug becomes:

$$FS = \frac{\gamma_t \, T A_b + \tau A_p}{\gamma_w \, H_b \, A_b} \tag{8}$$

where, τ is the mobilized shear A_b is the plug base area, A_p is the plug perimeter area and H_b is the piezometric head acting on the base of the cohesive plug. While a localized lower strength zone within the bottom plug could fail in shear before the entire plug does, calculations show that as the plug diameter becomes larger, the factor of safety becomes lower with all other factors unchanged. This means that the plug area based on the full plug diameter or width should generally be used.

Regarding selection of a perimeter shear value, consideration should be given to strength reductions caused by installation of the shaft wall elements and to short term (undrained) versus long-term (drained) effects. The maximum shear resistance of cohesive soils at the time of initial excavation is generally the undrained shear strength, S_u which is half of the unconfined compressive strength, Q_u. Installation of the shaft wall sheeting or other wall units such as slurry diaphragm walls may cause shear strength to be reduced. Suggested reduction factors, 2, for driven piles are given in Figure 18.7 of Peck et.al. (1974). With time as excess porewater pressures dissipate and thixotropy develops, cohesive soils generally regain their lost strength. This process is called "set-up." Set-up with time may be estimated by the methods given by Skov and Denver (1988) or Svinkin et al. (1994). Effects of slurry diaphragm wall installation on wall-soil interface shear strength are addressed by Xanthakos, (1979).

Cohesionless soil present below the base of the cohesive portion of the bottom plug does not contribute to plug shear and should not be considered. However, if cohesionless soil is present above the clay plug, it may be considered. The shear strength, τ, of cohesionless soil may be computed as:

$$\tau = \bar{\sigma} \tan \phi \tag{9}$$

Determination of $\bar{\sigma}_h$ should consider the reduction of vertical stress within the plug zone due to the excavation as well as lateral presses on the plug from load transfer through the earth retention system walls. The extent of pressures above at-rest values will be dependent on the earth retaining system stiffness and movements allowed during construction.

Febesh and Woolworth (1984) suggested that full shear may be considered if D≥1.5T. They recommend that no shear resistance be considered if D > 3T. For intermediate cases, shear, τ, is proportioned as:

$$\tau = 2 \, S_u \left[1 - \frac{D}{3 \, T} \right] \tag{10}$$

where, S_u is the average undrained shear strength at the time.

Febesh and Woolworth (1984) did not suggest what factor of safety should be used when shear is added to the plug weight.

Cohesive Plug Failure by Buckling or Tension Cracks. The risk of a cohesive bottom plug failure by buckling or due to development of tensile cracks within the upper portion of the aquiclude layer should also be considered. When the uplift pressure exceeds the weight of the plug and the horizontal compressive stress which resists exterior horizontal pressure on the plug, tensile stresses will develop within the upper portion of the plug. The stresses in an idealized cohesive bottom plug that is subjected to hydrostatic uplift pressure and horizontal pressure are shown in Figure 1. The net uplift pressure, σ_n is,

$$\sigma_n = \sigma_w - \sigma_v \tag{11}$$

where, σ_w is the hydrostatic pressure on the plug base, $\sigma_v = \tau T$ and T is the thickness of the cohesive portion of the bottom plug.

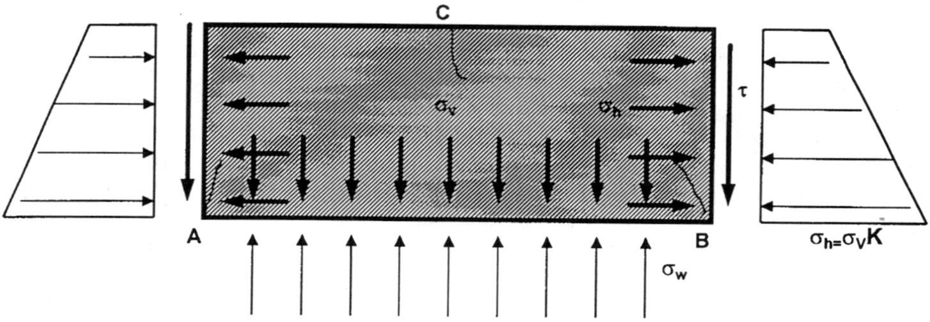

Figure 1. Pressures Acting on Cohesive Bottom Plug

The maximum moment and stress within the plug depend on the plug shape and fixity along the perimeter. For the case of long, narrow bottom plug, a two-dimensional analysis is reasonable. In this case the bottom plug may be analyzed as a unit width strip or beam. If we consider the ends of the beam to be fixed and transmit moments, then the maximum bending moment of the beam is at two end points, A and B (Graham and Godden, 1968):

$$M_A = M_B = -\frac{\sigma_n D^2}{12} \tag{12}$$

This means that for a bottom plug that is subjected to uplift pressure exceeding the pressure from the body mass of the plug, the largest tensile stresses will be at the base of the plug at each of the two ends. A smaller tensile stress will also develop at the middle of the upper surface of the plug. If the ends of the plug are considered pin connections instead of fixed, the maximum moment becomes the middle of the upper surface of the plug and is equal to:

$$M_c = \frac{\sigma_n D^2}{8} \tag{13}$$

The maximum outer fiber stress, σT in a beam of depth T is:

$$\sigma_T = \frac{MT}{2\,I} \tag{14}$$

where, I is the moment of inertia = $T^3/12$ for a unit width beam. If M_A from Eqn. 12 is substituted for M in Eqn. 14, the maximum tensile stress becomes,

$$\sigma_T = \frac{\sigma_n D^2}{2\,T^2} \tag{15}$$

Since lateral earth pressure also acts on the beam. the maximum tensile pressure will be the difference of the stress computed from Eqn. 15 and the horizontal earth pressure, σ_H. Goodman (1980) refers to this condition as a "clamped beam" and gives the following equation for maximum tensile stress:

$$\sigma_T = \frac{\sigma_n D^2}{2\,T^2} - \sigma_H \tag{16}$$

where $\sigma_H < \pi^2 ET^2/60D^2$ to limit σ_H to pressures which do not tend to buckle the beam from a high slenderness ratio.

If we solve for the net uplift pressure, σ_n, then for a long rectangular bottom plug,

$$\sigma_n = \frac{(\sigma_T + \sigma_H)\,2\,T^2}{D^2} \tag{17}$$

If the bottom plug is better modeled as a circular plate with a uniform surface pressure, σ_n, and is fixed around the perimeter, the maximum moment is (Young, 1989):

$$M = \frac{\sigma_n (D/2)^2}{8} = -\frac{\sigma_n D^2}{32} \tag{18}$$

The maximum fiber stress from bending is (Young, 1989):

$$\sigma_T = \frac{6M}{T^2} = \frac{3\sigma_n D^2}{16\,T^2} \tag{19}$$

After subtracting the horizontal earth pressure as discussed above,

$$\sigma_T = \frac{3\sigma_n D^2}{16\,T^2} - \sigma_H \tag{20}$$

If we solve for the net uplift pressure, σ_n, then for circular bottom plug,

$$\sigma_n = \frac{(\sigma_T + \sigma_H)\, 16\, T^2}{3\, D^2} \qquad (21)$$

The maximum net uplift pressure at the point when σT is maximum and tension cracks should develop may be computed from Eqn. 17 for a rectangular bottom plug or from Eqn. 21 for a circular bottom plug. The factor of safety against failure by tension cracking is then the ratio of σ_n maximum to σ_n actual.

The largest value for t which may be used is the maximum tensile strength of the clay. Generally, we do not count on tensile strength in soils and assume that σT = 0. Testing performed by Narain and Rawat (1970) found that the maximum tensile strength of an undrained compacted clay was in the range of 1/6 to 1/12 of its unconfined compressive strength, q_u, with the ratio highest for cohesive soils with lowest plasticity index. A ratio of 1/8 corresponds to clay with a plasticity index of 20 percent.

The maximum tensile strength, σ't, of a drained clay with a friction angle, ϕ', and a cohesion intercept, $\overline{c}$, is:

$$\sigma'_t = \frac{\overline{c}}{\tan \phi'} \qquad (22)$$

The maximum horizontal pressure at top of the cohesive plug is the passive earth pressure, σ_{HP}, is (Terzaghi et. al., 1996, p 248):

$$\sigma'_{HP} = \gamma' \, D\, N_{\phi'} + 2c \sqrt{N_{\phi}} \qquad (23)$$

where N'ϕ = $K_p = \tan^2(45° + \phi'/2)$. Passive pressure should only be used if inward movements of the walls are sufficient to mobilize it. Generally the horizontal stress in the plug will vary from the at-rest value before excavating to a maximum equal to the passive pressure. Using the preceding equations, a value for σ_m may be computed and inserted into Eqn. 17 or 21 to compute σ_n. Then, the maximum head, H_m, acting on the base of the cohesive plug prior to the development of tension cracks would be:

$$H_m = \frac{\sigma_w}{\gamma_w} = \frac{\sigma_n + \sigma_v}{\gamma_w} \qquad (24)$$

Then, the factor of safety against a bottom rupture by cracking or buckling would be:

$$FS = \frac{H_m}{H_b} \qquad (25)$$

where, H_b is the piezometric head acting on the base of the cohesive plug.

BOTTOM PLUG FAILURE CASE HISTORIES

Milwaukee NS-10 Junction Structure Shaft

During shaft excavation in 1988, the bottom plug of the NS-10 Junction Structure shaft failed when excess hydrostatic pressure caused the plug to suddenly rupture and blow-in at several locations. The blows or geysers grew to 1/2 to 1 m tall fountains of muddy and silty water. The bottom failure occurred after excavating in the dry to a depth of 12.5 m. Available information on soil, groundwater and excavation conditions allows back analysis of bottom stability with perimeter shear and to assess buckling and D/T ratios.

Shaft Conditions

The earth retaining structure consisted of interlocked steel sheet piles with internal steel ribs. The diameter at the inside face of the sheeting was 7.32 m. The steel sheet piles were CZ-114 sections with a total length of 18.3 m. The sheets were driven to a depth of 16.8 m below the ground surface. At the time of bottom blow-in, approximately half of the shaft had been excavated to a depth of 12.5 m and the remaining half to a depth of approximately 11.6 m. The base depth of the cohesive plug ranged from 14.8 to 15.0 m and averaged 14.9 m. The resulting cohesive plug thickness ranged from 2.4 to 3.4 m with an average of 2.9 m A plan view of the shaft conditions is shown in Figure 2.

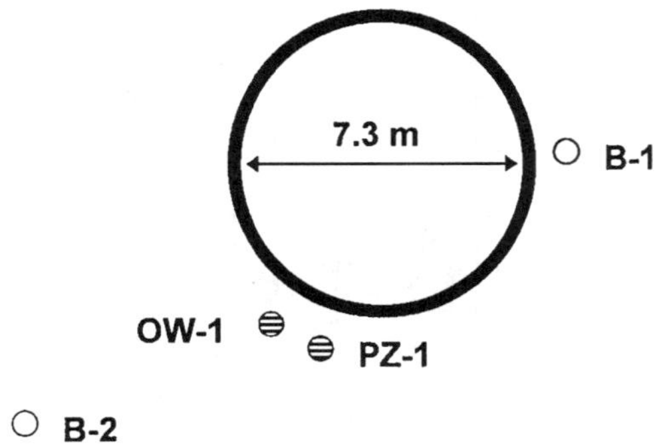

Figure 2. Plan View, NS-10 Junction Structure Shaft

Soil Conditions

Soil conditions at the Juncture Structure shaft were indicated by two soil borings which were located as shown in Figure 2. The soil conditions are shown on the profile in Figure 3. The properties of the soils as determined by laboratory testing are listed in Table 3. The cohesive soil plug at the time of failure was within the lower portion of the Estuarine Deposits. The two nearby borings indicated that the lower sub-stratum of the Estuarine Deposits was a silty clay with a Unified Soil Classification of CL, unconfined compressive strengths ranging from 48 to 172 kPa (averaging 104 kPa), water contents ranging from 22 to 30 percent, and a Plasticity Index average of 9 percent.

Table 3. Soil Properties at NS-10 Junction Structure Shaft

PROPERTY	UNIT	FILL	ESTUARINE	ALLUVIUM
Unconfined compressive strength, Q_u	kPa	-	48-240, avg. 76	-
Degree of Sensitivity, S_t	-	-	4	-
Effective Friction Angle, ϕ	deg.	30	32	30-40
Effective Cohesion, c	kPa	-	5	-
Young's Modulus, E	MPa	10-40	15-24	41-82
Poisson's Ratio, μ	-	0.3	0.4	0.3
Coeff. of Earth Pressure At-Rest, K_o	-	0.50	0.50	0.45
Saturated Unit Weight, γ_t	kN/m³	19.6	16.3	19.5-20.4
Effective Unit Weight, γ	kN/m³	9.8	6.5	9.7-10.6
Permeability, k	cm/sec	~1x10⁻³	1x10⁻⁴ to 1x10⁻⁶	1x10⁻² -1x10⁻³
Plasticity Index, I_p	%	-	3-12, avg. 9	-

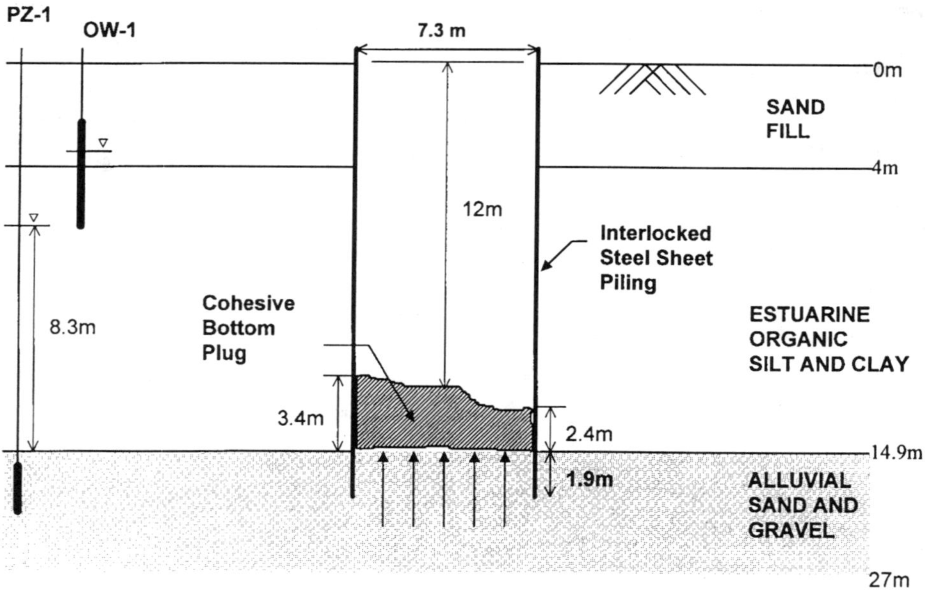

Figure 3. Junction Structure Shaft Cross-Section

Groundwater Conditions

Groundwater levels affecting the shaft were indicated by an observation well, OW-1, and a piezometer, PZ-1. These wells were located within 3 m of the shaft as shown in Figure 2. Approximate sensing zone locations are shown in Figure 3.

Normal water table depths prior to excavating ranged from 2.29 to 3.02 m. Just prior to the excavation bottom blow-in, OW-1 indicated a water table depth of 3.96 m. The lower level was probably due to seepage through sheet pile interlocks in the Fill zone.

Normal piezometric levels in the Alluvium ranged from 6.49 to 6.64 m. Just prior to the excavation bottom blow-in, PZ-1 indicated a level 6.61 m which is 8.3 m above the average level of the base of the plug. The Alluvium water levels had not fallen prior to the blow-in because upward seepage had not developed. The observations also show that upward seepage through the sheeting interlocks was not effective as a form of pressure relief.

Observations at Time of Bottom Blow-In

The shaft was excavated with a clam bucket and air spades. Excavation at depths below 4 m was started on 12/06/88. At the end of the shift on 12/07/88, the average bottom depth was at approximately 11.6 m. It was firm with no signs of "pumping" or quick conditions. During the morning of 12/08/88, approximately one-half of the bottom had been advanced to approximately 12.5 m when it began to rupture. At several locations ranging from 1 to 2 m away from the sheeting, cracks and blows developed. Fountains of muddy water protruded approximately 1/2 to 1 m above the bottom surface. No signs were observed of the bottom plug rising along the sheeting as would result from a perimeter shear failure. Workmen quickly dumped approximately a meter of gravel over the excavation bottom for surcharge and then flooded the shaft. This action prevented a more catastrophic bottom plug failure.

Analysis of Failure

Careful analysis of the shaft, soil and groundwater conditions suggests that this shaft likely failed by buckling of a relatively watertight, cohesive plug. Factors of safety for several conditions are discussed below.

First, a check of the factor of safety against piping through a possible channel or discontinuity was made. The maximum hydraulic gradient exists for shortest flow path. The shortest flow path extends from the lowest surface of the bottom plug to the bottom of the sheet pile cuttoff wall. The hydraulic gradient, is,

$$i = \frac{\delta H}{\delta L} = \frac{5.9}{4.3} = 1.37$$

Assuming that the unit weight of a sand filled channel is 18.8 kN/m^3, then the critical hydraulic gradient is,

$$i_c = \frac{\gamma'}{\gamma_w} = \frac{9.0}{9.8} = 0.92$$

The factor of safety against piping for this condition is approximately,

$$FS = \frac{0.92}{1.37} = 0.67$$

An even lower critical hydraulic gradient of 6.5/9.8 = 0.66 and factor of safety of 0.66/1.37 = 0.48 results if the unit weight of the organic silt is used. These low factors of safety against piping suggest that a channel or discontinuity blow was likely if a discontinuity was encountered. However, field observations did not reveal such a discontinuity nor the development of a boil or quick conditions. This case shows that high hydraulic gradients may exist across a continuous cohesive plug without development of a boil.

At time of failure, the ratio of plug diameter to average plug thickness was D/T = 7.3/2.4 = 3.0. If the suggestions of Febesh and Woolworth (1984) are followed, the plug should be evaluated with no shear since D/T ≥ 3. As previously stated, the factor of safety against cohesive plug heave neglecting shear is given by Eqn. 7. Substituting for conditions at the time of bottom plug failure results in a factor of safety of:

$$FS = \frac{16.3(2.9)}{9.8(8.3)} = 0.58$$

Since this factor of safety is much lower than 1.0, shear resistance must have been acting to resist failure. The maximum plug thickness, T for a factor of safety of 1 using this method is:

$$T = \frac{\gamma_w H_b}{\gamma_t} = \frac{9.8(8.3)}{16.3} = 5.0 \text{ m}$$

By this method, the recommended minimum bottom plug thickness would have been 5.0 m which is 2.1 m thicker than the average thickness existing at the time of failure.

The factor of safety against cohesive plug heave including perimeter shear was previously given as Eqn. 8. If the shear strength, τ, is assumed to be 24 kPa which is half the lowest range of unconfined compressive strengths measured for the lower sub-strata, then,

$$FS = \frac{16.3(2.9)\,\pi\,(7.32/2)^2 + 24\,\pi\,(7.32)\,2.9}{9.8(8.3)\,\pi\,(7.32/2)^2} = 1.05$$

Using this low shear strength value results in a factor of safety close 1.0 and thus may explain the bottom failure. The shear strength value is 24/52 = 0.46 or 46 percent of the average undrained shear strength measured for the bottom plug layer. Peck et. al. (1974) suggested reduction factors, α, ranging from 0.54 to 0.96 and averaging 0.8 for an average unconfined compressive strength of 104 kPa. For this soil, the smallest value of correlates best with that corresponding to the failure.

Bottom stability predictions were made prior to shaft excavating. A shear strength value of Su=19.1 kPa (80% of the lower limit) was assumed in a conservative attempt to consider strength reduction caused by sheet pile driving. Using Su = 19.1, a factor of safety of 0.95 results. Local experience with set-up in this organic silt deposit for tested bearing piles indicates that interface shear strengths in the range of 20 to 30 kPa are typical in this deposit after one to two weeks of set-up time.

Although the plug failure can be explained as an intact plug heaving by shear along the perimeter, actual observations suggest that the plug failed by the development of tension cracks that led to blows. Assuming undrained, $\phi = 0$ conditions, then the tensile strength of a clay with a plasticity index of 9 percent may be estimated as 1/10 of Q_u (Narain and Rawat, 1970). Substituting Q_u =104 kPa for the average unconfined compressive strength for the lower sub-stratum gives a tensile strength value:

$$\sigma_t = \frac{Q_u}{10} = \frac{104 \text{ kPa}}{10} \; 10.4 \text{ kPa}$$

The maximum horizontal earth pressure for passive conditions with $\phi = 0$ and using the average undrained shear strength according to Eqn. 23 is:

$$\sigma_{HP} = 2 \, Su = 2\,(52) = 104 \text{ kPa}$$

Substituting these values into Eqn 21 gives the maximum net uplift pressure when cracks should develop within the plug base:

$$\sigma_n = \frac{\left(\sigma_T + \sigma_H\right) 16 \, T^2}{3 \, D^2} = \frac{(34 + 269) \, 16 \, (2.1)^2}{3 \, (6.1)^2} = 192 \text{ kPa}$$

Substituting into Eqn. 24 to compute the maximum head before cracking gives:

$$H_m = \frac{192 + 22.0(2.1)}{9.8} = 24.4 \text{ m}$$

The actual water pressure head on the bottom of the clay plug at the time of failure was 8.3 m. The corresponding factor of safety against tension crack development is:

$$FS = \frac{24.4}{27.7} = 0.88$$

This result suggests that the plug should not have failed by buckling or cracking. However, If the tensile strength is assumed to be 0 as is normally done, and the horizontal earth pressure is assumed to be the estimated at-rest value prior to excavating, then t average is approximately 73 kPa. If instead of using the average plug thickness of 2.9 m, we use a thickness of 2.3 m for the thinner half, then the net uplift at the point of cracking becomes σ_n = 38.4 kPa and the corresponding maximum head, H, is 7.7 m. The resulting factor of safety is 0.9. Therefore, if the worst combination of plug conditions is assumed, the clay plug failure can be explained by buckling and cracking followed by a piping blow-in.

Milwaukee Jones Island Siphon Shaft

During shaft excavation in 1984, the bottom plug of the Jones Island Siphon shaft failed when excess hydrostatic pressure caused a cohesive bottom plug to suddenly rupture and blow-in . The bottom failure occurred after excavating in the dry to a depth of 32.6 m. Water and approximately 10 m^3 of soil accumulated with the shaft bottom before it was stabilized by partial flooding and surcharging with gravel. Available information on soil, groundwater and excavation conditions allows back analysis of bottom stability with perimeter shear and to assess buckling and D/T ratios.

Shaft Conditions

The earth retaining structure consisted of a frozen ground cofferdam extending to a depth of approximately 34.4 m. Below the frozen ground cofferdam, the earth retaining structure was intended to change to steel ribs and timber lagging. Dewatering wells were installed to depressurize glacial outwash deposits between cohesive till layers. However, the dewatering did not keep pace with shaft excavating. The resulting pressure relief was not adequate and the bottom ruptured and blew in before any ribs and lagging were installed.

The freeze pipes were set at a diameter of 7.9 m. At the time of excavating, the freezewall was approximately 1.2 to 1.5 m thick. As the shaft was being excavated, a permanent concrete lining with an inside diameter of 5.49 m was cast in lifts from the top down. When the bottom blew in, the lining had been cast to a depth of 30.2 m and the shaft had been excavated to a depth of 32.6 m. The base depth of the cohesive plug was 34.7 m. The resulting cohesive plug thickness was approximately 2.1 m. A plan view of the shaft conditions is shown in Figure 4.

Soil Conditions

Soil conditions at the Siphon shaft were indicated by the logs from two soil borings

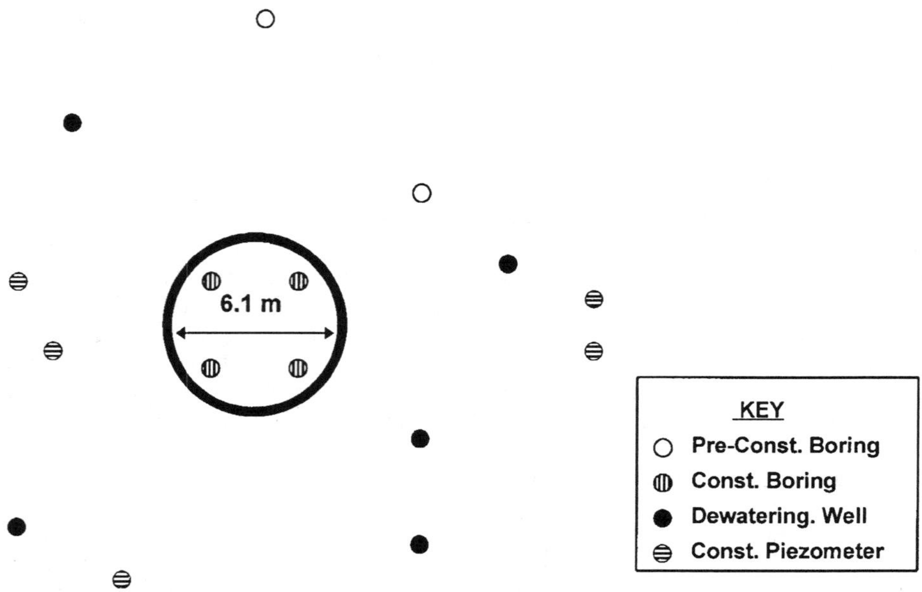

Figure 4. Plan View, Jones Island Siphon Shaft

prior to construction and from 17 additional borings and wells which were completed during construction. These boring and well locations are shown in Figure 4. The soil conditions are shown on the profile in Figure 5. The properties of the soils as determined by laboratory testing are listed in Table 4. The cohesive soil plug at the time of failure was within an overconsolidated silty clay glacial till deposit. The two pre-construction borings indicated that till had a Unified Soil Classification of CL, unconfined compressive strengths ranging from 240 to 350 kPa (averaging 306 kPa), water contents ranging from 15 to 19 percent, and a Plasticity Index averaging 17 percent.

Table 4. Soil Properties at Jones Island Siphon Shaft

PROPERTY	UNIT	FILL	ESTUARINE	ALLUVIUM	TILL
Unconfined comp. strength, Q_u	kPa	-	48-240, avg. 76	-	240-350
Degree of Sensitivity, S_t	-	-	4	-	2
Effective Friction Angle, ϕ	deg.	30	32	30-40	29
Effective Cohesion, c	kPa	-	5	-	32
Young's Modulus, E	MPa	10-40	15-25	40-80	55-90
Poisson's Ratio, μ	-	0.3	0.4	0.3	0.4
Coeff. of Earth Pressure At-Rest, K_o	-	0.50	0.50	0.45	0.8
Saturated Unit Weight, γ_t	kN/m^3	19.6	16.3	19.5-20.4	22.0
Effective Unit Weight, γ	kN/m^3	9.8	6.5	9.7-10.6	12.2
Permeability, k	cm/sec	$\sim 1x10^{-3}$	$1x10^{-4}$ to $1x10^{-6}$	$1x10^{-2}$ -$1x10^{-3}$	$1x10^{-5}$-$1x10^{-7}$
Plasticity Index, I_p	%	-	3-12, avg. 9	-	15-19, avg.17

Groundwater Conditions

Groundwater levels affecting the shaft were indicated by observation wells and piezometers with sensing zones within the alluvium and outwash deposits. These wells were located as shown in Figure 4. Approximate sensing zone depths of two of the pertinent wells are shown in Figure 5. Normal pre-construction piezometric levels in the Alluvium and Outwash ranged from 0.9 to 1.5 m. Construction dewatering from deep wells in the outwash had lowered the alluvial aquifer level to approximately 8.8 m just prior to the bottom blow-in. Pumping from deep wells had lowered the outwash aquifer level to approximately 11 m, however, two days prior to the blow-in, one of the deep wells was deactivated to allow a new well to be developed. As a result, water levels rose and were at approximately 7.0 m below ground surface at the time of the blow-in. This resulted in a head, H_b, of 27.7 m on the base of the cohesive plug.

Observations at Time of Bottom Blow-In

The shaft was excavated with air spades and a clam bucket. At the end of the day shift on 05/03/84, the average bottom depth was at approximately 31 m. It was firm with no signs of seepage, "pumping" or quick conditions. The bottom blow-in developed during the second shift, after excavating to a depth of 32.6 m. The blow was described as a rupture in the northeast quadrant that spewed soil and water into the shaft. Before remedial actions could be taken, approximately 10 m^3 of soil and 6 m of water had accumulated in the shaft bottom. No signs were observed of the bottom plug rising along the freezewall as would result from a perimeter shear failure. Workmen quickly dumped approximately 100 m^3 of gravel over the excavation bottom for surcharge and then flooded the shaft to a depth of 9.1 m. This action minimized further lost ground from outside the excavation.

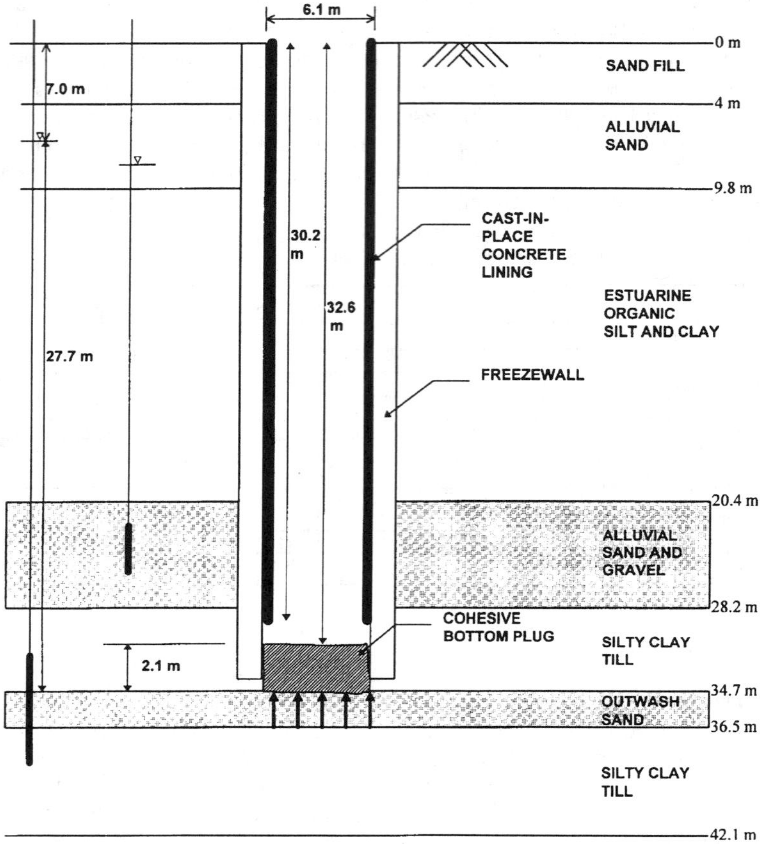

Figure 5. Jones Island Siphon Shaft Cross-Section

Analysis of Failure

Careful analysis of the shaft, soil and groundwater conditions suggests that this shaft likely failed by buckling of a relatively watertight, cohesive plug. Factors of safety for several conditions are discussed below.

First, a check of the factor of safety against piping within a channel or discontinuity was made. The maximum hydraulic gradient corresponds to shortest flow path which could extend through the cohesive bottom plug to the underlying outwash aquifer. This hydraulic gradient was extremely high:

$$i = \frac{\delta H}{\delta L} = \frac{25.6}{2.1} = 12.2$$

Based on an effective unit weight of 12.2 kN/m³ for the till, the critical hydraulic gradient is:

$$i_c = \frac{\gamma'}{\gamma_w} = \frac{12.2}{9.8} = 1.24$$

The factor of safety against piping for this condition is approximately:

$$FS = \frac{1.24}{12.2} = 0.10$$

This very low factor of safety against piping could not exist if a channel of discontinuity was present within the bottom plug. Field observations did not reveal such a discontinuity nor the development of a boil or quick conditions prior to the blow-in. If a continuous vertical channel or discontinuity existed, a pipe or blow would likely have developed sooner.

At time of failure, the ratio of plug diameter to average plug thickness was D/T = 6.1/2.1 = 2.9. If the suggestions of Febesh and Woolworth (1984) are followed, the plug would have been evaluated with very little shear since D/T ≈ 3. The factor of safety against cohesive plug heave neglecting shear is given by Eqn. 7. Substituting for conditions at the time of bottom plug failure results in a factor of safety of:

$$FS = \frac{22.0(2.1)}{9.8(27.7)} = 0.17$$

Since this factor of safety is much lower than 1.0, perimeter shear resistance must have been acting to resist failure. The plug thickness, T needed for a factor of safety of 1 using this method is:

$$T = \frac{\gamma_w H_b}{\gamma_t} = \frac{9.8(27.7)}{22.0} = 12.3 \text{ m}$$

By this method, the recommended minimum bottom plug thickness would have been 12.3 m which is 10.2 m thicker than the average thickness existing at the time of failure.

The factor of safety against cohesive plug heave including perimeter shear was previously given as Eqn. 8. The shear strength, τ, may be reasonably assumed to be 153 kPa which is half of the average unconfined compressive strength test results (120 to 175 kPa for the till at bottom plug level). Use of the average undrained strength is reasonable based on the small range of test values and because formation of a freezewall for ground support does not disturb the soil and reduce shear strength. The factor of safety against heave including perimeter shear after substituting into Eqn 8 is then,

$$FS = \frac{22.0(2.1)\,\pi\,(6.1/2)^2 + 153\,\pi\,(6.1)\,2.1}{9.8(27.7)\,\pi\,(6.1/2)^2} = 0.96$$

Using the average value of the shear strength in this case results in a factor of safety close to 1.0 and thus may explain the bottom failure.

Bottom stability predictions were not made by the authors prior to shaft excavating. The contractor's engineer apparently did not compute factors of safety either, The designer had anticipated little or no head on the bottom plug because dewatering of the outwash aquifers was planned and therefore they did not expect bottom instability problems.

Although the plug failure can be explained as an intact plug heaving by shear along the perimeter, actual observations suggest that the plug probably failed by the development of tension cracks that led to hydraulic fracturing and then a blow. The factor of safety against cracking or buckling of a circular cohesive plug with clamped perimeter conditions may be computed using Equations 21, 24 and 25. Assuming undrained, $\phi = 0$ conditions, the tensile strength of a clay with a plasticity index of 17 percent may be estimated as 1/9 of Q_u (Narain and Rawat, 1970). Substituting the average unconfined compressive strength for the lower sub-stratum of Q_u =306 kPa gives a tensile strength value:

$$\sigma_t = \frac{Q_u}{9} = \frac{306 \text{ kPa}}{9} \; 34 \text{ kPa}$$

For undrained conditions using Eqn 22, the drained tensile strength is:

$$\sigma'_t = \frac{\bar{c}}{\tan \phi'} = \frac{32 \text{ kPa}}{\tan 29^\circ} = 58 \text{ kPa}$$

Since the drained value is higher, the undrained shear strength value of 32 kPa appears to be a conservative estimate. We used 34 kPa, even though most practitioners commonly assume the tensile strength of soil to be zero.

The effective horizontal earth pressure, $\overline{\sigma}_h$, for this case may be taken as the at-rest value:

$$\bar{\sigma}_h = K_0 \, \bar{\sigma}_v = 0.8 \left[336.6 \text{ kPa} \right] = 269.3 \text{ kPa}$$

Substituting these values into Eqn. 21 gives the maximum net uplift pressure when cracks should develop within the plug base:

$$\sigma_n = \frac{(\sigma_T + \sigma_H)\, 16\, T^2}{3\, D^2} = \frac{(34 + 269)\, 16\, (2.1)^2}{3\, (6.1)^2} = 192 \text{ kPa}$$

Substituting into Eqn. 24 gives the maximum head before cracking:

$$H_m = \frac{192 + 22.0(2.1)}{9.8} = 24.4 \text{ m}$$

The water pressure head on the bottom of the clay plug at the time of failure was H_b=27.7 m. The corresponding factor of safety against tension crack development is then:

$$FS = \frac{24.4}{27.7} = 0.88$$

This result suggests that the plug failure may be explained by buckling or cracking followed by a piping blow-in. If the tensile strength is assumed to be 0 as is normally done, an even lower factor of safety would result.

Press Foundation Shaft

A rectangular shaft was required to construct a large press foundation within an alluvial plain along the Mississippi River. The shaft excavation was 10.1 x 17.1 m in plan and 13.9 m deep as shown in Figure 6. The earth retention system consisted of 0.76 m wide, 18.9 m deep slurry diaphragm walls that were installed in panels with semi-circular joints formed by using end stop pipes during concrete placement.

Soil Conditions

Two soil borings in the vicinity of the shaft indicated that the pertinent soils consisted of medium dense alluvial sand deposits separated by two horizontally continuous cohesive layers. The thicknesses and depths of these deposits relative to the shaft excavation are depicted in Figure 7.

Groundwater Conditions

The approximate water table level was at a depth of 6.1 m. Groundwater was to be controlled by a combination of cut off by the slurry diaphragm walls and by a pressure relief well from a 22.9 m deep well centered below the shaft. The design intent was to reduce the piezometric level within the lower sand to 9.1 m. However, the well was only installed to a depth of 16.8 m as shown in Figure 7. It was not effective at relieving pressure within the lower sand.

Field Observations

While excavating, the central well was used as a sump. Water levels in the sump were

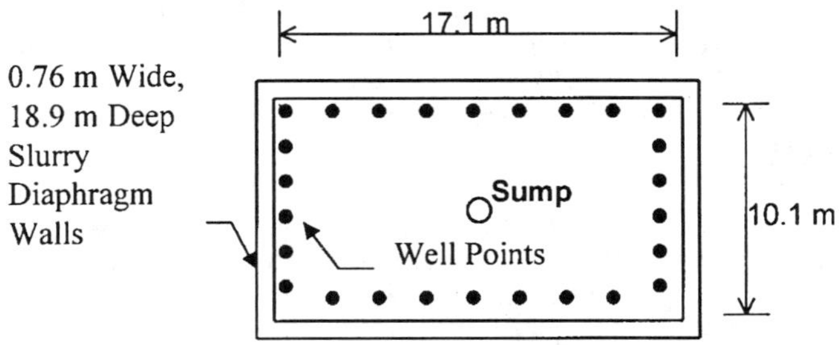

Figure 6. Plan of Press Foundation Shaft

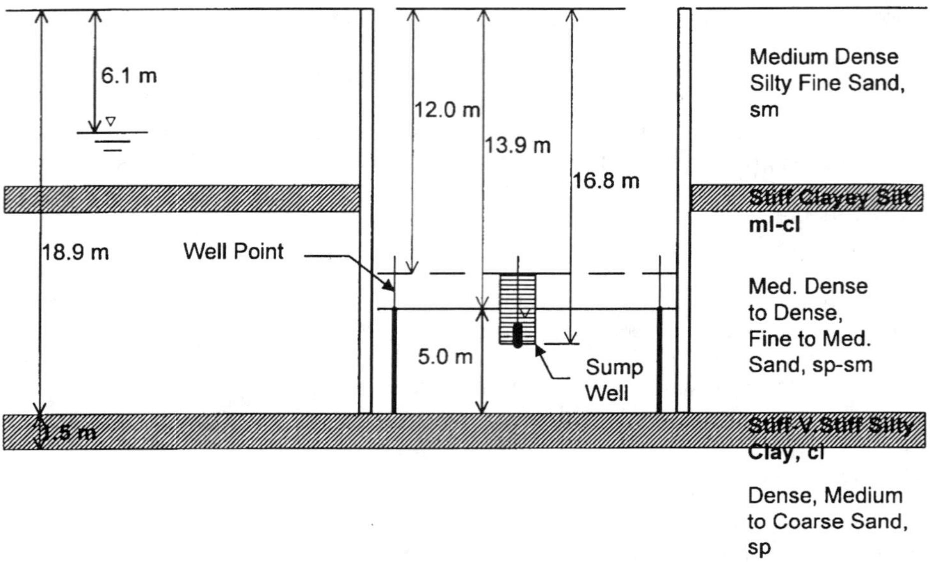

Figure 7. Press Foundation Shaft Cross-Section

generally lowered to within approximately 1 m of the sump bottom. This internal plug dewatering was found to be ineffective. When the excavation subgrade had reached approximately 12.0 m, minor boils were observed. To remedy the situation, 30 well points were installed around the inside perimeter as shown in Figures 6 and 7. The well points were successful in reducing water levels within the middle sand layer inside the slurry wall and allowing the excavation to resume.

As the excavation was being lowered, leaks through some of the slurry wall joints were observed. Cleaning and patching of these leaks were not successful in stopping the underlying joint inflows. In an attempt to depress the water level within the sand outside the slurry wall and below the bottom plug, deeper internal wells were planned. While drilling and as soon as the lower clay layer was punctured, a geyser (blow) developed through the well borehole. A flow rate ranging from 15 to 22 l/sec was estimated Considerable sand piped into the excavation with the water flow. The shaft was flooded with assistance from hydrant water. External dewatering wells were installed into the sand layers outside the shaft. These wells allowed the excavation work to be completed.

Analysis of Failure

Analyses of bottom stability and seepage conditions were made for various hydrogeology assumptions. NAVFACS DM7.1 charts (Anon., 1982) were used for the case of upward seepage through the bottom plug where a cutoff wall penetrates into sand of infinite depth. A factor of safety against piping of over 1.5 results for this case if the water table is lowered to at least 9.1 m by dewatering. If the water table is not drawn down, the factor of safety against piping and quick conditions drops to approximately 1.2.

Another analysis was made for the condition of cutoff walls penetrating into the lower clay layer. If perimeter shear is ignored, the factor of safety against heave is approximately 1.12, 0.98 and 0.88 for water table levels drawn down to 9.1 m. 7.6 m and unchanged at 6.1 m. If perimeter shear is added for the lower clay, the factors of safety against heave increase to 1.32, 1.16 and 1.04 for water table levels drawn down to 9.1 m. 7.6 m and unchanged at 6.1 m. These results show that pressure relief was needed to provide adequate bottom stability.

Two types of failure developed during excavation of the shaft. First, boils developed because the slurry diaphragm wall was not adequately watertight. A few open joints allowed a short seepage path through the walls with sufficient seepage pressure to result in boils and quick conditions. This consequence is predicable using the NAVFAC charts or the methods previously discussed in the Article entitled "Heave and Quick Conditions from Upward Seepage." This case shows that adequately watertight joints are critically important where a shaft with cutoff walls is surrounded by an aquifer that has not been dewatered.

The second type of failure was a blow-in through the cohesive bottom plug as a result of a discontinuity consisting of an advancing borehole. For this case, the critical hydraulic gradient as expressed by Eqn. 5 is:

$$i_c = \frac{\gamma'}{\gamma_w} = \frac{10.6}{9.8} = 1.1$$

The actual hydraulic gradient across the clay layer and sand bottom plug combined the was approximately 14.3/8.4m = 1.7. The resulting factor of safety against a piping blow after creating a borehole flow path was approximately 0.6 = 1.1/1.7. The blow should have been expected prior to attempting deep internal dewatering wells.

CONCLUSIONS

Methods of evaluating the stability of shaft bottom plugs were presented along with summaries of recommended factors of safety from a variety of references. The methods include heave by piping through granular soils, boils or blows through discontinuities within cohesive plugs, cohesive plug heave ignoring side shear, cohesive plug heave with side shear and cohesive plug buckling or cracking. Two bottom plug failure case histories from Milwaukee and one from along the Mississippi were analyzed using these methods to determine how well these methods accounted for the observed failures.

Based on case histories presented and upon our experience with dozens of other shafts at a variety of sites, the following conclusions and recommendations were reached:

1. Piping or Quick Conditions should always be evaluated. This is the primary mode of bottom instability for granular bottom plugs and is often the mode of instability for mixed granular and cohesive plugs and cohesive plugs with discontinuities. When the factor of safety against boils or quick conditions is less than 1.0, a risk of a bottom blow exists if a natural channel such as a granular inclusion or if a discontinuity such as an old pile or borehole exists within the plug. If the cohesive plug is continuous, the factor of safety against boils can be much less than 1.0. When bottom plug seepage is expected or very possible, we recommend a minimum

factor of safety of 1.5 against piping or quick conditions where the subgrade will be used to support settlement sensitive facilities and a minimum factor of safety of 1.25 for other cases. If the bottom plug is cohesive and not expected to have discontinuities or seepage channels, the factor of safety against piping may be less than 1.0.

2. Evaluation of the factor of safety against cohesive plug heave by simply comparing the weight of the plug to the upward hydrostatic pressure is generally conservative. Most cohesive bottom plug failures can not be explained by this method. If the method is used for cohesive bottom plugs within firm or stronger soils (Qu 50 KPa), we recommend a minimum factor of safety of 1.0. The cohesive plug consists of softer soils, we recommend using a minimum factor of safety of 1.2.

3. In most cases, we recommend that the factor of safety against heave of cohesive bottom plugs be determined with consideration of perimeter shear. The selected shear strength value needs to consider effects of disturbance by wall installation such as sheet pile driving. Experience indicates that where the soil is not disturbed, the average undrained shear strength may be used unless wide variations in shear strength have been measured. If the ratio of the plug diameter or width to thickness is approximately 3 or less, this method should provide a reliable indication of bottom stability. We recommend a minimum factor of safety of 1.3 against heave failure for general cases where perimeter shear is used. For cases where the subgrade will be used to support settlement sensitive facilities, we recommend a minimum factor of safety of 1.5.

4. Both the NS-10 Junction Structure and Jones Island Siphons shaft failures are explained by evaluating the factor of safety against development of tensile cracks or buckling. Analysis of a cohesive bottom plug as a clamped beam shows that the highest tensile stresses tend to develop within the plug base near the perimeter of the plug. Once cracking develops, hydraulic fracturing will generally lead to a blow of piping failure through the cracks. Whenever the factor of safety against heave is evaluated using perimeter shear, we recommend that the factor of safety against buckling be computed as well. We recommend a minimum factor of safety of 1.3 against buckling failure for general cases. For cases where the subgrade will be used to support settlement sensitive facilities, we recommend a minimum factor of safety of 1.5.

5. When the critical bottom condition will exist for a short period such as days to weeks, average undrained shear strength or undrained shear strength corresponding to reductions from wall installation may be used. If surcharge is placed on the bottom plug or underlying hydrostatic pressures are reduced within a short period, evaluations using drained shear strength parameters may not be necessary. Where the shaft bottom condition will exist for a longer period, the bottom stability should also be checked using drained shear strength parameters. The minimum factors of safety recommended above are also appropriate for evaluating stability using drained shear strength.

6. For shafts where cutoff walls are being used to create sufficiently long flow paths to reduce hydraulic gradients and seepage velocities to safe levels or where the

bottom plug contains granular soil overlying a lower cohesive layer, special attention is needed to cutoff wall joints. The joints do not need to be perfectly watertight. Interlocked sheet piling are seldom that watertight. However, the joints should be tight enough to preclude seepage pressures through leaks that would lead to piping or quick conditions. Where bad joints are found, external dewatering, grouting or other sealing methods such as ground freezing will likely be needed. Internal dewatering is often not an adequate remedy for leaking cutoff walls.

REFERENCES

Anon., 1982, *Soil Mechanics,* Design Manual 7.1, Department of Navy, Naval Facilities Engineering Command, Alexandria, VA., 7.1-263 to 7.1-271.

Anon., 1985, *Canadian Foundation Engineering Manual,* 2nd ed.,Meyerhof, G.G., ed., BiTech Publishers, Vancouver, pp 422-424.

Bowles, J.E., 1996, *Foundation Analysis and Design,* 5th ed., McGraw-Hill, New York, pp

Cedergren, H. R., 1967, *Seepage Drainage, and Flow Nets,* Wiley, New York, pp. 116-129, 225-227, 452-455.

Dismuke, T.D., 1975, "Cellular Structures and Braced Excavations," *Foundation Engineering Handbook,* 1st ed., Winterkorn, and Fang, ed., McGraw-Hill, New York, pp. 469-471.

Febesh, M., and Woolworth, R.S., 1984 "Temporary Earth Retaining Structures," *Handbook of Temporary Structures In Construction,* Ratay, R.T. ed., McGraw-Hill, New York, pp. 8-33 to 8-40.

Goodman, R.E., 1980, *Introduction to Rock Mechanics,* Wiley, New York, pp. 223-227.

Graham, J., and Godden, W.G., 1968, "Structural Analysis," *Structural Engineering Handbook,* Gaylord and Gaylord, ed., McGraw-Hill, New York, p. 1-29.

Peck, R.B. Hanson W.E., and Thornburn, T.T., 1974, *Foundation Engineering,* 2nd Ed., Wiley, New York, pp 51-55, 283.

Narain, J., and Rawat, P.C., 1970, "Tensile Strength of Compacted Soils," *Journal of Soil Mechanics and Foundations Division,* American Society of Civil Engineers, New York, Vol. 90, No., SM6, pp. 2185-2190.

Skov, R. and Denver, H., 1988, "Time Dependence of Bearing Capacity of Piles," *Application of Sress-Wave Theory to Piles,* BiTech Publishers, Vancouver, B.C., pp 879-888.

Svinkin, M.R., Morgano, C.M., and Morvant, M., 1994, "Pile Capacity as a Function of Time in Clayey and Sandy Soils," *Proceedings, Fifth International Conference on Exhibition on Piling and Deep Foundations,* Deep Foundations Institute, Englewoods Cliffs, New Jersey, pp 1.11.1-1.11.8.

Teng, W., 1962, *Foundation Design,* Wiley, New York, p. 104.

Terzaghi, K., Peck, R.B., and Mesri, G., 1996, *Soil Mechanics in Engineering Practice,* 3rd Ed., Wiley, New York, pp. 116-121, 213-223, 248-249, 285-286, 457-467.

Xanthakos, P.P., 1979, "Load Bearing Elements and Foundations," *Slurry Walls,* 2nd ed., McGraw-Hill, New York, pp. 375-411.

Young, W.C. 1989, "Flat Plates," *Formulas for Stress and Strain,* 6th. ed. Mc Graw-Hill, New York, pp. 391-406.

Chapter 24

INSTRUMENTATION FOR MONITORING GROUND MOVEMENTS: CENTRAL ARTERY/TUNNEL PROJECT, BOSTON, MASSACHUSETTS

S. Vaghar, D.J. Bobrow, and T.A. Marcotte
Bechtel/Parsons Brinckerhoff, Boston, Massachusetts

ABSTRACT

This paper discusses the geotechnical instrumentation program on the Central Artery/Tunnel Project, in Boston, Massachusetts. Particular attention is given to a state of the art Graphic Information System (GIS) application which has been developed for rapid analyses and reporting of data.

This application will see extensive use during the ongoing complex downtown tunnel construction. The excavations in soft ground require cuts of up to 120 feet deep and 250 feet wide. Deformation and groundwater monitoring instruments have been installed to monitor the stability of the excavation support systems and the abutting buildings.

The use of this application is illustrated with an example from an ongoing contract.

INTRODUCTION

The Central Artery/Tunnel (CA/T) Project is currently the largest government funded public works project in the United States. An existing elevated highway, constructed in the 1950s, is being phased out in favor of a tunnel alignment through downtown Boston, Massachusetts. Figure 1 shows the entire project location.

An integral element of the Project is the Ted Williams Tunnel; an immersed tube tunnel crossing of Boston harbor. The highway alignment leading to the Ted Williams Tunnel consists of cut and cover tunnels and boat sections, and jacked tunnels. The main portion of the Project, the Central Artery, is currently being constructed primarily by cut and cover and top down construction, utilizing diaphragm walls as the excavation support system.

The following gives a brief overview of the Project setting, within the context of which the instrumentation program has been designed and implemented. It also discusses Construction Contract C11A1 (Figure 1), which illustrates the use of the new instrumentation application system on the Project.

All elevations mentioned in this paper are referred to the Central Artery/Tunnel Project Datum (PD), which is 100 feet below the National Geodetic Vertical Datum (Mean Sea Level).

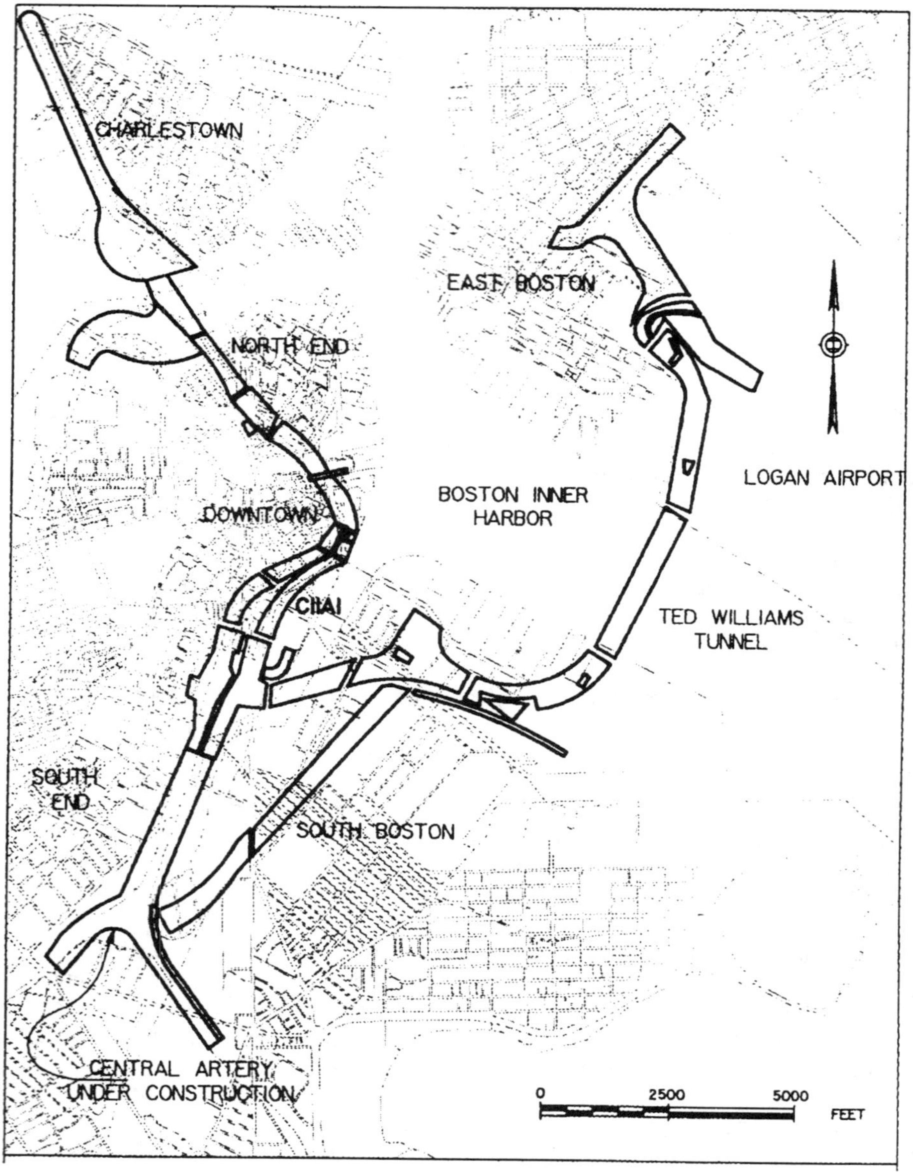

Figure 1. Project Location Plan

BACKGROUND

Geological Setting

Boston is in the geographic center of the Boston Basin; a triangular area with sedimentary, metamorphic and igneous rocks of late Precambrian and Cambrian age, overlain by glacial and post glacial soil deposits (Kaye, 1982).

An idealized subsurface geologic profile of the Project is shown in Figure 2.

A brief description of the strata follows.

Miscellaneous Fill: consists of loose to very dense granular and soft to hard cohesive soils with bricks, wood piles, ash and other miscellaneous debris.

Organic Deposits: consist of soft to medium stiff organic silts, interbedded with clay, sand, and peat. They are typically slightly overconsolidated.

Marine and Stream Deposits: these are medium dense to dense sands, which are generally discontinuous, and are rarely more than 10 to 15 ft thick.

Marine Clay: locally referred to as the "Boston Blue Clay", it is a soft to hard clay and silt with fine sand lenses. The clay is the most widespread surficial sedimentary deposit, reaching thicknesses of up to 120 ft in the project area.

Glaciomarine/Glaciolacustrine/Glaciofluvial Deposits: glaciomarine deposits consist of silt and clay strata interbedded with layers of sand and gravel and occasional boulders. Glaciolacustrine deposits consist of laminations of sand, silt and clay. It is a localized discontinuous unit. Glaciofluvial deposits consist of dense to very dense sands and gravels.

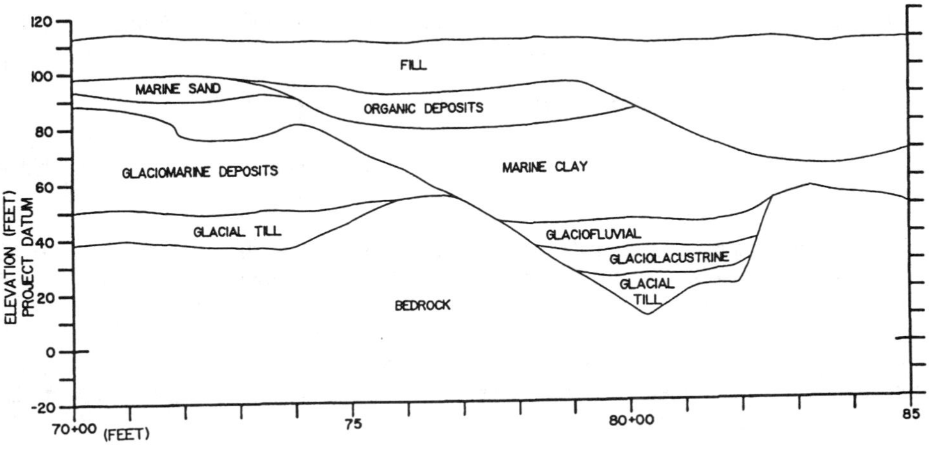

Figure 2. Idealized Project Subsurface Profile

Glacial Till Deposits: lodgement and ablation till deposits generally consist of a heterogeneous mixture of very dense sand, silt, and gravel, with some clay, cobbles and boulders.

Bedrock: the bedrock is the Cambridge Formation, a thinly bedded Argillite. The Argillite is intruded by igneous dikes and sills. Typically the upper 5 to 10 ft of the bedrock is severely to completely weathered. At some locations, the weathering extends to depths greater than 50 ft.

Site Development History

The Project site lies predominantly within an area that was once mostly covered by seawater, tidal marshes and estuaries. Most of these areas have been filled over the years, using clays dredged from the harbor, and sand and gravel from gravel pits and glacial drumlins, as well as building demolition rubble. The filling began in the late 1700s and generally ended in the 1960s (Ty, 1987).

Many of the buildings constructed during the 18th and 19th century consist of brick and wood frame structures supported on wood piles and cribbing or spread footings resting in the marine clay. Many of the older buildings and structures associated with railroads and other industries were subsequently demolished, with their pile and granite block foundations left in place, posing obstructions to this project.

Modern high-rise buildings along the alignment have deep foundations or reinforced concrete mats which are founded in the till or bedrock.

Groundwater Regime

The groundwater regime consists of two water tables; one perched above the marine clay deposit, and the other, a confined aquifer, located below in the bedrock. The perched aquifer phreatic surface is at approximate elevation 103 feet (PD) adjacent to the harbor and rises to elevation 108 feet (PD) inland. The confined aquifer piezometric elevations are typically between elevations 95 and 98 feet (PD).

INSTRUMENTATION PROGRAM

The geotechnical instrumentation program on the Project has been designed to achieve the following objectives:

1. Monitoring of ground and facilities deformations resulting from the construction, as a forewarning of unforeseen conditions which may impact the safety of the personnel, the public, and the construction work.

2. Monitoring of changes in groundwater patterns and possible resulting ground movements, which may adversely impact the abutting facilities, as well as monitoring of the facilities themselves.

3. Providing pre-construction baseline condition data for comparison with construction and post-construction data, to aid in the evaluation of construction related impacts.

Table 1. Typical Types of Instruments Used

Instrument Type	Parameter	Method
	Deformation	
Deformation Monitoring Point	Surface Vertical & Horizontal Deformation	Optical Survey
Convergence Gage	Convergence Across Excavation or Tunnel	Portable Mechanical Tape Extensometer
Utility Monitoring Point	Vertical Deformation of Utility	Optical Survey
Borros Point	Subsurface Settlement: Single Point	Optical Survey
Settlement Platform	Settlement of Original Ground Surface Below Fill	Optical Survey
Probe Extensometer	Subsurface Settlement: Multi-Point	Electrical Probe
Inclinometer	Subsurface Horizontal Deformation	Electrical Probe
Multi-Point Heave Gage	Heave Below Bottom of Excavation	Electrical Probe
Crack Monitor	Opening and Closing of Crack in a Structure	Portable Mechanical Gage or Fixed Grid Gage
Tiltmeter	Rotational Deformation	Plug-in Electrical Readout Unit
Liquid Level Gage	Surface Vertical Deformation	Datalogger
Electrolevel/Beam Sensor	Surface Vertical Deformation	Datalogger
	Groundwater Pressure	
Observation Well	Groudwater Level in Granular Fill	Electrical Probe
Vibrating Wire Piezometer	Groundwater Pressure in Soils	Plug-in Electrical Readout Unit
	Stress & Load in Temporary Supports	
Vibrating Wire Strain Gage	Strain on Surface of Steel (Used to Calculate Stress)	Datalogger or Plug in Electrical Readout Unit
Load Cell on Tieback	Load on Tieback	Plug-in Electrical Readout Unit
	Vibration	
Seismograph	Vibration	Automatic Reading

Table 1 lists the types of instruments used to monitor the performance of the construction operations and their effect on adjoining structures and the groundwater levels.

The construction contractors are responsible for purchasing, installing, and maintaining the majority of the instruments. The Management Consultant, Bechtel/Parsons Brinckerhoff, performs quality assurance of the contractor's work, as well as monitors and reports the data.

Figure 3 shows typical instrument installation details for selected instruments.

Contract C11A1

Many of the instruments described above have been used on Contract C11A1. It therefore presents a good opportunity to illustrate the use of the GIS application in the context of an ongoing contract. The location of Contract C11A1 is shown in Figure 1. The highway tunnel is a cut and cover tunnel approximately 1,900 feet long. The tunnel will be

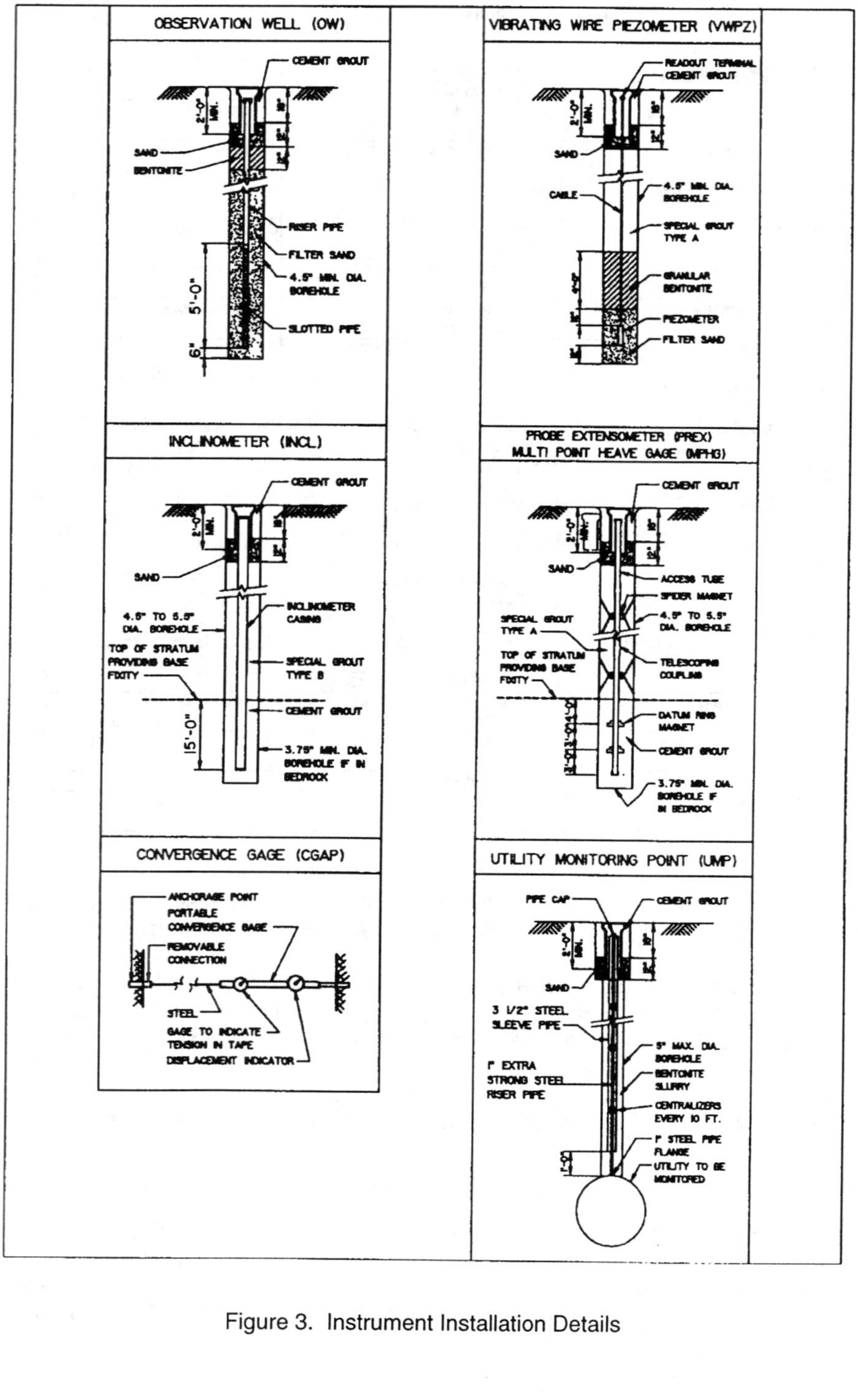

Figure 3. Instrument Installation Details

founded within the marine clay, till, and bedrock strata. The contract was awarded in February 1995 to Perini Corporation, and is ongoing as of the writing of this paper (January 1997).

The excavation support system consists of 3-foot thick soldier pile and tremie concrete diaphragm walls, founded in bedrock, which are constructed using slurry wall techniques. The contractor has used both a clam shell and a 'Hydrofraise', a rotary rock excavator commonly referred to as a milling machine, to excavate the 3-foot wide trenches down to a maximum depth of around 120 feet, under a bentonite slurry, after placement of the soldier piles. The trenches are then filled with tremie concrete and the walls are cross-braced with steel struts.

The concrete diaphragm walls are also intended to act as groundwater cutoff members. Each slurry wall panel is fitted with PVC sleeves which are used for grouting the bedrock underneath the wall for additional watertightness, if required.

PVC sleeves have been installed in some diaphragm wall panels to allow for the installation of inclinometers to measure horizontal deflection of the walls. The inclinometers have been installed in the walls every 150 feet along the alignment, opposite each other, on both sides of the excavation. Additional inclinometers have been installed behind the walls and at critical locations in front of some buildings.

Instrument clusters, including inclinometers, probe extensometers, piezometers and observation wells, have been installed behind the excavation support walls to monitor settlements in the soil, and groundwater patterns in the upper and lower aquifers above and below the Marine Clay stratum.

Survey monitoring points on the abutting buildings have been installed to monitor vertical and horizontal movements. Allowable (response) values have been defined by the designer in terms of angular distortion of the buildings.

DATA MANAGEMENT SYSTEM

Overview

The Geographic Information System (GIS) for the Project has been developed by the Project staff to improve the conventional methods of data processing and retrieval, which are too slow and cumbersome for the volume of instrumentation data generated. It is essential that the system functions as a true early warning mechanism, so that any required mitigating actions can be implemented rapidly. This automated system provides readily accessible data to the contractors, construction managers, and other interested parties.

The following is a description of the GIS application, and how it enhances the instrumentation program.

Data Collection

Selected instruments, typically building deformation monitoring points, have been installed prior to construction to provide baseline data. Most instruments are installed early during each contract. However, very few are effected by construction in the early phases of the contract. Therefore, this period of time is also used to collect baseline data. The frequency of monitoring during this phase of construction is typically monthly.

During construction, the instrument reading schedule is adjusted on a daily basis to monitor active or critical construction areas. Daily site observations are performed to aid in the evaluation of the data, as well as to pinpoint areas where additional monitoring may be required. Typically all instruments in active construction areas are monitored on a weekly to monthly basis. Daily or more frequent readings are taken in critical areas.

The Management Consultant currently collects the data and distributes it to all interested parties, including the construction contractors. Construction contracts require that data reports be provided to the construction contractors within 24 hours of the data collection.

Data are collected either manually by using readout units capable of electronic storage, or remotely via modem. Manual readings are recorded on forms which contain the last three observations. Observation wells, vibrating wire piezometers, tiltmeters, crack monitors, probe extensometers and multi-point heave gages (Table 1) are typically read and recorded manually. The inclusion of the previous data on the manual data form assists in the rapid detection of reader error.

Inclinometers, seismographs, vibrating wire strain gages and load cells are instruments which are read either manually, or using dataloggers, depending on the location. With the exception of seismographs, these instruments are read and the data are stored in a portable datalogger, or they are wired to a fixed datalogger via multiplexers, and accessed remotely with a computer and modem over telephone lines. Multiplexers are used to combine instruments to allow many instruments to be handled by a single datalogger.

Deformation monitoring points, borros points, utility monitoring points and convergence gages data are generally collected using a second order Class II digital level and bar coded rod. Optical survey data are collected and transferred to the database in electronic batch files.

Database Structure/Data Entry and Reduction

The GIS application allows a user to retrieve, display and generate output from textual and graphics databases. The textual data are stored in Oracle™, a relational database, while the graphics data such as mapping are stored in GDS™, CA/T Project CADD system. The Oracle™ database allows for quick and accurate data entry, storage, processing, and retrieval of data.

An important feature of a relational database management system is that it easily tracks relationships between records in different tables. These data are known as attribute data. Attribute data are indexed to allow rapid access, ensure that duplication of data does not occur, and correlate information from different tables.

A Structured Query Language (SQL) is used to communicate with the relational database. SQL is based upon English-like statements, and is used to query, insert, update, delete and modify the data. It allows the users to create the database tables and store and manipulate information in these tables, as well as maintain the database itself. It also inter-

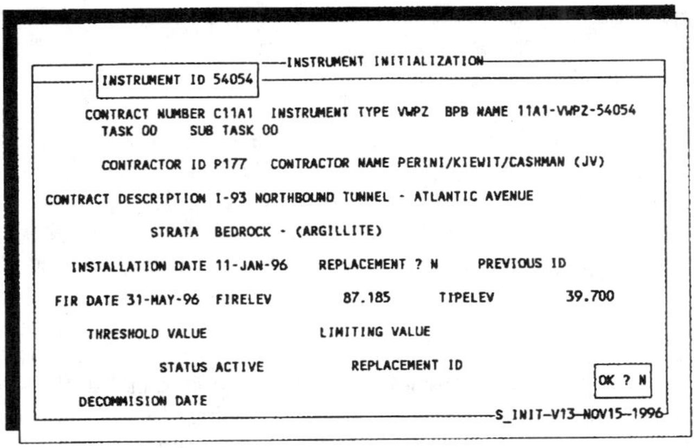
INSTRUMENT INITIALIZATION

INSTRUMENT ID 54054

CONTRACT NUMBER C11A1 INSTRUMENT TYPE VWPZ BPB NAME 11A1-VWPZ-54054
TASK 00 SUB TASK 00

CONTRACTOR ID P177 CONTRACTOR NAME PERINI/KIEWIT/CASHMAN (JV)

CONTRACT DESCRIPTION I-93 NORTHBOUND TUNNEL - ATLANTIC AVENUE

STRATA BEDROCK - (ARGILLITE)

INSTALLATION DATE 11-JAN-96 REPLACEMENT ? N PREVIOUS ID

FIR DATE 31-MAY-96 FIRELEV 87.185 TIPELEV 39.700

THRESHOLD VALUE LIMITING VALUE

STATUS ACTIVE REPLACEMENT ID

OK ? N

DECOMMISION DATE

S_INIT-V13-NOV15-1996

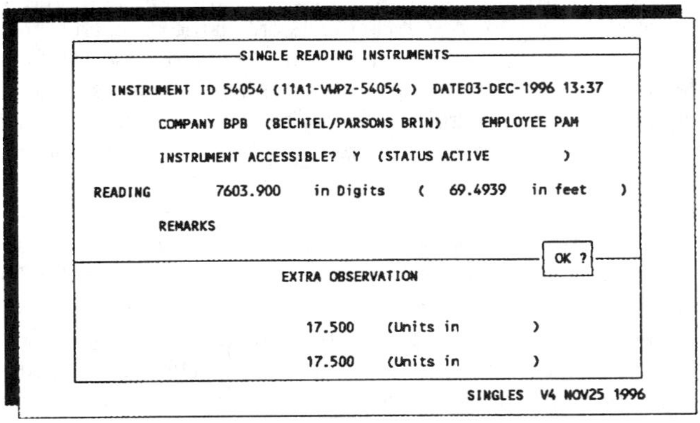
SINGLE READING INSTRUMENTS

INSTRUMENT ID 54054 (11A1-VWPZ-54054) DATE03-DEC-1996 13:37

COMPANY BPB (BECHTEL/PARSONS BRIN) EMPLOYEE PAM

INSTRUMENT ACCESSIBLE? Y (STATUS ACTIVE)

READING 7603.900 in Digits (69.4939 in feet)

REMARKS

OK ?

EXTRA OBSERVATION

17.500 (Units in)

17.500 (Units in)

SINGLES V4 NOV25 1996

Figure 4. Instrument Initialization and Observation Forms

faces with additional programming languages, thereby allowing for processing many sets of records at a given time.

Once the data are entered and checked, they are available for users across a Wide Area Network (WAN) with up to 1000 PCS linked to the network. Users unfamiliar with the location of instruments can access a GDS[tm] graphical interface for a spacial representation (location map) of the instruments. To safeguard the integrity of the data, triggers in the database restrict entries to a pre-defined range for each parameter. Daily backup protects against accidental loss of permanent data.

The database has controlled access, which limits access for general users to verified/ released information only, while authorized users have access to insert, update, and delete data. This controlled access allows users unfamiliar with the system to view data, without accidentally deleting permanent records.

Each instrument has an initialization file (Figure 4) which contains the instrument type, coordinates, installation date, stratum, pre-defined allowable (response) values, calibration factors, historical high and low elevations, etc.

With the exception of a few instruments such as inclinometers and seismographs, all the geotechnical and structural instrumentation observations are contained within the Oracle database. Required data entry parameters on the Observation form (Figure 4) include a 5-digit unique number, date and time, initials of the reader, accessibility of the instrument, the reading(s) and any comments. The forms require data entry in this order, and all fields must be completed.

All data initially entered into the database are tagged with a notation to indicate that the data are unchecked. The data are reviewed by an authorized instrumentation engineer for mathematical errors, and are compared to previous readings for accuracy before the tag is removed. Users without data entry privileges cannot view unchecked data, thus eliminating false alarms from data that have not been verified.

Electronic data which are to be transferred into the Oracle™ database are stored in ASCII comma delineated format. The ASCII file must contain the date, time and all the readings. A matching text file is used to define each parameter stored in the file for readings collected with a datalogger. Each line of the datalogger file represents a cycle of readings taken at a particular time for all the instruments wired to the datalogger. Use of the matching text file allows an unlimited number of cycles of readings to be transferred into the database at one time.

Oracle™ database stores and analyzes attribute data and generates reports. However, it does not incorporate spatially defined graphics such as map data. The data contained in the relational database can be linked to files containing graphics such as GDS™ files that contain the location of various geotechnical instruments. Using GIS technology, spatial analysis capabilities allow the users to map tabular data from the relational database, interactively query graphics for associated tabular data, and restrict tabular data to a geographic area defined in GDS™.

Oracle™ and GDS™ are integrated using a Graphical User Interface. Using the GDS™ module SQL*CAD™, customized menus and basic programs written in GDS™, the user is able to access and perform operations on both databases to generate customized maps, without an in-depth knowledge of either GDS™ or Oracle™.

The base mapping contains detailed graphics of roads, railroads, buildings, surface utilities, topography, and hydrographic features. The mapping is organized into contract areas. Using pull-down menus, a user can quickly display any desired area on the project. Once at the desired location, the base mapping can be displayed as well as the location of all instruments.

Data Reports & Evaluations

Sixty predefined tabular report formats have been created in order to expedite the report production process (Figure 5). Reports may be generated using a specific date, or may contain all readings regardless of the collection date. They may contain raw data or data reduced to engineering units. All the data for a construction contract can be included, or the data for only one instrument may be generated.

Instruments in overlapping contract areas can be included in more than one contract, ensuring the instruments needed from adjoining areas are included in the summary report for each contract.

Reports compare the reduced data to the pre-defined contract response values. If the response value is exceeded, the reading is automatically flagged.

Use of the GDS™ graphical database helps with the evaluation of the data. When a change in the data is identified, the user can locate the area geographically, and review instruments located adjacent to the affected instrument. A query can be performed to include only instruments within the same stratum to reduce the number of instruments shown on the screen. The user can zoom out to display an area as large as needed for the review.

Once the desired instrument types are displayed, an attribute query can be performed to display certain categories of instruments within the group, such as instruments read within the past week, instruments not read for one month, vibrating wire piezometers in bedrock, and instruments that exceed an allowable value. The user may also perform the above mentioned queries, and color code any instruments that meet the criteria, by using pull-down menus.

Tabular data reports can also be generated from the graphical database. The user defines an area using a square, polygon or radius from a center point. The Graphical User Interface links to the relational database to retrieve the data. The output is identical to reports printed using the Oracle™ database forms. For deformation data on buildings, the user can outline a building, and a report will be generated for that building. Typically the angular distortion is calculated between adjacent points on a building. Using the graphical database, a user can click on any two points on a building, and the angular distortion between the points will be calculated using the latest observations.

Figure 6 presents a typical GDS™ computer screen, which illustrates the base mapping and the location of vibrating wire piezometers on a portion of Contract C11A1. The lower part of the screen shows a report with the most current readings of a vibrating wire piezometer. The report was generated by simply 'clicking' on the selected piezometer.

Included in all reports sent to the Resident Engineer's office is a cover letter with an evaluation of the data. Any significant changes in readings are correlated with construction or natural activities and factors.

When instruments exceed the contract response values, a graph of the data accompanies the report. The Project uses Easyplot™ software to plot the instrumentation data. Oracle™ generates a text file containing all the data as well as default parameters used by Easyplot™. In addition, the Contract area and name, X and Y axis titles, and a legend with response values are included. Figure 7 shows, as an example, a plot of selected C11A1 vibrating wire piezometers when construction dewatering in August 1996 led to a drop in piezometric pressures in the lower aquifer.

Central Artery/Tunnel Project
Bechtel/Parsons Brinckerhoff
Date: 04-DEC-96 11:37:05

Generated by: VWPZ.sql

VIBRATING WIRE PIEZOMETER MONITORING REPORT (VWPZ)

Contract Area: C11A1
Contract Name: I-93 NORTHBOUND TUNNEL - ATLANTIC AVENUE
Company: PERINI/KIEWIT/CASHMAN (JV)
Checked By:

INST ID	B/PB ID	DATE (5)	PIEZO. ELEV. (ft.) (5)	CHANGE FROM HIST. LOW (ft.) (5)	DATE (6)	PIEZO. ELEV. (ft.) (6)	CHANGE FROM HIST. LOW (ft.) (6)	DATE (7)	PIEZO. ELEV. (ft.) (7)	CHANGE FROM HIST. LOW (ft.) (7)	DATE (8)	PIEZO. ELEV. (ft.) (8)	CHANGE FROM HIST. LOW (ft.) (8)	REMARKS
54043	11A1-VWPZ-54043	11/12/96	46.93	-35.61	11/15/96	47.15	-35.39	11/19/96	44.96	-37.58	11/22/96	44.26	-38.30	
54044	11A1-VWPZ-54044	11/12/96	36.90	-46.58	11/15/96	37.14	-46.34	11/19/96	34.93	-48.56	11/22/96	34.18	-49.31	
54045	11A1-VWPZ-54045	11/12/96	37.59	-44.29	11/15/96	37.81	-44.07	11/19/96	35.74	-46.14	11/22/96	35.01	-46.87	
54046	11A1-VWPZ-54046	11/12/96	38.77	-43.95	11/15/96	39.02	-43.71	11/19/96	36.96	-45.77	11/22/96	36.27	-46.45	
54047	11A1-VWPZ-54047	--------	-------	------	--------	-------	------	09/20/96	-------	------	11/22/96	-------	------	UNDER BIT PATCH
54050	11A1-VWPZ-54050	11/12/96	36.65	-48.36	11/15/96	36.93	-48.08	11/19/96	34.64	-50.37	11/22/96	33.89	-51.12	
54051	11A1-VWPZ-54051	11/12/96	37.74	-47.01	11/15/96	39.29	-45.45	11/19/96	38.10	-46.65	11/22/96	38.20	-46.54	
54052	11A1-VWPZ-54052	11/05/96	36.16	-43.88	11/15/96	33.46	-46.58	11/19/96	31.15	-48.89	11/22/96	30.44	-49.60	
54054	11A1-VWPZ-54054	11/12/96	69.94	-14.08	11/15/96	70.26	-13.76	11/19/96	69.59	-14.43	11/22/96	69.15	-14.87	
54056	11A1-VWPZ-54056	11/12/96	30.30	-51.70	11/15/96	30.62	-51.38	11/19/96	27.73	-54.27	11/22/96	28.66	-53.34	
54057	11A1-VWPZ-54057	11/12/96	40.40	-41.68	11/15/96	40.64	-41.42	11/19/96	38.97	-43.09	11/22/96	38.40	-43.66	
54060	11A1-VWPZ-54060	11/12/96	39.57	-42.86	11/15/96	39.74	-42.69	11/19/96	37.84	-44.59	11/22/96	37.12	-45.31	
54061	11A1-VWPZ-54061	11/12/96	36.79	-46.26	11/15/96	37.13	-45.93	11/19/96	35.20	-47.85	11/22/96	35.34	-47.72	
54265	11A1-VWPZ-54265	11/12/96	36.83	-50.27	11/15/96	37.20	-49.90	11/19/96	35.67	-51.43	11/22/96	35.55	-51.55	
54266	11A1-VWPZ-54266	11/12/96	36.77	-48.85	11/15/96	37.12	-48.50	11/19/96	35.03	-50.60	11/22/96	34.33	-51.29	
54267	11A1-VWPZ-54267	11/12/96	43.32	-42.76	11/15/96	44.06	-42.01	11/19/96	43.19	-42.88	11/22/96	43.01	-43.06	
54276	11A1-VWPZ-54276	11/12/96	37.21	-46.18	11/15/96	37.49	-45.91	11/19/96	35.30	-48.09	11/22/96	34.34	-49.05	
54277	11A1-VWPZ-54277	11/12/96	64.39	-24.74	11/15/96	64.89	-24.25	11/19/96	64.00	-25.13	11/22/96	63.71	-25.43	
53488	14C1-VWPZ-53488	11/15/96	70.08	-19.22	11/18/96	69.44	-19.86	11/19/96	68.88	-20.42	11/22/96	68.61	-20.69	

	THRESHOLD VALUE	LIMITING VALUE
VWPZ		

Figure 5. Instrumentation Data Report

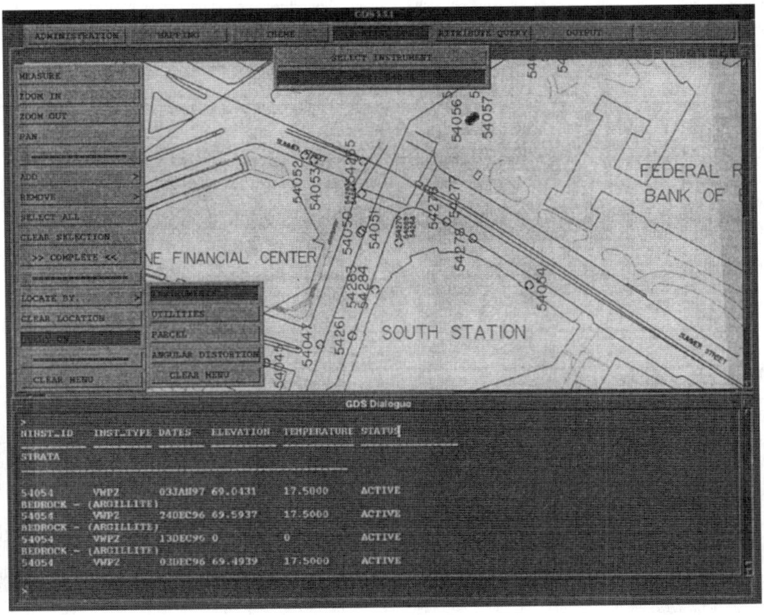

Figure 6. Instrument Data Accessed in GDS

The groundwater instruments serve as the first indicators of potential settlement. At Contract C11A1, groundwater data are reviewed in conjunction with probe extensometer data, which would indicate the onset of settlement at various depths beneath the site, as well as deformation monitoring data on the adjacent buildings. In the above case, deformation monitoring points indicated approximately 0.15 inches of settlement on several buildings. Much of this settlement was attributed to deformation of the till stratum as a result of groundwater depressurization.

Development of the GIS application has led to much streamlining of the data processing. Use of standardized forms has allowed users to quickly enter and retrieve data. Users have access to project wide information, as opposed to specific contracts only. Data collected for early contracts are used and accessed quickly for follow-on Contracts without any additional data entry. The flexibility of reporting formats suits the needs of construction for rapid processing and evaluation of data. On a typical contract such as C11A1, the users can generate reports, graphs/plots, and location plans; attach their site observations, evaluations and recommendations; and send, fax, or EMAIL the report to the field offices in one hour or less.

FUTURE DEVELOPMENTS

The GIS application described above has been under development and concurrent use since 1994. New features are continuously developed and added, as new needs are identified. Modules currently under development or testing include the following:

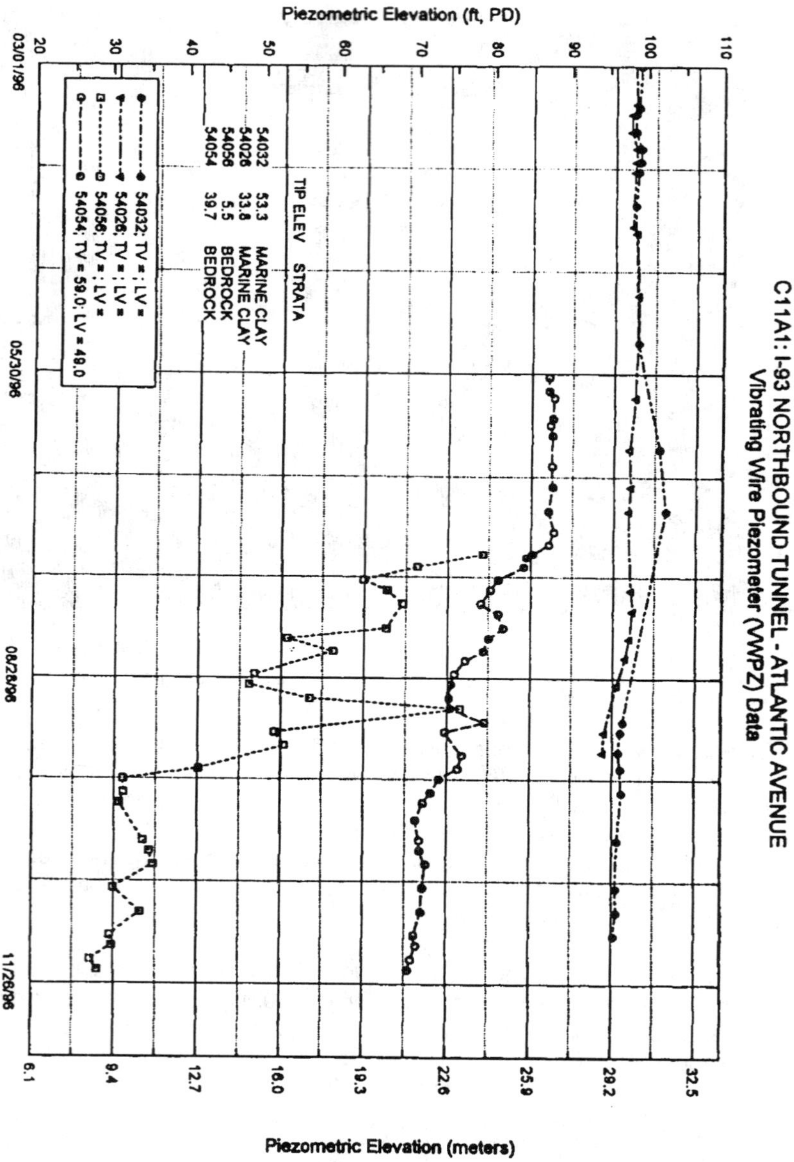

Figure 7. Instrumentation Data Plot

- A module which generates groundwater and settlement contour plans for different strata where instruments have been installed.

- A module which allows geological cross sections to be generated and superimposed on surface features profiles, including 3-D perspectives, at any desired location.

- A module which enables site photographs currently being taken with hand held digital cameras, as well as views generated by overhead video cameras, to be incorporated into the GDS^{tm} base mapping, and to be made available to the users, without a need to exit the GIS application.

SUMMARY

The instrumentation program at the Central Artery/Tunnel Project is, by necessity, large and complex. However, it is a flexible and responsible program which is designed to make monitoring information, along with evaluations, readily and rapidly available to the decision-makers. The brief account of the program given here is intended to demonstrate the value of employing current technology to achieve these enhancements on a large scale. However, with the increasing availability and decreasing cost of new systems, many of the features described here can be applied equally successfully to smaller projects.

ACKNOWLEDGMENT

The authors are indebted to the Massachusetts Highway Department for permission to publish the Project data.

REFERENCES

1. Kaye, C.A., 1982, “Bedrock and the Quaternary Geology of the Boston Area”, Geological Society of America, Reviews in Engineering Geology, Volume V.
2. Ty, R.K.S., 1987, “History and Characteristics of Man-made Fill in Boston and Cambridge”, MSc Thesis, Massachusetts Institute of Technology, pp. 132, 31 maps.

Chapter 25

GEOTECHNICAL EXPLORATION FOR A LAKE MEAD INTAKE

Richard F. Coon
CH2M Hill, Bellevue, WA

ABSTRACT

Six potential locations along the west shore of Lake Mead were evaluated to site a new water intake pumping station for the SNWA. The evaluations were based upon geotechnical exploration results and conceptual intake pumping station designs. The exploration included geologic mapping, geophysical testing, underwater ROV inspections and borings. Ground surface and underground chamber pumping station designs were considered. Sites at Saddle Island, Black Canyon and Hoover Dam are geotechnically suitable. A Saddle Island site was selected because it has the most favorable conditions for the intake pumping station and the aqueduct to the Las Vegas valley.

INTRODUCTION

The Southern Nevada Water Authority (SNWA) proposes to construct and operate a system of regional water supply facilities known as the Treatment and Transmission Facility (TTF). SNWA would use the TTF in conjunction with the existing Southern Nevada Water System. This existing system was constructed in the 1970's and today includes a raw water intake pumping station on Saddle Island, the Alfred Merritt Smith Water Teatment Facility adjacent to Saddle Island, three aqueducts leading to the west portal of River Mountains Tunnel No. 2, and a system of aqueducts and pipelines in the Las Vegas valley. The TTF will include a second intake pumping station, new aqueducts, booster pumping stations, a new water treatment facility, and the use of the existing River Mountains Tunnels No. 1.

This paper provides an overview of the studies and exploration performed over a period of several years to select the new intake pumping station site and the system of water conveyance to a new water treatment facility. Many factors were considered in the final selection of an intake system. Geotechnical conditions and the feasibility of a deep-water intake shaft and intake tunnel played a major role in the site selection process.

WATER INTAKE REQUIREMENTS

Early planning indicated that the ultimate capacity of the TTF system of approximately 2300 liters per day. The planning also indicated that the new intake should provide ultimate capacity at a Lake Mead water level as low as 305 m. above mean sealevel (amsl).

The alternative intake configurations evaluated are: (1) a subaqueous shaft with a deep tunnel and vertical turbine pumps extending to tunnel level; (2) a submerged pipe on the lake bottom with suction provided by pumps on the shoreline; or (3) submersible vertical pumps suspended from a bridge over the lake. The subaqueous shaft configuration was selected for this project. The submerged pipe and bridge configurations were found to be unsuitable because of technical problems with the large pumping rates required for this project, and because these types of installations have unavoidable aesthetic impacts which are undesirable in a National Park.

The subaqueous shaft could be constructed with either a blasted connection or a drilled connection to the lake bottom. The option of a blasted connection is commonly referred to as the Norwegian lake tap because is has been used on many projects in Norwegian fjords. The lake tap for the SNWS project is one of the first constructed by the drilling method.

SITES IDENTIFIED BY GEOLOGIC RECONNAISSANCE

Introduction

The area of limits of the new intake on are shown on Figure 1. The new intake must be on the west or north shore of Boulder Bay the area of Lake Mead above Hoover Dam. The west and north shoreline of Lake Mead from Hemenway Cove to Callville Bay northeast of the area shown in Figure 1, is underlain by Recent alluvial deposits, older partially cemented alluvial deposits, and poorly-indurated sedimentary units. There is no evidence of thick sequences of Tertiary-age volcanic rocks similiar to the River Mountains to the west or the well-indurated sedimentary rocks of the Frenchmen Mountains to the northwest. Along this 40 km shoreline the 305 m contour is typically more than two kilometers offshore. Geologic reconnisance indicated that this long shoreline provides no suitable intake sites. Saddle Island, Hemenway Cove, Black Canyon and Hoover Dam showed promise of having suitable geotechnical and water depth conditions.

Saddle Island Alternatives

Saddle Island is a small outcrop of Pre-Cambrian bedrock about three kilometers long in a north-south direction and about one half a kilometer wide. The bedrock of Saddle Island is older and dissimilar to the younger predominately sedimentary and volcanic rocks which surround it. There are, therefore, major geologic discontinuities on all sides of Saddle Island.

The island bedrock consists of two geologic zones or plates. The southern plate underlies the southern two-thirds of the island. It is composed of ampibolite and gneiss with a few light colored dikes and a number of shear zones. The southern plate is thought to be basement rock which was originally deep in the earth's crust and has been factured and changed by exposure to the ground surface conditions.

The northern plate underlies the northern one-third of island. It is composed of a wide range of metamorphic rocks and contains extensive intrusions of younger volcanic rocks. The northern plate is considerably more complex in rock type and structure than the southern plate and noteably poorer in quality from an engineering perspective. A fault zone, more

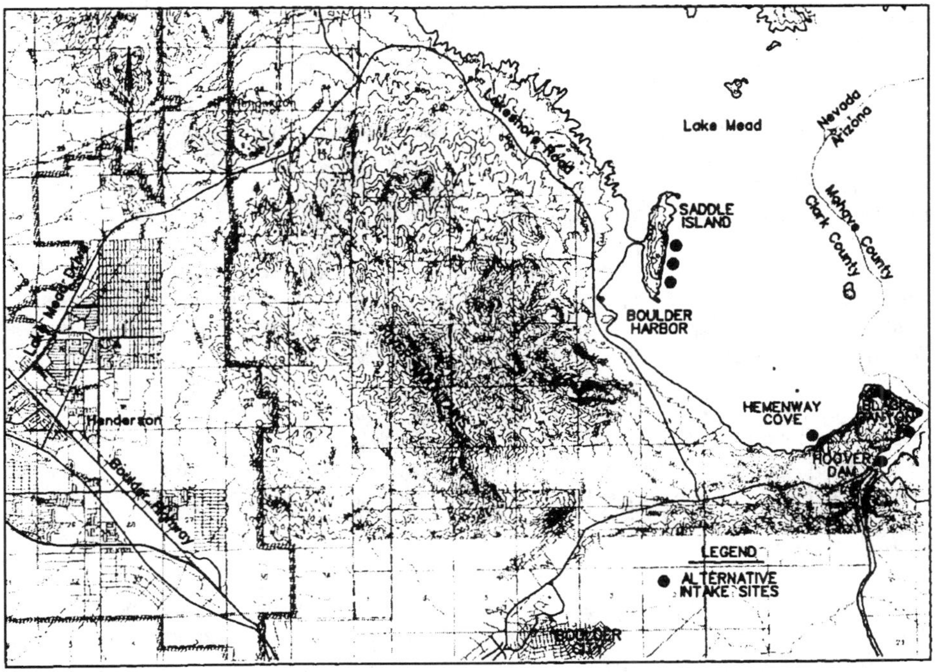

Figure 1. Alternative Intake Sites

than 30 m thick, separates the two plates and is thought to be evidence that the northern plate slide into position on the southern plate.

The existing water intake is in the southern plate. The surface intake pumping station occupies a small valley on the west side of the island and the intake shaft is located on an excavated bench part way down the the steep underwater slope on the east side of the island. The collar of the intake shaft is at elevation 318 m.

The east shore of Saddle Island appears to be underlain by metamorphic rock and the 305 m contour is typically a 100 to 200 m east of the shoreline. If the subsurface rock conditions are similar to those of the existing intake shaft and tunnel, the second intake could be located on Saddle Island. Geotechnical exploration for a new intake was generally in an area one kilometer long and 200 m wide. Geotechnical work described in the following section provided specific exploration targets within this area shown as the three sites on Figure 1.

Hemenway Cove Alternative

Tertiary volcanic rocks cropout on the south shore of Hemenway Cove. The lake bottom slopes at Hemenway Cove are steeper than the shoreline to the north suggesting that the underwater slopes may be underlain by rock instead of alluvium. The 305 m contour is also relatively close to shore. These site observations suggest the possiblity of an intake constructed in Tertiary bedrock with an intake shaft in Hemenway Cove.

Black Canyon Alternative

Hoover Dam is constructed on Tertiary volcanic rocks with the massive tuff of Hoover Dam as a principal rock in the dam abutments and the host rock for the four diversion tunnels. The diversion tunnels of Hoover Dam are exceptional tunnel projects as the 17 m tunnels were driven by drill-and-blast with essentially no initial ground support.

The Hoover Dam tuff has a limited outcrop area which includes portions of the Promonatory Point peninsula between Black Canyon and Hemenway Cove. A fault near the east end of Hemenway Cove separates the tuff of Hoover Dam on the east from the other Tertiary volcanics on the west.

The tuff of Hoover Dam is exposed in Black Canyon above Hoover Dam. A fifth intakealternative site is located in the west wall of Black Canyon a little more than a kilometer upstream of the dam. The intake would presumedly be a Norwegian type in the steep canyon wall with a long intake tunnel to an intake pumping station adjacent to Hemenway Cove.

Hoover Dam Alternative

The Hoover Dam penstocks in the Nevada abutment convey water to the turbines in the Nevada powerhouse. Under most operating conditions, water in excess of the generating requirements is returned to the Colorado River downstream of the powerhouse.

A hydraulic connection to the penstocks would provide an intake to the water of Lake Mead without underwater construction. The sixth intake alternative involves the penstock connection, and an underground pumping station near Hoover Dam.

GEOTECHNICAL EXPLORATION

Introduction

Geotechnical exploration of intake alternatives at Saddle Island, Hemenway Cove, and Black Canyon was conducted in several phases. The exploration included geologic mapping on land, and water-borne geophysical testing, underwater television inspection, and borings on land and from an anchored barge in Lake Mead.

The starting point was a literature review to obtain papers, maps, and photographs pertinent to the sites. This search revealed a large amount of information about the geotechnical conditions at the Hoover Dam site and the SNWS intake. There was very little information about the materials on the bottom of Lake Mead east of Saddle Island, Hemenway Cove or Black Canyon. The best available topographic mapping of Lake Mead had a contour interval of 12 m.

Pictures of the Saddle Island, Hemenway Cove, and upper Black Canyon prior to the formation of Lake Mead were obtained from unexpected sources. Aerial photographs taken in February 1935 were obtained from an archive collection of Fairchiled Aerials at Whittier College in California. These photographs were taken the month that Lake Mead started to form. These photographs showed valuable information about topography and drainage prior to the inundation by Lake Mead. Ground level pictures of the upstream portion of Black Canyon taken during dam construction were found in a National Park Service publication.

Geologic Mapping

Geologic maps on land were made on Saddle Island including mapping from a small boat along the steep eastern shoreline. Geologic maps were also made of the area at Hemenway Cove and the peninsula between Hemenway Cove and Hoover Dam. In each area, geologic maps by others provided a starting point with additional information added about lithology, structure, and rock quality.

Bathymetry

A preliminary bathymetric map of each intake site was developed photogrammetrically from the 1935 aerial photographs. The control for these maps was survey points near the shoreline identified on the ground and in the aerial photographs. These maps are less accurate than most photograpmmetric maps because of the poor print quality of the archive negatives. These maps of course could not show changes in topography due to geologic processes such as sedimentation and landslides in the last half century. However, a comparison of these photogrammetric maps and bathymetric maps might provide a measure of the rate of topographic change caused by these geologic processes.

Detailed baythmetric surveys were performed at Saddle Island, Hemenway Cove and Black Canyon. These surveys were concurrent with the ROV inspection which afforded an opportunity to compare location and depth measurements in the two programs. The bathymetric survey used a small pontoon boat as a work platform, a Geometrics, Inc GPS, and an Echotrak fathometer with both digital and analog chart output. Bathymetric maps for the three areas have a contour interval of 1.5 m. The bathymetric maps and the photogrammetrically prepared maps are reasonably similiar except that the bathymetric maps showed more topographic details. There was no detectable major changes in the underwater topography.

ROV Inspections

The Saddle Island, Hemenway Cove and Black Canyon intake sites were inspected by underwater television using a Phantom 500 HD/2 Remotely Operated Vehicle (ROV). The ROV is a tubular frame with three sets electrically-driven thrust propellers to provide underwater mobility, a instrument package, a television camera, and adjustable, high-intensity light. The ROV was connected to the boat by a 230 m umbilical.. The ROV, as equipped for inspection, was slightly buoyant. The propellers drive the ROV to submerge and control underwater location. Water depth and compass orientation of the ROV were provided by instrumentation on the ROV.

The location of the house boat, which served as the working platform for this survey, was determined by a global postioning satallite (GPS) system with a network of connected satellite signals. No local base station was necessary with this system. The location of the ROV in relation to the house boat was by TRACKER SYSTEM using sonic waves (sonar) from a pinger on the ROV, a directional receiver on the house boat, and gyroscopes in the ROV.

The ROV was typically submerged to the lake bottom and directed along a pre-designated track perpendicular to the shoreline. While in motion, the ROV could be approximately tracked. The ROV was set on the lake bottom at intervals of about 50 feet to allow a precise location determination. The television was operated continuously during submergence and the video record of this transmissions includes the compass orientation and water depth.

The television images from this exploration are high resolution and detailed where water column is clear. The camera was able to focus in on objects as small as the label on a submerged soft drink can. Visibility, however, was often poor because of flocks of sediment moving in the water column because of currents or a stirring of the lake bottom by the ROV propellers. Even under clear conditions, the field of view in the television record is typically limited 3 to 6 m. Perspectives, which would be easy to achieve above water, were hard to obtain in these records.

The ROV records provided the design team with a better appreciation of the variations in underwater slopes, the presence of small and large boulders and the ubiquitous layer of fine-grained lake bottom sediment. The ROV inspection provided information on the condition of the existing water intake and the quality of rock in the steep slopes at Saddle Island and the precipitous rock slopes in Black Canyon. The ROV inspection could not distinguish between thin and thick overburden conditions.

Geophysical Testing

The Saddle Island sites were explored by sub-bottom sonar sounding, magnetometer survey and refraction seismic testing. The first two tests were performed with instruments towed behind a slowly moving boat with instrument locations determined by GPS. The refraction seismic testing was performed from anchored small boats with location control by conventional survey from the shoreline.

The sub-bottom profiling was performed by a towed ORE GeoPulse acoutic gnerator operating at an energy level of 300 joules. The magnetometer survey was performed by a towed Geometric, model 866, proton magnetometer. The refraction seismic tests were performed with geophones floating is a fixed array and mutliple small shots at different locations on the lake bottom.

The sub-bottom profiles provide a weak and complex signature of the subsurface conditions apparently because of the high topographic relief and the relatively great thickness of lake bottom and pre-lake soil deposits. The magnetometer survey showed relatively uniform magnetic conditions except for a strong anomaly in the vicinity of a offshore cinder cone now submerged well below lake level. The anomaly may be a volcanic neck now filled with volcanic rock. the anomaly is of little direct interest because it is offshore of the alternative intake sites.

The refraction seismic testing provide well-defined velocity profiles perpendicular to shore in four possible intake sites. Two of the sites were not further explored because these velocity profiles showed the bedrock was covered by thick overburden. The remaining two sites, the north and center sites on Figure 1, were drilled because the overburden soils appeared to be thin and the bedrock velocities were encouragingly high.

Borings

Borings were made on land at Hemenway Cove and a one offshore location at Hemenway Cove and three offshore locations at Saddle Island. The offshore borings were made by a skid-drilling rig mounted on a Flexi-float barge. The water depth at the offshore boring sites ranged from 30 to 70 m.

The offshore boring at Hemenway Cove was performed first and the conditions there proved difficult. The anchorage area is entirely on a colluvial slope. Initially, four Danforth anchors were used with two anchors in water depth of less than 20 m and two anchors in a water depth of more than 100 m. This arrangement was unsuccessful as the barge moved several feet during drilling operations and broke the casing. Eventually, a six anchor pattern with reinforced concrete anchor blocks weighing 4,000 kg proved successful. Drilling at Saddle Island used the same six anchor pattern of concrete blocks. Nearshore anchorage in the steep rock slopes at one site was provided by rock anchors in shallow water installed by divers.

INTAKE SITE SELECTION

Saddle Island

The north and middle sites at Saddle Island were judged to be unsuitable for intake tunnel construction because the boring at these locations encountered soil-like zones. Tunnel construction at a depth of 80 to 100 m under potentially high hydrostatic pressures was expected to be high cost and to involve high risk.

The south intake site at Saddle Island was found to have good geotechnical conditions at the intake shaft and along the underwater portion of the intake tunnel. The one

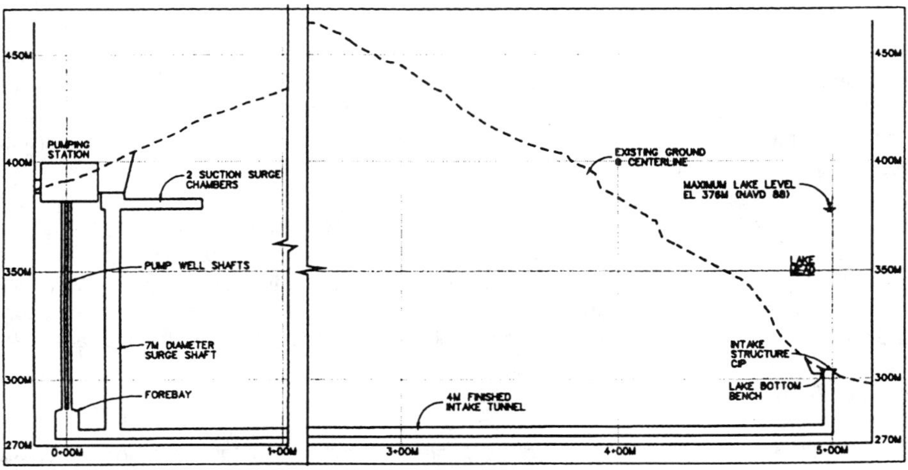

Figure 2. Intake Configuration for Saddle Island

potential problem with this location is a possible fault immediately west of the intake shaft site. In contrast with the north and middle sites at Saddle Island, this is clearly a geotechnically superior site.

The intake facility at the south end of Saddle Island could have the configuration shown in Figure 2. This configuration has a drilled intake shaft, a conventionally excavated intake tunnel, an underground forebay, multiple wells for turbine pump, an access/surge shaft, near surface surge facilities, and a ground surface pumping station.

An alternative with the pumping station sited in an underground chamber was also considered. Land borings showed that a chamber is geotechnically feasible. The chamber alternative would be a little more costly to construct than the ground surface alternatives and could involve complications relative to confined space operations. The surface facility and the south intake shaft site were selected as the Saddle Island alternative.

Hemenway Cove

Drilling on the shore in the Hemenway Cove area revealed that the Tertiary volcanics are of poor quality and contain soil-like layers to a depth of at least 100 m. These poor ground conditions well below the level of Lake Mead and adjacent to the shoreline would make construction of the tunnel and forebay of the Figure 2 configuration difficult and risky. A surface pumping station was judged to be infeasible at Hemenway Cove.

A second possibility for an intake at Hemenway Cove was the place the pumping station, intake tunnel and intake shaft in the tuff of Hoover Dam exposed onshore east of Hemenway Cove. A single deep offshore boring showed that lake is underlain by poor quality volcanics. This finding eliminated the option of the Hemenway Cove intake alternative.

Black Canyon

The alternative of a Black Canyon intake was considered using the underground chamber pumping station in the tuff of Hoover Dam on the east side of Hemenway Cove. This concept assumed that the intake would be developed in the very steep west wall of Black Canyon by means of the Norwegian method and that the intake tunnel would be constructed by TBM largely in the tuff of Hoover Dam.

This alternative was not selected because of major concerns about the geotechnical conditions, construction concerns, and operational concerns. This alternative would involved an unusually large geotechnical exploration program to verify the feasibility of a long intake tunnel and pumping station in chamber, construction times for this alternative could be long, and the facility would require access across the public field of view of Boulder Basin of Lake Mead.

Hoover Dam

The final intake alternative is a direct connection to the penstocks on the Nevada side of the Hoover Dam complex. Figure 3 shows a key element of this alternative a high head pumping station in an underground chamber. The chamber, shown in Figure 3, would be located at approximately the level of the powerhouses and just behind the west wall of Black

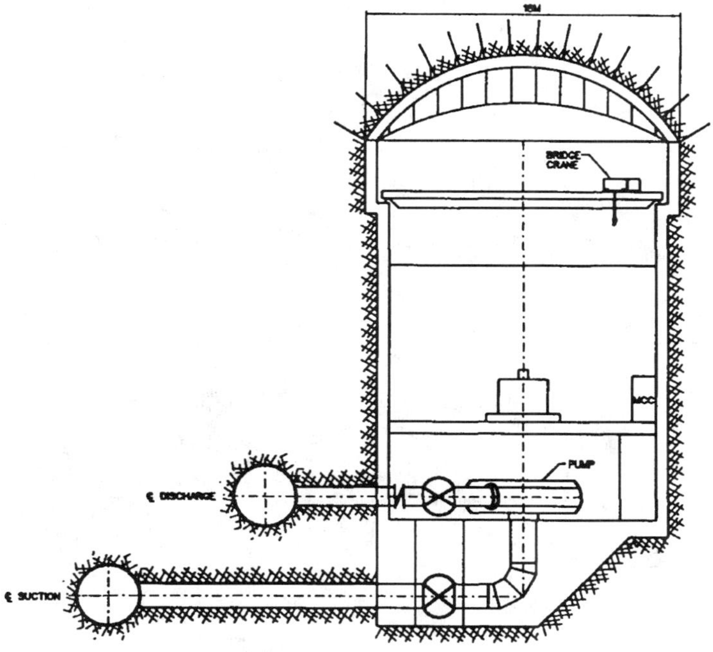

Figure 3. Chamber Pumping Station for Hoover Dam

Canyon downstream of Hoover Dam. Although borings have not been made in the chamber area, geologic mapping indicated a high probability that the tuff of Hoover Dam is present and the rock quality is adequate for the proposed construction. This alternative was judged to be geotechnically feasible.

FINAL INTAKE AND AQUEDUCT ALTERNATIVES

With the identification of two alternative intake sites there still remained a question of the site of future treatment facility and conveyance system alternatives. Routing studies and preliminary studies of the WTF sites were still in progress. The alternative routes are shown in Figure 4.

The Saddle Island intake alternative could serve the Foothills WTF by one of two aqueduct routes passing through the River Mountains in the existing River Mountains Tunnel No. 1. Water from the Saddle Island Intake could reach the River Mountains WTF by way of the Existing River Mountains and a pipeline west of River Mountains or through a new tunnel on the SI-5 route.

Water from the Hoover Dam intake site could reach the River Mountains WTF site by way of route HD-2 including a new tunnel in the River Mountains, or by route HD-8 involving still another new River Mountains tunnel.

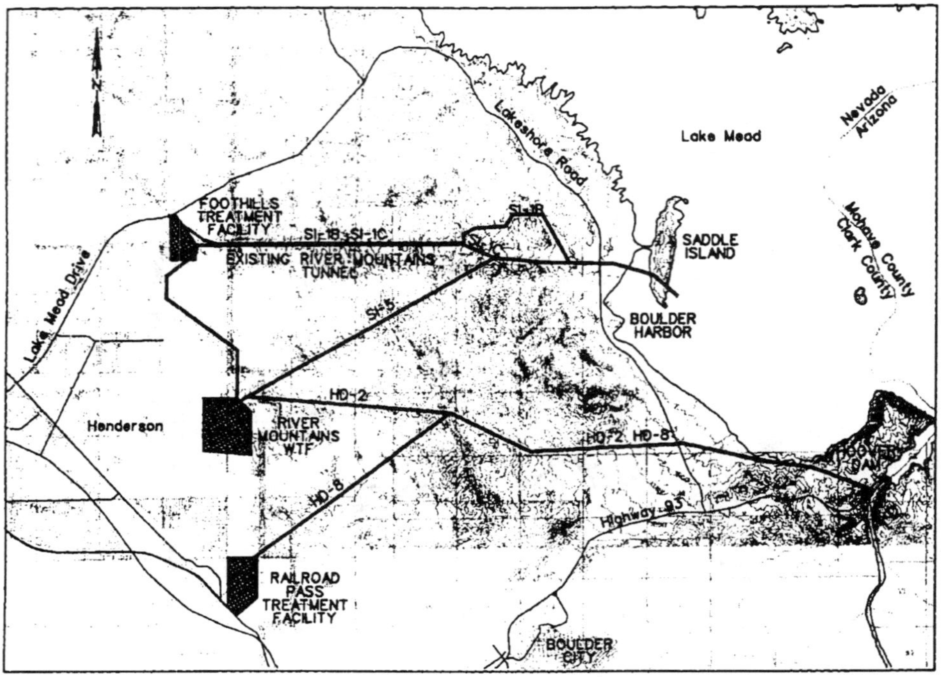

Figure 4. Final Intake and Aqueduct Alternatives

CONCLUSION

The Saddle Island intake with a surface intake pumping station, the SI-1C route, and the River Mountains WTF have been recommended for this project. This decision was made after consideration of the geotechnical risk for intake, tunnel and other construction; estimated construction cost along with consideration of the risk of potential changes during design and construction; and the operability of the new facilities along with the existing SNWS facilities.

ACKNOWLEDGMENTS

The work described in the paper was performed for the Southern Nevada Water Authority by the Joint Venture of Montgomery-Watson and CH2M Hill. Kleinfelder of Las Vegas provided geotechnical and drilling services; Inshore Divers of Pittsburg, CA provided ROV and diver services; Christianson-Boyle of Salt Lake City and Salisbury Drilling of Spokane, WA provided drilling services. The Boulder City, NV office of the Bureau of Reclamation provided valuable historic information and access to the restricted area around Hoover Dam.

Chapter 26

STRESS ANALYSIS AND DESIGN OF TUNNEL LININGS

Ching L. Wu[1] and Joseph Penzien[2]
[1]Bechtel Corporation, San Francisco, CA; [2]International Civil Engineering Consultants, Inc., Berkeley, CA

ABSTRACT

An analytical procedure is presented for evaluating the stresses in linings of bored tunnels caused by kinematic soil-lining interaction. Three cases of interaction as produced by the following conditions are treated: (1) relaxation of insitu soil stresses near a lining following its installation; (2) overburden pressure at the soil surface; and (3) free-field soil response resulting from an earthquake. Non-dimensional formulas and graphs resulting from parametric studies are presented for practical tunnel design applications. The procedure is applied separately to a steel lining and a concrete lining for a site located at the lower end of Market Street in San Francisco through which a BART (Bay Area Rapid Transit) tunnel passes.

INTRODUCTION

In 1964, Burns and Richard (Ref. 1) published a paper entitled, "Attenuation of Stresses for Buried Cylinders" which presented a procedure for evaluating the soil-structure interaction behavior on a cylinder produced by an overburden pressure at the soil surface. The development of this procedure treated the soil surrounding the cylinder in the form of a continuum and derived two-dimensional plane-strain solutions using Michell's series form of the Airy stress function. Later in 1972, Peck, Hendron, and Mohraz (Ref. 2) published their paper, "State of the Art of Soft-Ground Tunneling" which made use of the Burns/Richard procedure to determine the stresses in tunnel linings produced by the relaxation of insitu soil stresses near the linings following their installation in bored tunnels.

While the soil-lining interactions produced by overburden pressure at the soil surface and by relaxation of insitu soil stresses near a lining following its installation result in similar forms of stress patterns in the soil surrounding a lining, the intensities of these two interactions are different, even when the insitu soil stresses produced by the overburden pressure at the centerline elevation of the tunnel are equal to the corresponding stresses produced by gravity.

In Ref. 3, Penzien and Wu developed the theoretical formulation for the soil-lining interaction solutions for all three cases mentioned in the Abstract above, by application of classical theory of elasticity. The solutions for the case of the interaction produced by overburden pressure at the soil surface check very well with the solutions by Burns and Richard, although the derivation is significantly simplified. The solutions for the soil-lining interaction by relaxation of insitu soil stresses near a lining following its installation present a more realist condition for tunnel lining design. This approach also results in lower lining stress and reduces conservatism in the design. The solutions for the case of the interaction produced by free-field soil response due to earthquake enhance the traditional tunnel lining design for the important seismic conditions. This paper presents the formulas for the tunnel lining design in non-dimensional form for ease of practical design applications. Similar to Ref. 2, parametric studies were carried out and functional relations were presented in the form of graphs which have been found very useful in design.

DEFORMATIONS OF TUNNEL LINING FOLLOWING INSTALLATION

Kinematic Soil-Lining Interaction

When boring a cylindrical tunnel through relatively soft soil, a very rigid open-face shield is pressed forward longitudinally (*z* direction) into the soil, much like the action of a cookie cutter, transferring internal soil stresses to the outer surface of the shield. As the shield moves forward, a narrow segmental lining ring is installed directly behind the shield and grout is injected between the outer surface of the ring and the inner surface of the cylindrical cavity. It is intended that this grout fill any voids or gaps that may be present. This ring-installation operation is continually repeated as the shield moves forward with the adjacent lining rings being bolted together to secure them in place within the tunnel.

As the shield passes by a fixed point along the longitudinal axis of the tunnel, the soil pressure distribution is transferred to the lining. However, since the lining is quite flexible compared to the shield, the lining will deform under this distribution which, in turn, will modify the pressure distribution. This coupled behavior of cause and effect will be referred to as kinematic soil-lining interaction. The objective herein is to determine the stresses in the lining after full convergence of this interaction behavior following passage of the shield.

Deformations of a Cylindrical Tunnel Without Lining

Consider boring a cylindrical tunnel of diameter D into a soil region which is already under the gravity stress-state condition shown in Figs. 1 and 2 and assume no lining rings are installed following passage of the shield. As the shield moves forward, the surface of the cylindrical cavity at a particular cross-section will displace inward. To determine the pattern of this inward displacement around the entire circumference of the cavity, one must subtract the corresponding displacement pattern along the same circumference of an imaginary circle within the free-field soil (no cylindrical cavity present) caused by imposing the full gravity loading represented by the stress state in Fig. 2 from the corresponding displacement pattern around the circumference of a cylindrical cavity produced by the same imposed gravity loading. In determining these plane-strain ($\varepsilon_z = 0$) displacement patterns, it is convenient to separate the stress state into its dilatational and shear-type components. The above described subtraction of one displacement pattern from another can be carried out for each of these two stress-state components and then superposition can be used to obtain their combined patterns. These combined displacement patterns for the plane-stress condition can easily be converted to meet the plane-strain condition. Using the theory of elasticity (Ref. 4) and following the procedure described in Ref. 3, one can obtain the following relations:

$$\Delta D_{dc} = -\frac{D\overline{\sigma}_1}{2E_s}(1 + K_o)(1 + v_s)(1 - 2v_s)$$

$$\Delta D_{sc} = \pm\frac{D\overline{\sigma}_1}{2E_s}(1 - K_o)(1 + v_s)$$

$$\Delta D_{dt} = -\frac{D\overline{\sigma}_1}{E_s}(1 + K_o)(1 - v_s^2)$$

$$\Delta D_{st} = \pm\frac{2D\overline{\sigma}_1}{E_s}(1 - K_o)(1 - v_s^2) \tag{1}$$

$$\Delta D_d = -\frac{D\overline{\sigma}_1}{2E_s}(1 + K_o)(1 + v_s)$$

$$\Delta D_s = \pm\frac{D\overline{\sigma}_1}{2E_s}(1 - K_o)(1 + v_s)(3 - 4v_s)$$

Where D is the diameter of the lining, $\overline{\sigma}_1$ is the effective compressive soil stress, K_o is the effective lateral soil coefficient, E_s is Young's modulus and v_s is Poisson's ratio of the soil. Note that the subscripts *"d, "s", "c",* and *"t",* in these relations refer to dilational deformation, shear-type deformation, imaginary circle in free-field soil, and, tunnel, respectively.

Deformations of a Cylindrical Tunnel with Lining

Consider boring a cylindrical tunnel of diameter *D* into a soil region which is already under the gravity stress-state condition shown in Figs. 1 and 2 with lining rings being installed immediately following passage of the shield as described previously. Let us assume that the grouting operation is carried out with 100 percent effectiveness. As the shield moves forward, the surface of the cylindrical cavity at a particular cross-section will displace inward interacting with the lining. This interaction will cause the lining to deform into dilatational and shear-type (ovalling) modes which are radially compatible with the inner surface of the surrounding soil. The amplitudes of these modes depend upon the stiffness of the lining relative to those of the surrounding soil. If water is present in the soil, pore pressure will eventually build up on the outside surface of the lining after its installation. This pressure will affect the amount of soil-lining interaction in the dilatational mode.

Dilatational Deformation. The dilatational stiffness of the lining is defined as the intensity of uniform pressure required on its outer surface to produce a uniform inward ΔD equal to unity under plain-strain conditions. It is easily shown that this stiffness k_{dl} is given by

$$k_{dl} = \frac{2E_lA_l}{D^2(1 - v_l^2)} \tag{2}$$

in which E_l is Young's modulus of the lining material, v_l is its Poisson's ratio, and A_l is the cross-sectional area of a single ring segment divided by its longitudinal width. The subscript *dl* is used to indicate that this particular stiffness represents the dilatational mode displacement of the lining.

A dilatational stiffness of the soil outside the lining is defined as the intensity of a uniform pressure on the inner cylindrical surface of the soil cavity required to produce an outward ΔD equal to unity under plane-strain conditions. Previously it was shown that release of the dilatational pressure $\overline{\sigma}_1(1 + K_o)/2$ under this plane-strain condition resulted in an inward ΔD as given by the fifth of Eqs. (1); thus, the desired soil stiffness k_{ds} is given by setting ΔD_d equal to one in this equation, substituting k_{ds} for $-\overline{\sigma}_1(1 + K_o)/2$ on its right hand side, and solving for k_{ds} resulting in

$$k_{ds} = \frac{E_s}{D(1 + v_s)} \qquad (3)$$

in which subscript *ds* is used to indicate dilatational stiffness of the soil.

After full relaxation of the dilatational soil stresses in an unlined tunnel following passage of the shield, the cylindrical cavity will have a diameter equal to $D+\Delta D_d$; thus, in the case when unstressed lining rings of diameter D are installed, the return of pore pressure and the development of soil-lining interaction following construction will impose a uniform pressure at the soil/lining interface causing the lining to deform inward in a dilatational mode with diameter change Δ_{dl} and the surface of the soil cavity to deform outward in a dilatational mode with diameter change Δ_{ds}.

Satisfying the compatibility of these pressures at the soil/lining interface and satisfying compatibility of the soil and lining diameter changes, one can derive the following relation expressing the lining diameter change in the dilatational mode of deformation:

$$\Delta_{dl} = \frac{D\overline{\sigma}_1(1 + K_o)(1 + v_s)}{2E_s(1 + a_d)} + \frac{uD^2(1 - v_l)a_d}{2E_lA_l(1 + a_d)} \qquad (4)$$

in which a_d is a dimensionless soil/lining interaction parameter given by

$$a_d = \frac{2E_lA_l(1 + v_s)}{DE_s(1 - v_l^2)} \qquad (5)$$

Shear-Type (Ovalling) Deformation. The shear-type (ovalling) stiffness of the lining is defined as that intensity of pure shear-type pressures which will produce an outward ΔD in the transverse direction (x axis) and an inward ΔD in the vertical direction (y axis), both equal to unity under plane-strain conditions. It can be shown that this generalized stiffness is given by

$$k_{sl} = \frac{48E_lI_l}{D^4(1 - v_l^2)} \qquad (6)$$

in which I_l is the moment of inertia of a single ring-segment cross-section divided by its longitudinal width. The subscript *sl* indicates that this stiffness represents the shear-type modal deformation of the lining.

A shear-type (ovalling) generalized stiffness of the soil outside the lining is defined as the intensity of shear-type pressures which, if acting in the opposite directions on the inner surface of the cylindrical soil cavity, would produce an outward ΔD vertically and an inward ΔD transversely both equal to unity under the plane-strain condition. Previously it was shown that release of the shear-type pressures $\pm\overline{\sigma}_1(1 - K_o)/2$ under this plane-strain condition resulted in transverse and vertical diameter changes as given by the sixth of Eqs. (1); thus, the desired soil stiffness k_{ss} is given by setting ΔD_s equal unity in this equation, substituting k_{ss} for $\overline{\sigma}_1(1 - K_o)/2$ on its right-hand side, and solving for k_{ss} resulting in

$$k_{ss} = \frac{E_s}{D(1 + v_s)(3 - 4v_s)} \tag{7}$$

The subscript *ss* indicates that the stiffness represents the shear mode in soil.

After full relaxation of the shear-type soil stresses in an unlined tunnel following passage of the shield, the cylindrical cavity will oval with diameter changes as given by the sixth of Eqs. (1); thus, in this case when unstressed lining rings of diameter D are installed, the development of soil-lining interaction following construction will impose shear-type pressures at the soil/lining interface causing the lining to deform with an outward diameter change in the transverse direction and an inward diameter change in the vertical direction, both equal to Δ_{sl}, and the surface of the soil cavity to deform with an inward diameter change in the transverse direction and an outward diameter change in the vertical direction, both equal to Δ_{ss}. Satisfying compatibility of soil and lining diameter changes and of pressures at the soil/lining interface, similar to the procedure leading to Eq. (4) but with no pore pressure effect, one obtains

$$\Delta_{sl} = \frac{D\overline{\sigma}_1(1 - K_o)(1 + v_s)(3 - 4v_s)}{2E_s(a_s + 1)} \tag{8}$$

in which s_s is the dimensionless soil/lining interaction parameter defined by

$$a_s = \frac{48E_lI_l(1 + v_s)(3 - 4v_s)}{D^3E_s(1 - v_l^2)} \tag{9}$$

DEFORMATIONS OF TUNNEL LINING DUE TO CHANGE IN OVERBURDEN PRESSURE

Consider the deformations produced in a tunnel lining by the application of an overburden pressure p on the surface of the soil shown in Fig. 1. The principal stresses on an element at Pt. O will be

$$\sigma_y = -p; \ \sigma_x = \sigma_z = - K_o p \tag{10}$$

If no lining is present in the tunnel when the overburden pressure is applied, the dilatational and shear-type (ovalling) diameter changes under the plane-strain condition will be given by the third and fourth of Eq. (1), i.e.

$$\Delta D_d = \Delta D_{dt} = - \frac{D_p}{E_s} \quad (1 + K_o)(1 - v_s^2)$$

$$\Delta D_s = \Delta D_{st} = \frac{2D_p}{E_s}(1 - K_o)(1 - v_s^2) \quad (11)$$

Now assume an unstressed lining is in the tunnel when the overburden pressure p is applied to the surface of the soil causing soil/lining interaction to take place. This interaction will impose (1) a uniform pressure component at the soil/lining interface causing the lining to deform inward in a dilatational mode with diameter change Δ_{dl} and the surface of the soil cavity to deform outward in a dilatational mode with diameter change Δ_{ds}, and (2) a shear-type pressure component causing the lining to deform with an outward diameter change in the transverse direction and an inward diameter change in the vertical direction, both equal to Δ_{sl}, and the surface of the soil cavity to deform with an inward diameter change in the transverse direction and an outward diameter change in the vertical direction, both equal to Δ_{sf}.

Again, satisfying compatibility of pressures at the soil/lining interface and satisfying compatibility of the soil and lining diameter changes, the following relations can be derived expressing the lining diameter changes in the dilatational and shear-type (ovalling) modes of deformation, respectively:

$$\Delta_{dl} = \frac{Dp(1 + K_o)(1 - v_s^2)}{E_s(1 + a_d)} \quad (12)$$

$$\Delta_{sl} = \frac{2Dp(1 - K_o)(1 - v_s^2)}{E_s(1 + a_s)} \quad (13)$$

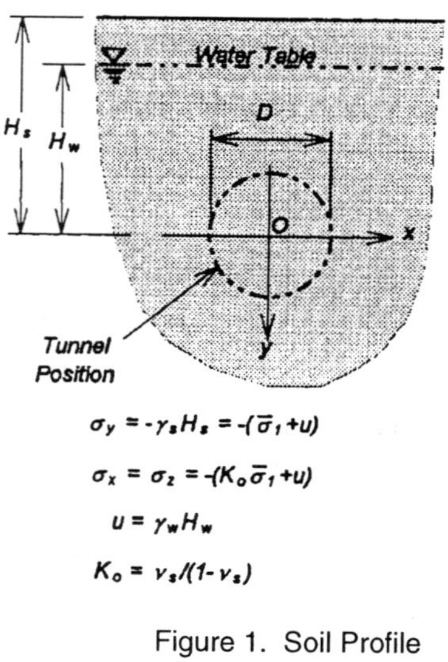

Figure 1. Soil Profile

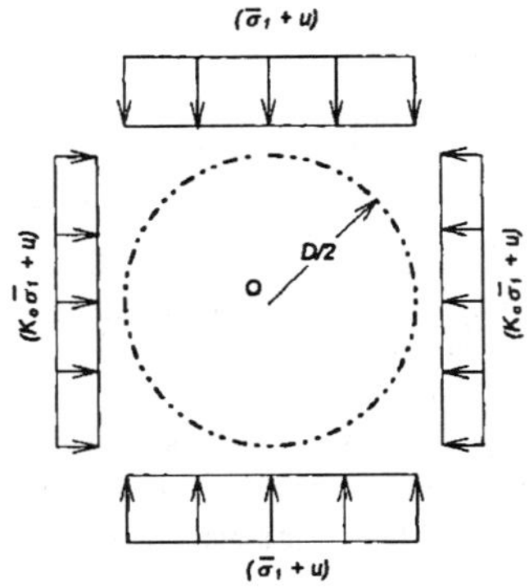

Figure 2. Free-field Soil Stress State Over Tunnel Diameter D

DEFORMATIONS OF TUNNEL LINING DUE TO SEISMIC GROUND RESPONSE

Consider the deformations produced in a tunnel lining due to its interaction with the free-field ground response produced during an earthquake. Let this ground response be in the critical transverse direction to the tunnel. From a standard site-response analysis, using the vertically-propagating shear-wave model, generate the distribution of the free-field ground displacement $u(y,t)$ using the coordinate system in Fig. 1. Then find the critical distribution $u(y,t_c)$ at time t_c which will produce maximum shear-type (ovalling) deformation in the tunnel lining. Assuming no significant change in the soil properties E_s and v_s over depth D at the intended tunnel location, the function $u(y,t_c)$ will usually be reasonably linear over the range $-D/2 < y < D/2$. In this case, the average free-field soil shear strain in the xy plane over this depth is given by the approximate relation

$$y_c = \frac{[u(-\frac{D}{2}),t_c] - [u(\frac{D}{2}, t_c)]}{D} \tag{14}$$

The corresponding state of stress over the depth D is of the shear-type producing shear and principal normal stresses equal to

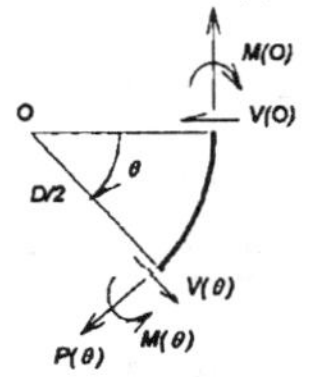

Figure 3. Sign Convention for Force Components in Lining

$$T = \sigma_1 = -\sigma_2 = \frac{y_c E_s}{2(1 + v_s)} \tag{15}$$

in which the shear modulus G_s is equal to $E_s/2(1+v_s)$ and should be assigned a value compatible with the critical shear-strain level y_c. These stresses will produce ovalling of an imaginary circle in the free-field soil region with major and minor axes in the ±45° directions.

If a lining is present in a tunnel which experiences the above ground deformation environment, its increase in diameter along the major axis at $\theta = -45°$ due to ovalling can be obtained by simply substituting $y_c E_s/2(1 + v_s)$ for $p(1 - K_o)/2$ in Eq. (13) resulting in the relation

$$\Delta_{sl} = \frac{2Dy_c(1 - v_s)}{(1 + a_s)} \tag{16}$$

The diameter along the minor axis at $\theta = 45°$ will decrease by this same amount.

FORCE COMPONENTS IN TUNNEL LINING

Adopting the sign convention shown in Fig. 3 for the internal force components $M(\theta)$, $P(\theta)$, and $V(\theta)$, one can establish that for the dilatational displacement mode

$$P_d(\theta) = \frac{E_l A_l \Delta_{dl}}{D(1 - v_l^2)} \tag{17}$$

$$M_d(\theta) = V(\theta) = 0$$

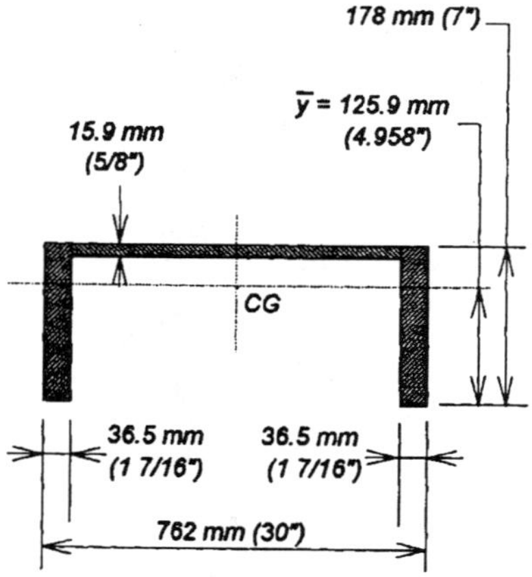

Figure 4. Cross-section of BART Steel Lining

for the shear-type (ovalling) displacement mode with its major axis at $\theta = 0°$

$$P_s(\theta) = \frac{24E_l A_l \Delta_{sl}}{D^3(1 - v_l^2)} \cos 2\theta$$

$$M_s(\theta) = \frac{6E_l I \Delta_{sl}}{D^2(1 - v_l^2)} \cos 2\theta \qquad (18)$$

$$V_s(\theta) = \frac{48E_l I \Delta_{sl}}{D^3(1 - v_l^2)} \cos 2 \sin\theta$$

and for the shear-type (ovalling) displacement mode with its major axis at $\theta = -45°$

$$P_s(\theta) = - \frac{24E_l I \Delta_{sl}}{D^3(1 - v_l^2)} \cos 2\left(\theta + \frac{\pi}{4}\right)$$

$$M_s(\theta) = - \frac{6E_l I \Delta_{sl}}{D^2(1 - v_l^2)} \cos 2\left(\theta + \frac{\pi}{4}\right) \qquad (19)$$

$$V_s(\theta) = - \frac{48E_l I \Delta_{sl}}{D^3(1 - v_l^2)} \cos 2\left(\theta + \frac{\pi}{4}\right) \sin 2\left(\theta + \frac{\pi}{4}\right)$$

Substituting Eqs. (4) and (12) separately into the first of Eqs. (17) yields the uniform internal force component $P(q)$ for the case of stress relaxation following installation of the lining and for the case of overburden pressure applied at ground surface, respectively; substituting Eqs. (8) and (13) separately into Eqs. (18) yields all three internal force components for the case of stress relaxation following installation of the lining and for the case of overburden pressure applied at ground surface, respectively; and substituting Eq. (16) into Eqs. (19) yields all three internal force components for the case of subjecting the lining to the seismic free-field ground-deformation environment.

Direct superposition can be used to find the combined internal force components for the three cases considered. In doing so, it is helpful to recognize that when combining components in Eqs. (18) with the corresponding components in Eqs. (19) the following identity can be used:

$$A \cos 2\theta + B \cos 2\left(\theta + \frac{\pi}{4}\right) = \sqrt{A^2 + B^2} \cos 2(\theta + \lambda) \qquad (20)$$

in which

$$\lambda = \frac{1}{2} \tan^{-1}\left(\frac{B}{A}\right) \qquad (21)$$

NORMALIZED TUNNEL LINING DESIGN FORMULAS

Compressibility Ratio and Flexibility Ratio

For soil-lining interaction studies, two key parameters that affect the lining behaviors, the compressibility ratio and flexibility ratio are defined as follows.

$$\text{Compressibility Ratio} \qquad c \equiv \frac{1}{a_d} = \frac{DE_s(1 - v_l^2)}{2E_l A_l(1 + v_s)}$$

$$\text{Flexibility Ratio} \qquad f \equiv \frac{1}{a_s} = \frac{D^3E_s(1 - v_l^2)}{48E_l I_l(1 + v_s)(3 - 4\,v_s)} \qquad (22)$$

The compressibility ratio is a measure of the extentional stiffness of the medium relative to that of the lining. The flexibility ratio is a measure of the flexibility stiffness of the medium relative to that of the lining. The larger the compressibility ratio or flexibility ratio, the more compressible or flexible the tunnel lining is relative to the soil medium.

While the definition of the compressibility ratio and flexibility ratio in Ref. 2 was based on the diameter strain of an imaginary circle in a uniform medium without the excavated tunnel as compared to the diameter strain of a lining ring subjected to pressure distribution shown in Fig. 1, the above definition of the compressibility ratio and flexibility ratio is based on the diameter strain of the excavated circular boundary of the medium. They represent a more realistic parameter for studying the interaction between the tunnel lining and the excavated soil medium.

The tunnel lining design formulas are normalized in non-dimensional form and summarized as follows.

Soil-Lining Interaction Produced by Relaxation of Insitu Soil

$$\frac{M_s}{(y_sH_s\text{-}y_wH_w)\,R^2} = -\frac{1}{4}(1 - K_o)\,\frac{1}{1+f} \qquad (23)$$

$$\frac{P_d}{(y_sH_s\text{-}y_wH_w)\,R} = -\frac{1}{2}(1 + K_o)\,\frac{1}{1+c}$$

$$\frac{P_s}{(y_sH_s\text{-}y_wH_w)\,R} = \mp\frac{1}{2}(1 - K_o)\,\frac{1}{1+f} \qquad \begin{array}{l}\text{—at Springline}\\ \text{+ at Crown}\end{array}$$

$$\frac{\Delta_{sl}/D}{(y_sH_s\text{-}y_wH_w)/E_s} = \pm\frac{1}{2}(1 - K_o)(1 + v_s)(3 - 4\,v_s)\,\frac{f}{1+f} \qquad \begin{array}{l}\text{+at Springline}\\ \text{—at Crown}\end{array}$$

Where y_s is the average unit weight of the soil over height H_s above the tunnel centerline, is the average unit weight of water, H_w is the height of the water table above the tunnel centerline, and R is the radius of the tunnel lining.

Soil-Lining Interaction Produced by Overburden Pressure

$$\frac{M_s}{(y_s H_s - y_w H_w) R^2} = -(1 - K_o) \frac{(1 - v_s)}{(3 - 4v_s)} \frac{1}{1 + f}$$

$$\frac{P_d}{(y_s H_s - y_w H_w) R^2} = -(1 + K_o)(1 - v_s) \frac{1}{1 + c} \qquad (24)$$

$$\frac{P_s}{(y_s H_s - y_w H_w) R} = \mp 2(1 - K_o) \frac{(1 - v_s)}{(3 - 4v_s)} \frac{1}{1 + f} \qquad \begin{array}{l} \text{— at Springline} \\ \text{+ at Crown} \end{array}$$

$$\frac{\Delta_{sl} D}{(y_s H_s - y_w H_w)/E_s} = \pm 2(1 - K_o)(1 - v_s^2) \frac{f}{1 + f} \qquad \begin{array}{l} \text{+ at Springline} \\ \text{— at Crown} \end{array}$$

Soil-Lining Interaction due to Pore Pressure

$$\frac{P_d}{y_w H_w R} = -\frac{1}{1 + c} \qquad (25)$$

$$\frac{\Delta_{dl}/D}{y_w H_w/E_s} = (1 + v_s) \frac{c}{1 + c}$$

Soil-Lining Interaction due to Seismic Ground Response

$$\frac{M_s}{y_c R^2 E_s} = -\frac{(1 - v_s)}{(1 + v_s)(3 - 4v_s)} \frac{f}{1 + f}$$

$$\frac{P_s}{y_c R E_s} = -\frac{2(1 - v_s)}{(1 + v_s)(3 - 4v_s)} \frac{f}{1 + f} \qquad (26)$$

$$\frac{\Delta_{sl}/D}{y_c} = 2(1 - v_s) \frac{f}{1 + f}$$

Comparison of Normalized Results with Ref.2

For the case of the soil-lining interaction produced by the overburden pressure, the normalized moment, thrust, and diameter change in the lining generated by the solutions derived herein compared very well with the solutions of the formulation by Peck (Ref. 2) based on the theoretical derivation by Burns and Richard (Ref. 1) as shown in Figures 5, 6, and 7 for a dry site condition where there is no pore pressure due to underground water. These results are almost identical if no slippage between the soil and lining is assumed,

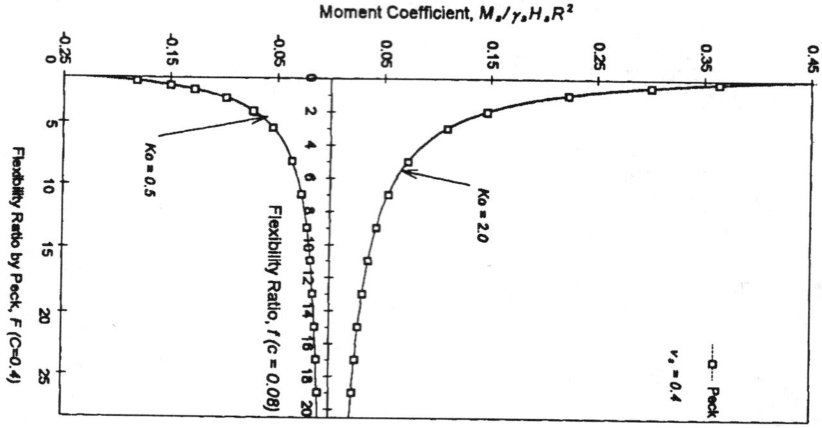

Figure 5, Moment due to Overburden Pressure (Comparison with Peck's Results)

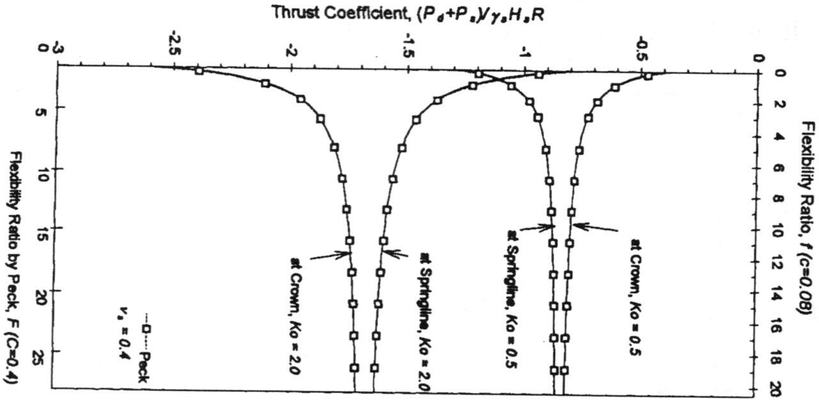

Figure 6, Thrust due to Overburden Pressure (Comparison with Peck's Results)

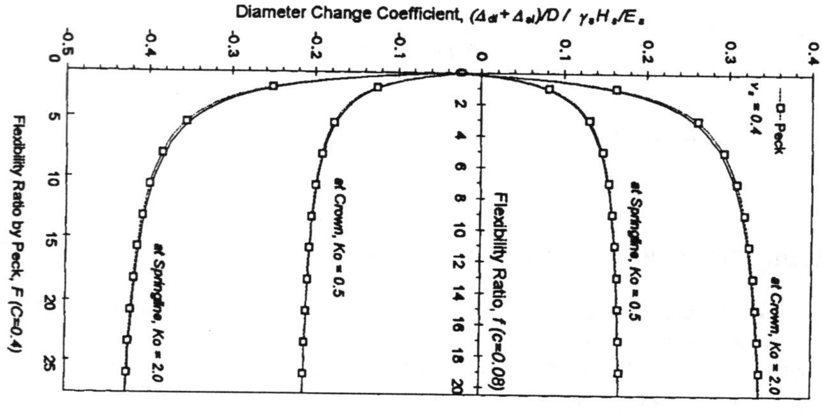

Figure 7, Diameter Change due to Overburden Pressure (Comparison with Peck's Results)

Figures 5, 6, and 7

which will affect the results for the thrust.

However, based on the solutions for the soil-lining interaction produced by relaxation of insitu soil stress near the lining which is the general case for tunnel lining design applica tion, the moment, thrust and diameter change in the lining is significantly reduced from those given by Ref. 2. Excluding the effects due to pore pressure these comparisons are shown in Figures 8, 9, and 10. For a rigid lining, the maximum bending moment in the lining is re duced by as much as 46% according to the solutions by relaxation of insitu soil stress as compared to the results yielded from the solution by overburden pressure. As the flexibility of the lining increases (*f* increases), the bending moment from both solutions decreases. By use of the solution by relaxation of insitu soil stress, the maximum thrust in the lining is reduced by 27% for the very rigid lining and 17% for the very flexible lining. Reduction for diameter change ranges from 45% to 39% for the K_o value selected. The comparisons indicate that the solutions for the soil-linings interaction produced by relaxation of insitu soil stress near the lining significantly reduce the calculated bending moment, thrust, and diam eter deformation of the lining. Using the formulas derived from this approach as de scribed herein significantly reduces the conservatism in the conventional tunnel lining design.

Figures 11 through 14 show the non-dimensional moment, thrust, and diameter defor mation due to the effective compressive soil pressure for seven values of K_o. These curves are intended for designers' use in design of the tunnel lining. K_o less than 1.0 represents normally consolidated soil where the horizontal insitu stress in the medium is a fraction of the vertical stress. K_o=1.0 is the special case of hydrostatic pressure distribution when the horizontal and vertical stresses are equal. For over-consolidated clay over the years, K_o is greater than 1.0 and the horizontal insitu stress in the medium is greater than the vertical stress.

Figure 11 gives the non-dimensional moment, and thrust component due to ovalling in the lining as a function of the flexibility ratio. The bending moment and the thrust component due to ovalling in the lining results from the shearing type (ovalling) deformation of the lining. As the flexibility ratio increases, the moment and thrust coefficients decrease. Although the definition of the flexibility ratio for this paper is based on the diameter strain of the excavated circular boundary rather than an imaginary circle by Ref. 2, the observations by Ref. 2 re main valid that for flexibility ratios less than 10, the moment and thrust coefficients decrease rapidly, and for the flexibility ratio greater than about 10, the lining behaves relatively flexible with respect to the medium. For flexibility ratio greater than 10, the moment and thrust coefficients are generally less than 1% and 2%, respectively, for normally consolidated soil, and less than 2% and 4%, respectively, for over-consolidated soil. The moment and thrust coefficient curves, as well as the diameter deformation curves discussed below, as functions of the flexibility ratio demonstrate the same general characteristics as those provided in Ref. 2, except that the magnitude of these coefficients is significantly reduced here.

Figure 12 provides the thrust coefficient resulting from the dilatational deformation of the lining versus the compressibility ratio. This thrust coefficient is relatively insensitive to the compressibility ratio and decreases slightly as the compressibility ratio increases. How ever, it increases markedly as the soil is more consolidated and the effective lateral soil coefficient K_o increases. The dilatational thrust component usually dominates over the thrust component resulting from ovalling lining deformation.

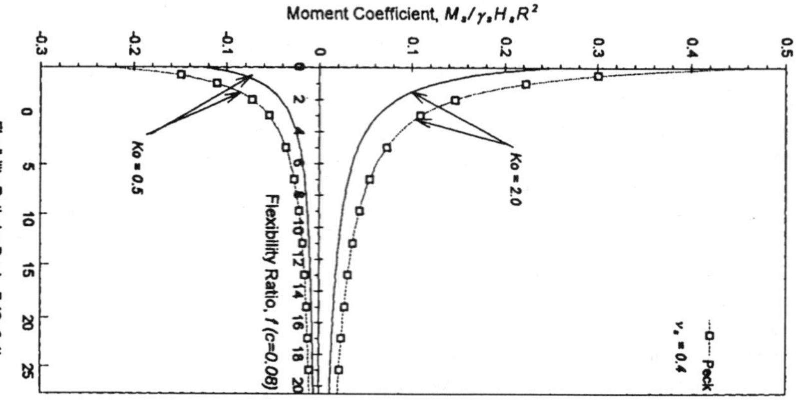

Figure 8, Moment due to Relaxation/Interaction (Comparison with Peck's Results based on Overburden Pressure)

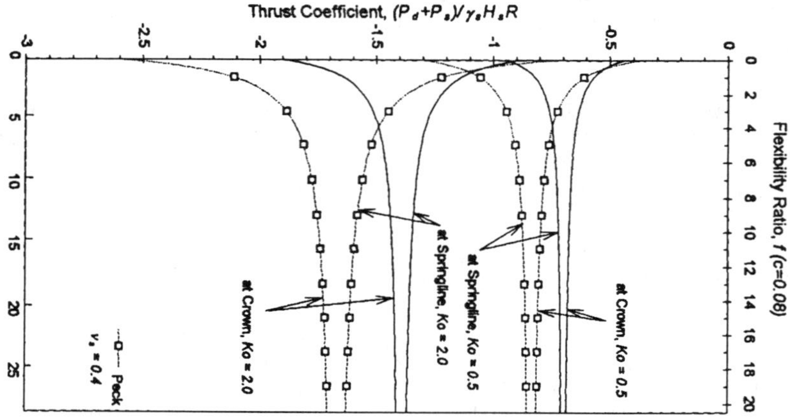

Figure 9, Thrust due to Relaxation/Interaction (Comparison with Peck's Results based on Overburden Pressure)

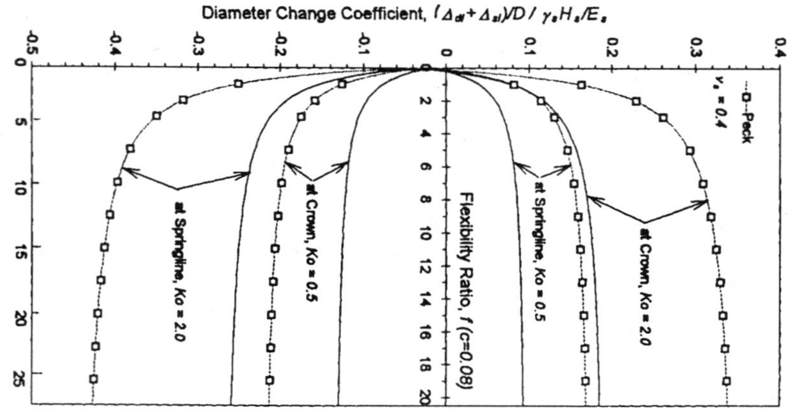

Figure 10, Diameter Change due to Relaxation/Interaction (Comparison with Peck's Results based on overburden Pressure)

Figures 8, 9, and 10

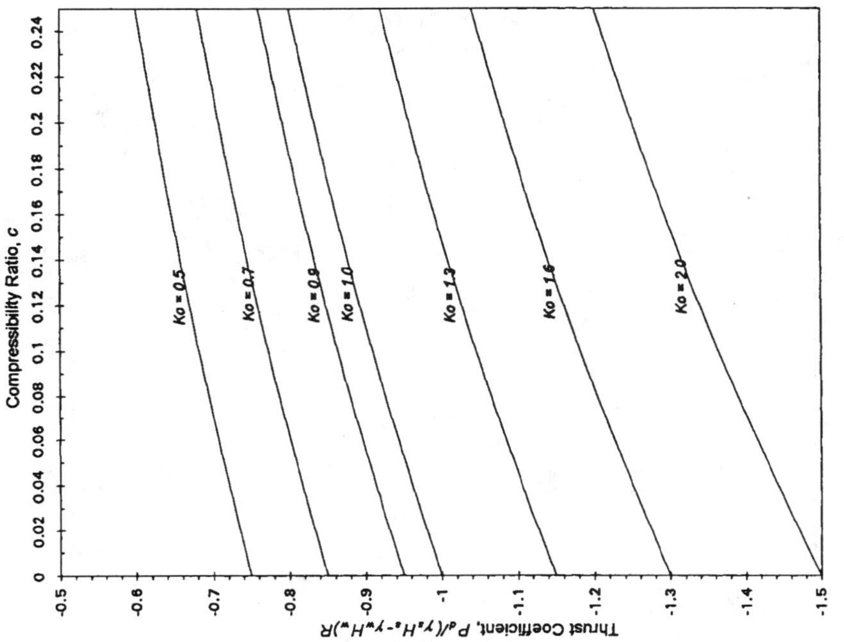

Figure 12, Thrust due to Effective Soil Pressure (Dilatation Mode)

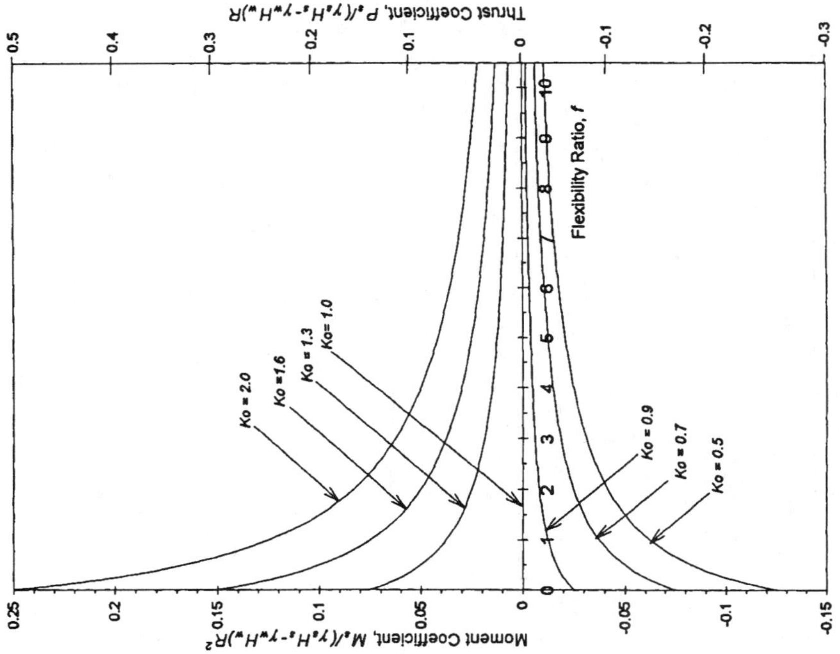

Figure 11, Moment and Thrust due to Effective Soil Pressure (Ovalling Mode)

Figures 11 and 12

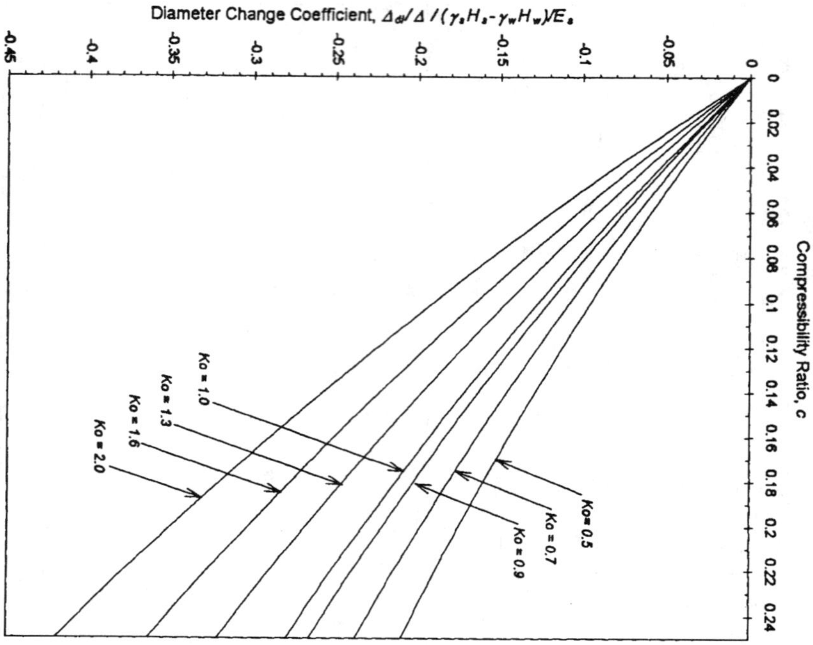

Figure 13, Diameter Change due to Effective Soil Pressure (Dilatational Mode)

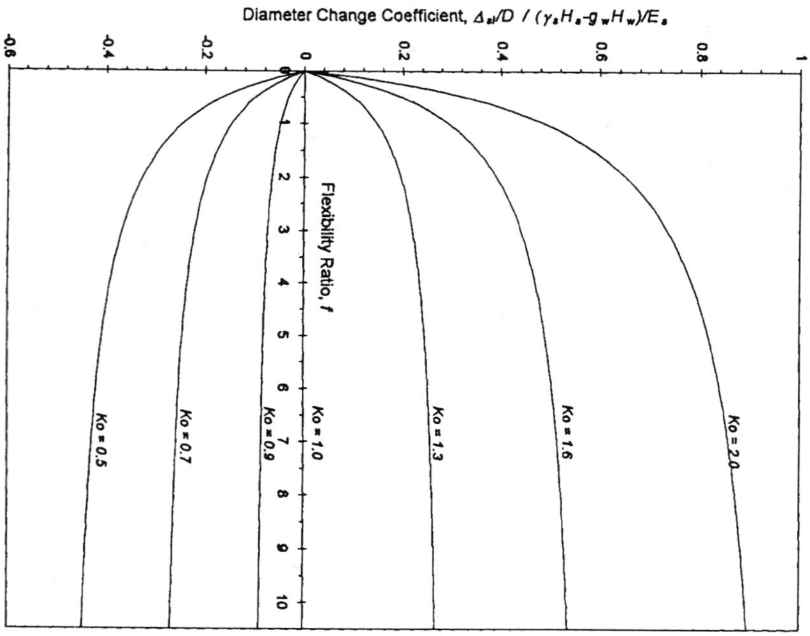

Figure 14, Diameter Change due to Effective Soil Pressure (Ovalling Mode)

Figures 13 and 14

Figure 13 gives the diameter change coefficient resulting from the dilatational deformation of the lining versus the compressibility ratio. This diameter change coefficient in creases almost linearly as the compressibility ratio increases. Figure 14 shows the diameter change coefficient resulting from the shear type (ovalling) deformation of the lining as a function of the flexibility ratio. For flexible linings, the diameter change coefficient increases rapidly as the flexibility ratio increases. For medium to rigid linings, the diameter change coefficient increases to reach a plateau. For flexibility ratio greater than 10, this diameter change coefficient is generally less than 0.4 for normally consolidated soil and less than 0.8 for over-consolidated soil. Normally, the dilatational component of the diameter change only amounts to a small fraction of the ovalling component.

Figures 15 and 16 provide the non-dimensional thrust and diameter change coeffi cients due to the pore pressure in the soil. These responses result from the dilatational deformation of the lining subject to uniform pore pressure in the soil due to the underground water. They are independent of the effective lateral soil coefficient K_o. The thrust coefficient decreases slightly as the compressibility ratio increases while the diameter change coefficient increases almost linearly as the compressibility ratio increases. When the ground water table is above the tunnel springline, these effects should be combined with the corresponding thrust and diameter change coefficients due to the effective stress to obtain the total static results for the lining design. The effects of pore pressure on the lining response depends on the level of the water table below the ground surface. When the ground water table is close to the surface, the thrust due to the pore pressure is on the same order of magnitude as the thrust due to effective soil stress. However, the diameter change due to the pore pressure is practically negligible.

The seismic induced bending moment, thrust and diameter change coefficients ver sus the flexibility ratio are represented in Figures 17 and 18 in non-dimensional form. These responses result from the shear type (ovalling) deformation of the lining subjected to seismic shear strain in the soil. They are also independent of K_o. Both the moment and thrust coefficients decrease rapidly as the flexibility ratio increases while the diameter change co efficient increases rapidly as the flexibility ratio increases. To account for the seismic effect, the seismically induced shear type (ovalling) thrust, moment and diameter deformation should be combined with the corresponding static response components. However, the seismic effect is induced by shear strain of the ground soil response, and the maximum responses occur at 45° from horizontal. Appropriate methods should be applied to combine the two sets of response vectors as shown in the examples below. The dilatational thrust and diam eter deformation due to soil and pore pressures are then superimposed with the correspond ing shear-type (ovalling) components to obtain the maximum thrust and diameter change. Finally, the maximum thrust and bending moment are applied to the critical locations of the tunnel lining to calculate the maximum combined tensile and compressive stresses in the lining.

APPLICATION OF THEORY

The previously developed theory will now be applied to two cases involving a concrete lining and a steel lining. Both case studies have been carried out for a site located at the lower end of Market Street in San Francisco where a BART (Bay Area Rapid Transit) tunnel passes through soft soil material (Recent Bay Mud). The conditions at this site as repre sented in Fig. 1 are the following:

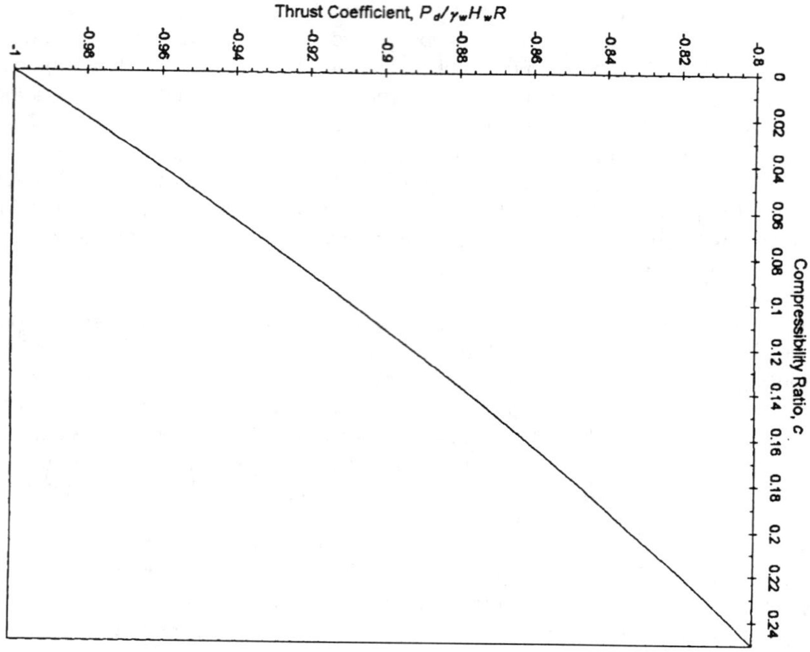

Figure 15, Thrust due to Pore Pressure (Dilatation Mode)

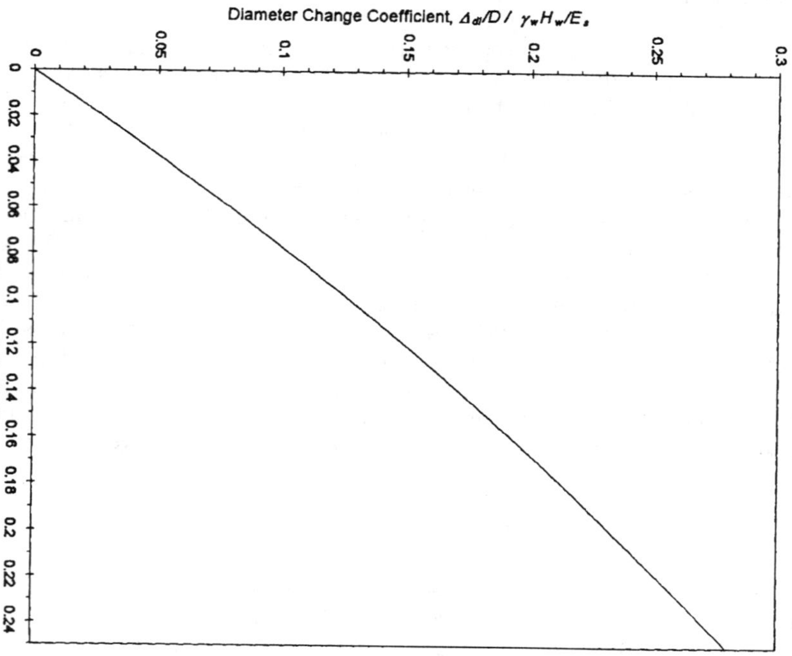

Figure 16, Diameter Change due to Pore Pressure (Dilatation Mode)

Figures 15 and 16

$E_s = 14.36 \times 10^6$ N/m^2 (300 k/ft^2); $K_o = 0.65$; $v_s = 0.394$; (a)

$y_s = 17.28 \times 10^3$ N/m^3 (0.110 k/ft^3); $H_s = 16.31$m (53.5 ft)

$y_w = 9.80 \times 10^3$ N/m^3 (0.0624 k/ft^3); $H_w = 13.26$m (43.5 ft)

The soil properties vary considerably at this site so the above numerical values for E_s, K_o, and v_s represent mean levels.

Case 1 Steel Lining

This case will consider the existing steel lining installed in the BART tunnel at the site described previously. It consists of segmental rings (six segments per ring) having the cross-section shown in Fig. 20. The properties of this lining are

$D = 5.49$m (18 ft); $E_l = 2.04 \times 10^{11}$ N/m^2 (29.6 x 10^6 psi); $v_l = 0.3$ (b)

$I_l = 9.63 \times 10^{-5}$ m^4/m (5.87 in^4/in); $A_l = 0.0314$ m^2/m (1.236 in^2/in)

Using these properties along with those given for the soil previously; obtaining the non-dimensional diameter change, thrust, and bending moment coefficients from Figures 10, 11, 12, 13, 14, 15, and 16, or by introducing their values appropriately in Eqs. (23) and (25), one obtains

$a_d = 249$; $a_s = 0.886$; $c = 0.004$; $f = 1.13$

$M_s(\theta = 0) = -0.0406\ (y_sH_s - y_wH_w)\ R^2 = -46.5 \times 10^3$ Nm/m (-10.4 kft/ft) (c)

$P_d(\theta) = -0.822\ (y_sH_s - y_wH_w)\ R - 0.996\ y_wH_wR = -69.8 \times 10^4$ N/m (-47.8 k/ft)

$P_s(\theta = 0) = -0.0812\ (y_sH_s - y_wH_w)\ R = -3.39 \times 10^4$ N/m (-2.32 k/ft)

$\Delta_{dl} = 0.046\ (y_sH_s - y_wH_w)\ D/E_s + 0.00557\ y_wH_wD/E_s = 0.54$ mm (0.0213 in.)

$\Delta_{sl} = 0.186\ (y_sH_s - y_wH_w)\ D/E_s = 10.8$mm (0.425 in.)

The maximum deformation and forces in the tunnel lining due to static loads, particularly the bending moment are significantly smaller than those obtained by formulations provided by Peck in Ref. 2 (with correction for the pore pressure by hoop stresses on the lining and the condition of no slippage between the lining and soil):

Static Responses	By Stress Relaxation	By Overburden Pressures (Ref. 2)
Max. Bending Moment	46.5 x 10^3 Nm/m (10.4kft/ft)	81.82 x 10^3 Nm/m (18.39 kft/ft)
Max. Thrust	73.2 x 10^4 N/m (50.10 k/ft)	83.10 N/m (56.94 k/ft)
Max. Diameter Change	11.4 mm (0.446 in.)	19.53 mm (0.769 in.)

The reduction of static responses for the tunnel lining design (bending moment reduced by 43%, thrust by 20% and diameter change by 42%) by use of the new formulation is shown on Figures 19, 20 and 21.

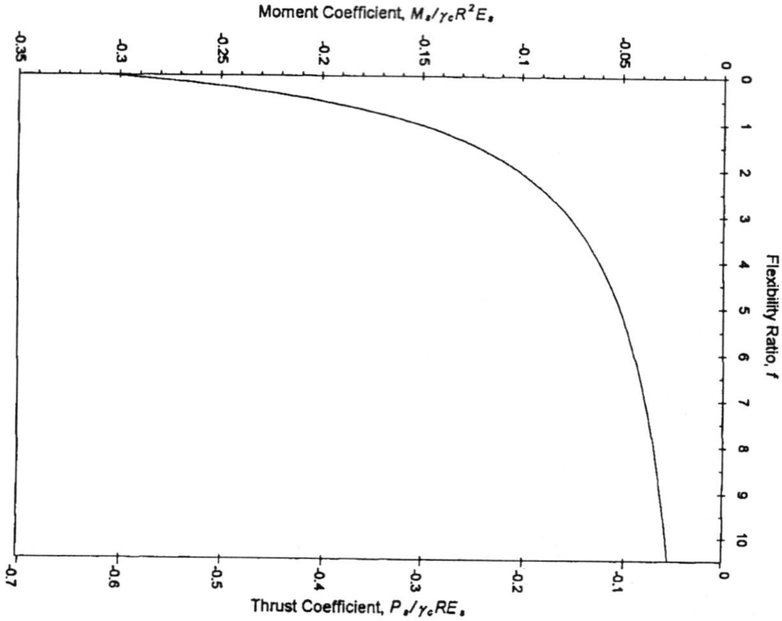

Figure 17, Moment due to Seismic Soil Strain (Ovalling Mode)

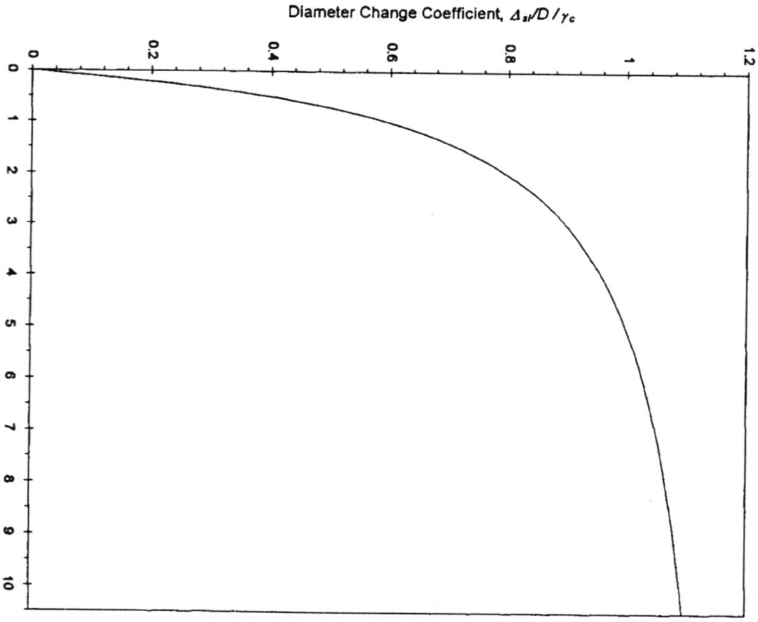

Figure 18, Diameter Change due to Seismic Soil Strain (Ovalling Mode)

Figures 17 and 18

At the above BART tunnel site, it has been estimated that the average free-field soil shear strain y_c could reach the level 0.0028 under maximum credible earthquake conditions (Ref. 5). Making use of this value, Figures 17, 18 and 19 or Eqs. (41) and (44) yield

$$M_s(\theta = -\frac{\pi}{4}) = -0.142\ y_c R^2 E_s = -42.9 \times 10^3 \text{ Nm/m } (-9.64 \text{ kft/ft})$$

$$P_s(\theta = -\frac{\pi}{4}) = -0.283\ y_c R E_s = -3.13 \times 10^4 \text{ N/m } (-2.13 \text{ k/ft}) \qquad \text{(d)}$$

$$\Delta_{sl} = 0.649\ y_c D = 9.98 \text{ mm } (0.392 \text{ in.})$$

The maximum combined force components in the ovalling mode, when combining Eq. (18) with the corresponding Eq. (19) in accordance with Eq. (20) are the following:

$$M_s(\theta + \lambda) = -[(46.5)^2 + (42.9)^2]^{1/2}\, 10^3 = -63.3 \times 10^3 \text{ Nm/m } (-14.2 \text{ kft/ft})$$

$$P_s(\theta + \lambda) = -[(3.39)^2 + (3.13)^2]^{1/2}\, 10^4 = -4.61 \times 10^4 \text{ N/m } (-3.16 \text{ k/ft}) \qquad \text{(e)}$$

$$\Delta_{sl}(\theta + \lambda) = [(10.8)^2 + (9.98)^2]^{1/2} = 14.7 \text{ mm } (0.578 \text{ in.})$$

The corresponding value of λ is $(1/2)\tan^{-1}(3.13/3.39) = 21.4°$. Adding $P_s(\theta + \lambda)$ to $P_d(\theta)$ gives the total combined axial force at $\theta = -21.4°$ equal to -74.4×10^4 N/m (-51.0 k/ft) where the moment equals -63.3×10^3 Nm/m (-14.2 kft/ft). Combination of the shear and dilatational diameter deformations gives the maximum diameter change of 15.3 mm (0.60 in.). Combining the resulting circumferential normal stresses produced by these components, one obtains the following stresses using the cross-section properties and dimensions shown in Fig. 26.

$$\sigma_o(\theta = -21.4°) = +10.5 \times 10^6 \text{ N/m}^2 \text{ } (+1{,}508 \text{ psi}) \qquad \text{(f)}$$

$$\sigma_i(\theta = -21.4°) = -106.4 \times 10^6 \text{ N/m}^2 \text{ } (-15{,}433 \text{ psi})$$

at the outer and inner fibers of the lining, respectively.

At $\theta = (-90° - 21.4°)$, $P_s(-111.4°) = +4.61 \times 10^4$ N/m (+3.16 k/ft) and $M_s(-111.4°) = +63.3 \times 10^3$ Nm/m (+14.3 k/ft).

Combining the normal stresses produced by $[P_s(-111.4°) + P_d(\theta)]$ and $M_s(-111.4°)$, one obtains

$$\sigma_o(\theta = -111.4°) = -54.9 \times 10^6 \text{ N/m}^2 \text{ } (-7{,}951 \text{ psi}) \qquad \text{(g)}$$

$$\sigma_i(\theta = -111.4°) = +61.0 \times 10^6 \text{ N/m}^2 \text{ } (+8{,}991 \text{ psi})$$

at the outer and inner fibers, respectively. The maximum compressive stress -106.4×10^6 N/m^2 (-15,433 psi) and the maximum tensile stress $+61.0 \times 10^6$ N/m^2 (+8,991 psi) are approximately 43 and 25 percent of the linings yield stress $+249 \times 10^6$ N/m^2 (36,000,000 psi), respectively.

From the above results, it is quite clear that the BART steel lining would perform quite satisfactorily under the above combined static and seismic loading conditions.

Case 2 Concrete Lining

Consider a concrete lining of diameter $D = 5.49$m (18 ft) and of thickness $t = 0.305$m (1 ft) having a compressive strength $f_c' = 34.47 \times 10^6$ N/m² (5,000 psi), a Young's modulus

$E_l = 27.77 \times 10^9$ N/m² (0.58 x 10⁶ k/ft²), and a Poisson's ratio $v_l = 0.17$. Using these values, the cross-section flexural and axial stiffnesses, per unit of longitudinal length, will be

$E_l I_l = 65.5 \times 10^6$ Nm (48.3 x 10³ kft) and $E_l A_l = 8.46 \times 10^9$ N/m (0.58 x 10⁶ k/ft). Using all of the above numerical values, one obtains the non-dimensional diameter change, thrust, and moment coefficients from Figures 10, 11, 12, 13, 14, 15 and 16, or from Eqs. (23 and (25) the following:

$a_d = 308;\ a_s = 2.71;\ c = 0.0032;\ f = 0.369$

$$M_s(\theta = 0) = -0.0639\,(y_s H_s - y_w H_w)\,R^2 = -73.1 \times 10^3 \text{ Nm/m } (-16.4 \text{ kft/ft}) \qquad \text{(h)}$$

$$P_d(\theta) = -0.822\,(y_s H_s - y_w H_w)\,R - 0.997\ y_w H_w R = -69.8 \times 10^4 \text{ N/m } (-47.8 \text{ k/ft})$$

$$P_s(\theta = 0) = -0.128\,(y_s H_s - y_w H_w)\,R = -5.33 \times 10^4 \text{ N/m } (-3.65 \text{ k/ft})$$

$$\Delta_{dl} = 0.00371\,(y_s H_s - y_w H_w) D/E_s + 0.00451\ y_w H_w D/E_s = 0.440 \text{ mm } (0.0173 \text{ in.})$$

$$\Delta_{sl} = 0.0937\,(y_s H_s - y_w H_w)\,D/E_s = 5.44 \text{ mm } (0.214 \text{ in.})$$

Again, the maximum deformation and forces in the tunnel lining due to static loads, particularly the bending moment are significantly smaller than those obtained by the formulation of soil-lining interaction by overburden pressure provided by Peck in Ref. 2 (with correction for the pore pressure by hoop stresses on the lining and the condition of no slippage between the lining and soil):

Static Responses	By Stress Relaxation	By Overburden Pressure (Ref. 2)
Max. Bending Moment	73.1 x 10³ Nm/m (161.4 kft/ft)	131.2 Nm/m (29.5 kft/ft)
Max. Thrust	75.1 x 10⁴ N/m (51.5 k/ft)	86.7 N/m (59.4 k/ft)
Max. Diameter Change	5.88 mm (0/.231 in.)	10.2 mm (0.401 in.)

The reduction of static responses for the tunnel lining design by use of the new formulation is shown on Figures 19, 20 and 21.

Making use of the value $y_c = 0.0028$ given previously; obtaining the non-dimensional coefficient from Figures 17, 18, and 19, or by use of Eq. (26), the seismically produced values are

$$M_s(\theta = -\frac{\pi}{4}) = -0.223\ y_c R^2 E_s = -67.5 \times 10^3 \text{ Nm/m } (-15.2 \text{ kft/ft})$$

$$P_s(\theta = -\frac{\pi}{4}) = -0.446\ y_c R E_s = -4.92 \times 10^4 \text{ N/m } (-3.37 \text{ k/ft}); \qquad \text{(i)}$$

$$\Delta_{sl} = 0.327\ y_c/D = 5.03 \text{ mm } (0.198 \text{ ft})$$

Notice that both the static and seismically produced maximum axial forces in the shear-type (ovalling) mode, i.e., $P_s(\theta = 0)$ and $P_s(\theta = -\pi/4)$, are both small (approximately 10 percent) when compared with the static axial force $P_d(\theta)$ produced in the dilatational mode, of which about 1/2 is produced by the dilatational interactive soil pressure with the other 1/2 produced by the pore pressure.

To find the maximum combined force components in the ovalling mode due to the static and seismic contributions, i.e., combining Eq. (18) with the corresponding Eq. (19), make use of Eq. (20) to obtain

$$M_s(\theta + \lambda) = -[(73.1)^2 + (67.5)^2]^{1/2}\, 10^3 = -99.5 \times 10^3 \text{ Nm/m } (-22.3 \text{ kft/ft})$$

$$P_s(\theta + \lambda) = -[(5.33)^2 + (4.92)^2]^{1/2}\, 10^4 = -7.25 \times 10^4 \text{ N/m } (-4.97 \text{ k/ft}) \qquad \text{(j)}$$

$$\Delta_{sl}(\theta + \lambda) = [(5.44)^2 + (5.03)^2]^{1/2} = 7.41 \text{ mm } (0.292 \text{ in.})$$

in which $\lambda = (1/2)\tan^{-1}(4.92/5.33) = 21.4°$ in accordance with Eq. (46). Adding $P_s(\theta + \lambda)$ to $P_d(\theta)$ gives the total combined axial force at $\lambda = -21.4°$ equal to -77.1×10^4 N/m (-52.8 k/ft) where the moment equals -99.5×10^3 Nm/m (-22.3 kft/ft). Combination of the shear and dilatational diameter deformations gives the maximum diameter change of 7.85 mm (0.31 in.). Combining the resulting circumferential normal stresses produced by these components, one obtains,

$$\sigma_o(\theta = -21.4°) = +3.90 \times 10^6 \text{ N/m}^2 = +565 \text{ psi} \qquad \text{(k)}$$

$$\sigma_i(\theta = -21.4°) = -8.96 \times 10^6 \text{ N/m}^2 = -1{,}298 \text{ psi}$$

at the outer and inner fibers, respectively. At $\theta = (-90° - 21.4°)$, $P_s(-111.4°) = +7.25 \times 10^4$ N/m (+4.97 k/ft) and $M_s(-111.4°) = +99.5 \times 10^3$ Nm/m (+22.3 kft/ft).

Combining the normal stresses produced by $[P_s(-111.4°) + P_d(\theta)]$ and $M_s(-111.4°)$ gives

$$\sigma_o(\theta = -111.4°) = -8.48 \times 10^6 \text{ N/m}^2 \ (-1{,}229 \text{ psi}) \qquad \text{(l)}$$

$$\sigma_i(\theta = -111.4°) = +4.37 \times 10^6 \text{ N/m}^2 \ (+634 \text{ psi})$$

at the outer an inner fibers, respectively. The maximum compressive stress of 8.96×10^6 N/m² (1,298 psi) at $\theta = -21.4°$ is only about 26 percent of f_c'; however, the maximum tensile stress of 4.37×10^6 N/m² (634 psi) at $\theta = -111.4°$ is about 30 percent above the tensile strength of the concrete; thus, cracking of the concrete is indicated. This situation calls for either thickening the lining to reduce the tensile stresses or providing reinforcement.

Consider the first option of increasing the thickness of the lining. Carrying out the same evaluations described above for discrete values of thickness t, the stresses in the inner fibers at the critical locations $\theta = (-\lambda - 90°)$ and $\theta = -\lambda$ can be obtained as functions of lining thickness. It is determined that a thickness of 0.0366 m (1.2 ft) is required to reduce the maximum tensile stress to 3.45×10^6 N/m² (500 psi) corresponding to the tensile strength of the concrete.

Case 3 Other Site Locations

For illustration purposes, consider a different site location at Los Angeles, California where the soil condition is largely a sand formation and the ground water table is low. Assuming the same diameter BART tunnels are to be constructed at the same depth and subjected to the same seismic ground response. The representative soil parameters are:

$E_s = 100.1 \times 10^6$ N/m² (2090 k/ft²); $K_o = 0.5$; $v_s = 0.4$;

$y_s = 18.9 \times 10^3$ N/m³ (0.120 k/ft³); $H_w = 0$; $y_c = 0.0028$

For the steel BART lining, the maximum compressive stress (-122.6×10^6 N/m² or -17,792 psi) and the maximum tensile stress (83.4×10^6 N/m² or 12,107 psi) are approximately 49 and 36 percent of the lining's yield stress ($+249 \times 10^6$ N/m or 36,000,000 psi), respectively. However, the concrete lining of 0.3048 m (1 ft) thickness would be grossly inadequate. It would require a thickness of 0.533 m (1.75 ft) to reduce the maximum tensile stress to the tensile strength of the concrete at 3.45×10^6 N/m² (500 psi).

Also for illustration purposes, examine another site location in Washington, D.C., where the soil condition is overly consolidated clay and the ground water table is slightly below surface. For this location assume the same diameter BART tunnels are to be constructed at the same depth and subjected to one-third of the seismic ground response as BART. The representative soil parameters are:

$E_s = 103.4 \times 10^6$ N/m² (2160 k/ft²); $K_o = 1.33$; $v_s = 0.4$;

$y_s = 18.9 \times 10^3$ N/m³ (0.120 k/ft³); $H_w = 3.048$m (10 ft); $y_c = 0.0010$

For the steel BART lining, the maximum compressive stress and the maximum tensile stress are well within the lining's yield stress. The concrete lining of 0.254 m (10 in.) thickness would also be adequate to maintain the maximum tensile stress within the tensile strength of the concrete.

DISCUSSION

The soil-lining interaction by relaxation of insitu soil stresses near the lining is developed in this paper based on a direct and realistic kinematic relationship between the tunnel lining and the surrounding soil. The resulting solutions lead to significantly reduced forces, stresses and deformations of the lining. Therefore, this new formulation reduces unnecessary conservatism and produce a more effective design for the tunnel lining than those procedures now commonly used in the industry.

With the increasing use of computers, numerical analysis for tunnel lining design is also becoming popular. However, tunnel lining design based on elasticity formulation remains as the fundamental approach and the starting point of the tunnel lining design. It is hoped that the solutions and graphs presented herewith will aid engineers in the design of tunnels as Ref. 2 has achieved during the last 25 years.

It should be emphasized that the analytical procedure developed herein for determining lining stresses following installation is based on the assumption of perfect construction methods, i.e., the lining is placed in position in a circular stress-free state and grouting outside the lining completely fills all voids present. Since a perfect installation of this type cannot be achieved, the stress-state of linings as actually installed and the deficiencies which can occur in grouting should be taken into consideration in the design process. As pointed out by Kuesel (Ref. 6), the steel lining rings installed in the BART tunnels were subjected to considerable ovalling due to elastic flexure and working of the joints under their own dead weight and additional distortions were introduced as the lining rings were bolted together.

Steel linings of the type used by BART are very ductile; thus, if large ovalling deformations should occur for whatever reasons, they will simply conform to the tunnel cavity while carrying the outside pressure distribution primarily in the dilatational mode, which has a large margin of safety. The stiffer concrete linings will, however, carry a larger share of the outside pressure in the ovalling mode producing relatively larger internal bending moments. From the previous case study using a concrete lining, it was shown that this type of lining can easily carry the imposed pressure distribution when designed with adequate thickness. Some reinforcement may be required, however, when the thickness is held to some limited, but practical, level.

REFERENCES

Burns, J. Q. and Richard, R. M., “Attenuation of Stresses for Buried Cylinders”, Proceedings of the Symposium on Soil-Structure Interaction, University of Arizona, Engineering Research Laboratory, Tucson, Arizona, September 1964.

Peck, R. B., Hendron, A. J. Jr. and Mohraz, B., “Sate of the Art of Soft-Ground Tunneling”, RETC Proceedings, Vol. 1, 1972.

Penzien, J. and Wu, C.L. “Stress in Lining of Bored Tunnels” under review for possible publication in International Journal of Earthquake Engineering and Structural Dynamics.

Timoshenko, S. and Goodier, J.N., “Theory of Elasticity”, Second Edition, McGraw Hill Book Company, Inc., 1951.

Wu, C.L. and Penzien, J., “Seismic Design of MUNI Metro Turnback Project”, Fifth U.S. National Conference on Earthquake Engineering, Vol. IV, 1994.

Kuesel, T. R., “Soft Ground Tunnels for the BART Project”, RETC Proceedings, Vol. 1, 1972.

Chapter 27

ANTICIPATED GROUND CONDITIONS FOR THE RIVERSIDE BADLANDS TUNNEL

R.R. Redd,[1] J.T. Waggoner,[2] J.S. Hill,[2]
S.J. Klein,[2] and D.B. Desai[3]
[1]Metropolitan Water District of Southern California; [2]Woodward-Clyde Consultants; and [3]Daniel, Mann, Johnson, and Mendenhall

ABSTRACT

The 13 km (8 mile) long Riverside Badlands Tunnel is planned for construction starting in 1998. Extensive geotechnical investigations indicate that the 3.7 m (12 ft) finished diameter tunnel will be excavated through soft ground, weak sedimentary rocks, and heavily fractured metamorphic rocks. Nearly 80 percent of the tunnel will be excavated beneath the groundwater level. Tunneling conditions are expected to range from raveling and overstressed in the sedimentary rocks, to blocky and seamy/crushed with some squeezing in the metamorphic rocks. Two alluvium reaches will require dewatering. The variable and interspersed ground conditions are expected to be a key factor in selection of tunneling methods, equipment, and initial support.

INTRODUCTION

The Metropolitan Water District of Southern California (Metropolitan) is constructing the Inland Feeder Project, which will convey raw water from the State of California's facilities in Devil Canyon, near San Bernardino, California, to Metropolitan's new Eastside Reservoir Project, near Hemet, California. The Riverside Badlands Tunnel is one segment of the Inland Feeder Project, and consists of an approximately 12,800 m (42,000 ft) long, 3.7 m (12 ft) finished diameter pressure tunnel. This tunnel is to be constructed through the Crafton Hills and the San Timoteo Badlands, in San Bernardino and Riverside Counties, as shown on Figure 1. The purpose of this paper is to describe the variable and challenging ground conditions that will be encountered in the Riverside Badlands Tunnel. It is noted that the Contract Documents for this project have not been finalized, and there could be some differences between the conditions and interpretations described in this paper, and the final Contract Documents. Accordingly, the conditions described in this paper are intended to be a brief overview and should not be relied upon for bidding purposes or construction, and the information presented in the final contract documents will supersede the information presented herein.

PROJECT DESCRIPTION

As shown on Figure 1, the Riverside Badlands Tunnel trends approximately north-south from a location east of the City of Redlands, in southern San Bernardino County, to just south of State Route 60 at the northeastern edge of the San Jacinto Valley, in northern

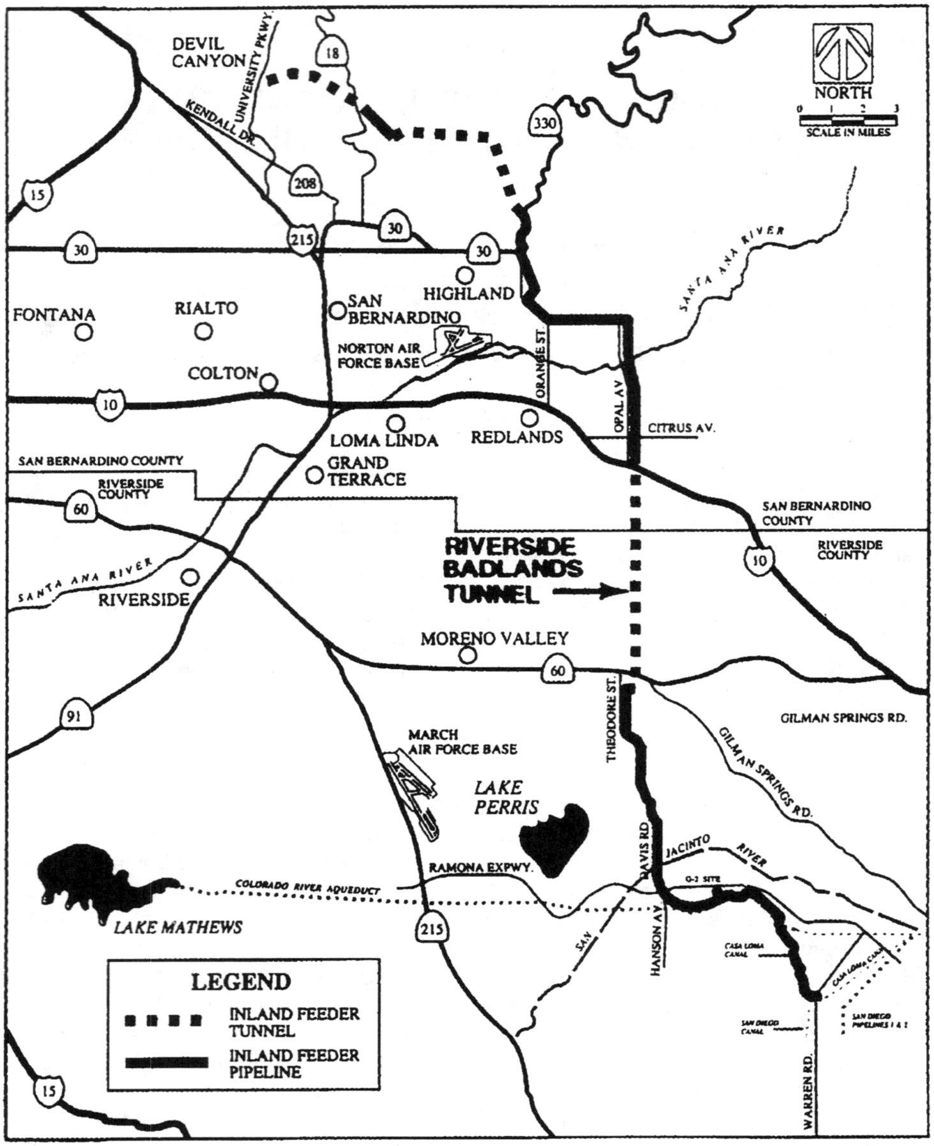

Figure 1. Site Location Map

Riverside County. Much of the tunnel extends beneath an area of dissected hills known as "The Badlands," from which the tunnel's name was derived. Topographic conditions over the tunnel are illustrated on Figure 2.

The construction contract for the tunnel will also include 200 to 250 m (700 to 800 ft) of 3.7 m (12 ft) diameter open cut welded steel pipeline at the north and south ends of the tunnel. A permanent access structure will also be constructed at each end of the tunnel.

When completed, the tunnel will convey a water flow up to about 0.07 cubic meters per second (1,000 cubic feet per second [cfs]). The static hydraulic grade line for the tunnel is at Elevation 1937 feet, which results in maximum static internal hydrostatic pressures ranging from about 640 Kpa (90 psi) at its north end to slightly less than 780 Kpa (110 psi) at the south end. Transient pressures during operation of the completed tunnel could increase the internal hydrostatic pressure up to about 890 KPa (125 psi).

Tunnel construction is to be staged from the north and south portals, with intermediate access available at the two optional shaft sites. One of these shaft sites is located on the north side of Live Oak Canyon Road, about 5 km (3 mi.) south of the Opal Avenue portal. The other optional shaft site is located on the south side of San Timoteo Canyon Road, about 5 km (3 mi.) north of the Gilman Springs Road portal.

The tunnel alignment consists of five tangent sections connected by three horizontal curves and two angle points. The angle points are at optional shaft sites, in Live Oak and San Timoteo Canyons, and can be replaced with 330 m (1000 ft) radius horizontal curves, if the tunnel construction contractor elects not to use the optional shaft sites for staging tunnel construction.

The tunnel profile has a slight southward downgrade. From the north portal (at Opal Avenue) to Live Oak Canyon, the downgrade is about 0.2 percent. From Live Oak Canyon to the south portal (at Gilman Springs Road), the downgrade is about 0.03 percent.

ANTICIPATED GEOLOGIC CONDITIONS

Field investigations performed to characterized ground and groundwater conditions along the tunnel alignment included: about 6,500 m (21,400 ft) of seismic refraction surveys; resistivity surveys at 10 locations; about 600 m (2,000 ft) of side-hill excavations; 200 m (700 ft) of trenches; about 3,400 m (11,000 ft) of borings; and several types of insitu testing in the borings. Laboratory tests were performed to evaluate geotechnical properties of the various ground types, and TBM boreability. Aquifer pumping tests were performed to develop parameters required for Contractor's development of a dewatering plan for alluvial tunnel reaches, and chemical analyses of groundwater samples were performed for evaluating water treatment and corrosion protection requirements.

The results of geological/geotechnical investigations indicate that the tunnel will encounter three principal types of ground: soft ground (alluvium); soft or weak sedimentary rock (the Upper, Middle, and Lower Members of the San Timoteo Formation); and stronger, but variably weathered, sheared, and fractured hard rock (metamorphic rocks). In addition, the alignment crosses several fault zones. With the exception of limited reaches near the two portals, the tunnel is below the groundwater table. The maximum groundwater head above the tunnel invert is about 100 m (330 ft). The expected distribution of ground conditions along the tunnel alignment is shown on Figure 2 (a geologic profile of the tunnel). Based on the available data, approximately 10 percent of the tunnel is expected to encounter alluvium, 66 percent the San Timoteo Formation, and 24 percent metamorphic rocks.

Mixed face tunneling conditions are expected at several locations where the tunnel

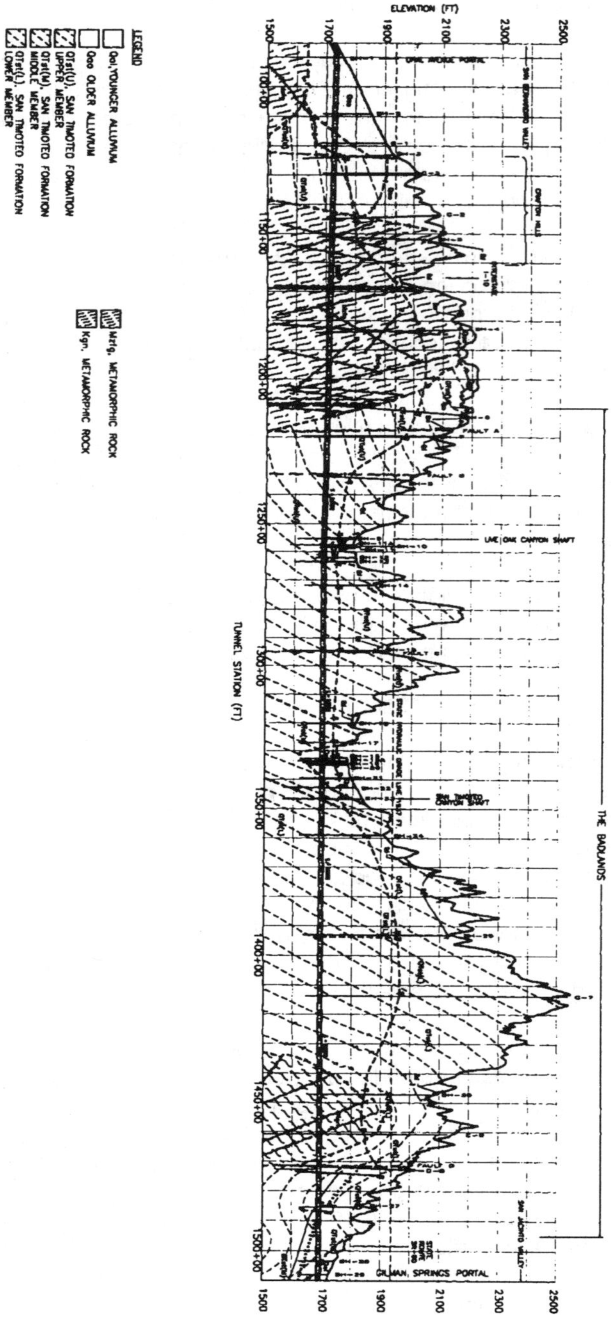

Figure 2. Tunnel Geologic Profile

excavation will transition from one type of ground to the next. In some cases, the visual appearance and character of the ground in adjacent reaches will be similar, particularly at the contact between the older alluvium and the youngest San Timoteo Formation, and thus the change from one reach to another may be difficult to determine.

For the purposes of describing the anticipated ground conditions, the tunnel alignment was divided into ten reaches, numbered from north to south, as shown on Figure 2. These reaches were established based on the principal type of ground (i.e. alluvium, San Timoteo Formation, or metamorphic rock) anticipated in each reach of the tunnel. The anticipated ground conditions for each of the three principal ground types are described below generally using Terzaghi's rock and soil classification terms (Proctor and White, 1946; Terzaghi, 1950) and as adapted by Deere et al. (1969).

Reaches in Alluvium

Alluvium is expected in Reaches 1, 5, and 7, as shown on Figure 2. The lengths of these tunnel reaches are: 760 m, 150 m, and 330 m (2,500 ft, 500 ft, and 1,100 ft), for Reaches 1, 5, and 7, respectively. The alluvium is considered soft ground for tunneling purposes. It generally consists of medium dense to very dense silty and clayey sands; medium dense to dense, poorly graded and well graded sands; and stiff to very stiff and, in some instances hard, sandy clays and silts. Some layers in the alluvium are gravelly, and boulders up to about 1 m (3 ft) in maximum dimension are anticipated to be present in Reaches 5 and 7.

In the alluvium, ground conditions above the groundwater level are expected to be running, fast raveling, slow raveling, or firm depending upon the soil type, grain size distribution, and the fines content. Below the groundwater table, flowing and cohesive running ground conditions are expected in the sand and silt deposits, and slow raveling ground conditions are expected in the clay deposits. Terzaghi indicates that flowing conditions should be expected in ground below the groundwater table with an effective grain size (D_{10}) greater than about 0.005 mm (Terzaghi, 1950). This criterion appears to apply to all of the sandy alluvial deposits for which grain size data was obtained. A refinement proposed by Deere et al. (1969) indicates that dense silty sand with a Uniformity Coefficient greater than about 6 can be considered cohesive running ground. Cohesive running ground generally behaves similarly to running or flowing ground, except that it has enough cohesion to allow for a brief period of raveling before it breaks down and runs (above the groundwater table) or flows (below the groundwater table) into the tunnel.

Reaches in the San Timoteo Formation

Five of the tunnel Reaches, totaling about 8,300 m (27,400 ft) of the tunnel, are expected to encounter the San Timoteo Formation (Figure 2). These Reaches and their lengths are:

Reach 2:	800 m	(2,700 ft)
Reach 4:	1,400 m	(4,500 ft)
Reach 6:	2,000 m	(6,600 ft)
Reach 8:	3,000 m	(9,800 ft)
Reach 10:	1,150 m	(3,800 ft)

The San Timoteo Formation consists of sandstone, conglomeratic sandstone, conglomerate, and siltstone/claystone. These "rocks" are typically uncemented to weakly cemented, weakly indurated, and friable, although some portions are moderately to strongly cemented. The general characteristics of this formation are described below.

Lithology

Based on geologic mapping by Shuler (1953) and the distribution of the constituent lithologies (i.e. sandstone, conglomerate, and siltstone/claystone), the San Timoteo Formation was subdivided into an Upper, Middle, and Lower Member. As shown on Figure 2, the Upper Member will be encountered in three Reaches (Reaches 2, 4, and 6) of the northern half of the tunnel. The Middle Member will be encountered in the vicinity of San Timoteo Canyon (in Reaches 6 and 8), and near the south end of the tunnel (in Reach 10). The Lower Member will be encountered along a 3,000 m (9,800 ft) reach south of San Timoteo Canyon (Reach 8).

All three members of the San Timoteo Formation contain a significant amount of sandstone. The Upper Member also contains large percentages of conglomerate and conglomeratic sandstone, but very little siltstone/claystone. The Middle Member tends to be more fine-grained, having a greater percentage of siltstone/claystone than the Upper and Lower Members, and a relatively low percentage of conglomeratic sandstone and conglomerate. Similar to the Upper Member, the Lower Member contains a significant percentage of sandstone, contains a large percentage of conglomeratic sandstone and conglomerate, and a low percentage of siltstone/claystone. The approximate distribution of lithologies within the three Members is summarized in Table 1 below.

Table 1. Summary of Lithologies Within the San Timoteo Formation

Formation Member	Approximate Distribution of Lithologies (in Percent)			
	Siltstone / Claystone	Sandstone	Conglomeratic Sandstone	Conglomerate
Upper	6%	39%	24%	31%
Middle	33%	57%	7%	3%
Lower	4%	60%	27%	9%

The clasts (i.e., pebble to boulder size particles) in the sandstone, conglomeratic sandstone, and conglomerate are typically embedded in a silty sand matrix and, based on the results of grain size analyses, it is estimated that about 95 percent of the clasts will be less than 0.15 m (6 in.) in diameter. The largest clasts observed in outcrop and in the borings were on the order of 0.6 m (2 ft), and boulders as large as 0.75 m (2 1/2 ft) in maximum dimension have been observed at one location as "float" on the ground surface.

Mineralogy

Based on petrographic analyses, quartz contents in the sandstone, siltstone, or claystone, range from about 8 to 20 percent. Feldspar minerals, rock fragments, and matrix materials in these same samples range from about 5 to 25 percent, 5 to 37 percent, and 38 to 63 percent, respectively.

X-ray diffraction tests indicated smectite, an expansive clay mineral, to be the dominant clay mineral present in samples of claystone from the San Timoteo Formation. Atterberg Limit data indicate liquid limit values range from about 26 to 45 percent and plasticity index values ranging from 5 to 24 for samples from borings in Reaches 2, 4, and 6. In Reaches 8 and 10, the claystone exhibits higher plasticity with liquid limit values ranging from 32 to 58 percent and plasticity index values ranging from 11 to 33.

Lithology of Clasts

Clasts from six large (270 to 540 kg [600 to 1200 lbs] each) samples of conglomerate or conglomeratic sandstone from the Upper and Lower members of the formation were classified based on visual observations of the specimens. The results of these observations are summarized in Table 2.

Table 2. Summary of Clast Lithologies

Formation Member	Approximate Distribution of Clasts by General Lithology (in Percent)			
	Granitic (1)	Volcanic (2)	Gneissic	Other (3)
Upper	1 to 35	1 to 10	35 to 95	1 to 20
Lower	65 to 85	5 to 35	5 to 35	1 to 2

(1) = Primarily granodiorite and tonalite; (2) = Primarily andesite and dacite; and (3) = Primarily quartzite and cemented conglomerate

Cementation

Moderately calcareous cemented or strongly calcareous cemented beds, lenses, or concretions within the San Timoteo Formation were encountered in most of the borings. One of the cemented layers was submitted for petrographic analysis. The cement (calcium carbonate) content of that sample was estimated at about 33 percent.

The moderately cemented or strongly cemented layers make up about 5 percent of the total length of core drilled in the San Timoteo Formation. While 92 percent of these moderately cemented and strongly cemented layers are less than 0.3 m (1 ft) thick, layers on the order of 3 to 5 m (10 to 15 ft) thick were encountered in the borings, and one layer about 15 m (50 ft) thick was encountered. Based on the apparent dip of the bedding with respect to the tunnel alignment, about 15 percent of the tunnel length within the San Timoteo Formation is expected to encounter moderately or strongly cemented layers or concretions.

The boring data also indicate that San Timoteo Formation contains uncemented zones of extremely weak, friable rock. The borings completed in Reach 2 encountered the largest amount of uncemented material, with over 30 percent of the core described as uncemented, extremely weak rock. Borings completed within Reaches 4, 6, 8, and 10 indicate less than 10 percent of uncemented, extremely weak rock.

Rock Strength

Due to the variable cementation and induration of this formation, the strength was found to vary widely. Available unconfined compressive strength test data for the San Timoteo

Formation are summarized in Table 3, according to lithology and Formation Member. For the samples strong enough to test, the unconfined compressive strength, based on 128 laboratory tests, was found to range from 0.05 to 14.7 MPa (7 to 2130 psi). Eleven of those tests were completed on samples from layers described as “moderately” or “strongly” cemented. The unconfined compressive strengths range from about 2.5 to 14.7 Mpa (350 to 2130 psi), averaging almost 5.5 Mpa (800 psi). It is noted that some of the core retrieved from the San Timoteo Formation (mostly from the Upper and Middle Members) was sufficiently lacking in cementation and induration that strength tests could not be performed. Such samples were typically conglomeratic sandstone or conglomerate with a low fines content. The variation of rock strength with lithology of this formation is summarized in Table 3.

Table 3. Summary of Unconfined Compressive Strength Results

Rock Type	Formation Member	No. of Tests	UCS [Mpa, (psi)] Minimum	Maximum *	Average UCS [Mpa, (psi)]**
Sandstone	Upper	19	0.05 (7)	1.4 (208)	0.6 (88)
Sandstone	Middle	14	0.07(10)	11(1,600)	1.3(185)
Sandstone	Upper	28	0.2 (24)	14.7 (2,130)	3.7 (544)
Conglomerate	Upper	8	0.14 (20)	0.4 (60)	0.3 (38)
Conglomerate	Middle	0	-	-	-
Comglomerate	Lower	2	0.5 (78)	0.75 (110)	-
Conglomerate Sandstone	Upper	3	0.06 (9)	0.14 (20)	0.1 (16)
Conglomerate Sandstone	Middle	0	-	-	-
Conglomerate Sandstone	Lower	4	0.8 (122)	3.8 (550)	-
Siltstone/Claystone	Middle	7	0.25 (36)	1.3 (183)	0.7 (102)
Siltstone/Claystone	Lower	20	0.06 (9)	3.0 (437)	1.0 (14)
Siltstone/Claystone	Lower	22	0.8 (117)	7.1 (1,030)	3.1 (457)

* Includes results of tests on cemented materials.
** Average values are listed if at least 5 test results are available.
UCS - Unconfined Compressive Strength.

As indicated in Table 3 above, the strength of the San Timoteo Formation generally depends on which member the sample came from, and its specific lithology. Overall, the Upper and Middle Members are weaker than the Lower Member, and generally exhibit unconfined compressive strengths less than about 1.0 Mpa (150 psi) (except for the “moderately” or “strongly” cemented layers). The strength of the Lower Member tends to be higher, averaging about 3.1 Mpa (450 psi) for siltstone/claystone and almost 3.8 Mpa (550 psi) for sandstone. The Lower Member also contains some zones with unconfined compressive strengths lower than 1.0 Mpa (150 psi).

The boundary between rock and soil, based on unconfined compressive strength, is usually considered to be about 1.0 Mpa (150 psi). Therefore, the San Timoteo Formation (especially the Upper and Middle Members) is considered a borderline soil/rock material. Accordingly, this formation is expected, from a tunneling perspective, to exhibit characteristics similar to both soft ground and weak rock.

Clast Strength

The strength of clasts in the San Timoteo Formation was evaluated by performing point load strength tests on conglomerate clasts from bulk surface samples and from core samples. The results of these tests indicate that the strength of the near surface clasts ranges from extremely weak to moderately strong, with point load indices ranging from 0.2 to 307 psi. The point load indices for clasts obtained from the core ranged from 9 to 1763 psi. This difference in strength is presumed to be a result of weathering of the near-surface clasts.

Using a factor of 24 to estimate the approximate unconfined compressive strength from the point load index data, the strength of the clasts at tunnel depth is expected to range from about 1.5 to 300 Mpa (220 to 42,000 psi).

Clast/Matrix Bond Strength

The extent to which the clasts in conglomerate and conglomeratic sandstone are held within the matrix was initially a concern in planning the exploratory drilling program, and may also be of significance in construction of the tunnel. With regard to core recovery from the borings, the main concern was that the clasts might be plucked from the matrix, and could roll around at the bottom of the hole, causing the core to be ground up and resulting in poor core recovery. This problem was not experienced. In fact, core recovery in the San Timoteo Formation was excellent, averaging almost 100 percent, with the clasts being easily cut by the diamond bit without disturbing the bond with the matrix. This high core recovery is attributed both to the bond strength of the matrix and careful drilling techniques that were employed.

Discontinuities

Relatively few joints were observed in the core samples obtained from the San Timoteo Formation. This is likely a result of the generally low strength of the rock that can deform plastically without fracturing in response to tectonic stresses or deformations.

Shears were observed in borings and in outcrops, and several are expected to be encountered in the tunnel within this formation, especially in the vicinity of the identified faults. As observed in the field, shears with narrow apertures generally do not appear to appreciably affect the overall character of the rock mass. More significant shears or shear zones, which will likely impact tunneling, are expected locally.

Estimated Ground Behavior

The behavior of the San Timoteo Formation in the tunnel excavation will depend, for the most part, on the ratio of the in situ overburden stress to the strength of the rock, and the lithology. In the tunnel reaches where the San Timoteo Formation is expected to be encountered, the cover ranges from about 15 to 260m (50 to 860 ft), resulting in an overburden stress ranging from about 0.3 to over 5.5 Mpa (50 to over 800 psi).

Deere et al. (1969) indicate that overstressed or yielding rock conditions will result in the development of a plastic zone around the tunnel excavation when the overload factor (i.e., the overburden stress divided by + of the unconfined compressive strength of the rock) is greater than 1. Ground movements in the tunnel are usually small, and controllable providing that this factor is less than about 4. Peck (1969) notes that yielding conditions that would significantly affect face stability and tunneling, do not usually develop unless the overload factor exceeds about 5 to 6. Table 4 summarizes estimated overload factors for the San Timoteo Formation Reaches.

Table 4. Estimated Overload Factors in San Timoteo Formation Reaches

Reach	Estimated Range of Overload Factor
4	0.8 to >10
6	0.6 to >10
8	0.6 to 8
10	2.4 to >10

The behavior of the San Timoteo Formation under yielding or overstressed conditions will vary depending on the lithology. According to Deere et al. (1969), frictional materials exhibit different behavior than frictionless materials, when subjected to overstressed conditions. For the San Timoteo Formation tunnel reaches, it is anticipated that the sandstone, conglomeratic sandstone, and conglomerate will tend to behave like frictional materials and the claystone will tend to behave like a frictionless material. The siltstone is anticipated to exhibit characteristics of both frictional and frictionless materials depending on amount of fines present, the plasticity characteristics of the fines, and the degree of cementation.

The behavior of frictional materials in the San Timoteo Formation (i.e., sandstone, conglomeratic sandstone, conglomerate, and siltstone) in an overstressed condition will depend on the amount of cohesion present and groundwater conditions (Deere et al., 1969). The amount of cohesion will depend on the degree of cementation, the amount of clay binder, or both. Above the groundwater level, the frictional materials have the potential to run if uncemented and cohesionless or, if cohesion is present, they have the potential to develop chunks, slabs, or spalls due to stress-induced fracturing that will lead to raveling conditions. Below the groundwater table, seepage forces from water flowing towards the tunnel excavation may overcome slight amounts of cementation or cohesion and cause the materials to flow into the tunnel opening (i.e. flowing ground). Where significant cementation or cohesion is present, frictional materials below the groundwater table will have the potential to exhibit firm to raveling behavior depending on the amount of cementation or cohesion present, although, it is more likely that these materials, when overstressed, will tend to fracture leading to slabbing (Deere et al., 1974). The stand-up time for such ground is usually very short, so that immediate full-perimeter ground support is often required.

The frictionless materials (i.e. claystone and siltstone) of the San Timoteo Formation will likely squeeze in an overstressed condition, when the overload factor exceeds 1. As encountered in the borings, siltstone/claystone layers occur throughout the San Timoteo Formation in beds that range from less than 0.3 m (1 ft) to about 6 m (20 ft) thick.

Another potential stability concern in frictionless materials is swelling potential. Due to the smectite present in the claystone and the Atterberg Limit test data there is a potential for the claystone to be swelling ground. Terzaghi and Peck (1967) indicate that materials with a plasticity index in the range of 10 to 35 have a medium swelling capacity. Based on the Atterberg Limit test data, it is anticipated that claystones encountered north of San Timoteo Canyon (i.e. Reaches 2, 4, and 6) will have a low swelling potential, whereas the claystones encountered south of San Timoteo Canyon (i.e. Reaches 8 and 10) will have a moderate swelling potential. In addition, the claystone/ siltstone are expected to exhibit slaking behavior when exposed to air or water for even short periods of time. The rock in all San Timoteo Formation tunnel reaches will soften and lose strength in the presence of standing water.

Reaches in Metamorphic Rock

Metamorphic rock will be encountered along Reach 3 and Reach 9, as shown in Figure 2 for distances of 2,150 m and 950 m (7,100 ft and 3,100 ft), respectively. In general, these metamorphic rocks are gneissic in composition. In addition, the northern reach (Reach 3) contains a variety of granitic and volcanic dikes.

Rock Strength

Two types of strength tests were performed: laboratory unconfined compressive strength tests; and point load index tests. The point load tests were completed to provide a larger number of tests, for the purposes of better defining the range in strength among the metamorphic rocks. It is noted that because of the highly fractured nature of the metamorphic rock, it was difficult to find suitable core samples for unconfined compression tests and many of the tests that could be performed, failed along foliation or pre-existing weakness planes. For this reason, the point load index strength data is considered to be more representative of the actual range of strength of the intact rock, than the strength range indicated by the unconfined compressive strength data. The laboratory unconfined strength tests were used to help "calibrate" the results of the point load tests. The results of the point load tests are summarized in Table 5 below, for both the north and south metamorphic rock reaches. Table 6 provides a summary of the laboratory unconfined compressive strength data.

Table 5. Summary of Point Load Strength for Metamorphic Rock

Rock Type	Metamorphic Rock Reach	Type of Failure (2)	Estimated UCS from Point Load Tests [Mpa, (psi)] (1)			
			No. of Tests	Minimum	Maximum	Average
Gneissic Rocks	Northern Reach (Reach 3)	F I	45 25	1.5 (200) 1.5 (200)	220 (32,100) 370 (53,400)	50 (7,300) 85 (12,400)
	Southern Reach (Reach 9)	F I	18 13	12 (1,700) 5 (700)	180 (26,000) 200 (29,200)	80 (11,500) 100 (14,300)
Dike Rocks	Northern Reach (Reach 3)	F I	4 7	10.5 (1,500) 1.5 (200)	80 (11,600) 140 (20,500)	N/A 45 (6,700)
	Southern Reach (Reach 9)	N/A	N/A	N/A	N/A	N/A

(1) Based on the assumption that unconfined compressive strength (UCS) is approximately 24 times the point load index in psi.(2) Failure type F was along foliation; failure type I was through intact rock.

Table 6. Summary of Laboratory UCS Strength for Metamorphic Rock

Rock Type	Metamorphic Rock Reach	Type of Failure (2)	Unconfined Compressive Strength UCS [Mpa, (psi)]			
			No. of Tests	Minimum	Maximum	Average
Gneissic Rocks	Northern Reach (Reach 3)	F	11	5.4 (780)	62.1 (9,000)	25.1 (3,640)
		I	11	29.5 (4,270)	164 (23,780)	73.3 (10,630)
	Southern Reach (Reach 9)	F	13	3.2 (470)	74.3 (10,780)	40.9 (5,930)
		I	3	4.8 (700)	50.5 (7,320)	-
Dike Rocks	Northern Reach (Reach 3)	F	3	18.5 (2,680)	113.2 (16,420)	-
		I	3	64.8 (9,400)	139.1 (20,170)	-
	Southern Reach (Reach 9)	N/A	N/A	N/A	N/A	-

(1) Failure type F was along foliation; failure type I was through intact rock.

With the exception of the upper weathered zones near the contacts with the San Timoteo Formation, and in fault/shear zones, the metamorphic rocks are generally moderately strong to strong (24 to 103 Mpa [3,500 to 15,000 psi]) based on the results of the unconfined compression tests; whereas the metamorphic rock appears to be strong to very strong (52 to 259 MPa [7,500 to 37,500 psi]) based on the results of the point load index data.

The average intact unconfined compressive strength of the metamorphic rock in Reaches 3 and 9 is estimated to be about 103 Mpa (15,000 psi), and the maximum rock strength is not expected to exceed 345 MPa (50,000 psi) in each reach. It should be recognized, however, that tunnel stability and ground support requirements are likely to be controlled more by the discontinuities in the rock mass not the average unconfined compressive strength of the rock which may be more important for evaluating the performance of tunneling excavation equipment.

Discontinuities

The behavior of the metamorphic rock mass will be greatly influenced by the discontinuities in the rock mass, i.e., fracture frequency, orientation, and physical characteristics of the discontinuities. According to fracture frequency data, the northern metamorphic rock mass is highly to intensely fractured with approximately 35 percent to 63 percent of the rock cored in the borings having greater than 10 fractures per 0.3 m (1 ft). The southern metamorphic rock mass is less intensely fractured, with approximately 5 to 10 percent of the rock cored in the borings having greater than 10 fractures per 0.3 m (1 ft).

In addition to fracture frequency, Rock Quality Designation (RQD) is often used to describe the quality of rock masses. Based on the RQD values from the boring logs, an assessment of RQD for all the rock core obtained in the northern and southern metamorphic rock reaches was made. The results are summarized in Table 7.

Table 7. Summary of "Average" RQD

Metamorphic Rock Reach	**Total Length Cored in Metamorphic Rock [m,(ft)]**	**Total Core Piece Length > 10 cm (4 in) [m,(ft)]**	**"Average" RQD (1)**
Northern (Reach 3)	398 (1307)	56 (183)	14%
Southern (Reach 9)	187 (612)	86 (281)	46%

1) For summary purposes, an "average" RQD for an entire boring and for a group of borings is taken to be the total

piece length of rock core that exceeded 10 cm (4 inches) in length, divided by the total length of rock core drilled.

Estimated Ground Conditions

It is anticipated that ground conditions in the two metamorphic rock tunnel reaches will be controlled by the degree of fracturing of the rock mass. Reach 9 is generally less fractured than Reach 3 (see Table 7), and ground conditions will be more favorable for this reach. Estimated ground conditions, in terms of Terzaghi's rock conditions (Proctor and White, 1946), for these two reaches are summarized in Table 8.

Table 8. Estimated Ground Conditions for Metamorphic Rock Tunnel Reaches

Tunnel Reach	Ground Condition (% of Reach)			
	Massive, Moderately Jointed	Moderately Blocky and Seamy	Very Blocky and Seamy	Crushed Rock
3	5	5	45	45
9	20	20	50	10

In Reach 3, clay gouge in fault and shear zones could be squeezing or swelling ground. It is anticipated that at least 25 percent of the crushed rock in this reach (10 percent of the total reach) will encounter squeezing/swelling ground and that the majority of this ground will be located in the vicinity of the Banning Fault. Considering the results of laboratory tests, squeezing ground behavior is considered more likely than swelling ground behavior.

POTENTIAL GEOLOGIC HAZARDS

Four potential geologic hazards were evaluated as part of final design for the tunnel. They are: seismic shaking, fault displacements, hazardous gas, and warm ground and groundwater temperatures. These potential hazards are discussed below.

Seismic Shaking

The proposed tunnel is situated in a relatively highly seismic area, between two major active faults: the San Andreas; and San Jacinto faults. A few other shorter and less active faults are in the vicinity of the tunnel. The potential levels of ground shaking from all these faults needs to be considered and accounted for in the design of temporary and permanent structures. Based on upper bound earthquake magnitudes selected for the San Andreas and San Jacinto faults, peak horizontal ground acceleration at the ground surface is estimated to range from 0.42g to 0.72g, and peak ground velocity (at the ground surface) is estimated to range from 0.5 to 1.4 m/s (1.8 to 4.6 ft/s). At tunnel depth, these values are estimated to range from 0.27g to 0.50g, and from 0.5 to 1.1 m/s (1.6 to 3.7 ft/s), respectively.

Fault Displacements

Based on the investigations completed, and an evaluation of available published information concerning the faults identified, several faults are considered to have a potential for future fault displacements as a result of an earthquake. Faults that cross the tunnel alignment and that should be considered to be active or potentially active include: the Reservoir

Canyon Fault; the Banning Fault; Faults "A", "B", and "D" (previously unnamed faults); the Chicken Hill Fault; and subsidiary splays of the San Jacinto fault. Estimates of potential fault displacements for these faults range from 0.15 m to 1m (+ ft to 3+ ft). The likelihood of fault displacements of these magnitudes occurring during tunnel construction is considered low. However, it should be recognized that this potential exists.

Hazardous Gas

The potential for encountering hazardous gas was investigated by monitoring the borings for explosive gas and hydrogen sulfide. This monitoring was performed during drilling, and at the same time some of the water level measurements were collected in piezometers. No explosive gases or hydrogen sulfide were detected during any of the monitoring that was performed. Nonetheless, because the San Timoteo Formation is of sedimentary origin, and because oil/gas is reported in the general region (Shuler, 1953), the possibility of encountering flammable gas or oil cannot be completely discounted. CAL/OSHA has assigned a classification of "potentially gassy with special conditions" to the tunnel. As a result, the construction contractors will need to plan to use equipment and construction methods that are consistent with this classification, and provide adequate ventilation capacity to reduce the possibility of hazardous gas concentrations being encountered in the tunnel.

Ground and Groundwater Temperatures

While there are no known hot water springs near the tunnel alignment, hot water is present along the San Jacinto fault near Gilman Hot Springs, located about 16 km (10 mi) southeast of the Gilman Springs Road portal. As a result, the potential for encountering warm or hot ground and groundwater during tunnel construction was investigated by collecting temperature measurements of groundwater in piezometers installed in the borings. With the exception of a tunnel reach from about Station 1380+00 to 1460+00, ground and groundwater temperatures are expected to range from about 18 to 24 degrees C (65 to 75 degrees F). Ground and groundwater temperatures from about Station 1380+00 to 1460+00 are expected to range from about 24 to 32 degrees C (75 to 90 degrees F), with the warmest tunnel reach occurring between about Station 1392+00 to 1424+00, the approximate area of highest ground cover. Given the depth of ground cover, a temperature of 32 degrees C (90 degrees F) appears to be about 6 degrees C (10 degrees F) higher than would be suggested by a typical geothermal gradient of about 0.6 degrees C (1 degree F) rise per 30 m (100 ft) in depth below the ground surface.

SUMMARY AND CONCLUSIONS

As discussed above, the approximate 13 km (8 mile) long Riverside Badlands Tunnel will be excavated through diverse and variable ground conditions, including soft ground, weak sedimentary rocks, and heavily fractured metamorphic rocks. In general, tunneling conditions are expected to range from overstressed sedimentary rocks, to blocky and seamy/crushed rock with some squeezing ground in the metamorphic rocks. Local running/flowing, swelling, slabbing, and slaking tunneling conditions are also anticipated. In addition, local warm ground and groundwater conditions are anticipated.

As shown on Figure 2 (the tunnel geologic profile), the ground types are interspersed along the tunnel such that tunnel excavation methods selected by the contractor will need to be versatile, or perhaps the contractor will elect to change excavation methods to suit the ground conditions.

The fact that nearly 80 percent of the ground will be excavated beneath the groundwater level adds to the construction challenges in building this project. Dewatering will be needed to control the excavation performance of the soft ground in the two alluvium reaches. In addition, high groundwater inflows are anticipated associated with two reaches of highly sheared and faulted ground. Probe holes are required in these two reaches, and pressure grouting will be performed as required, ahead of the tunnel excavation, to control groundwater inflows and also to improve the performance of the ground.

It appears that the Crafton Hills and Riverside Badlands present interesting and challenging geologic conditions for tunneling. In many respects, the successful completion of the project will hinge on the careful selection of appropriate tunnel excavation and tunnel support methods, and also to implement these methods effectively.

REFERENCES

Deere, D.U., Peck, R.B., Monsees, J.E., and Schmidt, B., 1969, "Design of Tunnel Liners and Support Systems," Report for U.S. Department of Transportation, OHSGT Contract 3-0152, No. PB 183 799, NTIS, Springfield, Virgina, 287p.

Deere, D.U., Merrit, A.H., and Cording, E.J., 1974, "Engineering Geology and Underground Construction," 2nd International Congress for the International Association of Engineering Geology, Sao Paulo, Brazil, August, VII-GRI.

Peck, R.B., 1969, "Deep Foundations and Tunneling in Soft Ground," Proc. 7th International Conference on Soil Mechanics and Foundation Engineering, Mexico City, State-of-the-Art Volume, pp. 225-290.

Proctor, R.V., and White, T.L., 1946, "Rock Tunneling with Steel Supports," with Karl Terzaghi, Commercial Shearing, Youngstown, Ohio, 296p.

Shuler, E.H., 1953, "Geology of a Portion of the San Timoteo Canyon Badlands near Beaumont, California," M.S. Thesis, Department of Geology, University of Southern California.

Terzaghi, K., 1950, "Geologic Aspects of Soft-Ground Tunneling," Chapter 11 in Applied Sedimentation, P.D. Trask, ed., John Wiley and Sons, New York, NY, pp. 193-209.

Terzaghi, K., and Peck, R.B., 1967, "Soil Mechanics in Engineering Practice, John Wiley & Sons, Inc. New York, NY, 729p.

9

Shafts & Raises

Chairmen
M. Kobler
UCM
Santa Rosa, CA

A.C. Abrams
Atkinson Construction Co.
Lakewood, CO

Chapter 28

MANHATTAN SHAFT, NEW YORK CITY WATER TUNNEL NO. 3

Edward (Ted) M. Dowey
New York City, Dept. of Environmental Protection

ABSTRACT

Shaft 26B of New York City's Third Water Tunnel was constructed by Frontier-Kemper Constructors, Inc. of Evansville, Indiana. The 176.8 m (580 ft) deep shaft will serve as the construction shaft for the Manhattan spur of Water Tunnel No. 3. Twelve m (40 ft) of overburden was supported by a rectangular cofferdam. This shaft was excavated in 13 months, through 164.6 m (540 ft) of rock, using 24,040 kg (26.5 tons) of explosives. Concrete lining cycles included the installation of stainless steel pipes embedded in the lining as the shaft advanced. A total of 635 m (2,084 linear feet) of stainless steel pipe was installed.

INTRODUCTION

New York City uses over 4.9 billion liters (1.3 billion gal) of water a day, 90% of which comes from the Delaware and Catskill watersheds and reservoirs west of the Hudson River. The water is conveyed from these collecting reservoirs to the balancing reservoirs of Kensico and Hillview via the Delaware and Catskill Aqueducts. From Hillview Reservoir the water is brought to the five boroughs via City Tunnel No. 1 and City Tunnel No. 2. City Tunnel No. 1 was completed in 1917, City Tunnel No. 2 in 1936. Both have been in continuous operation since their completion. The remaining 10% of the City's water supply is delivered to the City from the Croton system of reservoirs and aqueducts. (Figure 2)

The use of City Tunnel No. 3 will increase delivery capacity to meet growing demand, increase water pressure in the southern and eastern sections of the City farthest from Tunnels No. 1 and No. 2, and reduce peak flows which periodically overload Tunnels No. 1 and No. 2. The Third Water Tunnel will also permit dewatering, inspection, and possible rehabilitation of Tunnels No. 1 and No. 2 and will provide an alternate delivery system in the event of a problem with Tunnels No. 1 and No. 2.

Stage One of City Tunnel No. 3 consists of a 21.7 km (13.5 mi) of 7.3 m and 6.1 m (24 ft and 20 ft) diameter concrete-lined pressure tunnel plus three subterranean valve chambers. The Stage One Tunnel begins at Hillview Reservoir in Yonkers, N.Y. (as do Tunnels No. 1 and No. 2), proceeds to the massive Van Cortlandt Valve Chamber, continues south into the Bronx, then under the Harlem River into Manhattan and south to the Central Park Valve Chamber. The tunnel continues east under the East River to the Roosevelt Island Valve Chamber and into Queens where it extends for one mile. The tunnel is 122 to 244 m (400 to 800 ft) below the surface. Along the route of the Stage One tunnel are 14 riser shafts which

Figure 1. The Manhattan Shaft, Empire State Building and Muck Pile — Shaft is Behind Muck Pile

will bring water to the surface and to the distribution mains. Stage One is complete and waiting final disinfecting before being put into service.

Stage Two consists of the Brooklyn/Queens tunnel and the Manhattan tunnel. An 8 km (5 mi) long section of the Brooklyn / Queens tunnel is 4.8 m (16 ft) in diameter and is being constructed by Shiavone Construction Co., J. F. Shea, a Joint Venture. The excavation, done by tunnel boring machine, is complete. The other 8 km (5 mi) long 6.1 m (20 ft) diameter section in Queens is being constructed by Grow/Perini/Skanska J.V. The tunnel boring machine is presently excavating.

The Manhattan Shaft, Shaft 26B, will serve as the construction shaft for the Manhattan tunnel. The Manhattan tunnel will run from the Central Park Valve Chamber down the west side of Manhattan, to Shaft 26B, and south toward the World Trade Center to an as yet undetermined terminus. Construction of the Manhattan tunnel is scheduled to begin in 1997.

PROJECT DESCRIPTION

The Contract

Contract 538A for Shaft 26B of Water Tunnel No. 3 was designed and the construction supervised by New York City Department of Environmental Protection. The contract was awarded to Fontier-Kemper Constructors, Inc. for $24,396,810.00 on April 8, 1994. The contract is for the construction of a 176.8 m (580 ft) deep concrete lined shaft. The construction site is located on the west side of Manhattan in New York City northwest of the intersection of 30th St. and 10th Ave. (Figure 1) Included in the contract is the fabrication of over 634,900 kg, (1.4 million lbs) of 1.2 m (48 in) stainless steel pipe to be used in the Stage 2 shafts in Brooklyn and Queens. The pipe is not covered in this paper.

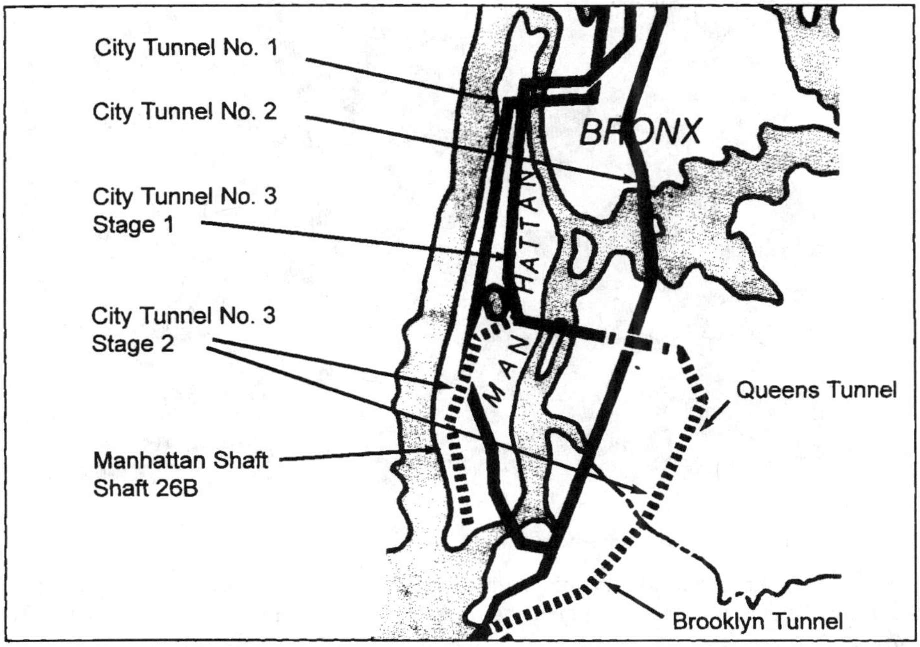

Figure 2. Location of City Water Tunnels 1, 2, and 3

Manhattan Shaft: Multiple Purposes, Complex Geometry

The Manhattan Shaft will serve as the construction shaft for the Manhattan tunnel. The Manhattan tunnel is planned to be constructed in two parts, first north to Central Park, and then under a second contract, the southern section. The 6.1 m (20 ft) diameter of the shaft is to accommodate the construction of the tunnel. The Manhattan tunnel is 3.05 m (10 ft) finished diameter. Following the completion of the north and south sections of the Manhattan tunnel the shaft will serve as a distribution shaft. In the event that the tunnel has to be inspected the Manhattan tunnel will be dewatered from Shaft 26B using the pipes installed in the concrete lining under this contract.

The large upper shaft accommodates the components of the distribution shaft and the dewatering pipes which determines its geometry. While only the perimeter walls of the shaft are constructed under this contract, the geometry of the excavation and geometry of the concrete lining are determined by the final use of the shaft, a distribution/dewatering shaft. The major features of the distribution shaft are the distribution chamber, anchorage for the valves in the distribution chamber, the riser valve chamber, and anchorage for the riser valves.

The distribution chamber, from elevation +14.25 to -11.00, is for connections to the street water mains. (All elevations are given in feet above (+) or below (-) mean sea level at Sandy Hook NJ of the United States Coast and Geodetic Survey Datum as per the contract.) From elevation -11.00 to -107.00 the shaft will contain two 1.2 m (48 inch) diameter pipes that serve to anchor the valves in the distribution chamber. The riser valve chamber will exist from elevation -107.00 to -133.00. (The riser valves will function as emergency shut-off valves

Figure 3. Elevation +3.0 down to the Work Deck at Elevation -37.5

in the event of a problem with the distribution chamber.) From elevation -133.00 to -237.00 two 1.2 m (48 inch) pipes will be installed to serve as anchorage for the riser valves. The lower shaft, from elevation -237.00 to -560.00, will serve as a waterway.

The shaft cross section at the top, elevation +14.25 to elevation to -11.00, is a circle 9.7 m (32 ft) in diameter but with two 0.9 m (3 ft) straight segments on the north and south forming a race track with geometry. (Figures 4 and 5) This is the shape of the excavation in the overburden that has to be supported. The concrete shaft walls are 0.4 m (16 inches) thick . The walls contain one mat of reinforcing bars around the perimeter except at four locations where a double mat is installed. These four sections support the beams that support the cover. The shaft walls in this section will be removed when the distribution chamber is built under a subsequent contract.

The second shaft cross section is from elevation -11.00 in the overburden to elevation -37.5 in the rock. (Figures 6 and 7) The excavation cross section is the same race track section. However the inner face of the concrete lining is a 7.9 m (26 ft) diameter circle. This section is reinforced with an inner, middle, and outer reinforcing bar mats. The shaft walls range from 0.9 m (3 ft) to 1.8 m (6 ft) thick crossing the boundary from soil to rock. The reason for the variation in thickness of the walls is the four sets of stainless steel pipes embedded in the lining on the east side (10 degrees south of east). Keys must be formed in the concrete. The keys are placed on 2.3 m (7 ft 9 in) centers up and down the shaft. They will serve to lock in the refill concrete around the 1.2 m (48 in) pipe that will anchor the distribution chamber valves.

The third shaft cross section extends from -37.5 to -156.0. (Figure 8) This section is entirely in rock. The race track section for excavation is gone and the concrete shaft walls are reduced to 0.43 m (17 in) in thickness. The excavation continues to accommodate the stainless steel pipe on the east. This "pipe chase" continues to the bottom of the shaft. The concrete lining of the shaft varies in this section. The keys to lock in the refill concrete are the

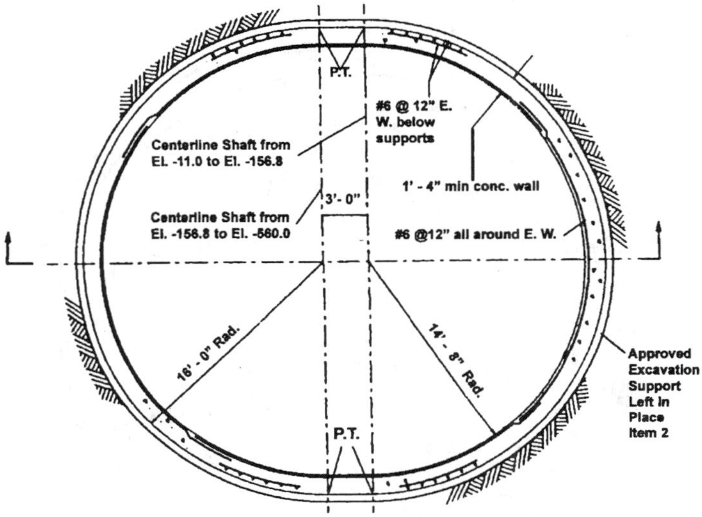

Figure 4. Shaft El. +14.25 to El. -11.00

same from -37.5 to -100.0. A large 0.45 m (1 ft 6 in) deep key runs around the perimeter of the shaft from elevation -102.0 to -107.0. This large key will support a reinforced concrete beam that supports the refill and the 1.2 m (48 in) pipe. The riser valve chamber, with reinforced walls, extends from -107.0 to -133.0. A 0.15 m (6 in) deep key runs around the bottom of the shaft from elevation -133.0 to -135.0.

The fourth and lowest section of the shaft, from -156 to the bottom at elevation -560 is a 6.1 m (20 ft) diameter waterway. (Figure 9) The centerline of this section of the shaft is 1 m (3 ft) to the east of the upper shaft. At the transition from the upper section to the lower section, at elevation -156.0, the east walls are tangent. The displacement of the centerline of

Figure 5. Workdeck at Elevation -11.0

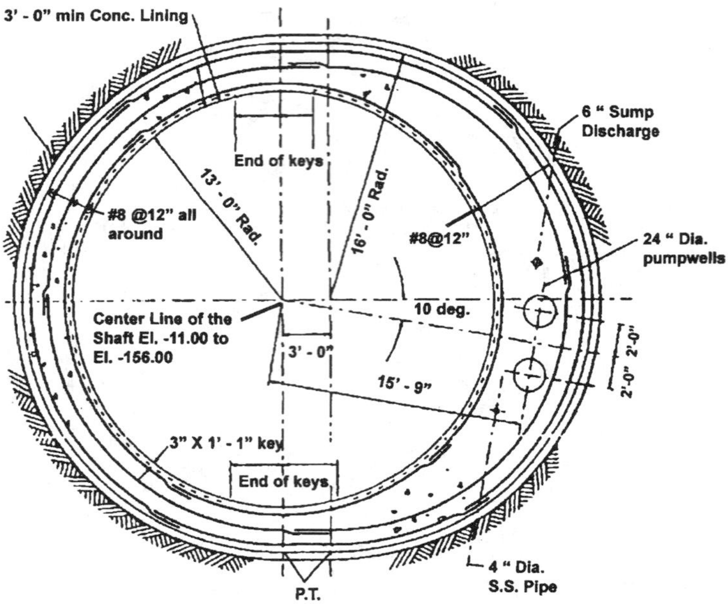

Figure 6. Cross Section of Shaft El. -11.00 to -37.50

the lower shaft east from the upper shaft presents logistical challenges for working in the shaft. The concrete shaft walls are 0.43 m (17 in) thick and unreinforced

Excavation and Rock Support

The initial shaft excavation is required by the contract to proceed from the surface,

Figure 7. Shaft from El. -11.00 to El. -90.00 (+/-), Rock Excavation Below

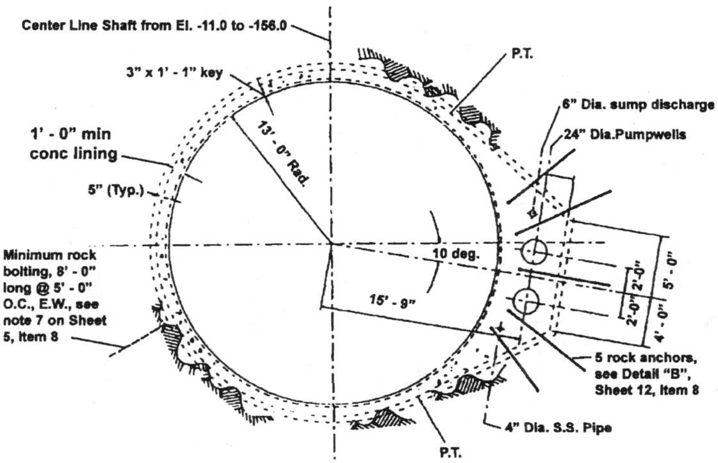

Figure 8. Shaft Cross Section El. -37.5 to El. -156.0

elevation +18.5, through approximately forty feet of overburden and 20 feet into rock to elevation -42.5. The contractor may select the method of ground support for the overburden, subject to approval by the NYC DEP. The concrete lining must then be placed from elevation -37.5 to elevation -11.00. Thereafter excavation in rock may proceed for approximately 30.5 m (100 ft) prior to the required placement of the concrete lining. The excavation cycles of 30.5 m (100 ft) shall proceed to the end of the concrete lining at elevation -530.

The contract calls for the excavated portions of the shaft to be supported with 2.4 m (8 ft) rock bolts with straps on a 1.5 by 1.5 m (5 by 5 ft) pattern with wire mesh over the rock surface.

Complicating these excavation and concrete lining sequences are four sets of stainless steel pipes which must be placed in the shaft prior to each concrete lining segment in order to be incorporated into the lining.

Dewatering Stainless Steel Pipes

This shaft will serve as the dewatering shaft for the Manhattan tunnel in the event that the tunnel must be inspected once it is in service. The flow of water would be shut off at the Central Park Valve Chamber and the water in the Manhattan tunnel pumped out through the stainless steel pipes embedded in the lining here. The Manhattan tunnel slopes down to a low point at the bottom of Shaft 26B allowing the Manhattan tunnel to be drained from this point.

For the purposes of dewatering, four sets of pipe are embedded in the shaft lining. The pipes are one 10 cm (4 in), one 15 cm (6 in) and two 61 cm (24 in) pipes, all stainless steel. In order to install the pipes the excavation cycles and the pipes lengths must coincide. The contract allowed the contractor to fabricate/order various lengths of pipe so that the excavation cycles could match the pipe lengths.

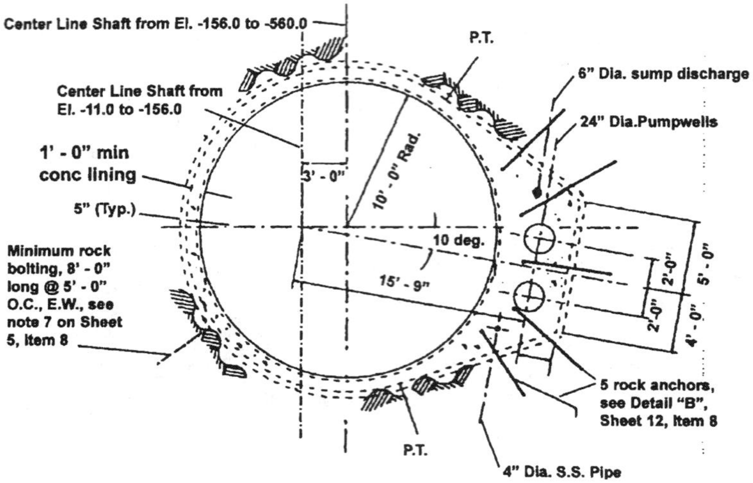

Figure 9. Shaft Cross Section El. -156.0 to El. -560.0

The pipe must be installed after each section is excavated but before each section is lined with concrete. Logistics dictate that the pipe must be installed beneath and under the previously lined section.

URBAN ENVIRONMENT

Shaft 26B is located north of the northwest corner of 30th St. and 10th Ave. in Manhattan. The shaft is located at the east end of a 244 m (800 ft) long site, 4.5 m (15 ft) from the sidewalk on 10th Ave. The site is covered with 20 cm (8 in) unreinforced concrete slabs. The shaft site was occupied by an LIRR /Amtrak joint venture building which was removed. Twenty-four thousand and forty kg (26.5 tons) of explosives were used on the job making good community relations essential. Structures in the neighborhood to the north, northeast, and east required continuous blast monitoring and notification.

Clockwise from the north are the following structures:

Eighteen m (60 ft) to the north of the shaft excavation, is 24 hour per day LIRR train repair facility where entire trains are elevated several feet on hydraulic lifts in order to access the undercarriage.

Also to the north, 91m (300 ft) from the shaft are the two primary Amtrak Northeast Corridor tracks running east from New Jersey under the Hudson River to Pennsylvania Station. In addition there is a third rail tunnel which turns north and travels up the west side of Manhattan and up the Hudson River.

To the northeast is a 14 story building built over the LIRR and Amtrak tracks. This building houses a skating rink on the top floor, corporate residents with sensitive instruments and computers, and a major metropolitan daily newspaper.

Under 10th Ave., 15.2 m (50 ft) from the shaft, is a 0.5 m (20 inch) diameter 1880's bell and spigot cast iron gas main with a joint every 4 m (12 ft). Twenty-one m (70 ft) from the centerline of the shaft, in the street is a 0.5 m (20 in) high pressure gas main. Both gas mains are in service and operated by Con Edison.

Across 10th Ave from the site is a business housed in a brick structure and an adjacent 5 story walk up built in the 1880's with residents in 3 of the 20 apartments.

To the southeast of 30th St. is the Morgan Post Office Facility, another 24 hour a day operation. To the south and bordering the site for its entire length is an abandoned and contaminated one story brick structure where homeless people lived. Columns through the brick structure support an abandoned overhead Conrail track.

SHAFT CONSTRUCTION

Site Preparation

Work began on the site in late August of 1994. Preconstruction surveys were conducted of surrounding buildings. A base line subsidence survey was established. Initial site preparations included the demolishing of the LIRR/Amtrak building, cleaning up the site, and installing fencing. The homeless in the building south of the site were removed from the building. They were offered options for relocation by the NYC Mayor's Office for Outreach and MTA police. The building was sealed. Utilities were brought into the site. The offices, change houses, and maintenance shop areas were all constructed.

The Overburden Excavation

Frontier-Kemper Constructors, Inc. chose to support the overburden with a rectangular cofferdam of sheeting and wales. (Figure 3) The support and excavation of the overburden was subcontracted to River Pile and Foundation Co., Inc. The driving of the sheeting started in October 1994, with a Manitowac 3500 crane and an FMC Kink Belt HC-218A truck crane with a Tunkers D4030 vibratory hammer. Later the sheets were driven with a McKiernan-Terry Crop 11B-3 hammer. The sheets were PLZ -23 with some PLZ-25 sheets substituted. The sheeting was driven in the shape of a rectangle, 12.6 by 11.4 m (41.33 ft by 37.33 ft) with 20 sheets on the north and south sides and 18 on the east and west sides. The geometry of the shaft to be enclosed by the sheeting is the shape of a race track described earlier. (Figure 4) However to attempt to drive the sheeting in the same shape as the shaft might result in problems with closure and the sheeting would be more difficult to brace.

In addition to the shaft excavation, River Pile and Foundation Co. Inc., also drove H piles, for Frontier-Kemper Constructors Inc. They drove 11 H piles for the shaft cover/deck, 12 piles for the muck bin and 3 for the "holster" for the drill jumbo for Frontier-Kemper Constructors Inc. A Vulcan 1A hammer was used.

Some difficulty was encountered because part of the perimeter of the excavation where the sheeting was to be driven consisted of an unpenetratable fill. The fill in this area was shot rock. Sections of the shaft had to be pre-excavated to a depth of 3 to 3.6 m (10 to 12 ft) to get below the rock fill. The sheets were driven to rock, a depth of about forty feet. There was some difficulty in seating some of the sheets. Two of the sheets in the southwest corner split apart which eventually allowed the inflow of some ground and impinged on the third brace. The average elevation of the bedrock was -24.3. The elevation of the bedrock in the northeast corner of the shaft dipped to -32.0 and the east side averaged an elevation of -27.4.

With completion of the sheeting installation, mid November 1994, excavation started with a Manitowac 3500 crane and clam shell to the depth of the first brace at elevation +8.0. Waler beams with corner braces were then welded together in place in the shaft. The water table was encountered as expected at elevation +4.0. The ground consisted mostly of a red/brown sand.

Excavation and bracing proceeded for three cycles with braces for the cofferdam at elevations +8.0, -3.0, and -13.0. Part of the third wale required an alteration to accommodate the two sheets that split apart during the installation. A partial fourth brace was installed on the east side with extensions along the north and south sides at elevation -20.0 to brace the sheeting and accommodate the lower bedrock elevation.

With the completion of the cofferdam and prior to rock excavation, Frontier-Kemper Constructors, Inc. placed a concrete curb along the bottom of the sheeting, covering the rock/overburden border. There was some water infiltration through the cofferdam prior to blasting and through fractures in the rock subsequent to blasting.

Rock Excavation

Blasting Parameters. The contract contains a limit on the magnitude of the vibrations, the noise produced as a result of blasting, and the hours of blasting. The maximum peak particle velocity allowed at the nearest surface structure is a conservative 1.27 cm/sec (0.5 inches per second). The noise produced by the blast must not exceed 140 db, 7.6 m (25 feet) from the shaft or 95 dbA (on the A scale) at the nearest residence. There is a limit on the pounds of explosives per delay which ranges from 1.9 to 3.1 kg (4.2 to 6.9 lb) per delay depending on the compliance with vibration and noise criteria. There is also an addendum to the contract requiring the contractor to follow the regulations of the New York City Fire Department who oversee the use of explosives in the City. The explosive used must be approved for use in New York City by the Fire Department. The use of electric or non-electric detonators is directed by the Fire Department. The type and location of the explosive magazines is approved by the Fire Department. Fire Department inspectors are present for each blast. Blasting is restricted to the hours between 8 AM and 7 PM. Extensions to 10 PM may be allowed on a case by case basis.

Blast Pattern. The blasting pattern utilized by Frontier-Kemper Constructors Inc. was designed to blast the entire face of the shaft in one shot while adhering to the contract limits on noise and vibration. The blast pattern for a typical section of the shaft used 110 holes each 4.76 cm (1 7/8 in) diameter and 3.05 m (10 ft) deep. The blast holes were drilled in 5 concentric circles surrounding the central "burn" holes. The holes were spaced 0.9 m by 0.9 m (3 by 3 feet) but with a 0.75 by 0.75 m (30 by 30 in) spacing on the perimeter holes. The depth of the holes, the size of the face, and the desired volume of rock to be excavated per shot, coupled with the contract limits on vibration and noise required that each hole be detonated individually.

The Ensign-Bickford Company manufactured the Primadet (tm) Nonelectric Delay Detonators used to meet these requirements. A series of 4 MS (millisecond) and 17 LP (long period) delays were used in the blast holes producing 21 discrete detonation periods. The "in hole" delays used in conjunction with 17 MS and 42 MS trunk (surface) delays staggered each set of 21 detonation periods up to 8 times resulting in a possible 168 delay periods. While the physical wiring of the delays resulted in a blast pattern that looked like a pie with slices, the actual sequence of detonation proceeded from the center of the shaft to the perimeter. While the longest LP delay was 9.6 seconds, the blast duration from surface initiation to the final reverberations lasted close to 15 seconds.

The explosive used was Austin Powder Company's Emulex 520A emulsion. The explosive is approved by the NYC Fire Department for use in the city. The explosives were delivered to the site by Explo Powder Inc. On the site were 2 magazines, one for explosive storage and one for detonator/delay storage. The blast system is designed to effectively break the rock while providing for a large degree of safety for all involved in transporting the explosives and detonators and for loading at the bottom of a deep shaft. The blasting system performed well in breaking the rock and did so safely.

Blast Monitoring. Rock excavation started in February, 1995. Flyers were posted to alert the neighborhood and the Community Board was notified of the impending blasting. Due to the size of the blasts, with 363 kg (800 lb) of explosives, the duration of the blasts, 15 seconds, and the number of blasts, 85, arrangements were made with the neighbors regarding notification prior to each blast. As a result four calls were made to local businesses, Con Ed, and the LIRR facility notifying them of each impending blast.

Vibrations were monitored for each blast at 4 sites. The vibration monitoring sites were the LIRR rail workshop, the 14 story building to the northeast, in the business east of the site on 10th Ave. and the 5 story walkup tenement . The vibrations did not exceeded the allowable limits for any of the 85 blasts.

Con Edison monitored for gas leaks in the cast iron main in 10th Ave. prior to and after each blast. About halfway through the project, Con Ed placed a PVC sleeve in the pipe. A whistle was blown prior to each blast and the traffic on 10th Ave. was stopped for the duration of the blast. The shaft deck served to contain the flyrock.

In addition the DEP measured blast sound pressures at the nearest residence. The requirement of less than 95 decibels on the "A" scale was initially hard to reach. Various sound attenuation measures were taken to reduce the noise. A wooden fence with styrofoam insulation was constructed inside the site fence on 10th Ave. Concrete blankets were draped over additional sections of the fence.

Excavation Cycle

The holes for blasting were drilled by a hydraulic-electric shaft drill jumbo constructed by Frontier-Kemper Constructors Inc. and Canon Industries for the project. The drill is MSHA approved for use in mines and contains its own fire suppression system. The jumbo stands on a single central leg and is braced by three horizontal pistons. The two booms with Canon drills are operated from two additional booms containing the controls for the drills. The drills are hydraulically fed and produce faster drilling rates than the traditional air powered feeds. At well over 18,000 kg (20 tons) it was the heaviest lift for the Manitowac crane servicing the shaft.

Explosives were typically delivered first thing in the morning. The loading of explosives normally took up the morning of the day shift. The blast holes were loaded and clipped together for the proper detonation sequence. The shaft was evacuated, covered, and blasted.

Rock support was installed from the muck pile around the entire perimeter of the shaft. The rock support consisted of a pattern of 2.4 m (8 ft) long Atlas Copco Swellex rock bolts on a 1.5 by 1.5 m (5 by 5 ft) pattern with straps running vertically and horizontally. Chain link wire mesh was installed over the rock surface and attached to the Swellex rock bolts with Swellex washers. The chain link generally was installed 1.5 to 3.0 m (5 to 10 ft) above the face to reduce damage during blasting.

Mucking was done with one of two diesel powered Menzi muckers. The muckers had 0.2 m^3 (0.25 cu yd) buckets. They filled a 4.6 m3 (6 cu yd) bucket which was hoisted to the surface by the Manitowac 4100 crane and dumped into the muck bin on the surface. The final clean up prior to drilling was done with an air and water "blow pipe" to allow inspection of the bottom.

A total of 85 blasts using over 24,040 kg (26.5 tons) of explosives were used. The drill, blast, rock support, and muck cycles took 4 shifts resulting in blasting every other day. The blasting started February of 1995 and continued until mid-March of 1996. Excavation cycles for each 30.5 m (100 ft) took 16 blasts (average) including ancillary blasts for "tights", uneven rock surfaces and a "break in" period at the start of excavation limiting the blasts. After the fifth excavation/concrete cycle an additional four blasts were required for the last 30 feet of unlined shaft.

As required by the contract one blast was initiated below the level required for the setting of the stainless steel pipe and the concrete forms. This blast was partially mucked out to the elevation required for setting the pipes and concrete forms. Subsequent to the placing of the concrete lining this blast was mucked out creating some space between the pipes and the concrete lining and the next rock face to blasted. Protective steel covers were installed on the stainless steel pipes and a blast mat was draped over the pipe for protection from the blast.

Geology and Ground Water

The surface rock of central Manhattan has been reclassified as the Hartlandt Formation according to the 1994 United States Geological Survey map by Dr. Charles Baskerville. According to the map the Manhattan Schist underlies the Hartlandt Formation in the vicinity of Shaft 26B. The DEP has recovered rock cores from two boreholes 15 m and 24 m (50 and 80 feet) from the shaft. These were available to contractors bidding on the job in the pre-bid exhibit. The cores indicated that large sections of the shaft were in a fine grained igneous rock (aplite) while the rest was in a schist. Indeed only about 40 percent of the rock in the shaft consisted of schist. The aplite comprised 60 percent of the rock. The aplite was found to blast well. In addition to the schist and aplite the rock was penetrated by numerous coarse grained feldspathic granitic pegmatites. Some of the larger boulder sized specimens of the pegmatite had workers, inspectors, and visitors filling the trunks of their cars with rocks.

The first 30.5 m (100 ft) of rock excavation of the shaft started with a schist striking N30E and dipping steeply to the west. However the aplite entered the shaft from the east and occupied the entire shaft by elevation -120.0. The aplite continued uninterrupted for 61 m (200 ft). Another 9.1 m (30 ft) of schist was encountered with the same contact orientation. However below elevation -400.0, the orientations of contacts between the schist and aplite became totally unpredictable and variable.

In mid-October, 1995, while drilling for blasting at elevation -368.00, a small geyser of water erupted from a blast hole. The water inflow, later measured to be on the order of 490 liters per minute (130 gpm), necessitated the removal of the drill jumbo, mobilization of some larger pumps and a subsequent grouting operation. With the City's approval, the contractor drilled holes 19.8 m (65 ft) deep around the perimeter of the shaft and in the center of the shaft. The idea was to impregnate the fractures in the rock with grout to reduce the inflow. The grouting reduced the inflow to less than 75 liters per minute (20 gpm). Subsequent excavation revealed grout filled joints in the rock. While the inflow was reduced, the shaft remained uncomfortably wet throughout its depth.

Logistics in the Shaft

The shaft cover sat on 11 piles seated on rock. The shaft cover consisted of the personnel hoist system (on a moveable bridge), and three winches for the suspension of the concrete forms and the work decks. It also served as a cover during blasting and secured the shaft during off hours.

The shaft was accessed with a personnel cage throughout the construction of the shaft. The personnel hoist consisted of a "cage", headframe, and bridge fabricated by EFCO Inc. The hoist and drum for the personnel cage was manufactured by Timberland. The personnel cage traveled along two guide/safety ropes to the counter weights at the bottom. The counter weights could be raised or lowered from the hoist house to accommodate blasting and work at different elevations. Two sets of safety dogs were mounted on the top of the cage. The dogs were tested each month and the ropes were electromagnetically tested every three months. The entire assembly was set on a moveable bridge. The bridge could move over the shaft to close the top, or south off the shaft to allow the shaft to be serviced by the crane without obstructions. The headframe for the personnel cage was repositioned to access the two different diameter portions of the shaft.

A flexible winch system was needed to access the shaft with two different center lines. Three New Era winches with 30 hp motors were attached to the beams of the shaft cover. Two of the winches were repositioned in order to service the lower section of the shaft which has a smaller diameter and different centerline. The winches were outfitted with safetys that could be set to trip when designated work loads were reached. Another set of safetys monitored the rotation of the winch drums for excessive speed. The winches had an electric brake on the drive motor and an air powered band brake on the drum.

Two personnel work decks, of matching large and small diameters were used to access the shaft for all the major operations. As a redundant/independent suspension system, they were outfitted with an independent set of four static lines, in addition to the three winch cables. The static lines were attached to the same bolts in the concrete that had supported the forms.

Frontier-Kemper Constructors, Inc. had two sets of shaft concrete forms for the two diameters. They were assembled and disassembled at the bottom of the shaft for each concrete cycle. They were 9.1 m (30 ft) high and 6.1 and 7.9 m (20 and 26 ft) in diameter. A typical pour sequence consisted of four pours, starting from the lowest and continuing up to the closure pour meeting the previously placed lining. The forms were suspended from the winches after the first pour of the sequence and raised for the subsequent pour. The forms were then set on bolts embedded in the concrete. The concrete forms had work decks at the top and the bottom of the 9.1 vertical m (30 vertical ft) and catwalks at interim levels.

Pipe installation

The stainless steel pipe installed in the shaft was procured by Frontier-Kemper Constructors, Inc. from the Nooter Corporation of St. Louis.

The stainless steel pipes were installed in the shaft using two pipe handling tools designed and built by Frontier-Kemper Constructors, Inc. (Figure 10) The tools were handled by the crane servicing the shaft. The tools had to be able to pick the pipes up from the horizontal position on the surface, lower them down the shaft, and place them 0.9 m (3 ft) laterally under the previously lined section or under the previous pipe. The tools, about 2.4 m (8 ft) long, had a swiveling yoke on one end and counterweights on the other end. The tool was suspended from the crane using four slings and one adjustable link to accommodate the weight redistribution that occurred when the pipe was removed. The yoke attached to the pipe allowed the insertion of the bolts into the flanges while installing the pipe.

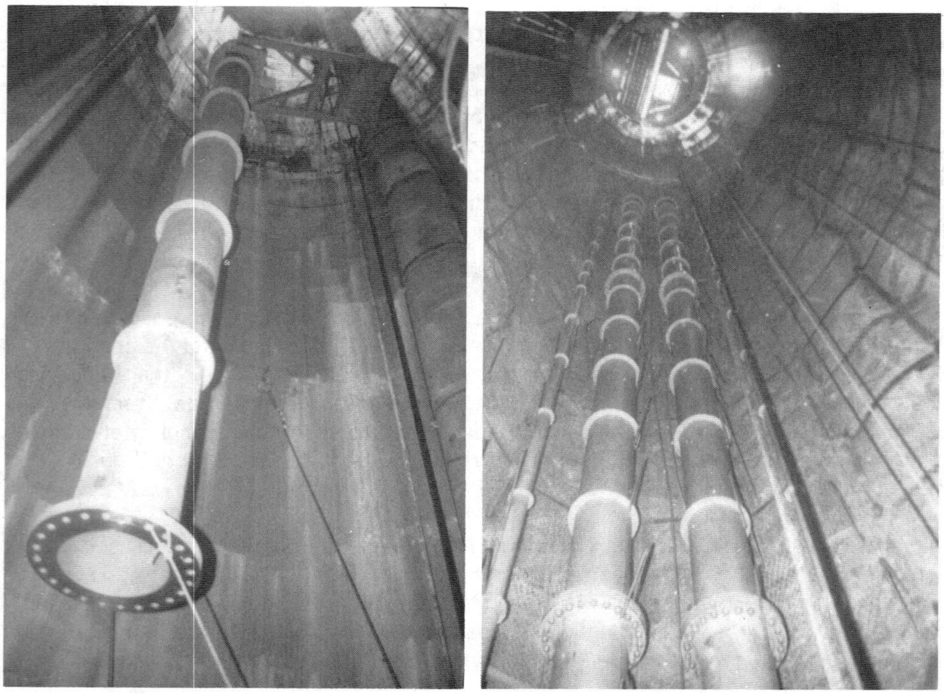

Figure 10. Lowering the Pipe (left); Installed Pipes (right)

Each 30.5 m (100 ft) of excavation would typically contain two sets of 0.6 m (24 in) pipes, one set of 15 cm (6 in) pipes, and one set of 10 cm (4 in) pipes, of lengths 10.6, 10.6, and 9.1 m (35, 35, and 29.75 ft). (Figure 10) The total weight of pipe was about 16,780 kg (37,000 lb) for the of 30.5 m (100 ft). Each joint of the 0.6 m (24 in) pipe had to be cleaned, two "O" rings glued into place, bolts installed in the specified order, torqued to the required amount and air tested for one hour, all while working off the hanging work deck in the shaft. The pipes were surveyed for plumbness and braced at the bottom, top of the concrete form, and at two other points higher in the shaft. The bottom 0.6 m (2 ft) of the pipe was blocked out so it would not be encased in concrete and would be accessible for connection to subsequent pipes.

CONCRETE LINING

Concrete Specifications

The contract specifies that the concrete for the lining be 31 MPa (4500 psi) in 28 days. The cement is a Portland cement of ASTM Designation C150-89, Type II, low alkali that is described in full for nine pages in the contract. The concrete contains sand, and aggregates of 0.6 to 1.9 cm (1/4 to 3/4 in), and 1.9 to 3.8 cm (3/4 to 1-1/2 in). The concrete has a minimum of 6.5 bags of Portland cement per cubic yard, and a maximum water content of 20.8 liters (5.5 gal) per bag of cement. The slump is between 2.54 and 10.16 cm (1 and 4 in). All elements of the concrete are inspected by the DEP's Quality Assurance and Quality Control Divisions. The cement plant, the cleanliness of the silo's and barges and trucks, the strength size and gradation of the sand and aggregates, the design mixes and trial batches, and numerous other inspection items are described over 22 pages of the contract.

Inspectors tested the concrete for slump, unit weight, air content, and temperature in the field. They also prepared cylinders for compressive strength testing at one week and one month.

Shaft Preparation for Concrete

When a blasting cycle has finished, the shaft must be cleaned up and a work deck installed in order to access the shaft walls. Prior to the placement of the concrete lining in the shaft, the muck from the last blast must be leveled at the correct elevation, the shaft excavation must be cleaned and surveyed for tights, rock anchors in the pipe chase installed, stainless steel pipes installed and plumbed, and water inflows contained or diverted. The muck is leveled at the elevation for the first pour and the concrete forms are constructed in place on the muck, following shaft preparation. The cleaning of the shaft includes removing all loose rock from behind the wire mesh that covers the shaft and washing the rock surface. Hooked reinforcing bars on a 1.5 by 1.5 m (5 by 5 ft) pattern are grouted into the rock in the area of the stainless steel pipes. The four sets of stainless steel pipes are installed. Water inflows are located and a hole drilled into the rock to localize the inflow at the source. A pipe is inserted into the hole so the water flows through the pipe. After the forms are installed the pipes are brought through the forms and the water allowed to flow. These operations are performed from the hanging work deck in two or three passes up and down the shaft.

When the shaft is clean and the pipes installed, the concrete forms are constructed on the muck at the bottom of the shaft. The construction of the forms is followed by the installation of a bottom deck, catwalks, ladders, a top deck, hoppers and “elephant trunks” behind the forms, and a large rotating hopper. A small sump in the muck keeps the shaft water below the concrete placement level.

The Concrete Pours

Five concrete cycles occurred averaging four pours per cycle (with one cycle of three and one of five). A total of 3,893 cubic meters (5,093 cu yd) was poured including 254 m^3 (332 cu yd) of additional change order concrete. The pours occurred between February 24, 1995 and February 22, 1996. The lining extended from elevation +14.25 down to -530.00.

The concrete was delivered to the site by truck from Ferrara Bros. plant in Queens. The concrete was delivered down the shaft by the crane and a 0.9 m^3 (2.5 cu yd) bucket. For the first cycle the concrete bucket could feed the hoppers mounted on the top of the forms directly. The concrete for the second cycle was delivered through drop pipes embedded in the lining of the first cycle. Logistics dictated that the drop pipes could only be used for one cycle. For the remaining 13 pours (3 cycles) the concrete bucket was emptied into a rotating hopper assembly that delivered the concrete to hoppers mounted on the top of the forms. The rotating hopper would hang from the winch cables during the pours. The concrete forms and work decks were bolted to the shaft lining below. From the hoppers on top of the forms the concrete went into removable “elephant trunks” behind the forms. The pours generally took eight to ten hours. The concrete was vibrated by hand held vibrators through doors in the concrete forms.

Grouting

The contract specified that the water inflow into the shaft be no more than 7.5 liters per minute (2 gpm). After a concrete cycle the shaft was generally grouted to reduce or stop the flow of water into the shaft. The work deck was placed at the bottom of the shaft and the shaft was grouted from the bottom up. The concrete had reached its 85 percent of its final strength and could withstand grouting pressures. Frontier-Kemper brought in their grout pump with two mixing barrels and a Moyno type screw pump. Initially a pressure of 0.7 MPa (100 psi) was used to fill any voids behind the lining. A second grouting pass used pressures of 2.06 MPa (300 psi). The grouting program used the same cement as specified for the concrete.

Extensive cement grouting brought the water inflow into the shaft close to but not to the allowed 7.5 liters per minute (2 gpm). As a last resort chemical grout, an item in the contract, was used to stop the small inflows. The chemical grout used was de neef Hydro Active Grout Sealfoam NF manufactured by de neef Construction Chemicals, Inc.

Covering the Shaft

The temporary shaft cover consists of 4 main steel beams sitting in recesses at the top of the shaft walls. Twenty-one precast concrete slabs are placed on the shaft walls and the steel beams. The cover is then covered with clean fill. The slabs have galvanized fittings used for lifting. The fittings are recessed when not in use.

The fabrication and painting of the steel was subcontracted by Frontier-Kemper Constructors, Inc. to Seibel Modern MFG & Welding Corp. The construction of the pre-cast concrete slabs was subcontracted to The Fort Miller Co., Inc. The Quality Assurance Division of the NYC DEP inspected the physical plant of and approved the work of each of the subcontractors. The beams and the slabs were delivered to the site and installed by Frontier-Kemper Constructors, Inc.

The sheeting, from the cofferdam, was cut off at elevation +14.25 at the same elevation as the top of the shaft walls. The beams were installed and the slabs placed on the beams. The cover was then buried under 0.9 to 1.2 m (3 to 4 ft) of clean fill, leaving virtually no trace of the shaft visible.

At the end of construction of the shaft a precast concrete covers is place on steel beams that span the shaft. The covers are then covered with 1 to 1.3 m (3 to 4 ft) of clean fill.

CONCLUSION

Frontier-Kemper Constructor, Inc. successfully completed the construction of the shaft in 21 of the allotted 30 months. New York City now has the Manhattan shaft ready for the construction of the Manhattan tunnel.

Chapter 29

SINKING CAISSONS AS AN EFFECTIVE MEANS OF CONSTRUCTING SHAFTS

F.J. Klingler,[1] K.M. Swaffar,[1] and J.C. Neyer,[1] and R. Hausmann[2]
[1]NTH Consultants, Ltd.; [2]Walbridge Aldinger Co.

ABSTRACT

This paper presents the results of a study of case histories of projects on which sinking caissons were used to construct shafts for tunnel construction and other underground works. Design methodologies are reviewed with respect to the impact of soil and groundwater conditions on the performance and constructability of sinking caissons. Factors which have led to success or failure are reviewed and evaluated, and recommendations for future investigation, design and construction activities are presented. Cost data for five recent projects are reviewed and compared to the cost of shaft installation by other means.

INTRODUCTION

A sinking caisson consists of a circular, square, or rectangular concrete wall structure that is initially formed at the ground surface and then sunk into place through the underlying soils, using a systematic approach of adding successive lifts of concrete to the top of the structure walls, while removing soil from the interior of the caisson. This process is continued until the required depth of excavation is achieved, after which a structural base slab is typically installed. Once sunk into position, the caisson serves the dual purposes of providing the necessary earth support during construction, as well as providing the walls of the final structure.

The technology of using sinking caissons was originally developed in Europe and brought to North America in the mid 1800's. The technology was first used in North America for the construction of bridge foundations for crossings of the Mississippi River in St. Louis and the East River in New York. These early caissons involved the use of compressed air to mitigate ground water infiltration at the excavation bottom. Excavation proceeded with small construction equipment or by hand, and muck was evacuated from the compressed caisson chamber through pressure relief pipes. The use of more modern construction equipment and materials, as well as better knowledge of soil conditions and their behavior, has enabled the sinking caisson to become an important option for making excavations into and through difficult ground conditions.

The authors combined experience in the last decade includes design and construction of more than two dozen sinking caissons. These have typically included projects where sinking caissons were judged to be the most practical means of construction because of concerns related to either bottom stability or potential lateral ground movements. Ground conditions through which these caissons have been constructed include both predominantly granular as well as predominantly cohesive deposits. This study examines five recent projects where sinking caissons were used in favor of other systems to construct shafts through soft and very soft clay strata.

DESIGN METHODOLOGY

The phrase "difficult ground conditions" can mean many things to many people. With respect to shaft construction, difficult ground conditions are typically those in which conventional means of earth support such as the use of internally braced soldier piles and lagging or steel sheet piling are not appropriate. Factors which often contribute to conventional systems or approaches not being feasible within predominantly clay soils include: high lateral earth pressures; strict ground movement limitations; instability of the excavation bottom as a result of weak clayey soils; the potential for hydrostatic uplift of the shaft bottom due to underlying artesian pressures; or the potential for ground water inflows from isolated granular layers. Factors which sometimes contribute to conventional systems or approaches not being feasible in predominantly granular soils include strict ground movement limitations and the potential for uncontrollable groundwater inflows into the shaft.

For many of these situations, a number of available construction techniques are suitable and have proven to be effective. These techniques include the use of diaphragm wall and tangent pile support wall systems, in-place soil mixing, or soil freezing. With such earth retention systems, the purpose of the system is only to support the earth and adjacent facilities until the final structure is built inside. The sinking caisson method of construction, on the other hand, will provide temporary earth support during construction, and can also be used as the permanent structure.

Sinking Caissons to Support High Lateral Earth Pressures

Many shaft construction methods exist which are appropriate for supporting high lateral earth loads. However, where lateral loads become very high, internal bracing may become large and difficult to install, and may obstruct the contractor's access to the shaft. In addition, where overload factors exceed eight, clay soils may be subject to squeezing or sudden collapse (Broms and Bennermark, 1967). In these cases, sinking caissons may be an effective alternative means of ground support. The caisson wall is typically sized to insure sufficient weight to sink the structure, which usually results in a wall thickness greater than necessary to support lateral earth loads.

Sinking Caissons to Reduce Ground Movements

Limitations on ground movements may also preclude the use of conventional means of earth support for shafts. Work by D'Appolonia (1971) suggests that for clay with shear strength less than 2000 psf, lateral movement of sheet pile shaft walls can be expected to be about 2 percent of the height of the wall, while lateral movement of rigid wall systems can be expected to be about 0.2 percent of the height of the wall. Data from studies by Peck (1969) suggests that vertical settlement adjacent to shafts may range from one-half to twice the magnitude of lateral wall movement. These correlations compare well with ground movement data observed by the authors for flexible braced shafts and rigid shafts in the Detroit area. As a result of concern regarding ground movements, rigid ground support systems (such as sinking caissons) have been specified for several projects in the Detroit area.

Sinking Caissons to Prevent Bottom Instability

Initial shaft design efforts in clay soils typically include analysis to determine the factor of safety for bottom stability using the given shaft depth and diameter, either by methods after Terzaghi (1943) or Bjerrum and Eide (1956). Current industry practice requires that where the factor of safety against bottom instability is less than 1.3, measures must be taken to resist ground movement below the base of the excavation. Conventional means of earth support such as internally braced steel sheet pile shafts can be designed to extend below the base of excavation, but are generally not capable of providing sufficient resistance to below-invert lateral loads where the factor of safety against bottom instability approaches 1.0. In these cases, use of a rigid earth support system (such as a sinking caisson) extending below the base of the excavation is often necessary to provide the required factor of safety against bottom instability.

Sinking Caissons to Control Hydrostatic Uplift And/or Groundwater Inflow

Sinking caissons may also be suitable for conditions where high hydrostatic pressures exist in formations below the base of the shaft, or where the shaft is to be installed through granular soils below the water table. In each case, the caisson can be sunk while maintaining the water level inside the caisson to balance outside or underlying hydrostatic forces. Soil is excavated from inside the caisson using a clamshell, and the base slab is either placed under water, or placed after lowering the groundwater pressures below the base by means of dewatering wells or other methods.

SINKING CAISSON CASE HISTORIES

Five case histories of recent projects in which sinking caissons were used to construct shafts through difficult ground conditions were reviewed by the authors with respect to: geotechnical and hydrogeologic conditions; unique project requirements; other shaft construction systems considered; why the sinking caisson method of construction was chosen; and the outcome of the construction effort. The following case histories demonstrate the ability to successfully use the sinking caisson method of construction for varied types of ground conditions where conventional earth support systems were not considered to be practical and/or economical.

Detroit Wastewater Treatment Plant Pumping Station 2A Project

The Detroit Water and Sewerage Department's Pumping Station 2A project in southwest Detroit included construction of two sinking caisson shaft structures. The shaft required for the main pumping station was 38.4 meters (126 feet) inside diameter and approximately 21 meters (70 feet) deep, while the nearby connecting shaft was 9.1 meters (30 feet) inside diameter and approximately 15.2 meters (50 feet) deep. The structures were designed by Metcalf and Eddy, Inc., and constructed by Walbridge-Aldinger, Inc. Construction on the project was begun in 1990 and completed in 1994.

The Pump Station 2A project was constructed on the grounds of Detroit's Wastewater

Treatment Plant, within close proximity to numerous existing structures and underground facilities. Protection of these facilities, many of which were critical to the operation of the existing plant, was dependent on controlling construction-related ground movements.

Ground conditions at the site consisted of surficial fill and a relatively thin desiccated clay layer, underlain by thick deposits of very soft silty clay with shear strength as low as 17.2 ka (360 psf). The clay deposits were underlain by a relatively thin layer of very dense glacial till (locally termed hardpan), which was further underlain by fractured dolomitic limestone containing artesian groundwater.

Several methods of ground support for construction of the pump station and connecting structure were considered during the design process, including circular freeze walls, diaphragm walls, and sinking caissons. Ground freezing was considered undesirable due to local contractors' unfamiliarity with this method, its expected expense, and questions regarding potential for ground movements and associated utility disruption occurring during thawing. Diaphragm walls were also considered to be undesirable due to lack of local experience. The size and depth of the structures, combined with poor soil conditions, very low ground movement tolerances, and much local experience with the sinking caisson technique, made sinking caissons the method of choice for installing the pump station and nearby junction chamber structures.

Preparations for installation of both caissons included pre-excavating the ground surface approximately 3 meters (10 feet). In addition, a three stage rock grouting program was performed for the pumping station caisson to prevent hydrogen sulfide laden groundwater from entering the excavation as the caisson approached the hardpan and bedrock layers. Due to the presence of artesian water, these proceedures would have been required for the pumping station regardless of the system used. At the connecting shaft location, an approximately 3.6 meter (12 foot) clay separation was expected between the caisson bottom and the hardpan aquifer, and as such, artesian water infiltration was not expected and rock grouting was not specified.

The pumping station caisson was constructed with seven foot thick concrete walls which were placed in five lifts (including the caisson shoe). Using mass placement methods, the contractor placed each lift as a single pour, without vertical construction joints. Type II cement was used to combat the potential effects of dissolved sulfides and sulfates in the artesian groundwater. During excavation and sinking, a soil plug ranging in height from 3 to 6.4 meters (10 to 21 feet) above the tip of the shoe was required in order to maintain a stable excavation bottom through the very soft clay soils. Two levels of bentonite slurry injection pipes were cast into the caisson walls to allow for lubrication of the exterior during sinking. However, the contractor was able to control the sinking adequately by discrete excavation of the interior of the caisson, and bentonite injection was not necessary. Prior to placement of the last lift of the caisson walls, steel pipe landing piles were driven around the perimeter of the caisson to ensure that it would be plumb and level when it landed. After landing the caisson with the shoe approximately 0.6 to 1.8 meters (2 to 6 feet) above the sloping rock surface, the caisson shoe was underpinned to the bedrock surface. Because of overload factors as high as 16, the soft clay below the shoe could not be safely exposed. As such, a frozen soil wall was installed along the outside of the shoe to prevent soil squeezing and/or soil flow below the shoe, while the caisson was underpinned by excavating approximately 3 meter (10 foot) long zones of soft clay and hardpan, and filling with concrete. The rock

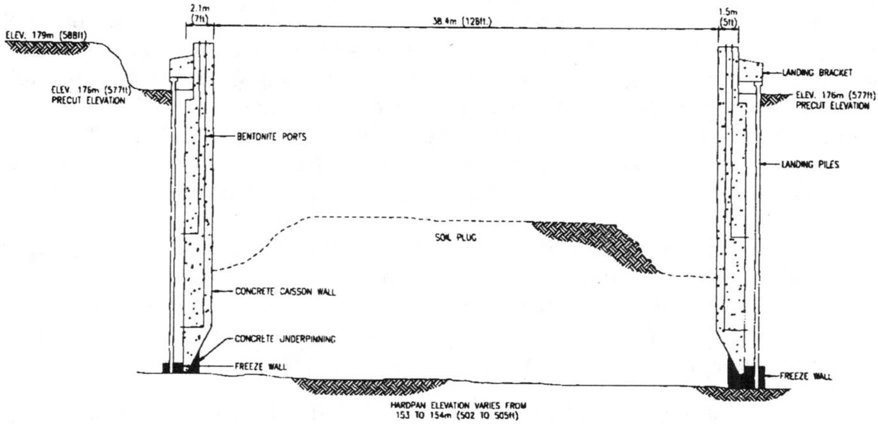

Figure 1. Pumping Station 2A Caisson

grouting program for the pumping station caisson proved to be a success, with one leaky area of the bedrock producing only about one gallon per minute of seepage. Construction time for the pumping station caisson, from shoe construction to landing the caisson, was approximately 32 weeks. Underpinning the shoe and placement of the base slab took approximately 8 additional weeks. An independent estimate by the contractor indicated that a braced steel sheeting shaft would have taken twice as long to construct.

The connecting shaft caisson was constructed with 1.5 meter (5 foot) thick walls which were placed in five lifts including the shoe. Type II concrete was also used for the junction chamber caisson. As with the pumping station caisson, excavation and sinking were completed by maintaining a soil plug within the caisson walls to maintain bottom stability, and steel pipe landing piles were driven around the perimeter of the caisson to ensure that it would be plumb and level when it landed. After sinking the caisson to the required depth, an artesian spring developed at the shaft invert. This flow was apparently the result of the landing piles penetrating the clay separation barrier and the underlying ungrouted artesian aquifer. In an effort to maintain a stable excavation bottom, the shaft was flooded. The caisson base slab was later installed underwater, using tremie concrete placement methods. Construction time for sinking this caisson, from shoe construction to landing, was approximately 22 weeks, with approximately 2 additional weeks required to install the base slab. However, several months elapsed in between the caisson landing and the base slab installation, due to the artesian flow into the shaft and the subsequent shaft flooding required to maintain a stable bottom.

Rouge Steel Scale Pit

Rouge Steel Company, located in Dearborn, Michigan, began work on a new continuous casting operation in 1987. The new facility required construction of a rectangular mill scale pit, 37.2 meters (122 feet) long, 14 meters (46 feet) wide, and 15.2 meters (50 feet) deep. The structure was designed by Lockwood Greene Engineers and NTH Consultants, Ltd, and constructed by Walbridge-Aldinger Company.

Ground conditions at the site of the new scale pit were very similar to those at the nearby Detroit Wastewater Treatment Plant. Factors important for the performance of the earth retention system included controlling bottom stability, and minimizing lateral movements and their impacts on adjacent structures. Several methods of installing the scale pit were considered, including an internally braced steel sheet pile cofferdam, soldier pile and lagging systems, and frozen ground techniques. The poor ground conditions, together with the use of the caisson walls as part of the permanent structure, made the sinking caisson option the least time consuming and the most cost effective in the contractor's view.

Prior to constructing the shoe for the scale pit caisson, the area was precut approximately 4.6 meters (15 feet) below the prevailing ground surface, and steel pipe landing piles were driven. The excavation bottom was expected to be within thick clay deposits approximately 6.1 meters (20 feet) above the water bearing hardpan and rock layers. As such, groundwater control measures such as rock grouting were not required.

The 0.9 meter (3 foot) thick caisson walls were placed in 4 lifts including the shoe. Type III concrete was used, since groundwater (and dissolved hydrogen sulfide) infiltration was not expected to be a significant problem. Bottom instability was prevented during sinking the caisson by maintaining a soil plug within the caisson interior. A 1.8 meter (6 foot) thick concrete base slab was installed after the caisson was sunk to design elevation.

Construction time for sinking the caisson and installing the base slab was approximately 11 weeks. This compared to an estimate of 20 weeks to install an internally braced cofferdam and construct the scale pit walls and slab within.

Redford CSO Pumping Station and Swirl Concentrator Caissons

The Redford Combined Sewage Overflow (CSO) retention facility includes a 18.3 meter (60 foot) inside diameter reinforced concrete pump station and a 10.7 meter (35 foot) inside diameter reinforced concrete swirl concentrator, constructed within 2.4 meters (8 feet) of each other. Both of these facilities were constructed as sinking caissons with completed base slabs requiring excavation to depths on the order of 16.8 meters (55 feet). These structures were designed by Wade-Trim/ Associates, Inc., for the owner, Charter Township of Redford. The project was constructed by Walbridge Aldinger Company.

The original contract documents as well as subsequent contractor test borings indicated soil conditions consisting of approximately 6 feet of surficial sand underlain by thick deposits of medium consistency silty clay, further underlain by compact predominantly granular layers, then shale bedrock. The soils below the base of the caisson excavations were expected to consist of silt, sand ,and hardpan layers under artesian groundwater pressure. Because these layers were expected to be very difficult to dewater, the contract documents required that hydrostatic uplifts forces be balanced during construction by flooding the caissons during excavation. In addition, the design called for dome shaped caisson base slabs to be placed under water using tremie concrete placement methods. The purpose of the domed base slab was to carry permanent hydrostatic loads and eliminate the need for rock anchors within the base slab.

Recognizing the difficultly and expense which would be associated with performing the excavation and base slab construction under water, the contractor proposed to develop

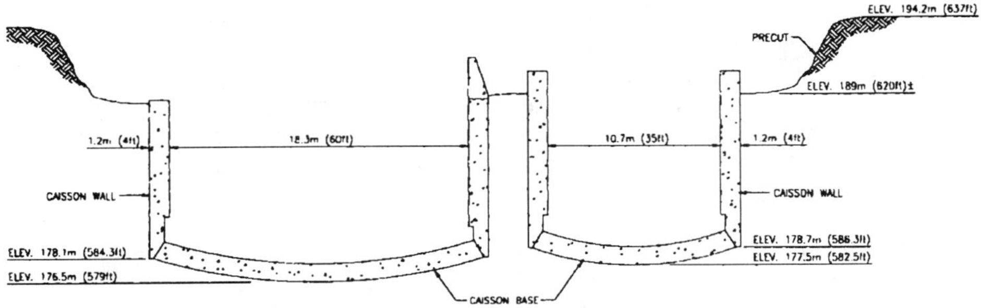

Figure 2. Redford CSO Shafts

a dewatering plan which would allow construction in the dry. A supplemental geotechnical investigation at the site helped define the presence and extent of the aquifer below the caissons. Based on grain size analysis of the soil samples, it was determined that the aquifer was dewaterable. A dewatering plan was then developed which was subsequently accepted by the owner's engineer.

In order to minimize the number and size of wells, a groundwater pumping test was conducted to further define the hydrogeologic parameters of the aquifer and possibly allow a reduction in the number of dewatering wells estimated to be required based on the grain size analysis. The results of this test allowed modification of the dewatering plan to include four, 0.9 meter (3 foot) diameter gravel packed wells.

The walls of the caissons were constructed in four lifts, including the shoe. Using the modified dewatering plan, the excavation was completed in the dry and the domed base slabs were placed without incident. The caissons were grouted in place by injecting grout through the bentonite injection pipes. Construction time was approximately 20 weeks for the 18.3 meter (60 foot) diameter caisson, and 15 weeks for the 10.7 meter (35 foot) diameter structure, not including the base slab construction. In each case, these slabs required approximately 3 additional weeks to construct.

River Rouge CSO Basin Demonstration Project

The City of River Rouge, Michigan has undertaken a major construction project to control the release of combined sewage into the Rouge River. The major component of the project is a large diameter concrete reinforced sinking caisson which will serve as the main combined sewage detention facility. The caisson has seven foot thick walls, with an outside diameter of 45.4 meters (149 feet) and a depth of approximately 22.9 meters (75 feet). This project was begun in the spring of 1996, and is planned to be completed in winter 1999. The project was designed by Sigma Associates, Inc., and is being constructed by Walbridge - Aldinger Company.

Soil conditions at the site consisted of approximately 2.4 meters (8 feet) of surficial sand and debris fill, underlain by an approximately 3 meter (10 foot) thick desiccated clay layer. The fill and sand layers typically contained perched groundwater. The desiccated clay is underlain by an extensive deposit of soft silty clay, which is further underlain by the hardpan till layer at a depth of about 20.7 meters (68 feet). The bedrock underlying the hardpan consists of dolomitic limestone, which is typically highly fractured within the upper 1 to 3 meters (3 to 10 feet). Artesian pressure within the hardpan and bedrock was measured at approximately 1.5 meters (5 feet) above the ground surface. The hardpan at this site is considered relatively impermeable, although the fractured bedrock is expected to produce high volumes of groundwater if it is exposed. Where encountered, artesian groundwater is expected to contain high concentrations of dissolved hydrogen sulfide, and moderate concentrations of methane gas.

Prior to constructing the caisson, the project site was pre-cut approximately 2.1 to 3 meters (7 to 10 feet) in order to prevent difficulties in sinking the caisson through the upper fill and desiccated clay layers. Perched groundwater was controlled by pumping from open sumps. A four stage rock grouting program was then undertaken to seal the fractured bedrock and to minimize infiltration of groundwater during excavation. Additionally, a geotechnical instrumentation program was undertaken with the purpose of protecting existing adjacent structures.

The caisson wall construction is planned to be competed in six lifts including the shoe. The excavation from the center of the caisson has been conducted in the dry, using a clamshell and smaller earthmoving equipment working within the caisson. At this writing, the caisson is close to its final elevation, and is planned to be completed within five weeks. Following sinking of the caisson walls to the desired depth, the clay zone below and around the perimeter of the shoe is planned to be subject to ground freezing to prevent soil squeezing below the shoe prior to base slab placement. Rock anchors will then be installed from the rock surface, and post tensioned in place following base slab placement.

The caisson construction is planned to be completed in approximately 28 weeks, not including the base slab construction. The base slab construction (including rock anchor installation) is expected to be completed in an additional 8 weeks.

Northwest Lateral Pipeline

Construction for the Costal Water Authority and City of Houston's Northwest Lateral Pipeline began in 1988 and represented an integral step in the expansion of the city's water supply system. The original design called for approximately 3350 meters (11,000 feet) of tunnel crossing two major bodies of water: the Houston Ship Channel and Greens Bayou. After award of the contract to the Greenfield Merriman Joint Venture, the joint venture retained NTH Consultants to redesign the water conveyance system under a value engineering provision in the contract. As part of the redesign, the tunnel depth was decreased and sinking caissons were used to construct "siphon" type extensions below the Houston Ship Channel and Greens Bayou.

At the Houston Ship Channel crossing, octagonal shaped shafts, which were specified in the contract documents to be either diaphragm walls or freeze walls, were redesigned to be round sinking caissons. This modification allowed a reduction in the wall thickness and reinforcement required, and made sinking caissons the economical choice. At the Greens Bayou crossing, round sinking caissons were designed to replace long sloping tunnel sections which extended below the Bayou. As a result of this design, the majority of the tunnel was installed at relatively shallow depth, while only the portions of tunnel directly below the

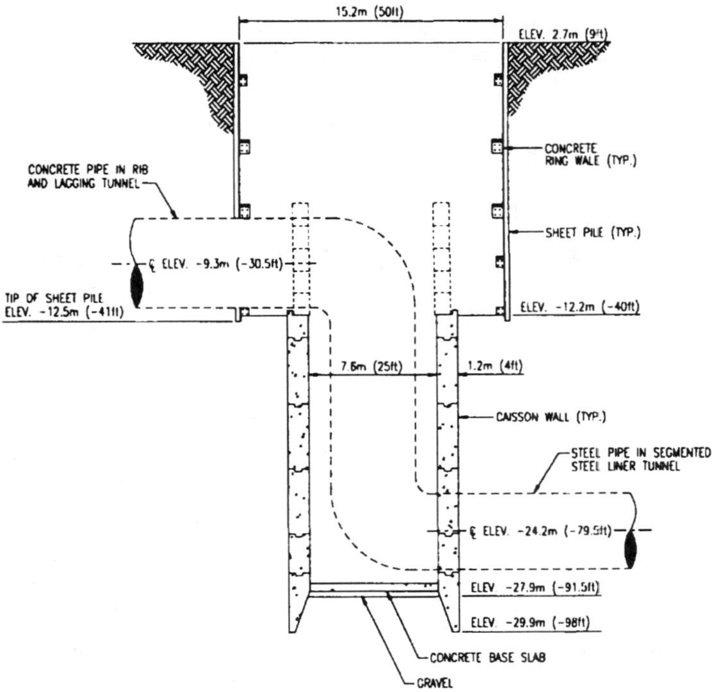

Figure 3. Greens Bayou Shaft

bodies of water were mined at the deeper depth. The decreased tunnel depth through most of the tunnel alignment allowed the use of steel rib and lagging primary tunnel liner rather than steel liner plate as called for in the original design. In addition, steep tunnel grades which were called for in the original design were eliminated.

The four shafts used to construct the siphon extensions were each constructed in similar ground conditions. Subsurface soils of each shaft locations generally consisted of medium to stiff clay with interbedded water bearing zones of fine sand and silt, underlain at depths of 21 to 30 meters (70 to 100 feet) by generally soft clay. Where groundwater was present within the water bearing layers, the potentiometric level was about 3 to 4.6 meters (10 to 15 feet) below ground surface.

The shaft designs at the Greens Bayou and Houston Ship Channel crossings called for two stage shaft construction, consisting of an upper sheet pile supported shaft and a lower concrete sinking caisson shaft. The upper steel sheet pile shafts were 15.3 meters (50 feet) deep, and were supported internally by concrete ring wales. The lower sinking caisson portions of each shaft were 7.6 meters (25 feet) in internal diameter with 1.2 meter (4 foot) thick walls. The shafts extended to overall depths of approximately 33.5 meters (110 feet) below ground surface. It should be noted that only one of the four sinking caisson structures was designed to be incorporated into the permanent structure. The remaining three caissons were used only as temporary structures, and were abandoned in place.

Table 1. Cost Comparisons for Sinking Caissons
Vs. Other Shaft Construction Techniques

Project/Shaft	Total Cost (Sinking Caisson Method)	Cost per Unit Volume for Caisson Method	Comments
PS2A - Pumping Station PS2A - Conn. Shaft	\$5,080,000 (caisson walls) \$2,213,000 (base slab w/ freeze & underpin) Not Available	\$842/m³ (\$643/cy) \$816/m³ (\$624/cy) Not Available	1. Costs for steel sheeting and bracing were quoted at \$2 million higher than sinking caisson. 2. Costs for freeze wall were quoted at \$3 million higher than sinking caisson. 3. At least 6 months were saved over alternative methods.
Rouge Steel Scale Pit	\$2,065,000 (caisson walls) \$301,000 (base slab)	\$943/m³ (\$721/cy) \$322/m³ (\$246/cy)	1. Savings of \$0.7 million gained over use of braced sheet pile shaft construction.
Redford CSO (60'dia) Redford CSO (35'dia)	\$337,000 (caisson walls) \$152,000 (base slab) \$173,000 (caisson walls) \$49,000 (base slab)	\$388/m³ (\$297/cy) \$387/m³ (\$296/cy) \$322/m³ (\$246/cy) \$323/m³ (\$247/cy)	
Rouge CSO Basin	\$4,717,000 (caisson walls) \$635,000 (base slab)	\$801/m³ (\$612/cy) \$177/m³ (\$135/cy)	1. Savings of \$2.5 million gained over use of braced sheet pile shaft construction. 2. Construction time for caisson estimated to be half that of braced sheeting.
NW Lat Pipe (all shafts)	Not Available	Not Available	

Control of ground movements during construction was critical due to the presence of an active railroad line which was located within 6.1 meters (20 feet) of the edge of one of the shafts. Borehole extensometers were used to measure ground movement in this area. Within the sinking caissons, bottom instability was typically prevented by maintaining a soil "plug" within the shaft bottom. Within one of the Greens Bayou shafts however, flooding the caisson was also required during sinking to balance hydrostatic pressures and groundwater seepage into the shaft. Typically, bentonite was injected at the shoe during excavation to reduce friction and adhesion on the outside of the caisson, and to aid in sinking. Once the caissons were advanced to the required depth, they were grouted in place by injecting grout through the bentonite injection pipes. The average time to construct each of the shafts was approximately 3 to 4 weeks for the upper sheet pile stage and 10 to 12 weeks for the lower sinking caisson (including the base slab).

COST COMPARISONS

Costs for constructing the sinking caissons discussed above were summarized and compared to other construction methods which were considered. All costs were adjusted to 1996 dollars. Based on our review of cost data for these projects, the overall cost for sinking caissons in place (excluding base slab construction) was found to be approximately $795 per cubic meter ($610 per cubic yard), which can be equated to approximately $1450 per square meter ($135 per square foot) of exposed shaft wall. However, it should be noted that unit area costs varied more widely than unit volume costs due primarily to differences in the wall thicknesses. Other methods which were considered for the above projects ranged from $1940 to $2150 per square meter ($180 to $200 per square foot) of wall, including the cost of the final concrete shaft walls in place. A summary of cost data for the above discussed projects is presented in Table 1.

CONSIDERATIONS FOR DESIGNERS

Several considerations are apparent based on the author's experience in the design and construction of shafts using sinking caissons. These are discussed below:

1) Sinking caissons are often cost effective when utilized as both the construction shaft and final structure. The volume of concrete used for the caisson walls will typically be greater than the volume of concrete required for the walls of the final structure, as a consequence of the weight required to sink the caisson walls. As a rough comparison of construction costs between sinking caisson construction and temporary earth support with an internally constructed structure, the preliminary design may compare the cost of the additional concrete required in the caisson walls (above that amount which would be required for the final structure), to the cost of a temporary earth support system. The designer should be aware, however, that the majority of caisson wall construction costs are from the labor for vertical form work, rather than the bulk materials. As such, the costs for thickened caisson walls do not increase the overall shaft costs incrementally.

2) Sinking caissons are highly effective in controlling ground movements adjacent to constructed shafts. Based on empirical studies of ground movement by Peck and D'Appolonia, as well as project experience from the above case histories, settlement adjacent to sinking caisson shafts is typically on the order of 0.2 percent of the depth of the shaft. This compares to estimates by Peck of 1 percent to 2 percent of the shaft depth for flexible earth

support systems such as internally braced steel sheet piles.

3) The design of sinking caissons in clay soils is highly sensitive to adhesion and friction of the soil surrounding the caisson walls, as well as the shear strength of soils below the base. Methods of evaluation of basal stability (Bjerrum and Eide, and Terzaghi), suggest that failure zones will extend to a depth below the caisson shoe approximately equal to 0.7 times the shaft width. As such, geotechnical investigations should be planned to gather data to depths equal to a minimum depth corresponding to one shaft width below the planned final shaft invert. Where bottom stability concerns require utilization of a soil plug within the caisson to control bottom heave during excavation, geotechnical investigations should extend to a depth below the caisson shoe of at least 0.7 times the width of the shaft. Soil testing should include unconfined compression strength evaluations or unconsolidated undrained triaxial tests of relatively undisturbed Shelby tube samples. For soft and very soft clay, in-situ vane shear testing should also be considered. Such samples and/or in-situ testing should be procured and/or performed at minimum intervals of approximately 15 feet. Where artesian groundwater is encountered within test borings, a piezometer should be installed to allow for design provisions to prevent hydrostatic uplift of the shaft base.

4) The use of sinking caisson techniques is often most successful and most cost effective where ground conditions are considered very poor. In the words of a contractor's tradesman: "The worse the clay is, the better these sinking caissons work!" It is significant that in most of the above cited cases, sinking of the caisson was controlled by discrete excavation, without using the bentonite lubrication systems which were installed.

5) Sinking caissons have proved to be an effective means of construction for both owner-designed and contractor-designed shafts. Advantages to owner-designed shafts include: ability to control ground movements (by specifying construction means such as sinking caissons); ability to incorporate the sinking caisson into the permanent structure; speed of construction in appropriate conditions; and the ability to obtain competitive bids for the specific type of shaft construction desired. The primary advantage of allowing the contractor to design shafts is that creative and innovative solutions can be proposed by the contractor, and the owner can share in the cost savings which result from contractor ingenuity.

BIBLIOGRAPHY

Bjerrum, L. And Eide, O., 1956, "Stability of Strutted Excavation in Clay," Geotechnique, Vol. 6, pp. 32-47.

Broms, B.B. and Bennermark, H., 1967, "Stability of Clay at Vertical Openings," Journal of the Soil Mechanics and Foundations Division, Proceedings, American Society of Civil Engineers, Vol. 93, No. 1, Jan., pp 71-94.

D'Appolonia, D.J., 1971, "Effects of Foundation Construction on Nearby Structures," Proceedings of the 4th Panamerican Conference on Soil Mechanics and Foundation Engineering, Vol. 1, pp. 189-236.

Ivor-Smith, D.J., et al., 1989, "Deep Soft Ground Tunneling Under Houston Ship Channel and Greens Bayou," Proceedings, Rapid Excavation and Tunneling Conference, June, 1989, Pond, R.A., and Kenny, P.B., ed., pp. 230-243.

Olden, J.M., and DiPonio, J.J., 1989, "Northwest Lateral Progress," World Tunneling and Subsurface Excavation, Vol. 2, No. 6, Aug., pp. 310-316.

Peck, R.B., 1969, "Deep Excavations and Tunneling in Soft Ground, State-of-the-Art Report," Proceedings of the 7th ICSMFE, Mexico City, State-of-the-Art Volume, pp 225-290.

Porter, T.G., et al., 1993, "Shaft Excavation in Soft Clay by Caisson Construction," Retaining Structures, Clayton, C.R.I., ed., Thomas Telford, London, pp 816-825.

Terzaghi, K., 1943, Theoretical Soil Mechanics, John Wiley & Sons, Inc., New York, New York.

Chapter 30

RENOVATION OF THE INLAND SYSTEM

Christopher D. Orlandi,[1] Thomas P. Kwiatkowski,[1] Len Cobb,[1]
and Judith H. Jones[2]
[1]Jenny Engineering Corporation; [2]Christy/Cobb, Inc.Christy/Cobb, Inc.

The Inland System supplies raw water to the industrial and domestic users in the City of Birmingham, Alabama. The Inland System, which is nearly 60 years old, consists of an approximately 57.9m (190 feet) high earthen dam across the Blackburn Fork of the Black Warrior River, a 39.6m (130 feet) high reinforced concrete intake structure, a 1524mm (60 inch) diameter Dresser coupled steel transmission pipeline, and various distribution lines extending into Birmingham. The Inland System was owned, operated and maintained by the Industrial Water Board (IWB) of the City of Birmingham, an agency of the city government, from 1958 until 1994. In 1994, the City of Birmingham consolidated all water and distribution functions under the Birmingham Water Works Board (BWWB).

In 1988, the IWB determined that the Inland System had reached a point where its reliability could not be assured without an in-depth evaluation and renovation. The IWB engaged Christy/Cobb, Inc. (CCI) of Birmingham as its consultant for the renovation work. CCI evaluated the Inland System and submitted a report to the IWB in April, 1992. The report contained several recommendations for capital improvements to the system including increasing the emergency spillway capacity, reconstructing the fuse plug, relocating public access to the lake away from the dam area, constructing a new intake structure, and constructing a tunnel through the dam abutment to deliver the water from the proposed intake structure to the existing 1524mm (60 inch) diameter supply line. The new intake structure and tunnel would provide for a future maximum demand flow rate as well as provide bypass and level control capabilities. CCI was authorized in 1992 to proceed with the engineering design for the renovation of the Inland System with emphasis on the spillway and intake structure.

PROJECT DESCRIPTION

The major component of the renovation of the Inland System was the new intake structure and tunnel through the dam abutment. CCI engaged Jenny Engineering Corporation (JEC) of Springfield, New Jersey as a consultant for the tunnel design. JEC was responsible for the design of the tunnel, the connecting shaft between the intake structure and tunnel, and the steel pipe final lining, while CCI was responsible for the intake structure and the connection to the existing supply pipeline as well as overall project management. CCI and JEC evaluated the project site in order to select the location for the new intake structure and set the alignment for the tunnel. A factor in the evaluation process was that the level of the lake could not be lowered to construct the new intake structure since the supply of water from the Inland System had to maintained. Therefore, a location was required which was

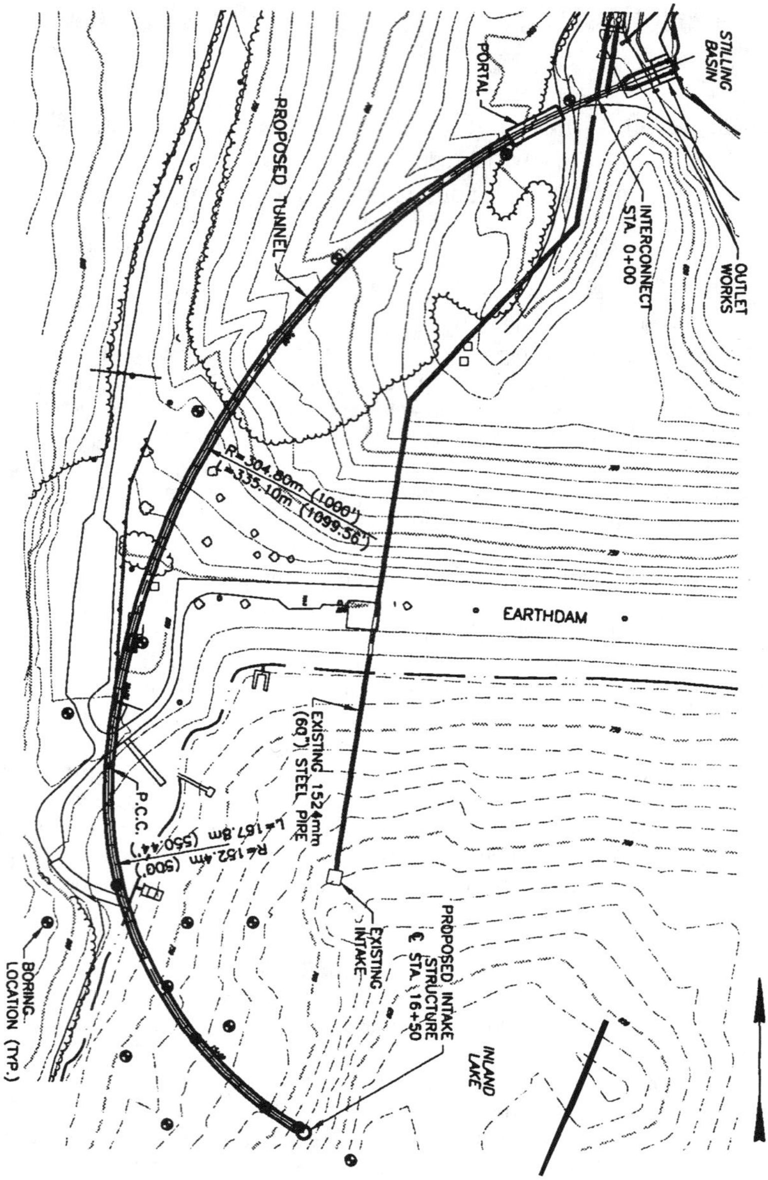

Figure 1. Tunnel Alignment

either land based or within the lake itself. Another factor in the evaluation was the requirement for the new intake structure to be capable of drawing water from varying depths in order to satisfy water quality concerns over the life of the structure. It was determined that due to the slope of the lake bottom and the topography around the dam, a land based structure was not feasible. A location in the lake was then selected.

The location selected for the new intake was approximately 182.9m (600 feet) south of the dam and approximately 98.4m (300 feet) south of the existing intake structure. The water depth at the new intake location was approximately 31.7m (104 feet) to the lake bottom. The lake bottom at this location was sloping which required underwater drilling and blasting in order to create a flat base area for the new intake.

The tunnel alignment was set after the location of the new intake shaft was established. It was determined that the tunnel alignment would originate at the interconnection point of the existing supply line and terminate at the centerline of the new intake shaft. The alignment was laid out in a continual curve which could be constructed by either tunnel boring machine (TBM) or drill and blast excavation methods (Figure 1). The initial curve consisted of a 304.9m (1000 ft.) radius for a length of 335.1m (1099.56 ft) to a point of compound curvature (PCC). From the PCC the alignment followed a 152.4m (500 ft.) radius curve for 167.8m (550.44 ft.) to the centerline of the new intake shaft. Station 0+00 was subsequently located at the interconnection point of the existing supply line and the shaft centerline was located at Station 16+50. Based on the topography at the supply line side of the dam, JEC determined that a portal would be the optimum method for starting the tunnel. The portal was located between Station 0+50 and Station 1+20 on undeveloped watershed land.

The tunnel elevation was based on several factors including the rock quality found during the geotechnical investigation, the elevation of the stilling basin at the downstream side of the dam, the elevation of the existing supply pipe, and the elevation of the bottom of the new intake shaft. The design team wanted to ensure that sufficient rock cover was maintained over the crown of the tunnel under the lake, while also minimizing the amount of rock excavation required for the pipe trench between the tunnel portal and the existing supply pipe and the stilling basin. The final design established a tunnel invert elevation of 609.16 at Station 1+20 (the face of the portal) and an invert elevation of 615.25 at Station 16+50, with a pipe slope of approximately 0.4% (Figure 2). The tunnel depth below the ground surface varied along the alignment. From the portal to the top of the dam the ground elevation increased with cover above the crown varying from 10.7m (35 ft.) to 54.9m (180 ft.). From the lake edge to the new intake shaft, however, the cover between the crown of the tunnel and the lake bottom decreased from approximately 48.8m (160 ft.) to approximately 7.9m (26 ft.).

GEOLOGIC CONDITIONS AND GEOTECHNICAL INVESTIGATION

Inland Lake is located in Blount County, Alabama, which is in the northern part of the state (Figure 3). The regional geology of the area consists of the Pottsville Formation of Pennsylvanian age which consists of a wedge-shaped body of sediments containing fragments of older rocks that thicken in a southwestward direction. The Pottsville Formation is exposed primarily in the Black Warrior Basin. It is the southern most of a series of Pennsylvanian age basins within the Appalachian Plateau. The Pottsville Formation in the Black

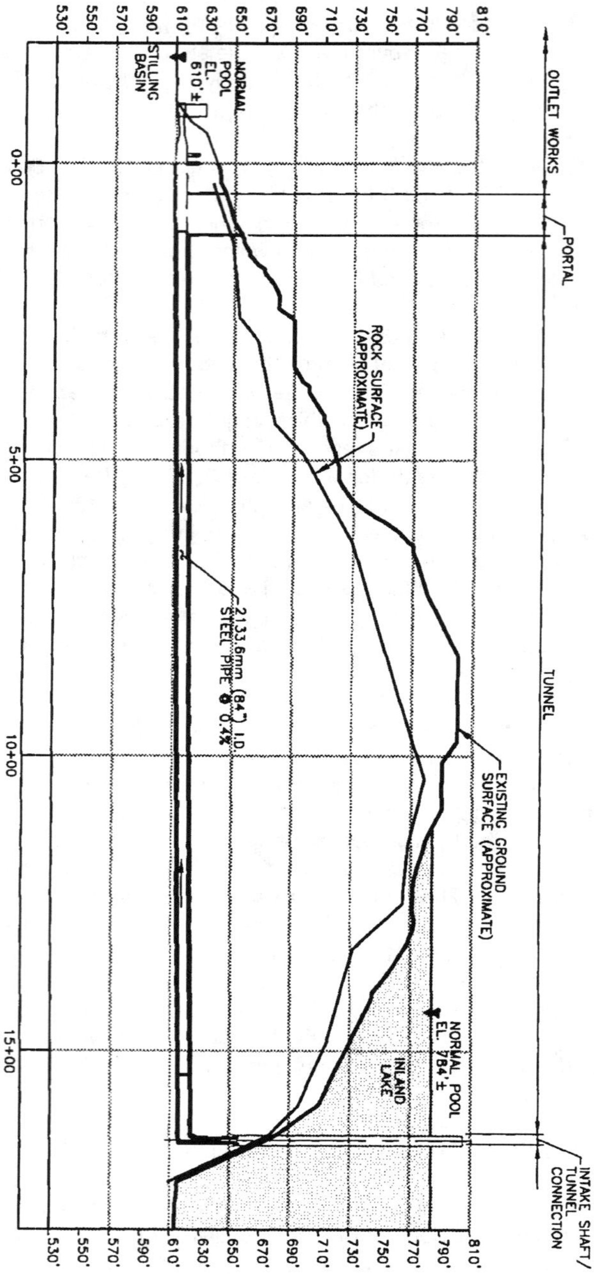

Figure 2. Tunnel Profile

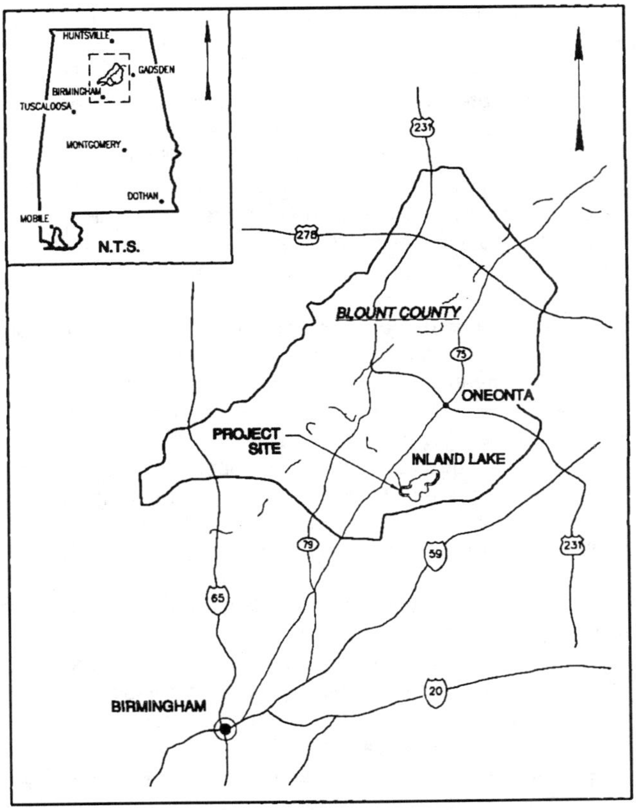

Figure 3. Location Plan

Warrior Basin consists of over 1828.8 m (6,000 feet) of sandstone, siltstone, shale and coal which overlies carbonates of Bangor Limestone of Upper Mississippian age.

A geotechnical investigation program consisting of 18 vertical exploratory borings was conducted for the project. Eight borings were made along the proposed alignment on land and an additional ten borings were performed in the lake in order to provide detailed design information for the tunnel and intake shaft sections of the project. The boring spacing generally ranged from 30.5m to 76.2m (100 to 250 feet) along the land portion of the tunnel alignment, and from 10.7m to 42.7m (35 to 140 feet) for the water borings. The borings and associated tests were performed by Law Engineering of Birmingham as contracted by the IWB.

The borings along the alignment were generally drilled two to three tunnel diameters below the proposed tunnel invert. The varied spacing of the borings was utilized in an attempt to define the geologic conditions at the project site, including the presence or likelihood of any anomalies and representative discontinuity types. The depths of the land borings varied along the alignment due to the local topography. They ranged from 10.7m to 56.1m

(35 to 184 feet) in depth, while the water borings ranged from 33.5m to 60.4m (110 to 198 feet) below the water surface.

The boring program indicated that the stratigraphy of the project area generally consisted of an overburden of fill material overlaying a thick sequence of sedimentary strata of the Pottsville Formation of Pennsylvanian age. The fill material varied in thickness from 3.0m to 9.1m (10 to 30 feet) and was composed of coarse sand, clayey sand and inorganic clay, sand and silt mixes. The Pottsville Formation consists of a basal orthoquartzitic, conglomerate sandstone and an upper irregularly alternating sequence of shale, sandstone and siltstone, randomly punctuated by thin coal beds.

The principal rock types along the tunnel corridor were sandstone and shale with minor subzones of siltstone. The sandstone unit extended throughout the length of the tunnel and continued well below the invert elevation. The sandstone texture ranged from fine to coarse grained, thin to very thick, and evenly bedded. Some beds exhibited cross bedding and contained ripple laminations. The sandstones were tan to greenish gray on fresh surfaces, but weathered to various shades of brown. The upper shale sequences were generally silty, medium to dark gray or greenish, well indurated, but very fissile, and contained both small and large randomly dispersed siderite nodules. The beds ranged from thin to moderately thick and were generally cross bedded containing ripple laminations.

Data from the geotechnical investigation and testing programs indicated that the proposed tunnel excavation would encounter sandstone throughout the entire length of the alignment. Rock quality along the tunnel route was expected to be good to excellent with some variation in compressive strength values. Test results classified the sandstone as moderately hard to hard, although some boring logs described zones of very hard material. The sandstone beds along the tunnel alignment between the land borings were typically thick to very thick (generally blocky) and contained some closely to moderately jointed zones. The boring logs for the water borings, however, described the bedding as thinly laminated and close jointed. This was interpreted as indicating that the rock structure along the tunnel alignment could be expected to be seamy, leading to slabbing and overbreak in the tunnel crown.

Dominant tectonic joints trending north 53 degrees west at spacing ranging from 4.6m to 6.1m (15 to 20 feet) apart were found to be well developed throughout the shale and sandstone strata, and commonly dipped 87 degrees in a northeast direction, although variations in dip and strike of five degrees were observed at some joints. Large scale relief joints oriented north 10 to 20 degrees east dip toward the valley floor at about 70 degrees. Joint spacing on the relief joints generally ranged from 0.6m to 1.2m (2 to 4 feet) apart. The sandstone joint patterns, based upon observed exposures, were more closely spaced, irregular, with offset steps to continuations in the underlying beds. These joints were frequently open and controlled the flow of groundwater.

Groundwater infiltration into the tunnel was expected to be mainly through joints and fractures in the rock. The boring program confirmed a direct interconnection between joints when, while one land and one water boring were being drilled, water was observed to be flowing from an observation well installed in one of the previously drilled land borings. The land and water borings were located approximately 30.5m and 121.9m (100 and 400 feet) respectively from the observation well. This was interpreted as indicating that the joint/

fractures encountered in the tunnel could be linked to the lake, thereby providing a continuous recharge source.

As a result of the finding of the potential link between the joint/fractures and the lake, the contract documents required a heading probe hole to be drilled ahead of the tunnel excavation in an attempt to intersect any water bearing seams prior to the tunnel excavation face reaching the seam location. If infiltration was visible through the heading probe hole, cut-off grouting operations could be initiated to control the inflows. The documents also required that the intake shaft/tunnel connection be pregrouted prior to any excavation taking place. The purpose of the pregrouting was to inject grout into any joints or fractures that were present in order to minimize infiltration of water into the shaft from the lake.

TUNNEL AND SHAFT DESIGN

Tunnel

The tunnel was designed to allow for TBM or drill and blast methods of excavation. The cross section of the tunnel for the TBM excavation consisted of a minimum 2.9m (9'-6") circular opening with initial supports consisting of either pattern rock dowel reinforcement or steel ribs, depending on the rock conditions. The drill and blast opening was a 2.9m (9'-6") wide by 2.8m (9'-3") high horseshoe with initial support requirements similar to the TBM excavation. The openings were sized to permit the installation of the 2133.6mm (84 inch) diameter steel pipe final lining (Figure 4). A heading probe hole was to be drilled ahead of the tunnel excavation in order to intersect any water bearing seams. If water infiltration was visible through the heading probe hole, cut-off grouting operations were to be initiated to control inflows. A temporary bulkhead was specified to be constructed in the tunnel at Station 15+40 prior to excavating the intake shaft/tunnel connection. The temporary bulkhead was specified in order to guard against flooding the tunnel due to the construction of the shaft between the intake structure and the tunnel. It was believed that blasting could open joints or fractures in the rock which could be connected to the lake bottom.

The steel pipe final lining was designed to withstand both the internal water pressure as well as anticipated external pressure from 27.4m (90 feet) of hydrostatic head of the lake. The external pressure controlled the design since there was a possibility of the concentrated hydrostatic pressures causing local buckling of the pipe. The pipe design was similar to that used for steel penstocks where the differential shrinkage between the steel pipe and the concrete fill of the annulus around the pipe was calculated to determined a gap ratio through which the pipe would be subject to the external hydrostatic pressures. It was determined that a 12.7mm (1/2 inch) pipe thickness was adequate for the project conditions.

Intake Shaft/Tunnel Connection

The connection between the intake structure and the tunnel was designed as a 2.9m (9.5 feet) diameter circular shaft (Figure 5). The depth between the rock at the bottom of the structure and the tunnel invert was approximately 10.7m (35 feet). The lower 5.2m (17 feet) of the shaft was designed to be over excavated between the shaft wall and the tunnel crown in order to accommodate the 90 degree steel pipe elbow. The initial supports for the connecting shaft consisted of pattern rock dowels and 76.2mm (3 inches) of shotcrete with welded wire reinforcement installed around the circumference of the shaft.

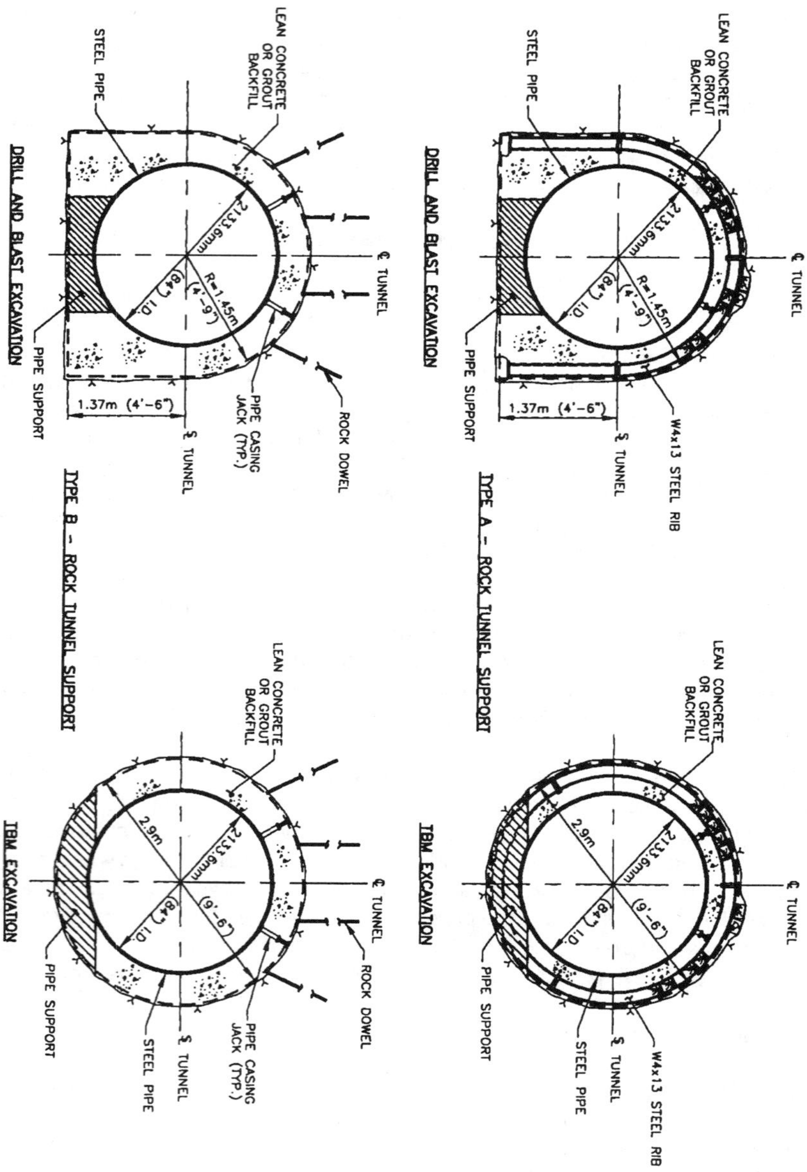

Figure 4. Tunnel Types

The construction of the connecting shaft between the new intake structure and the tunnel was a concern due to the possible connection between rock joints and the lake bottom which could cause a flood in the tunnel. Therefore, the specifications required that the rock area from the bottom of the intake structure to 1.5m (5 feet) below the tunnel invert be pregrouted prior to any excavation.

Intake Structure

The intake shaft was designed as a precast concrete segmental structure. It was anticipated that the concrete segments would be cast on-site and then transported out to the shaft location in the lake and lowered to the bottom where a flat base area would have been excavated out of the sloping lake bottom. The bottom 4.6m (15 feet) of the shaft would be sealed with tremie concrete placed around and in the structure. After the segments were all placed, the entire structure would be post-tensioned using rock anchors drilled and grouted into the rock to a depth of approximately 21.3m (70 feet) below the bottom of the structure.

The intake structure consisted of approximately 22 segments (Figure 6). Each segment was generally 2.1m (7 feet) high with an inside diameter of 4.9m (16 feet) and an outside diameter of 6.1m (20 feet). 1.5m (5 feet) long by 0.6m (2 feet) thick wingwalls were located at the quarter points of the structure. The rock anchors were located through conduits placed in the wingwalls. Trash racks for the four sluice gates were also attached to the outside of the wingwalls. The minimum concrete strength required for the segments was 34473.8 kPa (5,000 psi).

CONSTRUCTION

The project was issued for bids in September, 1993. The project was awarded in December 1993 to McNally Tunneling Corporation of Westlake, Ohio for approximately $12 million with a Notice to Proceed issued in February of 1994.

Intake Structure

The underwater drilling and blasting and the construction of the intake structure were carried out by Explosives and Diving Services, Inc. of Clarksville, Florida and The Industrial Company Southeast, Inc. (TIC) of Savannah, Georgia, respectively, under subcontract to McNally. The drilling, blasting and excavation work was performed from barges anchored by cables which were attached to points on land. The cables were connected to a hoist on the barges which was used to maintain the position of the barges over the intake structure location. Approximately sixty 101.4mm (4 inch) diameter presplit holes and sixty 101.4mm (4 inch) diameter production holes were drilled within the 10.7m (35 feet) by 15.2m (50 feet) base area for the intake structure. The hole depths varied from approximately 12.2m (40 feet) to 2.9m (9.5 feet). The total volume of rock excavated was approximately 1124.6cu. m (1470 cy). The excavated rock was placed downslope from the base area on the lake bottom. Drill placement and excavation quality was monitored by divers.

When the blasting for the base area was completed, construction of the intake structure was started. TIC had proposed to cast the segments and construct the intake structure on the barges used to excavate the base area. Their proposal was to construct a frame with jacks on which each segment of the structure would be cast in succession from the bottom

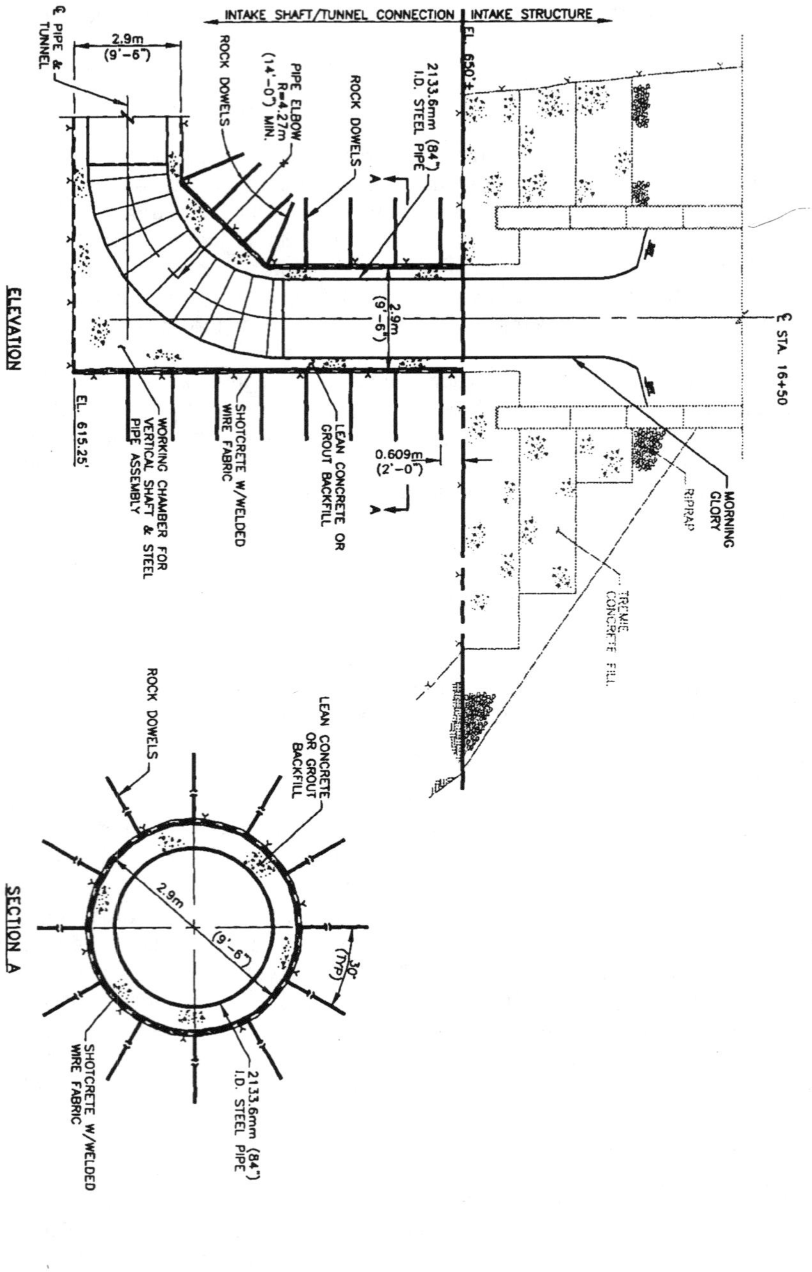

Figure 5. Intake Shaft/Tunnel Connection

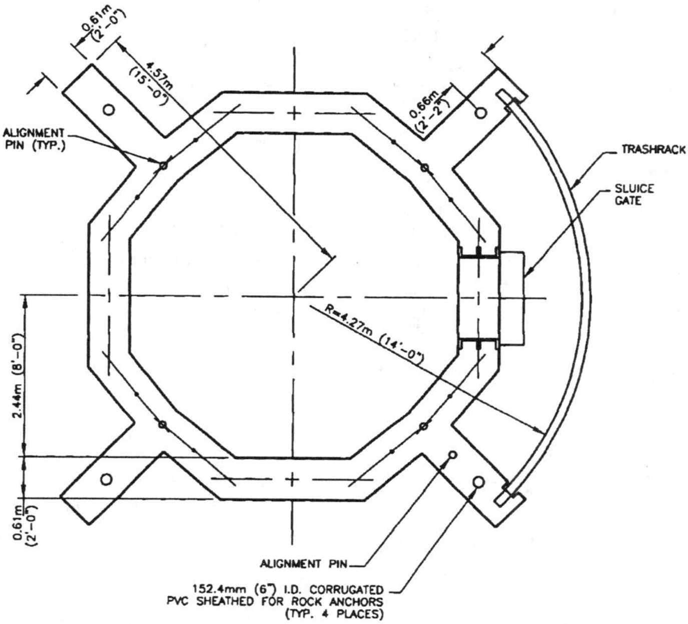

Figure 6. Intake Structure Segment

up. The jacks would hold the segment until the forms were able to be removed. Once the forms were removed, the segment would be lowered by the jacking system and supported by cables. The forms would then be placed above the previous segment and the next segment would be cast. The cycle of casting and lowering the segments would be repeated until the structure reached the lake bottom. The design team was concerned about the barges supporting the suspended weight of the entire intake structure, especially if any movement was to occur due to boat traffic on the lake. TIC provided documentation to support their proposal and the proposal was accepted by the design team.

The structure reached the base area at the bottom of the lake in March of 1995. Tremie concrete was placed around the structure to create a seal between it and the rock. A steel dome with a valve had been placed inside the lower segment during the casting operations. It was intended as an additional seal for the bottom of the structure during the placement of the tremie concrete and after the shaft was dewatered. The four rock anchors were drilled and installed through the sleeves in the wingwalls following the placement of the tremie concrete. Each of the rock anchors were tensioned to 622.8kN (140,000 lbs) in opposite pairs in order to resist the buoyant uplift forces exerted on the structure. The intake structure was then dewatered so that the pregrouting operations could be conducted from within it.

Tunnel

McNally had decided to construct the tunnel using drill and blast methods. Portal excavation was begun on September 7, 1994 and tunneling was started on October 18,

1994. The portal was ramped so that Load-Haul-Dump (LHD) equipment could be used for muck removal operations. Closely spaced presplit holes were used for the portal excavation to protect the 1524mm (60 inch) supply pipe which was only 5.2m (17 feet) away. The portal was successfully excavated without any damage to the supply pipe.

The tunnel excavation was a 2.9m (9.5 feet) by 2.9m (9.5 feet) horseshoe. Approximately thirty-two 38.1mm (1.5 inch) diameter holes were drilled for each tunnel round using a Minejack single-boom hydraulic jumbo. The drill holes varied in depth between 2.4m (8 feet) and 3.0m (10 feet). As the tunnel excavation progressed, pattern rock dowels were installed in the crown behind the drill jumbo. Due to the horizontal bedding of the sandstone, overbreak in the tunnel generally varied between 152.4mm (6 inches) and 304.8mm (12 inches) and the crown was usually a flat slab instead of an arch shape.

A horizontal clay seam located near the tunnel crown was discovered during the start of tunneling operations. This clay seam was visible for the initial 21.3m (70 feet) of the tunnel excavation and caused a large amount of overbreak at the tunnel crown. The clay seam and the tunnel excavation diverged at approximate Station 1+60. The contractor installed five W4x13 steel ribs with timber lagging on 1.2m (4 foot) centers to support the rock between the tunnel crown and the clay seam. These were the only steel ribs installed in the tunnel. Additional steel straps and wood blocking were also installed between the portal face (Station 1+20) and the steel ribs in order to further support the rock.

The tunnel excavation continued from October 1994 to late March 1995. The excavation encountered minor water inflows through vertical joints along the tunnel alignment, but no cutoff grouting was required. Tunneling was halted after reaching station 15+90. McNally had proposed that a rock plug be left in place between the shaft connection and the tunnel in lieu of constructing a bulkhead as required in the contract documents. CCI and JEC accepted the proposal with the stipulation that a probe hole be drilled between the tunnel and the shaft location in order to determine the possibility of water inflows. A 76.2mm (3 inch) diameter probe hole was drilled through a valve and plate assembly bolted to the face of the tunnel heading. No significant water was intercepted by the probe hole.

Intake Shaft/Tunnel Connection

McNally began the pregrouting operations of the rock below the intake structure on August 16, 1995. The drilling and the grouting operations were all performed from the bottom of the intake structure. Access for personnel and the drilling and grouting equipment to the invert of the intake structure was by a crane located on the barges. The top of the intake structure was above the cab level of the crane, so a spotter was located on a temporary platform constructed at the top of the structure. The spotter maintained communication between the crane operator and the personnel in the shaft and directed the crane operator during the lowering and raising of equipment and the mancage.

Initially, five pregrout holes were drilled through the steel dome at the bottom of the structure. Four holes were located at the quarter points between the wingwall locations. The fifth hole was located at the center of the shaft. Grout was injected through the initial holes until there was no more take. An additional eight holes were then drilled around the inside perimeter of the shaft and grouted. Approximately 18292 kg (40326 lbs) of cement were used for the pregrouting operations. When the pregrouting operations were completed, the steel dome was cut away with a torch and removed by the crane from the intake structure.

Drill holes for the shaft excavation were started in late September of 1995. The contractor's blasting plan called for line drilled holes at the shaft perimeter and approximately twenty-eight 50.8mm (2 inch) diameter production holes drilled in two concentric circles within the line drilled holes. A 127mm (5 inch) diameter cut hole was drilled at the center of the shaft. Drilling was completed and the first shaft round was fired on October 2, 1995. A total of 18 shots were made for the shaft excavation. Muck from each shot was removed by a clamshell bucket and placed in the lake away from the intake structure. The contractor used a spotter within the intake structure to direct the crane operator during the mucking operations since the operator and the spotter at the top of the structure could not see the invert of the shaft. A shelter constructed out of a W36 beam was placed on the ledge between the intake structure and the shaft to protect the spotter. The last amount of muck for each shot was hand loaded into a skip bucket since the clamshell could not effectively remove all the muck. The shaft excavation reached the tunnel invert elevation on October 26, 1995. The contractor proposed that initial supports not be installed in the shaft since the rock appeared to be sound, water inflows were minimal, and it was intended to place and grout the vertical steel pipe and elbow assembly within the shaft as soon as the tunnel excavation was completed. The design team accepted the proposal and the owner received a credit for the work not performed.

The final 15.2m (50 feet) of tunnel was excavated between November 13, 1995 and November 16, 1995 using hand drills to drill the last rounds for blasting. After the tunnel was broken through to the shaft, the crown area was excavated for the pipe elbow assembly. Pipe installation was begun in December, 1995. The elbow assembly was lowered in one piece by the crane through the intake structure and set in-place on supports at the tunnel/shaft invert. One 6.6m (20 feet) section of vertical pipe was also lowered through the intake structure and welded to the top of the elbow. The last vertical section of pipe was the morning glory. It was a 6.6m (20 feet) section of pipe with a transition at one end. The transition was required in order to reduce the entrance losses between the 4.9m (16 feet) diameter intake structure and the 2133.6mm (84 inch) diameter pipe. The maximum diameter of the transition was approximately 2775mm (109 inches). Grout was placed in the annulus between the steel pipe and the rock and concrete fill was placed in the space between the intake structure walls and the morning glory.

Pipe-in-Tunnel

After the pipe was placed, welded and grouted in the shaft between the intake structure and the tunnel, pipe placement within the tunnel was begun. Concrete fill had been placed at the tunnel invert and rail was set for the pipe traveler and locomotive. Two lengths of pipe were used due to the curvature of the tunnel. Pipe sections from the elbow to the point of compound curvature (PCC) were 6.6m (20 feet) in length and pipe sections from the PCC to the portal were 13.1m (40 feet) in length. Each pipe section was lowered by a crane at the portal and placed on the pipe traveler. The locomotive pushed the pipe section on the traveler through the tunnel to the preceding pipe section. The pipe was fitted to the preceding pipe with alignment adjustments being made by hydraulic jacks on the traveler. Steel supports were placed under the pipe and the traveler and locomotive went back to the portal for the next pipe section. Welders immediately started to weld the pipe joints after each section was installed, and other personnel placed tie-downs on the pipe to restrain it from the uplift forces due to grouting of the annulus. The tie-downs for the pipe consisted of cables stretched across the top of the pipe and attached to turn buckles grouted into the rock in the invert on each side of the pipe. The tie-downs were spaced approximately 1.6m (5 feet) apart for the length of the tunnel. The contract documents required that a maximum of 66m (200 feet) of pipe could be installed before the annulus around the pipe was grouted. This requirement was intended to provide for filling the annular void around the pipe by limiting the distance that grout would be pumped.

The backfill grout was placed using a slick line installed at the crown of the pipe and extending to the previous grout placement. A temporary bulkhead was constructed at the end of each run of pipe so that the entire annulus would be filled with the grout. As the grout filled the annulus to the crown of the pipe, the slick line was withdrawn. After the grout set, the temporary bulkhead was removed and the next run of pipe was installed. Voids were discovered over the top of the pipe after the backfill grouting operations were completed. These voids were contact grouted through grout holes located in the crown of the pipe.

The pipe installation in the tunnel was completed to the portal in February 1996. The pipe was extended from the portal to the existing 1524mm (60 inch) supply line and the connection was made in April 1996. The Inland System is currently supplying water to the City of Birmingham utilizing the new intake structure and pipe.

SUMMARY

The renovation of the Inland System was important in order to ensure a reliable water source for the City of Birmingham, Alabama. The design of the major components of the renovation work, the new intake structure and the tunnel through the dam abutment, was intended to provide the owner with a suitable replacement for the existing facilities while also providing for future anticipated demands on the System. The construction of the complex and critical project was successfully completed and the new facilities are supplying water to the City of Birmingham.

Chapter 31

BARRICK MEIKLE MINE SHAFT CONSTRUCTION

Cherie M. Tilley
Project Manager, Dynatec Mining Contractors, Elko, Nevada

ABSTRACT

Dynatec Mining Contractors (Dynatec) entered into a contract with Barrick Goldstrike Mines (Barrick) to develop the deep gold ore reserves at the Meikle Mine north of Carlin, Nevada.

The Ventilation and Production Shafts were started in early 1994 and were completed in December, 1995. The mine development and mine construction programs were completed in November, 1996.

This paper describes the Meikle Mine development program.

INTRODUCTION

In March of 1994, Dynatec began sinking shaft and mine development work at the new Meikle Mine owned by Barrick Incorporated.

The Meikle Mine is on Barrick Goldstrike property that covers 2,800 hectares (7,000 acres). The Goldstrike property is located on the Carlin gold trend, a 80 kilometer (50 mile) long band of major gold deposits in north-central Nevada.

Prior to the decision to proceed with the development of the Meikle Mine, Barrick conducted extensive exploration drilling which developed a reserve of 205 metric tons (6.6 million ounces) of gold with an average grade of 23.3 grams/metric ton (0.68 ounces/ton).

This valuable gold reserve prompted a decision to proceed with construction. The estimated cost for the mine development program was $180 million dollars (US). The new mine included special challenges of high rock temperatures, 60 degrees Celsius (140 degrees Fahrenheit) water and voids or vugs that range from fist size to very large caverns. Barrick installed the largest mine refrigeration system in North America to cool the Ventilation air and lower the underground temperature.

Barrick also dewatered the Meikle Mine development area with deep well turbine mine dewatering pumps. This system was so effective that within the entire scope of work conducted by Dynatec not a single shaft dewatering pump was used.

The initial work performed by Dynatec was called the Pre-sink phase. During this phase of the work both shafts were sunk to approximately 21.3 m (70 ft). This excavation allowed the sinking stages or galloways to be assembled on the surface and lowered into each shaft prior to erection of the headframe and surface construction. After completing the Pre-sink phase, Dynatec suspended their work until the surface contractor erected the headframe. Dynatec Engineering Inc. designed the shaft sinking components of the work. Dynatec site construction personnel worked with the surface contractor to construct the sinking apparatus within the permanent headframe.

The Production Shaft and Ventilation Shaft are 115 m (350 ft) apart. The shafts are oriented in an east west position. The Production Shaft is due west from the Ventilation Shaft.

VENTILATION SHAFT

The Ventilation Shaft was designed for 600,000 c.f.m. of air volume. It was also designed as an equipment lowering shaft for lowering mine equipment in one piece using a Timberline Hoist with 90,000 lbs of line pull.

The Ventilation Shaft is a 4.88 m (16 ft) diameter shaft. The shaft is concrete lined with a designed nominal 0.31 m (1 ft) thickness (See figure 2). It is sunk to a depth of 402.5 m (1320 ft). The shaft has four stations with a temporary loading pocket constructed on the bottom 1225 level. The shaft was sunk using rope guides for the crosshead and sinking bucket guidance. The Ventilation Shaft headframe is 34 m (110 ft) high. The Ventilation Shaft was sunk using a Denver Vulcan 2.7m (100 inch) diameter single drum hoist powered by two 447 KW (600 HP) motors with 6 ton muck buckets. This hoist will be relocated to the Production Shaft as the final service hoist.

Meikle Mine Site

Cavern on 1075 Level

Barrick and Dynatec management made a decision to start the shaft excavation utilizing a Webster 2000 SS roadheader. The decision to use the Webster roadheader was due to the type of ground to be excavated. The upper part of the shaft is located within the Carlin and Vinini formations (See figure 1). These two formations are mudstones, siltstones and softer rock types. The roadheader is a style designed for shaft sinking. The boom is bent 90 degrees down for better digging of the bottom. It also has no muck handling system. A Cryderman mucker was used for mucking. The Webster was able to cut the rock without any problems. The problem with using the roadheader was two fold. A very high overhead protection canopy and too small of a shaft excavation area. Because of the height of the unit the excavation open ahead of the permanent concrete lining was substantial. The mud and siltstones required 1.5 m (5 ft) split sets instead of the planned 0.9 m (3 ft) split sets to adequately support the open ground. The shaft bottom at approximately 5.6 m (18.5 ft) in diameter, was too tight a work space to install the long split sets. Therefore, crews were having to setup and move the roadheader three times during the ground support cycle. During this time no other excavation activity was possible. The ground support activity required stopping every 1.8 m (6 ft) of advance. The time to install the ground support became too slow for an acceptable shaft sinking advance rate. After trying several things to improve the situation and not succeeding the decision was made to suspend use of the Webster.

The crews then started using a benching method with sinking hammers to advance the shaft bottom. A two boom pneumatic Tamrock jumbo was constructed on site and used in the lower part of the shaft. The shaft sinking cycle consisted of drilling and blasting, mucking with a Herman Cryderman into a 6 ton shaft sinking bucket, pouring a 20 foot section of concrete lining, and installing service lines to support the work.

In the lower part of the shaft at approximately 360 m (1180 ft), two large caverns were encountered. One of these caverns was in the shaft pillar and was filled with 965 cubic yards of concrete. The second cavern occurred in a location that was to be excavated for a station and nothing further was done.

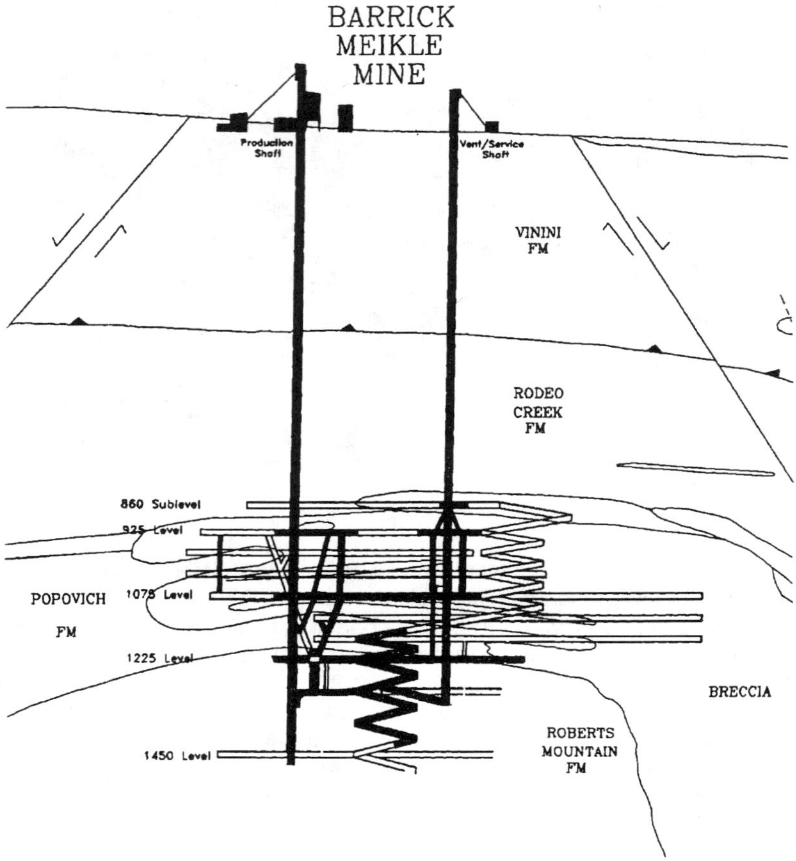

Figure 1

The Ventilation Shaft levels were excavated at 860, 928, 1078, and 1228. There is a 0.9 m (3 ft) difference in sill levels between the two shafts. The water drainage is from the Production Shaft toward the Ventilation Shaft at each main level. In order to handle development muck, a muck raise, 3.1 m (10 ft) in diameter was raise bored from the 1225 level to the 925 level at the Ventilation Shaft. A crosscut drift was driven to this raise on the 1075 level so that all the material from level and ramp excavation could be hoisted from the 1225 level. A 120 cubic foot temporary loading pocket and skip were installed to handle development muck. This allowed mine development to proceed while the Production Shaft and permanent crushing and muck handling system were being completed.

During sinking activities the Ventilation Shaft had four systems of escape from the bottom. First was the sinking bucket. Second was the Cryderman winch. The Cryderman was set up so that personnel could get from the Cryderman carriage to the sub collar when it was brought to surface. The third system of escape was the galloway. The galloway could be brought from the bottom to the sub collar. The fourth system of escape was an emergency escape hoist and capsule. This hoist could be powered by line power or an emergency generator. The emergency escape hoist was setup with its own landing in the sub collar and a capsule which traveled independent of any other conveyance guiding system in the shaft.

PRODUCTION SHAFT

The Production Shaft is designed to hoist 3000 T.P.D. but has an ultimate capacity of 6000 T.P.D. from the 1330 loading pocket.

Loading Ventilation Shaft Round

VENTILATION SHAFT

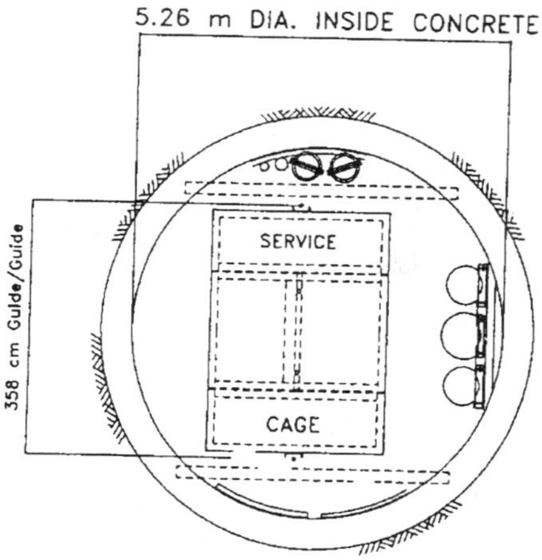

PLAN VIEW

Figure 2

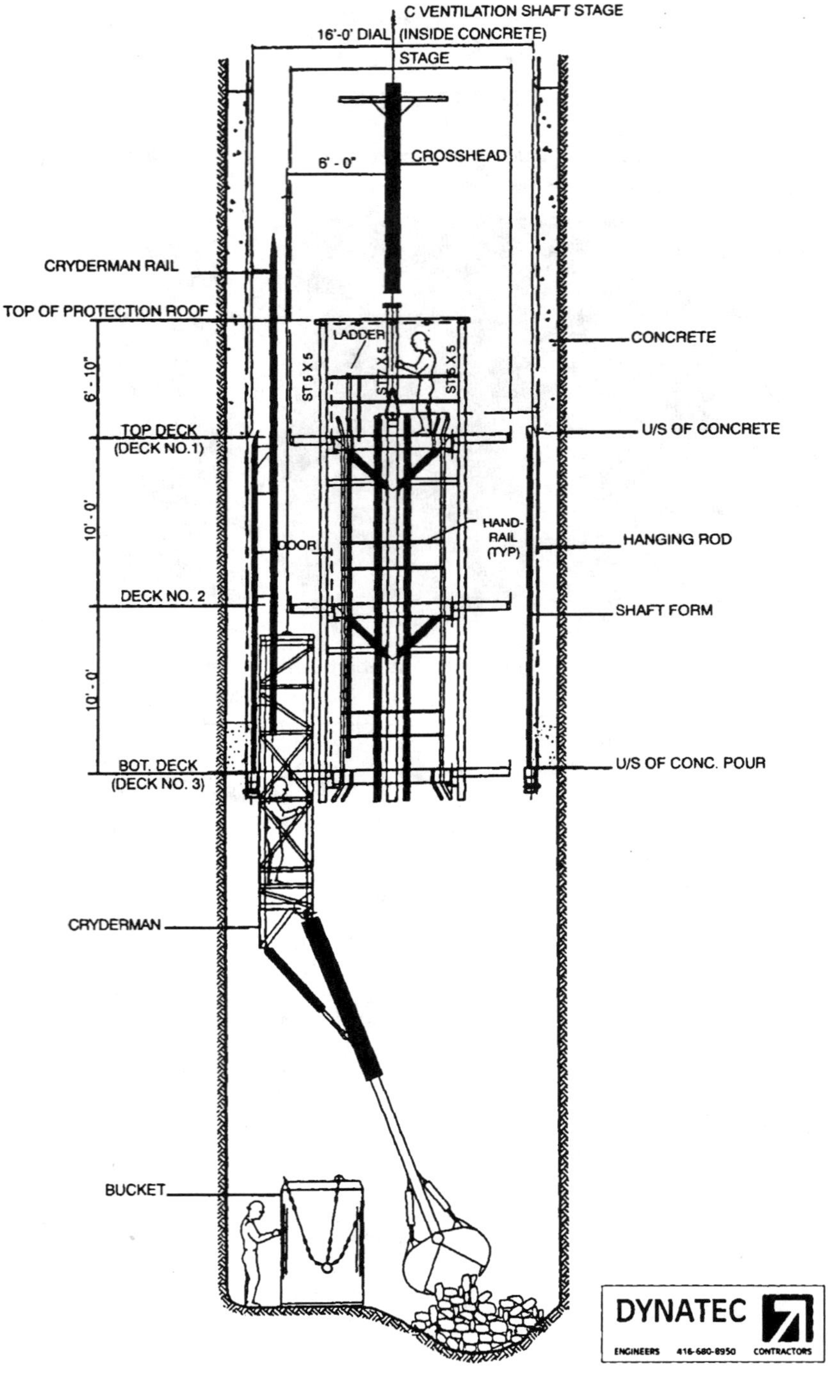
C VENTILATION SHAFT STAGE
16'-0' DIAL (INSIDE CONCRETE)
STAGE
CROSSHEAD
6' - 0"
CRYDERMAN RAIL
TOP OF PROTECTION ROOF
LADDER
ST 5 X 5
ST 7 X 5
ST 5 X 5
CONCRETE
6' - 10"
TOP DECK
(DECK NO.1)
U/S OF CONCRETE
10' - 0'
HAND-
RAIL
(TYP)
DOOR
HANGING ROD
DECK NO. 2
SHAFT FORM
10' - 0'
BOT. DECK
(DECK NO. 3)
U/S OF CONC. POUR
CRYDERMAN
BUCKET
DYNATEC
ENGINEERS 416-680-8950 CONTRACTORS

Typical Dynatec Mining Style Galloway

The Production Shaft is a 5.49 m (18 ft) diameter concrete lined shaft. The concrete has a nominal 0.30 m (1 ft) thickness. The Production Shaft is constructed with steel shaft sets every 4.57 m (15 ft) to support the major shaft conveyances. The shaft conveyances are intended to transport personnel and material within the mine (see Figure 3). The shaft in the permanent mode contains, two muck skip compartments, a service cage compartment, a Mary Ann cage compartment, and a compartment for mine services lines. Wood guides were used for all conveyances.

The Production hoist is a Canadian Ingersoll-Rand 3.9 m (144 inch) diameter double drum, single clutch hoist. It has a 933 KW (1250 HP) motor. The headframe is 55 m (180 ft) high. This headframe was designed to this height in order to accommodate the storage bins and loading of 172 mt (190 ton) trucks. The height of a truck is approximately 7.3 m (24 ft). Shaft sinking was done using the permanent hoist and headframe.

SINKING METHOD

The Production Shaft utilized the drilling technology described in a paper presented at the 1991 Rapid Excavation and Tunneling conference in Seattle, Washington. The paper

Looking Down the Production Shaft

PRODUCTION SHAFT

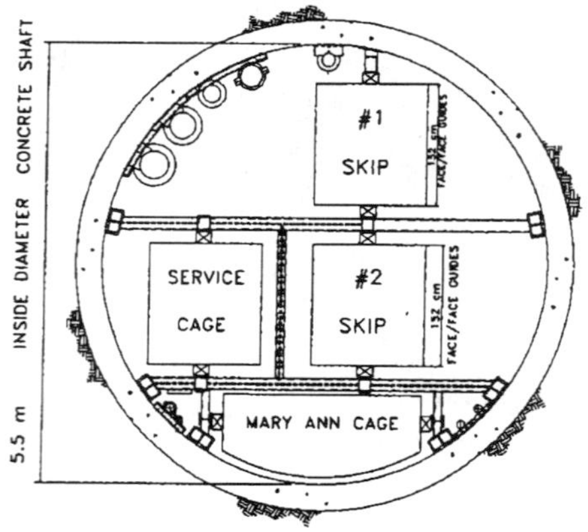

PLAN VIEW

Figure 4

was titled "New Technology in Shaft Sinking—The Craig Project". The authors of this paper are W. R. Dengler, Chairman, Dynatec International Ltd.; W. M. Shaver, President, Dynatec Mining Ltd.; and R. LeTourneau; Craig project manager, Dynatec Mining Ltd.

The drilling procedure was to lower the two boom Tamrock pneumatic jumbo from inside the galloway to just below the bottom deck of the galloway. Then the galloway was lowered to its drilling position at the shaft bottom. Drillers then began to drill the periphery holes of the round. While drilling the periphery holes, an ITH (In The Hole Hammer) drill was then lowered from the surface through the galloways bucket wells and setup to drill a large diameter cut hole. The rounds were drilled to a depth of 3.7 m (12 ft) using the jumbo. The ITH drill would overdrill by 0.6 m (2 ft) to help ensure full depth breakage of the round. The rounds were full bottom rounds and the large diameter cut hole location alternated to avoid any safety problems with the previous round. A typical round contained 65-4.76 cm (1 7/8 inch) holes. The 20.3 cm (8 inch) large diameter cut hole was typically drilled 4.27 m (14 ft) in depth. When the large diameter cut hole was completed, the ITH drill was removed and hoisted to surface. The two boom Tamrock pneumatic jumbo would then complete the round. After the round was completely drilled out the shaft jumbo was retracted back into its staged position in the galloway. The shaft crew would then load and blast the round. The only pumping that was required during shaft sinking was to remove the water produced by drilling. This water was removed by pumping into the sinking bucket with a double diaphragm air pump and hoisted to surface.

The drilling at the Meikle project differed from the Craig project in that Meikle utilized two Tamrock pneumatic drills instead of the electric hydraulic drills used at Craig. The decision to use pneumatic drills was due to the softer rock types of northern Nevada versus the harder rock types of the Canadian Shield.

When the smoke cleared from the shaft, the crew would descend into the shaft, wet down the muck pile, lower the galloway, and lower the Herman Cryderman for the mucking part of the cycle.

The original plan was to utilize two Cryderman shaft muckers. However, at Meikle because of the shallow shaft, speed of the hoist, and use of a spare bucket to change buckets in the bottom, the mucking was conducted with only one Cryderman for the entire shaft. The sinking buckets were 7 ton buckets. When mucking of a round was complete the crew would position the galloway for a concrete pour. The Production Shaft pours were 4.6

Concrete Forms

m (15 ft) in height. This height was to accommodate the center to center spacing of shaft steel sets. The day crew was scheduled to install steel, place concrete lining, and install services (compressed air, water, power cable, and ventilation duct). Drilling, mucking, ground support, and blasting were scheduled for the two back shifts. The concrete pours were normally 4.6 m (15 ft) to 7.3 m (24 ft) above the bottom of the shaft to allow room for the shaft jumbo to drill to the outside edge of the concrete lining. Concrete pours required scribing and screen to be placed at the bottom to seal the forms. This bottom section of concrete form is 0.9 m (3 ft) in height and is called the curb or A ring. The curb would be lowered into place, hung on hanging rods, lined, leveled, sealed, and then poured. When the curb was poured, the remaining 3.7 m (12 ft) of forms or matcher was then lowered into place. The match form would be connected to the curb. The match form was then poured. Concrete pours typically required 55 cubic meters (72 cubic yards) of concrete. The thickness of the lining was designed to be 30.5 cm (12 inches). Because of the ground conditions, which required rockbolting and screening throughout the shaft, the lining thickness was much greater, [49.5 cm (19.5 inches)]. When the concrete pour was completed the shaft bottom excavation would resume. At times, the concrete pour would carry over from day shift to swing shift. After a concrete pour, the next day shift was usually installing steel. Sometimes due to delays or the need to reduce the bottom depth from a previous pour there would be pours back to back. When installing shaft steel sets, the crew would place all parts required for two or three sets on the collar. When the shaft steel sets were installed the crew then extended the service lines. The service lines would be connected to support brackets which were previously installed with shaft sets. Often during the work cycle things would get out of balance, i.e. drilling problems, bad ground, rock bolting, late concrete trucks, problems with hoist, etc. Crews would adjust their activities to accommodate these interruptions and get back into rhythm as soon as possible.

EXCAVATION OF STATIONS AND LOADING POCKETS

Shaft Stations

Shaft stations were developed by stopping the shaft concrete lining in the Ventilation Shaft 8.8 m (29 ft) above the permanent station sill elevation and 8.2 m (27 ft) in the Production Shaft. As the excavation of the shaft continued, the crews would flare out the shaft or widen the entire circumference of the shaft on a 45 degree angle. As the shaft was mucked out, the crews would install resin bolts at various length depending on the actual ground conditions. The bolts installed were #7 rebar or dywidag from 2.4 to 4.26 m (8 to 14 ft) in length and spaced on a 1.2 m by 1.2 m (4 by 4 ft) pattern. After the resin bolts were installed the ground was shotcreted. The shotcrete was nominally 15.2 to 20.3 cm (6 to 8 inches) thick in the station areas.

The station brows were angled down to the 4.8 m (16 ft) height of the drift back. The drifts extended from both shafts, with one advancing east and one west. This was the case on all levels except the 860 Ventilation Shaft level. The Ventilation Shaft 860 level station only goes to the east.

At the Production Shaft, the station drifts were driven 15.2 m (50 ft) to the west at a width and height of 4.8 m (16 ft). They were driven 15.2 m (50 ft) to the east at a width of 6.0 m (20 ft) and a height of 4.8 m (16 ft). These wider drifts were to connect with the other shaft.

The Ventilation Shaft station drifts were all driven to the east 15.2 m (50 ft). The west drifts from the Ventilation Shaft were driven to approximately 27.4 m (90 ft) to allow for excavation of the muck pass crosscut. After the construction of the temporary loading pocket the three main levels were driven to connect to the east drifts of the Production Shaft. There was a crosscut drift driven to the muck raise on each main level.

The excavation of the stations was conducted using the shaft jumbo and Cryderman mucker as much as practical. Station breakout work outside the reach of the shaft equipment utilized jacklegs, a single boom Gardner Denver jumbo (which would pass through the bucket wells in the galloway), and Cat 931 track loaders.

Loading Pockets

Dynatec constructed and used three loading pockets during its work developing the Meikle Mine. The first loading pocket was the temporary pocket located on the 1225 level of the Ventilation Shaft. This pocket was constructed on the east side of the 1225 station sill. The pocket had a hopper that set above the station sill which was filled by a LHD loader.

Below the 1225 station sill was a measuring flask and one chute gate. The sinking bucket was replaced with a muck skip at 7.3 mt (8 ton) which operated on rope guides. The galloway was lowered to the area below the chute gate to serve as a weight on the guide ropes to provide the proper tension.

The second loading pocket constructed was the 1181 temporary loading pocket in the Production Shaft. The 1181 loading pocket area was excavated and supported from the shaft bottom. One double size chute throat was driven up at 45 degrees from the shaft for a distance of 9.4 m (31 ft). Crews drove two raise rounds up both the north and south muck passes from the top of the 2 compartment chute. The north and south raises connect the

1225 Shaft Station

Ventilation Shaft 1225 Level

Production Shaft 1075 Level

1181 loading pocket to the 1075 north and south truck dumping positions. The shaft was then deepened. The measuring bin and chute gate locations were excavated and ground support was installed as the shaft excavation continued down.

The third pocket constructed was the 1330 permanent loading pocket. With the shaft bottom at the 1300 level, the level was driven approximately 15.2 m (50 ft) to the end of the electrical room in an easterly direction. The conveyor gallery was driven in a northeasterly direction off of the electrical room drift for 11.58 m (38 ft). The shaft was then deepened and the loading pocket chute area was excavated and supported down to the 1331 level. Then the shaft was completed to its final depth of 451 m (1480 ft) and shaft steel furnished to 424 m (1390 ft).

Shaft crews then went back and installed all steel on the 1181 pocket. When the pocket was operational, the remaining 22.8 m (75 ft) for both the north and south raises was driven. The 1181 pocket then became the major muck handling system. Once the 1181 loading pocket was finished, the crew moved to the 1330 loading pocket. The structural steel and loading bins were installed. When the pocket was operational, crews then drove the conveyor gallery to its end. A short drift was driven from the end of the conveyor gallery to the 1450 development ramp.

MINE DEVELOPMENT

The initial plan was to develop the main levels when the Ventilation Shaft reached that level. This plan was changed by placing the priority on shaft sinking first and level development second. To facilitate the later level development, a 3.05 m (10 ft) muck raise was bored from the 1225 level to the 925 level. The top portion of this raise, 925 level to 1075 level, would serve later as the sand silo for the backfill plant. A temporary loading pocket was designed and constructed on the 1225 Ventilation Shaft station level. This pocket was designed to handle at least 1089 mt (1200 ton) per day.

With the completion of Ventilation Shaft sinking, raise boring, the muck pass and the temporary loading pocket operational, Barrick crews took over the development on the 925 level. As Dynatec completed sufficient work on the 1075 level Barrick crews were added and assumed the continuing stope development activity. The 1225 level, South Meikle drift diamond drilling positions, and ramp excavation both up and down was completed by Dynatec on Sept. 1, 1996. At this time Barrick took over all underground mining and stope development. By the end of 1996, a total of 2577 m (8452 ft) of development drifting work had been completed by Dynatec.

Dynatec equipment for conducting the level and ramp excavation included:

TYPE OF EQUIPMENT	UNITS
3 Boom pneumatic Tamrock Jumbo	3
1 Boom pneumatic Gardner Denver Jumbo	1
Tamrock EJC 210 LHD	3
Tamrock EJC JS285 LHD	2
Teledyne Scissor lift truck	3
Reed M-20 concrete pump	1
Reed 3080B concrete pump	1
Reed Lova 16-4 shotcrete machine	3
Austin Western grader	1
Kubota KX101 backhoe/breaker	1
Kubota 410 forklift	1

MINE CONSTRUCTION

The Dynatec scope of work for mine construction included a temporary muck handling system, permanent underground crusher station, permanent muck handling system, a un-

1225 Level

derground truck shop, underground fuel storage/handling system, refuge stations, wash bay, offices, electric substations, and a pump station.

The temporary muck handling system has two skip loading pockets on the 1181 level, two truck dumps with grizzlies and hydraulic rock breakers on the 1075 level, and one truck dump on the 925 level.

The permanent underground crushing system on the 1225 level consists of a Ross chain feeder control, a vibrating feeder, a 92 cm (3 ft) by 122 cm (4 ft) Kue-Ken crusher, a

Looking Up the Production Shaft

14' Diameter Raise Bore Cutter Head

Crusher Discharge

conveyor, and two 6 m (20 ft) diameter ore and waste bins with truck dumps, grizzlies, and hydraulic rock breakers. The crusher weighs 48 metric tons (53 tons). It is 244 cm (8 ft) wide, 335 cm (11 ft) tall, and 396 cm (13 ft) long.

Dynatec personnel lowered and moved the jaw crusher (53 tons) in one piece from surface down the Ventilation Shaft and across the 1225 level to the crusher station. The crusher was then jacked up 3.05 m (10 ft) moved to the north end of the station and temporarily supported. The crusher support steel was installed and the crusher was moved forward into its permanent position. Once the crusher was in place, the vibrating feeder and

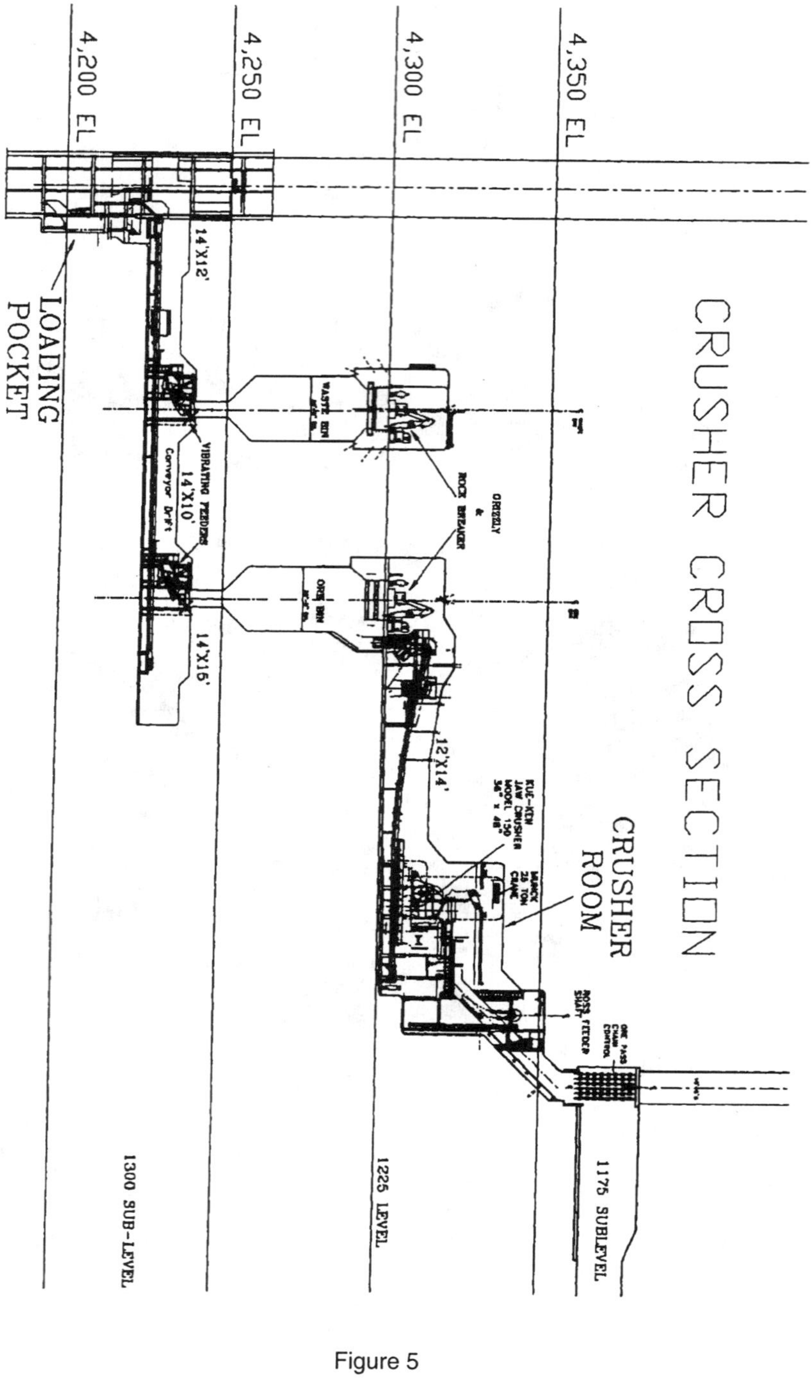
CRUSHER CROSS SECTION
4,350 EL
4,300 EL
4,250 EL
4,200 EL
LOADING POCKET
14'X12'
VIBRATING FEEDERS
14'X10'
Conveyor Drift
14'X16'
GRIZZLY
&
ROCK BREAKER
WASTE BIN
ORE BIN
12'X14'
KUE-KEN
JAW CRUSHER
MODEL 150
36" x 48"
MUNCK
25 TON
CRANE
CRUSHER ROOM
ROSS FEEDER
1300 SUB-LEVEL
1225 LEVEL
1175 SUBLEVEL

Figure 5

Crusher Station

Ross feeder were installed. A 25 ton Munck overhead crane was installed over the crusher station.

The permanent muck handling system consists of two skip loading pockets at 1330, a loading pocket feed conveyor on 1300 level, the two 6.1 m (20 ft) ore and waste bins, the crushing system, a truck dump on the 1175 with two sets of chain controls, and additional truck dumps on 1225, 1125, 1075, and 925. This crushing/handling/loading system is designed to produce a maximum of 544 mt (600 tons) an hour.

The 1225 underground truck shop has three truck bays and one welding shop. It has a 9.1 mt (10 ton) overhead Kone crane in the main shop. An Equipment wash bay and a oil/water separator were constructed near the truck shop.

Conveyor System

Truck Shop

Ore/Waste Pass

The underground fuel storage/handling system is located on the 1075 level next to the Ventilation Shaft. The fuel bay has a 13,250 liters (3,500 gallons) enclosed fuel tank with additional drum storage area. A containment area sufficient to handle the total fuel storage was constructed. The fuel is transferred underground through a pipe line in the Ventilation Shaft.

Dynatec raise bored holes for material (sand and gravel) storage, ventilation, muck passes, pipe passageways, and pilot holes for slashing the large muck bins. Dynatec raised bored 13 holes from 1.2 m (42 inch) to 4.59 m (14 ft) in diameter. Dynatec used a Robbins

DYNATEC MINING
MEIKLE MINE

DAYS

	PROD SHAFT	VENT SHAFT	DEVELOPMENT	CONSTRUCTION	OVERALL
BID	332	291	271	202	413
BID WITH CHANGE	369	343	230	277	805
ACTUAL	324	320	229	238	743
SCHEDULE GAIN	45	23	1	39	62

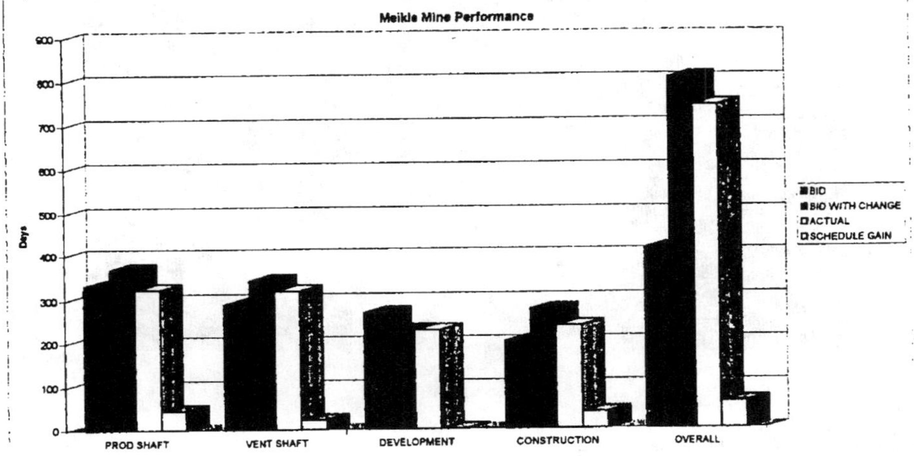

RBM 7SP machine on all the holes except two. Two 1.6 m (5 ft) raises were pulled with a Subterranean 009 Raise Drill.

SCHEDULE

The original scope of work was modified to take advantage of opportunities that became apparent as the work progressed. Barrick needed to start their mining crews at the earliest prudent time. The following schedule reflects the changes by using the units of time in the original contract. The primary project goal was to produce a certain amount of gold by the end of December, 1996. This call for ounces was met.

SUMMARY

In summary, the Meikle Mine Project was on schedule with the adjustments for additional work. The mine met its goal of required ounces produced at the end of the 4th quarter 1996. The project was conducted over a period from March 1994 to December 1996 or 33 months by Dynatec with only 4 lost time accidents. During the shaft sinking phase, Dynatec had only one lost time accident. During the life of the project only 2 lost time accidents occurred underground. This accident performance is a tribute to the safety awareness and attitude of the employees, supervision, Barrick personnel, and with full support Dynatec and Barrick management. At the Meikle Mine development project both Barrick and Dynatec senior management played a proactive role in project management. When accidents occurred senior management became involved in a positive or supportive way in the accident

1075 Caverns

investigations. Senior management insisted on and helped in establishing safe guards to prevent recurrences of accidents. The safety performance support by senior management was always firm and supportive. The quality of work received high praise from the owner and is something that Dynatec is proud of accomplishing. During the work in 1996, Barrick crews were brought on and took over the excavation work in stages. In September, 1996, Barrick assumed all underground excavation. This changing of responsibilities required a lot of co-operation from Dynatec and Barrick personnel.

ACKNOWLEDGMENTS

I would like to express my appreciation to John Newman, Shaft Engineer; Bob Stevenson, Shaft Superintendent, Terry Tilley, Project Engineer, and Jamie Nelson; of Dynatec Mining Contractors for their assistance in preparing this paper.

A special thanks goes to Barrick Meikle personnel, Steve Long, Mine Superintendent; John Breen, Project Engineer; and Frank Brown, Engineering Technician; who provided slides, drawings, and information to complete this report.

REFERENCES

Dengler, W. R.; Shaver, W. M.; and LeTourneau, R.; 1991-" New Technology In Shaft Sinking- Craig Project" for R.E.T.C.

10

Future/Special Projects

Chairmen
P. Freeland
Granite Construction Co.
Watsonville, CA

B. Kaub
Granite Construction Co.
Watsonville, CA

Chapter 32

3D ANIMATION IN TUNNELING

Jarrett E. Carlson,[1] Anthony D. Songer,[2] James E. Diekmann[3]
[1]Engineer for Kiewit Construction, Underground Division, Oxon Hill, Maryland;
[2]Assistant Professor of Civil Engineering, University of Colorado, Boulder, Colorado;
[3]Professor of Civil Engineering, University of Colorado, Boulder, Colorado

ABSTRACT

Computer Aided Design (CAD) systems are currently recognized as a necessary tool for the management of projects. Currently, CAD is used for TBM and conveyor setups, concrete lift design and overall constructability reviews. Typically, such CAD models exist only in 2 dimensions. The advent of more powerful computers, enhanced CAD programs and after market animation packages provides tremendous potential to the Tunneling industry. In particular, effective application of these technologies enhances cost, time and quality in field operations.

INTRODUCTION

Recent advances in computer technology allow the integration of Computer Aided Design (CAD) and scheduling at the personal computer level. 3D animation occurs through the linking of CAD objects with exterior databases that change with time. The databases can range from a scheduling program, procurement spreadsheet or even an in house cost control system.

Such integration provides planners, designers and constructors with the tool to visually analyze a design with a proposed schedule and "see" the impact of changes in the design or schedule. Advantages of 3D animation include aiding designer visualization and evaluation of the design's functionality, assisting engineering in identifying constructability issues, eliminating costly models, and providing models to construction planners early in the design process.

This paper describes the application areas for 3D animation models at the field level, the required granularity to achieve quality visualization, and the benefits to visualizing tunnel projects prior to construction. A case study is described to illustrate the benefits of 3D animation for the Tunneling industry.

APPLICATION AREAS OF 3D ANIMATION

Visualizing or "seeing" the as-planned construction process provides opportunities for improving activities traditionally performed using 2D drawings. These include estimating, scheduling, constructability reviews, lift and formwork design, rebar detailing, miscellaneous steel and traffic control. Additionally, clearances, tolerances, equipment layout including rail, haul roads, batch plants also stem from the 2D contract drawings. Surveyors and engineers finally take derivatives of all of these drawings and then create as-builts. As-planned and as-built drawings often form the basis of claims and historical records of the company.

3D animation provides additional applications for the Tunneling industry not offered by or extended from 2D. These include schedule criticality, equipment layout, and ventilation measurements. Objects in the 3D model are highlighted as per a critical float value. Managers then view the "criticality" of the project with respect to time to gain an overall view of the problem areas visually. Another application integrates the 3D model to the submittal database allowing a project view with respect to planned and actual engineering requirements.

Equipment layout is another application of 3D animation. The minimization of movement of a resource such as a crane or form can be visually maximized using 3D animation.

Other possibilities for viewing data over time can stem from measurements of ventilation or gas levels. A display of air flows and/or gas concentrations over a project with many cross-passages, shafts and portals would lead to insights on problem areas of ventilation, gas pockets and dissipation rates.

Although 3D animation holds great promise for the Tunneling community it is not a panacea for unlimited success. Strategic barriers to successful implementation and functional concerns of appropriate level of detail in the 3D animation must be addressed.

APPROPRIATE LEVEL OF DETAIL

One question users of 3D animation must ask is what level of detail or granularity is appropriate in the 3D animation to optimize its use. Results from a research study at the University of Colorado identified four variables critical for developing appropriate granularity. These include; time increment, user viewpoint, schedule granularity, and model granularity.

Time increment is the frequency by which the animation occurs. For example, if a time increment is one week, there is little to be gained by modeling items which take one shift to install. Conversely, if the time increment is one day and the schedule activities are rounded to the week, simulation time is wasted.

User viewpoint is the spatial position of the person viewing the animation with respect to the 3D model. There is no point modeling small items when the viewpoint within the simulation is a significant distance from the small item modeled.

Schedule granularity addresses the issue of balancing the number of activities modeled compared to the number of physical objects modeled. There is little point in simulating thousands of objects with only several activities. A one to one ratio is ideal, but not always achievable. Some schedule items may have more than one object and one object might have more than one schedule activity.

Model granularity is the level of detail included in the 3D model. There is a level of detail inherent in the physical objects. This refers to the amount of effort placed in creation of the model and how the elements in the CAD package are grouped. For example, a W-section is usually modeled by three planes, rather than the twelve planes required to include

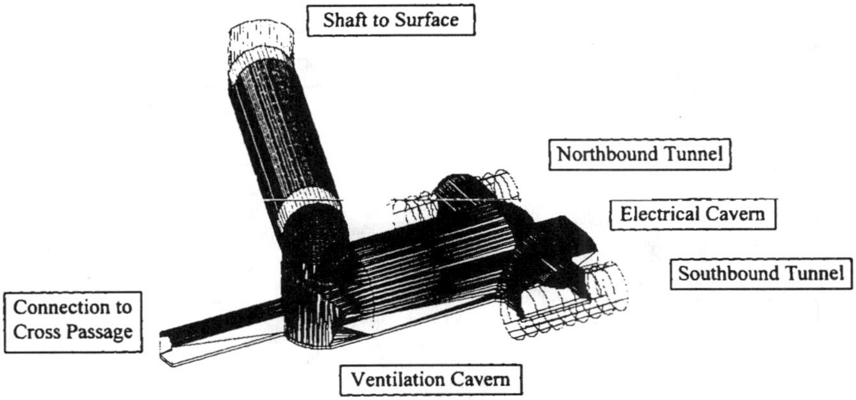

Figure 1. Willis Ventilation Structure Key Plan

the thickness of the flanges and web.

Study results demonstrate that simulation time increment and the user viewpoint within the model can both be easily adjusted to maximize the visualization of a given user.

Resolving conflict of schedule and model granularity is critical for successful use of 3D animation. Typically, the level of detail required to effectively visualize a problem differs from the level of detail given in engineering drawings, and this level can be different than the level driven by the work breakdown structure. Therefore, organizations implementing 3D animation must consider the issues of granularity early in the design phases of the project. 3D modelers must consult with planners to minimize incongruencies in granularity.

THE CASE STUDY

The sample project used for this illustration was the Dallas Area Rapid Transit's NC-1B contract for the local underground light rail system. The general contractor on the project was The S. A. Healy, Company, of McCook, Illinois. Healy had successfully utilized 3D drawings of many of the complex design configurations encountered on the project, primarily for the purposes of clearances, formwork design, sequencing of lifts, and fostering communication between engineering and superintendents. Many areas of the project did not merit the time and effort expended into creating the electronic drawings, but there were several areas of difficult geometry that were dealt with efficiently through the use of 3D drawings. One specific area that did merit the drafting of 3D drawings was the Willis ventilation structure and associated emergency exit, Figure 1.

A simple experiment was developed through simulating an excavation sequence for the tunnel level access passages, the control room, the vent cavern and the connection to the shaft. Any related drawings were referenced and included in the 3D model, such as rock bolt installations, invert and lining concrete and the instrumentation. This model was then animated as per the original approved schedule. Because of the integration of all of the contract drawings in a time based format, it was easy for the control group to identify two

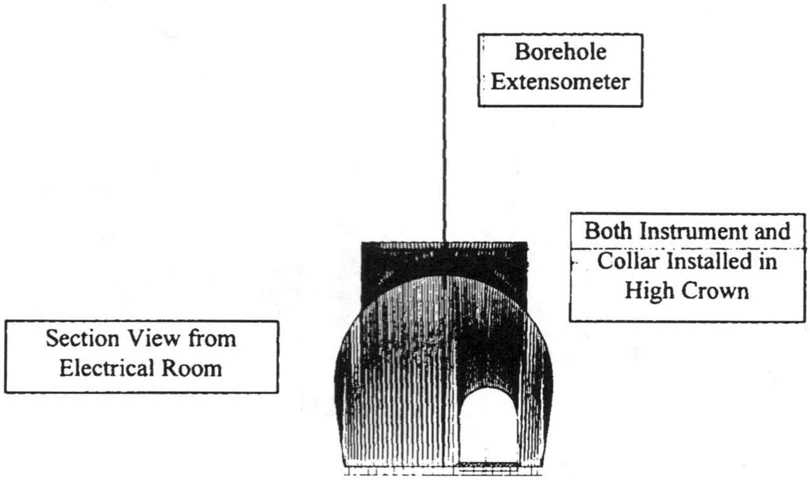

Figure 2. Section View Showing Height Problems

related problems. The original schedule had called for a full face advance through the vent cavern, which left the road header at a height disadvantage for excavating the collar and first few feet of the vertical shaft to connect to the cross adit. This height problem was due to the lack of a bench from which the mining could be conducted. Also identified was a conflict between the equipment available to install the roof borehole extensometer and the same height requirement. The extensometer drilling and installation required equipment that could not reach the crown of the cavern. A short delay was encountered in the production, mucking was temporarily reversed to provide the bench and mining was resumed upon installation of the equipment. This instrument was also the result of a change request by the engineer

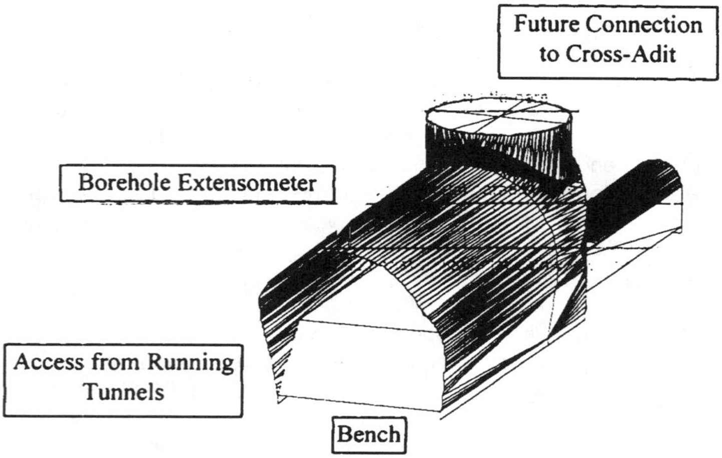

Figure 3. Isometric of Actual Construction

near the time of excavation. The 3D animation would alleviate minor problems such as these due to last minute changes. Figure 1 shows a slide from the original simulation after both excavation and installation of the instrument. Figure 2 shows a slide from the simulation after the top heading was advanced and the instrument installed and collar excavated.

Although the problems identified here are simple, they are easily fixed in the office. The rework adds man-hours without production, and increases the possibility of an accident. These instances can add up to a significant portion of the project's budget. As the instrument was a change order, the use of the visualization would improve negotiating parties understanding of the price for the change order.

CONCLUSION

This paper has suggested approaches by which 3D animations can be implemented effectively in a construction environment by addressing the keys to model creation.

The discussion also describes what types of tasks should be considered in this perspective, as those tasks that do not inherently exist in four dimensions have little to benefit from modeling.

A simple case study was included to demonstrate how even the simplest of models would have benefited a genuine situation. Although individual gains from modeling are small, the simplicity and cumulative effect of such gains are great.

Also, these animations have a secondary effect of improving safety and quality. Safety is a continuing priority in the industry and these types of models can be used for decreasing accidents and tort defenses. New quality techniques on the horizon, such as ISO 9000, actually require modeling of a structure prior to onset of construction. Tunnel modeling will exist in an electronic format due to the impracticality of physical models.

Finally, the software community has reached a level of adaptability not seen in even the recent past. These 3D animation systems can be integrated into existing corporate structures with no changes required of existing systems. Indeed, everything is in place for tunnel contractors to effectively implement an electronic system for which projects can be modeled through time.

ACKNOWLEDGMENTS

The authors would like to offer thanks to the employees of The S. A. Healy Company of McCook, Illinois, specifically Herb Middleton, Project Manager, and Vladimir Goubanov, the forward thinking tunnel engineer responsible for creating the three dimensional drawings for utilization in the animations. Also, the authors would like to offer thanks to Maryland based Jacobus Technology for providing the cutting edge object oriented software to the University of Colorado at Boulder. In particular, Mr. Cleveland and Mr. Reck for making the software available to research institutions and Mr. Simons for the training seminars.

REFERENCES

Griffis, F. H., Hogan, D., Li, W., August 1995, "An Analysis of the Impacts of Using Three

Dimensional Computer Models in the Management of Construction," Draft Report to the Construction Industry Institute, Columbia University, New York, New York.

Nielsen, R. W., Stover, A. K., Spring/Summer 1993, "The Use of Simulation Software for a Power Plant Construction Project," Civil Engineering Practice, pp. 65-72.

Alciatore, D. G., O'Connor, J. T. Dharwadkar, P., September 1991, "A Survey of Graphical Simulation in Construction: Software, Usage and Application," Final Report to the Construction Industry Institute, University of Texas at Austin, Texas.

Mahoney, J. J., Tatum, C. B., September 1994, "Construction Site Applications of CAD," ASCE Journal of Construction Engineering and Management, Vol. 120, No. 3.

Kuprenas, J. A., Um, J. T., Booth, C. E., 1995, "Use of CAD Three Dimensional Modeling as a Constructability Review," Computing in Civil Engineering, Proceedings of the Second Congress Held in Conjunction with A/E/C Systems '95, pp. 195-201, Atlanta, GA.

Chapter 33

FACE STABILISATION IN CLOSED SHIELD TUNNELLING

G. Anagnostou and K. Kovári
Swiss Federal Institute of Technology, Zurich, Switzerland

ABSTRACT

Slurry and EPB-shield tunnelling have been successfully applied worldwide in recent years. Under extremely unfavourable geological conditions, however, face instabilities may occur. The present paper aims at a better understanding of the mechanics of face failure. For the case of *slurry shield*, attention will be paid to the time-dependent effects associated with the infiltration of slurry into the ground ahead of the face. Related topics, such as the stand-up time and the effect of the excavation advance rate, will be discussed. In the case of an *EPB-shield*, a distinction will be drawn between fluid-pressure and effective pressure in the chamber. The effective pressure can be visualized as a "grain to grain" contact pressure between the muck and the ground at the face. The water pressure in the chamber reduces the hydraulic head gradient in the ground and, consequently, the seepage forces acting in front of the face. The face is thus stabilized both by the direct support of the pressurized muck and by the reduction of the seepage forces in the ground. The paper includes normalized diagrams which allow the assessment of tunnel face stability.

INTRODUCTION

Tunnel construction in saturated soils is being carried out ever more frequently using closed shields, which allow the control of surface settlement and limit the risk of tunnel face failure through the continuous support of the face during excavation (Stack 1982). Slurry shields use a bentonite suspension as supporting fluid (Fig. 1a). By means of an air cushion, the fluid pressure may be accurately applied and maintained without fluctuations (Babendererde 1991). Earth pressure shields provide continuous support of the tunnel face using freshly-excavated soil (Fig. 1b), which under pressure completely fills-up the work chamber (Fujita 1981). The supporting pressure is achieved through control of the incoming and outgoing materials in the chamber, i.e. through regulation of the screw-conveyor rotation and of the excavation advance rate.

SLURRY-MODE OF OPERATION

The computational model

Face stability in homogeneous ground will be assessed by considering the limit equilibrium of a wedge loaded by a prismatic body (Fig. 2). The critical inclination ω of the slip surface ABFE is determined iteratively (minimization of safety factor). This sliding mecha-

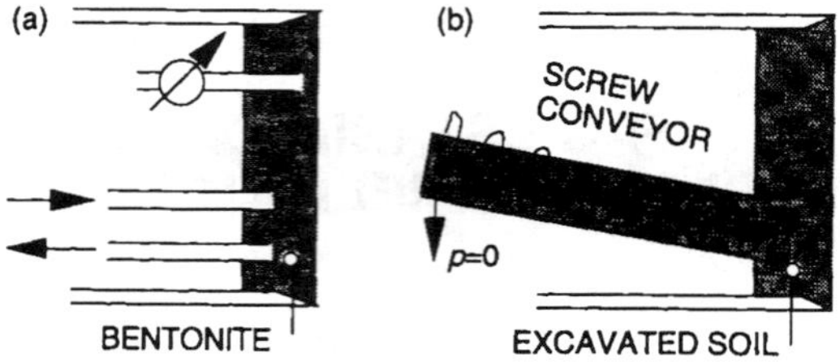

Figure 1. Closed-shield Types: (a) Slurry Shield, (b) EPB-shirld

nism was proposed by Horn (1961) and it takes into account the formation of slip surfaces, which may frequently be observed during the failure of tunnel faces in shallow tunnels. Mohr-Coulomb failure condition and drained conditions will be assumed, i.e., all computations are carried out in terms of effective stresses. The load of the prism is computed based upon silo-theory (Janssen 1895). Besides this load, the submerged weight of the wedge, the normal and shear forces along the slip surfaces ADE, BCF and ABFE, as well as the supporting force of the slurry are introduced into the equilibrium equations.

The calculation of the support force of the slurry deserves especial attention. To prevent a seepage flow towards the excavation face, the slurry pressure must be higher than the ground water pressure in the soil. Due to this pressure difference Δp (Fig. 3), the slurry infiltrates into the soil. The stabilizing force of a slurry depends essentially on the depth of infiltration of the slurry into the ground: the less the slurry infiltrates, the greater the support force will be (cf. Müller- Kirchenbauer 1972, Xanthakos 1979).

Under optimum operational conditions, a filter-cake forms on the tunnel face, acting like a membrane and inhibiting the infiltration of the suspension into the ground. The support

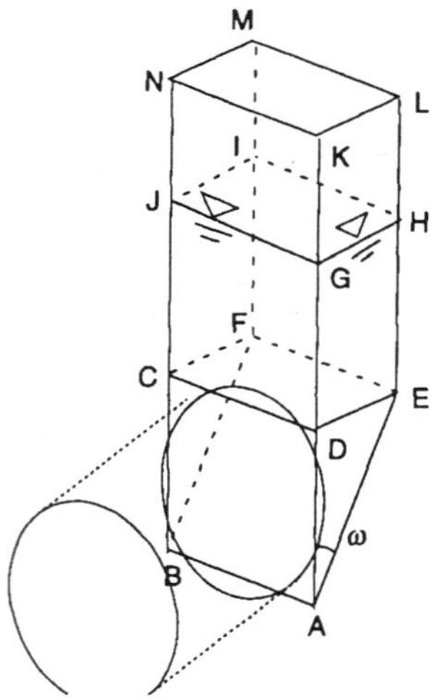

Figure 2. Sliding Mechanism (after Horn, 1961)

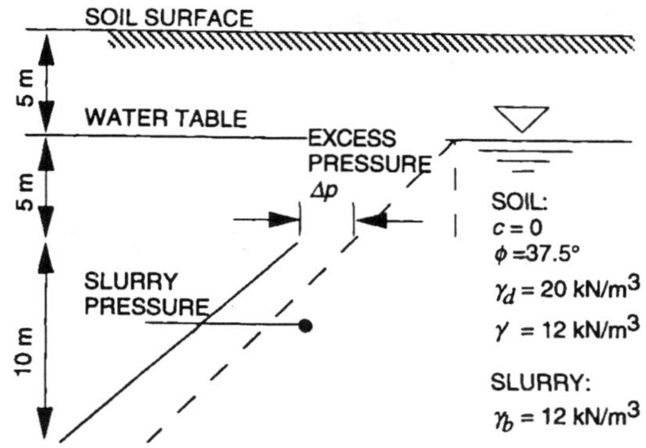

Figure 3. Data of the Numerical Example

force results from the difference in hydrostatic pressure between the slurry and the ground water (so-called "membrane-model"). A filter-cake is formed when the suspension contains aggregated solid matter which is filtered out at the beginning of the slurry infiltration. Often, suspended material is present in the slurry due to mixing with the excavated soil; aggregates may be added to the suspension when required - for example, in a uniform, coarse and poorly-graded ground.

For soils with exceptionally high permeability or when the shear resistance of the slurry is low, the bentonite will penetrate into the ground to a certain degree. The final distance of infiltration e_{max} can be estimated by means of the following empirical formulae (DIN 4126):

$$e_{max} = \frac{\Delta p d_{10}}{2\tau_f} \tag{1}$$

where Δp, d_{10} and τ_f denote the excess slurry pressure, the characteristic grain diameter of the soil and the yield strength of the slurry, respectively. Accordingly, the extent of slurry infiltration is governed by the finer soil particle fraction. The yield strength τ_f of the suspension depends essentially on the bentonite concentration. When the slurry infiltrates into the soil, it exerts a mass force, which is equal to the pressure gradient (Anagnostou and Kovári 1994). The support force of the slurry is calculated by integrating the mass forces over the suspension-saturated zone of the wedge.

Numerical Example

The effect of slurry infiltration on the face stability will be demonstrated by a numerical example concerning a tunnel in a cohesionless ground. For given geometric data, shear parameters and unit weights (see Fig. 3), the safety factor depends on the excess pressure Δp and on the infiltration depth, i.e. (see eq. 1), on the chracteristic grain size d_{10}. of the soil and on the yield strength τ_f of the slurry.

It is, therefore, possible to represent the safety factor as a function of the characteristic grain size d_{10} (Fig. 4). On the abscissa of this diagram, the grain size range for gravels and sands is given as well; a d_{10}-value of 0.60 mm, for example, characterizes a poorly-graded

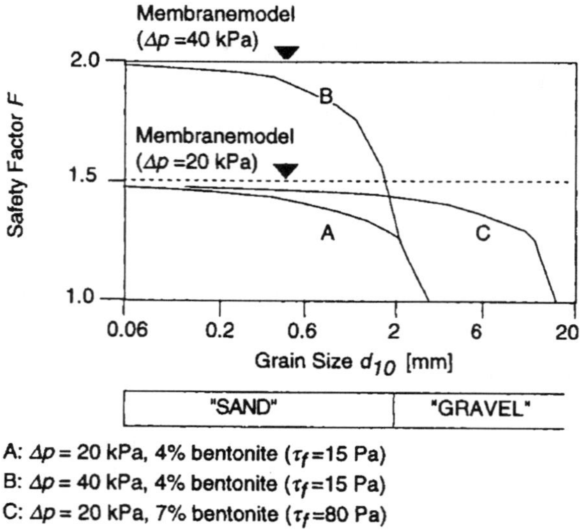

Figure 4. Safety Factor as a Function of Characteristic Grain Siez d_{10}.

medium sand with a small silt or clay fraction. The curves A, B and C in Fig. 4 correspond to different values of slurry yield strength τ_f (i.e., to different bentonite concentrations) and of excess pressure Δp. These two parameters can be directly controlled by the engineer in practice.

Consider curve A. In a soil with a d_{10}-value which is smaller than the grain size of medium sand, the deviation from the membrane model (safety factor =1.50) is negligible, i.e., the slurry acts as though the face were sealed. Within the range of coarse sand (0.60-2.00 mm), the safety factor experiences a steep decrease. It should be noted, however, that soils with a d_{10}-value higher than 0.60 mm occur rather infrequently. At a characteristic grain size in the magnitude of fine gravel, the safety factor becomes equal to 1; in a coarser subsoil, heading failure will occur.

The question then arises as to what extent the stability of the tunnel face can be increased by raising the excess pressure or the bentonite concentration. Comparison of curves A and B (Fig. 4) shows that raising the excess pressure causes an increase in safety, but only in fine-grained soils. When the characteristic grain size is bigger than approximately 2 mm (i.e., poorly graded gravel without a fine-grained fraction), increasing the fluid pressure will only cause further infiltration and fluid loss. The deeper infiltration of slurry into the ground represents a safety risk that cannot be compensated by raising the fluid pressure. In extremely coarse and poorly-graded soils, an important technique for stabilizing the face - the control of the slurry pressure - becomes ineffective.

However, by selecting a higher bentonite content, the grain-size range of soils that can be supported by a bentonite slurry becomes, in this example, larger by an order of magnitude (compare curves A and C in Fig. 4). Note that the bentonite content has little effect when the soil is fine-grained. Consequently, in a fine-grained soil, a slurry with a low concentration of bentonite might also be used. Such a suspension is favourable due to the easier separa-

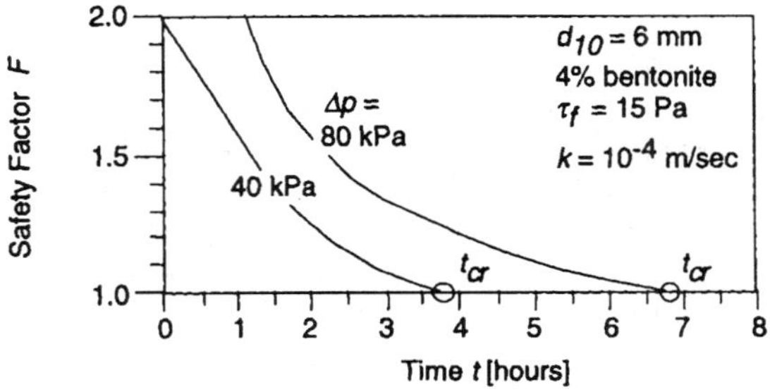

Figure 5. Safety Factor as a Function of Time *t*

tion and better handling of the excavated material.

Time-dependent effects

During an excavation standstill, the infiltration distance increases gradually over the course of time. Since the support force decreases with increasing infiltration distance, the safety factor will decrease gradually. Quantification of this effect is based upon the calculation of the time-development of the infiltration depth. Related details are given elsewhere (Anagnostou and Kovári 1994).

Fig. 5 shows the safety factor as a function of time for the example of Fig. 3. At t=0 the infiltration depth is zero, and the safety factor obtains its maximum value of the membrane model (approximately 2 for Δp=40 kPa). The initial safety margin vanishes after a critical time-period t_{cr}, i.e. at t_{cr} the limit equilibrium is achieved. Accordingly, t_{cr} represents the stand-up time of the slurry-supported tunnel face. Obviously, permeability has a decisive influence on the stand-up time. The lower the permeability is, the slower the infiltration and the consequent loss of stability will take place. By a dimensional analysis, one can verify that the stand-up time is proportional to the reciprocal value of permeability. This relationship is represented by a straight line in a double logarithmic plot (Fig. 6). In this example, the tunnel face will remain stable during an excavation standstill of up to several hours, provided that the permeability is lower than 10^{-4} m/sec. In highly-permeable ground (e.g., k=10^{-2} m/sec),

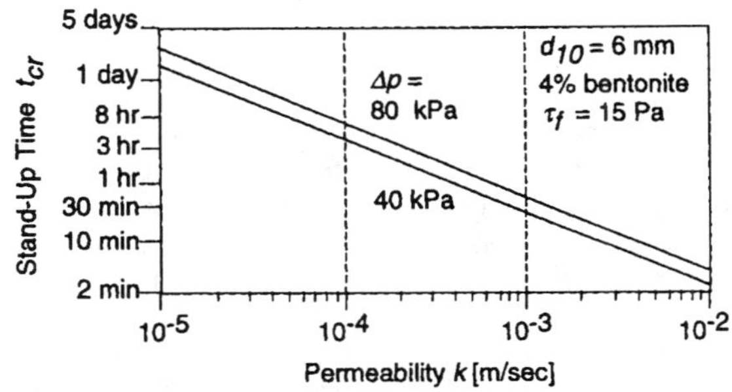

Figure 6. Stand-up Time t_{cr} as a Function of Permeability *k*

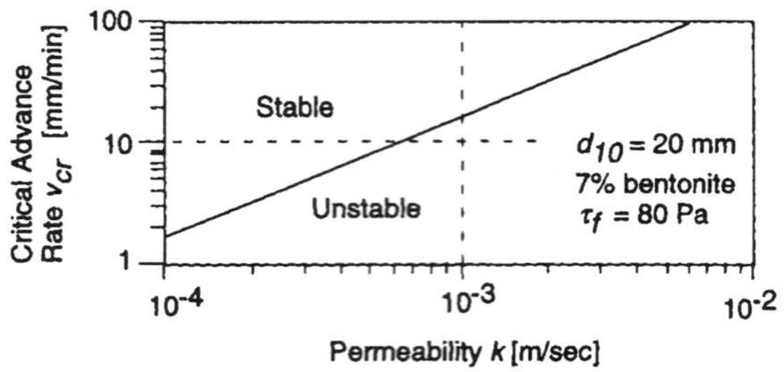

Figure 7. Critical Advance Rate v_{cr} as a Function of Permeability k

face instability will occur within few minutes.

In order that the face remains stable (safety factor ≥1), the support force should not underpass a minimum value, and, consequently, the infiltration distance should not exceed a critical maximum value e_{cr}. During excavation standstill in a coarse-grained and poorly-graded soil, the infiltration distance may reach this critical value after certain time (Fig. 5). During continuous excavation, however, the infiltration of slurry takes place simultaneously with the removal of ground at the face, i.e. the infiltration is partially compensated by the excavation. Continuous excavation is, therefore, more favourable than excavation stand-stills.

Details on the analytical determination of the effect of advance rate on the slurry infiltration can be found in Anagnostou and Kovári (1994). The infiltration distance is governed by the ratio v/k of advance rate to soil permeability. The higher the advance rate, the less the slurry penetrates into the ground, and, consequently, the higher the support force and the safety factor will be. In particular, if the advance rate is higher than a critical value v_{cr}, the critical infiltration depth will not be reached, i.e. the face will remain stable. Fig. 7 shows the critical advance rate as a function of permeability for the example of Fig. 3. Accordingly, in a soil with a permeability of, e.g., 10^{-3} m/sec, the advance rate should amount to at least 17 mm/min. At higher advance rates a safety margin will be present; at lower advance rates, the tunnel face becomes unstable.

EPB-MODE OF OPERATION

The computational model

The assumed sliding mechanism is the same as previously. However, as the work chamber is filled with excavated soil under pressure, a distinction must be drawn between the total and effective stress acting upon the face. Only the effective normal stress can be denoted as actual support pressure on the excavation face. This will be termed “effective support pressure” and denoted in the following text by s' (Fig. 8).

What effects does the pore water pressure, i.e. the piezometric head h_F in the work chamber have from a stability point of view ? If it is lower than the piezometric head h_o in the undisturbed state, then the groundwater will seep through the tunnel face. So seepage forces f will act towards the tunnel face and could endanger its stability. Determination of the seepage forces calls for a three-dimensional numerical seepage-flow analysis. The numerical model is described elsewhere (Anagnostou and Kovári 1996). Fig. 9 shows typical results (contour-lines of the piezometric head in the plane of symmetry). The increasing density of the potential lines close to the tunnel face indicates an increasing value of the seepage forces. As the seepage forces are oriented perpendicularly to the potential lines, the resultant seepage force acting on the wedge slopes slightly downward, while that in the prism above is practically vertical. The destabilizing effect of the seepage forces acting on the wedge is thus clearly apparent. An approximately horizontal load is exercised on the wedge, while the vertical load from the prism is simultaneously increased. The first effect is taken into account by introducing the seepage forces acting within the wedge into its equilibrium equations. The second effect requires a modification of the silo-theory. Computational details are given in the above-cited publication.

The necessary effective support pressure

At limit equilibrium, the effective support pressure s' depends on the tunnel diameter D, on the overburden H, on the piezometric head in the chamber h_F, on the elevation of the water table h_o, on the shear strength parameters c and ϕ, on the submerged unit weight γ' (for the soil beneath the water table) and on the dry unit weight γ_d (for the soil above the water table). With a dimensional analysis and by taking into account the linearity of the equilibrium and failure equations, one obtains the following general form of the limit equilibrium condition (Anagnostou and Kovári 1996):

$$s' = F_o \gamma' D - F_1 c + F_2 \gamma' \Delta h - F_3 c \frac{\Delta h}{D} \tag{2}$$

where F_o to F_3 are dimensionless coefficients that depend on the friction angle ϕ, on the geometric parameters H/D and $(h_o\text{-}D)/D$, and on the ratio of the dry to the submerged unit weight γ_d/γ'. Fig. 10 shows these coefficients as functions of the friction angle ϕ. The diagrams have been computed numerically. The curves cover the relevant range of the geometric parameters H/D and $(h_o\text{-}D)/D$. The ratio γ_d/γ' was taken to 1.60, a good enough approximation for practical purposes. Eq. (2) together with the diagrams in Fig. 10 provide a

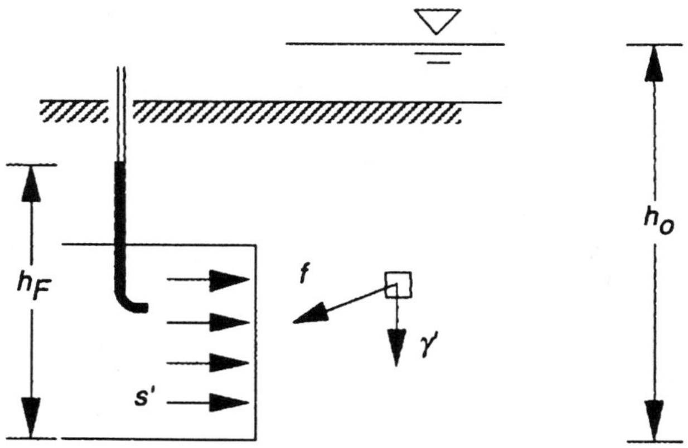

Figure 8. Seepage Force f and Effective Support Pressure s'

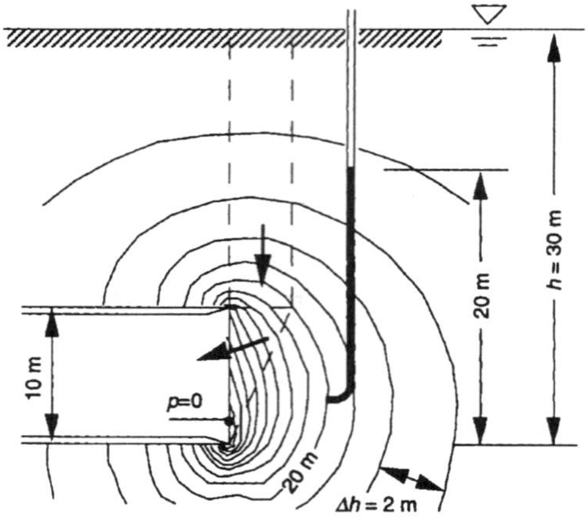

Figure 9. Numerically Computed Contourlines of Piezometric head h

simple but powerful tool to characterize quantitatively the face stability at EPB-operation mode.

Example

The Great Belt project involves two single-lane undersea tunnels (ø 8.7 m) bored by four EPB tunnel boring machines. The tunnels cross glacial tills and fissured marls. The upper till is a preconsolidated undisturbed soil with up to 20% clay. The lower till is an extremely heterogeneous material containing irregular sand lenses, gravels and glacial boulders. Due to the high porewater pressures (up to 4 bar in the tills), extensive drainage works have been carried out. This so-called Moses project reduced the seepage forces, simplified cross passage constructions and made entry into the working chamber possible at low air pressure (Biggart *et al.* 1993).

In order to provide a better understanding of face stability conditions, statical calculations have been performed based on the model described above (Kovári and Anagnostou 1994). In the following, the interaction between the main factors will be demonstrated by means of a numerical example concerning a tunnel section in the tills with an overburden and a sea depth of 30 m and 10 m, respectively.

By using eq. (2) and the nomograms of Fig. 10 (with ϕ=32.5°and γ'=13 kN/m^3), one obtains the limit equlibrium relation between the effective support pressure s', the head difference $(h_o - h_F)$ and the cohesion c (Fig. 11).

According to Fig. 11, the compensation of the water pressure (i.e., h_o=h_F) suffices for face stability even when the ground shows a cohesion c as low as 10 kPa (point A in Fig. 11). If the difference between the water level h_o and the head h_F in the chamber for some reasons is high, a high effective support pressure s' is required to ensure face stability. For example, in the extreme case of atmospheric pressure in the chamber (and for a cohesion c of 30 kPa - a reasonable value for the tills), the necessary effective support pressure amounts to 160

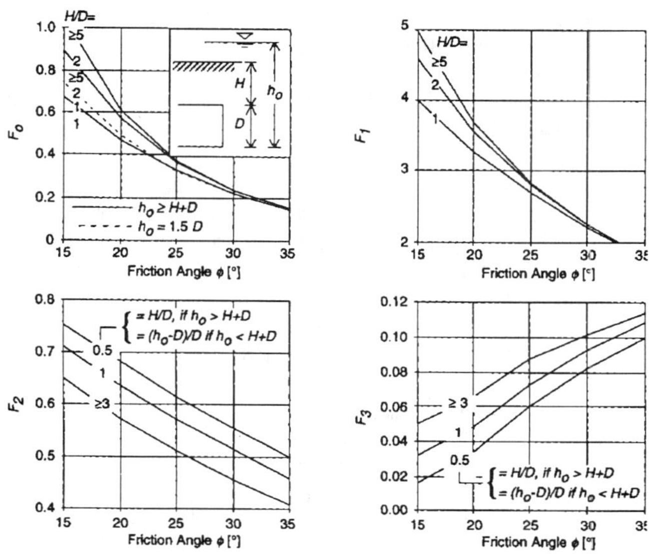

Figure 10. Nomograms for the Dimensionless Coefficients

kPa (point B in Fig. 11).

A high effective support pressure, however, has considerable operational disadvantages such as excessive wear of the cutter head, high torque, or arching of the muck at the entrance to the screw conveyor. The head difference between chamber and ground should, therefore, be kept as low as possible. In this example, when reducing the head difference by 20 m, the required effective support pressure *s'* decreases from approximately 160 kPa to 60

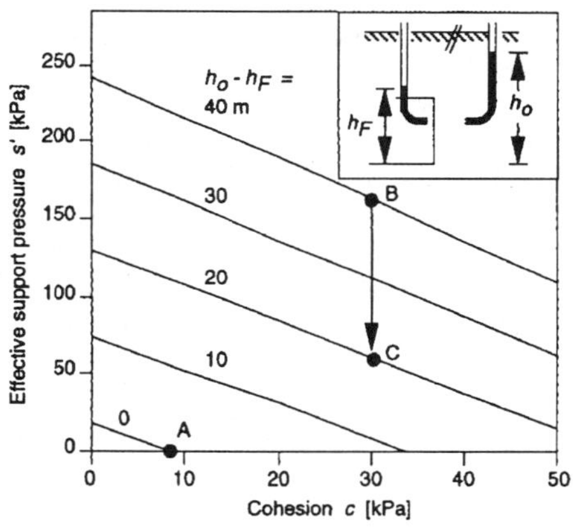

Figure 11. Effective Support Pressure *s'* as a Function of Cohesion *c* and of Hydraulic Head Difference (h_o-h_F)

kPa (point C). A low head difference can be maintained either by decreasing the pore water pressure in the ground (by pump operation) or by maintaining a sufficient head h_F in the chamber. The latter depends upon the ground properties, the properties of the muck and machine-specific details. An overview of existing techniques can be found elsewhere (Anagnostou and Kovári 1996).

Fig. 11 shows, furthermore, that, in a cohesionless ground, the same reduction in the required effective support pressure s' of approx. 56 kPa can be achieved either by reducing the head difference by 10 m or by increasing the ground cohesion c by approximately 28 kPa. In other words: The effects of $\Delta s' \cong 56$ kPa, $\Delta(h_o - h_F) = 10$ m and $\Delta c \cong 28$ kPa in stabilizing the tunnel face are statically equivalent.

CLOSING REMARKS

The results of the parametric studies presented are in accordance with experience. Even when - in the absence of laboratory or field investigations - the data necessary for a stability analysis are incomplete, the computations may be useful, for they improve understanding of the complex interrelations involved and, therefore, enable a better assessment of the residual risks or - in the case of face instabilities - of the effectiveness of various countermeasures.

REFERENCES

Anagnostou, G. and K. Kovári 1994, "The Face Stability of Slurry-shield-driven Tunnels", Tunnelling and Underground Space Technology, 9, No. 2, pp. 165-174.

Anagnostou, G. and K. Kovári 1996, "Face Stability Conditions with Earth Pressure Balanced Shields", Tunnelling and Underground Space Technology, 11, No. 2, pp. 165-173.

Babendererde, S., 1991, "Kritische Betrachtungen zum Einsatz von Hydroschilden", Int. Symp. Sicherheit und Risiken bei Untertagebauwerken (eds. R. Fechtig and K. Kovári), pp. 47-51 (in German).

Biggart, A.P., J.-P. Rivier and R. Sternath, 1993, "Storebælt Railway Tunnel - Construction", Proc. Int. Symp. on Technology of Bored Tunnels under Deep Waterways, Copenhagen, Denmark, pp. 63-93.

DIN-4126, 1986, "Ortbeton-Schlitzwände. Konstruktion und Ausführung" (in German).

Fujita, K., 1981, "Use of Slurry and Earth-Pressure-Balance Shield in Japan", Int. Congress on Tunnelling, Düsseldorf, Tunnel 81, vol. 1, pp. 383-406.

Horn, M., 1961, "Horizontaler Erddruck auf senkrechte Abschlussflächen von Tunneln", Landeskonferenz der ungarischen Tiefbauindustrie, Budapest (in German).

Janssen, H. A., 1895, "Versuche über Getreidedruck in Silozellen", Zeitschrift des Vereins deutscher Ingenieure, XXXIX, No. 35, pp. 1045-1049 (in German).

Kovári, K. and G. Anagnostou, 1994, "Storebælt Railway Tunnel - Assessment of face stability", Report prepared for the MT-Group, Consortium for Storebælt Crossing.

Müller-Kirchenbauer, H., 1972, "Stability of slurry trenches", Proc. 5th europ. conf. SMFE, Madrid, vol. I, pp. 543-553.

Stack, B., 1982, Handbook of Mining and Tunnelling Machinery, Chichester a.o., John Wiley and Sons.

Xanthakos, P., 1979, Slurry Walls, p. 622, New York a.o., McGraw-Hill.

Chapter 34

MASSACHUSETTS BAY TRANSPORTATION AUTHORITY CONCRETE IMMERSED TUBE TUNNEL CROSSING, FORT POINT CHANNEL

Thomas P. Kwiatkowski[1] and John S. Anderson[2]
[1]Jenny Engineering Corp.; [2]Frederic R. Harris, Inc.

ABSTRACT

The Massachusetts Bay Transportation Authority (MBTA) has undertaken an underground transit project to extend public transportation service from downtown Boston to the South Boston Piers area. This new system, known as the Transitway, will connect existing subway lines to this area which is comprised of approximately 300 acres, in light of the potential for 21 million square feet of development by 2010.

This system includes construction of a concrete immersed tube tunnel crossing of Fort Point Channel. The crossing will be accomplished in the highly congested setting of downtown Boston and will be located just to the north of the famous Boston Tea Party ship exhibit and amid major Central Artery Construction activity. The most recent concrete tube constructed in the United States was in 1962. The Fort Point Channel Crossing will employ current design and construction methods commonly utilized in the Netherlands, Australia, Hong Kong and elsewhere. This paper will present the design philosophy, planning, and evaluation of alternative crossing methods which led to the selection of the concrete immersed tube tunnel and construction techniques envisioned for this particular application.

INTRODUCTION

The Massachusetts Bay Transportation Authority (MBTA) provides public transportation in the greater Boston metropolitan area via surface bus routes, commuter rail and surface/subsurface transit lines. When put into service, the Transitway will be the fifth operating subsurface transit line and will connect the downtown to the South Boston Piers district. The four existing lines extend from the Central Business District (CBD) to points north, south and west. The Piers district, located immediately east of the CBD, and separated by the Fort Point Channel, has historically been served by surface buses. It is expected that this former rail yard and industrial area will be developed over the next 15 to 20 years as office/commercial space generating approximately 18,000 peak hour and 90,000 daily one-way trips by 2010. Major developments completed or commenced to date that support this expectation include a new Federal Courthouse, an interchange on I-90 (part of the Central Artery/Ted Williams Tunnel project), and a hotel/conference facility adding to the existing World Trade Center complex at Commonwealth Pier.

In anticipation of this development and associated transportation demand, the MBTA proposed the Transitway Project in 1987. An aggressive planning phase was pursued with

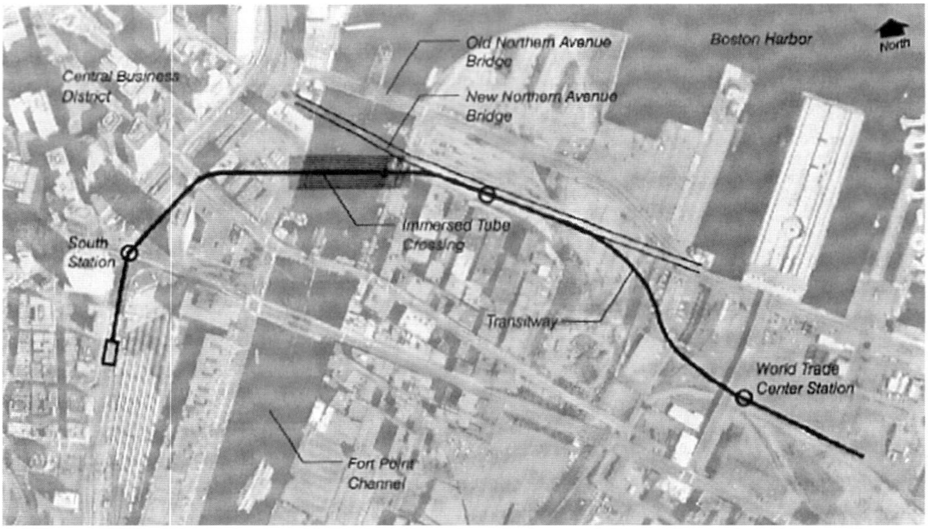

Figure 1. Transitway Alignment in Initial Phase

a Feasibility Study, DEIR/S and FEIR/S documents completed and a Record of Decision received in early 1994. The project is proposed to be developed in two phases. The initial phase will construct approximately 1,280 m (3,900 feet) of alignment (tunnel and boat) connecting with the existing Red Line Subway service at a reconstructed South Station and extending to the easterly terminus in South Boston. See Figure 1. The full-build phase will extend the tunnel westerly from South Station to the Boston Common where it will connect with the existing Green Line at Boylston Station. In addition, this full-build segment of tunnel will connect to the Orange Line at Chinatown Station. During the time that the initial phase alignment is operated, Transitway vehicles will be reversed by means of a turn-around loop located below Atlantic Avenue to the south of South Station. The initial phase has a Capital Cost Estimate of $350 - 400 million (1993 $) and the Full Build Capital Cost Estimate is $550 - 600 million.

The typical tunnel cross-section consists of two travel-ways 3.93 m (12 feet) wide and with a minimum vertical clearance of 5 m (15.25 feet). The proposed tunnel is typically constructed of cast-in-place concrete and the alignment and cross section will accommodate both rubber tired articulated buses and Light Rail Vehicles which could be introduced in the future as ridership increases. A substantial portion of Transitway tunnel to the west of Fort Point Channel is to be constructed within the excavation created for the Central Artery highway tunnel, thereby saving significant construction costs and minimizing utility relocations and disruptions to traffic relative to an independent construction effort.

The Transitway alignment traverses a heavily developed urban area which presents formidable engineering challenges. These include reconstruction of the South Station Red Line Station within the limits of Dewey Square while maintaining existing Red Line transit operations and surface traffic, tunneling beneath an occupied, pile-supported historic building and crossing the Fort Point Channel. It is the latter endeavor that is the topic of this paper.

PROJECT AREA

The Fort Point Channel is located to the southeast of the Financial District of Boston and southwest of Boston Inner Harbor. Both shorelines of the channel are defined by granite seawalls and timber bulkhead structures and wharfs. The westerly shore line was established in the colonial period and resulted from incremental filling with native soil material from Fort Hill, miscellaneous fill and debris from fire ruins. The easterly shore line was established in the late 1800's and resulted from an extensive landfill of tidal marsh to support industrial expansion and a rail yard.

Fort Point Channel is approximately 1.7 km (5,500 feet) long extending from the Roxbury Canal Conduit to the Old Northern Avenue Bridge near the junction with the Boston Inner Harbor. Water depths in the area of the proposed crossing range from 3.28 to 8.2 m (10 to 25 feet) at Mean Low Water (MLW) with a tidal range of approximately 3.28 m (10 feet). The width of the channel in this area is approximately 197 m (600 feet).

The upland properties abutting the channel on the westerly side in the vicinity of the Transitway alignment include (south to north) the Russia Wharf Buildings (three pile supported, steel framed, seven story structures used for office/commercial space and listed on the National Register of Historic Places), the former Boston Edison Substation site (where the Massachusetts Highway Department proposes to construct a Ventilation Building for the Central Artery highway tunnel and BECo a commercial development inclusive of subsurface parking facilities) and 270 Atlantic Avenue (caisson supported, 12-story, steel-framed office building). All existing and proposed development extends to the edge of the channel. The Transitway alignment enters the channel from the former BECo site.

The upland properties abutting the channel on the easterly side in the vicinity of the Transitway alignment include (south to north), the Children's Museum/Computer Museum facility (inclusive of a floating restaurant and proposed floating exhibit space), an undeveloped parcel to be obtained by the MBTA (which was formerly occupied by the Victoria Station restaurant) and the New Northern Avenue Bridge with approach from New Northern Avenue. Sleeper Street runs parallel to the channel approximately 50 m (165 feet) to the east of the channel. The Transitway alignment enters the channel from the former Victoria Station site.

The existing and proposed development precludes the use of a local casting basin for construction of the tubes and presents the need for extensive detailed planning and coordination to optimize available site access.

There are seven bridges crossing Fort Point Channel. Two are located between the Transitway alignment and Boston Inner Harbor. They are the New Northern Avenue Bridge, a fixed span highway bridge (5.3 m clearance at Mean High Water) and the Old Northern Avenue Bridge, a truss swing span highway bridge (26 m horizontal clearance when open). The alignments of the New Northern Avenue Bridge and the Transitway are skewed and the distance separating these structures ranges from 92 m (280 feet) on the west side of the channel to 16.5 m (50 feet) on the east side.

The granite bulkheads on the east and west sides of the channel in this area are considered historic resources requiring that, to the extent they are disturbed, they be restored to existing appearance and function following Transitway construction.

There are four commercial water users in or near Fort Point Channel. They include The Gillette Company, the New England Aquarium, James Hook Lobster Co. and the Barking Crab Restaurant. The two latter users are in the immediate vicinity of the project and will require mitigation to prevent suspended solids from entering the intakes during dredging.

GEOTECHNICAL CONDITIONS

A subsurface exploration program consisting of 13 geotechnical borings and six environmental borings was performed along the proposed alignment of the Transitway Channel crossing. Two existing borings from other projects were also used to develop soil profiles along the alignment.

The subsurface exploration program included installation of piezometers, field permeability tests, field vane shear tests, and recovery of undisturbed tube samples for laboratory testing. The laboratory testing program included standard soil index tests, chemical analyses, one-dimensional consolidation tests, and triaxial tests.

Subsurface Conditions

Subsurface profiles developed from the geotechnical investigation program are presented in Figure 2. The soil and rock strata encountered are described below in order of increasing depth.

Fill (F) - A layer of miscellaneous fill approximately 6.6 to 8.2 m (20 to 25 feet) thick is present where the Tunnel alignment crosses land areas adjacent to the Fort Point Channel on the westerly edge. Similarly, at the easterly edge, the fill ranges in thickness from 1.6 to 4.9 m(5 to 15 feet). The fill contains primarily of fine to coarse sand with some gravel and silt and occasional cobbles and boulders. Foundation remnants such as concrete floor slabs, granite blocks and timber piles, along with other types of construction debris(bricks, steel, paving materials) are likely to be present in the fill soils.

Organic Soils (O) - Organic silt and clay with some sand were deposited in tidal marsh and estuarine environments. Adjacent to Fort Point Channel, fill material was placed on the organic material. In Fort Point Channel, the bottom is comprised of organic soils which range in thickness from 1.6 to 8.2 m (5 to 25 feet) along the Transitway Channel crossing. The organic soils are known to be contaminated from historic industrial sources, oil spills, urban runoff and combined sewer overflow effluent.

Marine Clay (C) - A deposit of marine clay (commonly known as the Boston Blue Clay) lies beneath the organic soils. The clay ranges from 6.6 to 13.1 m (20 to 40 feet) thick within the westerly half of the Transitway alignment with the thinner stratum dimension near the midpoint of the crossing and increases up to 24.6 m (75 feet) thick in the easterly half of the alignment. The upper part of the clay deposit has been overconsolidated and desiccated. The clay generally becomes softer with increasing depth as the overconsolidation due to desiccation decreases. In some areas, the upper 3.3 to 4.9 m (10 to 15 feet) of the clay consists of a very stiff to hard desiccated crust, which is often distinguished by a yellow-brown color resulting from oxidation. Below this desiccated crust, the clay is generally me-

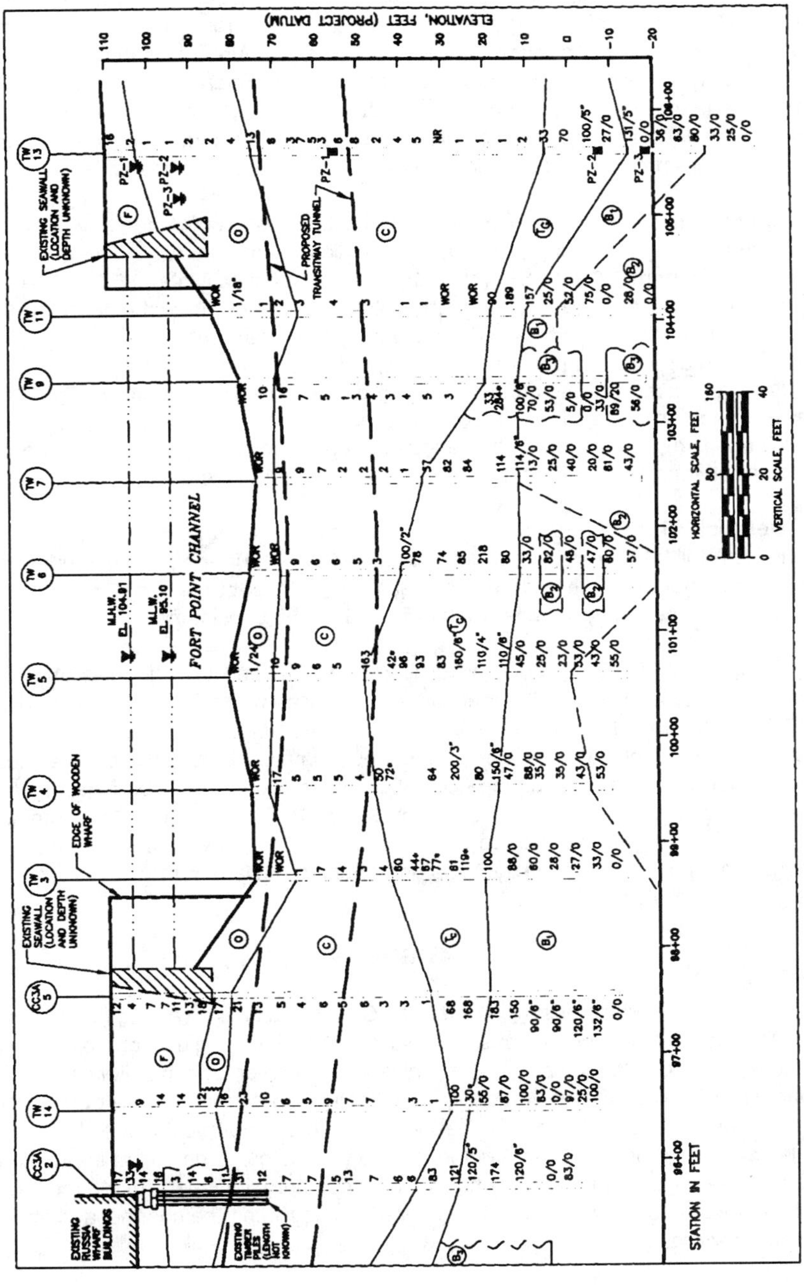

Figure 2. Subsurface Profile Along Alignment Centerline Sta. 93+00 to 107+00

dium stiff to stiff. The deepest portion of the clay at the east end of the alignment is soft to medium stiff. An extensive program of field and laboratory testing was performed to determine the undrained shear strength of the clay. The shear strength is necessary to evaluate bottom stability in the construction methods considered. Shear strengths were determined from field shear vane tests, laboratory shear vane tests, conventional unconsolidated-undrained triaxial compression tests, and undrained triaxial compression and extension tests on samples anisotropically consolidated to the in-situ effective stresses.

Glacial Till (T_c and T_g) - A glacial till deposit consisting of dense to very dense sand, gravel, silt and clay in varying proportions underlies the marine clay. The thickness of the glacial till varies from 3.3 to 11.5 m (10 to 35 feet) and is typically about 6.6 to 9.8 m (20 to 30 feet) thick in the central and western portion of the alignment where the bottom of the tunnel bears on or near the till surface. The glacial till in the central and western portion of the alignment contains a high percentage of silt and clay and has relatively low permeability. This material is described as cohesive till (T_c) on the subsurface profiles. The glacial till in the eastern portion of the alignment, which is described as granular till (T_g), contains less silt and clay and is more pervious. The borings performed for the project and previous borings in the area encountered only occasional cobbles or boulders in the glacial till.

Bedrock (B1, B2 and B3) - The bedrock along the tunnel alignment is weathered Cambridge Argillite. The rock stratum shown on the subsurface profiles has been divided into sub-units that reflect the degrees of weathering. Most of the upper portion of the bedrock has been completely weathered (kaollinized) to the consistency of a stiff to hard clay.

Groundwater - The ground water level in the fill and organic soils above the clay reflects the water level in the Channel and is typically between Elevation 100 and Elevation 105 (Project Datum is 100 feet below Mean Sea Level, NGVD 1929). The piezometric level measured in the glacial till in two borings is approximately Elevation 94. The piezometric levels measured in the rock in boring AC12-4 and several other CA/T borings on the wet side of the Channel are in the range of Elevation 90 to 92. The depressed piezometric levels (below mean sea level) in the deep strata below the clay may be due to the presence of permanent deep underdrainage systems and ongoing dewatering for construction projects in the general vicinity.

ALIGNMENT

The Transitway crossing of Fort Point Channel occurs on a tangent section of alignment. This tangent section crosses the channel at a skew angle of approximately 25 degrees to a perpendicular crossing. The total length of immersed tube within the limits of the Channel is 230m (700 feet). Vertically, the crossing represents the low point in the entire alignment and includes a 131m (400 feet) vertical curve. Approaching the channel crossing from the west the gradient is -3.25 percent. The vertical curve begins at approximately the midpoint of Tube No. 1 (the westerly tube) and ends approximately 14m (43 feet) easterly of the joint between Tube Nos. 2 and 3. The gradient departing the channel and approaching the easterly shoreline is +1.77 percent.

A condition of the regulatory permits for this facility requires that the channel bottom be restored to its original condition and that no portion of the tunnel project above the existing bottom. This represents one of several factors controlling the Transitway profile in this area.

The bottom elevation of the tunnel ranges from Elevation 52.65 at the westerly end to Elevation 45.2 at the low point to Elevation 49.97 at the easterly end. The channel mudline elevation ranges from Elevation 75 to Elevation 85 with adjacent upland existing grade at approximately El. 110.

A low point sump and pump station is located within Tube No. 2.

FORT POINT CHANNEL CROSSING

During the conceptual design phase of this project, a Type Study was conducted to determine the most economical and practical method to cross Fort Point Channel. The Type Study was completed in late 1993 and recommended a concrete immersed tube be utilized for the Channel crossing.

Consideration was given at that time to prepare Construction Documents for both the concrete immersed tube and the cofferdam construction method and take alternate bids for the Channel crossing. However, it was determined to select the most economical and practical crossing method based on the Type Study and prepare final documents for that method.

Final design began in early 1995 and has progressed just beyond the 60% stage. Sixty percent (60%) design documents have been submitted to the MBTA. However, the immersed tube design was placed on hold until it was known what impact design modifications and studies being conducted to determine the optimum method to construct the Transitway beneath the Russia Wharf buildings would have on the crossing. Final design is expected to restart early in 1997.

The following sections describe the Type Study conducted for this project followed by specific design details developed to date for the concrete immersed tube crossing.

TYPE STUDY

The Type Study evaluated two methods of construction for the Fort Point Channel Crossing. One is the open cut cofferdam and the other is the immersed tube method. The horizontal and vertical alignment for each crossing was the same. The immersed tube will be discussed first, followed by the open cofferdam.

For the immersed tube crossing, two types of tubes were considered, a steel shell and concrete tube. A steel shell tube includes as its basic elements a steel shell which forms a watertight membrane and in combination with the reinforced concrete interior lining and concrete which surrounds the shell, provides the necessary structural strength for the finished tunnel. A concrete tube consists of a reinforced concrete box section enclosed within a waterproofing system to provide strength and serviceability requirements.

Certain factors regarding the crossing will be the same whether the crossing is accomplished by concrete or fabricated of steel.

These factors include the following:

- o Each cross section will have two tracks/lanes, one for each direction of travel.

- The tube tunnel will extend to just behind each existing seawall and terminate in cofferdams where a transition structure will connect the tube sections to cut and cover tunnel box sections. Construction procedures for the cofferdams will be similar to the procedures for the open cut cofferdam option.

- The crossing will consist of three sections of tube. Two sections will be 73m (240 feet) long and the other 79m (260) feet long. The 79m (260 foot) section will be located in the middle of the Channel. Constructing the crossing in three stages will allow a majority of the Channel to remain open throughout the construction.

  An evaluation was conducted to determine whether two tubes, each 112.8m (370 feet) long, or three tubes of the lengths mentioned above, were best suited for the crossing. In the two tube option, horizontal and vertical curves in the Transitway alignment resulted in reduced lateral clearances as the tube passed through the old Northern Avenue Bridge and reduced vertical clearances as the tube passed beneath the new Northern Avenue Bridge. Therefore, it was determined, to reduce the difficulty in floating the tubes into place, three tubes would be more practical for this crossing.

- Steel sheet piling will be installed along the lateral limits of the trench. The purpose of the steel sheet piling is to limit the width of excavation. Generally in immersed tube construction, the trench for the tube is dredged permitting the sides of the excavation to reach its natural slope resulting in a relatively wide excavation. Since the organic sediments in the bottom of the Channel are known to be contaminated, a cost comparison was conducted to determine whether it was more economical to install the sheet piling or dispose of the contaminated material. The results of this analysis indicate the steel sheet piling is the more economical choice. In addition, reducing the volume of organic sediment to be removed and the time spent dredging reduces the likelihood of fouling the intakes of commercial water users.

Steel Shell and Concrete Tube Alternatives

The present trend in the United States is to employ steel shell tubes for immersed tube construction. In Europe, concrete tubes are more widely used. There is no clear cut reason why this is the case. For this crossing, each method of construction was evaluated with respect to the conditions encountered on this project. The following is a summary and comparison of some of the advantages and disadvantages of each type of construction.

Construction Duration

Steel shell construction is generally of a shorter duration than concrete tube construction. This results from the need to construct a casting basin near or adjacent to the site for a concrete tube. However, for this project given the relatively short tube length of 243m (740 feet) and the presence of underutilized dry docks greater than 295m (900 feet) in length, the possibility exists that a nearby shipyard or dry dock can be utilized as a casting basin. Therefore, the overall project duration can be reduced.

Waterproofing

Waterproofing of the structure is inherent in steel shell construction since the shell acts as the waterproofing membrane. In concrete tube construction, a waterproofing membrane is added to the outside of the structure. This is generally not as preferable as having the waterproofing built in as part of the structure since there is greater potential for an attached membrane to be damaged.

Foundation

Steel shell tubes generally utilize a screeded foundation course, while the concrete tube utilizes a pumped or jetted foundation. A screeded foundation course is usually formed when foundation material such as coarse sand or well graded gravel is placed in the trench and leveled. The leveling is performed using a heavy screed which may require several passes to achieve the proper shape. All this work is performed from a barge with winches used to move the screed.

A jetted foundation involves the placing of the immersed tube on temporary supports in the trench and then hydraulically placing sand between the tube and the ground. The jetting is done to fill the void under the tube and give it uniform bearing.

Shape and Layout

The steel shell provides the structural advantage of a circular shape, however, it requires greater interior and exterior structure dimensions to provide an equivalent rectangular clearance envelope. The rectangular concrete tube is able to closely follow the shape of the clearance envelope resulting in a smaller overall structure. For this project, the overall exterior dimensions of the concrete tube are approximately 12.8m (42 feet) wide by 7.6m (25 feet) high and the steel tube is 14.3m (47 feet) wide and 9.5m (31 feet) high.

Production of Units

Steel shells are fabricated in steps in a manner similar to an assembly line process. At each step a certain operation is performed beginning with the cutting of struts and plates to the assembly of 6.1m (20 feet) long circular segments which are then connected to make 60-122m (200-400 feet) long tube sections. Concrete is placed and the tube launched from the dry dock. The tube is then transported to an outfitting location where final concrete is placed prior to sinking of the segments.

As each shell is completed and launched another tube is then assembled. This permits steel shells to be launched on a regular basis, as required by the construction schedule.

Conversely, concrete tubes are generally constructed in batches depending on casting basin size and the quantity of formwork available. Therefore, it is envisioned, the three tubes required for this project will be constructed concurrently. This also meets the need for the long curing durations required for concrete.

ON SITE FABRICATION

Fabrication of immersed tubes near the crossing site is generally considered where sufficient room is present to allow for the construction of a dry dock or casting basin. This dock must be of sufficient size to accommodate tubes 60-90m (200 - 300 feet) long, 12-15m (40 to 50 feet) wide and 6-9m (20 to 30 feet) high. Additional at grade space is required for construction staging, material storage, offices and parking.

An evaluation of the project site indicates construction areas will be limited and generally congested. The western side of the project is bordered by the BECO substation site, the future site of the Central Artery vent building, Russia Wharf and Harbor Plaza (270 Atlantic Avenue). This site is very congested and does not contain sufficient area for a casting basin.

On the eastern side of the Channel, beyond the seawall, there is an area approximately 42m (140 feet) wide by 91m (300 feet) long, approximately one acre. This area is currently occupied by the Victoria Station Restaurant and a parking lot and is much smaller than the three to six acres necessary for a casting basin or dry dock. This limited area would create a cramped construction work area and would not be able to accommodate the fabrication of even one tube section of the proposed length at a time. Therefore, this site is not considered adequate for a casting basin.

Alternative casting basins sites were evaluated in proximity to the site at Fan Pier and at the site of the Central Artery crossing of the Channel. The Fan Pier site appeared to have sufficient area available for construction of a dry dock and the staging needed for tube construction. This location is adjacent to the site and would require the floating of the tubes through the Northern Avenue Bridges. This site was not considered since there would be a major loss of parking and future plans included the construction of a new courthouse at this site. The courthouse is currently under construction.

The Central Artery's Seaport Access Road Channel Crossing crosses Fort Point Channel from the Dorchester Avenue Bridge to Mt. Washington Avenue. This project includes the construction of a casting basin adjacent to the crossing for the concrete tubes. The basin will be constructed on the Gillette site. The basin will be removed upon completion of tube construction and a cut and cover tunnel will be constructed in the remaining excavation.

Once construction starts for the Seaport Access Road Channel Crossing, the casting basin area will always be occupied by some part of the Central Artery project. Therefore, the only possibility of utilizing the Central Artery casting basin for the Transitway project is to assume the Transitway would be constructed far in advance of the Seaport project. Therefore, this scenario is not possible.

OFF SITE FABRICATION

Off site fabrication is generally accomplished at an existing shipyard facility since these sites are sufficiently larger and contain equipment and skilled labor to complete the project. As part of this type study, contact was made with a number of shipyards to determine their capability for fabrication, outfitting and/or tube construction.

The Thomas Register for 1992 indicates that shipyards are presently operating in virtually every state along the Atlantic seaboard from Maryland to Maine. The following sites were researched as potential fabrication sites:

Port Deposit, Maryland
Bath, Maine
Sparrows Point, Maryland
Quincy, Massachusetts
South Boston, Massachusetts

Preliminary investigations indicate that shipyards are available to fabricate the steel shell and possibly the concrete tubes. The determination of which shipyard would be selected by the successful bidder will be based on such items as shipyard past performance, fabrication, transportation cost and availability of materials as determined at the time of bidding and construction.

OPEN CUT COFFERDAM CROSSING METHODS

As mentioned previously, the open cut cofferdam method was also evaluated during the Type Study phase of this project. This method involves the use of sheet piling internally braced to create an open work space in the Channel, within which a cast-in-place concrete box would be constructed. Using this method, the crossing of the Channel would be performed in the dry and would be accomplished in two stages. Two stage construction would allow half of the Channel to remain open at all times.

Stage I

The first stage would begin on land and extend approximately halfway into the Channel. This would avoid more than half of the Channel being closed at any one time to allow boat traffic to pass as well as the tide, which varies up to 3m (10 feet). The width of the cofferdam varies up to a maximum of 16.1m (53 feet) and the depth varies from 14.6m (48 feet) on land to 21.6m (71 feet) in the Channel. The sheeting would extend 2.7m (9 feet) above mean high water based on a flood elevation of 111.

For the Stage I cofferdam, all sheet piles would be driven deep enough to obtain toe embedment in the glacial till. The borings indicate that the top of the glacial till varies from 0 to 6m (0 to 20 feet) below the tunnel invert in this section.

For the open water portion of the cofferdam, it is anticipated that the upper level bracing would utilize prefabricated bracing frames consisting of at least two brace levels connected vertically to form self-supporting frames. The bracing frames would serve as the template for driving the sheet piles. The cofferdam would be dewatered after the sheet piles are installed along the perimeter of the bracing frames. Excavation below the Channel bottom and installation of the lower brace levels would then be performed in the dry. For the portion of the cofferdam on land, excavation and installation of bracing would be performed in the dry, proceeding downward from the ground surface, after the open water portion of the cofferdam is dewatered.

Since the borings indicate that glacial till is present relatively close to the bottom of Stage I excavation, soil anchors would be utilized to support the lower portion of the steel sheet piling. This would permit the concrete box structure to be cast without interference of the lower cross bracing.

After the Stage I tunnel section is completed, a sheet pile wall would be installed around the end of the completed tunnel section within the confines of the existing cofferdam. The remainder of the existing sheet pile cofferdam would be flooded in stages and cross bracing removed as the water level rises. The Stage I sheet piling would then be cut off at the Channel bottom and removed. This now opens the west side of the Channel.

Stage II

Sheet piling for the Stage II cofferdam would be installed. The Stage II cofferdam would require a different construction procedure than defined for Stage I due to the large thickness of soft clay below the tunnel.

Installation of the sheet piles and upper brace levels would be similar to the Stage I cofferdam, except that dewatering inside the cofferdam would be limited to the level required to excavate and install the upper brace levels in the land portion of the cofferdam. Excavation of the lower portion of the cofferdam would be performed in the wet to maintain stability of the clay in the bottom of the cofferdam. Bracing for the lower portion of the cofferdam would be suspended below the upper brace levels and lowered into position as excavation proceeds. A tremie concrete base slab will be poured under water prior to dewatering the cofferdam. The base slab will brace the toe of the sheet piles and to help resist base heave in the soft clay. Relief holes will be drilled through the tremie concrete slab to prevent hydrostatic uplift.

CONSTRUCTION SEQUENCING, SCHEDULING AND COST

For the Type Study, the anticipated construction procedures were utilized to estimate durations for each major item of work and to develop statements of probable construction costs in 1993 dollars. The sequence and schedule were developed with the intention of minimizing construction durations in order to limit the impacts of construction on the public and industry along the Channel.

Steel Shell Immersed Tube Method

Construction Sequence - The sequence of construction within limits of the Transitway crossing would consist of four stages as described below. Prior to work beginning in the Channel for Stage 1, general site mobilization will be underway as well as mobilization of the steel tube fabrication plant.

Stage 1 - Guides for sheet pile driving would be installed followed by the installation of the steel sheets surrounding the entire excavation. Within the limits of the East and West Work Shafts, internal cross bracing would also be installed.

Stage 2 - Stage 2 consists of dredging across the entire Channel within the sheet piling and placing the screeded foundation course.

Stage 3 - Following completion of foundation course, the steel tubes would be transported, one at a time from a nearby outfitting dock to the Channel. Once in the Channel, a laying barge would lower the tube into position.

The tubes will be placed from east to west. Beginning with Tube No. 3 and ending with Tube No. 1. This sequence is dictated by a conflict between the tube and the cross bracing designed within the limits of the East Work Shaft. The cross bracing prohibits the tube from being floated directly above the excavated trench and then lowered into position. The cross bracing is required because of the poor soil conditions surrounding the shaft.

Therefore, to place Tube No. 3 the tube would have to be lowered below the bracing elevation and then "slid" into position over the excavation and lowered into place.

After installation of Tube No. 3, Tube No. 2 would be lowered into place and connected to Tube No. 3 with the aid of coupling jacks and divers. This process will then be repeated for Tube No. 1.

Stage 4 - Sheet piling would be installed transversely to form the East and West Work Shafts, respectively. Locking backfill, ordinary backfill and rip rap if required, would then be placed around and over the tube between the work shafts. Within the limits of the work shafts the transition structures would be constructed followed by placement of backfill.

Construction Schedule and Cost - The estimated overall construction contract duration for the steel tube crossing of the Channel is 23 months with an anticipated cost of $30,400,000.

Concrete Tube Method

Construction Sequence - The construction sequence for the concrete tube crossing is essentially the same as that for the steel shell tube with one exception.

Concrete Tube Nos. 1 and 3 would be fabricated with the transition section on the appropriate end of the tube. This would eliminate the need for an East and West Work Shaft as mentioned in Stage 4 above. Therefore, following placement of the three tubes, locking backfill, ordinary backfill and rip rap, if required, would be placed.

Construction Schedule and Cost

The estimated duration of the concrete immersed tube crossing is 19 months with an anticipated cost of $22,400,000.

Cofferdam Method

Construction Sequence - The construction of the crossing by cofferdam methods would be performed in two stages in order to permit approximately one-half of the Channel to remain open at all times for passage of boats and tidal fluctuations.

Stage I

As with the immersed tube crossings, the first item of work would be the construction and installation of templates for driving the sheet piling. Following template construction, the steel sheet piles would be driven around the limits of excavation. Cross bracing would then be installed as the water level is lowered within the cofferdam. Once dewatered, excavation of soil would continue to the foundation elevation with tiebacks being installed as the excavation progresses. The cast-in-place reinforced concrete box would then be constructed and prior to backfilling, a steel sheet pile wall would be installed across the cofferdam. This wall would serve as the western limit of the Stage 2 cofferdam. The steel sheet piling would then be removed, to reopen the Channel.

Stage 2

Steel sheet piles would be driven to surround the remaining excavation. The previous east end wall of Stage 1 would be removed. The water level would be lowered within the cofferdam and cross bracing installed. Unlike Stage 1, tie-backs cannot be utilized in Stage 2. The soft clay on the east side of the Channel does not provide sufficient strength to anchor the tie-back. Therefore, cross bracing would be utilized to support the entire excavation. Following excavation support installation, the concrete box would be constructed and backfilled as specified.

Construction Schedule and Cost

The estimated duration of the cofferdam crossing is 30 months with an anticipated cost of $25,900,000.

EVALUATION

Steel Shell Tube Method

Advantages - The steel shell tube has been utilized on 19 of the 21 immersed tube crossings in the United States since 1910 and was recently utilized on the Third Harbor Tunnel Crossing for the Central Artery. This current and past construction history has produced Contractors experienced in steel tube construction as well as experienced skilled labor.

The steel shell tube also has an advantage that it is designed to be transported over the open ocean and for great distances if required. This increases the number of potential fabrication sites and often results in greater competition between bidders.

Since steel tubes are generally fabricated at an existing shipyard facility, tube fabrication proceeds concurrently with dredging operations. This results in a shorter overall construction duration than that of the cofferdam method. The construction schedules indicate the duration of the steel tube crossing is 22 months, approximately 8 months less than that of the cofferdam. However, it is 5 months longer than the concrete tube crossing. The reasons for the shorter concrete tube contract duration will be discussed below.

Disadvantages - The disadvantage of the steel tube is its cost. The steel tube is the most expensive alternative with a probable construction cost of $30,400,000, which is $4,500,000 above the cofferdam method and $8,000,000 above the concrete tube. This higher cost is essentially due to the fact that the steel tube in addition to the massive fabricated steel shell, has the same volume of concrete as both the concrete tube and cast-in-place box. The cost of the fabricated steel shell greatly increases the cost. To a lesser degree, the exterior dimensions of the steel tube are larger than the concrete tube and cast-in-place box which also results in additional costs for excavation, support and backfill.

Concrete Tube Method

Advantages - The concrete tube has two significant advantages over the other two methods studied for this crossing. It has the least probable construction cost and the shortest construction duration. The probable construction cost of $22,400,000 is $3,500,000 less than the cofferdam method and $8,000,000 less than steel tube method. The reasons for the lower cost compared to the steel tube method have been discussed above.

The second advantage of the concrete tube crossing method is that it possesses the shortest construction schedule of the three crossing methods studied. The concrete tube duration is projected at 17 months. Five months shorter than the steel tube crossing and 13 months shorter than the cofferdam crossing. The reasons for the longer duration of the cofferdam method will be discussed below.

It should be noted the construction schedule and cost are based on the assumption that the concrete tubes will be fabricated in existing dry docks close to Fort Point Channel. This assumption is substantiated by the presence of several existing dry docks in the area each of which could accommodate the three tubes being cast simultaneously. In addition, this will facilitate shipping of the tube sections in fair weather and calm seas in order to minimize impacts of shipping on the tubes.

Disadvantages - Of the 21 immersed tube crossings constructed in the U.S. from 1910 to 1986 only two were constructed of reinforced concrete. Both were two lane road tunnels constructed in Oakland, California. One was constructed in 1928 and the other in 1962. This type of construction, although not familiar to this country, is common in Europe and as mentioned previously, utilizes conventional construction techniques and will be used for the Central Artery immersed tube crossing of Fort Point Channel.

Since the materials and construction techniques utilized are conventional operations, it appears the reason concrete tubes have not been widely used in the U.S. is due to the lack of land available adjacent to the project site for construction of a casting basin. Many crossings involve 12 to 20 tube sections, 91-122m (300-400 feet) long and require a large casting basin. In urban locals where immersed tubes are generally constructed, such land is not available.

Cofferdam Method

Advantages - The cofferdam method does permit essentially all work to be accomplished in open air. This provides greater opportunity for quality control and monitoring of such items as excavation, foundation preparation and backfilling.

Also, the cofferdam method does provide the most positive barrier for protection against siltation of the Channel waters during construction. The interlocking steel sheet piling will confine sedimentation within the cofferdam whereas, in both the immersed tube options, silt

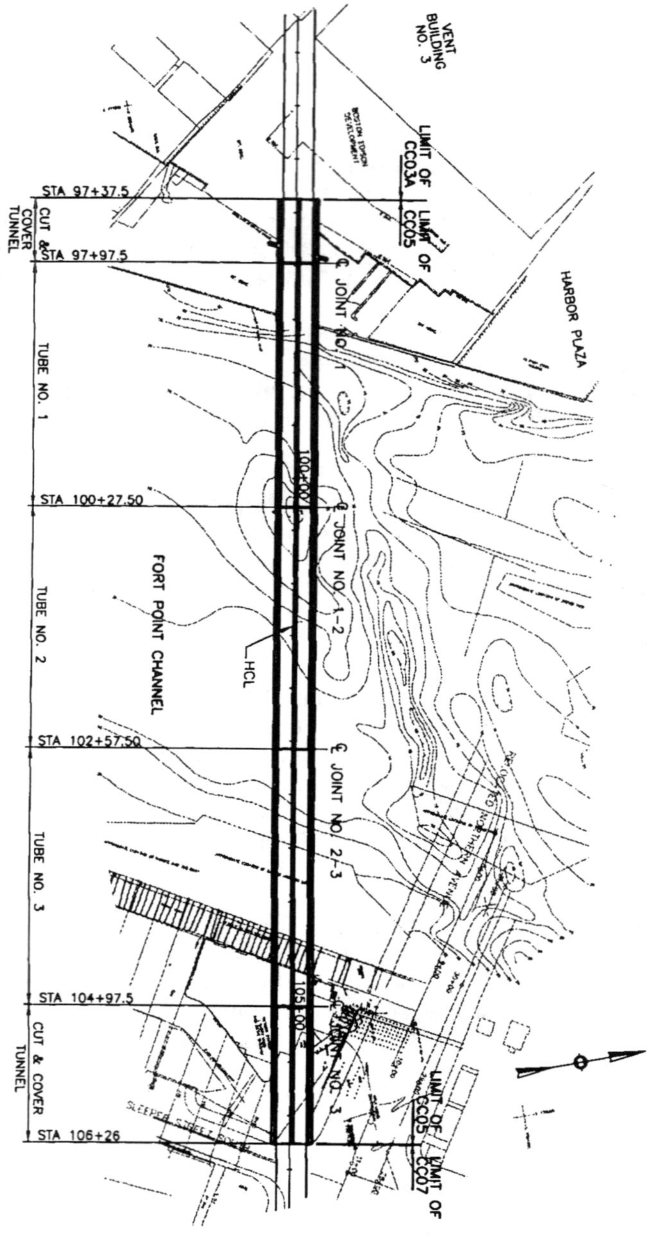

Figure 3. Location Plan

curtains will be utilized which requires maintenance and protection against damage.

Disadvantages - The cofferdam crossing has the longest duration of the three alternatives studied. The duration is 30 months, eight months longer than the steel tube and 13 months longer than the concrete tube is due to the fact that no cast-in-place concrete construction can begin until the steel sheet pile cofferdam is in place. Conversely, in the tube crossings, casting of the concrete tube structure is concurrent with dredging and sheeting operations eliminating a substantial period of time from the construction schedule.

RECOMMENDATION AND SELECTION

Based on the available information the steel tube was eliminated from consideration based on its relatively high cost. Of the remaining two methods, the concrete immersed tube had some distinct advantages.

It is the least expensive method, $3.5 million less than the cofferdam method and the concrete tube's construction duration of 19 months is 11 months shorter than that of the cofferdam. This shorter duration will greatly reduce the impact to the general public, tourists, traffic and industry along the Channel.

Therefore, it was recommended the concrete immersed tube method be utilized for the Transitway crossing of Fort Point Channel. This recommendation was subsequently approved by the MBTA and final design of the concrete immersed tube crossing of Fort Point Channel began in early 1994.

CONCRETE IMMERSED TUBE

As mentioned previously, following the submission of 60% design documents to the MBTA, the project has been placed on hold due to design modifications and studies to determine the most economical and practical method to construct the Transitway beneath Russia Wharf. It is anticipated, final design will resume early in 1997. The following is a brief summary of design details developed to date for the immersed tube crossing of Fort Point Channel.

TUBE DETAILS

During the Type Study, the immersed tube alignment contained two horizontal curves along with a single vertical curve. During the final design stage of the project, the Transitway alignment has been modified resulting in the elimination of both horizontal curves. The current horizontal alignment (Figure 3) for the three tube elements is a tangent alignment. The vertical alignment is indicated in Figure 4. The alignment changes which have been implemented since the Type Study have simplified construction and facilitated a more economical structure.

Typical Cross Section

The concrete tube cross section is indicated in Figure 5. The overall structural tube is 7m (23'-3") high by 11.3 (37'-0") wide. The top slab is .7m (2'-3") thick with a bottom slab .8m (2'-9"). The exterior walls are .75m (2'-6") thick and the interior wall is .6m (2'-0").

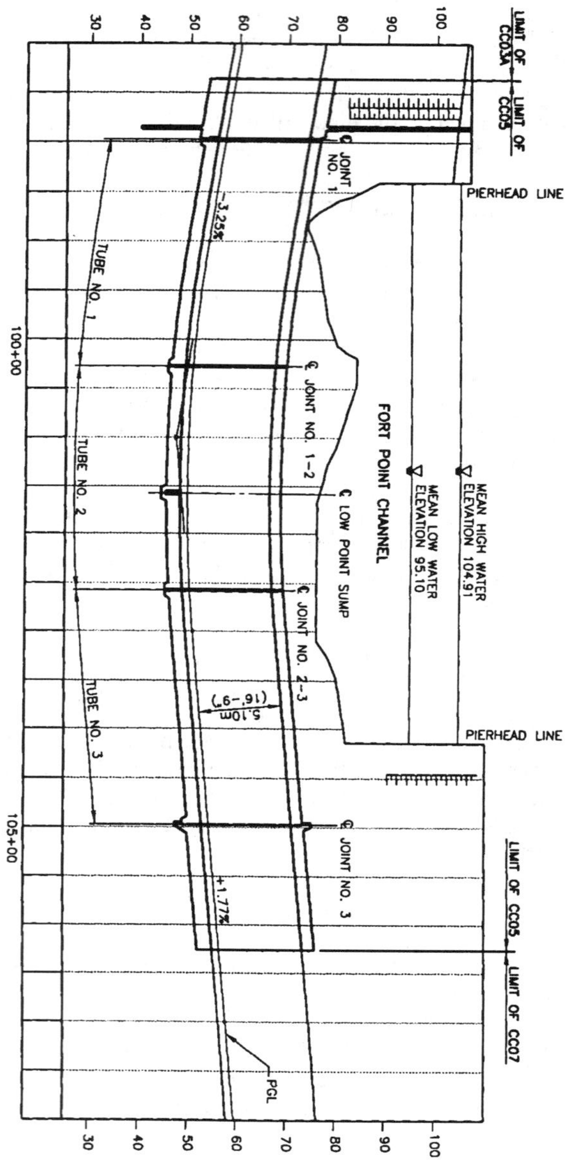

Figure 4. Profile

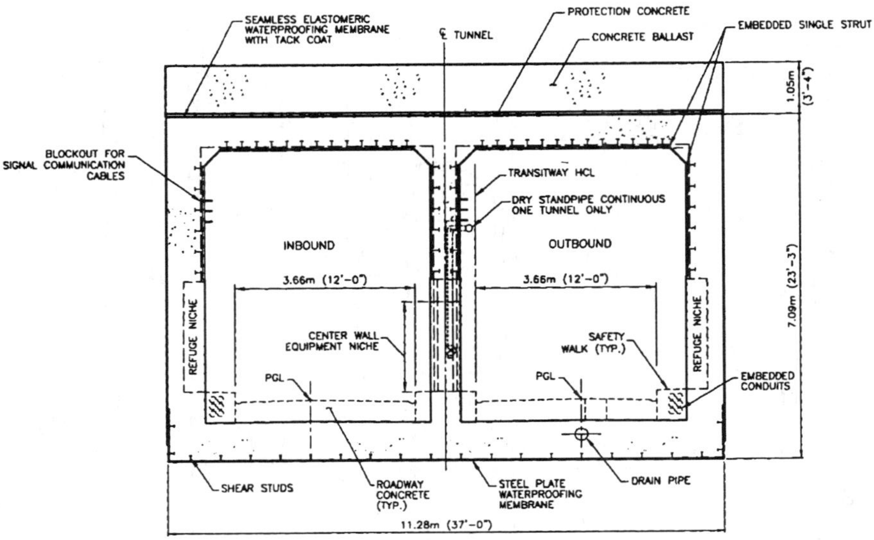

Figure 5. Typical Tunnel Section

It should be noted, ballast concrete will be placed on top of the immersed tube sections after sinking. The immersed tube sections had to be designed light enough to float with sufficient free board to permit sailing of the tube sections into Fort Point Channel. Therefore, to obtain a sufficient factor of safety against flotation of the final structure, ballast concrete is required.

System wide design criteria required numerous opening and block-outs within the tube walls. See Figure 6. The .75m (2'-6") exterior walls receive .45m (1'-6") deep refuge niches every 6m (20') and the interior wall in addition to having refuge niches every 6m (20') on each side, receives standpipe niches through the wall as well as personnel cross passages. These numerous niches within the structure walls have prompted detailed structural evaluations and have contributed to the decision to utilize a waterproofing system on the tube elements.

Waterproofing System

The waterproofing system recommended for use on this project consists of a 7.6mm (1/4") steel plate on the bottom of the tube sections and a liquid applied elastomeric membrane on the walls and top slab. In addition to its waterproofing abilities, the steel plate on the bottom slab was selected for its resistance to damage during shipment and placement.

The tube elements will be transported through relatively shallow water, subject to 3m (10') tides, with narrow clearances. Therefore, the possibility does exist of the tube scraping or striking the channel bottom. Also, the use of steel sheet piling to support the excavated area, presents another obstacle the tunnel elements must clear. The steel membrane pro-

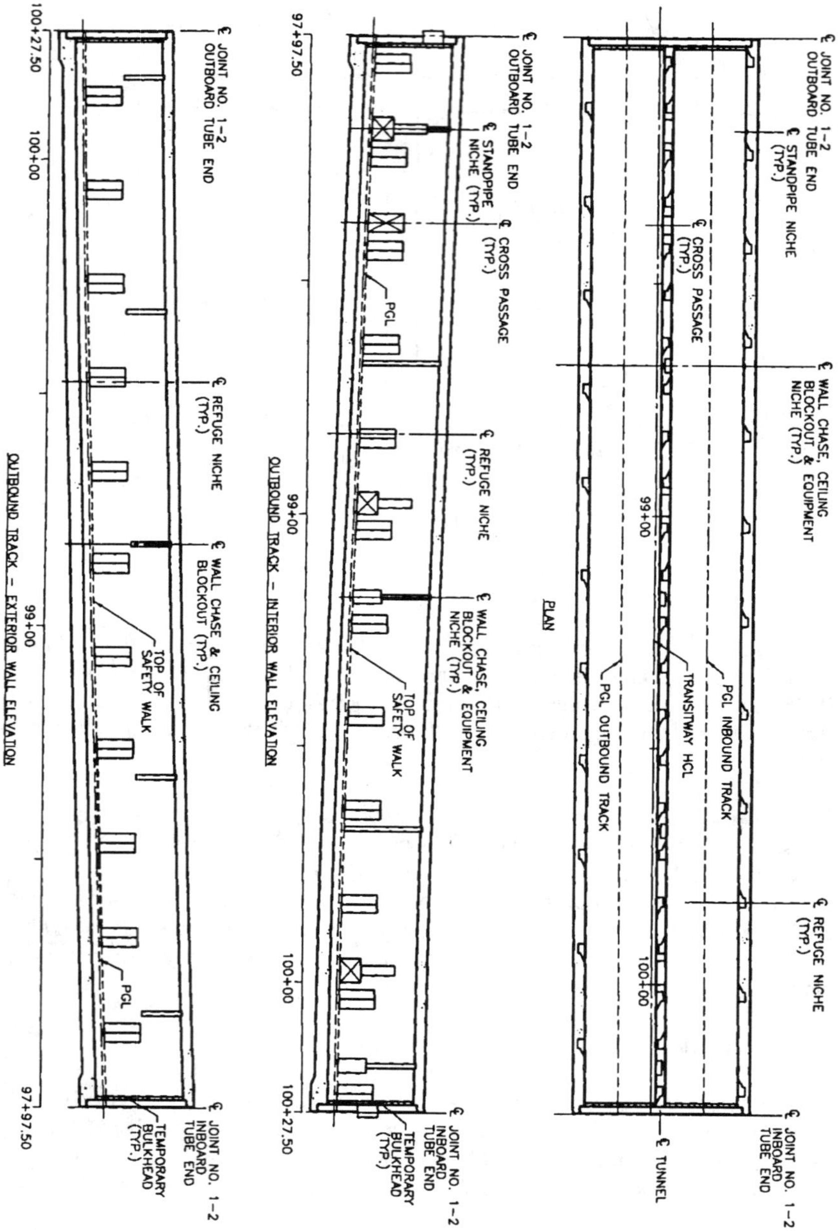

Figure 6. Tube Plan & Elevations

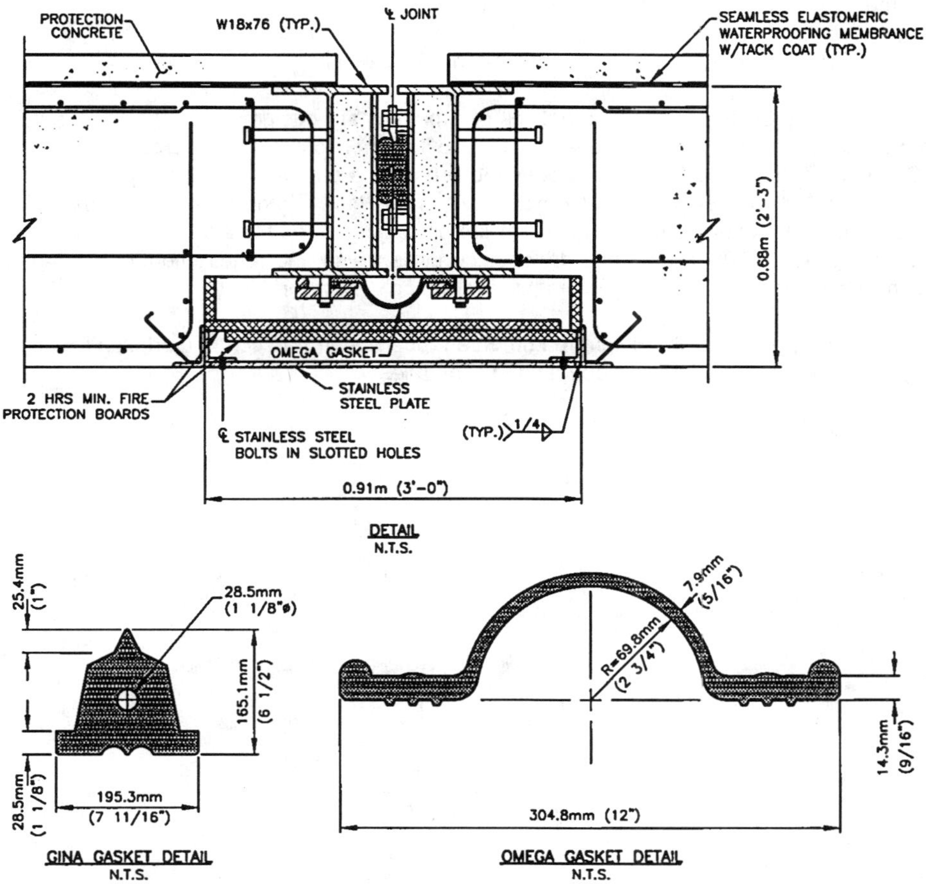

Figure 7. Gasket Details

vides the most positive resistance to damage.

The method of cooling concrete during curing to control cracking and utilizing no external waterproofing system was evaluated for this project. Due to structural and practical considerations, it was determined not to be an optimum system for this project.

Gasket System

The tube element will receive both a Gina and Omega gasket to prevent water infiltration at the tube joints. See Figure 7. The GIna gasket is bolted to the inboard end of each tube section before flotation and serves to seal the joint during closure. Following final joint closure, the Omega gasket is installed to provide a permanent joint seal. The joint will be designed to permit rotation between the tube elements but transverse and vertical movements will not be permitted. This system has been utilized on numerous steel and concrete immersed tube projects throughout the world.

SUMMARY

A comparison of three alternatives of a preliminary design level concluded that the concrete immersed tube tunnel provides the most economical method of crossing Fort Point Channel. The project construction cost and schedule are impacted favorably by the existence of dry dock facilities in close proximity to Fort Point Channel within Boston Harbor. Several dry docks are sufficiently large to permit casting of the three tube elements, a total length of 243m (700 feet) in one dry docking cycle. Also, the requirement of casting the concrete immersed tube is a good match for Boston's labor force.

The selection of the concrete immersed tube option has contributed significantly to maintaining the overall project budget and schedule.

Chapter 35

SUCCESSFUL PROJECT TEAM INTEGRATION FOR RIVERSIDE BADLANDS TUNNEL

James G. Gallanes[1] and Drupad B. Desai[2]

[1]Principal Engineer and Project Manager for Inland Feeder Project, Metropolitan Water District of Southern California; [2]Vice President, Technical Director —Underground Structures, Daniel, Mann, Johnson, & Mendenhall

ABSTRACT

A key component of the Inland Feeder Project is the Riverside Badlands Tunnel, a 7.9-mile stretch subjected to high internal and external hydraulic pressures passing beneath highly eroded hills in a semirural environment. The tunnel presented significant design and management challenges to the project team. This paper describes the formula for project success used to manage the design of this complex pressure tunnel, from team member selection to the completion of all contract documents, on schedule and below budget. The project team dealt with sensitive regional and local community issues and won the trust of the community. The team included participation by senior management and recognized experts from inception to finish. The team worked diligently to build quality into design and adopted a claims avoidance approach to management and preparation of documents.

INTRODUCTION

The Metropolitan Water District of Southern California (Metropolitan) is planning to construct a series of pressurized, 12-foot finished diameter water conveyance tunnels and connecting pipelines for the Inland Feeder Project. One tunnel segment of the Inland Feeder Project has been designated the Riverside Badlands Tunnel. As shown in Figure 1, the alignment of this segment extends from near the community of Redlands in San Bernardino County to San Jacinto Valley in Riverside County. Metropolitan retained DMJM and Woodward-Clyde Consultants (the DMJM team) to provide geologic/geotechnical and tunnel design consulting services for the Riverside Badlands Tunnel.

PROJECT DESCRIPTION

The Riverside Badlands Tunnel is a 41,720-foot-long, 12-foot finished diameter pressure tunnel that will be excavated under as much as 850 feet of cover, with over 80 percent of its length below the groundwater table. It is envisioned that construction will be staged from portals located at each end of the tunnel (Opal Avenue and Gilman Springs Road), and, at the Contractor's option, from temporary access shafts located in Live Oak and San Timoteo canyons.

The geologic/geotechnical studies included field investigations and laboratory tests to characterize ground and groundwater conditions along the alignment; tunnel boring machine (TBM) boreability tests for estimating TBM advance rates; aquifer pump tests to de-

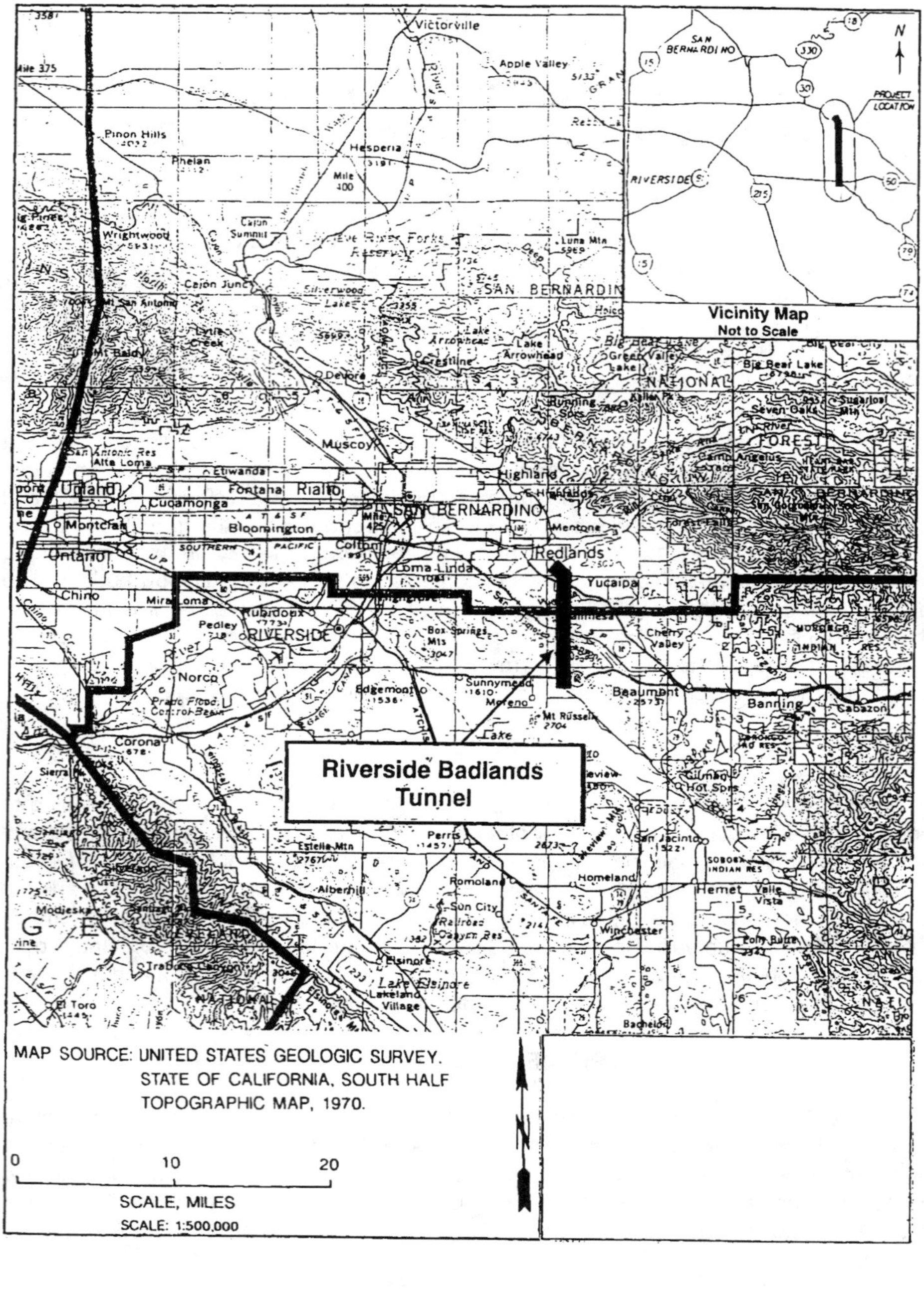

Figure 1. Location Map

velop parameters required for the Contractor's development of a dewatering plan for alluvial tunnel reaches; and chemical analyses of groundwater samples for evaluating water treatment and corrosion protection requirements. (See Figure 2)

Key considerations in the selection of the "Final" profile included transitions to open-cut pipelines at each end of the tunnel; maintaining adequate cover at the portals for turning under; and maintaining adequate cover at the Yucaipa Creek and San Timoteo Creek crossings to avoid exposure due to potential long-term scour. Management of wastewater from tunnel inflows, dewatering wells, construction activities, and stormwater runoff will be an important consideration for the Contractor's layouts.

Approximately 80 percent of the tunnel will be excavated below the groundwater table. Groundwater control is considered critical for excavations through the alluvium reaches. The most feasible methods of groundwater control appear to be dewatering wells and/or jet grouting. Probing and pre-excavation grouting ahead of the tunnel heading are required to mitigate the anticipated flush flows at several fault zones.

The final lining designs are based on the ground characterizations and loading and performance criteria developed for the project. The designs take into consideration hydraulic and operational criteria, rock loads, external and internal hydrostatic pressures (including potential shunt flows and operational transients), construction loads, and the estimated characteristics of the design basis earthquake. The seismic design parameters were developed from maximum design earthquakes on the San Andreas and San Jacinto faults using deterministic methods and median values. Based on performance and design criteria considerations, the acceptable final lining alternatives are either (1) welded steel pipe (WSP) or (2) prestressed concrete cylinder pipe (PCCP) throughout.

ROLES AND RESPONSIBILITY

Fundamental to the success of the project team was a conscious effort by both MWD and the DMJM team in clearly defining roles and responsibilities of all the team players. First and foremost was the recognition of the fact that Inland Feeder Project was a one-of-a-kind mega project, undertaken by Metropolitan with the clear desire to use the latest technology and management tools to implement the program. MWD also recognized that the last major underground construction project carried out by Metropolitan staff was almost 20 years ago, and they should be prepared to modify and adopt new design management practices.

During the negotiation process, Metropolitan staff and the DMJM team discussed in detail the conventional practices in place at MWD in administering the Consultant Contracts and the roles and responsibilities of both entities. During this phase, DMJM introduced several new concepts of design management based on their experience with managing mega projects. Metropolitan reviewed these concepts with respect to their management practices. After considerable discussion, a detailed listing of activities required to fulfill the project needs was developed, along with a list of assumptions regarding the respective responsibilities of MWD and DMJM. A detailed workplan breakdown was developed to establish timely response requirements.

Metropolitan introduced a "Board of Consultants" to the project team. This Board's role was to advise the project team during the development of the project and act as an

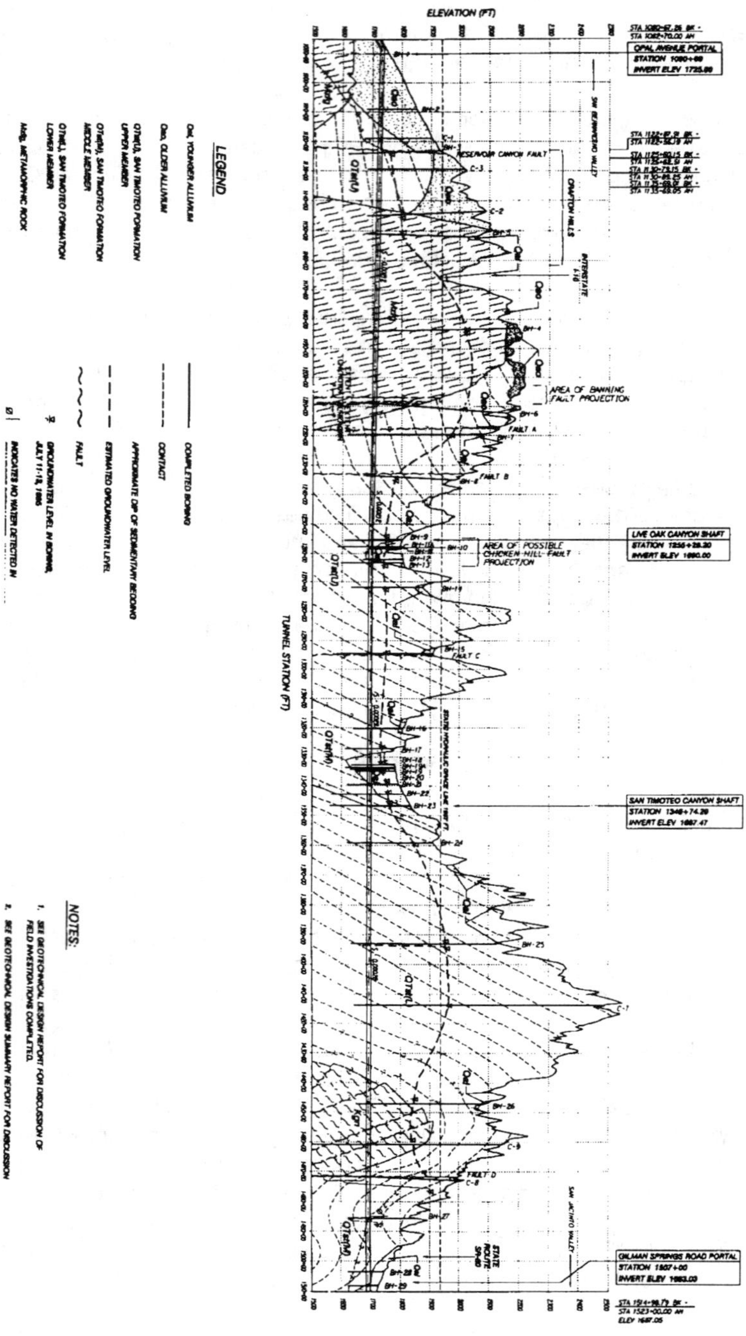

Figure 2. Tunnel Profile

extension of Metropolitan staff to report to MWD top management. The Board was a great success. They were more than a technical advisory board, traditionally used on other mega projects. They assisted the project team in project development phase and design management phase by providing underground expertise at all stages of the project, including consultant selection process.

Numerous subconsultants, including technical specialists, minority and women-owned business organizations, field and office staff, and special laboratory test organizations, participated in the program. Their participation accomplished a variety of complex tasks on this project and helped Metropolitan meet its outreach program goals. The DMJM team established a Corporate Oversight Committee to act as in-house experts and to be available to the DMJM team and Metropolitan management group at regular intervals throughout the life of the project.

During the early phases of negotiations, we jointly decided to work on the principle of matrix management, a concept which was relatively new for Metropolitan staff. They recognized that it would be cost-effective and quality responsive for the project tasks at hand. Metropolitan's willingness to adopt a new management style made project management integration much easier for all members, enabling them to execute their assignment in a timely and efficient manner.

TEAMWORK

Teamwork is a highly respected concept in our industry. Partnering sessions are held to smooth out the rough edges amongst the team players, sometimes with the aid of a special facilitator hired just for this task. On the Riverside Badlands Tunnel project, this was accomplished without any fanfare by the project team leaders.

All of the team players—on the Metropolitan side as well as the DMJM side—understood that we were working together to make the Inland Feeder Project a great success. Project team leaders from both MWD and the DMJM team directed the design and management teams. The key ingredients to success of design teams can be described as follows:

- Timely decision-making process
- Proven methods to resolve technical issues
- Value-engineered approach to design
- Participation of industry-recognized experts and technical specialists throughout the design process
- Quality built into every step of the design
- Design excellence not a goal but a requirement

Another major component of the success formula was the management approach from the inception of the project design to its finish. Metropolitan recognized the need for strong management leadership to achieve the goal of combining technical talent with tactful implementation of the project assignment. It was essential that top management of both Metropolitan and the DMJM team be involved and be visible to project team managers. This action was a great confidence builder for all members. We are sure that their active involvement provided necessary force to produce timely input to policy decisions. Another major factor was that the DMJM team was able to make available group managers who were also technical experts and were actively involved in project planning and the design process. The key ingredients to the successful management process can be described as follows:

- Active participation of senior management
- Metropolitan's willingness to go the extra mile to earn the trust of the community
- Claims avoidance approach to management — insist on quality product at all levels of execution at all times
- Be flexible and cooperative
- Be willing to make hard choices, including adopting innovative solutions to advance the project goals
- Establishment of realistic milestone schedules

No public works project is without budget limitations. Inland Feeder is no exception. During the negotiation phase, we established detailed cost data for all the activities, along with detail scope statements for each task, including recognition for uncertainties and unknowns. Metropolitan and DMJM team diligently worked to maintain control on the budget. The tools used included the following:

- Regular progress meetings with production staff and design managers
- Detailed reviews of budgets and percentage completion of deliverables on a monthly basis with key staff
- A biweekly review of schedule status, issues, and action lists with key staff
- A detailed project status briefing and involvement of top management in the budget maintenance process throughout the life of the project

QUALITY IN DESIGN

The common objective of any Owner/Designer team is to obtain high-quality facilities through the use of good planning, good design, good engineering, and good management.

It was the DMJM team's responsibility, through proper communication at different levels of execution, to meet this goal. They set out to produce quality documents that would enable construction bids within the client's budget and deliver a high-quality facility without minimum field changes—preferably at an optimum cost.

This was achieved on the Riverside Badlands Tunnel project by establishing the criteria and expectations of the finished installations based on Metropolitan's vast experience with similar projects, input from the Board of Consultants and technical specialists, and experience of the DMJM team members on similar assignments. A written document was developed and circulated amongst all staff to ensure clear understanding regarding their commitment and establish quality acceptance levels for this project.

This process was further strengthened by regular biweekly discussions with the Board of Consultants, to iron out any misunderstandings or avoid any misinterpretations of the governing design criteria approved by Metropolitan. In some instances, the design team was allowed to obtain critical additional geotechnical and structural design and field data on an as-needed basis, without overextending the budget, to improve the quality of design.

Timely policy decisions were vital in order not to impact quality of work. This became evident during preparation of project specifications. Metropolitan wanted to produce project-wide specifications, while the DMJM team was keen on producing project specifications that met their needs. A considerable effort was expended on both sides, and open communications between staff were key to this effort.

On large, complex projects, we always face extremely tight schedules and budgets. The complexity of the Riverside Badlands Tunnel project required the skills of many different disciplines and experts within each team organization. To achieve quality in the end product, it is of paramount importance that each team member know what needs to be done and also respect others' contribution.

Producing quality in design documents is not easy. It requires management leadership, setting standards of performance, and establishing quality criteria when the project is scoped. For the Riverside Badlands Tunnel project, we developed a Quality Control manual once the scope of work was completely developed, and each person on the team was required to read it. The group managers reviewed with the staff periodically to see that the manual was being followed.

The net result of the effort was

- Meeting the project objective of producing a good set of plans and specifications.
- With a quality design, Metropolitan can expect the Contractors to develop a more realistic and competitive bid, with a minimum of field changes, thereby better meeting MWD's expectation of building a cost-effective facility.

CONCLUSIONS

A key component of successful team integration was the willingness of all the team members to work together. Metropolitan and DMJM team management were able to create an atmosphere of cooperation by establishing a well-organized plan describing roles and responsibilities for all the parties involved. The team was able to involve recognized experts in a manner that resulted in solving difficult design and management issues in a timely manner. This created a win-win atmosphere amongst all team players, and the result is a set of quality documents that everyone is proud to be associated with.

ACKNOWLEDGMENTS

We wish to acknowledge the many people and organizations without whose diligent effort this very successful project completion would not have been possible. In particular, we would like to recognize the leadership and assistance of the following:

Metropolitan Water District: Gary Snyder, Gary Hazel, Wally Liu, Ron Redd, Tobin Tellers, Rob Bell, John Narvaiz, Rob Lyons, Meang Kour Hassell, Penny Lew, Mark Van Gessel, Mike McReynolds, Jay Arabshahi

Board of Consultants: Dr. Tor Brekke, Dr. Gregg Korbin, Mr. Eugene Murphy, Mr. Richard Balcerzak, Dr. Clarence Allen, Mr. Ron Proctor, Mr. Mike Kobler

DMJM Team: John Somerville, Ray Holdsworth, Charles Rendall, Tom Freeman, John Bischoff, Steve Klein, Jean Suter-Hill, John Waggoner, Sohail Khan, Steve Fowler, David Cregger, Dr. C.T. Chu, Zoran Batchko, Richard Erickson, Paul Womack, Afsaneh House, Robert Sistrunk, Joe Sperry

We would also like to thank the administrative and engineering field and office support staff of both organizations for providing excellent help throughout the life of the project.

REFERENCES

Daniel, Mann, Johnson, & Mendenhall and Woodward-Clyde Consultants, 1996, "Final Design Report forRiverside Badlands Tunnel, Inland Feeder Project," June 14, prepared for Metropolitan Water District of Southern California.

Daniel, Mann, Johnson, & Mendenhall and Woodward-Clyde Consultants, 1996, "Geotechnical Data Report

for Riverside Badlands Tunnel, Inland Feeder Project," June 14, prepared for Metropolitan Water District of Southern California.

Desai, D.B., 1996, Proceedings, North American Tunneling, Volume 2.

Fox, Arthur J., Jr., and Cornell, Holly A., 1984, "Quality in the Constructed Project," ASCE Proceedings

11

International Projects I

Chairman
C. Linden
Obayashi Corp.
No. Hollywood, CA

Chapter 36

WORLD'S LARGEST SLURRY SHIELD TUNNELING REPORT IN TRANS-TOKYO BAY HIGHWAY CONSTRUCTION

Tuneyoshi Funasaki,[1] Yasutaka Izumi,[1]
Makoto Kanai,[2] and Keizou Miki[2]
[1]Trans-Tokyo Bay Highway Corporation; [2]Obayashi Corporation Tokyo Bay Tunnel, Kisarazu South Joint Venture

INTRODUCTION

The Tokyo metropolitan region, inclusive of the three neighboring prefectures (Kanagawa, Chiba and Saitama), has the highest concentration of population and industry in Japan; with an area of 13,500 km^2 or only 3.6 percent of the national land, but a population of 31.8 million or 26 percent of the national total, and a Gross Regional Product (GRP) of 146 trillion yen (US$1.33 trillion) or 36 percent of the overall GDP of Japan as of 1990. Tokyo Bay, embraced by the metropolitan region as shown in Figure 1 (a), (b), is open to the Pacific Ocean as a marine gateway and is ringed by large cities and coastal industrial zones together with port facilities, all constituting a concentration of prime importance for socio-economic activities.

The Trans-Tokyo Bay Highway (TTB highway), which crosses Tokyo Bay and provides a direct transportation link between the west and east sides of the bay, has long been a sort of "dream-come-true" project in Japan. As shown in Figure 2, linked to the Tokyo Bay Shore Highway, Metropolitan Inter-city Expressway, Tokyo Outer Ring Road, Tateyama Expressway, and others, it will play an important role as an integral part of the regional highway network to help promote spatial redistribution and multipolarization of urban functions, thus strengthening cooperation between cities and restructuring the metropolitan region.

The TTB Highway is a project of 15.1 km in length designed as a toll highway to cross the bay by tunnel and bridge, and will require ten years to complete at a total cost of more than 1,438 billion yen (US$13.077 billion).

Following more than 20 years of feasibility and environmental assessment studies, construction of this giant offshore highway project commenced in May 1989 as the first large-scale highway project with private-sector participation under the prevailing Road Law in Japan, and is now steadily proceeding toward the goal of completion in fiscal 1997.

This report outlines the entire TTB Highway project, including the waterproofing measures against high water pressure and the rapid construction method applied in the construction of shield tunnel sections.

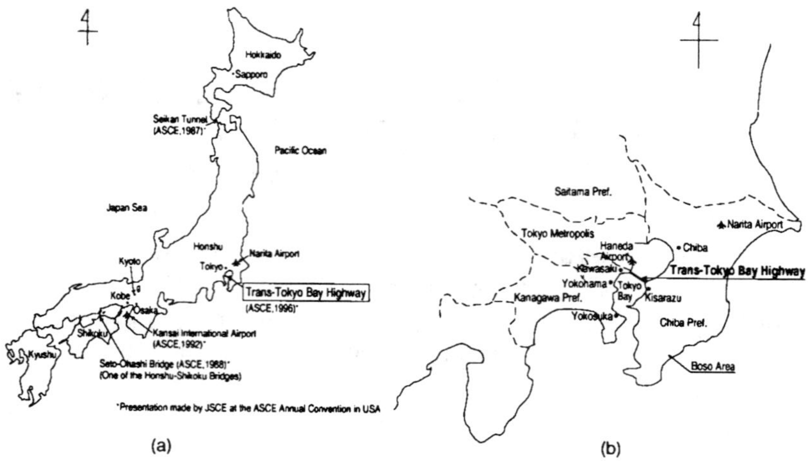

Figure 1. Location of Trans-Tokyo Bay Highway

OUTLINE OF THE HIGHWAY STRUCTURES

During the preliminary study of the highway, various structural options for tunnel and bridge were evaluated before commencement of the project. The final design was determined on the basis of the following reasons:

1) To ensure the safety of aircraft taking off from and landing at the Tokyo International Airport at Haneda, airspace restrictions were imposed in the area between the Ukishima Access and the Kawasaki Man-made Island (Kawasaki Island). This posed problems for building a long-span bridge.

2) Shipping traffic surveys indicated that of the some 1,400 vessels crossing the planned line of the highway, about 90 percent were concentrated in a 10-km section on the Kawasaki side. This meant that a bridge spanning the bay could produce adverse effects in terms of maritime safety and vessels trying to seek refuge in stormy weather.

3) Soft subsoil conditions extending to considerable depths posed difficulties in selecting a suitably engineered and cost effective foundation.

The general plan of the highway is illustrated in Figure 3, and having a highway length of 15.1 km in total, of which the marine section is 14.3 km. At the Kawasaki end, the land section of the highway joins the Tokyo Bay Shore Highway at the Ukishima Interchange. At its Kisarazu end, it is linked to the Tateyama Expressway by the TTB Highway Connector Road (7.1 km). The land sections at the Kawasaki and Kisarazu ends are approximately 0.3 and 0.6 km respectively. An outline of the project under the terms of the project license is given in Table 1 below.

The highway is being constructed as a four-lane facility: two lanes in each direction. For this, two parallel tunnels are being driven in the undersea section. However, in order to

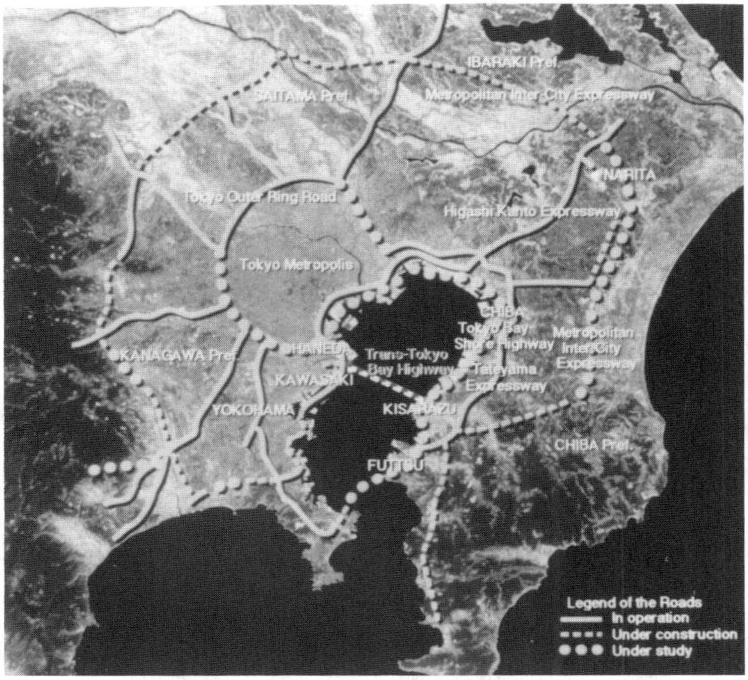

Figure 2. Metropolitan Highway Network

accommodate the expected future increase of traffic, the dimensions of major structures such as the man-made islands and the bridge have been determined to make future widening feasible.

Marine Section

The structural components of the marine section of the highway will consist of a shield-driven undersea tunnel, approximately 10 km in length, for the western two-thirds of the highway where shipping traffic is heavy; and a bridge about 4.4 km in length for the remaining section. Near the middle of the tunnels, a ventilation facility has been provided in the form of the Kawasaki Island, and where the tunnel and bridge meet, the Kisarazu Man-made Island (Kisarazu Island) is being constructed. The Kisarazu Island will serve as a rest area for users of the highway. Both islands have also been used as launching bases for the shield machines driving the tunnel. Figure 4 shows the highway structures of the marine section.

Land Section

The land section on the Kawasaki side serves as an approach section to the tunnel from the interchange, and has structural components of a large U-shaped retaining wall and a box culvert. Whereas the section on the Kisarazu side is an embankment structure, where a tollgate is being constructed. Sandy muck excavated during the shield tunnel work was effectively used to build an approximately 11-m high embankment.

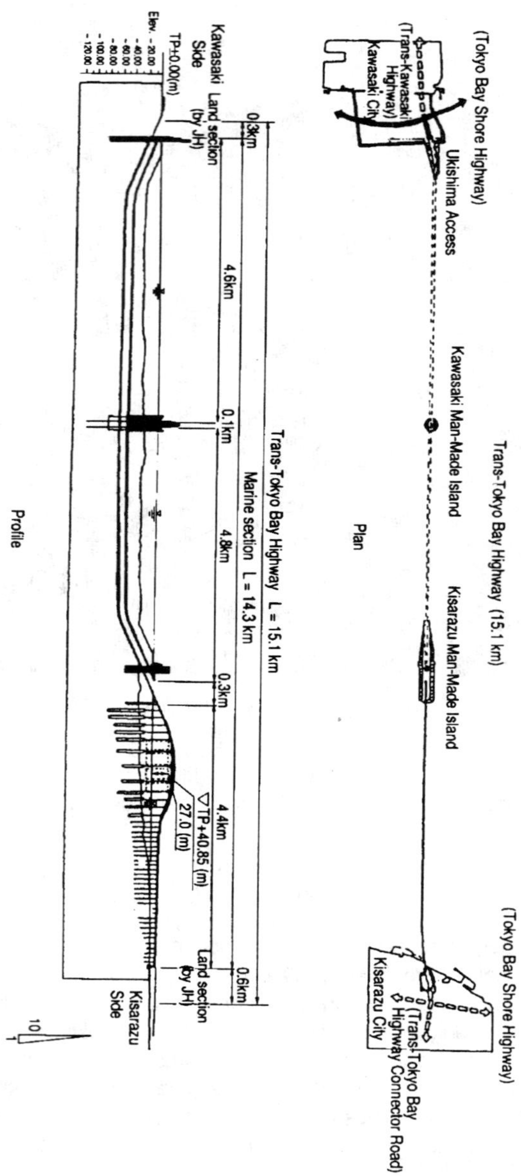

Figure 3. General Plan/Profile of Trans -Tokyo Bay Highway

Table 1. Outline of the Project under the Terms of the Project License

Name of highway	Trans-Tokyo Bay Highway
Official route designation	National route 409
Origin and destination	From Ukishima, Kawasaki City To Nakajima, Kisarazu City
Length of highway	15.1 km (Marine section 14.3km)
Number of lanes	Dual 2-lane (6 lanes at final stage)
Design speed	80 kilometers per hour
Forecast traffic volume	33,000 veh/day (in the first year of operation)
Year of completion	Fiscal 1997
Total project cost	¥1,438.4 billion (US$13,077 million)

Exchange rate : US$1 = JPN¥110

Table 1. Outline of the Project under the Terms of the Project License

CONSTRUCTION SCHEDULE

Figure 5 shows the construction schedule for the highway. Construction began in 1989, and is scheduled for completion within fiscal 1997.

OUTLINE OF SHIELD TUNNELS

Outline

The tunnel section of the TTB Highway is about 10 km long, or two-thirds of the total length of the 15.1-km highway. It runs from the Ukishima Access on the Kawasaki coast, sloping down through a sloped section into the seabed, then passing through a level section and finally up through a second sloped section, to the Kisarazu Island. Constructed midway along the level section between the Ukishima Access and the Kisarazu Island (at a point about 5 km offshore from the Ukishima Access) is the Kawasaki Island, where ventilation facilities are located. The horizontal alignment of the tunnel is basically straight. The sloped sections have been designed with a 4 percent grade in order to shorten the sections as much as possible. The level seabed section has a 1.0 D-thick overburden (where D is the tunnel outer diameter), so that the combined weight of the overburden and tunnel structure will counteract buoyancy forces.

Figure 6 shows the basic tunnel cross section. The upper section of each tunnel is 10.5 m wide and accommodates the roadway, comprising the left-hand shoulder (2.5 m wide), carriageway (dual 3.5-m traffic lanes) and the right-hand shoulder (1.0 m wide). The lower half of the tunnel consists of a central service and evacuation passage and two flanking sections containing telecommunications, electric power lines and other facilities. Ventilation is provided by a longitudinal system with electrostatic precipitator installed on the ceiling of the tunnel.

(a) Ukishima Access

(b) ShieldTunnels

Figure 4. Highway Structures of Marine Section

The wider left-hand shoulder for the entire section of the tunnel (2.5 m in width) will serve a continuous lay-by and enable drivers to evacuate from the carriageways when an emergency arises. In addition, the evacuation passage prepared under the carriageway slab does not necessitate construction of evacuation cross passages connecting two tunnel tubes, thereby saving construction cost and time.

Basic Plan

For the basic structure of the tunnel, a comparative study was conducted between a submerged tunnel and a shield tunnel and the shield tunnel was finally adopted for the highway based on the following reasons:

1) A submerged tunnel requires dredging a trench in the seabed, backfilling, and towing and sinking the tunnel tube sections, all of which would necessitate placing prolonged restrictions on shipping traffic. This was not a good choice for a congested shipping environment.

2) With technology for shield tunneling established in the late 1970s and the method

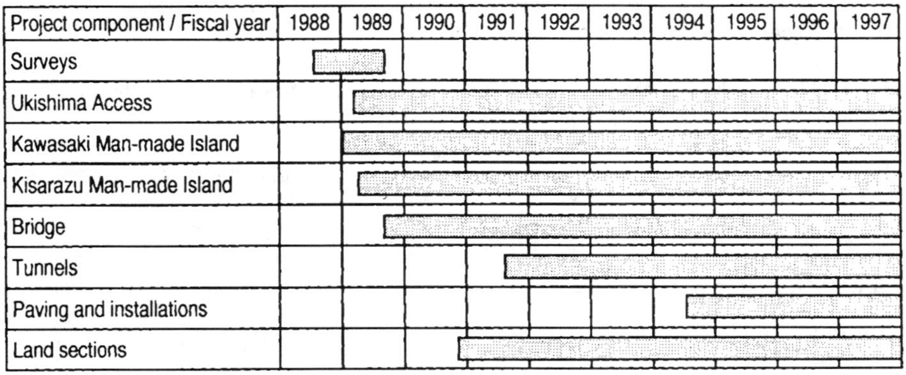

Figure 5. Construction Schedule

since proven in the construction of long, large diameter tunnels under high water pressure, shield tunneling was also determined to be applicable to the execution of the TTB Highway project.

3) Although the tunnels will initially carry four traffic lanes, plans are for future expansion of the roadway to six lanes in order to cope with future traffic demand increase. Shield tunneling solution can easily meet the requirements for such an extension by construction of an additional tunnel.

Tunnel Design

Since the tunnels would be driven under the severe conditions outlined below, careful

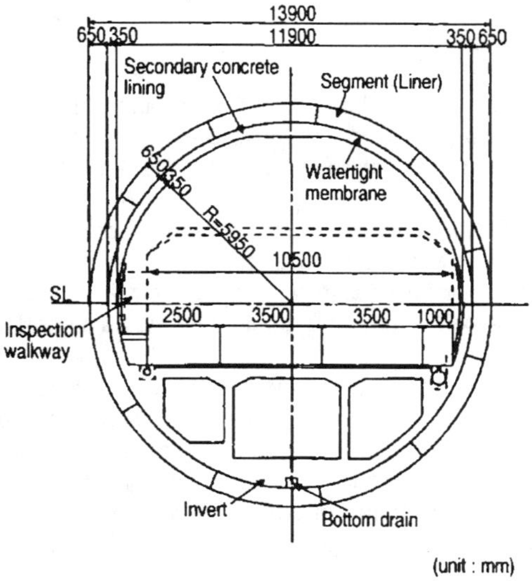

Figure 6. Tunnel Cross Section and Traffic Lanes

and extensive deliberation went into the design and execution work, including meetings of a committee that included advisory experts.

1) The roadway configuration would necessitate construction of a shield tunnel having a large cross section (13.9 m in diameter), and requiring long-distance tunneling operations.

2) The tunnels would be driven through soft ground, and in a seismically active area.

3) Since the water depth at the site is high (the maximum head would reach 50 to 60 m), and because there is no impermeable layer, the tunnels would be subject to extremely high water pressure (0.49 to 0.59 MPa).

4) The tunnels would experience the corrosive effects of salts contained in the sea water.

5) The structure being a twin tunnel spaced 0.5 to 1.0 D along the sloping sections and 1.0 D along the level seabed section could cause construction work on one tunnel very easily affecting the other. Shield tunnel cross effects had to be considered in the tunnel design.

6) The tunnels being long, several shield machines would have to advance concurrently from the vertical shafts at the Ukishima Access, Kawasaki Island and Kisarazu Island in order to shorten the construction period. Underground connections would therefore be made below the seabed and under high water pressure.

7) To enhance both safety and work precision, there would be a need to automate various operations such as excavation and assembly of segmental ring liners.

EXCAVATION OF SHIELD TUNNELS

Measures to Enhance Durability

With respect to tunnel waterproofing and corrosion control, the following measures were adopted to support the various watertight measures used (ranging from backfill grouting to membrane waterproofing) against any possibility of leakage:

1) Even with plastic (as opposed to elastic) materials, considered the most reliable choice for a backfill grout, there would be a change in the coefficient of permeability of the grout in the long term. It was therefore decided, that the backfill grout could not be expected to provide semi-permanent waterproofing, but could only act as a short-term water stop.

2) For the watertight gaskets provided along the outer perimeter of each liner segment, a hydro-expansive seal that swells when it becomes wet was chosen. A sealer was selected whose hydro-expansive component would not dissolve easily, and this would be installed in strips 35 mm or more in width. This would satisfy waterproofing require-ments during both construction work and actual operation. A lubricant would also be applied to prevent mechanical damage to the watertight gaskets during segmental ring assembly.

3) Damage protective gaskets carrying the same amount of resilience as the water-

tight gaskets would be installed along the inner perimeter of each segment to prevent segment damage during installation.

4) Normal portland cement containing 50 percent by weight of blast-furnace slag would be used to produce the reinforced concrete segments. This would give the concrete a coefficient of permeability of about one-tenth that of normal portland cement concrete, thereby increasing its resistance against saltwater corrosion.

5) Each segmental ring was provided with four backfill grout holes. Around the perimeter of each hole, a butyl-based rubber seal would be set. When concrete was placed, this would react with the concrete to produce hydro-expansion and bonding.

6) Steel materials would be protected with concrete as much as possible to place them in an alkaline environment. Allowances were made for a certain amount of corrosion in those steel materials which were unavoidably exposed. The segment bolts, however, were given an anti-rust coating with semipermanent durability.

7) A watertight membrane would be laid between the primary and secondary liners in order to lead a water leak down to the bottom for drainage.

Rapid Construction of the Shield Tunnels

The Shield Tunnel sections of the Trans-Tokyo Bay Highway are being constructed by means of the Pressurized Slurry Shield method. This employs a tightly sealed shield in which pressurized slurry maintains a stabilizing force against the working face while the cutters excavate the soil and the shield machine moves forward. The particular characteristic of this method is that while the slurry is forced against the working face, excavated material is pumped up in slurry form to a treatment plant above ground. In the treatment plant, the constituent sand is separated out from the slurry, which is the returned by pump to the working face, after having its density and specific gravity adjusted.

With the object of opening the tunnel for use as soon as possible, a number of measures were adopted for the excavation and the lining operations. Those adopted at the planning stage, those related to the installations, and those related to the execution of the work are here outlined separately.

Measures adopted at the planning stage

1) The tunnel was divided into eight sections, each one 3,000 m or less in length. This would ensure that if by any chance one of the tunneling machines should suffer a breakdown and be unable to continue boring, the machine coming from the opposite direction to meet it would be able to complete that section.

2) Temporary facilities were designed to permit excavation and lining to be carried out simultaneously, and a process introduced by which the lining could be completed without interruption when the excavation operations came to an end.

3) To reduce the amount of time taken to install the lining, the liner rings were divided into eleven equal segments and installed automatically.

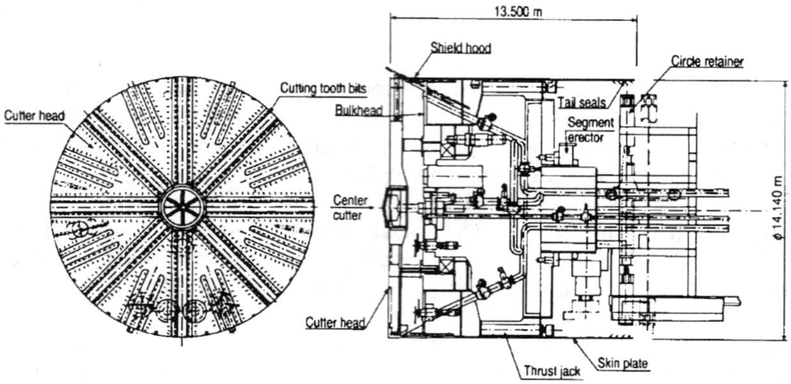

Figure 7. Slurry Shield

Measures Taken Regarding the Installations

SHIELD MACHINES

(1) Standard specifications

Twin parallel tunnels are being driven between the Ukishima Access and the Kawasaki Island, and between the Kawasaki and Kisarazu Islands, making a total of eight work sections. Tunneling involves eight shield machines, each starting from one of the vertical access shafts at the Ukishima Access, Kawasaki Island and Kisarazu Island, to eventually connect below the seabed.

Figure 7 shows the front view and cross section of the shield machine, and Figure 8 shows a shield machine assembled at the fabricating plant.

① Basic configuration

a.	Bore diameter:	14.14 m
b.	Shield machine length:	13.5 m
c.	Total weight:	29.4 to 31.4 MN

② Driving machinery

a.	Total thrust:	235 MN (48 jacks with 4.9 MN capacity)
b.	Jack stroke:	2,550 mm

③ Cutting head

a. Cutting teeth: Pin-jointed cutting tooth bits are being used to facilitate any necessary replacement. The cutter head is mounted also with leading bits, peripheral bit-abrasion protection bits, peripheral lateral-protection bits, and abrasion-sensor dummy bits.

Figure 8. Shield Machine

b. Cutter head: The cutter head is operated by electric motors with a speed of rotation ranging from 0.39 to 0.45 min^{-1}.

④ Tail seal: The mechanical shield is equipped with four lows of wire-brush tail seals to improve abrasion resistance, durability and water pressure resistance of the tail seals.

⑤ Auxiliaries: The shield machine is also equipped with a circle retainer, a man lock, backfill grout injectors, and equipment for tunnel connection.

(2) Special specifications for the shield machine

(A) Special specifications for underground tunnel connections

There was a choice between using one shield machine to excavate each section between vertical access shafts, and using two, one from either end of each section. As the result of an overall evaluation that considered, for example, the work schedule, durability of the shield machine, reliability of execution and safety, the latter option was chosen. This meant that the TTB Highway, with its two tunnels for eastbound and westbound traffic, required a total of four underground tunnel connections to be made.

After comparing three alternative methods; ①freezing method combined with center cutter retraction; ② shield hood extension method; ③connection ring method, it was decided to adopt a method which had already been used with considerable success, one combining ground freezing with retraction of the shield's center cutter (①). In this method, retracting the center cutter of each shield machine allows two shields driving from opposite directions to be brought very close to one another. As a result, it is possible to minimize the size of the freezing zone.

It was determined that for connecting two shield machines midway in the tunnel section, the machines would be brought opposite each other, and then the surrounding natural ground frozen by means of ground freezing pipes extended from the first-arrived shield machine.

To perform tunnel connections in this way, the shield machines were equipped with the following instruments:

① Center cutter retractor

In order to expedite the ground freezing process, the shields were equipped with a center cutter retractor to allow the cutting heads of two opposing shields to come within about 300 mm of each other.

② Instrumentation to confirm relative shield positions

When two shield machines were brought within 50 m of each other, a horizontal bore of about 300 mm diameter was driven between the two machines to measure the relative position and direction (attitude angle).

③ Freezing pipes and pipe insertion jacks

The shield machines were equipped with freezing pipes along the inner surface of the shield hood skin plate and the cutter head perimeter plate. In addition, the shield hood skin plate of the body of the shield machine was given 48 holes to permit radial insertion of freezing pipes.

(B) Special specifications for automatic segment assembly

Segment erection work was automated for greater work safety, labor economy, assembly precision and work efficiency. Reducing primary lining erection time was a particularly important issue in constructing the shield tunnels. Automation permitted the time required to assemble one segmental ring to be reduced to less than 110 minutes.

Features of liner segments and the shield machine's segment erector necessary for robotic assembly are listed below, many of which were developed specifically for the TTB Highway project:

① Segments
- a. Bolt dislodgement prevention devices
- b. Division of liner ring into equal segments to shorten erection time
- c. Furnishing of sensing grooves for accurate segment position detection
- d. Special metal parts for mechanical gripping of segments

② Segment erector
- a. Use of a mechanism and control system to enable high positioning accuracy
- b. Adoption of automatic bolt fastening devices
- c. Installation of bolt delivery mechanism inside segments
- d. Nut and washer feed mechanism.

SLURRY TREATMENT FACILITIES

Slurry treatment facilities were installed on top of the buffer structure on the man-made island. The facilities have a large processing capacity, being designed to handle approximately 1,500 cubic meters of muck (in natural ground) per tunnel each day. Muck separated from the slurry-muck is hauled by belt conveyors to a barge to be carried off site. The designed transport capacity is approximately 16 and 20 cubic meters per minute, respectively, for slurry feed and slurry-muck discharge. At the slurry treatment plant, spoil components are separated from the slurry mixture. Sand gravel and cobbles are removed by a primary separator, whereas silt and clay are first dewatered by filter presses, and then treated with the addition of unslaked lime (0.29 kN per cubic meters) and carried to barges by means of belt conveyors for secondary processing.

CONCRETE MIXING FACILITY

The surface equipment used in placing the secondary tunnel lining includes a concrete mixing barge. The cement and aggregate for the concrete is delivered by surface vessels. Ready mixed concrete is pumped under pressure to hoppers at the bottom of the shaft by concrete pumps on board the mixing barge.

TRANSPORT FACILITY INSIDE TUNNEL

The tunnels of the TTB Highway are provided with in-tunnel transport facilities including rails because they: (a) are large diameter and therefore require haulage of large quantities and extremely heavy loads of materials and equipment; (b) require long-distance tunneling, with correspondingly great transporting distances; and (c) will have their primary and secondary tunnel liners installed more or less simultaneously.

BACKUP CARS OF THE SHIELD MACHINE

The shield machine is followed by nine backup cars. They are an automatic segment-transport and bolt-retightening car, a pipe-extension car and the following cars for placing secondary lining:

1) Bottom half of tunnel; a car for sheeting water-proof membranes, and that for assembling steel bars for secondary lining, side-wall traveling forms, slab traveling forms.

2) Top half of tunnel; a car for sheeting waterproof membrane, and that for assembling steel bars, roof arch travelling forms.

Measures Taken Regarding Execution of the Work

1) Excavation was carried out on six days of the week, with Sundays given to maintenance of the machines.

2) Each day's boring was done in two shifts, both of ten hours' duration.

3) Excavated soil was used for the toll gate embankments or as covering material for industrial waste disposal sites. It was removed in such a way as to avoid interference with

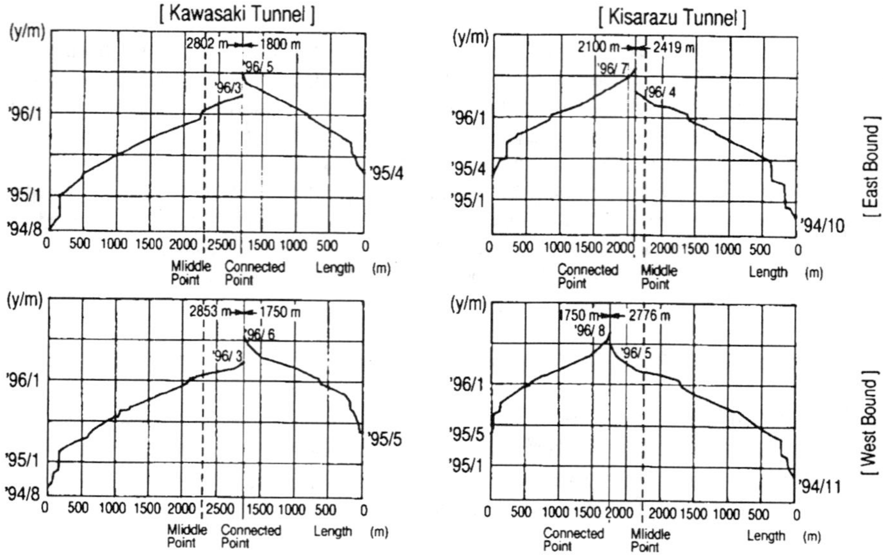

Figure 9. Tunneling Construction Process

the rate of shield progress.

Process Applied in the Excavation Work

1) As shown in Figure 9, the eight sections of the shield-excavated tunnel, measuring from 1,750 m to 2,853 m, were scheduled to be finished in 13 to 19 months. Disregarding modifications to schedules, excavation was able to advance in average cycles of 150 m or more. The best monthly progress recorded at any one of the eight tunneling sites was 310.5 m.

Lining work, from the invert to the crown arch, was done in 15 m cycles and progressed at an average monthly rate of 160 m.

12

NATM

Chairman
J. McDonald
Traylor/Obayashi Corp. J.V.
San Ysidro, CA

Chapter 37

SUCCESSFUL SOFT GROUND NATM APPROACH ON METRO SECTION E4B IN WASHINGTON, D.C.

Galyn Rippentrop[1] and Peter Kolitsch[2]
[1]Kiewit/Kenny, AJV; [2]Kolitsch Consulting, Inc.

INTRODUCTION

On April 14, 1994, the joint venture of Kiewit Construction Company and Kenny Construction Company was awarded a 76 million dollar contract for the construction of the New Hampshire Avenue Tunnels in Washington D.C. The Notice to Proceed was given on May 16, 1994 and scheduled completion is September 1997. This project is one of five contracts for the Mid-City E Line as part of the Washington Metropolitan Area Transit Authority (WMATA) Greenbelt Route. Once completed, the Mid-City E Line subway tunnels will connect service between the U Street-Cardozo and Fort Totten stations (See Figure 1).

The New Hampshire Avenue Tunnels project consists of approximately 6800 feet of twin single track subway tunnels, three vertical shafts, associated utility work and comprehensive restoration work. Section E4b involved two methods of tunneling along the project alignment. The first method consisted of the well known soft ground open face shield driven tunnel and the second method consisted of tunnels driven using the New Austrian Tunneling Method (NATM) .

The NATM section of twin tunnels begins at the Buchanan Vent Shaft and proceeds approximately 2630 feet under Rock Creek Cemetery, including the Farragut Emergency Access Shaft and connects approximately 390 feet beyond to the existing Fort Totten running tunnels (See Figure 2). Special considerations had to be given to the portion of the tunnel alignment which passed beneath Rock Creek Cemetery. The analyses and subsequent selection of the tunnel support system were based on anticipated geological conditions and the sensitive tunnel alignment.

The New Hampshire Avenue Tunnels project was the second soft ground NATM project in Washington D.C. The first, by no coincidence, was completed at the adjacent Fort Totten Station and Running Tunnels Project by Mergentime/Hochtief, AJV. Currently, on a nearby WMATA jobsite, the J.F. Shea Construction Company has started a third NATM soft ground tunnel.

Tunnel support systems were studied with respect to three tunneling methods:

- A one-pass system consisting of a bolted precast concrete segmental lining
- A two-pass system consisting of an unbolted precast concrete segmen

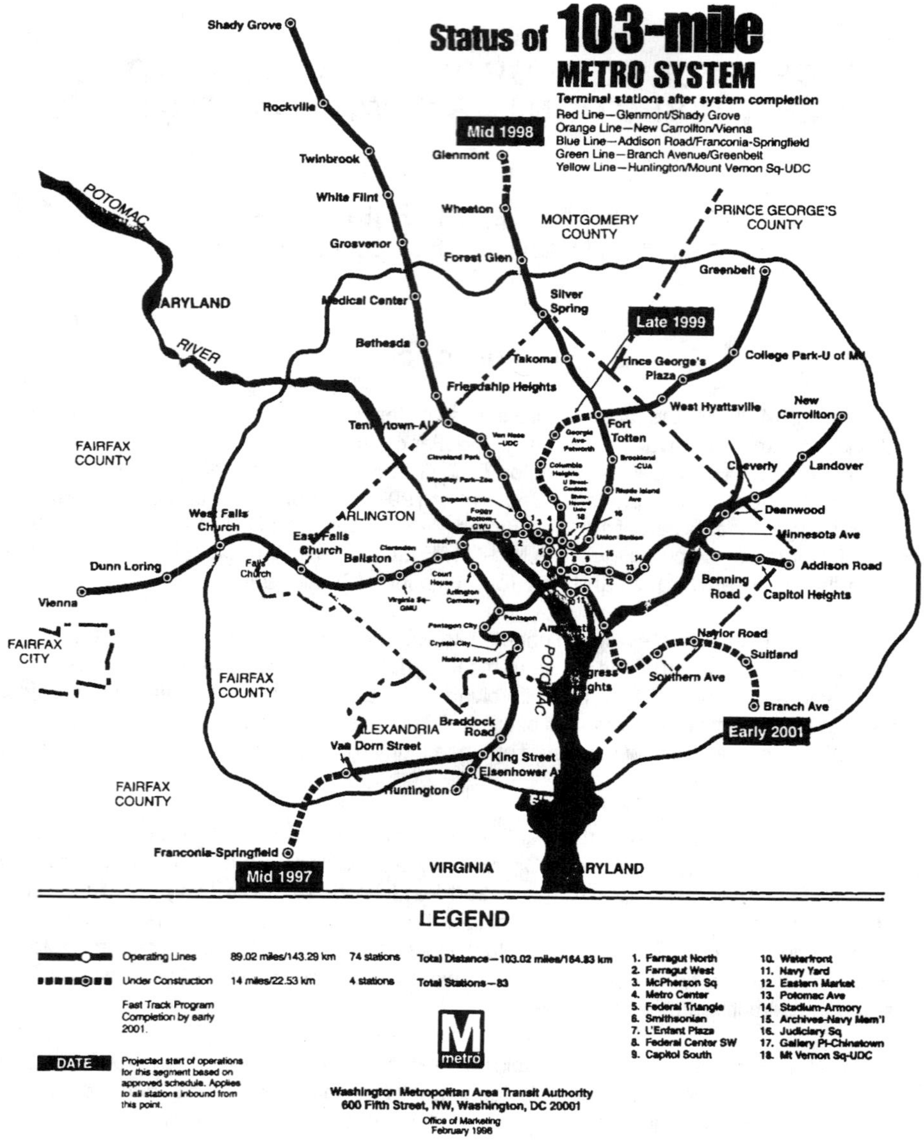
Status of 103-mile
METRO SYSTEM
Terminal stations after system completion
Red Line—Glenmont/Shady Grove
Orange Line—New Carrollton/Vienna
Blue Line—Addison Road/Franconia-Springfield
Green Line—Branch Avenue/Greenbelt
Yellow Line—Huntington/Mount Vernon Sq-UDC
Mid 1998
Late 1999
Early 2001
Mid 1997
Shady Grove
Rockville
Twinbrook
White Flint
Grosvenor
Bethesda
Friendship Heights
Glenmont
Wheaton
Forest Glen
Silver Spring
Takoma
Fort Totten
Greenbelt
College Park-U of Md
West Hyattsville
New Carrollton
Landover
Cheverly
Deanwood
Minnesota Ave
Addison Road
Capitol Heights
Benning Road
Naylor Road
Suitland
Southern Ave
Branch Ave
Vienna
Dunn Loring
Ballston
Braddock Road
King Street
Van Dorn Street
Huntington
Franconia-Springfield
MONTGOMERY COUNTY
PRINCE GEORGE'S COUNTY
FAIRFAX COUNTY
FAIRFAX CITY
ARLINGTON
VIRGINIA
POTOMAC
RIVER
LEGEND
Operating Lines 89.02 miles/143.29 km 74 stations Total Distance—103.02 miles/164.83 km
Under Construction 14 miles/22.53 km 4 stations Total Stations—83
Fast Track Program Completion by early 2001.
DATE Projected start of operations for this segment based on approved schedule. Applies to all stations inbound from this point.
1. Farragut North
2. Farragut West
3. McPherson Sq
4. Metro Center
5. Federal Triangle
6. Smithsonian
7. L'Enfant Plaza
8. Federal Center SW
9. Capitol South
10. Waterfront
11. Navy Yard
12. Eastern Market
13. Potomac Ave
14. Stadium-Armory
15. Archives-Navy Mem'l
16. Judiciary Sq
17. Gallery Pl-Chinatown
18. Mt Vernon Sq-UDC
metro
Washington Metropolitan Area Transit Authority
600 Fifth Street, NW, Washington, DC 20001
Office of Marketing
February 1996

Figure 1

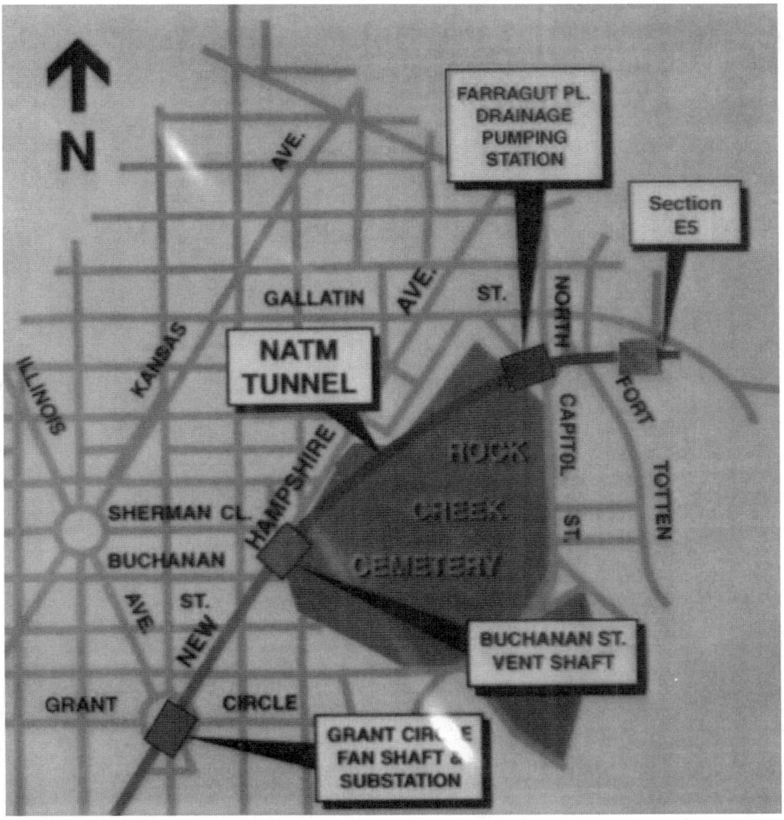

Figure 2

tal lining with a cast in-place concrete final lining

- A two-pass system consisting of a NATM reinforced shotcrete initial l ining and a cast-in-place concrete final lining

After careful consideration of the various tunneling methods available, WMATA selected the two-pass tunnel support system, which specified the NATM method of initial support followed by a concrete liner (See Figure 3). The design for the E4b project was done by Rummel, Klepper & Kahl Consulting Engineers. R, K&K provided a special team to develop the NATM design for this portion of the contract.

NATM USAGE & BACKGROUND

The New Austrian Tunneling Method, also called the flexible shotcrete method, has been used in Europe for decades. NATM typically is applied in large diameter tunnels in technically stressed rock, or shallow soft ground tunnels of varying geology. The basic philosophy of NATM is to retain the initial strength of the surrounding ground in a multi-heading excavation by constructing a load carrying arch.

New Hampshire Avenue Tunnels Project
WMATA - Section E4b Washington, D.C.

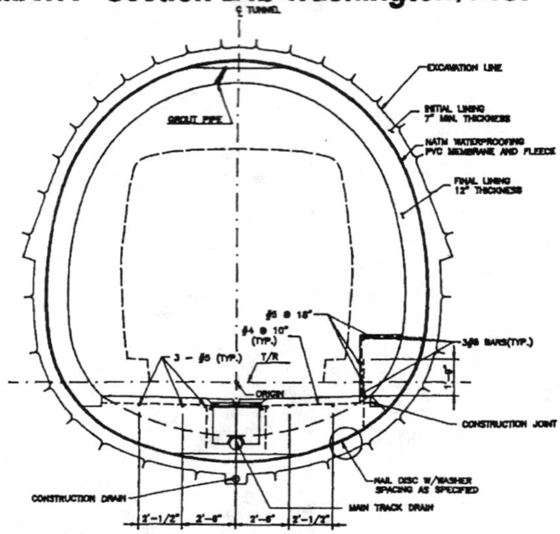

TYPICAL CROSS-SECTION
NATM TWO PASS LINING SYSTEM

Figure 3

The key features of the NATM construction technique include:

Sequential excavation and support of the tunnel

- Excavation sequences and face areas that can be varied (e.g. top heading, bench and invert)
- The primary support is provided by shotcrete in combination with steel mesh, girders, rockbolts or spiling
- Monitoring the behavior of the primary lining

For tunnels in urban areas limiting settlement is of paramount importance to avoid damage to overlying structures. To achieve the essential limitations of settlement, the following principal measures must be undertaken:

1) Excavation stages must be sufficiently short, both in terms of dimen sions and duration.
2) Completion of primary support must not be delayed. In particular, closure of the shotcreted ring needs to be completed as soon as possible.

The NATM method provides the contractor with the most flexibility to control ground subsidence by preserving the self-bearing capacity of the in-situ soils and reducing ground relaxation. The excavated face is virtually unobstructed allowing direct observation of soil conditions and ground losses, as well as permitting access for roof pre-support measures and dewatering to be performed at the face. Depending on the soil conditions encountered, initial support measures can be adjusted to control the face. The owners of the Rock Creek Cemetery imposed stringent requirements for surface settlement and allowed limited access within the property. Therefore, the selection of the NATM method was considered essential for a successful completion.

The Farragut Emergency Access Shaft, which was included within the NATM portion of the Section E4b project was also specified to be supported using NATM techniques. This method was incorporated to maintain ground strength and to provide an inseparable contact in the form of a closed elliptical shape. Also, the compatibility of a homogenous structure of shaft and tunnel provided additional strength and stability. Another benefit during construction is that the composite structure of shotcrete, lattice girders and welded wire fabric preclude the need for internal bracing.

GEOLOGY

The District of Columbia is located in the transitional area between two major physiographic provinces of the eastern United States. The Piedmont Province to the north and the Coastal Plain Province to the south. The E4b project is located southeast of this area, which is commonly referred to as the Fall Line Zone. The Coastal Plain Province is characterized predominately by soils from the Cretaceous Age, which consist of interbedded layers of clays and sands with occasional gravel and boulders.

On this contract, two major strata were encountered during the site explorations the clays of Stratum P1 and the sands of Stratum P2. These soils have been further divided into two substrata clays and three substrata sands. These distinctions are important as they relate to tunneling construction methods and start-up time. The descriptions of the soil characteristics are as follows:

A. STRATUM P1 (CN), PLASTIC CLAY

A stiff to hard, red, brown, and gray plastic clay frequently slickenedsided, meaning smoothly polished from past slippage. Preconsolidated stresses range from 14 to 16 ksf., with shear strength ranging from 1 to 9 ksf. Generally, the P1 clay can maintain an extended period of stand-up time.

B. STRATUM P1 (CL), SANDY CLAY

A stiff to hard, gray, silty or sandy clay with some fine cemented sands. Shear strengths have a more narrow range than the plastic clays of Stratum P1, but typically in the upper half of P1 materials, therefore yielding similar stand-up time.

C. STRATUM P2 SILTY AND CLAYEY SAND

An interlayered compact to very compact, light gray, brown or tan silty or clayey sand that can contain sand lenses. The sand can be cemented in layers ranging from a fraction of an inch to several feet thick. The test borings revealed high blow counts and low penetration indicating higher shear strength and very good stand-up time.

D. STRATUM P2 (SP-SM) SAND, TRACE SILT, CLAY

A very compact, gray, fine to medium sand with trace silt and cemented sand layers. The combination of relatively clean, cohesionless sand with cemented sand lenses has been proven to be distinctly unfavorable in tunneling construction with poor stand-up time.

E. STRATUM P2 (G) SAND, SOME GRAVEL

A very compact, light gray and brown, fine to coarse gravel and sands that occur most frequently between clays and overlying sands. Stand-up time is considered very poor.

F. STRATUM P3 SILTY CLAY AND CLAYEY SAND

Consists of hard, gray-green or gray-blue, silty or sandy clay, sandy silt and silty or clayey fine sand with occasional gravel or boulder zones. According to the boring logs, this material was to be expected at or near the tunnel invert and exhibits very good stand-up time.

Another important geologic consideration is ground water. Detailed measurements of ground water levels have been obtained during site explorations performed over the past 22 years. The ground water table ranged from zero to four feet above tunnel invert. Therefore, dewatering wells were specified by contract to provide adequate drawdown below tunnel invert. Perched water was encountered within the P2 sands on top of the P1 clay layers. These perched water zones may have been fed by infiltrating rainfall or leaking utilities.

METHODOLOGY

The concept of NATM construction makes it imperative that careful attention be given to the behavior of the soil during the execution of the work. The lining itself interacts with the surrounding soils, which are considered part of the lining. NATM excavation does not require any special equipment, but must be performed in a smooth and continuos fashion.

The installation of the initial lining, which immediately follows the excavation, must be completed before advancing to the next cycle. The initial lining is composed of lattice girders , welded wire fabric, pre-support (sheets or spiling, etc., if required)and shotcrete. NATM construction requires a minimum stand-up time for the excavated soil, otherwise additional pre-support measures must be taken. Therefore, particular attention must be given to the sealing shotcrete that is placed on the exposed ground surface immediately after excavation.

A. FARRAGUT SHAFT

The Farragut Shaft is an elliptical shaped vertical shaft with the major axis approximately 60 feet in length and the minor axis approximately 43 feet in width. The shaft was excavated and supported to a total depth of 72 feet. The initial lining wall thickness was one foot for the first 32 feet and increased to two feet thick for the remaining 36 feet (See Figure 4). For the Farragut Shaft, a cycle consisted of an eight segment lattice girder installed one segment at a time and in 3 ft. 4 in. deep increments (rounds).

The sequence of construction for the installation of one segment in a ring for the Farragut

New Hampshire Avenue Tunnels Project
WMATA - Section E4b Washington, D.C.

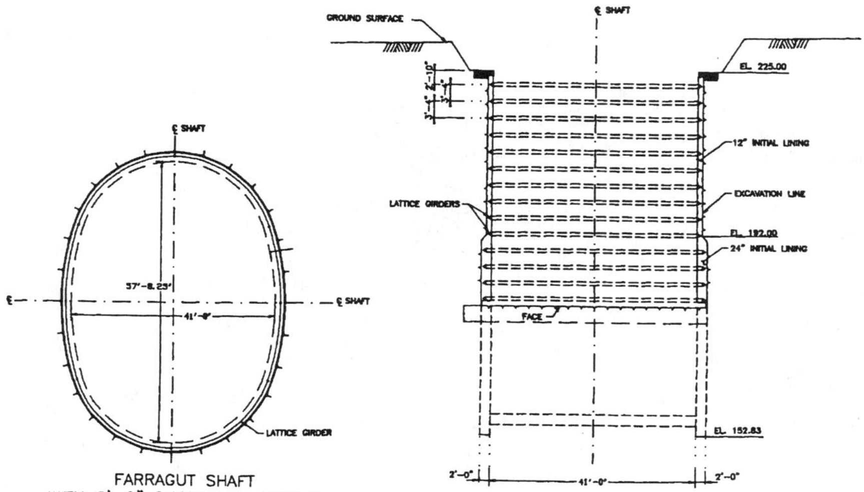

FARRAGUT SHAFT WITH 2'–0" SHOTCRETE SECTION
PLAN VIEW

FARRAGUT SHAFT PROFILE

Figure 4

Shaft was specified as follows:

1. Excavate one segment width by 3 ft. 4 in. maximum depth.
2. Support excavated surface immediately with 1 inch thick sealing shotcrete.
3. Install first layer of welded wire fabric 4x4-W4.0 x W4.0.
4. Install lattice girder.
5. Install shotcrete to cover lattice girder.
6. Install second layer of welded wire fabric 4x4- W4.0 x W4.0.
7. Apply final layer of 1 inch thick shotcrete.

The remaining seven segments were installed in this manner all the way around the shaft to close the ring and complete the cycle.

The construction sequence at the Farragut Shaft included the interface with the four tunnel openings. The lattice girders were installed continuous through the tunnel opening, but only sealing shotcrete was used in the tunnel openings. In addition, the tunnel openings contained load bearing apron rings, necessary to transfer the soil loading from the overlying shaft support elements.

B. RUNNING TUNNELS

The tunnel profile in the NATM section resemble an egg shape with an arched top half

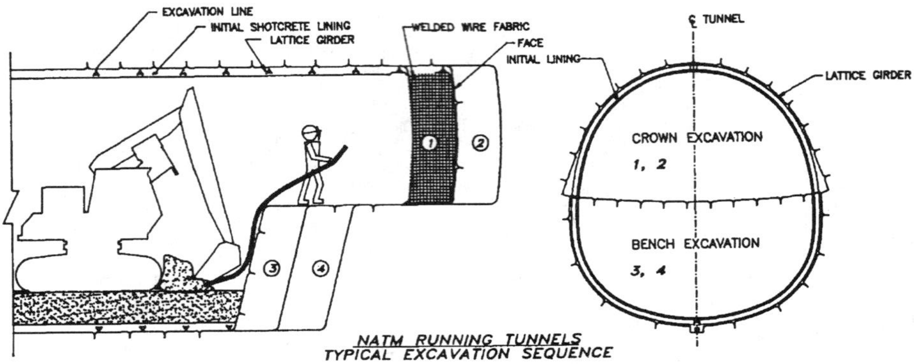

Figure 5

and a more circular bottom half. The dimensions were approximately 19 feet high and 17 feet wide. The excavation cycle for the initial lining was limited to 3 feet 4 inch rounds. For the running tunnels, a ring or round consisted of a four segment lattice girder. Once completely installed to form a closed ring, the initial lining maintained the stability of the whole formation. The excavation progressed in stages. The first stage was a top heading, followed by a bench in one round increments. The top heading always preceded the bench by at least two rounds (See Figure 5).

The distinct difference between the NATM shaft excavation & support and the NATM tunnel excavation & support was that the tunnel specifications requires at least one type of pre-support. Some type of roof pre-support was required for every round in the tunnel to increase the stand-up time of the excavated soils. The different types of pre-support that were specified on this contract are as follows:

1. <u>Chemical Grouting</u>

Chemical grouting was to consist of injecting chemical solutions such as sodium silicate into the groutable soil, with the intent to transform the soil into a cohesive mass with increased strength and reduced permeability. Chemical grouting was to extend at least two diameters ahead of the face covering the area above the tunnel crown. A constant 20 foot grout zone ahead of the face would then be maintained to minimize leakage of water into the tunnel and provide a stable face. This method involved the time consuming interruption of tunneling excavation to perform intermittent grouting operations, consequently, a different method was selected. The use of directional drilling and chemical grouting will be described later in this brief.

2. Forepoling Sheets

Forepoling plates consisted of interlocking metal plates driven or jacked above the lattice girder for approximately six feet into the ground ahead of the tunnel face. Forepoling plates were required prior to each round of tunnel excavation where specified.

3. Forepoling Pipes

Forepoling pipes consisted of 8 to 15 foot long, 1 1/2 inch diameter perforated steel pipe installed at the tunnel heading in predrilled holes. Following the placement of the pipe, cement grout was injected through the perforations of the pipe into the ground to form a cylindrical mass of stabilized soil around each grout pipe. The procedure was repeated for every round with 8 foot long forepoling pipes or every second round with 15 foot long forepoling pipes of tunnel excavation where specified.

4. Face Shotcrete

Face shotcrete was applied in the same manner as sealing shotcrete. Face shotcrete was required when work at the face was interrupted for more than four hours, and would consist of 1 to 4 inches of shotcrete.

The sequence for the construction of one round of the top heading was specified as follows:

a. Install roof pre-support as specified and local support as required.
b. Excavate one round at 3 foot 4 inch depth.
c. Install sealing shotcrete immediately after excavation with a 1 inch thickness or as required to stabilize the face.
d. Install two sections of lattice girders.
e. Install five inches of shotcrete.
f. Install one layer of welded wire fabric 6 x 6- W6.0 x W6.0.
g. Apply final one inch layer of shotcrete.

The sequence for the construction of one round of the bench was specified as follows:

a. Install local support as required.
b. Excavate one round at 3 foot 4 inch depth.
c. Support excavated surface immediately with 1 inch thicksealing shotcrete.
d. Install two sections of lattice girders.
e. Install five inches of shotcrete.
f. Install one layer of welded wire fabric 6x6- W6.0 x W6.0.
g. Apply final layer of one inch thick shotcrete.

C. INSTRUMENTATION

Instrument measurements have to be considered as an integral parts of the New Austrian Tunneling Method. Therefore, an instrumentation program was designed for monitoring soil movements and loadings. The final decision for the type of support is primarily based on measurement results. No other method of excavation and support relies so heavily on correct evaluation of the bearing capacity of the in-situ ground as NATM. These measurements are considered of great importance because of their support and economic implications.

Instrumentation that was installed during the NATM construction consisted of thefollowing:

1. Roof Leveling Points

The survey of roof movements required a hanger device that was installed in the initial shotcrete lining. Daily measurements were made with a rod and level. Roof leveling points were spaced 70 to 100 feet as specified.

2. Convergence Points

Convergence points consisted of eye bolts fixed in the shotcrete initial lining at the tunnel crown and both sides at spring line. Measurements were made with a tape extensometer, an instrument that applies a constant tape tension and reads to 0.001 of an inch. Readings were made once per shift during excavation, then the results are compared to initial readings to determine tunnel deformations.

Roof leveling and convergence measurements are of significant importance as they provide immediate evaluation of the initial lining support performance. In the case of severe movement, remedial action can be performed in the field at the same time of reading. Results are typically presented in graphical format for identification of convergence trends.

Based on the interpretation of roof leveling and convergence points, the engineer can specify changes in the support method procedure.

SCHEDULING

The excavation was proposed to start from the Farragut Shaft with two crews per heading, three shifts per day and five days a week. These crews would first drive the north tunnels, towards Ft. Totten, and then drive south towards the Buchanan Shaft. Another two crews would start from the Buchanan Shaft and drive toward Farragut.

This would allow enough time for the tunnel concrete to be completed to meet the tunnel completion interim milestone. However, this required the purchase of two complete excavation set-ups and the supervision to man them. It was decided to excavate the tunnels entirely from Farragut Shaft, using a single plant set-up with the same crews. By concentrating all efforts in one place, this approach reduced equipment and supervision cost. The risk was that if our scheduled production was not attainable, overtime would have to be utilized to make up the schedule.

As tunneling advanced from Farragut, towards Buchanan Shaft, our continued ability to pump the shotcrete up to 1400 feet in the tunnel ultimately initiated one last change in our excavation plan.

Once excavation passed the midpoint between Farragut and Buchanan Shafts the entire NATM setup was moved to Buchanan Shaft and mining operations were completed using Buchanan Shaft to service the NATM excavation. This allowed two very important operations to occur:

Shotcrete could continue to be pumped directly from surface and the requirement to transport shotcrete through the tunnel with mixer trucks was no longer necessary.

Farragut Shaft could be dedicated entirely to the tunnel concrete operations thus eliminating the logistical problems of utilizing a single shaft to perform excavation and concrete operations concurrently.

EXCAVATION & SUPPORT OPERATIONS

The excavation and initial support procedures for NATM as specified were augmented by Kiewit/Kenny with respect to the actual geological soil conditions as encountered on site. For on site consulting, a NATM specialist, Kolitsch Consulting Inc. was hired during the excavation and support phases.

A. MOBILIZATION

In preparation for the excavation and support of the Farragut Shaft, Kiewit/Kenny combined the mobilization efforts with the relocation of conflicting utilities in extremely close quarters in a wedged shaped lot less than 1000 square feet. Using the adjacent city row house, which had been purchased by WMATA foroffices and storage, work proceeded amidst heavy traffic in a densely populated city block.

The major operations for the mobilization of the Farragut Shaft and tunnel excavation were as follows:

- Set-up volumetric batch plant for wet mix shotcrete with 15-50 cy/hr variableoutput.
- Set-up Reed concrete pump with 20 cy/hr. output and three inch victaulic piping to two inch flexible hose at the nozzle.
- Set-up 150 ton Northwest crane with 100 foot boom and 10 cy muck skip.
- Install covered sand and aggregate bins, approximately 50 cy each.

Install U-shaped muck bin with embedded soldier piles and lagging. The dimensions were: 30 ft. (L) x 18 ft. (W) x 19 ft.(H). The muck bin capacity were approximately 380 CY which handled nearly 10 rounds of tunnel excavation.

Install compressors, electrical equipment and temporary utilities.

B. SHAFT EXCAVATION

The excavation for the Farragut Shaft began on September 14, 1994. The first five rings did not encounter any problems related to soil or water conditions and were completed on September 21, 1994. However, the excavation of Ring No. 6, perched water of approximately 6 GPM made the support in the Statum P2 sands (SP-SM) very difficult. The stand-up time was basically down to zero, so forepoling sheets had to be driven as a pre-support. The forepoling sheets were nine inch wide interlocking metal plates driven with a special backhoe attachment. The length varied from four to eight feet allowing the crew to securely anchor the forepoling sheets and contain the raveling P2 sands.

During the excavation for Ring No. 7, the perched water increased up to 11 GPM. Additional dewatering together with forepoling sheets and deck mats for the backhoe to sit on were required to keep the excavation advancing. At Ring No. 10, soil conditions improved and forepoling sheets were only used locally. Below Ring No. 10 a 2 inch perforated pipe was installed around the perimeter of the elliptical shaft to dewater the lining and collect the water.

The shaft excavation continued down to Ring No.16, at which time the apron rings for the tunnel heading openings were installed down to spring line. The apron rings consisted of a box shaped circular beam that was prefabricated and shotcreted in six overlapping sections.

During excavation of the shaft, the groundwater was measured to be at Elevation 162.50, approximately ten feet above invert. Consequently, a 12 vacuum lance dewatering system was installed within Ring No.15 at Elevation 173.00 and produced 50 to 60 GPM during the remainder of the shaft excavation and during initial tunnel excavation in all four headings.

The manpower and equipment that were incorporated for the excavation of the Farragut Shaft were as follows:

Manpower:	**Equipment:**
1 - Site Superintendent	1 - Northwest 190D - 150 Ton
1 - Walker x 2 Shifts	1 - Caterpillar 966 - Loader
1 - Foreman x 2 Shifts	1 - Liebherr R902 - Backhoe
4 - Laborer x 2 Shifts	1 - RPM Volumetric Batch Plant
3 - Operators x 2 Shifts	1 - Reed 3000 - Shotcrete Pump
1 - Mechanic x 2 Shifts	

C. RUNNING TUNNELS

Prior to excavation of the tunnels, significant changes were initiated, particularly with the requirement for the running tunnel roof pre-support specification.

1. Pre-Support Changes - Directional Drilling:

Pre-support proposed and performed by Hayward Baker, as an alternate method to the contract specifications, substituted the drilling and grouting from the tunnel face with directional drilling and grouting from the shaft. The significant advantage gained by the directional drilling and grouting method was that the excavation cycles could proceed uninterrupted, as opposed to the original specification which required grouting be performed every 20 feet from within the heading.

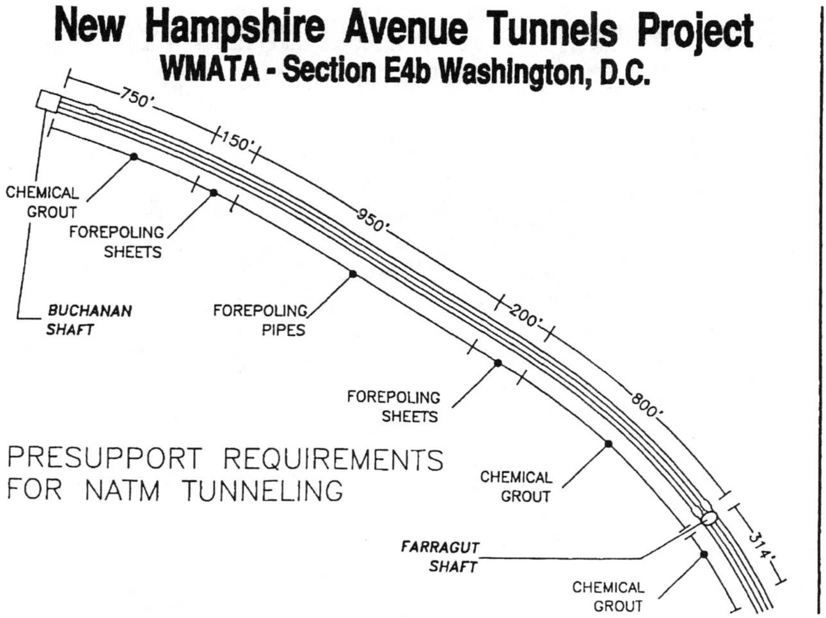

Figure 6

Directional drilling and grouting was performed as follows:

a. General

Chemical grouting was required per the contact in both the inbound and outbound tunnels for the following sections:

1) From Farragut Shaft 314 foot towards FortTottenStation.
2) From Farragut Shaft 800 foot towards Buchanan Shaft
3) From Buchanan Shaft 530 ft. towards Farragut Shaft.

The directional drilling was performed in the north direction towards Fort Totten Station from Farragut Shaft. The directional drilling was performed in the south direction towards Buchanan Shaft from within specially designed drilling chambers. This was planned to allow for the tunnel excavation to begin in the northern direction, while directional drilling was performed simultaneously in the southern direction from within the chamber section (See Figure 6).

b. Drilling Procedure

Hayward Baker subcontracted with Underground Research, Inc. (URI) for the drilling phase. URI specialized in horizontal directional drilling with projects that required a high degree of navigational and steering control. To augment the navigational devices furnished with the directional drilling system, Scientific Drilling, Inc. (SDI) surveyed each borehole with

New Hampshire Avenue Tunnels Project
WMATA - Section E4b Washington, D.C.

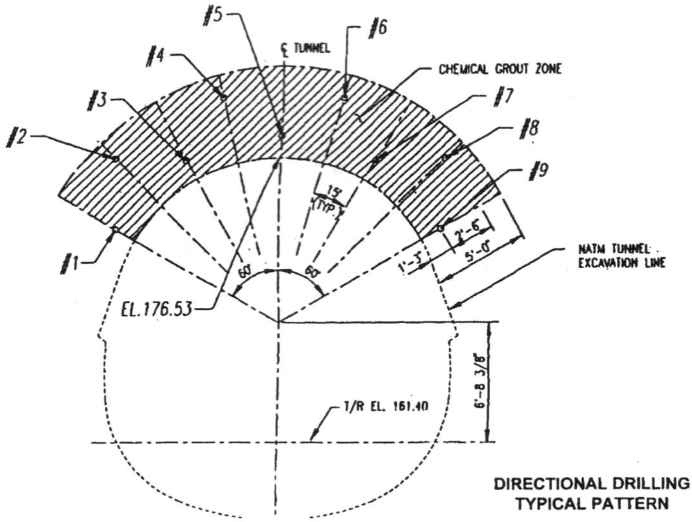

Figure 7

a high precision survey instrument. SDI is regarded throughout the oil well drilling industry as the premier borehole survey instrument manufacturer. The SDI instrument utilized highly sensitive gyros to measure horizontal and vertical direction changes with a tolerance of 0.2 percent in 1000 feet of drilled length. Directional information was fed into a computer at interval distances that calculated the angular direction change and a three dimensional offset from the previous location. The computer continually updated the incremental data to provide both a tabular and graphical plot of the borehole trajectory.

Nine boreholes were drilled parallel above each tunnel heading to provide the specified grout cover (See Figure 7). The boreholes were then as-built with the SDI survey instrument at which time adjustments were made to the chemical grout injection volumes for any alignment deviations. Tube-a-manchette grout pipes were installed within the boreholes in 10 foot sections of 2 inch ID Schedule 40 steel pipe with packers at 2 foot intervals. By utilizing the steel pipe rather than plastic pipe, it was extremely unlikely that a sleeve would be rendered unusable because the pipe could be mechanically cleaned. In addition higher grouting pressures were obtainable.

Installation of the tube-a-manchette pipes was carried out with the URI model HRD 750 drill rig modified for close quarters drilling. The HRD 750 is capable of 150,000 pounds of push and pullback, and up to 40,000 foot pounds of torque. The rig was capable of approximately eight feet of vertical lift in order to reach all holes without the need for re-cribbing or other support bracing. Because the rig weighed only 9 tons, sufficient reaction force was applied to effectively carry out the drilling. The pipe sections were threaded end to end with pipe couplings, thereby maintaining strength and sleeve port spacing continuity throughout the length of the bore.

The heart of the directional drilling system was the steerable bit. Due to the variations of soils encountered, the drill bit had to be capable of maintaining the desired alignment while compensating for changing soil conditions. To efficiently penetrate widely varying soil conditions, URI developed a proprietary off-set drill bit which utilized pulsed jets of drilling fluid that were timed to the drill string rotation to effect the steering in difficult soils. In softer soils, the off-set drill bit was steered by selecting the rotation corresponding to the desired direction and give a sustained push with no rotation. For example, a right turn was achieved with the off-set drill bit in the 3 o'clock position. To maintain a straight alignment, a high rotation speed was used with a moderate push. In all cases, a low flow of bentonite slurry was forced through the drill steel to cool the bit and to lubricate the casings.

The navigational device used during the drilling was the Accu-Nav. The Accu-Nav was mounted directly behind the off-set drill bit and provided the driller with continuous directional information. The tilt and inclination sensors were gravity controlled and accurately assessed grade. However, due to the interference from the steel tube-a-manchette pipes, the magnetic alignment sensor was rendered inoperable.

Therefore, at predetermined increments, the drilling was stopped and the SDI survey instrument was inserted and provided a more precise assessment of grade and alignment. Initially, the steerable drill bit was advanced in 20 foot increments before inserting the SDI survey instrument, but quickly increased to 70 ft. increments as the drillers' skill improved.

The drilling was accomplished using a duplex type system with a steerable inner drill string and an outer "wash-over" casing installed afterwards. The inner drill steel was a 3.5 inch OD API drill rod with 2-3/8 in. ID to allow for insertion of the SDI survey instrument. The "wash-over" casing was 4.5 inch OD and provided sufficient annular space between the casings. During the installation of the "wash-over" casing, bentonite slurry was pumped and flowed back to the shaft in the annular space between the casings carrying the drill cuttings, thereby eliminating the possibility of soil contamination. Upon successful completion of the installation of the inner drill string, the "wash-over" casing was drilled to full depth at which time the inner drill casing was removed and reused.

Then the tube-a-manchette pipe sections were installed within the "wash-over" casing, capped and filled with a cement-bentonite slurry until the theoretical pipe volume was injected. As the outer casing was removed, additional cement-bentonite was pumped to compensate for the removed casing volume. After the withdrawal of the outer casing, the shaft wall was patched with a rapid set, non-shrink hydraulic cement to seal the tube-a-manchette in the bored hole.

c. Chemical Grouting

The chemical grout injection work was carried out utilizing equipment and techniques developed by Hayward Baker specifically for horizontal grouting projects. The chemical grout was a composition of sodium silicate, organic reactant, water and enhancer and has been successfully used by Hayward Baker on various projects over the past several years.

For conveying the grout mixture to the individual injection sleeves, Hayward Baker utilized high pressure polyethylene grout pipes mounted on large hose reels. The hose reels were constructed with multiple high pressure swivels and individual hoses to transfer three components to the packer assemble. One hose carried the grout mixture, one hose carried an anti-freeze solution for the inflation of the packers and the third line carried water to the free side of the packer assembly to prevent build-up of grout which might leak past the packer. This multiple hose reel assembly was sufficient enough in length to grout the farthest grout pipe without splicing.

Each packer assembly had two packer glands separated 20 inches by a slotted center tube, to ensure that only one sleeve port could be grouted at a time. Once a packer assembly was centered on the desired sleeve port by manually pulling the device, a burst of high pressure water was applied to open the sleeve. After the sleeve had been opened, grout injection began with a continuous monitoring of the injection pressures and flowrates throughout the grouting operation. Grouting proceeded until the specified pressure was reached or a specified volume was achieved. Depending on the sleeve port location and line loss of the injection tubes, grout pressures ranged from 130 to 220 psi. Grout quantities exceeded 750,000 gallons for the 3,300 feet of tunnel and nearly 30,000 feet of grout pipe.

2. Tunnel Excavation

The NATM tunnel excavation and support was performed in accordance with the specified procedures as described above.

The sequence for the excavation and support of the NATM tunnels was as follows.

a. Excavate and support Outbound Tunnel from Farragut Shaft towards Fort Totten Station.
b. Excavate and support Inbound Tunnel from Farragut Shaft towards Fort Totten Station, after the Outbound Tunnel had progressed 100 feet, as specified.
c. Excavate and support Outbound Tunnel from Farragut Shaft towards Buchanan Shaft.
d. Excavate and support Inbound Tunnel from Farragut Shaft towards Buchanan Shaft, after the Outbound Tunnel had progressed 100 feet as specified.

The manpower and equipment that were used for the NATM excavation of the running tunnels included:

1) Top side and common service:

Manpower:	Equipment:
1 - Site Superintendent	1 - Northwest 190D - 150 TN
1 - Laborer x 3 Shifts/Day	1 - Caterpillar 966 - Loader
4 - Operators x 3 Shifts/Day	2 - Reed 3000-Shotcrete Pump
1 - Mechanic x 3 Shifts/Day	1 - RPM Volumetric Batch Plant

2) Two Tunnel heading crews (Inbound & Outbound Tunnels)

Manpower Per Heading:	Equipment Per Heading:
1/2- Walker x 3 Shifts/Day	1/2 - 953 Cat Track Loader
1 - Foreman x 3 Shifts/Day	1 - EJC 130 Mucker- 31/2 cy
3 - Miners x 3 Shifts/Day	1 - Liebherr R900-Backhoe
2 - Operators x 3 Shifts/Day	1/2 - Liebherr R902-Backhoe (with drill assembly)

CONCLUSION

The main objective during tunneling under the Rock Creek Cemetery was to ensure that no perceptible ground settlement would occur and it was for this reason that WMATA picked the New Austrian Tunneling Method to excavate this portion of their underground metro system.

As the NATM tunneling operations advanced their were many “tried and true” methods used, but there were also various new and innovative ideas introduced into the excavation plan.

Everyone involved with the NATM operation (i.e. crews, supervision WMATA and consultants worked together to produce a quality product in a safe and timely manner. It is this type of cooperation and input that is vital in making any project successful and was definitely one of the main factors in the successful completion of the New Hampshire Avenue NATM Tunnels.

Chapter 38

EIGHT HEADING ROCK NATM

Larry H. Heflin,[1] James M. Marquardt,[2] and John H.N. Yen[3]
[1]Design Coordinator, WMATA; [2]Project Manager, J.F. Shea Co., Inc.;
[3]Project Manager, Washington Metropolitan Area Transit Authority

INTRODUCTION

On February 17, 1995, WMATA Contractor J.F. Shea holed through the last segment of rock tunnels of the planned 103-Mile Rapid Transit System for the Washington Metropolitan Area Transit Authority. Through the joined efforts and cooperation between the owner and the contractor, the project was completed on time and within budget. The Glenmont extension is on schedule to open in Mid-1998. The complete 103-mile system is scheduled to open in 2001.

WMATA's earlier 19 miles of rock tunnels had been done since 1976 ? By drill and blast with heavy horseshoe steel sets, sets with shotcreting over wire-fabric by TBM and NATM. All rock tunnels were primarily in metamorphic gneiss and schist, part of an east coast lithology that outcrops in a band running from below Richmond, Va. To above New York City.

The 5,400 ft. long twin six meter tunnels of the Glenmont line had been earlier designed in 1979 by URS of NYC using Jenny Engineering of Pittsburgh, Pa., as a tunnel design specialist subconsultant. Jenny had provided two alternative designs, one conventional drill and blast horseshoe and one for Tunnel Boring Machine. Funding considerations had put the design on the shelf and it was clear that the state of the art had led to improved design methods in the intervening years. Before URS was tasked to update their design in 1992, the Authority reviewed its earlier hard-rock experiences in light of the funding mandate to complete each segment on a limited budget and an expedited schedule. Many factors considered and many opinions were in leading to a final decision to drop the earlier designs and redesign for a single designated means of tunneling. That method was the New Austrian Tunneling Method.

The NATM had been used earlier on WMATA for both soft-ground and rock tunnels. In the public's eyes the NATM had become inextricably linked with the use of PVC's and filter material enveloping or partially enveloping the underground structures to give the holy grail of tunnels, a water-tight structure. Although this link of waterproofing methods with excavation and support methods was essentially severed, a general feeling of their interdependence no doubt helped lead to acceptance of the NATM decision because of the waterproofing design that accompanied it. The Authority decided to drop the TBM and conventional options and provide bid documents for NATM only.

URS asked LAW Engineering of Atlanta to provide the NATM design of the tunnels and associated shafts.

DESIGN

The initial lining was designed with the aid of the ABBACCUS finite element computer program. The program input included soil and rock parameters and the excavation shapes. Based on the data available, the following approach was used to develop the Ground Types:

- Geotechnical information provided with the boring logs was evaluated to determine rock mass classifications. The Rock Mass Rating (RMR) and the Tunneling Quality Index (Q) were evaluated following procedures reproduced in Chapter 2 of Underground Excavations in Rock, 1982, by Hoek and Brown.
- Ground types and support requirements were divided into four categories:
    - The initial lining as required by the NATM Tunnels included those measures necessary for maintaining the inherent strength of the ground surrounding the tunnel opening while preventing unnecessary loosening and enhancing the stress redistribution process.
    - Ground Type I: was confined to the Quartz Mica Schist to Gneiss lithology. Core recovery's and RQD's generally ranged from 90% to 100% and 75% to 90%, respectively. Q values range from about 1.5 to 5 while RMR values generally varied from 40 to 56. This means relatively sound rock, moderately jointed and general rock mass behavior stable during construction.
    - Ground Type II: traversed the Quartz Mica Schist to Gneiss and the Quartz Hornblende Schist (Amphibolite) lithologic units. Core recoveries and RQD's generally ranged from 80% to 95% and 50% to 75%, respectively. The general values for Q and RMR ranged from 0.3 to 5 and 24 to 50, respectively. This means rock with closely spaced joints and possibly intensely sheared zones. General rock mass behavior is stable, but stand-up time is less than one day. Local rock loosening and fallouts, possibly, if not supported immediately after excavation.
    - Ground Type III: was found within the Amphibolite and Quartz Mica Schist to Gneiss lithologic units. The poorer quality rock encountered in the contact zone between the Amphibolite and the Schist to Gneiss was included in Ground Type III. Core recoveries and RQD's generally ranged from 60% to 95% and 30% to 50%, respectively. The general values for Q and RMR ranged from 0.005 to 0.2 and 5 to 25, respectively. This means rock and weathered rock with closely spaced joints and sheared zones probable. Rock mass behavior may be unstable and rock loosening and fallouts are possible if not supported immediately after excavation. Rebar spilling and benched headings may be required where weathered rock zones are present.
    - Ground Type IV: was the mixed face condition where decomposed and/or weathered rock are encountered along with weathered rock and/or highly jointed to moderately jointed materials. This means the mixed face condition where soil, weathered rock and rock are encountered. Ground behavior may be unstable if not supported immediately after excavation. Pipe spilling and benched headings will be required.

Based on this criteria, the percentage of tunnel by ground type expected and found:

	EXPECTED	ACTUALLY ENCOUNTERED
Type I	20%	52%

Type II	50%	25%
Type III	25%	18%
Type IV	5%	5%

Access to the job included the portal and two shafts, one at Mason and one at Henderson, access was a maximum of some 2200 feet apart, spaced in reference to National Fire Protection Association codes and evacuation of passengers from the subway in the event of emergency. See Figure 1:

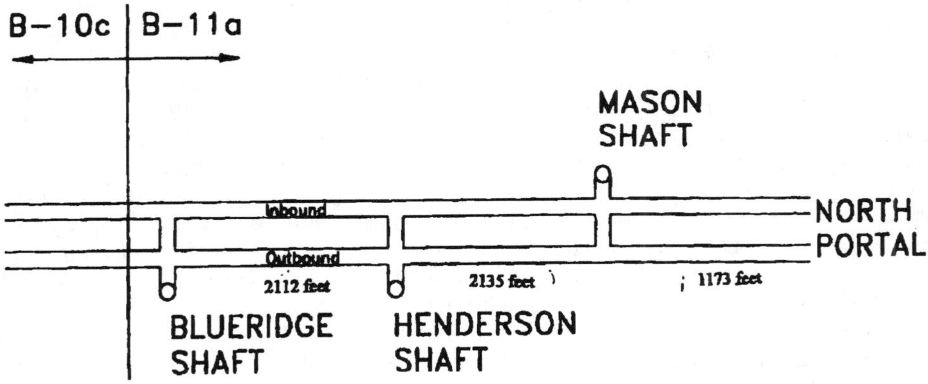

Figure 1. Pre-construction

The bid documents were sent out in April 1993 with NATM as the only acceptable method. During the bid period potential contractors examined the available materials, core samples, construction area and expressed their general enthusiasm for the NATM: or asked why a Tunnel Boring Machine Option was not made available and if a TBM Value Engineering Change Proposal could be suggested by the winning contractor. The contract does provide for a VECP which can be submitted after the Authority has accepted a contractor.

The NATM had been selected primarily because of three reasons:

- The Glenmont Tunnels had only one portal area for access and the Mason shaft, a logical point for TBM entry, was small and with a limited surrounding work area. TBM entry from the North Portal area would have required a more specialized machine to handle the mixed face conditions.
- The end of the tunnel opposite the portals was a break-through to a working tunnel in service for the Metro system... any TBM arriving at this point would be forced to back out or, some thought, be turned around in a room excavated especially for this purpose. In a project where each single tunnel was only some 5,400. feet in length, our estimate indicated that TBM mobilization plus the possibility of a turn around room would increase the cost beyond a reasonable bid.
- Earlier experiences with (admittedly earlier generations of) TBM's that had caused delay for repair or having become temporarily stuck led to a concern that lost work-days could wreak a political disaster on the delicate balancing of funding for future projects.

Bids were opened June 9, 1993 with valid bids in a 20%, 50 to 60 million dollar range. The low bidder was J. F. Shea Co., Inc. of California, with portions of the work, including the cut and cover section, to be done by the George Hyman Company of Washington, D.C as a major subcontractor. Both contractors had records of successful completion of earlier Metro constructions and excellent records in the construction community.

J. F. Shea was given a Notice to Proceed on August 17, 1993, and promptly submitted a Value Engineering Change Proposal to work with a TBM. Shea felt much of the work was clearly a TBM job. They had a machine available for renovating which could be on the site in a matter of months and promised to finish the work 3 months early and split a two million dollar savings with the Authority. They cited the environmental and community benefits of an essentially blast-free tunnel job and the comparative safety of TBM operations versus drill and blast shotcrete. Shea had done drill and blast excavation for tunnels and stations in earlier WMATA projects, and had TBM experience on the Metro Shady Grove line construction. The engineers and lawyers of the Authority went back to their estimates, and restudied their reasons for having chosen NATM. In discussions, Shea stressed that the anticipated greater advance rate for a TBM could more than make up for any potential delay should there be a breakdown or work stoppage for any reason. Although both Shea and the Authority acknowledged that the shallow portion of the job and the north end leading to the cut and cover area from the Mason shaft might be handled with NATM because of the varying nature of the softer ground, the Authority felt that several headings in NATM would always provide continuing progress, at least in some if not all headings.

With a tight schedule forcing a timely response the Authority stuck to its original convictions and rejected the VECP offer. Shea was instructed to continue its work under the original bid documents and complete the tunnels and shafts using the New Austrian Tunneling Method.

CONSTRUCTION

The Glenmont tunnels were twin tunnels each over one mile in length (1.3 miles each), separated by a 20' pillar generally in Micaceous Gneiss and Schist with a average compressive strength generally ranging from 5,000 to 25,000 psi with a high test result of 30,400 psi.. The North Portal floor was some 60 feet below ground from which the tunnel ran south essentially level at an .035% downhill grade for the first 500 feet and then continued south at a 4% downward grade for the next 3200 feet. The last 2500 feet, to its mating with the tail-track of the Wheaton Station was again essentially level, a .035% downhill slope. At its deepest the tunnels are nearly 220' deep with ground water remarkably level staying at a general elevation of 430', giving us a maximum head of about 190'. The surface was immediately adjacent to a busy State Highway, Georgia Avenue, with a high vehicle per day usage. The Wheaton South end of the tunnel was also the site of the Blueridge Fan Shaft, an in-service ventilation structure.

Because of the tightness of the metamorphic rock, water-flows were not expected to be severe and no preconstruction dewatering was required. Waterproofing was achievd using the PVC membrane lining system used on earlier WMATA tunnels done with the NATM. The shotcrete lining was covered with a "Fleece Drainage Fabric" which was in turn covered with a heat-sealed watertight Synthetic Membrane. Water flows towards the tunnel were thus intercepted and channeled to enclosed drains on each side of the tunnel. A Geotechnical Basis of Design and Construction Report had been prepared by (JEC) Jenny Engineering

Corporation for printing with the specifications and numerous sub-surface reports of investigation had been prepared by Muser Rutledge Consulting Engineers of NYC. The Authority's conservative estimate of the work was that of the over 10,000 feet of tunnel, some 2000 feet would be in the Type I, or best and most easily supported: allowing unlimited length of round and minimal support of a 4" initial shotcrete coat with occasional 10' rock bolts for local support. The estimate was very conservative with over half the job being done as Type I, an acknowledgment in part, of the careful excavation techniques of the contractor.

Type II ground was supported with 4' of shotcrete, rockbolts and the addition of lattice girders as needed and welded wire fabric and occasional rebar spilling.

Type III had additional requirements. It required 6" of shotcrete, welded wire fabric, lattice girders at 4' to 8' spacing, 10' rockbolts, 12' rebar spilling and a maximum round length of 4' to 8' was permitted.

Type IV, for the most difficult ground required a heading and bench, had the type III requirements and added 10' pipe spilling, 2" diameter. Only five percent of Type IV was expected. Five percent was found.

The Authority anticipated excavation to be done in 10 months and the entire job in 36 months. With Shea's cooperation with WMATA Resident Engineer Dave Shilling, the schedule was met.

The Mason street shaft had a limited work area and was a 19 foot diameter NATM shaft 95 feet deep. The Henderson Shaft was 180 feet deep and 32 feet in diameter . The main construction work area at Henderson was considered adequate at about two acres. A later requirement for more extensive treating of groundwater required the area to hold retention basins as well as support the construction. Minimal mucking was done from the Mason Shaft and the Henderson shaft area was outfitted with an American 9260 125 ton crawler crane with a 16 cu. yd. muck box. Mucking was limited by work-hour restrictions because of the urban environment and 10 yard dump trucks transported the muck offsite in a work day limited to 7 a.m. to 5 p.m. each day.

A further restriction on the work was a need to hold noise below a 75 dbl. level during the day and 55 at night. Shea accomplished this by installing the contract-required wooden construction fence 12 feet high with a three-foot inward-facing top lip and backed by foam-board. The interior of the crane engine compartment was lined with insulation bating, as were the walls of the shotcrete batch plant located at Henderson. The muck-bucket cables were covered with rubber hose for additional noise dampening. Compressors were electric and enclosed. Despite the presence of a nursing home directly across a residential street, and the presence of neighbors eager to express their annoyance at sporadic noises, the noise requirements were met and a letter of commendation was received from the County Noise Enforcement Officer.

Light levels were similarly restricted by aimed shades preventing light-spillage to the nursing home and neighbors.

THE TUNNELS

To meet the tight schedule Shea began from the Henderson Shaft Adit with the inbound,(most distant) westerly tunnel first, excavating towards the Mason Shaft. Shea mobilized two mobile rubber tire hydraulic drill jumbos. Shortly thereafter, headings in the second tunnel began in two directions. Rubber tire loaders were used as platforms for the placement of lattice girders, wire mesh and shotcreting operations. Shotcrete was almost entirely dry-mix and produced 20% to 25% rebound and reached the required 5000 psi strength. Some 5 cubic yards of shotcrete per hour were placed. An average round was 10' in type I and II ground, 6' in type III and IV.

About a month later, the North Portal tunnels started in generally mixed face materials heading slightly down-grade with parallel tunnels maintained at a 150' blasting face separation as required by contract.

From the Henderson Shaft the remaining headings started with three shift operation. In the meantime, excavation had begun on the Mason shaft. Upon its completion heading were advanced in both directions in each of the two tunnels, giving us, for a short time, a total of ten NATM headings being worked at the same time. From six headings working three hour shifts of 8 hours each, the total sustained advance from all faces peaked at 85' feet per day with an average of 52' per day with a three shift operation. For a brief few days, headings were advanced from the Mason Shaft to yield a peak advance of some 100 feet per day from a total of ten headings.

The work was done with two crews of drillers and two crews of muckers and shotcreters, the Henderson shaft area allowing greater flexibility in managing the work. Slow going in the western tunnel between the North Portal and the Mason Shaft required setting Stage IV supports for a distance of some 250 feet, the only lengthy area of “most-difficult ground” encountered. It was here that our only severe overbreak occurred, a fallout of some 17 cubic yards that stopped at a natural rock joint intersection long enough to be strengthened with shotcrete. Later the hole was filled with concrete during the tunnel lining operation.

The final lining was plain (unreinforced) cast in place concrete (shotcrete was used in the adits with their varying internal dimensions) cast with a minimum 10" thickness. The final lining was designed with “feedback” loads from the initial lining and water loading of 16 feet at the crown and zero at the (drained) invert.

CONCLUSION

Both the contractor and the Authority feel this was a good job. Partnering was used, but an anticipated Disputes Review Board was never initiated. Shea would still have preferred to have used its TBM, but the Authority feels it got its tunnels on time and within budget... just what it asked for.

Chapter 39

LONDON BRIDGE STATION, JUBILEE LINE EXTENSION

Kurt Zeidler[1] and Vojtech Gall[2]
[1]Technical Director and [2]Vice President, Dr. G. Sauer Corporation

ABSTRACT

London Underground Limited are currently constructing an extension of the Jubilee Line. The approximately 16-km long Jubilee Line Extension Project connects the existing Green Park Underground Station with the Canary Wharf area and Greenwich. Being a part of the project, the London Bridge Station (Contract 104) was built using a combination of the New Austrian Tunneling Method (NATM) with Compensation Grouting. NATM was successfully applied in softground and urban areas under restrictive requirements regarding settlement limitation. Information on the flexibility of the method and the influence of Compensation Grouting on ground and tunnels are given.

INTRODUCTION

London Underground Limited (LUL) are currently constructing an extension to the existing Jubilee Line underground metro system. The new metro line will connect the existing Green Park Station in Central London with Stratfort in the East of London. The 16 km long extension - approximately seventy-five percent of which are aligned underground - is subdivided in eleven principal construction contracts. The project comprises the new construction and adoption works for twelve metro stations and running tunnels linking the stations. The construction works started in 1994 and the line is expected to start operation early 1998.

The majority of the underground schemes is being constructed using traditional cut and cover and shield machine methods. The mined tunnels are lined with cast iron (SGI) or precast concrete segements. There are two contracts, where the New Austrian Tunnelling Method (NATM) is employed to construct substantial sections of the scheme: Contract 102, Waterloo Station and Contract 104, London Bridge Station. The alternative bids of the Contractors and their consultants convinced the Client of the benefits of the NATM in terms of economic savings on time, labour and materials.

On 21 October 1994 the NATM station tunnels being under construction for the Heathrow Express Rail Link collapsed. As a consequence, all NATM works were suspended in the UK. After approximately 7 months review and discussion period a status of 'non objection against NATM' was reached and NATM works could be restarted.

This paper reports on the main aspects of the construction works for the new London

Bridge Station scheme and the implications of the Heathrow event on the design and construction.

THE LONDON BRIDGE STATION

General

The civil contract works for the London Birdge Station commenced end of 1993. The works are being carried out by the Costain Taylor Woodrow Joint Venture as the main contractor. The main site installation and the Main Access Shaft is located at Red Cross Way, Central London.

Ground Conditions

The majority of the new tunnels of the London Bridge Station were constructed in London Clay, an over-consolidated, tertiary clay deposit with varying silt and sand content. The invert region of the Station Platform Tunnels and the Central Concourse Tunnel cut into the upper parts of the Woolwhich and Reading Beds, which are over-consolidated silty, sandy clays. The Blackheath Beds, a highly consolidated, normally water saturated sand band between the London Clay and the Woolwich and Reading Beds was not encountered apart from one isolated little occurance in the invert area of the Central Concourse Tunnel.

Young gravel deposits of fluviatile origin and man made fill is located above the London Clay. The gravels are water saturated, whereas the Clay appears to be underdrained. Only the contact zone between the granular deposits and the London Clay and some sand/ silt pockets incapsulated within Lonon Clay are soft and water bearing. Little, locally limited water seapage through some fissures was observed during the excavation. Damp spots or minor water seapage through the shotcrete lining was observed after typically three days to one month. The ground cover above the tunnels' roof varyies between 25 and 32 m.

Buildings and Subsurface Structures

Contract 104 is the most complex scheme of the Jubilee Line Extension Project with regards to its location beneath the vast brick structure of the existing London Bridge Railway Station, other historic listed and sensitive surface structures, as well as Borrow High Street, that is a major traffic corridor underlain by countless service ducts for water, gas and electric power. The London Bridge Railway Station is said to be the world's largest brick structure. Its foundations could not be fully investigated and their geometry and conditions are only vaguely known. The response to ground movements by the brick structures could only be conservatively estimated and somewhat accurate modeling was not feasible due to the lack of detailed information on the structure.

Furthermore, old SGI-lined tunnels of the City and South London Line and the Norhern Line had to be undermined. A thorrough survey of the existing underground structures has been carried out as to their structural integrity and the conditions of the linings at the preconstruction stage. However, limited information was available on the behavior of old SGI-lined tunnels when new NATM underground openings are constructed in very close proximity to them.

Due to the above described circumstances, very stringent limits to surface settlements and ground deflections have been imposed on the contract by British Rail, represented by Rail Track, as well as service operators and the property owners. Maximum allowable settlement values and maximum angular distortion related to protected buildings have been established. A settlement reducing measure, i.e. Compensation Grouting, had to be activated, before the maximum value had been reached, to avoid damage of the surface structure. The maximum value for the settlement has been defined with 25 mm while the maximum angular distortion was not allowed to exceed 1/1000.

Apart from requiring the use of settlement reducing excavation and support procedures, these contract requirements also demanded that the grouting system for the Compensation Grouting had to be in place and ready for operation before the zone of influence by the underground construction works reached the relevant building(s).

Photo 1. Brick Foundations for Railway Viaducts at London Bridge Station

General Arrangement

The following description depicts the main sections of the station scheme as it was planned to be constructed before the Heathrow event. The changes to the layout and design after the Heathrow event are described in a later section.

Contract 104 comprises three main construction parts: First, the construction of the Jubilee Line London Bridge Station including several hundred meters of running tunnels, one bifurcation ('Step Plate Junction'), platform tunnels, one Central Concourse Tunnel, Ventilation Tunnels and machine rooms involved the majority of the NATM excavation activities. The construction of a new Platform Tunnel for the Northern Line including the running tunnels to the junctions with the existing running tunnels and the junctions themselves form the second part of the contract. The main access shaft at Red Cross Way, two trial tunnels and various access tunnels constitute the third part.

The Step Plate Junction with its total length of approximately 62 m is located at the contract boundary to Contract 103 (west of Contract 104) encompassing the junction between the Westbound Running Tunnel and the Cross Over Tunnel which leads to the East Bound Tunnel in Contract 103. The excavation cross section area varies from approximately 22 m² to 120 m². The East- and Westbound Running Tunnels with a total length of approximately 500 m lead from the contract boundary Contract 103 and Step Plate Junction respectively in easterly direction toward the Jubilee Line London Bridge Station Platform Tunnels. The excavation cross section area of the running tunnels is approximately 22 m². Close to the Step Plate Junction the Cable Tunnel links the Jubilee Line Eastbound and Westbound Running Tunnels. It is approximately 23 m long and has an excavation cross section area of approximately 40 m². The Jubilee Line Station Platform Tunnels (east- and westbound) are approximately 140 m long each and have an excavation cross section area of approximately 50 m². The Jubilee Line Central Concourse is located between the Station Platform Tunnels and is approximately 80 m long. It has an excavation cross section area of approximately 60 m². It was planned to connect the platform tunnels to the Central Concourse Tunnel with 8 Cross Passages. An 'Interlocking Machine Room' (IMR)at the west-end of the Station and a 'Machine Chamber' (MC) near the east-end of the station were designed to provide sufficient space for accommodating electro-mechanical equipment. During the course of the re-design of the station after the Heathrow event, the IMR and the station tunnels have been substantially re-developed, whereas the MC has been abandoned.

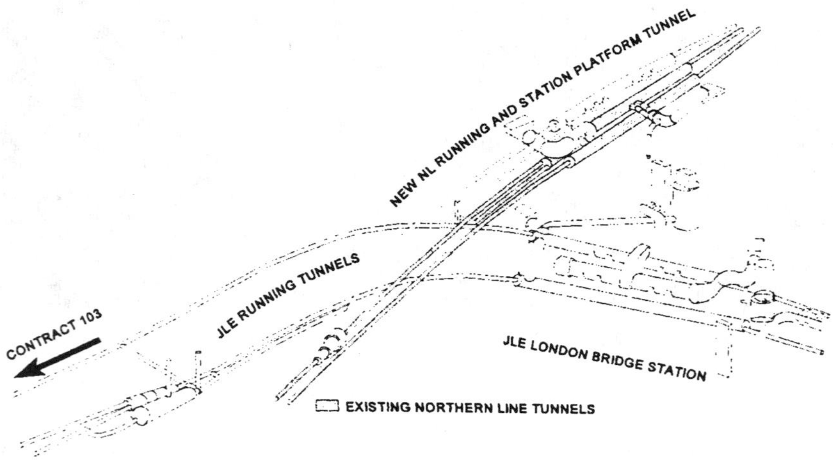

Figure 1. 3-Dimensional Sketch of the London Bridge Station Scheme
Based on Sketch by JLEP.

Escalators and connection tunnels lead the passengers to the existing Northern Line Station and to the surface.

Before the construction works for the actual station could commence, two trial tunnels had to be constructed. To access the Trial Tunnels and the subsequent construction work faces a 12 m OD Main Access Shaft has been constructed close to the west end of the contract. One 11.3 m wide tunnel was constructed from the Main Access Shaft toward west to test the ground response to the excavation of a large diameter tunnel and Compensation Grouting. This Trial Tunnel is approximately 45 m long and was used later as access to Contract 103 and as muck tunnel. Above the second half of the 11.3 m Trial Tunnel trial Compensation Grouting was carried out to investigate the ground and shotcrete lining response to the hydro fracturing, explore the heave effect at the surface and to calibrate the control system. A large amount of monitoring instruments were installed to monitor the ground structure interaction and the ground response during the construction of the Trial Tunnels (Kimmance, J.P. and Allen, R. 1996). The second trial tunnel has an excavation span of 5.3 m and is approximately 50 m long. It was driven from the Main Access Shaft toward east to simulate the tunneling situation equivalent to that of running tunnels. During a later stage of the construction works, it served as a section of the access tunnel to the construction faces of the Northern Line works.

Contract 104 also includes a safety upgrade for operation of the existing Northern Line Station. An access tunnel leads from the end of the 5.3 m Trial Tunnel to the work faces for the new Northern Line (NL) tunnels and joins in close proximity to the existing NL Southbound Northern Line Tunnel with the new NL Southbound Northern Line Running Tunnel. The latter links the new Northern Line Platform Tunnel to the existing NL Southbound Running Tunnel. The NL Running Tunnel is approximately 214 m long and has an excavation cross section area of approximately 17 m^2. The new NL Platform Tunnel is approximately 114 m long and has an excavation cross section area of approximately 50 m^2. Cross Passages connect the new Northern Line Platform Tunnel to the existing Northern Line Southbound Platform Tunnel, which will be converted into a Concourse Tunnel later. A lift shaft, interchange tunnels and escalator shafts connect the upgraded Northern Line Station to the Jubilee Line Station and the surface.

NATM Works

At the time of awarding the contract, the Jubilee Line Extension (JLE) Project Team restrained the use of NATM to the deep level tunnels above which sufficient clay cover to the water saturated gravels is maintained. It was allowed the construct the Running Tunnels, Platform Tunnels and Concourse Tunnel as well as the Ventilation Tunnels using the NATM. Escalator shafts, access and lift shafts were only permitted to be built using NATM to a level, where a minimum clearance to the clay - gravel interface of 2 m could be demonstrated.

Compensation Grouting above NATM Tunnels

As described above, the contract works had to be carried out beneath sensitive and historic surface buildings. To protect these buildings from damage caused by undue surface settlements, an extensive Compensation Grouting system has been developed during the project design period and further refined by the Contractor during the preparation period for the contract execution. The Joint venture contracted GKN Keller-Soilcrete to carry out the design and execution of the grouting works.

The former running tunnels of the City and South London Line were used as access to grouting adits. From these grouting adits and from several grouting shafts, which were sunk from the surface, an array of sleeve pipes were installed, through which the grouting material was injected into the London Clay. By fracturing the clay and forcing the grouting material

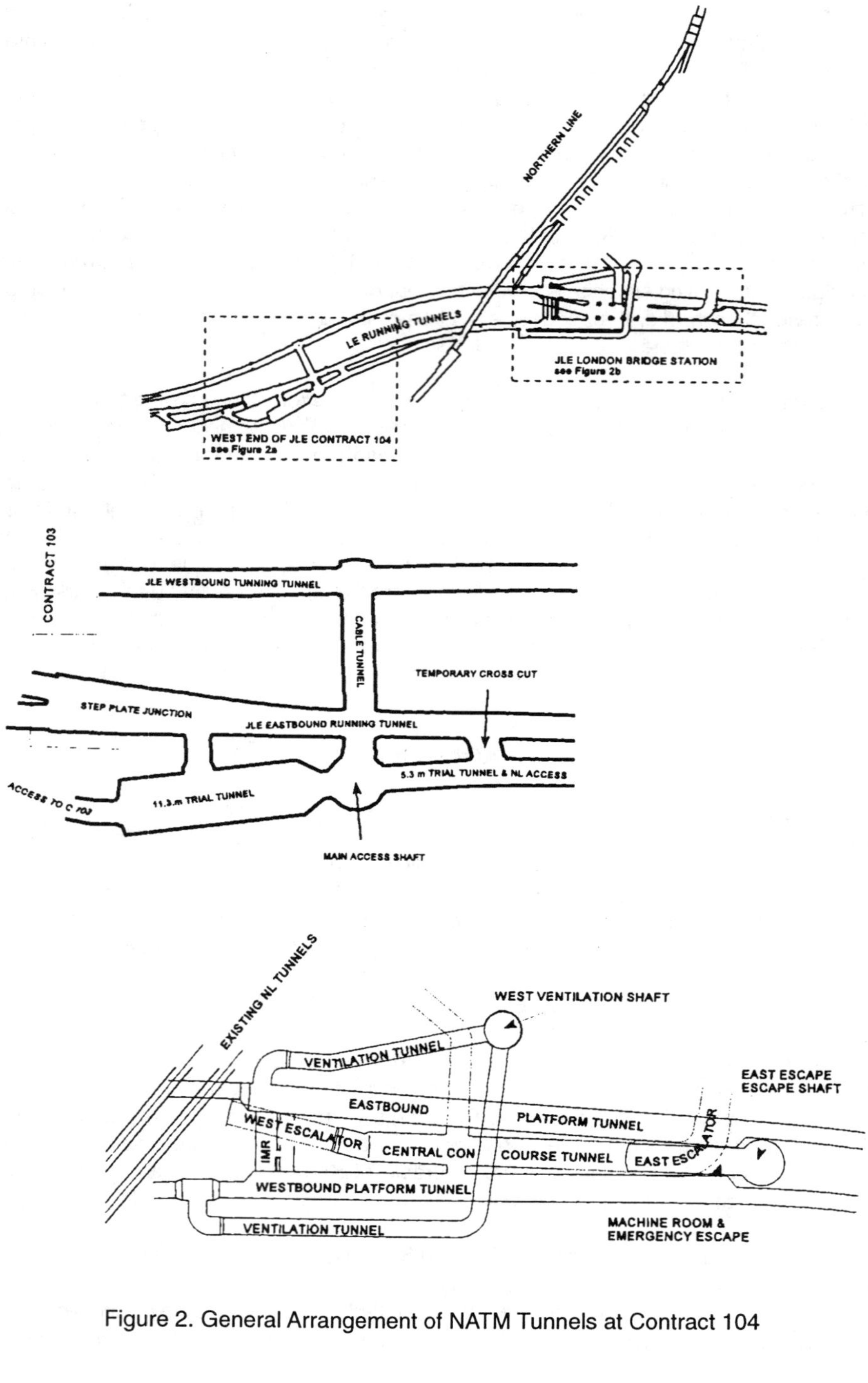

Figure 2. General Arrangement of NATM Tunnels at Contract 104

into the ground, soil is displaced and settlements compensated. A highly sophisticated control and pumping system ensures that the injection, displacement and subsequently the surface heave takes place in a controlled manner.

The clearances between the tunnels' roof and the grouting pipes ranges between 8 and 2 m, being the smallest above the central station area.

NATM EXCAVATION FINISHED BEFORE THE HEATHROW EVENT

General

Substantial parts of the NATM excavation and support works were successfully finished before the station tunnels at Heathrow collapsed. Subsequently, all NATM works in the UK were suspended. The following text describes the works finished by that date and the experience gained with applying the method in London Clay in an urban area.

Excavation of all tunnels was carried out in sequences either top heading, bench and invert, or half face drifts for the larger span tunnels. Lengths of excavation round was typically limited to a maximum of 1m. However, on a number of tunnels the invert excavation length was increased to 2m.

Trial Tunnels

Both, the 11.3 m and the 5.3 m span Trial Tunnel were successfully completed.

The 11.3 m Trial Tunnel was excavated from a previously formed smaller shaft eye funneling up to full cross section size over a transition of approximately 3 m length. For the excavation and support of the full Trial Tunnel cross section the half face drift method had been used. The initial support is formed by 300 mm thick welded wire fabric (WWF) reinforced shotcrete and lattice girders.

The 5.3 m Trial Tunnel was also excavated from a previously formed shaft break out eye. The excavation and support followed the top heading and bench/invert sequence, where the bench and invert was excavated and supported in one operation step. The initial support is built up by 150 mm thick WWF reinforced shotcrete and lattice girders.

Cable Tunnel and Jubilee Line Running Tunnels

The Cable Tunnel was driven from the third break out eye in the Main Access shaft. The excavation was carried out in top heading, bench and invert drifts with an initial lining comprising 200 mm WWF and lattice girder reinforced shotcrete. The tunnel was constructed beyond the JLE Eastbound Running Tunnel and ended there with a short stub. For the construction of the Running Tunnels break out eyes were formed, from which the construction of the JLE Running Tunnels commenced.

Both the Eastbound and Westbound Running Tunnels were constructing following the same excavation and support sequence as the 5.3 m Trial tunnel.

The Eastbound Running Tunnel was progressed to the contract boundary Contract

103 whereas the tube of the Westbound Running Tunnel reached from the Step Plate Junction to a place some 30 meters short of the east end of the Station Platform Tunnel. On the way there the existing shotcrete lined new NL Southbound Running Tunnel and the live SGI lined old NL Southbound Running Tunnels had to be undercrossed with a maximum clearance of 0.5 m.

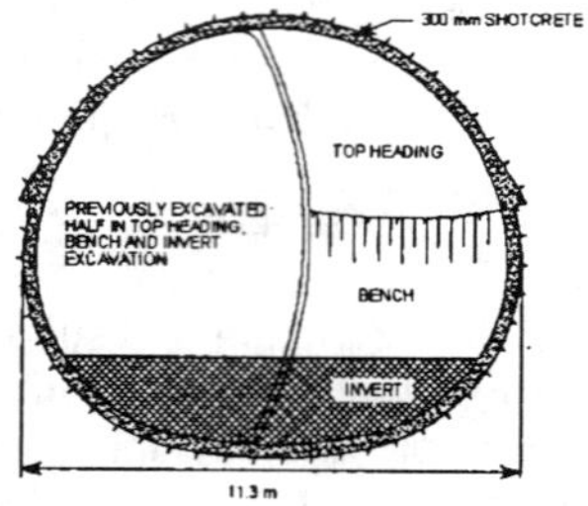

Figure 3a. Excavation and Support of 11.3 m Trial Tunnel

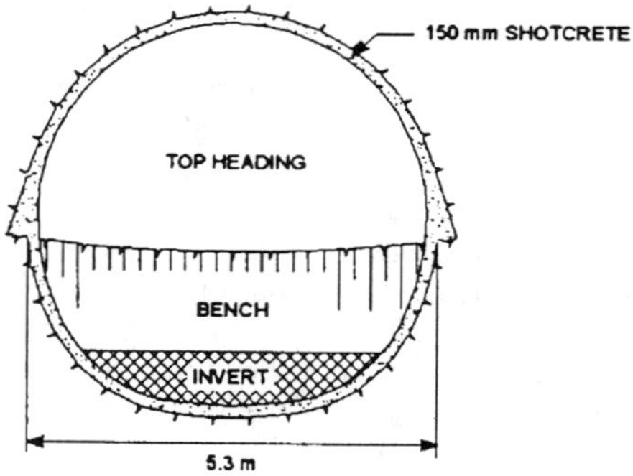

Figure 3b. Excavation and Support of 5.3 m Trial Tunnel

Jubilee Line Step Plate Junction

The Step Plate Junction was constructed starting from the Running Tunnel funneling up to its maximum width of approximately 12 m. The excavation and support sequence started with top heading, bench and invert drifts and changed after approximately 30 m to the half face drift method, as it had been previously carried out for the construction of the 11.3 m Trial Tunnel.

The initial support is formed by 200 mm WWF reinforced shotcrete at the small end. The shotcrete thickness was increased to 300 after 30 m, catering for the increasing tunnel span.

The left half of the Step Plate Junction was constructed to the headwall at its east end whereas the right half was approximately 50 % completed when the NATM works were suspended.

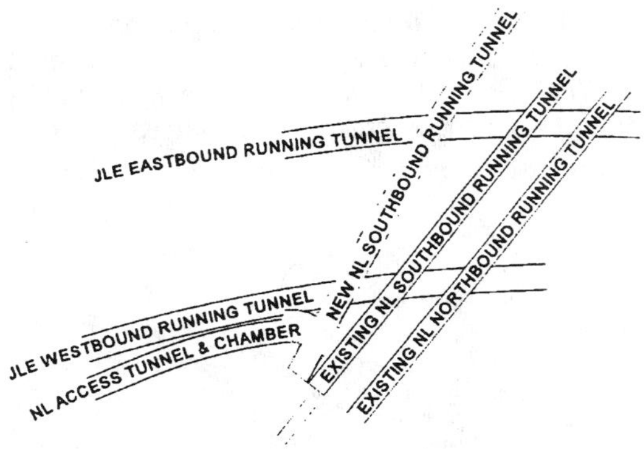

Figure 4. Detailed Plan View of the Undercutting Tunnels

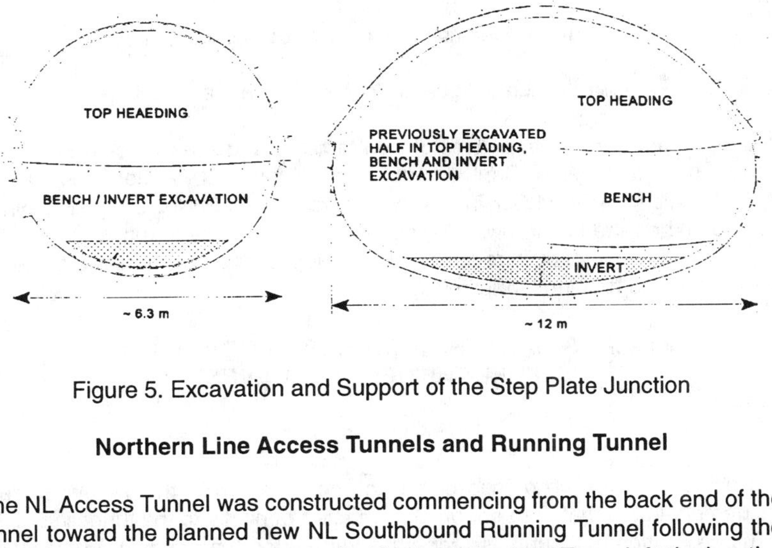

Figure 5. Excavation and Support of the Step Plate Junction

Northern Line Access Tunnels and Running Tunnel

The NL Access Tunnel was constructed commencing from the back end of the 5.3 m Trial Tunnel toward the planned new NL Southbound Running Tunnel following the same excavation and support sequence as applied for the 5.3 m Trial Tunnel. At the junction to the planned new NL Running Tunnel a junction chamber was built in close proximity to the old SGI lined NL Running Tunnel (approximately 0.5 m clearance). The excavation area of this chamber has been split into top, bench and invert heading. It is known that the NL running tunnels have undergone a substantial refurbishment and have been enlarged to fit larger trains through in earlier days. Therefore, the excavation area of the access chamber was split into two halves, which were excavated subsequently to cater for any unforeseen ground condition during the last three meters of excavation.

Northern Line Station Platform Tunnel

Commencing with a short transition at the end of the new NL Southbound Running

Tunnel the new NL Station Platform Tunnel was enlarged to the full 8 m span. Excavation was carried out following top heading, bench and invert excavation. For the initial support 200 mm shotcrete with WWF and lattice girders reinforcement was applied. Approximately 70 m of the platform tunnel were successfully constructed when the NATM works were suspended.

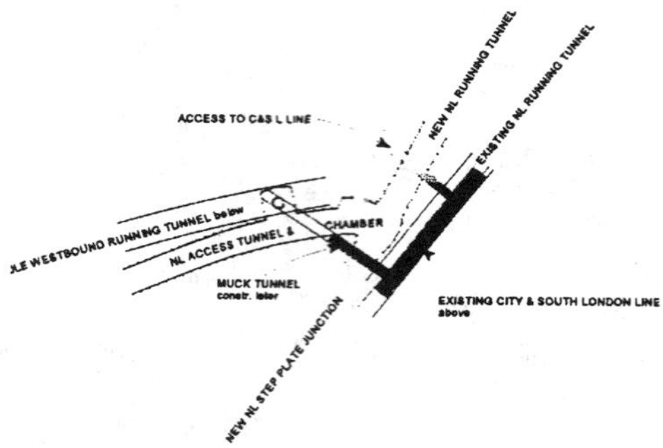

Figure 6. Situation at the NL Access Chamber

Jubilee Line Access Tunnel to Contract 103

In order to access the work faces of Contract 103 and allow muck transport from the shield driven running tunnels in Contract 103, an access tunnel was constructed commencing from the back end of the 11.3 m Trial Tunnel toward the JLE Westbound Running Tunnel Contract 103. When the Heathrow collapse occurred, the access tunnel and the junction chamber to the running tunnel were successfully finished, lined with shotcrete and ready for the break out into the running tunnel toward east.

FINDINGS FROM THE NATM EXCAVATION FINISHED BEFORE THE HEATHROW EVENT

General

The overall experience gained during the construction of the various underground openings demonstrated that the London Clay reacted benignly to the tunneling activities at London Bridge Station, although its properties were such that it could be excavated with standard tunnel excavators. Ground deflections and overbreak stayed within limited ranges. The monitoring results demonstrate, that an adequate design, fast excavation, prompt application of the support measures and early support ring closure provided sufficient safety and support capacity to restrict overbreak, undue ground deflection and loads and, consequently, surface settlement. Also, the correct implication of the design, good workmanship and adequate construction supervision ensured the success at this contract.

Trial Tunnels

During the construction of both Trial Tunnels London Clay with low silt and sand content was encountered. The ground showed numerous, randomly orientated fissures and appeared rather blocky and hard. Excavation was carried out using excavators. No significant water ingress could be observed during the excavation period. The ground provided sufficient stand-up-time such that one meter long rounds were excavated and the support was safely installed without overbreak. Overbreak in this material generally occurs along disconitinuities in the form of small wedges and blocks. The ground response, lining deflections and surface settlement ranged within the predicted limits.

The maximum surface settlement above the first, not Compensation Grouting-treated part of the 11.3 m Trial Tunnel was 30 mm. This value has to be taken with some caution, since the 12 m OD Main Access Shaft is located in close proximity of the measurement point and some influence by late ground movements originating from the shaft construction cannot be totally excluded. Maximum settlements were predicted with 35 mm based on a face volume loss of 1.2% during the tunnel construction. Size and shape of the settlement trough matched well with the predictions. The total convergence of the tunnel lining ranged between 0 and 10 mm.

Unexpected deformation of a short section of the temporary shotcrete middle wall was detected early and remedial actions initiated immediately. Mining activity at the left half of the tunnel was halted and excavation and support application in the right half of the tunnel resumed to eliminate the effected section of the temporary shotcrete middle wall. Investigation in the reason for the undue deformation revealed that the temporary middle wall had not been constructed to the shape required by the design. As a consequence, a strict procedure for profile control was introduced.

The maximum surface settlement above the 5.3 m Trial Tunnel was 24 mm. However, this value appears to be influenced by the close proximity of the measuring point to the Main Access Shaft, since more distant settlement points above the 5.3 m Trial Tunnel and the NL Access Tunnel show settlements ranging between 14 and 17 mm. The tunnel convergence ranged between 0 and 5 mm.

Jubilee Line Cable Tunnel and Running Tunnels

The construction of the JLE Running Tunnels commenced after the NL Access Tunnel and a part of the new NL Southbound Running Tunnel were finished.

The construction of the Cable Tunnel and the JLE Running Tunnels progressed in a similar successful manner as the Trial Tunnels. Surface settlement and tunnel deformation matched well with the predictions and the experience made during the 5.3 m Trial Tunnel construction.

When tunnels were constructed in close proximity to existing underground structures, interactions between these tunnels were observed (Kimmance, J.P. et al, 1996)

During the construction of the JLE Westbound Running Tunnel parallel to the 5.3 m Trial Tunnel in a clear distance of approximately 5 m some influence on the previously constructed Trial Tunnel was observed.

For example at chainage 46 of the 5.3 m Trial Tunnel a significant increase of the horizontal distance between the sidewalls below axis level was detected whereas the diagonal cord toward the running tunnel (DD4) showed convergence during the same period.

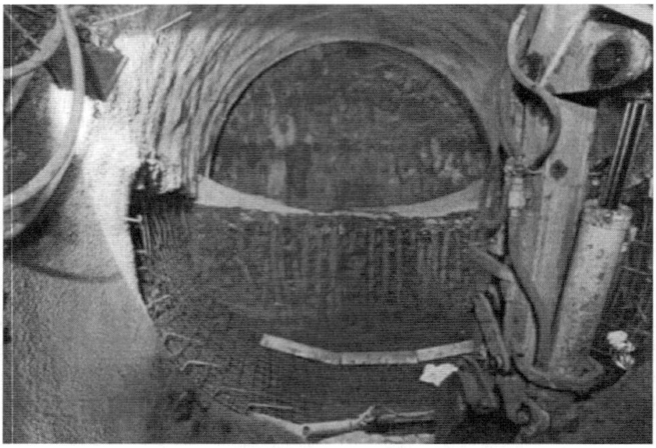

Photo 2. Excavation of the JLE Running Tunnel

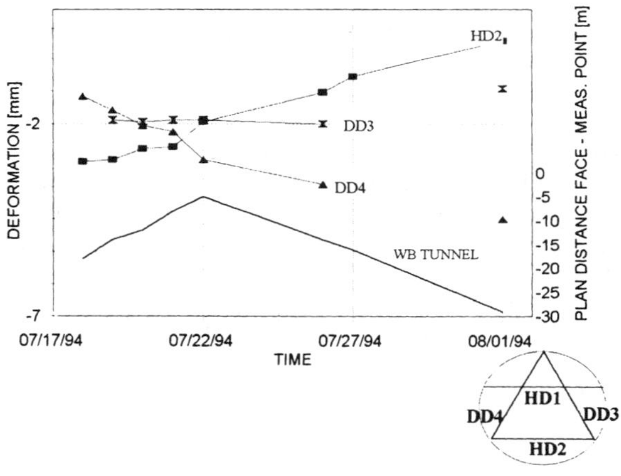

Figure 7. Deformation Trends at 5.3 m Trial Tunnel,
Chainage 46 During the Approach of the JLE Westbound Running Tunnel

The NL Access Tunnel rises up in level to meet the level of the planned new NL Running Tunnel, while the JLE Running Tunnels stay level and underpass the NL tunnels. When the JLE Westbound Running Tunnel approached the previously built new NL Southbound Running Tunnel a 'divergence' was observed. Once the JLE Running Tunnel undercut the new NL tunnel with a clearance of approximately 0.5 m and progressed farther, the reading results indicated a convergence in horizontal and diagonal directions more than compensating the previous increase. Subsequently the lower horizontal cord (HD2) exhibits a gradual divergence until deformations apparently ceased.

Shortly after the above event the construction of the JLE Westbound Running Tunnel undercut the existing SGI lined NL Running Tunnels. Deformation of the SGI lining was observed, as summarized in the table following.

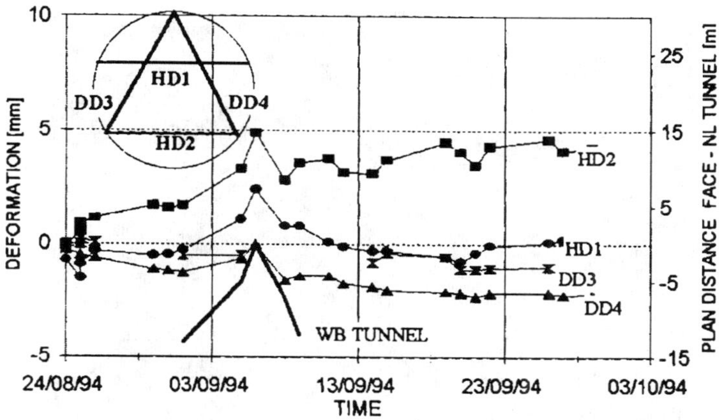

Figure 8. Deformation Trend of the New NL Southbound Running Tunnel While the Excavation of JLE Westbound Running Tunnel Under Crosses

Table 1. Summary of deflections of the SGI lined NL Running Tunnels During Undercrossing of the JLE Westbound Running Tunnel

Existing Tunnels	Total deformation when JLE Westbound undercut
Existing NL Running Tunnel (SGI lined)	
Deformation	
Horizontal	–4 mm
Vertical	8 mm
Settlement	9 mm
Existing NL Running Tunnel Northbound (SGI lined)	
Deformation	
Horizontal	–4 mm
Vertical	2 mm
Settlement	15 mm

Jubilee Line Step Plate Junction

The Step Plate junction was excavated as described in the above until NATM works were suspended and the completion of right half of the tunnel had to be ceased. A temporary headwall was installed to support the top, bench and invert heading face as well as to stabilize the temporary shotcrete middle wall.

Where the tunnel was constructed beneath a brick railway viaduct, Compensation Grouting was employed to minimize the settlement impact on the sensitive structure.

The construction performed without any problem to that date. The monitoring readings indicated deformations ranging well within the values that were anticipated during the de-

Photo 3. The 'half' finished Plate Step Junction

sign. Deformation monitoring devices were installed in order to watch the behavior of the temporary face support during the period of stand-still. Monitoring readings indicated stable conditions. The appearance of wet stains on the shotcrete support indicated the gradual reinstatement of the groundwater/porewater pressure regime and the influence of Compensation Grouting on the pore/fissure water.

Deformation of the previously constructed 11.3 m Trial Tunnel due to the construction of the Step Plate Junction stayed with maximum 8 mm horizontal deflection and maximum 12 mm roof settlement well below the expected values.

Northern Line Access Tunnels and Running Tunnel

Tunneling for the NL Access Tunnel and the new Southbound Running Tunnel progressed according to conditions expected. Overbreak and ground deflection stayed within the predicted values. The presence of old SGI lined running tunnels of the City and South London Line above the new NL Running Tunnel did not have any detrimental influence on the ground behavior.

Approaching the alignment of the planned new NL Running Tunnel, the Access Tunnel had to be enlarged to form a 'wrap around' chamber, from which adequate break out eyes could be formed for the subsequent construction of the NL Running Tunnel. The headwall of this chamber was constructed 0.5 m clear of the existing SGI lined NL Running Tunnel. The ground close to the existing SGI lining was as competent as London Clay previously encountered and did not produce significant overbreak. Some grout seams stemming from the grouting activities for the NL enlargement were found in fissures.

The SGI lining was equipped with a series of deformation monitoring pins, which were regularly read during and after the approach of the junction chamber. The roof leveling points settled continuously during the construction period of the chamber. The measuring point

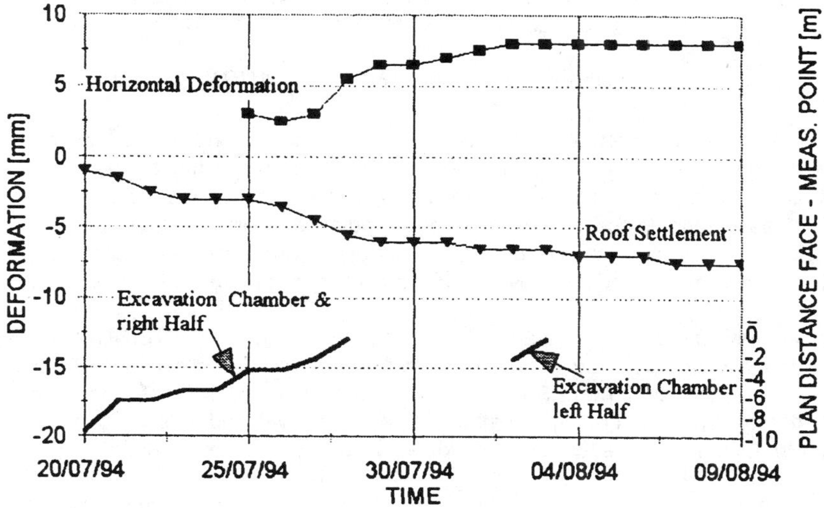

Figure 9. Deformation Trends of SGI lined NL Southbound Running Tunnel While the Access Tunnel/chamber is Being Constructed

closest to the chamber head wall indicates the maximum displacement (approximately 8 mm). The horizontal deformation followed basically a similar pattern indicating divergence with the maximum of approximately 8 mm.

The excavation works for the NL Running Tunnel induced deformation of the adjacent existing SGI lined NL Running Tunnel of approximately 4 to 6 mm in horizontal direction and roof settlements of approximately 5 to 7 mm (Kimmance, J.P. et al, 1996).

Northern Line Station Platform Tunnel

Approximately 50 % of the platform tunnel was constructed without any difficulty. The ground/support monitoring values stayed well below the anticipated values. The presence of the old SGI lined NL in close proximity did not significantly influence the ground behavior.

Deformation of the SGI lining support for the existing NL Platform Tunnel stayed below the tolerable limits.

Jubilee Line Access Tunnel to Contract 103

Construction of the adit and the junction chamber progressed without difficulties although the alignment of the tunnel follows a tight curve radius. During the break out from the head wall of the 11.3 m Trial Tunnel the ground was found not to be substantially softened by the previous Trial Tunnel Construction. Marginally more overbreak was observed over the initial few meters of tunnel construction.

AMENDMENTS TO THE GENERAL ARRANGEMENT AND DESIGN POST HEATHROW EVENT

General

The suspension period of the NATM works of almost 7 months was used for a review of the design, safety measures and general arrangement of the station tunnels. The latter effort aimed for a simplification of the layout in terms of the number of parallel and over/under crossing tunnels and, therefore, to reduce potential interaction between adjacent tunnels. Furthermore, an extensive discussion about the application and applicability of the NATM was carried out and a series of additional safety requirements were introduced.

A review of the design, construction works, monitoring data and quality test results of the shotcrete lining as well as an audit of the safety procedures in place at Contract 104 was carried out by the Health and Safety Executive (HSE). The results of these reviews and audits demonstrated that NATM construction performed well and no significant risk or even danger to the safety of the tunnels had existed at Contract 104 at any time.

Although the principles of the design of the NATM works were not changed and basically similar excavation and support techniques were used after the resumption of the NATM works, strict audits and more onerous load cases were implemented on the design. This lead subsequently to a design of a more rigid shotcrete support lining.

The Client instructed the Contractor to make provisions to change the construction method from NATM to the traditional tunneling method using a shield and cast iron segmental linings, since it was uncertain when and if at all the NATM works could resume. When the re-design for the NATM works was started, the cast iron segments were ordered and in production. Therefore, the SGI lining had to be incorporated in the NATM design as permanent support for the Station Platform and Central Concourse Tunnels.

Jubilee Line

<u>Access to Contract 103</u>: The decision to switch the construction method to the traditional tunneling method with hand mining and SGI linings required the backfill of the junction

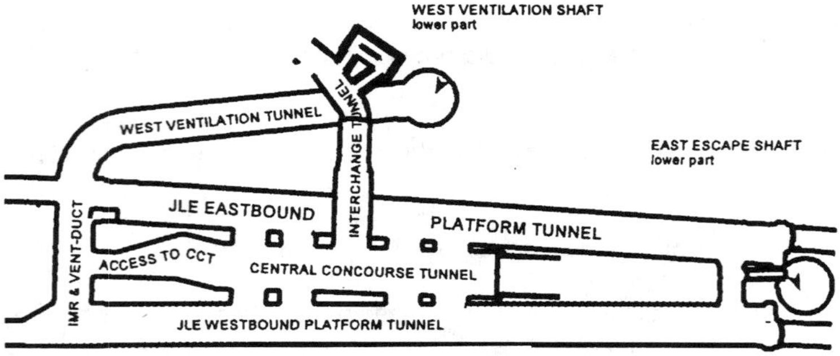

Figure 10. New General Arrangement of London Bridge

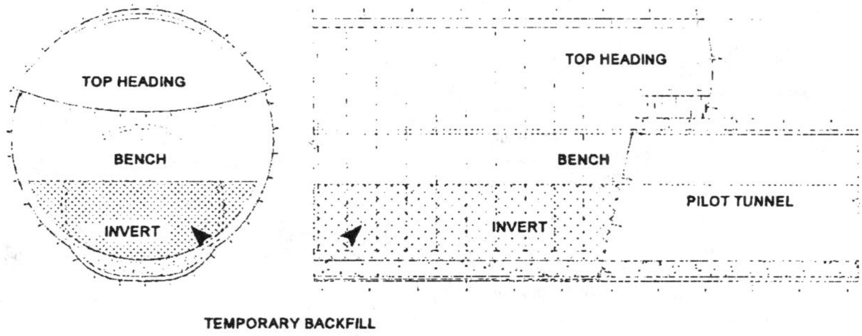

Figure 11. Excavation and Support Sequence for the Station Platform Tunnels

chamber with fill concrete and a re-alignment of the access tunnel to overcome the difficulties in terms of curve radii.

Station Tunnels: The first step toward a resumption of NATM works was made, when the permit for the construction of tunnels with running tunnel size was given. Consequently, Pilot Tunnels were driven along the planned Station Platform Tunnels, which provided not only an early access to the construction face of the JLE Running Tunnels toward Contract 105, but also additional information on the ground conditions and face support for the subsequent construction of the Station Platform Tunnels.

The Pilot Tunnels were finished before the enlargement to the full Station Platform Tunnels size commenced. Each Pilot Tunnel a concrete cradle was installed in the invert to form a sound, pre-constructed invert for the subsequently built platform tunnel. The Pilot Tunnels were supported by 200 mm reinforced shotcrete with lattice girders.

During the period of the westbound Pilot Tunnel construction the decision was made to resume NATM works based on a re-design. The Station Platform Tunnels are fully circular with an excavation diameter of 8.6 m. The initial support comprise 300 mm thick shotcrete reinforced with WWF and lattice girders. Excavation was carried out with excavators and a roadheader following a top heading, bench and invert excavation procedure.

Interlocking Machine Room (IMR): To reduce the number of tunnels running parallel and over crossing each other it was decided to combine the ventilation for both the westbound and eastbound tunnels in one scheme at the west end of the station. This required an enlargement of the IMR in height to accommodate a ventilation duct which leads the air from the westbound across the eastbound into the combined West Ventilation Tunnel. Furthermore, the previous 'Dead End Cross Passage', a small emergency cross passage at the west end of the station, was incorporated into the IMR. The IMR extended beyond the eastbound tunnels to allow the construction of an adequate break out eye for the subsequent construction of the eastbound Station Platform Tunnel.

The re-designed IMR has an excavation width of 8.0 m and a height of 8.6 m and was excavated following the standard top heading, bench and invert excavation sequence. 300 mm thick reinforced shotcrete and lattice girders form the initial support.

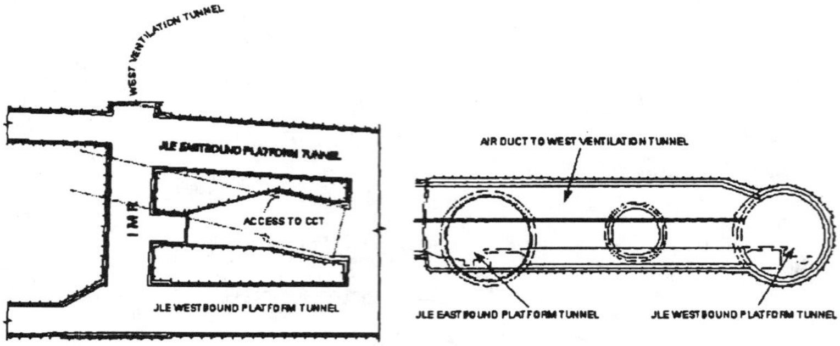

Figure 12. Plan and Longitudinal Section of the IMR

Ventilation Tunnels: Due to the above described combination of ventilation tunnels, a larger West Ventilation Tunnel had to be designed to provide the clearance area required for two horizontally separated air ducts.

The excavation area of the Ventilation Tunnel with 6.6 m width and 6.1 m height was split into the standard top heading, bench and invert excavation. The initial support comprised 200 mm reinforced shotcrete and lattice girders.

The construction of the East Ventilation Tunnel was based on a similar design but followed a far more complicated alignment. The East Ventilation Tunnel will lead the air from the JLE east and westbound Running Tunnel at the east end of the station above the JLE westbound Running Tunnel and, subsequently, down slope into the East Ventilation Shaft. Gradients of up to 15°, curve radii of minimum 7.5 m had to be handled during the construction and complicated construction sequences for the junctions to the JLE Running Tunnels had to be executed.

Central Concourse Tunnel (CCT): In order to access the work face for the Central

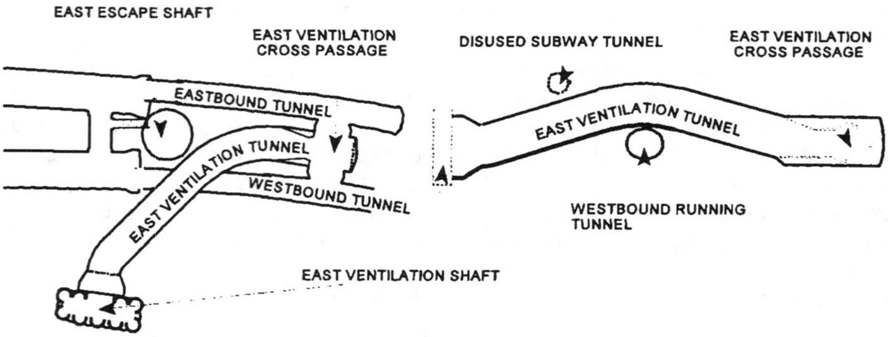

Figure 13. Plan and Longitudinal Section of the East Ventilation Tunnel

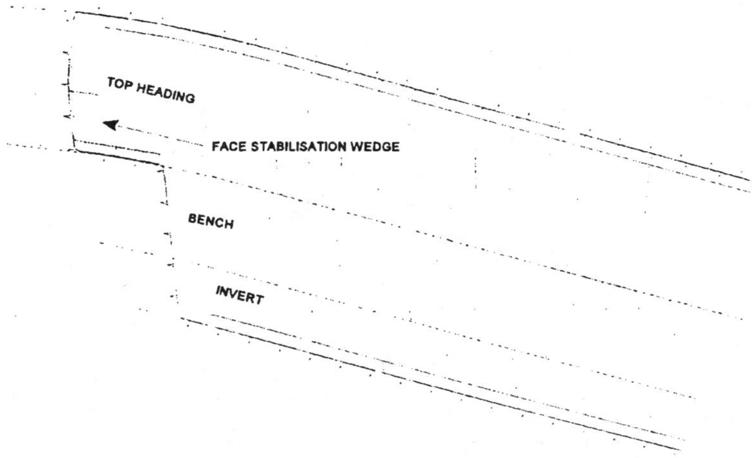

Figure 14. Excavation Sequence for the East Ventilation Tunnel

Concourse Tunnel an access tunnel had to be broken out from the side wall of the IMR. This access tunnel was located and designed such that some parts of the underground space gained by this opening could be used for the lower parts of the East Escalator. The access tunnel started with a running tunnel size and funneled up (trumpet) to the full CCT size. Before the trumpet was built, the running tunnel size access was continued to the planned east end of the CCT. Starting from there, a 6.0 m excavation diameter pilot tunnel was driven to the planned east end of the CCT and supported with steel fibre reinforced shotcrete and steel arches. By finishing the pilot tunnel the works for the enlargement up to the CCT could commence.

The re-designed Central Concourse Tunnel has an excavation diameter of approximately 12 m to encompass the 10 m ID SGI lining including the jamb frames for the cross passage openings. The excavation area was split into top heading, two bench and one invert excavation. No provisions for the 7 m wide cross passages linking to the Station Platform Tunnels were made during the construction of the CCT. These were constructed after completion of the 400 mm thick, reinforced shotcrete lining of the CCT.

Northern Line

After the suspension of the NATM works, all NL works were changed to the traditional tunneling method with hand/excavator mining and SGI linings.

Compensation Grouting above NATM Tunnels

A general exclusion zone for Compensation Grouting above any progressing tunnel face was introduced.

In order to save time and costs, the Ownert decided to install the arrays of horizontal grouting pipes above the central station area from within the existing tunnels of City and

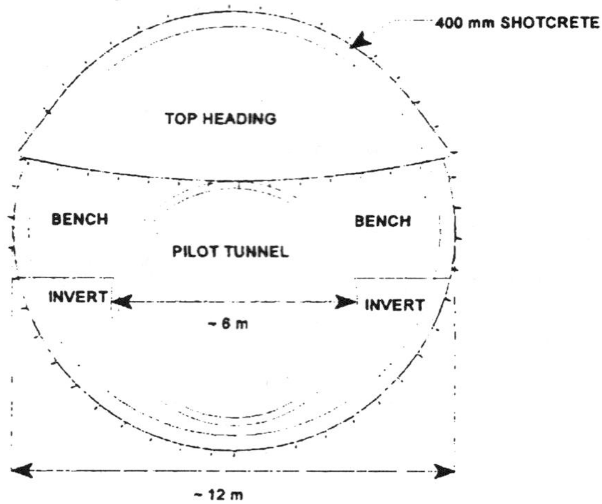

Figure 15. Excavation Sequence for the Central Concourse Tunnel

South London Line. Therefore, the sleeve pipes were located approximately 2 m above the roof of the Central Concourse Tunnel. This lead to the development of a specific exclusion zone which reached 14 m ahead and 6 m behind the leading edge of the closed invert of CCT. The exclusion zone was limited in width to 4 m from each centerline of a Station Platform Tunnel. Furthermore, a strict control procedure was introduced for the period of CCT construction, such as continuous observation of that section of the CCT was carried out by an experienced NATM tunnel engineer, above which Compensation Grouting was being performed. The decisions of this person were supported by the regular monitoring reading results and visual inspection of the tunnel linings. The person had to stop Compensation Grouting as soon as any undue occurrence was observed.

The Contractor decided to implement a 'trigger value' of 5 mm settlement. Having reached this value, Compensation Grouting was started to counteract the settlement.

FINDINGS FROM THE NATM WORKS CARRIED OUT POST HEATHROW EVENT

Restart of Jubilee line Step Plate Junction

The Step Plate Junction was finished based on the pre-Heathrow-event design. The ground ahead the face appeared slightly softened due to an increase of the water content, but provided sufficient stand-up-time to safely excavate and support the individual rounds. No additional support measures were required for the construction.

Jubilee Line Pilot and Running Tunnels

During the construction of the Pilot and Running Tunnels basically similar conditions were found as during the previous construction activities. The ground became more blocky toward the east end of the contract approaching a fault zone. This lead to increased wedge failures, mainly occurring in the lower part of the face (Groves, P.N. and Morgan, S., 1997). The ground at the tunnel perimeter was kept sufficiently under compression, such that wedge failures did not occur in the roof or side wall area. The increasing silt/sand content of the ground and some silt lenses were combined with small groundwater inflow at locally confined areas and discrete fissures.

Tunnel deformations were similar to those previously experienced at similar tunnels at Contract 104.

Monitoring readings taken at the NL tunnels indicate similar interaction as experienced during the construction of the JLE Westbound Running Tunnel.

Jubilee Line Station Platform Tunnels

For the enlargement from the Pilot Tunnels to Station Platform Tunnels backhoes and a road header was used. The shell of the Pilot Tunnel was removed during the procedure of bench excavation. The Pilot Tunnel, being located in the face, formed on one hand an obstacle for the excavation. In particular during the bench excavation, it reduced the room to maneuver at the face significantly. On the other hand, it reduced the ground surface area in the face significantly and contributed, therefore, to the face stability. Undue deformation or any sign of distress of the Pilot Tunnel shotcrete shell ahead the advancing pilot tunnel face was not observed during the construction period.

Tunnel deformation of the shotcrete shell for the Station Platform Tunnels stayed well below the predicted values.

Surface settlement speed above the tunnel markedly increased when the advancing tunnel face was within a zone of approximately 18 m plan distance from the monitored point. Also a significant delay of the cease of surface settlements in these areas was detected, where the ground was treated with Compensation Grouting. Compensation Grouting was able to compensate the surface settlement induced by subsurface activities and to keep the settlement to the stipulated values. It was also noticed that during a period of stoppage of tunneling surface settlement decelerate but they do not cease, although the tunnels were adequately supported and a temporary shotcrete face support was installed. The face was stepped, consistent with the excavation sequence. Only the fully domed headwall at the end of the tunnel lead to a prompt halt of the surface settlements in ungrouted areas (Zeidler, K. et al, 1997).

Jubilee Line Central Concourse Tunnel (CCT)

The Station Platform Tunnels were lined with SGI linings prior to the commencement of the enlargement for the CCT.

The ground conditions encountered during the construction of the CCT were very similar to those described above. Only at the west end of the CCT a of soft, water saturated silt and sand layer was excavated in the invert area. This material appeared soft or even running when disturbed. The geological feature - most likely a relict channel feature within the Woolwhich and Reading Beds overlain by Blackheath sands and localized limestone depositions- was locally limited and did not require a change of the overall support sequence.

For the subsequent 3 meters the invert closure was installed earlier than required by the design.

The CCT shotcrete lining was equipped with 10 convergence monitoring cross sections. Also, the SGI linings within the Station Platform Tunnels were monitored as to their deformation due to interaction between the tunnels and equipped with convergence sections in 5 m intervals.

Where any ground remained between the CCT and the Station Platform Tunnels, the rather slender ground pillars did not indicate any sign of distress and did not develop undue strain or cause any significant load upon the shotcrete lining. They appeared to carry only minimal ground load, what indicates that a load carrying ground arch spanning across the entire station scheme (more than 29 m) had been generated ahead the advancing face. Monitoring data seem to support this since absolute measurements within the Station Platform Tunnels indicate that the extreme lateral sidewalls tended to deflect inwards as opposed to the sidewalls toward the CCT. The platform tunnels generally squat slightly. This trend is not surprising, since the ground pillar and the ground toward the advancing excavation must be softer and therefore provide less bedding to the platform tunnels' sidewall than the 'infinite' ground at the lateral sides of the station. However, the ground pillars did not display observable yielding, neither during the 'open' unsupported period nor later behind the shotcrete support generating increasing ground loads on the lining (CCT and platform tunnels). Being confined between the linings the ground would gain strength, if it would be re-loaded, and subsequently, it would increasingly transfer ground loads onto these linings, which would be reflected in a reversal of the deformation trend of the linings. However, this trend reversal was not observed during several weeks after construction.

The shotcrete lining of the CCT showed small deformation during the period of construction only, which converged to zero, once the shotcrete support ring was closed. Stress cells indicated a confinement stress level within the shotcrete shell, which lied far below the characteristic strength required by the design.

Table 3. Summary of Convergence Data from Station Tunnels during Construction of CCT

Tunnel Measured Cord	HD1 [mm]	D1 [mm]	D2 [mm]	HD2 [mm]	D3 [mm]	D4 [mm]	D5 [mm]	D6 [mm]
Westb. Platform Tunnel maximum average maximum	 - 2.9 - 0.9	 - 1.3 - 0.4	 - 13.1 - 5.2	-	-	-	-	
Central Conc. Tunnel average	0.8	0.7	0.2	3.6	2.2	1.1	2.3	4.0
Eastb. Platform Tunnel maximum average maximum	 - 1.9 - 0.4	 - 10.7 - 4.9	 -1.7 - 0.6	-	-	-	-	

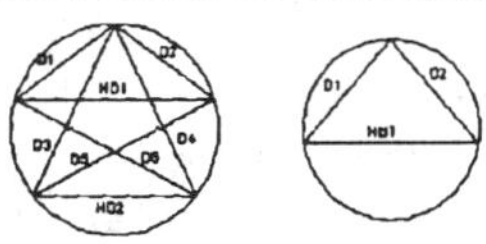

Photo 4. Construction of the Central Concourse Tunnel

Jubilee Line Cross Passages

The cross passages between the station tunnels were finished after the completion of the NATM works for the CCT and did not induce any tunnel deformation of any tunnel or surface settlements.

Jubilee Line Ventilation Tunnels

The West Ventilation Tunnel was constructed without any significant influence on the eastbound Station Platform Tunnel. Ground behavior and lining deformation was very similar to that experienced during the construction of the Interlocking Machine Room.

The more blocky ground facies and the complexity of the East Ventilation scheme involved increased attention to the design and its execution. Complicated lattice girder geometries resulted from the gradient of the tunnel and the requirement to keep the tunnel face vertical rather than perpendicular to the tunnel axis. The complex procedure for setting out the lattice girder on site extended the period between excavation and installation of the shotcrete support lining. A layer of 50 mm thick unreinforced flash shotcrete provided sufficient short term support to increase the stand-up time of the ground.

However, surface settlements induced by the construction of the East Ventilation Tunnel were not excessive and did not require additional grouting measures.

CONCLUSION

Experiences gained during the construction of the Jubilee Line London Bridge metro Station have demonstrated that

- adequate design,
- appropriate application of the design,
- good workmanship and

- skilled as well as experienced construction site supervision on both, the Client's and the Contractor's side

are essential for the successful execution of a contract.

Furthermore, NATM does not involve more risks in tunnel construction than any other modern tunneling method, given above principles are observed.

During construction interactions between adjacent tunnels are existing, but an adequately designed shotcrete support shell is able to handle these influences. An adequate design requires the thorough assessment of all likely ground conditions and risks, as well as adverse site conditions such as varying material and equipment quality.

Observations made during the construction of London Bridge Station in London Clay have tought that slender ground pillars between adjacent large diameter underground openings do not carry significant ground loads during the construction period. A load-carrying ground arch spans over the entire scheme that transfers ground loads to the lateral sides of the scheme. As applicable for single tunnels and for systems of parallel tunnels the shedding of the ground load also takes place ahead of the face, so that a three-dimensional, almost spheriodal space of reduced ground loads is created. This 'ground self-support' ahead the advancing tunnel face is the reason for the freshly excavated round always being located under the protection of the support provided by the ground itself.

REFERENCES

Kimmance, J.P. et al, 1996: Observations of deformations created in existing tunnels by adjacent and cross cutting exavations. In: Proc. Int. Symp. Geotechnical Aspects of Underground Construction in Soft Ground, Mair & Taylor (eds), p 702 -712

Kimmance, J. P. and Allen, R., 1996: The NATM and Compensation Grouting Trial at Redcross Way. In: Proc. Int. Symp. Geotechnical Aspects of Underground Construction in Soft Ground, Mair&Taylor (eds)

Way, A. 1996: JLE Contract 104: all change at London Bridge. In: Tunnels & Tunnelling, November 1996, p 52 - 56

Groves, P.N. and Morgan, S., 1997: Engineering consequences of ground conditions on NATM works at London Bridge, Jubilee Line Extension. Proc. For the Eighth International Symposium, Tunnelling '97, in print.

Zeidler, K. et al, 1997: Surface Settlement due to the Construction of the London Bridge Station Tunnels by NATM. In: Proc. For International Tunnelling Conference 1997, Vienna, in print

Chapter 40

HEATHROW EXPRESS — NEW RAIL TUNNELS UNDER LONDON'S MAIN AIRPORT

A.P. Deane,[1] A.G. Myers,[2] and G.C. Tipper[3]
[1]Director, Tunnel Division, Mott MacDonald, U.K.; [2]Construction Director, Balfour Beatty Civil Engineering; [3]Construction Manager, Heathrow Express

SCHEME DESIGN

London's Heathrow Airport, as the world's busiest international airport, is missing one vital aspect - a dedicated rail link which connects directly to the city. All this is about to change with the construction of Heathrow Express - a fast, high quality rail service between the airport and the centre of London.

It was recognised in the mid 1980s that if Heathrow was to accommodate the forecast growth in passenger numbers and the resulting increase in road traffic, a dedicated airport rail service would be vital.

The main access into the centre of Heathrow is via a road tunnel with a direct link to the M4 motorway. The airport has introduced a number of traffic management schemes, but these can only achieve so much benefit. Growth in air passenger traffic is predicted to continue at an average of 4% per annum well into the next century and it was clear that significant investment was needed. Heathrow Express at a cost of over £350 million will provide this when it starts its service in June 1998.

The business case for the scheme is based on a ridership of around 15% of air passenger traffic. The initial Heathrow Express service will be based on four trains per hour, up to 8 car train lengths, providing a high speed non-stop service between the airport and central London in just 16 minutes.

Initially a joint venture between the British Railways Board and Heathrow Airport, by mutual consent of the two parties, Heathrow Airport recently purchased the British Railways Board share making this one of the largest private sector rail projects currently under construction in the UK.

Leaving Paddington, the main rail terminus in West London, Heathrow Express trains will travel 12 miles on the newly electrified line to airport junction, north of the airport. At this point they will leave the mainline to enter a 5 mile tunnel section that runs underneath the heart of the airport.

The railway companies are responsible for track work, signalling and electrification of the 12 mile main line section including a grade separated junction over the main tracks for the Great Western Railway which links London and South Wales. Heathrow Airport is re-

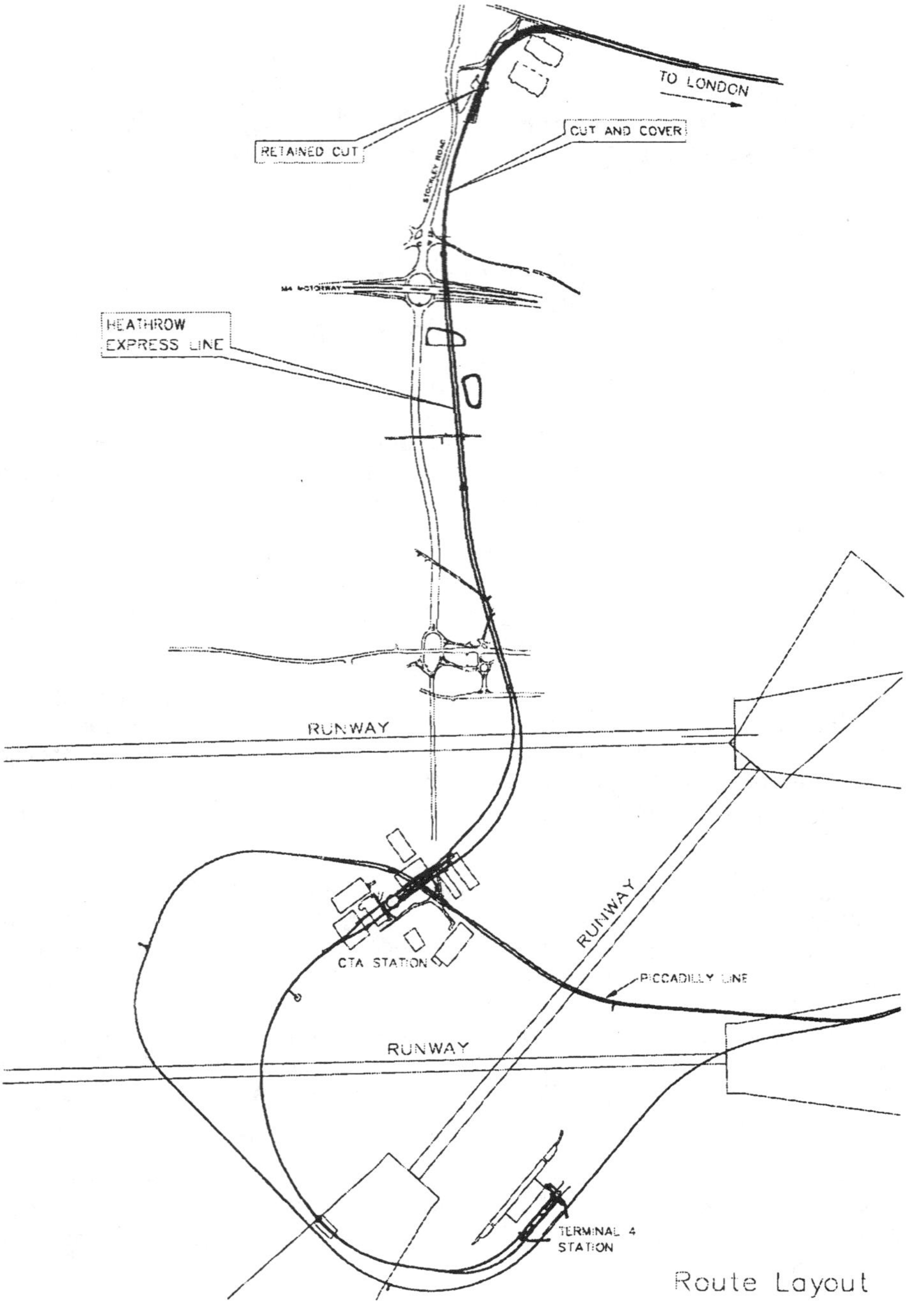
TO LONDON
CUT AND COVER
RETAINED CUT
HEATHROW
EXPRESS LINE
RUNWAY
RUNWAY
CTA STATION
PICCADILLY LINE
RUNWAY
TERMINAL 4
STATION
Route Layout

Figure 1.

sponsible for the 5 mile underground section from airport junction to Terminal 4 via the central area of the airport.

Heathrow Airport set up their own project team to manage the construction of the new works.They in turn appointed Mott MacDonald as their lead designer.

This part of the route commences with a 500m. long section carrying the tracks in a twin cell cut and cover reinforced concrete box which passes through an old urban refuse landfill site last used in the seventies. The line then changes to bored tunnel to go under Shepiston Lane, a local road carrying heavy commuter traffic and with sensitive services in the verges either side.

Approximately 100m south of Shepiston Lane there is the first emergency intervention shefl. From here twin 5.7m. diameter running tunnels take the track into the Central Terminal Area (CTA) with two further intermediate intervention shafts along the route. The first of the two major stations is constructed under the buildings and roads of the CTA.

South of this station there are major junction tunnels to allow the railway to be extended to the future Terminal 5 west of the airport. Then the two running tunnels converge and a single tunnel runs to a station at Terminal 4 on the south side of the airport with two further intermediate shefls on this length.

The geology over the airport area is very uniform. In outline the succession is approximately two metres of made ground overlying 4 to 6 metres of water-bearing gravel. Beneath the gravel is approximately 50 metres depth of London clay. Whilst the top 1.5 metres of the clay is weathered, below this the clay is a uniform stiff gray-brown material with undrained shear strength increasing with depth. At the depth of most of the Heathrow Express tunnels the strength is approximately 70 kN/m^2. As well as being relatively strong and uniform the clay has a very low permeability making it a very attractive tunnelling material.

TUNNEL DESIGN

The design for the running tunnels uses an expanded precast concrete lining of six segments plus a shorter wedge key which is pushed into place by a ram on the tunnelling machine expanding the lining against the ground. The key is situated at tunnel axis level to reduce the problems which can arise in filling a key void in the crown.

Although considered, cut and cover construction for the stations was rejected early in the design process because of the disruption it would cause to airport operations.The stations were, therefore, constructed in bored tunnel.

The CTA station shown in figure 2 has two 220m long plafforms with a central 300m long concourse providing access via iifls and escalators to Terminal 1 at the north end and to Terminal 3 at the south. Ventilation shefls are provided at each end of the station together with intermediate emergency escape shelves.

The Temminal 4 station has two 220m plafform tunnels with an 80m concourse tunnel and similar provisions for access and ventilation.

Both station layouts were principally determined by the need to provide good connections to the airport terminals, but there were also major restrictions on where shafts could be situated in the highly built up airport area.

Initial planning of the station complexes was based on the use of segmental linings for the tunnels. As part of the constructability and cost reviews carried out during the design process, the use of primary shotcrete support, often referred to as NATM, was considered in the more complex areas. Although this system has a track record in sofl ground, there had been very limited use in the United Kingdom.

However detailed study of tunnels in Frankfurt, Tokyo, Sao Paulo and elsewhere suggested that NATM was a feasible option for Heathrow. Together with the technical review, a cost study was undertaken which showed that substantial savings could be achieved in the time for construction and therefore in the costs of the station complexes by the adoption of NATM in place of the segmental linings.

The main advantages of the NATM support system are the ability to construct non-circular tunnels, the flexibility to allow changes in diameter to be made relatively easily, and the three dimensional action of the lining which assists in the construction of the larger openings between tunnels which are often required by the operators of modern railways. Also the construction plant required is more standard and therefore more easily available than the machines used for large segmentally lined tunnels.

To support the decision to utilise NATM on this large scale it was decided in 1992 to construct a trial tunnel using the technique. A 100m long tunnel with the same cross section as the station plafform tunnels was excavated. It was positioned along the line of the running tunnel between CTA and Terminal 4. This trial tunnel was extensively instrumented, particularly with regard to surface and subsurface ground movements. No significant problems were identified in its construction and surface settlements were similar to or less than those that would have been expected from a similar size of segmentally lined tunnel.

Even with the use of NATM construction of the large tunnels proposed at Heathrow would inevitably cause ground movements which could affect the airport infrastructure above. Prior to the award of the construction contract a review was undertaken of every structure and utility above the area of the tunnels. Allowable settlements and distortions were established for each item. These were set out in the construction contract to ensure that potential damage was minimised.

To ensure that these specified settlements were not exceeded, it was decided to employ compensation grouting in a number of areas. This technique consists of predrilling arrays of tube-a-manchette grout holes above the area to be excavated. Before tunnelling stafts, grout is injected under pressure until a small amount of heave is observed on the structure requiring protection. As tunnelling proceeds further grout is injected to compensate for the setleement taking place and ensuring that the movement of the structure remains within the predetermined limits.

CONSTRUCTION CONTRACT

The construction of the railway was split into a number of packages with the tunnelling contract including the stations awarded to Balfour Beatly Civil Engineering with a start date of February 1994. The remainder of this paper focuses on this contract.

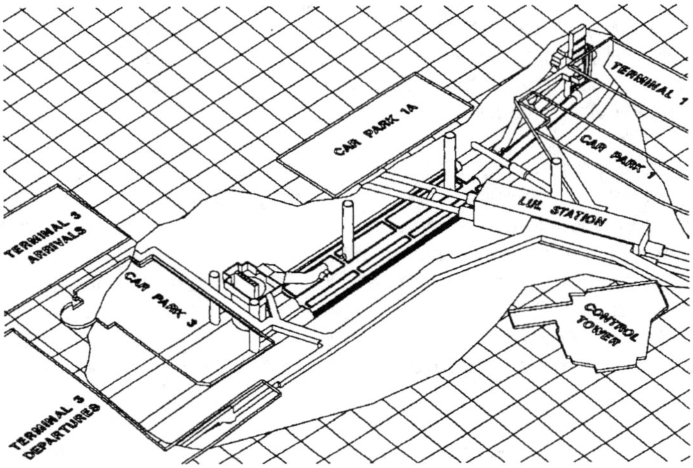

Figure 2.

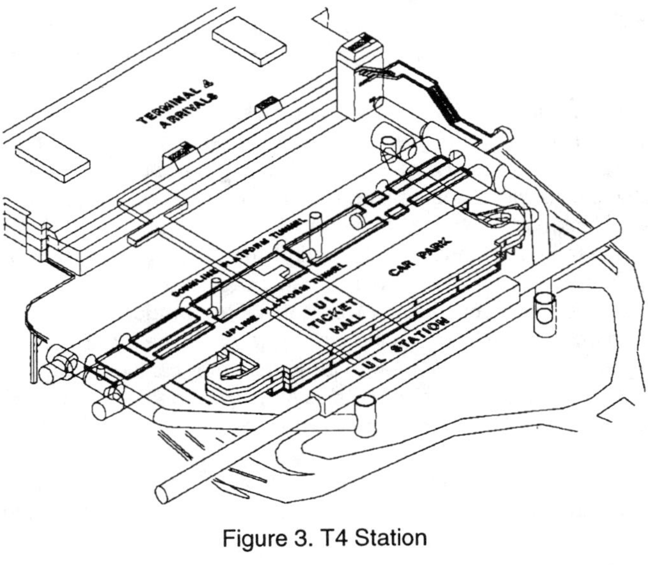

Figure 3. T4 Station

A main work site was available near the north end of the works. Here Balfour Beatty elected to combine two circular escape shafts into a single elliptical shotcrete lined shaft from which the main running tunnels to the CTA were driven. A conveyor system was then installed from the shaft bottom across Shepiston Lane to transfer the clay excavated from the tunnels to the cut and cover site to be used as backfill. The running tunnels were driven sequentially using an open face tunnelling machine. Advance rates of over 180 metres per week were regularly achieved with a maximum of 40m in one day.

An area was available for a construction site at the CTA, which had been occupied previously by the airport fuel companies. A 20m diameter segmental lined shaft was constructed on this site under an advance contract. This alowed an early start to the excavation of the concourse and plafform tunnels for the station. As noted above, these were constructed using NATM techniques.

The design approach for the NATM tunnels was based on a two layer system; a shotcrete primary support with an intemal insitu concrete lining. The design assumed that long term loads on the tunnels would be carried only by the insitu concrete lining and the shotcrete would, therefore, be 'temporary', although left in place at the end of the construction period rather than removed. For shotcrete lined tunnels design and construction are part of a single integrated process. It was therefore decided that the contractor would be responsible for the design of the temporary shotcrete with the client's designer retaining responsibility for the permanent insitu lining.To carry out the design of the NATM the contractor retained Geoconsult of Salzburg.

On the night of 20th/21st October 1994, there was a major collapse of the three 9m wide shotcrete lined tunnels being constructed for the CTA station adjacent to the 20m access shaft.

A short time prior to the colapse significant spelling and cracking of the shotcrete was observed by the people working underground and the tunnels were evacuated. Due to the prompt action by well trained staff on duty, nobody was injured by the failure.

The collapse resulted in a large depression at the ground surface, almost totally within the construction site boundary. However, until the full extent of the failure could be confirmed, some roads and a section of the London Underground Line which serves the airport were temporarily closed with some consequent minor disruption to the airport. A significant length of tunnel, which had been excavated and lined using the same NATM techniques as the failed section, remained standing. In the period immediately following the collapse, however, the cause of the failure was not clear. Given the possibility of very major disruption to the airport if further failures took place, it was considered prudent to ensure the stability of the remaining shotcrete lined tunnels in the area. This was done by filling them with concrete through existing access points and through large diameter casings installed by a piling rig. The method adopted was to form 'bulkheads' of structural concrete at points close to where the tunnels had failed and then fill the remainder of the tunnels with foam concrete.

Immediately following the collapse the U.K. Government's Health and Safety Executive, requested a voluntary suspension of NATM tunnelling in the U.K. until all parties could be satisfied as to the safety of the system.

THE CHALLENGE

The project then faced a number of challenges, particularly

- to plan a way forward for the construction; specifically to establish procedures which would satisfy all parties concerned as to the safety of the construction methods to be used,

- to establish the financial basis on which construction was to proceed,
- to reduce to an absolute minimum any delays to the project opening date.

The team set up to review the construction included representatives from the client Heathrow Express, the contractor Balfour Beatty, the designers Mott MacDonald and the Project Insurers. A full review of all aspects of the work was carried out and the following were the principal conclusions of the team

- the station at CTA should be constructed essentially where it was originally planned. However, the shotcrete lined tunnels would generally be replaced by segmentally lined tunnels. This decision was driven principally by programme requirements. The CTA station had always been on the critical path for the project. Given that the cause of the collapse had not been established it was possible that there could be a significant delay while all parties were satisfied as to the use of NATM in this sensitive area.
- the area of the main collapse consisted of highly disturbed clay where tunnelling would be extremely difficult. It was therefore decided to surround this area of disturbance with a 60m diameter cofferdam constructed from secant piles and remove all the disturbed ground within it.
- in other major areas where NATM support was originally proposed, principally Terminal 4 station and the three large junction or "tumout" tunnels, shotcrete would continue to be used. There would, however, be a full review of the NATM design and construction and, if necessary, modifications made.

These decisions required a significant amount of revised design which would need to be carried out while construction was continuing. In order to achieve the opening of the railway as quickly as possible, there was a need for al parties involved to co-operate. The structure of the contract was, therefore, modified to form a single team, where the senior manager from the contractor, the client and the designer reported to the project director, forming a single team within the project. The revised agreement was based on a tight construction programme and it was, therefore, essential that the designer was a part of the team to ensure that there was no delay in the supply of the design information through what hac /ecome essentially a "fast-track" programme.

As noted above NATM had been chosen originally because of its economy and its flexibility in complex areas. These advantages still existed. Coupled with this, in the Terminal 4 station the two main plafform tunnels were already under construction using this method. To change to the only realistic alternative of segmental construction would have involved major delays and the excavation of large chambers for tunnelling machines with the potential to cause settlement and possibly damage to the Terminal buildings above them.

Having decided to proceed with shotcrete support, a full review of the design and construction methods was undetaken including extensive risk analyses and planning of contingency measures. In some areas, for example, the major turnout and crossover tunnels where tracks merge, these reviews confirmed that NATM was a safer option than a conventional segmental approach which would have required handbuilt rings up to 15m. diameter.

With the Health and Safety Executive investigation on-going and with the recent high profile failure at Munich, as well as the Heathrow colapse, there were potential problems and delays in proposing to continue with shotcrete lining for the tunnel primary support. As well as the HSE, a number of parties including London Underground, the airport operators and the project insurers all had a major interest in any proposal to use a tunnelling method over which questions had been raised.

It was essential therefore that the design and construction procedures should be clear and auditable. The design of the NATM tunnels was carried out using hand calculations supported by extensive numerical analysis. The approach was to confirm that for the predicted range of loading conditions, and allowing for ground-lining interaction, the support system satisfied the required axial and bending moment capacities during the various stages of construction.

For the majority of the tunnels two dimensional analyses were carried out using the FLAC finite difference program. The clay elements were modelled to follow a variable stiffness/strain law which depends on applied load stress history and boundary conditions. This ground model is supported not only by extensive testing of the clay but also by back-analysis of a number of instrumented tunnels constructed in London.

The construction sequence both for the individual tunnels and the tunnel complexes was included in the analysis. To allow for the three-dimensional behaviour that occurs as the tunnel is excavated a Hypothetical Modulus of Elasticity (HME) was used for the shotcrete. This approach utilizes a low modulus for the shotcrete during the analysis of the first stages of the construction sequence. The HME is derived from the back-analysis of measurements on previous tunnels. The Heathrow trial tunnel had been extensively instrumented and provided high quality information for the derivation of the HME employed for the main design.

For the major tunnel junctions three dimensional modelling was employed using the ADINA finite element package. However because there are only limited measurements taken from previous tunnels to confimm the shorttemm three dimensional behaviour of the shotcrete these analyses were used only as a back-up to simplified hand calculations of the stress distribution in the junction, and not as a main design tool.

As well as the normal ground loadings the linings, including the partially complete shotcrete linings, were checked for loadings due to compensation grouting. Aithough theoretically the grouting pressure cannot be greater than the overburden bad the effect of local grouting was found to be significant. Limits on grouting pressures and volumes were put into the site operating procedures.

For the Terminal 4 station where three tunnels were to built in close proximity an exceptional load case of the effect on the adjacent tunnels of the collapse of one tunnel was also considered.

The review of the design resulted in the following changes to the NATM which have the effect of providing a fully engineered design to load factors usually associated with permanent works

- the cross-sections were made more nearly circular, particularly in the invert;
- lattice girders were used around the whole section rather than just in the top half of the tunnel;
- load factors for full overburden were increased from 1.0 to 1.4.

In parallel with the design check all the construction procedures were reviewed.This resuted in the following improvements

- retraining took place of all the staff and operatives involved in the NATM construction together with the introduction of detailed check sheets and hold points in the construction process to ensure that the highest quality was maintained at all times
- senior design staff were located on site both to assist in training and quality control for the NATM and to deal immediately with any design questions which arose.

Initially the review of the designs for the NATM tunnels was to be carried out by Geoconsult based on these revised criteria and followed by independent checks by a team of designers from Mott MacDonald. These checks included all aspects of the design and construction sequence including extensive studies on the stability of the excavated face. As the work progressed it was decided that a more rational approach would be for Mott MacDonald to take responsibility for the design with Geoconsult acting as their subconsultant.

NATM tunnelling restarted in September 1995 at Temminal 4 station, with the other main areas following at agreed intervals over the next six months. All tunnel excavation was complete in February 1997.

The architectural concept for the passenger tunnels in the Terminal 4 station was to leave the insitu concrete finish exposed. The design originally utilized unreinforced insitu concrete linings wherever possible with a waterproof membrane between the concrete and the shotcrete to prevent water staining in the event of local cracking.

However the revised design parameters resulted in a construction sequence for this station where the two platform tunnels needed to have their final insitu linings installed before the concourse was excavated between them. In turn this required the lining to be reinforced because of the deformations which were predicted to occur as the concourse was excavated. Use of a waterproof membrane meant that this reinforcement needed to be self supporting over 9m spans. To achieve this a scheme utilising fabricated lattice girders within the reinforcement cage was developed.

The three major junction tunnels have been constructed using the same techniques. These caverns are of constantly varying cross-section and at up to 15m wide are the largest tunnels built to date in London Clay. Extensive monitoring has demonstrated that the structures behaved as predicted in the design. Stresses and movements in the linings and the ground were within acceptable levels and stabilised rapidly following the construction of any particular closed section.

Aithough in areas of the station which will be used by passengers on foot the NATM tunnels have a secondary lining of insitu concrete, in other locations, particularly the ventilation tunnels and the large junction tunnels which have a continuously varying cross-section a permanent lining of shotcrete has been used. A wet shotcrete mix utilizing steel fibre reinforcement and bw levels of additives was developed and tested. This has been applied using a robot arm mounted on an excavator to produce linings up to 400mm thick.

In the area of the main colapse it was therefore decided to construct a 60m diameter circular shaft enclosing the majority of the collapsed area. The temporary walls of the shaft were formed of bored cast insitu piles. The piles were at approximately one metre centres

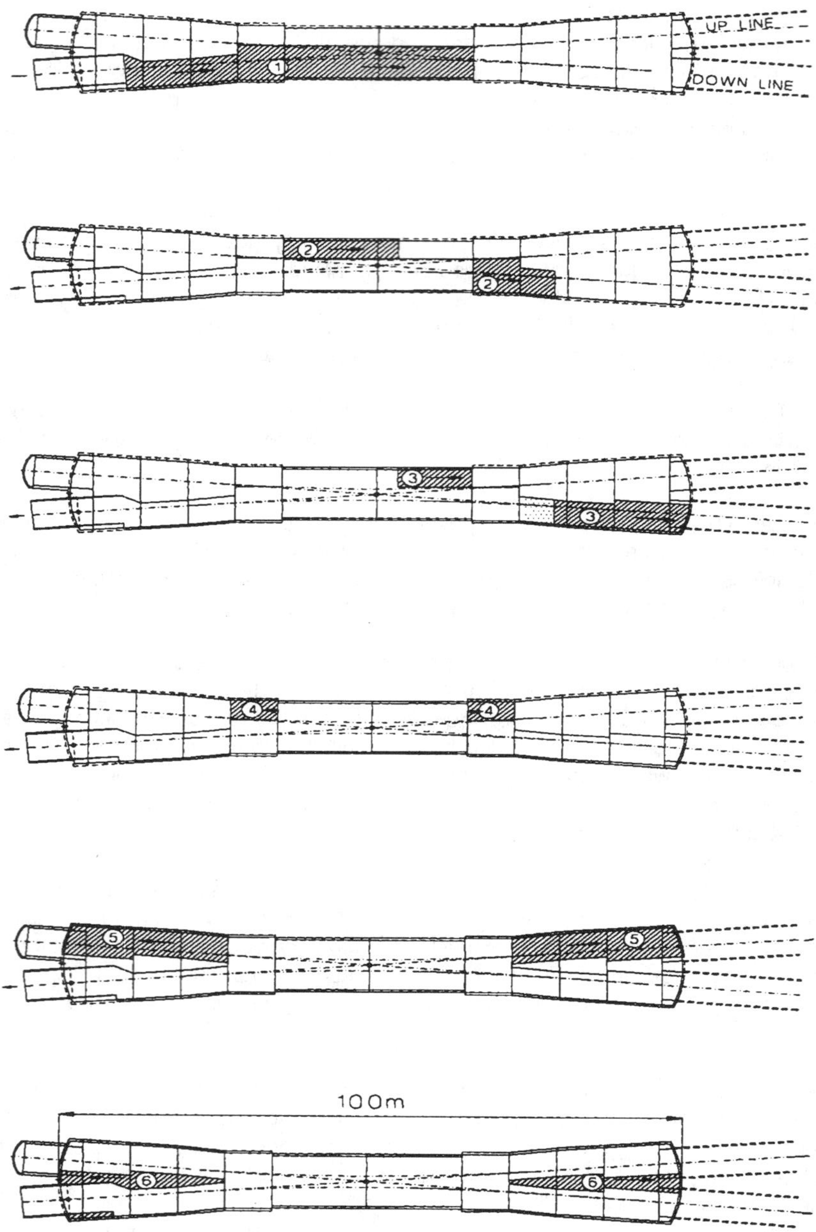

Figure 4. Sequence of Construction of the Crossover Tunnel Restart of Construction

around the perimeter. For the first 20m the piles were 1.2m in diameter resulting in a secant wall. Below this level the diameter was reduced to 0.9m producing contiguous piles for a further 20m. As well as providing a structural wall to allow excavation to proceed safely the piles were to provide a cut-off to prevent water from the gravels entering the excavation.

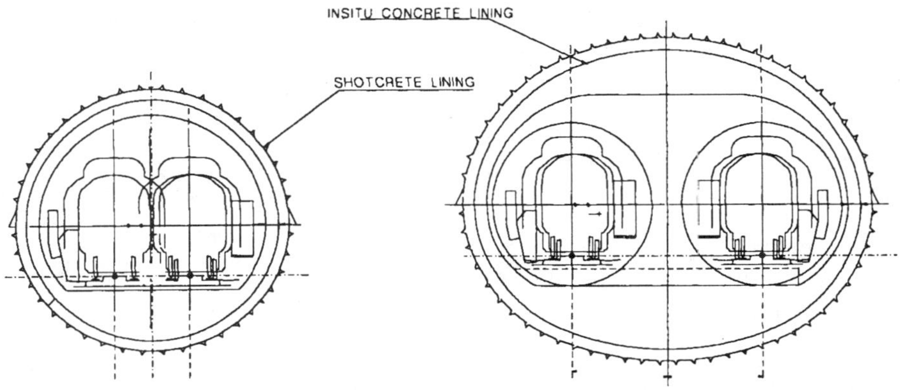

Figure 5. Narrowest and Widest Sections of Crossover Tunnel

Although an extensive site investigation of the collapsed area was undertaken there were major uncertainties as to how much support this ground would give to the piled wall. The method of construction chosen therefore was to excavate each 1.2m stage of the 60m. shaft and cast the 1m thick wall in a complete ring before the next excavation step was started. An extremely efficient construction system was developed for this work and very quickly the programme reduced from the original estimate of a 7 day cycle to less than 4 days for each lift.

The majorily of the remainder of the station at CTA was built using segmental linings. Nine metre diameter shields were installed in access shafts north of the cofferdam to construct the station tunnels. The contractor's staff were fully involved in the design of these machines and when a second machine was ordered after the start of the first drive no modifications required to the design. The length of concourse tunnel which had been excavated prior to the collapse and then filled with concrete was re-excavated using pneumatic breakers and lined with cast iron segments. Two temporary shafts were sunk to provide access for this work. Because of a constricted site one of these shafts was started at ground level at 9.5m diameter and belled out to 14m at tunnel level using a shotcrete lining.

An innovative form of temporary works was used to construct the openings in the segmental plafform and concourse tunnels. Instead of connecting the temporary lintel beams to the lining to allow the segments for the opening to be taken out, a ring of shotcrete was sprayed inside the final lining against a hessian layer to facilitate later removal. This scheme significantly simplified the construction of the openings.

The construction is still meeting the ambitious targets set after the tunnel collapse with the first train timetabled to run in June 1998 only six months later then the original programme. This achievement is due largely to the spirit of co-operation of all the staff working on the site coupled with the commitment of the client, contractor and designers to the unique "single team" concept.

13

International Projects II

Chairman
T.P. Smirnoff
Parsons Brinckerhoff Quade & Douglas
Los Angeles, CA

Chapter 41

DESIGN AND CONSTRUCTION OF THE NEW TUNNEL AT GENTING SEMPAH, MALAYSIA

S.J. Porter,[1] M.T. McRae,[2] and R. Wright[3]
[1]Hyder Consultants, Australia; [2]Jacobs Associates, El Segundo;
[3]Transfield Tunnelling, Malaysia

ABSTRACT

The new tunnel at Genting Sempah is part of a privately funded infrastructure project that involves duplicating the existing highway extending from the Gombok Toll Plaza near Kuala Lumpur to Karak, in Malaysia. The 60 km long tollway includes an 800 m long, two-lane highway tunnel with excavated dimensions of 10.5 m wide by 8 m high. The majority of the tunnel extends through slightly to moderately weathered, moderately to highly jointed quartz porphyry. Excavation of the tunnel was performed using drill and blast methods and the primary tunnel support consists of rock dowels and shotcrete except for short sections adjacent to the portals where steel sets were utilized. A portion of the tunnel encountered sheared and altered material that was successfully supported using rock dowels and shotcrete. The final lining for the tunnel consists of mesh and fiber reinforced shotcrete. The flexible nature of the primary support and shotcrete final lining resulted in considerable cost savings and expedient construction methods.

INTRODUCTION

The Kuala Lumpur-Karak Highway is located northeast of Kuala Lumpur in central Malaysia. An existing two lane highway is being duplicated by private investors under a Build-Own-Operate-Transfer (BOOT) contract with the Malaysian Government.

The Kuala Lumpur-Karak Highway is approximately 60 km in length. Mid-length it passes through the Genting Highlands - an area famous in that part of Asia for its casino and resorts as well as its natural beauty. The highway crosses the highlands at Genting Sempah ("Sempah" meaning saddle). A two lane bi-directional tunnel, approximately 900 m in length passes through the saddle. As the highway is being duplicated to provide two lanes in each direction it has been necessary to construct a second tunnel, parallel to the existing alignment through the saddle. Refer to Figure 1 for a site plan. At the time of preparation of this paper the excavation and lining of the tunnel were complete and fit-out was in progress.

MTD Prime, a Malaysian public company has the "Concession" from the Government to build and operate the toll road. They will own the toll road during the "Concession Period", after which time ownership will transfer to the Government. MTD Construction has a contract with MTD Prime to design and construct the highway duplication. MTD Construction have in turn contracted Transfield Constructions, an Australian company with locally owned operations in Malaysia, to design and construct the new tunnel at Genting Sempah due to their expertise in tunneling. The construction cost of the tunnel is approximately U.S. $12 million.

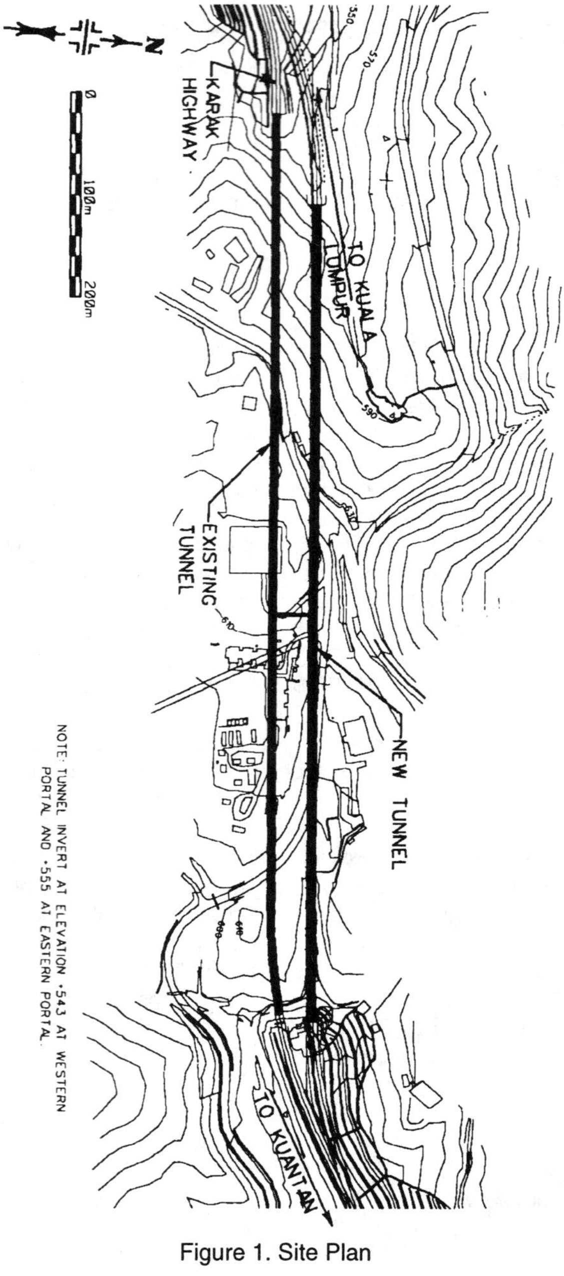

Figure 1. Site Plan

Transfield engaged Hyder Consulting in Australia to undertake the design of the tunnel, including the geotechnical investigations, temporary and permanent tunnel support and mechanical and electrical services. Jacobs Associates provided specialist expertise in the design of the tunnel support.

GEOLOGY

Geologic Setting

The tunnel is located below the Genting Sempah Pass, a low point of the Genting Chain which is a major range of mountains trending from the northwest to southeast and delineating the watersheds of the Gombak River to the west and the Kenyol River to the east. The slopes in the vicinity of the tunnel are very steep with gradients ranging from 30 to 70 percent. There are few natural outcrops in the area of the tunnel due to the 2 to 5 m cover of lateritic soils over much of the area. The combination of steep slopes and a continuous soil cover makes the characterization of the rock mass quite difficult.

The majority of the rock within the region has undergone tectonic deformation associated with the Genting thrust belt, the major fault zone of the area. Most of the faults are low angle thrust faults developed in response to east-west lateral compression.

Geologic Units

The great majority of the tunnel extends through a quartz porphyry that has also been described as a rhyolite porphyry and includes porphyritic rhyolite, rhyolite, rhyodacite, dacite and rhyolitic tuff breccia. Generally the rock is moderately to highly jointed and most joints have been recemented with secondary minerals, mainly calcite, kaolin and chlorite, and subordinately with quartz. Generally the rock encountered at tunnel depth is slightly to moderately weathered. However, the weathering profile is highly variable and the highly fractured and faulted material tends to be more highly weathered. The weathering, alteration, shearing and close jointing has largely destroyed the original texture of the parent rock. The three rock classes, defined during the geologic characterization, are summarized below:

- Class A - Slightly altered/fresh rock
- Class B - Altered rock, zones disturbed by weathering and faulting
- Class C - Highly altered, cataclastic to mylonitic rock

The unconfined compressive strength of the samples of intact rock tested during the geological investigation for the original tunnel ranged from 24 to 240 MPa.

A seismic refraction survey was performed during the geotechnical investigation for the new tunnel and included surface seismic refraction, surface-to-borehole, borehole-to-surface, and cross-hole seismic methods. Typically the recorded compression wave velocities close to tunnel depth are in the range of 3500 to 5000 m/sec. Lower velocity materials were generally limited to depths less than 25 m.

Geologic Structure

The predominant joint sets within the rock mass are subvertical, followed by those inclined from 31° to 60° with a third set having an inclination of less than 31°. The average spacing of the subvertical joints is approximately 50 mm, 100 to 200 mm in the intermediate system, and 300 to 900 mm in the subhorizontal set. Typically the subvertical joints are undulose with thin coatings of calcite. The intermediate joint set are typically rough planar to irregular, and commonly stepped with some porous zones. The subhorizontal set can generally be characterized as clean, uncoated, planar and smooth.

Groundwater

In general the groundwater level is relatively close to the ground surface. Reportedly, water inflows into the existing tunnel were significant during construction but have been successfully controlled in the completed tunnel.

Rock Mass Classification

The Norwegian Geotechnical Institutes Q-System (Barton et. al., 1974) was used to characterize the rock mass during the geologic investigation. The Q values were assessed based on inspection of the rock core obtained from the exploratory drilling. Q was estimated to range between 0.01 to slightly greater than 0.1 which corresponds to extremely poor rock using the Q-systems descriptive terminology.

DESIGN

General Arrangement of Tunnel

The alignment of the new tunnel was chosen to take advantage of the natural landform, allowing its length to be minimized to approximately 800 m (approximately 100 m shorter than the existing tunnel). The tunnel profile is a typical horseshoe shape with a finished width of approximately 10 m and a height of approximately 8 m (see Figure 2). Near the eastern portal the new tunnel is located as close as 17 m to the existing tunnel with the separation increasing rapidly away from the portal to a maximum of 30 m.

The finished tunnel will incorporate two 3.75 m lanes, shoulders on each side, and a 650 mm wide emergency walkway. Architectural wall panels will be provided on each side of the tunnel. The tunnel services will incorporate ventilation, lighting, traffic controls, communications, fire emergency services, drainage and a central monitoring and control system.

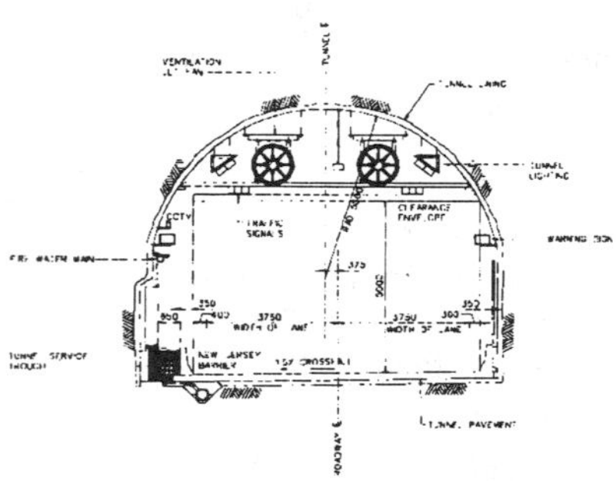

Figure 2. Tunnel Cross Section

Design of Tunnel Support

The design of the tunnel support was developed using empirical methods (Terzaghi, 1946, Wickham, Tiedemann, and Skinner, 1974, Bieniawski, 1993, and Barton et. al., 1974) and the past experience of the designers. Based on observations made during a visit to the site, it was surmised that many of the discontinuities within the rock mass are healed, and these healed discontinuities were "broken" during the drilling and sampling process such that the assessment of the rock mass based on inspection of the core was misleading. This conclusion was confirmed by the results of the seismic refraction survey. Grimstaad and Barton (1993) provide the following relationship between compression wave velocity and the Q-value:

$$q = 10^{\frac{(V_p - 3500)}{1000}}$$

where V_p is the compression wave velocity in m/sec.

Using a conservative estimate for V_p of 3000 m/sec provides a Q value of 0.3 which is significantly higher than the upper end of the estimated range for Q (i.e., 0.1) developed during the geotechnical investigation.

Four tunnel "support types" were developed for the range of rock conditions anticipated along the tunnel and general rock mass descriptions were defined to aid in the selection of the appropriate support type. These descriptions are presented below in Table 1.

Table 1. Description of Rock Conditions for Different Support Types

Support Type	General Rock Description
1	Fresh to slightly weathered (altered) rock, with tight, rough to slightly rough, moderately to closely spaced joints, slightly weathered walls.
2	Fresh to moderately weathered (altered) rock with tight, slightly rough, closely spaced joints, moderately weathered walls.
3	Moderately weathered (altered) rock, with coated, very closely spaced joints, highly weathered walls; rock includes numerous sheared zones.
4	Crushed and sheared, moderately weathered (altered) rock, with gouge zones

Support Types 1 to 3 utilize rock dowels and shotcrete for primary tunnel support. The rock dowels are 3.3 m long, untensioned and full column grouted with cement grout. The longitudinal spacing between the dowels ranges from 1.5 to 2 m and the circumferential spacing ranges from 1.25 to 1.5 m. Steel sets or lattice girders and shotcrete were the specified primary support measures for Type 4 ground and the contractor elected to use steel sets. The sets consist of 200UC60 sections. Shotcrete was specified to be applied as necessary to control raveling in Type 1 ground, whereas a minimum shotcrete thickness of 50, 100, and 100 mm was specified for Types 2, 3, and 4 ground, respectively. All primary shotcrete was specified to be reinforced with collated fibrillated polypropylene (CFP) fibers to reduce rebound and improve the flexural toughness, and increase the fatigue and impact resistance. Fiber reinforcement also helps reduce shrinkage cracking which, in turn, reduces the permeability of the shotcrete. A reinforced shotcrete wallplate was designed and installed at the base of the top heading in Type 4 ground.

The final lining for the tunnel consists of mesh (F41) and fiber (CFP) reinforced shotcrete, 125 to 150 mm thick. It was considered that the two types of reinforcing would optimize the performance of the lining as the mesh provides a quantifiable flexural strength and the fiber reinforcement provides the benefits listed above. Drainage for the tunnel is provided by strip drains, spaced at 5 m, installed behind the layer of primary shotcrete. The layout of the tunnel support for Type 3 ground, the most prevalent along the tunnel, is shown on Figure 3.

CONSTRUCTION

The tunnel was excavated by drill and blast methods. Excavation was carried out using a heading and bench method due to the size of the tunnel and generally poor rock conditions.

A heading from the east, approximately 500 m in length, was constructed first due to the accessibility of this portal area. After development of the western portal, involving a large open cut excavation, a western heading commenced. At this time, advancement of the eastern heading was stopped and the bench removal operation undertaken. Following the completion of the bench excavation, the final lining was installed in the eastern heading. After breakthrough, the bench removal and final lining operations were completed in the western heading.

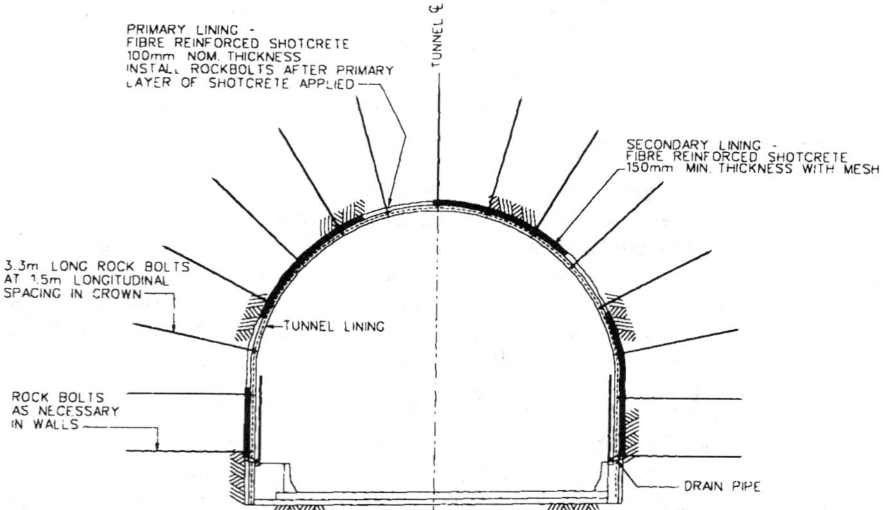

Figure 3. Tunnel Support Measures in Type 3 Ground

Excavation/Primary Support Sequence

The excavation and primary support sequence was typically as follows:

(i) drill face;
(ii) charge face;

(iii) blast and dissipate smoke by ventilation;
(iv) scale loose material from the crown and sidewalls;
(v) remove muck from face;
(vi) apply shotcrete to crown and walls; and
(vii) install rock dowels.

Face Drilling and Blasting

Drilling of the face was carried out using an Atlas Copco H135 drill rig with two drill booms and a man basket. The heading drilling comprised approximately 80 holes in a burn cut pattern. The length of drilled hole was varied according to the ground conditions. The holes were charged using locally produced Emulite cartridge explosives and non-electric detonators. Lighter charges were placed in the perimeter holes to control overbreak.

As the separation to the existing tunnel is only 17 m at the eastern portal and 30 m maximum, care was taken to control blast vibration levels at the existing tunnel. Vibrations were monitored in the existing tunnel adjacent to each blast. Blasting vibration was successfully controlled and the recorded vibrations were consistently within the specified maximum peak particle velocity of 50 mm/s and no damage to the existing tunnel was observed. As a precaution, the existing tunnel was cleared of traffic prior to each blast.

Mucking Operations

Mucking was carried out using a Caterpillar 966F rubber tired loader and articulated dump trucks. The mucking operation included scaling of all loose material from the tunnel crown and walls. Air quality was monitored regularly to ensure diesel fumes were being effectively removed by the ventilation equipment.

Figure 4. Tunnel Mucking Operation in Top Heading

Shotcrete

Wet mix shotcrete was batched at an on-site batch plant. It was applied to the walls and roof of the tunnel using an Alive 285 shotcrete machine and Alive 305 remote boom. The shotcrete was applied immediately after mucking. Where considered necessary for stability, however, the boom allowed application of shotcrete over the top of the in-place muck pile.

Rock Dowels

The Type 3 support employed for the majority of the tunnel required installation of eleven 3.3 m long rock dowels over the crown of the tunnel and in the side walls as necessary. After drilling the pattern of holes, the dowels were installed with prefixed grout tubes and centralizers. The collars were sealed and the dowels grouted using an MAI pump.

Quality Control for Tunnel Support

The Project was undertaken as a “Quality Assurance” (QA) contract and, hence, the responsibility for quality control was totally with the Tunnel Contractor. This responsibility included the selection of the appropriate tunnel support type based on the encountered geotechnical conditions. During the early stages of the project, inspectors and “check engineers” were employed by the Client to overview the work of the Contractor. However, in the later stages of the project, the Client had full confidence in the Contractors quality control methods and these additional quality assurance personnel were no longer required. This resulted in immediate cost savings as well as providing the Contractor additional flexibility in planning the sequence of construction activities which helped in completing the project early and within budget.

The following were the principle quality control testing measures for the tunnel support specified by the designers and implemented by the Contractor:

- material testing for concrete and shotcrete such as cube and slump tests;
- test panel field trials for shotcrete;
- cores for thickness measurement and strength testing for the final shotcrete lining;
- Schmidt hammer testing of the final shotcrete lining;
- pull-out tests on rock dowels at commencement of work and tensile load tests of the rock dowels at a rate of approximately 2 percent during tunneling; and
- convergence monitoring.

Excavation and Support Performance

As noted earlier, the predominant support type used was Type 3. Type 4 support was employed near the portals and approximately 100 m of Type 2 support was used. The maximum advance rate achieved was 36 m per week in the heading and 65 m per week in the bench.

The primary tunnel support measures used for the great majority of the tunnel, consisting of rock dowels and shotcrete, proved to be very effective apart from two minor collapses in the tunnel crown that occurred shortly after installation of the rock dowels and shotcrete. These occurred in the time before the support had reached a significant proportion of its design strength. The use of accelerated grout with the dowels may have provided earlier support and alleviated this problem. Resin grouted dowels were not used because obtaining the required bond with resin cartridges may have proved difficult to achieve consistently in the ground conditions encountered and because of corrosion considerations.

Convergence was measured using survey equipment at regular intervals along the tunnel and at locations with poor ground conditions selected by both the client and the contractor. No appreciable convergence was measured and there was no evidence of stress related cracking in either the primary or final shotcrete linings.

The shotcrete final lining proved to be an economical alternative to a cast-in-place concrete lining as the surface profile of the shotcrete can follow the excavated profile quite closely, resulting in considerable savings in materials alone.

Tunnel Fit-out

Civil works commenced in the eastern heading following completion of excavation and lining. Blinding concrete was placed in the invert to minimize deterioration of the tunnel floor and to ensure an even thickness of the overlying no-fines concrete drainage layer. The tunnel drainage system was constructed and then the no-fines concrete placed and roller compacted. The pavement, consisting of 230 mm of reinforced concrete, was then constructed. Finally, the precast concrete new jersey barriers, cable ducts, pulling pits and walkway were installed.

At the time of preparation of this paper the mechanical and electrical services fit-out was in progress. This work is being undertaken by Siemens under a subcontract to Transfield. Following behind the mechanical and electrical works will be the final construction activity - the installation of the architectural wall panels. The tunnel operating and emergency services will then be tested and commissioned and the new tunnel at Genting Sempah will be ready to receive traffic.

CONCLUSIONS

The design process used for the new tunnel at Genting Sempah illustrates several important points. First, it shows that with today's communication technology it is possible to perform a fast-track design remote from the construction site - in this case with parts of the design performed on two continents. Second, the design and successful completion of the tunnel validates the use of empirical design methods for tunnels with diameters of approximately 10 m. Finally, this project highlights the importance of using a number of exploratory methods to characterize a rock mass and showed that borings alone, especially in a rock mass with healed discontinuities, can provide an unfavorable and perhaps misleading view of rock mass quality.

The construction of the tunnel was completed ahead of schedule and within budget and is another example of the efficiency and economic benefits of using rock dowels and shotcrete for permanent tunnel support. The quality assurance contract used for the project was successful and indicates that in many cases it may be beneficial to assign some, if not all, quality assurance responsibility to the Contractor on a design-build project.

BIBLIOGRAPHY

Barton, N.R., Lien, R. and Lunde, J., 1974, Engineering Classification of Rock Masses for the Design of Tunnel Support, Rock Mech. 6, 189-239.

Grimstaad, E. and Barton, N., 1993, Updating the Q-System for NMT. Proc. Int. Symp. on Sprayed Concrete - Modern Use of Wet Mix Sprayed Concrete for Underground Support, Fagernes, Oslo.

Lang, T.A., 1961, Theory and Practice of Rockbolting, Tans. American Inst. Min. Engrs, 220, 333-348.

Terzaghi, K., 1946, Rock Defects and Loads on Tunnel Supports. In Rock Tunneling with Steel Supports, (eds R.V. Proctor and T.L. White) 1, 17-99. Youngstown, OH: Commercial Shearing and Stamping Company.

Wickham, G.E., Tiedemann, H.R., and Skinner, E.H., 1972, Support Determination Based on Geological Predictions, In Proc. North American Rapid Excavation and Tunneling Conf., Chicago, 43-64.

Chapter 42

PRIMARY SUPPORT DESIGN FOR LARGE SPAN TUNNELS IN WEAK ROCK: THE MELBOURNE CITY LINK TUNNELS

M.T. McRae,[1] C. Wilson,[2] S.J. Porter,[3] and B. Hutchison[4]
[1]Jacobs Associates, El Segundo; [2]Wilson and Pass, South Africa;
[3]Hyder Consultants, Australia; [4]Transfield-Obayashi Joint Venture, Australia

ABSTRACT

The Melbourne City Link project includes two three-lane highway tunnels, 1.6 and 3.4 km long, respectively, which are both approximately 16 meters wide and 9 meters high. The tunnels will be excavated through interbedded, moderately to highly jointed siltstones and sandstones of the Silurian Age Melbourne Mudstone formation over the major parts of their lengths. The strength of the intact rock varies from less than 5 MPa to greater than 70 MPa. Based on the available information regarding the performance of large span tunnels in weak rock, it was considered that the commonly applied empirical design guidelines for tunnel support, based on rock mass classification systems, should be augmented by more explicit design methods. The support design was, therefore, validated by analytical procedures, including numerical analyses, which are described in this paper.

INTRODUCTION

The Melbourne City Link Project will connect two existing major freeways in the Melbourne road network and so will provide a much needed bypass of the CBD for commercial transport and private users alike. The project involves construction or upgrade of approximately 22 km of twin 3 or 4 lane expressway - a significant proportion of which will be on viaduct or in tunnels. The tunnels are being constructed beneath the highly sensitive Botanical Gardens, Yarra River, sports facilities and residential areas where surface road construction would not be acceptable. Refer to Figure 1 for a site plan.

The Project is being delivered under a Build-Own-Operate-Transfer ("BOOT") arrangement with the Victorian State Government. Transurban, a publicly traded company on the Australian Stock Exchange, won by tender the "Concession" from the Victorian Government to build, own and operate the Melbourne City Link (MCL) as a toll road for a period of 34 years. After this period, ownership will return to the State. Transurban has contracted the Transfield Obayashi Joint Venture (TOJV) to design and construct the Project. The construction cost of the MCL Project is approximately A$1.2billion (US$960 million).

TOJV engaged the Hyder/CMP Joint Venture (a joint venture of the consulting engineers Hyder Consulting and CMPS&F) to undertake the design of the Southern Link component of MCL which includes the tunnels. The scope of work for the driven tunnels component includes the geotechnical investigations, temporary and permanent tunnel support, civil works and mechanical and electrical services. Jacobs Associates of San Francisco are providing specialist expertise in the design of the temporary and permanent tunnel support. Geotechnical investigations have been undertaken by Golder Associates.

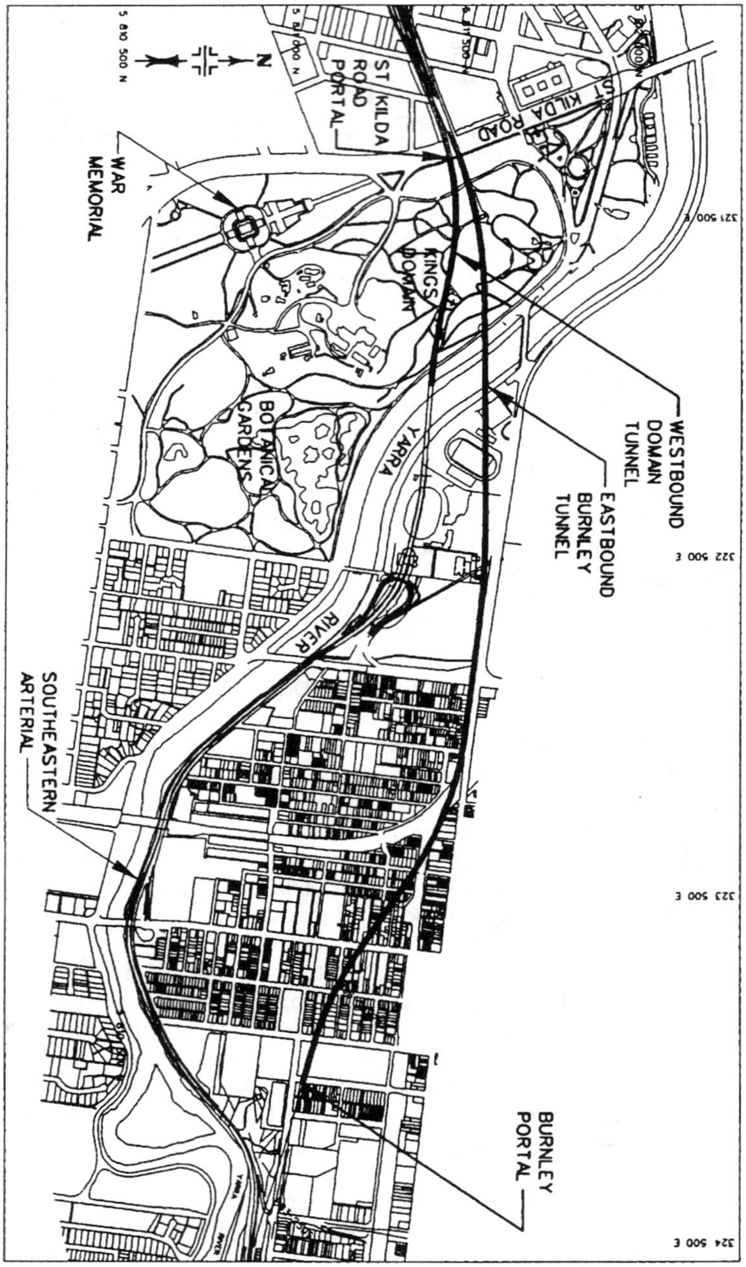

Figure 1. Site Plan

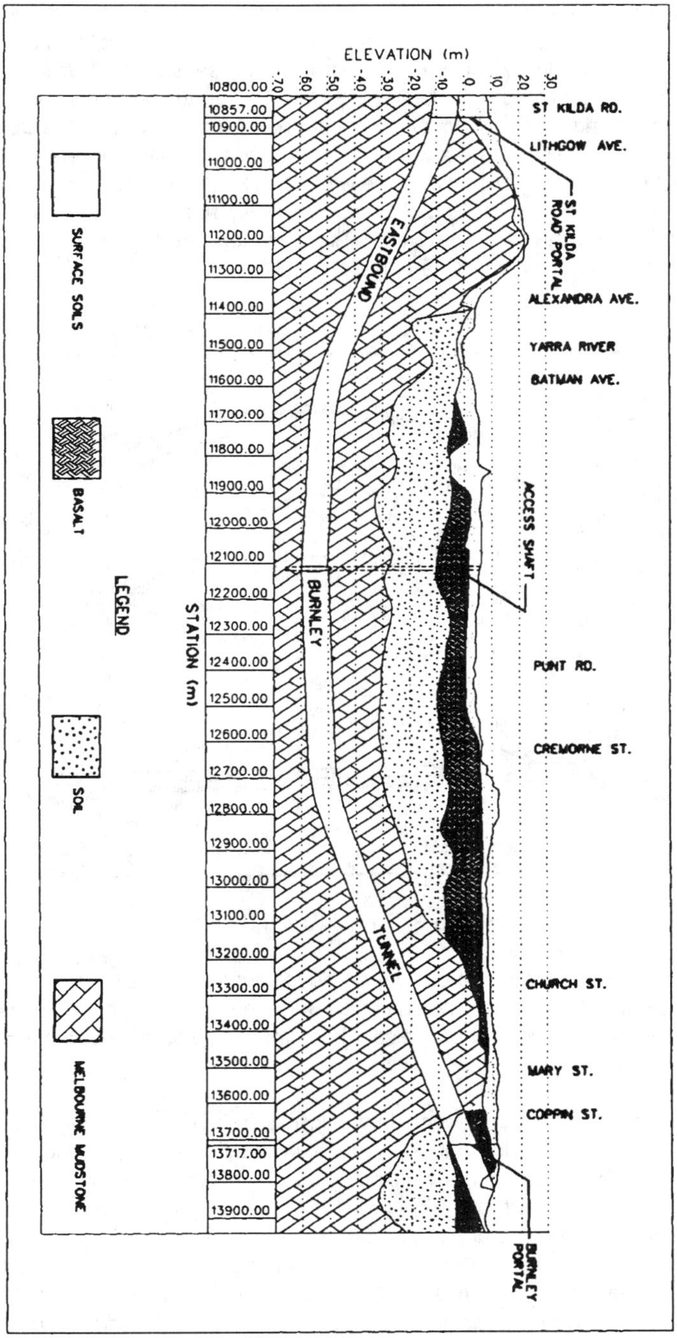

Figure 2. Long Section of the Eastbound Burnley Tunnel

Two tunnels are under construction. The westbound "Domain" tunnel will be approximately 1.6 km long, comprising approximately 0.7 km of driven tunnel and 0.9 km of cut and cover tunnel (including the shallow river crossing). The eastbound "Burnley" tunnel will be approximately 3.4 km in length, including 2.9 km of driven tunnel (including tunneling deep beneath the river) and 0.5 km of cut and cover tunnel. A long section of the Burnley tunnel is included as Figure 2. The tunnels will accommodate three traffic lanes each. The driven tunnels will have a modified ellipsoid shape with approximate excavated dimensions of 16 m wide by 9 m high (see Figure 3). A shaft of approximately 10 m in diameter is under construction mid-length along the eastbound tunnel to provide construction access and a supply shaft for the ventilation system of the completed tunnel.

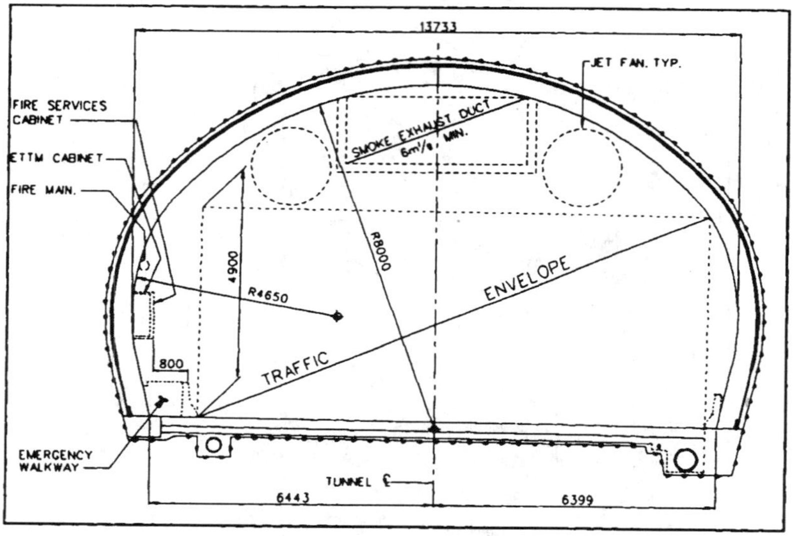

Figure 3. Tunnel Cross Section

At the time of preparation of this paper the excavation of the eastbound driven tunnel was underway from one heading, the shaft was under construction, and various elements of the cut and cover works were in progress. There will be up to five headings for the driven tunnels at the peak of construction activity.

The design of large span tunnels in weak rock has progressed significantly over the last 15 to 20 years. The Poatina underground power station and the Drakensburg pumped storage scheme, constructed in the 1960's and 1970's with spans of 13.7 and 16.7 meters, respectively, were two of the pioneering projects for large span openings in relatively weak sedimentary rock (Sharpe, et. al., 1984). Currently there are over 250 km of tunnel being constructed in the weak sedimentary rock of Taiwan and the span of many of these tunnels approaches the span of the openings used at both Poatina and Drakensburg (Chang, 1996). Clearly, significant experience has been gained in the area of the design and construction of large span openings in weak rock. Advanced numerical methods that allow detailed analysis of underground excavations have also been developed over the last 20 years. However, even with the expanding empirical database and the improved analytical methods, the design of large span openings in weak rock still offers significant design challenges as evidenced by the excessive convergence - up to 1.2 meters - that occurred over several hundred meters of the Mucha Tunnel in Taiwan (Chang et. al., 1996).

Empirical tunnel design methods based on rock mass classification schemes in common use today - the Norwegian Geotechnical Institutes (NGI) Q-System and the Rock Mass Rating (RMR) method - provide a starting point for designing tunnel support. However, these empirical methods should be supplemented by analytical methods, especially in the design of large span openings in weak rock. The wealth of experience obtained from the weak rock tunnels and underground openings in Taiwan indicates that the standard empirical design methods have some limitations when used for the design of large span openings in weak rock (Hou, 1996).

A design approach involving both empirical and analytical methods is recommended by Barton (1996), Hoek (1996a), and Bieniawski (1993). This basic methodology was adapted for the design of the MCL tunnels and extended to include a site specific classification scheme which could be used during construction to select appropriate support measures. This paper provides an overview of the tunnel design philosophy and methodology used for the more prevalent conditions where the tunnels are entirely in rock and the rock is fresh to highly weathered.

GEOLOGY

Geologic Setting

The MCL tunnels are to be driven through interbedded marine sediments of Silurian age, referred to as the Melbourne Mudstone (the Dargile Formation). These sediments have been folded and faulted and occasionally intruded by igneous rocks. Near the portals, and in the higher ground, these rocks have undergone various degrees of weathering. In the central and eastern project areas, a 200 m wide Tertiary valley (the Jolimont Valley) was cut through the rock, essentially stripping away the weathered rock material at the base of the valley.

This valley was subsequently filled by a variety of volcanic rocks and sediments. At the eastern portal of the east bound Burnley Tunnel, the tunnel passes through this buried valley and encounters mixed face conditions consisting of basalt flows overlying silts, clays, and colluvium (see Figure 2). The ventilation and construction access shaft, located approximately halfway along the Burnley Tunnel, encountered similar conditions, including a major sand and gravel aquifer below the basalt.

Geologic Units

The Melbourne Mudstone formation consists of interbedded siltstone and sandstone with siltstone comprising approximately 70 percent of the rock mass. The siltstone beds are typically 100 mm to 800 mm thick. The unconfined compressive strength of the intact siltstone varies from over 70 MPa where it is fresh, to less than 5 MPa where the siltstone is highly to completely weathered. The unconfined compressive strength of the more typical rock ranges from 10 to 50 MPa. The sandstone generally occurs in beds between 10 mm and 100 mm thick, although beds up to 1 m thick are not uncommon. The unconfined compressive strength of the sandstone is similar to the strength of the siltstone, although there have been occasional tests indicating strengths up to 249 MPa.

The Melbourne Mudstone is relatively durable and is not highly susceptible to slaking, softening or swelling upon excavation and exposure.

Quartz or feldspar porphyry or lamprophere dykes and sills are common in the Melbourne Mudstone. These igneous instrusives can vary greatly in their properties but can be typically characterized according to the width of the particular feature, as follows:

- high to very high strength rock (15 to 160 MPa), of variable widths, up to 20 m wide.
- low to medium strength , highly weathered rock (1 to 15 MPa), generally less than 10 m wide.
- extremely weathered rock (stiff to very stiff clays) less than 2 m wide; or at the margins of the wider dykes as crushed or extremely weathered seams 0.5 m to 2 m wide.

The dykes are often near vertical and are generally oriented in a north-south direction. It is common for these dykes to deviate along relatively steep bedding, to become sills. Locally these dykes have been encountered in swarms.

Geologic Structure

The Melbourne Mudstone has been folded into a series of synclines and anticlines, with the fold axes dipping gently to the north and south. The tunnels are aligned near perpendicular to the fold axes and, therefore, for the most part will be cutting across the strike of the beds. Local experience has shown that minor folds can occur between major fold axes, resulting in rapid, localized changes in bedding orientation.

Several major fault zones occur in the Melbourne area, although no known major fault zones cross the tunnel alignments. There are, however, several major lineations in the project area which may be associated with fault zones. Cored holes have shown the existence of highly fractured and sheared zones, and crushed seams, often parallel to bedding. Most of the individual sheared and crushed seams are on the order of 10 to 100 mm thick. The sheared zones are zones of closely spaced fractures which are often smoothed and polished. The crushed seams in fresh to slightly weathered rock consist of angular granular rock fragments (sand or gravel size with some silt and clay). In the more weathered rock, the crushed seams consist of gravely clay or clay of low plasticity. In fault zones mapped at the surface, or encountered in the cored holes, these sheared and crushed seams have been spaced at 0.1 to 1 m apart with the overall fault zones being typically less than 6 m wide. At least one major low dipping fault, 20 m wide, has been identified which could influence the tunneling for a significant distance. Three other potential fault zones have been tentatively identified.

Seven joint sets have been identified in the Melbourne Mudstone but in general only three are well developed at any one location, with only minor development of the other sets. The most common joint set is that parallel to bedding, with the joints having lengths up to 10 m, spacings of 0.1 to 0.3 m, and rough and irregular surface textures. The other joint sets typically have lengths of 0.3 to 3 m, spacings of 0.1 to 0.3 m and are also rough, but generally planar.

The joints in the siltstone are generally tightly closed except in some surface outcrops. In the sandstone the joints are often slightly open (less than 2 mm) and a correlation to higher permeabilities has been observed. Most joints are limonite stained and pyrite occurs in the joints of fresh to slightly weathered rock. In the more weathered rock clay coatings are common.

Groundwater Conditions

The project area is near sea level and the tunnels pass under the Yarra River (see Figure 1). Groundwater levels are, therefore, near surface over most of the tunnel lengths. At the deepest section of the Burnley Tunnel, there will be approximately 55 m of head at the crown of the tunnel. In this region there will be typically 15 to 20 m of rock cover to a major sand and gravel aquifer (see Figure 2). The proximity of this aquifer has necessitated careful attention to water control aspects in the temporary and permanent condition.

DESIGN PROCEDURES

Design Methodology

The design program for the MCL tunnels progressed through three phases. A preliminary design was developed during the bid period that allowed costing of various alternative alignments, tunnel support measures, and final lining alternatives. The tunnel support design developed during this phase of the project relied almost entirely on empirical methods (Terzaghi, 1946, Wickham, Tiedemann, and Shimmer, 1972, Grimstaad and Barton, 1993, and Bieniawski, 1993) supplemented by limited analyses to investigate specific anticipated design conditions. Stage 1 design saw the extension and refinement of the preliminary design and involved developing alternative approaches to tunnel support. During this phase, the Contractor evaluated the various alternatives from both price and schedule considerations and chose a preferred alternative. Detailed working drawings for the preferred alternative were developed during the Stage II design period. More detailed analytical methods, including numerical analyses, were used during both Stage I and Stage II design, however, the previously referenced empirical methods, local tunnel experience, experience in similar geological environments, and the specific experience of the tunnel designers was always used to check the results of the detailed analytical studies.

Due to the short design period, geological investigations and design tasks were performed concurrently. This schedule necessitated clear communication between the project team and occasionally design modifications were required as more geological information became available. The design program was established such that the key areas (i.e., the access shaft) were designed early in the program to allow the start of construction prior to the completion of the overall design.

Design Philosophy and Assumptions

The three basic temporary support measures utilized for the MCL tunnels include:

1. rock reinforcement consisting of rockbolts and fiber reinforced shotcrete in ground where squeezing is not anticipated and where bolts can achieve adequate anchorage,
2. steel sets and shotcrete where squeezing is anticipated or where rockbolts cannot

achieve adequate anchorage, and

3. targeted cable anchors where large rock wedges cannot be adequately stabilized by the rockbolts in measure 1. above.

All rockbolts will be tensioned to 120 to 200 kN depending on the size and length of the bolt. Cable anchors will not be tensioned, though they will always be used in conjunction with tensioned bolts. There is not a general consensus in the field of tunnel design with regards to the tensioning of rockbolts. Some authors indicate that tensioning helps preserve the integrity and strength of the rock mass and minimizes bed separation in sedimentary rock (Sharpe et. al., 1984) while others indicate that tensioned bolts are most effective in controlling block falls in hard rock (Hoek, 1996a). However, in many cases experience has shown that untensioned bolts are effective provided they are installed close to the face. The decision was made to tension the bolts to maintain the integrity of the rock, prevent bed separation resulting in flexural deformation where gently inclined beds span the tunnel excavation, and also to "test" the bolts to confirm adequate anchorage is available to carry anticipated rock loads (i.e., quality control).

The modified ellipsoid shape of the tunnels, as shown on Figure 3, was chosen because it efficiently accommodates the traffic envelope and various appurtenances and is efficient for roadheader excavation and tunnel support. The curved side walls were chosen over straight walls because they: provide a more efficient shape for the steel sets used for temporary support in areas of squeezing ground; provide a more efficient shape for the design of the unreinforced concrete final lining; accommodate tunnel services niche cabinets; and eliminate sharp corners in the profile and, hence, are easier to excavate by roadheader.

Control of groundwater by grouting prior to excavation, such that the water does not impair the installation of the support measures or significantly increase the loads on the support is another key element of the overall design philosophy.

Excavation of the tunnels will be by Mitsui S300 roadheaders. The excavation sequence will typically involve a two drift heading followed by full width excavation of the bench. In areas of poor ground the Contractor has the option to excavate a full width heading in short advances (1.0 to 1.5 m) to allow installation of the full top arch of the sets or alternatively to excavate the heading using two drifts using a temporary central prop. In areas of squeezing ground, excavation for a curved or flat invert strut may be required during the benching operation. Some parts of the tunnels will require a "tanked" or undrained final lining designed to prevent water inflow into the tunnels. These areas will also require excavation for a curved invert to resist the hydrostatic head which will develop. This invert is not required for stability of the excavation prior to construction of the lining and, therefore, may be excavated as a separate exercise at a later stage in the construction sequence.

Long term tunnel support will be provided by an unreinforced cast-in-place concrete lining, 300 to 450 mm thick. It is assumed the rockbolts will provide no long term support capacity. However, a proportion of the steel sets and their shotcrete encasement has been considered in the final lining design for the poorer ground areas.

In view of the lengths of the tunnels and the anticipated geological conditions, it was expected that ground conditions would vary considerably and unpredictably along the lengths of the tunnels. As primary support requirements would, therefore, also vary, it was considered impractical to develop and specify support for specific segments of the tunnels. Primary support was instead designed to suit a number of Typical Geological Conditions, each of which was defined by specification of relevant geological characteristics, and were not associated at the outset with particular segments of the tunnels. These characteristics were :

- unconfined comnressive strength of the intact rock fragments.
- orientation of the bedding in relation to the tunnel axis.
- spacing of fractures or joints.
- condition along bedding planes and fractures.

Estimates were made of the proportions of the lengths of the various segments of the tunnels along which each of the defined Typical Geological Conditions might exist. The locations where instances might occur was not, however, a factor in the design of the primary support for each Condition. It was instead an essential aspect of the design that applicable Typical Geological Conditions would be identified during excavation, and that the appropriate primary support design required within the corresponding segment of the tunnels would be confirmed accordingly at that time.

To assist in determining the appropriate support type at any location and in monitoring the performance of the various support types the headings will be mapped by experienced geologists on a routine basis. This mapping will be most intensive for the forward headings. Probe holes will be maintained a minimum of 5 m in advance of the forward heading face to identify water bearing features as well as areas of poor ground. At the commencement of tunneling three probe holes from the forward heading are proposed. The need for the three holes may be reviewed during construction. The probe holes will assist in assessing expected conditions ahead of the face and, hence, the timing of changes in support types.

The general design philosophy and methodology used for the MCL tunnels - using a combination of empirical and analytical methods with specific support requirements determined during construction - is similar to that described as the Norwegian Method of Tunneling (NMT, Barton, 1996). However, there are several key differences. First, long term support is provided by a cast-in-place concrete lining, whereas, the NMT relies on the rockbolts and fiber reinforced shotcrete for permanent support. The decision to utilize a cast-in-place final lining was based on long-term maintenance and serviceability requirements as well as economic considerations. Steel sets and shotcrete will be used in the worst ground conditions that are anticipated along the MCL tunnels while the NMT favors the use of reinforced shotcrete arches where heavy support is required. The selection of the appropriate support type for different sections of the MCL tunnels will utilize a site-specific ground classification scheme rather than one of the more general classification schemes to aid in the selection of the appropriate support measures.

Empirical Design Methods

As discussed above, empirical design methods (Terzaghi, 1946, Lang, 1961, Wickham, Tiedemann, and Shimmer, 1972, Grimstaad and Barton, 1993, and Bieniawski, 1993) and a review of specific precedent experience in similar geological conditions formed the basis for the design of the temporary tunnel support during the early stages of the project. These methods were also used later in the design phase in conjunction with analytical methods to develop the final design for tunnel support. The Q-System (Grimstaad and Barton, 1993) was used more than any other method because it is more applicable to large span openings and it provides design guidelines for rockbolts and fiber reinforced shotcrete which are the predominant temporary support measures used for the MCL tunnels.

It is anticipated that Q will vary between 0.7 to 7 for the majority of the rock encountered along the tunnels. Poorer ground conditions will be encountered close to the portals and in the major fault zones and it is estimated that the Q value in these areas could be as low as 0.02. The Q value of the weathered dyke material is estimated to be less than 0.01.

Analytical Design Methods

The three different modes of instability considered during the design process and which are associated with distinct Typical Geological Conditions are :

Block instability - falling and sliding of blocks of rock which are separated from the balance of the rock mass by bedding planes and other discontinuities. A potential for this mechanism to occur is expected to exist throughout the tunnels. The potential for major wedge failures, however, is expected to occur infrequently.

- Squeezing - overstress in shear or compression of zones within the rock mass, due to the altered stress state around the tunnels caused by their excavation. This mechanism is expected to affect only a few highly weathered fault zones.
- Bed flexure - thinly laminated beds of rock are prone to flexural deformation when spanning over an underground excavation especially if the bedding planes are weak and slip between the beds is possible. Based on the anticipated orientation, spacing, and strength properties of the bedding plane discontinuities, excessive flexural deformation of the rock mass is not expected to occur over a significant length of tunnel.

Block Instability. If an underground excavation is not supported, then blocks of rock may fall or slide from the roof and sides where fractures exist through the rock mass which divide it into discrete blocks, slabs, or wedges which are smaller than the excavation dimensions. Such fractures may exist prior to excavation, or may develop due to the effect of the excavation on stresses in the surrounding rock mass.

Loss of such blocks from the excavation profile may lead to one of the following situations :

a) A "natural stable profile" may develop in "ordered" rock masses, which would be located some depth beyond the excavated profile, despite the absence of support. Primary support may, in this case, be designed to suspend all of those blocks, slabs and wedges below the natural stable profile which may otherwise potentially fall or slide, by means of rockbolts which are anchored beyond the stable profile.

b) Continuous raveling may take place in "disordered" rock masses, in which fractures are closely spaced and randomly oriented, until the enlarged excavation finally completely fills with bulked loose rock, or until a sinkhole breaks through to the ground surface. Primary support may be designed for such conditions to either develop or to maintain shear strength within the loose, or potentially loose rock, such that individual blocks, slabs or wedges do not become detached. A number of design procedures and/or theories have been proposed in this regard, including those of Terzaghi (1946), Lang (1961), Rabcewicz (1969), and Goodman and Shi (1985).

The latter theories may also be applied to design of support in ordered rock masses, where they may result in more economical design, if the natural stable profile is thought to be located excessively deep beyond the excavated profile, and if less support is necessary to render the potentially loose material self supporting, than may be required to suspend it.

The bulk of the rock mass for the MCL tunnels is expected to be ordered. The primary support for such Typical Geological Conditions has, therefore, been designed to support the appropriate estimated depths of potentially loose material.

Material in fault zones which has been affected by advanced weathering and intense fracturing is, on the other hand, expected to be effectively disordered. Primary support was accordingly designed for such conditions to be able to provide sufficient confining pressures to restrain deformation of the rock.

Two factors determine the dimensions of naturally stable profiles in ordered rock masses:

a) The orientation, continuity, and condition of discontinuities, including fractures and bedding planes. These characteristics determine the location around the profile and maximum depth of potential wedges which might fall or slide into the excavation.
b) The magnitude and direction of stresses around the excavation. Such stresses may limit the depth of potentially unstable wedges by "locking them in" if the wedges are deep enough to extend into this zone of higher stress outside the unstable material.

The location and depth of the apex of a zone of potentially loose material can in principle be determined by analysis of the orientations of the discontinuities (Goodman, 1989) and the distribution of stresses around the excavation as determined by numerical analysis. In practice, however, the orientations of discontinuities can seldom be confidently anticipated. In the design of primary support for the Melbourne City Link tunnels only two alternative conditions were recognized in this respect :

a) Bedding dip near normal to the tunnel axis was considered to be "unfavorable", in that this condition might define the deepest potentially unstable wedges of rock, and
b) Bedding dip near parallel to the tunnel axis was considered to be "favorable".

It was assumed for simplicity for Condition b) that the maximum depth of potentially loose material might exist anywhere around the roof and upper sides of the excavation. Primary support was, therefore, designed to be symmetrical for these cases, and capable of suspending an entire "envelope" of potentially unstable rock of that depth, without regard for where the actual apex might be located.

For Condition a), the primary support design consisted of two components :

- a symmetrical pattern of support to cater for other shallower zones of potentially loose rock around the excavation.
- "targeted" supplementary support, which might not be symmetrically laid out, which was intended to suspend all potentially loose material under bedding planes which are near tangent to the excavation profile.

Squeezing. Squeezing ground conditions occur when the increased stresses that develop in response to tunnel excavation result in failure and plastic deformation of the ground around the tunnel excavation. The tendency for squeezing is, therefore, related, primarily, to two factors - the in situ stresses and the strength of the rock mass. A secondary factor which also controls the tendency for squeezing is the width of the zone of weak material. Generally, squeezing ground is associated with excavations at considerable depths. As the maximum depth (to the crown) of the MCL tunnels is approximately 60 meters, it might be concluded that squeezing conditions will not develop. Empirical methods suggest that squeezing may occur when the overburden, in meters, exceeds $350Q^{1/3}$ (Grimstaad and Barton, 1993). In the case of the MCL tunnels, this relationship indicates that squeezing could occur in the fault zones where Q is estimated to be 0.02 if the overburden depth exceeds 95 meters, again suggesting that major squeezing behavior is unlikely. Numerical analyses were performed to simulate the behavior of the tunnel excavation within a fault zone with maximum overburden to investigate the potential for squeezing. The results of these analyses are discussed below.

Bed flexure. Bending of relatively flat lying beds is another potential mode of deformation for the rock surrounding the MCL tunnels. Control of bending deformation requires prevention of excessive slip between the beds and reinforcing a sufficient thickness of the rock such that the resulting "laminated beam" can effectively span the excavation (Sharpe et. al., 1984). Hand calculations and numerical analyses were performed to investigate the potential for excessive deformation associated with bending behavior.

Design of Support Elements

Rockbolt Design. Rockbolts were designed to suspend the entire mass of the potentially unstable or loosened rock, and to be anchored beyond the naturally stable profile, with a suitable factor of safety (not less than 1.5). The load capacities of rockbolts were, therefore, determined in relation to their spacing and estimated maximum depths of the envelope of loosened rock applicable to the given Typical Geological Condition.

The lengths of rockbolts were specified to be not less than the estimated maximum depths of the loosened envelope, plus anchorage lengths necessary to assure the load capacity of the rockbolt. Anchorage lengths were designed to suit unit pull out strengths which the given rock was confidently expected to exhibit. Such anchorage lengths were estimated to vary between 0.8 m and 2.5 m, again to ensure a factor of safety of not less than 1.5. Based on a review of the results of extensive anchor and pile tests in the Melbourne Mudstone, it was assumed that the design ultimate bond stress (between the grout and the rock) would be not less than 10 percent of the unconfined compressive strength in the fresher rock and 20 percent of the unconfined compressive strength in the highly to completely weathered rock. The rockbolts were specified to be fully encapsulated with resin grout, with a shorter set time resin used in the anchor zone to allow the bolts to be pretensioned.

Based upon the above procedures, patterns of rockbolts were specified for each of the various Typical Geological Conditions, examples of which are shown on Figure 4.

Cable Anchors. In the very poorest ground, and in rock where bedding orientation is unfavorable, cable bolts are also specified, the lengths of which vary between 7.5 m and 9 m (see Figure 4). The required length of the cable anchors was determined based on simple wedge analyses (Hoek and Brown, 1980) facilitated by the program Unwedge (Rock Engineering Group, 1992). The cable anchors were specified to be cement grouted, and not pretensioned, although it was envisaged that they would generally be utilized in conjunction with pretensioned rockbolts.

Steel Sets. In the poor ground conditions where squeezing conditions are anticipated or where rockbolts or cable anchors cannot develop sufficient anchorage, steel sets and shotcrete will be used for temporary tunnel support. The use of steel sets was favored over other means of passive support (i.e., lattice girders and shotcrete) because of the ductility of steel sets and the ability of steel sets to provide immediate support. Close to the portals where overburden stresses are very low, the sets were designed to support the full overburden. At greater depths the design loads were estimated using empirical methods (Terzaghi, 1946) and the results from the numerical analyses.

Shotcrete. Application of shotcrete soon after excavation has been specified for all ground conditions in order to prevent raveling and to minimize deterioration of the rock matrix. In the better ground conditions, however, the distance from the face has been relaxed to some extent in comparison to more fractured geological conditions which are more prone to raveling.

Relatively thin shotcrete layers have been specified (50 to 100 mm) for areas other than those supported by steel sets, in which case the sets will be fully encased. For the typical rockbolt supported areas, the primary role of shotcrete will be to act as "lagging" between rock bolts. Whilst the ability of thin layers of shotcrete to act as a shell or arch is recognized, it has not been considered as a means to reduce the required capacity of the rock reinforcement to support unstable material. This is because such ability can be adversely affected by ground deformations which result in overstress and deterioration of the thin shotcrete layer. To the extent which it occurs, therefore, the shell behavior of the shotcrete will provide an additional margin of safety.

Other Support Measures. In specific instances, other support measures will be implemented in the MCL tunnels. These measures include spiling, invert struts and arches, and various methods to stabilize the face of the tunnel in poor ground conditions. Two types of spiling will be utilized - conventional rebar spiling, and pipe spiling, 12 m long and 114 mm in diameter. Invert struts and/or concrete invert arches will be used in poor ground conditions where excessive convergence or squeezing occurs. Maintaining the stability of the face is the responsibility of the Contractor and several possible methods have been identified, including:

- fiberglass rockbolts.
- buttressing, by maintaining a sloping face.

Numerical Analyses

The computer program FLAC (Itasca, 1995) was used to perform the numerical analyses of the primary support for the Melbourne City Link tunnels. The numerical analyses were not used on their own to develop specific support designs, but were rather used to: (1) confirm the basic design assumptions, (2) identify important behavioral mechanisms of the rock mass around the tunnel excavation, (3) estimate the loads in the tunnel support elements, and (4) estimate the deformations around the tunnel excavation. The numerical analyses were considered one of the design tools in the “tool box” of the available design methods. Two basic models were developed to represent the range of geologic profiles that will be encountered along the tunnel alignment. These baseline models were then modified to simulate the anticipated range of materials and corresponding support arrangements along the tunnel alignment. Numerous parametric studies were performed to evaluate the sensitivity of the predicted results to variations in particular parameters and these parametric studies were invaluable in gaining insight into the overall behavior of the tunnel excavations.

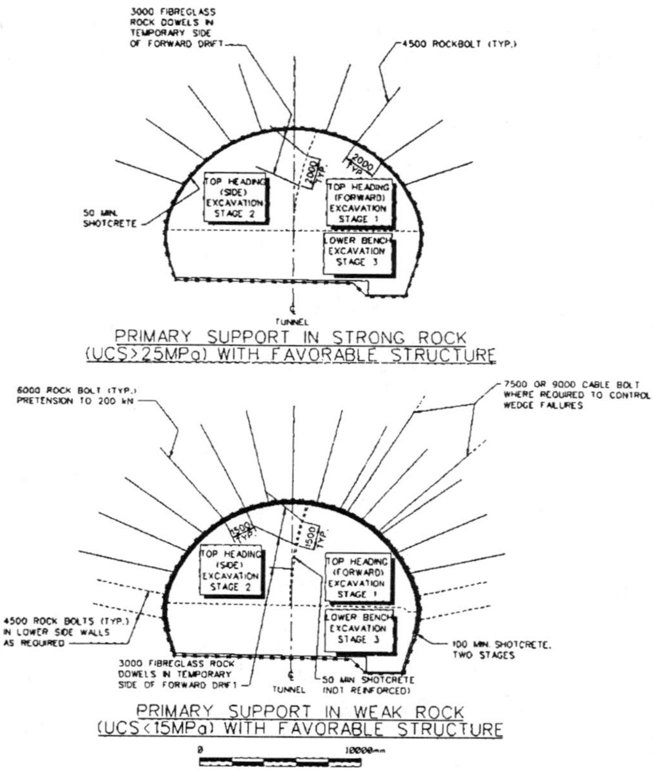

Figure 4. Examples of Rock Bolt Support

The rock mass was generally modeled as an isotropic elastic-perfectly plastic continuum with a Mohr-Coulomb failure criterion. This approach is considered valid for modeling a jointed rock mass where: the spacing of the joints is small compared to the size of the opening; there are a sufficient number of joint sets to assure isotropic strength properties for the rock mass; and when the strength and/or orientation of particular discontinuities or one joint set are not such that they dominate the behavior of the rock mass (Hoek, 1996b). Barton (1996) proposes that continuum models are valid for Q values less that 0.1 and for massive rock where the spacing of discontinuities is much greater than the scale of the opening and recommends discontinuum models for other cases. However, a realistic application of a discontinuum model presupposes a thorough knowledge of the orientation, extent, location and strength properties of the discontinuities. Given the sensitivity of discontinuum models to the properties of the discontinuities and the difficulty in accurately characterizing the location and strength properties of the discontinuities, a continuum approach, with the effects of the discontinuities accounted for implicitly in the selection of the continuum parameters, was adopted. It is recognized that in the better quality rock such a continuum approach may not completely capture the deformation modes of the rock mass, but provided the strength properties of the continuum model accurately reflect the influence of the discontinuities, the continuum approach will allow the basic behavior of the tunnel excavations to be evaluated.

The strength and deformability parameters of the rock mass were estimated considering the strength of the intact rock, the degree of jointing, the overall structure of the rock mass, and the condition of the discontinuities (Hoek, Kaiser, and Bawden, 1995). The bedding plane shears, prevalent throughout the Melbourne Mudstone, were considered to have the greatest influence on the behavior of the tunnel excavation. Therefore, several analyses were performed to investigate the effect of the bedding plane shears on the behavior of the tunnel excavation by modeling the bedding plane shears explicitly. The results of these analyses indicate that the predicted behavior of the models using explicitly modeled shears is comparable to the behavior predicted by the models where the effects of the bedding plane shears are accounted for implicitly in the selection of the continuum parameters, provided the spacing and strength properties of the bedding plane shears are not significantly below the anticipated values.

The results of two sets of analyses, providing an overview of the use of the numerical analyses, are discussed below. Both of these analyses used the base model developed to simulate the subsurface profile for the deep section of the Burnley Tunnel extending below the old river valley (see Figure 2). One of the analyses modeled the rock mass as slightly weathered to fresh siltstone with a fracture spacing of greater than 100 mm and with some sheared bedding plane surfaces. The second analyses modeled the rock mass as crushed, sheared, and moderately to highly weathered to simulate one of the fault zones that are anticipated along this reach of tunnel. These two conditions illustrate the wide range of rock properties that are anticipated for the sections of tunnel extending through the Melbourne Mudstone. The material properties used to characterize the slightly weathered to fresh siltstone and the fault zone material are presented below in Table 1.

Table 1. Estimated Rock Mass Properties

Rock mass description	GSI	UCS, MPa	m_i	Friction Angle	Cohesion MPa	Dilation Angle	Modulus of Deformation, MPa
Slightly weathered to fresh siltstone	40	25	12	48	0.2	6	2800
Fault zone	20	5	8	20	0.1	0	400

The term GSI referred to in Table 1 is the Geological Strength Index as defined by Hoek, et. al. (1995) and is related to RMR; m_i is an empirical constant used in the Hoek-Brown failure criterion (Hoek et. al., 1995); and UCS refers to the unconfined compressive strength of the intact pieces of rock.

The primary tunnel support modeled for the section of tunnel extending through the slightly weathered to fresh siltstone consists of eleven 4.5 meter long, 32 mm, fully grouted rockbolts in the crown of the tunnel, spaced at 2 meter centers along the tunnel together with 50 mm of shotcrete. The FLAC model simulated the recommended excavation sequence for this ground condition - two top headings and a bench - and the rockbolts were "installed" at the appropriate excavation stage. The primary tunnel support modeled for the major fault zones consists of 250UC89 steel sets placed at 1 m centers. The excavation sequence modeled for the sections of tunnel supported by steel sets involves a single top heading and a bench. The major support elements were modeled explicitly. The rockbolts were modeled using one dimensional cable elements coupled to the rock mass with the appropriate stiffness for the shear coupling springs to simulate a fully grouted bolt. The steel sets were modeled as beam elements, two-dimensional elements that can sustain bending, attached to the nodes of the rock mass elements. No attempt was made to model the thin shotcrete layer (50 to 100 mm) that will be used in conjunction with the rockbolts in the slightly weathered to fresh siltstone because the rockbolts are considered to be the major support element responsible for stabilizing the tunnel excavation; and the thin shotcrete layer is considered to serve as lagging between the bolts. This approach is conservative and eliminates the complexity of accurately modeling a thin layer of stiff, high strength material immediately adjacent to the weaker and more deformable rock mass.

The results of the analyses for the section of the tunnel extending through the slightly weathered to fresh siltstone are summarized on Figures 5 to 7. Figure 5 shows the predicted extent of the yield zone around the tunnel. The yield or "plastic" zone includes all zones where the stresses within the zone exceeded the yield criterion at any point during the calculation process including those zones with stresses currently below the yield surface. Because of the solution scheme used by FLAC, the yield zone should not be interpreted as a zone of "failed" rock that has to be fully supported, but rather as an indication of the extent of possible loosening of the rock mass. As shown on Figure 5, the rockbolts extend through the loosened rock mass and have a significant length anchored in the "intact" rock mass.

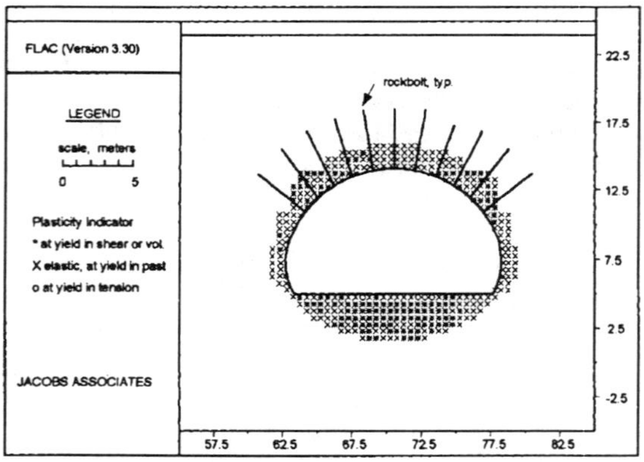

Figure 5. Yield Zone around Tunnel in Slightly Weathered Rock

Figure 6 shows the predicted displacements around the tunnel excavation. The maximum predicted displacement is 11 mm in the crown of the tunnel. It is recognized that the actual measured displacements in the tunnel could locally exceed the predicted values because the continuum model used to simulate the rock mass may not completely capture localized deformation modes within the rock. During the design process the predicted displacements were not considered to provide an absolute estimate of tunnel convergence, but rather as an indication of whether the specified support would be sufficient to limit displacements to relatively small values and as an indication of the dominant behavioral mechanism within the rock mass. Parametric studies were also performed to investigate the relative sensitivity of the predicted displacements to variations in the properties of the rock mass and the assumed support type.

Empirical methods are available to estimate the deformation that may occur in response to tunnel excavation (Grimstaad and Barton, 1993). However, these methods are very approximate and the predicted deformation varies over an order of magnitude. Therefore, numerical methods provide a more accurate method for estimating deformation, provided the properties of the rock mass can be estimated with a reasonable degree of accuracy and the model has the capability of simulating the dominant behavioral modes of the rock mass.

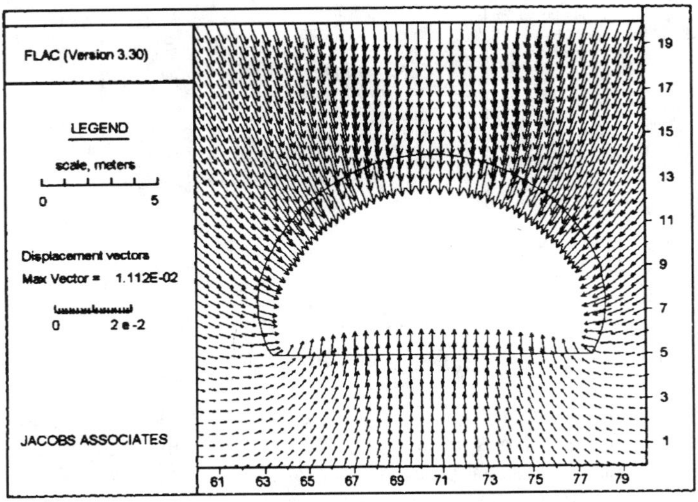

Figure 6. Displacements around Tunnel in Slightly Weathered Rock

Figure 7 shows the predicted axial loads in the rockbolts. The calculated loads have to be factored up to account for the longitudinal spacing of the bolts, but even after accounting for the spacing the calculated loads in the bolts are well below the yield load.

Figures 8 through 10 summarize the results of the numerical analyses for the section of the Burnley Tunnel extending through a fault zone. The initial analyses involved developing a ground reaction curve for the rock mass using a series of uniform radial support pressures acting around the entire perimeter of the tunnel excavation. The ground reaction curve is shown on Figure 8. As shown on this figure, the support pressure continuously decreases with increasing crown displacement because the constitutive model used to simulate the rock mass is elastic-perfectly plastic.

A support/ground equilibrium curve was then developed by simulating the installation of the support (i.e., steel sets) at different points along the ground reaction curve and simultaneously removing the support pressure. This analytical procedure is considered to reasonably approximate the actual stress redistribution that occurs in response to tunneling. The resulting ground/support equilibrium curve is shown on Figure 8 to be distinct from the ground reaction curve. This can be attributed to the fact that the modeled support has to undergo some displacement prior to developing a support pressure whereas the ground reaction curve is developed using a series of instantaneously applied and infinitely flexible pressures. Using this same reasoning, it can be concluded that a different ground/support equilibrium curve will be developed for each different support type depending on its axial and flexural stiffness and this was confirmed by varying the stiffness of the modeled support.

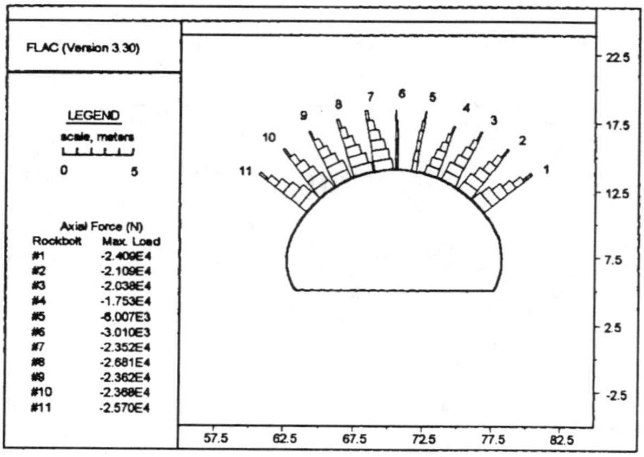

Figure 7. Calculated Rockbolt Loads

The support reaction curves shown on Figure 8 do not exhibit the characteristic initial soft response of a steel set (Hoek and Brown, 1980) because the sets are attached to the nodes of the rock mass zones. Although this approach is likely to artificially stiffen the response of the modeled steel sets to some degree, if is not considered completely unrealistic because the sets will be fully blocked and fully encapsulated with shotcrete.

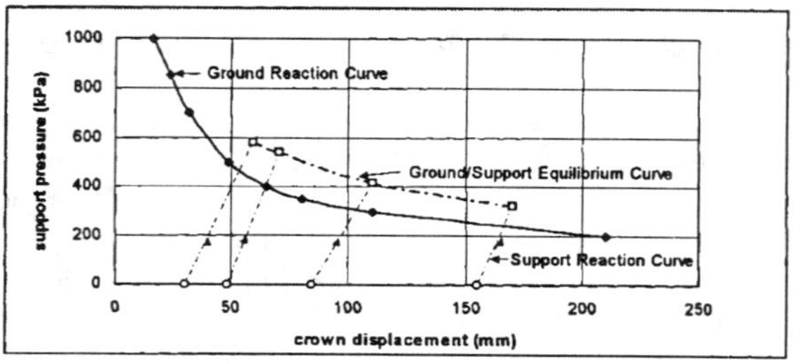

Figure 8. Ground/Support Interaction Curves

The analytical approach described above allows the design engineer to evaluate the interaction of the ground and support using well established tunnel design theory - the ground reaction curve. Numerical modeling allows the development of ground reaction curves for heterogeneous rock masses with complex constitutive relationships which is not possible with any closed-form solution. And the ground reaction curve and ground/support equilibrium curve provide a great deal of insight into the problem of efficient and effective support design. The ground/support equilibrium curve can be used to estimate the amount of tunnel convergence that must be allowed for to reduce the support pressures to acceptable values. The results of the analyses indicate that the loads developed in the steel sets are highly sensitive to convergence until approximately 75 mm of convergence has occurred. Beyond 75 mm of convergence, the predicted support loads are not highly sensitive to the amount of convergence.

The numerical analyses indicate that some squeezing ground behavior can be expected in the fault zones and that heave of the tunnel invert is an important factor in the overall deformation behavior within the rock mass. Very large convergence at the crown is predicted by the models unless heave is controlled. Heave was controlled in the numerical models by simulating a concrete invert up to 1 m thick. The predicted displacement pattern around the tunnel in the fault zone is shown on Figure 9. Some tendency for heave of the invert is apparent even though the model includes a concrete invert. An indication of the large volume of rock that is overstressed around the tunnel excavation is provided by Figure 10 which shows the predicted plastic or yield zone around the tunnel.

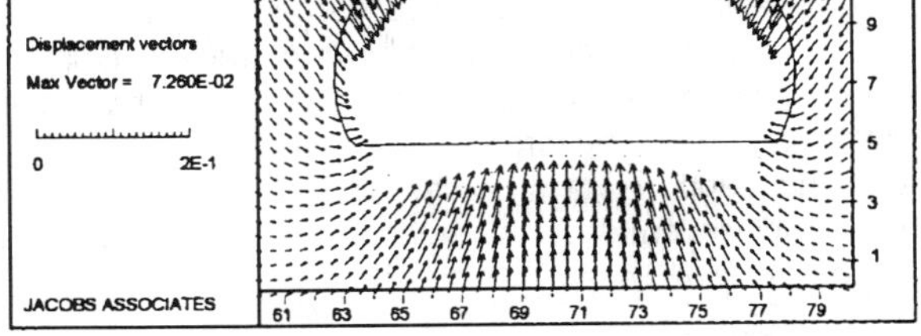

Figure 9. Predicted Displacements in a Major Fault Zone

All of the numerical analyses described above were performed prior to the start of tunnel construction and as such are based on an assessment of the rock properties from the geotechnical investigation. It is recognized that estimation of rock properties is not an exact science and it is intended to refine the material properties and the numerical models based on the observations and results of the instrumentation installed during the construction of the tunnels. Back analysis of the tunnel behavior will allow not only the refinement of the numerical models but also the refinement of the design for the various support elements.

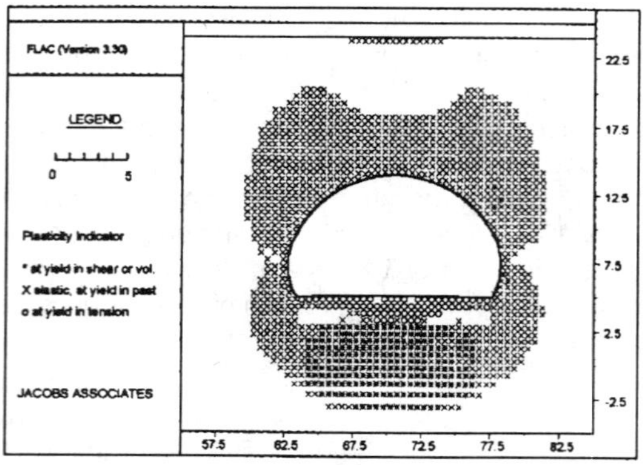

Figure 10. Yield Zone around Tunnel in Fault Zone

MONITORING PROGRAM

Instrumentation will be installed to:

1. monitor the deformation of the rock mass around the tunnel and confirm satisfactory performance of the tunnel support, and
2. determine the effects of the tunnel construction on the groundwater regime

The instrumentation program includes:

1. convergence monitoring points, with convergence measured using both tape extensometers (for the shaft and confined excavations) and total station (x,y,z) survey methods for the main drives.
2. multi-point extensometers (surface installations near portals and underground installations in the main drives).
3. slope inclinometers (for monitoring squeezing in mixed face conditions).
4. piezometers (for groundwater drawdown measurement).

As described above, estimates of possible convergence of the tunnel walls and crown were developed using numerical methods. On the basis of these estimates, allowances are made to determine the minimum overall excavated dimensions for the tunnels. It has been specified that where the measured convergence exceeds 75 percent of the recommended allowances, then additional support would be designed and installed, to limit further deformation in the rock mass and prevent encroachment of rock or primary support elements into the final lining envelope.

CONCLUSIONS

The use of empirical methods alone for the design of tunnel support is a reasonable approach for relatively small tunnels (say less than 10 m) in hard rock. However, these empirical methods should be supplemented by analytical methods for large tunnels, especially in weak rock, where experience indicates that the empirical methods have some limitations. The design of the MCL tunnels used a combination of empirical and analytical methods ranging from hand calculations to sophisticated numerical analysis. The empirical methods and less complex analytical methods were used to develop the basic support designs in the preliminary design phase. Analytical methods were used to develop support designs, later in the design program, when the rock conditions had been sufficiently characterized. During this stage the empirical methods and numerical analyses were used to check the designs and investigate the behavior of the supported tunnel excavations. Generally, it was found that the support designs developed using the analytical methods were comparable to the support measures recommended by the Q-System.

Numerical modeling provides insight into several important areas, including:

- the extent of the zone of disturbed or loosened rock that develops around a tunnel excavation in response to the stress redistribution that accompanies the excavation process.
- predominant deformation modes.
- an indication of whether the modeled support is adequate to control convergence.
- the loads developed within the simulated support elements.

Numerical modeling can also be used to develop ground reaction and ground/support equilibrium relationships that provide a good basis for: 1) understanding the behavior of the tunnel excavation, and 2) developing economical tunnel support designs.

Monitoring the performance of the tunnel excavations during construction is an important requirement for optimizing the design of the tunnel support. The information obtained from the monitoring program can be used to evaluate the overall behavior of the tunnel excavations and tunnel support, and allows calibration of the numerical models. Analytical methods can then be used confidently to refine the support designs and construction procedures which may potentially result in considerable savings, in both cost and schedule.

BIBLIOGRAPHY

Barton, N., 1996, Investigation, Design and Support of Major Road Tunnels in Jointed Rock using NMT Principles, Proc. of IX Australian Tunneling Conf., Sydney, 145-160.

Bieniawski, Z.T., 1993, Classification of Rock Masses for Engineering: The RMR System and Future Trends, Comprehensive Rock Engineering, J.A. Hudson ed., Vol. 3, Pergamon Press, New York, 553-573.

Chang, C.T., Lee, M.C., and Hou, P.C., 1996, Design of Twin-tube Tunnel through Soft Rock, Draft paper submitted to Conf. on Geotech. Aspects of Underground Construction in Soft Ground, London.

Chang, C.T., 1996, Vice President, Sinotech Engineering Consultants Ltd., Taiwan, Personal Communication.

Goodman, R.E. and Shi, G.H., 1985, Block Theory and its Applications in Rock Engineering, Englewood Cliffs, N.J.: Prentice Hall.

Goodman, R.E., 1989, Introduction to Rock Mechanics, John Wiley and Sons, New York.

Grimstaad, E. and Barton, N., 1993, Updating the Q-System for NMT, Proc. Int. Symp. on Sprayed Concrete - Modern Use of Wet Mix Sprayed Concrete for Underground Support, Fagernes, Oslo.

Hoek, E., 1996a, Review of Geotechnical Aspects, Melbourne City Link Tunnel Project, Melbourne, Australia, Project Report.

Hoek, E., 1996b, Consulting Engineer, Personal Communication.

Hoek, E., Kaiser, P.K., and Bawden, W.F., 1995, Support of Underground Excavations in

Hard Rock, A.A. Balkema.

Hoek, E., and Brown, E.T., 1980, Underground Excavations in Rock, London: Instn Min. Metall.

Hou, P.C., 1996, Deputy Manager/Senior Engineering Geologist, Sinotech Engineering Consultants, Inc., Taiwan, Personal Communication.

Itasca Consulting Group, 1995, FLAC Fast Lagrangian Analysis of Continua, Version 3.3, Itasca Consulting Group, Minneapolis, MN.

Lang, T.A., 1961, Theory and Practice of Rockbolting, Tans. American Inst. Min. Engrs, 220, 333-348.

Nelson, A., Porter, S., and Wilson, C., 1996, The Melbourne City Link Tunnels: Some Preliminary Design and Construction Concepts, IX Australian Tunneling Conference, August, Sydney.

Rabcewicz, L.V., 1969, Stability of Tunnels under Rockload, Water Power 21 (6-8), 225-229, 266-273, 297-304.

Rock Engineering Group, University of Toronto, 1992, Unwedge Version 2.22, University of Minnesota.

Sharpe, J.C., Endersbee, L.A., and Mellors, T.W., 1984, Design and Observed Performance of Permanent Cavern Excavations in Weak, Bedded Strata, Design and Performance of Underground Excavations: Construction, ISRM Symp., Cambridge, E.T. Brown and J.A. Hudson eds., 493-507.

Terzaghi, K., 1946, Rock Defects and Loads on Tunnel Supports. In Rock Tunneling with Steel Supports, (eds R.V. Proctor and T.L. White) 1, 17-99. Youngstown, OH: Commercial Shearing and Stamping Company.

Wickham, G.E., Tiedemann, H.R., and Skinner, E.H., 1972, Support Determination Based on Geological Predictions, In Proc. North American Rapid Excavation and Tunneling Conf., Chicago, 43-64.

Chapter 43

BUILDING THE UNDERGROUND RIVER TO SERVICE BUENOS AIRES

Bernie W. Martin,[1] Raymond Leray,[2] and Christian Criado[3]
[1]Director, Transportation and Infrastructure Klohn-Crippen Consultants Limited Toronto, Canada; [2]Deputy Manager of Infrastructure, Aguas Argentinas S.A., Buenos Aires, Argentina; [3]Director of Engineering Design Safege Montgomery Watson Buenos Aires, Argentina

ABSTRACT

Aguas Argentinas,[1] a consortium managed by Lyonnaise des Eaux of France and operator of the world's biggest privatized water supply and waste water collection system for the City of Buenos Aires, a city of 9 million people, was faced with the daunting task of providing new service to 1.6 million more people in the first five years of their 30 year concession.

Aguas Argentinas elected to use the Negotiated Compressed Process - NCP - for the fast track implementation of a 15000 meter, 3,5 meter diameter tunnel, a key component of the infrastructure master plan. NCP was developed by Klohn Crippen for the $200 million international railway tunnel between Canada and the United States. The conceptual design of the project was performed by Safege Montgomery-Watson, who engaged Klohn Crippen to assist with the NCP, the preparation of the tender documents, and implementation of the project during construction. The process included the purchase by the client Aguas Argentinas of two sophisticated 4 metre diameter earth pressure balance tunnel boring machines from Herrenknecht of Germany. Construction of the tunnel is being performed by the consortium Wayss & Freytag (Germany) and Sollazzo Hnos S.A. (Argentina).

The water tunnel is being built with a one pass, bolted segmented gasketed reinforced concrete liner.

The purpose of this paper is to describe a transferable methodology for this innovative contracting procedure, the project organization and quality control plan for this important world class tunnel project which is designed to service one of Latin America's largest metropolitan areas.

1.0 INTRODUCTION

At least 170 million people in urban areas lack a source of potable water near their homes, and the water supplied near their homes is often polluted. Concern for the deterioration of urban infrastructure and inefficient delivery of services has become, in Latin America, a major constraint on export growth. With 73% of its population living in towns and cities, Latin America is the most urbanized region in the developing world.

In 1990, Latin America contained four of the world's ten largest cities. The region has forty urban areas with more than one million inhabitants.

Most infrastructure entails large and lumpy capital costs that take decades to be amortized. Infrastructure developers and certain institutional investors share a long-term outlook. Environmental sustainable development does not mean foregoing convenient and efficient infrastructure. However, it does require efficient management, motivated employees and allocative and technical efficiency of municipal governments. This has lead to consider the role of the private sector in the delivery of municipal services. Clarity in the division of responsibilities for the delivery of services between the private and public sector would appear to be an essential condition of any reform in the structure of urban service delivery. Inefficient public sector monopolies are widely planned for the ineffective infrastructure provision. What emerges is the necessity of a private-public partnership where the guiding principal is to allocate the management and ownership of assets based on the comparative advantages of operating and investment efficiency.

Where the private sector has been invited to participate, Buenos Aires is but one example; improvements in service quality and staff productivity have followed.

The Buenos Aires system had fallen into disrepair as the government had stopped spending money on the system in the 1950's.

2.0 EXISTING SYSTEM

Buenos Aires, the capital of Argentina, is one of the largest cities in the southern hemisphere. The population of the metropolitan area is close to 9 million people. In 1993, only about 70% of that number was connected to the water supply network, with about 58% connected to the sewerage system. The remainder of the population was getting water from individual wells or privately owned water supply systems, developed out of necessity. The central system consists of:

- 2 water intakes from Rio de la Plata;
- 2 filtration water treatment plants - General San Martin with a capacity of 3.0 Mm^3/day (790 mgd) and General Belgrano with 1.0 Mm^3/day (264 mgd);
- 77 km (48 mi) of man-made underground rivers of "rios subterraneos" - tunnels ranging in depth from 5 to 30 m (16 to 98 ft), with diameters from 2.6 to 4.6 m (8 to 14 ft.);
- 9 pumping plants ranging in capacity from 340,000 m^3/day to 540,000 m^3/day (90 to 140 mgd);
- 338 water wells with a total combined capacity of 310,000 m^3/day (80 mgd) - only 270,000 m^3/day (71 mgd) currently available;
- 11,300 km (7,000 mi) of pipes, ranging in diameter from 75 mm to 2,100 mm (3 to 82 in), and
- 21 elevated water tanks with 278,000 m^3 (73 mg) of available storage - only 4 currently in use.

3.0 THE PRIVATIZATION PROCESS

To provide the necessary capital to expand and maintain the system, the government decided to privatize the water and wastewater services for Buenos Aires. An international proposal call was prepared by Sir William Halcrow & Partners in February 1992. In May of

1993, Aguas Argentinas was successful and became the operator of the world's largest privatized water and wastewater utility. Another joint venture company, comprised of Safege - Paris, France, and Montgomery Watson - Pasadena, California, is providing engineering services for the planning, design, and construction of the massive expansion and rehabilitation program required to fulfil the following contractual obligations on the water side of the system.

Aguas Argentinas elected to use the Negotiated Compressed Process -NCP- for the fast track implementation of a 15000 meter, 3.5 meter diameter tunnel, a key component of the $7 billion infrastructure master plan. NCP was developed by Klohn-Crippen for the $200 million international railway tunnel between Canada and the United States. The conceptual design of the project was performed by Safege Montgomery Watson, who engaged Klohn-Crippen of Toronto, Canada, to assist with the NCP, the preparation of the tender documents, and the review and analysis of the tenders received from the prequalified contractors, and manage the implementation and construction inspection of the project.

4.0 GEOLOGY

The geology of Buenos Aires, in the upper levels, is pleistocene deposits. These deposits include silty and clayey soils preconsolidated by desiccation with layers or lenses of cemented soils by calcareous impregnation. The total thickness of these sedimentary deposits is about 25 to 45 m; this part of the profile is the locally so-called "Formacion Pampeano". Underlying these formations, the "Puelches", fine and dense sands that correspond to the Pliocene Period into the dense sands there is a layer or stratum of plastic clayey silt or silty clay, a very plastic soil approximately 10 m thick. This green or blue clay is very stiff and generally normally consolidated under loads imposed by the Pampeano Average. SPT values vary from 27 to 42 which would indicate a very stiff to hard consistency.

5.0 THE PROJECT

The project consists of 15000 metres of 3.5 metre inside diameter soft ground tunnel built with one pass bolted gasketed concrete segment liners. Five access shafts, two of which will be converted into permanent water pumping once the mining is completed. As there is an intermediate target date of March 1998 to deliver water, approximately half the tunnel has to be completed by this date.

The access shaft/permanent pumping stations (P5 and P3) are each approximately 35 metres deep. Table 1 summarizes the accumulative chainage and internal dimensions of each shaft. It is noted that a shaft can be one cell or a cluster of up to 4 cells.

The molds for the segments are being produced by CBE of France with the following tolerances:

- Segment thickness ±2.0 mm
- Segment width ±0.3 mm
- Depth of the gasket slot ±0.5 mm
- Width of the gasket slot ±0.0 mm
- Error in location of bolt/dowel slots ±1.0 mm

Table 1

Shaft Number	Accumulative Chainage (metres)	Size (metres)
P2	21	1 cell, 7.6 metres dia.
P1	3597	4 cells, 11 metres dia.
P5	7794	2 cells, 11 metres dia. 1 cell, 12 metres dia.
P4	11 683	1 cell, 11 metres dia.
P3	15 316	2 cells, 11 metres dia. 2 cells, 12 metres dia.

The gasket is a Phoenix type M38527 designed to resist a maximum hydrostratic pressure of 3.0 bar and an internal water pressure of 1.5. A "tee-joint" test was performed with gaps and displacements exceeding project specifications and the phoenix gasket remained water-tight to pressures beyond 20.0 bar.

To manage the risk of variable geological conditions, stiff clays to sands with a high ground water table and for Occupational Health and Safety reasons, a "state-of-the-art" Earth Pressure Balance Tunnel Boring machine, manufactured by Herrenknecht of Germany, was selected. Chapter 6.0 describes the contracting process.

6.0 THE CONTRACTING PROCESS

There are four different contractual processes possible for the implementation of this project. These four methods are:

Traditional Competitive Bidding Process	(TCBP)
Design Build Process	(DBP)
Accelerated Bidding Process	(ABP)
Negotiated Compressed Process	(NCP)

The Negotiated Compressed Process, NCP, is a relatively new concept for construction contracting which can be considered as a variation of the European and American systems combined. It requires the involvement of all parties (client, engineer, contractor) in the selection and design of essential aspects of the project, which if pre-ordered and manufactured prior to selection of the main tunnelling contractor, could reduce the overall project schedule.

The concept was developed by Klohn-Crippen Consulting Ltd. (formerly Klohn Leonoff) for the two hundred million dollar CN intermodal tunnel between Canada and the United States. The NCP was adopted for the project. The key to this option was the pre-ordering of the TBM.

A key element of the NCP was the prequalification process for the general contractors, subcontractors and TBM suppliers. It was decided for the project that the segmental liners would be the responsibility of the general contractors.

Ads were placed in relevant trade journals. To prequalify, general contractors had to meet the following criteria:

1. Have a net worth of $150,000,000;
2. Have annual sales for the last 3 years of at least $400,000,000;
3. Have technical experience as follows:
   (i) Technical leader individually or of a joint venture with a project cost of at least $300,000,000;
   (ii) Technical leader individually or of a joint venture for a similar type of project of at least $100,000,000;
   (iii) Built in the last five years a tunnel using an earth pressure balance TBM, with reinforced concrete segments with a minimum diameter of 3.00 m and a length of at least 5 km, with access shafts and pumping stations.

All subcontractors had to have a net worth of at least $10,000,000, and relevant experience in similar municipal engineering works.

The following is the list of prequalified contractors:

1. Spie Batignolles, France;
2. Entrecanales Y Tavora S.A., Spain;
3. Wayss Y Freytag A.G., Germany;
4. Hochtief A.G., Germany
5. Bilfinger + Berger, Germany;
6. Ing. Civiles Asociados, Mexico;
7. Grupo Mexicano de Desarrollo, Mexico/Brasil;
8. Dickerhof & Widmann, Germany;
9. Impregilo, Italy.

Another element of NCP was the "Technology Agreement". The purpose of the Technology Agreement was to save harmless, release and indemnify AA from and against any and all claims, demands attributable to the design, manufacture, delivery, assembly and performance of the TBM. Only those contractors who maintained their pre-qualification status and who executed the Technology Agreement were allowed to bid the project.

Similarly, TBM suppliers were prequalified and technical information requested. Four suppliers responded:

1. Lovat, Canada
2. N.F.M., France/Japan
3. Herrenknecht/Germary
4. FCB, France/Japan

Each supplier was required to supply a list of similar type machines, the capacity of the fabrication plant and details of the machine. This technical information included:

1. General layout of machine;
2. Head and plenun;
3. Method of confinement and extraction;
4. TBM propulsion;
5. Method of erection of concrete segment;
6. Guidance system/pilotine;
7. Communication system;
8. Hydraulic systems;
9. Probine of soils in advance;
10. Method of soil treatment in advance of the TBM;
11. Trailing gear;
12. Grouting system;
13. Ventilation;
14. Fire protection.

TBM suppliers were required to supply performance criteria such as the speed of excavation, erection segments and grouting of annular ring.

Herrenknecht of Germany was awarded the TBM supply contract on June 23, 1995.

Letters to prequalified general contractors went soon afterwards to participate as a member of the technical committee to review the project parameters and geology. A meeting was held in Buenos Aires on the 26th and 27th of July which introduced the project to the technical committee which consisted of the prequalified contractors and presented an overview of the regional geology and the detailed geotechnical investigations. Another meeting of the prequalified contractors was held at the Herrenknecht plant in Germany on August 2, 1995. Tenders were received October 26, 1995.

A contract was signed with Wayss y Freytag, Sollazo HNDS., S.A., U.T.E., the 18th of April 1996. The commence work order was issued April 19, 1996. The anticipated completion date is April 19, 1999.

The construction documents also included the requirement for the contractor to prepare:

1. A quality control plan for construction.
2. An occupational health and safety plan.
3. A detailed critical path program using Primavera software, including monthly cash flow projections connected to detailed cost breakdown divided into pay items.
4. Detailed construction designs and drawings including all engineering surveys for the Owner's review.
5. A program for monitoring the "quality" of the muck derived from the excavation of the tunnels.
6. A detailed monitoring plan to verify and analyze the impacts if only on the settlement of adjacent structures.

7.0 PROJECT MANAGEMENT

Project management was a crucial element for the timely delivery of the project in order that Aguas Argentinas fulfil its obligation to the state. An intermediate target date was included in the contract documents to permit approximately half the tunnel project to be put into service by February of 1998, to service an additional 1.6 million customers. The liquidated damages therefore were correspondingly high to cover the opportunity costs.

The specifications also included a requirement that within 28 days of signing the contract a detailed construction schedule with all the activities connected to an anticipated cash flow for “early” and “late” activities using PRIMAVERA software be provided.

The engineer used a computer program to electronically manage the flow of documents expected to number in the thousands of pages by the end of the project. Minutes of meetings, submittals, and change orders were all prepared using a code of keywords developed at the beginning of the project. All key staff members of the construction management team were trained to use the software and the keyword system.

8.0 CONCLUSIONS

Aguas Argentinas expect to save a minimum of 12 to 15 months in the implementation of the NCP. The general conditions of the contract are those developed by the International Federation of Consulting Engineers (FIDIC).

9.0 REFERENCES

Cisic, S., Martin, L., de Clermont-Tonnere, E., Belley, J.C., R., 1995, “How to Model the Biggest Privatized Water Supply System in the World - Case Study of Buenos Aires”, C.P.F.

Martin, B.W., MacLennan, D.G., 1992, “St. Clair River Tunnel ... Past, Present, Future”, 10th Annual Canadian Tunnelling Conference, Banff, Canada.

Martin, B.W., 1995, “Management of Urban Infrastructure - Recapitalization and Equilibrium Model”, Master of Engineering Thesis, Carleton University.

[1] Aguas Argentinas is a joint venture company comprised of Lyonnaise des Eaux (France), Sociedad Comercial del Plata S.A. (Argentina), Sociedad General de Aguas de Barcelona (Spain), Meller S.A. (Argentina) and Banco de Galicia y Buenos Aires S.A. (Argentina), Compagnie Générale des Eaux (France), CFI Corporación Financiera Internacional (Washington), Anglian Water (Great Britain), and employees of Aguas Argentinas. The principal shareholder and the operator of the system is Lyonnaise des Eaux.

Chapter 44

SLURRY SHIELD TUNNELS ON THE CAIRO METRO

Donald P. Richards,[1] Patrick Ramond,[2] and Martin Herrenknecht[3]
[1]Parsons Brinckerhoff International, Cairo, Egypt; [2]Campenon Bernard SGE, Paris, France; [3]Herrenknecht GmbH, Schwanau, Germany

ABSTRACT

Phase 1 of the second line of the Cairo Metro includes approximately 5.83km, of double track tunnels excavated in water bearing sand of the Nile flood plain. Closed face bentonite slurry shield tunnel boring machines 9.43m in diameter were used. State-of-the-art technology was used in TBM operations and guidance, with surface settlements averaging less than 14mm, and alignment deviations averaging less than 25mm. Production rates up to 580m/month were achieved, with average production of 4.8m/shift, 13.4m/day and 78m/week. The completed tunnels, lined with precast concrete segments, are completely watertight under 20m of water pressure.

INTRODUCTION

Cairo is a huge sprawling city. The largest city in Africa, the Middle East and the Arab world, with a population estimated at 18 million. It is situated at the head of the Nile delta occupying a position at the crossroads of Africa, the Middle East, and the Mediterranean Basin and has a history stretching back 5000 years. Although Cairo's population has boomed in recent years, from 3.5 million in 1960 to its present day figure, the city has been a great metropolis for many centuries with Medieval historians calling the city "Mother of the World" and "the Greatest Metropolis in the Universe".

Historically, tunnels in the Cairo area have been limited to the rock plateaus on the sides of the Nile Valley and associated with tombs, cave dwellings and the like. It is only recently that bored tunneling has been carried out within Cairo and this is predominately associated with the Cairo Wastewater Scheme on the east bank of the Nile where soft ground tunneling using compressed air, earth pressure balance, and slurry shields has been successfully carried out (Flint & Foreman, 1991; Flint, 1992).

FACILITIES DESCRIPTION

As Cairo began its rapid expansion in the 1960's and 1970's the need for a mass transportation system was perceived. Studies were carried out between 1970 and 1974 which led to the concept of the Greater Cairo Metro System. The plan calls for a Regional Line and two Urban Lines as shown in Figure 1 which will be the heart of the city's mass transportation system.

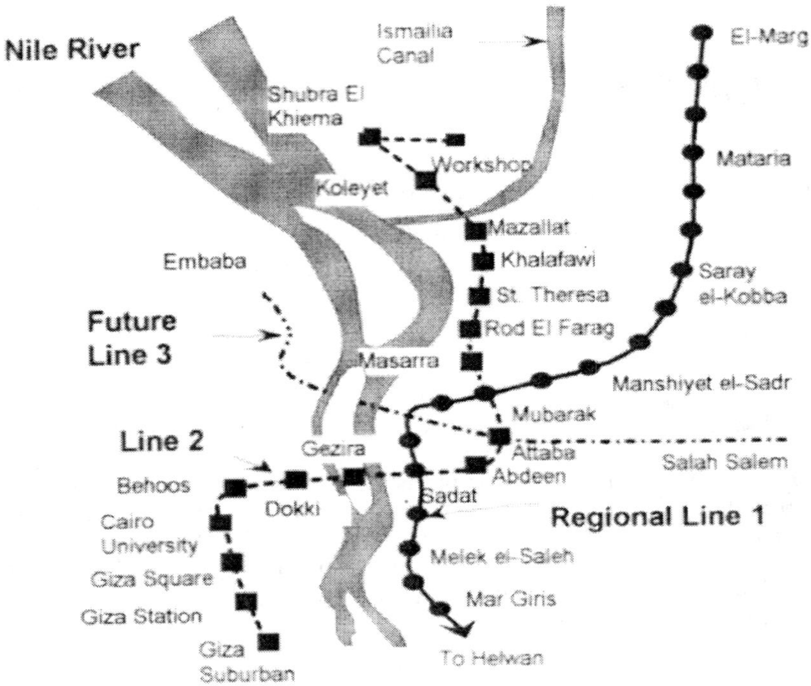

Figure 1. Map of Cairo Metro

The Regional Line (Line 1) was completed in 1989 and is the first metro in Africa. It is 42.5km long with about 4.5km underground through the heart of the city, built using cut and cover construction methods, as noted by Abdel Salam (1990,1991). The second line of the Cairo Metro, Phase 1, which is discussed in detail herein, has been described in general by Abdel Salam (1992). As in most metro construction, the route traverses a heavily congested area of high population density, and generally follows existing streets in order to minimize interferences with adjacent structures.

Line 2 was planned to be constructed and put into operation in two phases. Phase 1, running from Shubra El Kheima north of the city down into the heart of the central business district, intersects with Line 1 at Mubarak Station, and then terminates at Sadat Station which again will form an interchange with the existing Line 1. The northern portion of Line 2 from Shoubra El Kheima to Mubarak opened for service ahead of schedule in October 1996. The section from Mubarak to Sadat is schedule for opening in the summer of 1997. Phase 2 will extend this line across to the west bank of the Nile and terminate at Giza Suburban Station south of the city.

Line 2, Phase 1 includes a 2.5km long above ground section and an 8.5km underground section shown in Figure 2. The underground section can be summarized as follows:

Cut and Cover Tunnel	1.4 km	Cut and Cover Stations	7
Bored Tunnel	5.9 km	Existing Stations Enlarged	2

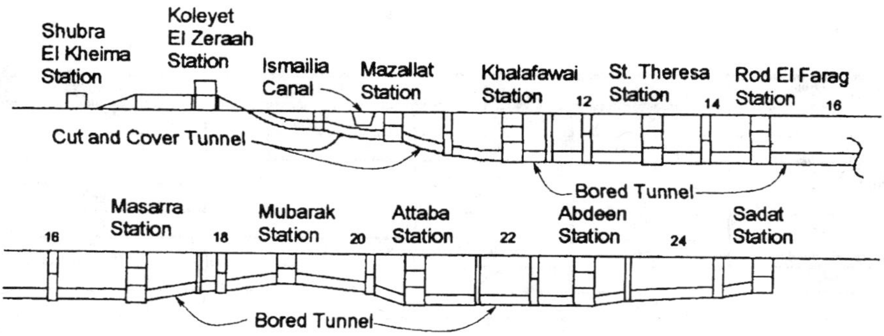

Figure 2. Line 2 Phase 1 Vertical Alignment

The system is a twin track heavy metro with both tracks in a single tunnel and side platforms at the station. As noted by Abdel Salam (1992), a single tunnel with double tracks was chosen to minimize construction impact upon adjacent existing facilities, and as a cost saving compared to twin parallel tube single track tunnels.

The Organization responsible for the planning design and construction of the metro system is the National Authority for Tunnels (NAT), a division of the Ministry of Transport. NAT was assisted by the Greater Cairo Metro Consultants, a joint venture of Parsons Brinckerhoff International, Electrowatt Engineering Services and Sabbour Associates. The civil works for design and construction of the 8.5km underground section was awarded as one contract to a consortium led by Interinfra and included Campenon Bernard-SGE, Dragages, Bouygues, Dumez, GTM, Soletanche-Bachy Egypt, Spie Batignolles and Arab Contractors.

SITE CONDITIONS

The geology of Cairo has been outlined in considerable detail by Shata (1988). He has noted that regionally, Cairo is underlain by tertiary sedimentary rocks and quaternary soils, both underlain by older basement rocks. The project area lies totally within the geomorphic unit known as the young alluvial plain which represents the majority of the lowland portion of the Nile Valley in the Cairo vicinity (See Richards and Burchell, 1994, for more details on geology). The ground and ground water conditions for Metro construction were governed by the Nile River. The Pleistocene age sediments in the alluvial plain are generally fairly consistent with depth, but vary somewhat laterally as a result of the long history of river meanders, and alternate cycles of sedimentation and erosion before the construction of Aswan High Dam in Upper Egypt in the 1960's. These sediments are approximately 60-90 meters thick in the Cairo area. The ground water level varies seasonally with the level of the Nile River, with the water level low in the winter when releases from Aswan High Dam are at a minimum. In addition, it varies annually with the river level as some years have a higher flow rate than others depending upon precipitation and run off in central Africa in the watershed of the upper Nile.

Generally the soil profile consists of a surficial fill layer, which varies in thickness (but is normally not more than a few meters thick) underlain by a natural deposit of stiff, relatively

Table 1. Geotechnical Parameters

SOIL TYPE	FILL	CLAY	SISAND	SAND
γ_T (KN/m^3)	18.0	18.5	19.0	20.0
K_o	.58 - .66	.66	.50	.38 - .40
E' (MPa)	3-7	4-11	25-45	50-160
v'	0.40	0.40	0.35	0.30
C' (KPa)	0	0	0	0
ϕ' (Degrees)	20	20	30	33-38
C_u (KPa)	50	75	-	-
E_u (MPa)	3-8	4-13	-	-
v_u	0.40	0.50	-	-

massive clay, which also varies in thickness (but is usually about 6 meters thick). This clay layer is underlain by a silty sand transition zone followed by an extensive deposit of sand which extends downward beneath the limits for Metro construction. It is generally poorly graded at any particular depth. The sands frequently contain lenses of silts and clays or gravels, and infrequently contain layers of cobbles and sometimes boulders at depths up to about 30m or so. Typical design parameters for the subsurface soil deposits as reported by Hamza (1993) are noted below in Table 1.

TUNNEL LINING DESIGN

The original bid documents required a precast concrete segmented tunnel lining. The tunnel structure was required by specification to be 8.30m inside diameter with a 40cm thickness, producing an external diameter of 9.10m. During the contract negotiations, an increase in the tunnel alignment tolerances of 25mm on the radius was approved, provided that the internal radius of the lining be increased accordingly in order to maintain the required structure gauge clearances. The final internal diameter for design was then adjusted to 8.35m with an external diameter of 9.15m, maintaining the 40cm thickness, and providing for a total construction tolerance of 125mm on the radius. The bid documents required a segment width of 1.20m , but this was increased to 1.50m during contract negotiations.

The precast concrete segmented tunnel lining was designed for the worst case conditions anticipated, considering applied loads due to overburden soil, external groundwater pressure, loads from adjacent or overlying structures, and all handling, erection and jacking loads. Full consideration was given to soil structure interaction, accounting for variations in the deformation characteristics of the surrounding soil. The detailed analysis of the tunnel was performed using a finite element program which accounted for soil deformations at the face and around the tunnel. Final deformations of the completed structure were estimated, considering the relatively strict requirements for the inside clearances between the rolling stock and the inside face of the tunnel lining and/or permanent utilities or appurtenant structures fixed to the tunnel wall.

The tunnel lining consisted of bolted precast concrete segments, with a single tunnel ring composed of seven segments plus a key. The lining ring was tapered to allow lining of the tunnel with a single ring design for any of the designed vertical, horizontal, or compound curves down to a radius of 150m, by simply rotating the position of the key (thinnest section of the taper) to achieve the desired overall curvature. The taper was 90 mm per ring, evenly divided between opposite faces. The segments were all cast using 64 molds (eight for each of the eight segments) at Egyptian Company for Precast Concrete. Production started on 23 February 1994 and was completed on 10 January 1996. A total of 3886 acceptable rings (31088 segments) were cast.

Tunnel Watertightness

The tunnel was required by specification to be watertight according to the following criteria:

Segments:	Class 1	=	Completely dry;i.e.nomoisture.
Joints :	Class 2	=	Substantially dry; i.e. slight moisture

The bid documents required watertightness to be achieved by the following methods, each of which is shown graphically in Figure 3, and each is discussed in more detail in the following paragraphs.

Elastomeric Gasket. The elastomeric gasket required by specification was to fit into a prepared groove around the perimeter of each segment, and a Phoenix brand gasket was approved for use. The gasket was fitted to all segments installed in the permanent tunnel lining.

It should be noted that the Contractor decided to supplement the required watertightness measures by adding a hydrophilic seal behind the elastomeric gasket. The brand name of this hydro expansive elastomeric strip was Aquaprene C 7 x 20, manufactured by Sanyo Chemical Industry Co. and distributed by Fosroc. This sealant material was fitted to all permanent segments on two faces: one circumferential joint and one longitudinal joint. This was expected to substantially increase the watertightness based on previous experience as noted by Burchell and Ellis (1987).

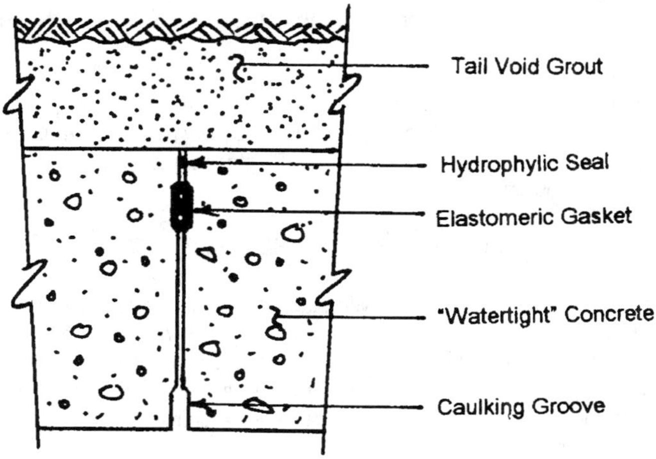

Figure 3. Tunnel Watertightness

Bolting. Galvanized bolts with a rubber grometted washer were proposed and approved for installation. All bolts installed were left permanently in place, contrary to the Contractor's request to remove them. Installation procedures required retightening the bolts in ring "n" after starting the shove for ring "n+1" in order to compensate for bolt loosening resulting from gasket compression due to the axial thrust developed during shoving of the shield.

Grouting. The TBM was designed to include automatic tail void grouting as the machine advanced. Injection was pressure controlled by upper and lower limit sensors, with the limits adjustable. The injection pipes were embedded in the tailskin of the machine at eight locations around the perimeter.

The tail void grout used was a lime and silica fume based mortar. This mortar provided strengths of about 1.0-1.2 MPa in 28 days, and had setting times of about 12-16 hours. This long setting time prevented set up of the grout in the grout mixing tank, pump, and injection piping system during short stoppages for ring erection, repairs, etc. The grout was injected at a pressure of about 3.5 bars. The pressure was always in excess of the hydrostatic water pressure in order to allow a certain amount of the mixture water to bleed into the surrounding ground, leaving a more dense (and higher strength) mortar material than in the laboratory specimens.

It should be noted that the injection quality plays a large part in the overall quality of the finished tunnel. It is a dominant factor in both short and long term conditions. In the short term, settlement control is largely dependent upon proper injection parameters. It is essential that the injection system allow for immediate adjustment of injection parameters such as volumes and pressures. In the long term, injection quality is critical to maintaining the position and shape of the erected lining ring. It should also be noted that complete and homogeneous injection of the tail void must be achieved since this mortar constitutes the first barrier against ground water seepage through the tunnel lining.

Caulking. The tunnel lining joints between individual segments were required by specification to be caulked in a groove cast into the four sides of each segment at the inside surface as shown in Figure 3. The specified watertightness of the tunnel was achieved without this caulking.

Watertightness Additives. Several different concrete admixtures were used during segment casting to improve the watertightness of the concrete segments themselves as structural units. During construction, no evidence of water migration through the uncracked concrete segments was observed.

TUNNEL BORING MACHINE SELECTION

The water bearing sandy soils of the Nile sediments limit the tunnel construction methods to full face tunneling machines. Compressed air tunneling with such a large face area is not feasible due to the difference in water pressure from the crown to invert of the tunnel plus the fact that air losses would be extremely high in the permeable ground. The selection of tunneling machines therefore fell to a decision between a slurry machine or an earth pressure balance (EPB) machine.

Broadly speaking, slurry machines are designed primarily for sandy or gravelly soils (see Maidl et.al. 1996). As long as a bentonite cake can be formed, it is possible to maintain face stability by applying a counter balance pressure. Slurry machines can be used in clayey ground but the slurry treatment process may require more time and be more costly if the clay disaggregates in the slurry fluid. EPBs are preferred to slurry machines when the grain size of the soil decreases to silts and clays. Low permeability clayey silty layers can safely be kept stable by applying an earth pressure. Other parameters to be considered when selecting a tunnel machine include items such as site surface area capacity for treatment plant with fine grained soils, discharge regulations, cost of additives for EPBs etc.

Soil conditions anticipated for the shield bored tunnel were generally the deeper sand deposit, although the finer grained transition layer upper deposits were expected locally where they dipped down to tunnel grade or where the tunnel rises in elevation to enter the stations. It was therefore critical to tunnel excavation progress to have the slurry treatment plant sized appropriately to "desand" the muck/slurry at a rate such that the excavation progress was not delayed. It was anticipated that the fines content (percent passing 74 microns sieve size) would vary from approximately 7 or 8 percent (as encountered in the tunnel face) in the deep sands to about 60 percent or more locally in the overlying finer grained deposits. An overall "average" fines content of about 8 to 12 percent was expected. The gradation of the sand and the overlying transition zone are shown in Figure 4.

Two bentonite slurry shields built by Herrenknecht of Germany were selected to drive the tunnels. Two machines instead of one were chosen in order to satisfy the construction schedule requirements. Details of the TBM design have been given by Herrenknecht (1994), and are outlined briefly below.

Machine Characteristics

The tunnel boring machine shield was 9.43m in diameter and the cutterhead was 9.45m. The shield was 9.5m (including cutterhead) in length. It was composed of three basic components; the cutterhead, the main shield body, and the tailskin. The shield weight was 4500 KN (459 T). The cutterhead weight was 930 KN (95T), and the cutterhead drive weight was 460 KN (47 T) for a total shield weight of 5890 KN (600 T). Table 2 summarized the various machine parameters.

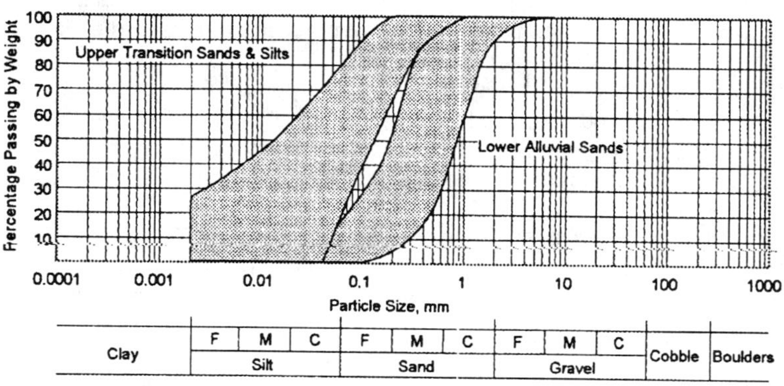

Figure 4. Grain Size Distribution

Table 1. TBM Operational Parameters

OPERATIONAL PARAMETER	UNITS	INSTALLED CAPACITY	NORMAL RANGE OF OPERATION
TOTAL INSTALLED POWER	kw (hp)	2450 (3264)	-
ROTARY DRIVE POWER	kw (hp)	1000 (1360)	-
ROTARY POWER/FACE AREA	kw/m²	14.3	-
TORQUE	T-M	770	440
ROTATION SPEED	rpm	2.25	1 - 1.5
ADVANCE RATE	mm/min	80	25-45
AGITATOR SPEED	rpm	0-60	50
NO. OF JACKS	pairs	15	all
TOTAL THRUST	tons	6000	2000 - 2500
SLURRY PUMP POWER	kw (hp)	2190 (2978)	-
SLURRY PUMP FEED	m³/hr	1500	900 -1100
SLURRY PUMP RETURN	m³/hr	1500	1100 - 1300
BENTONITE FACE PRESSURE	bars	3.0	1.5 - 2.5
GROUT INJECTION CAPACITY	m³/hr	40	15-20
GROUT INJECTION PRESSURE	bars	> 40	3.0 - 3.5
DRAINAGE	m³/hr	60	As required

This cutterhead was adjustable in position independent of the shield body, by hydraulic rams with a maximum extension of 40cm, which could be varied around the circumference. This capability provided a skew of the cutterhead allowing 4cm of overcut. The tailskin was extendable at the articulation joint a maximum of 25cm, again variable around the circumference.

The cutterhead was capable of operation in either direction of rotation, with cutter teeth arranged accordingly. The cutterhead was initially equipped with 280 drag bit type cutter teeth and 21 dual 38cm diameter roller disc cutters as shown in Figure 5. However, after the initial tunnel drive in Lot 12, (See Figure 2 for definition of tunnel drive Lot nos.) nine of the central roller disc pairs were removed to provide more open area in the cutterhead face thereby allowing better material flow into the cutting chamber. At the perimeter of the cutterhead were three gage cutters adjustable between 0 and 20mm, controlled manually for specific overcutting.

The design openings in the cutterhead face allowed passage of material up to 200mm in size. The machine was not equipped with a crusher to break this material size into smaller pieces in order to minimize potential blockages of the slurry pumps and piping. Inside the cutting chamber were located two counter-rotating agitator wheels to keep the excavated solids in suspension until removal by the slurry suction pumps. Face pressure of the bentonite slurry was provided and controlled by an air bubble chamber built into the body of the shield. The shield was equipped with a two chamber air lock to allow up to four workmen to enter the cutting chamber to repair or replace machine components as required.

The shield used an electro-mechanical erector for segment installation, having 6 degrees of freedom and a lifting capacity of eight tons. The erector used a vacuum pick up system, with the vacuum seals arranged to provide seals for both the larger regular segments and the smaller key segment.

Grouting of the annular void around the tunnel lining was accomplished by four sets of dual (normal and spare) injection pipes embedded in the tailskin of the shield. The normal injection points were located at the 2, 4, 8, and 10 o'clock positions. Back-up or spare

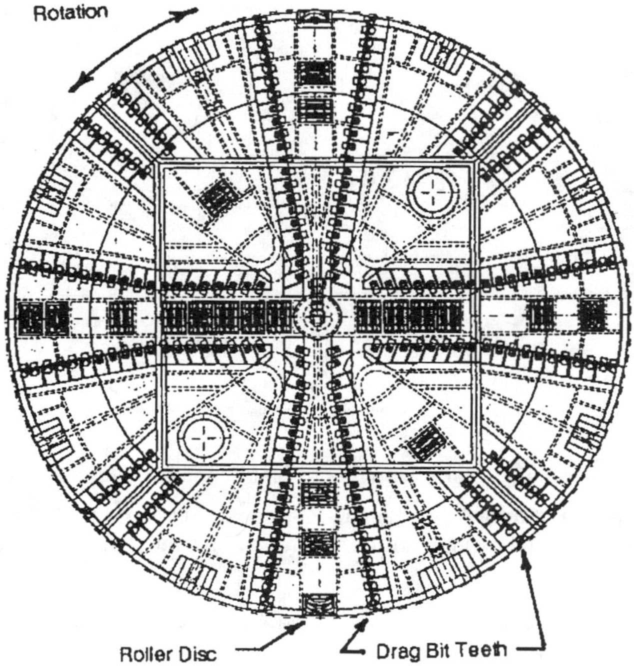

Figure 5. TBM Cutterhead Layout

injection ports were located at approximately the 3,6,9 and 12 o'clock positions. The injection pipes were equipped with pressure sensors to control automatic tail void grouting as the shield advanced, as discussed previously.

The tailskin was equipped with a flat plate seal at the rear, backed up by two wire brush seals. A paper pulp based mastic material, CONDAT WR 89, was continuously injected between the tail seals to provide a supplemental seal against ground water and tail void grout ingress around the perimeter of the erected lining. Sealant consumption averaged approximately 4.9 liters (6.9kg) per linear meter of bored tunnel.

Backup Equipment

The Tunnel Boring Machine also included four trailers which were pulled behind the advancing shield. Each trailer's equipment is briefly described below.

Trailer 1. Upper deck: Hydraulic motors for TBM cutterhead, thrust jacks, erector, agitator,etc.
Lower deck: Segment handling, slurry by-pass circuit, emergency drainage pump, grease pump for tailskin.

Trailer 2. Upper deck: Control cabin, mortar tank and pumps, electrical controls for slurry pump, drainage pump.

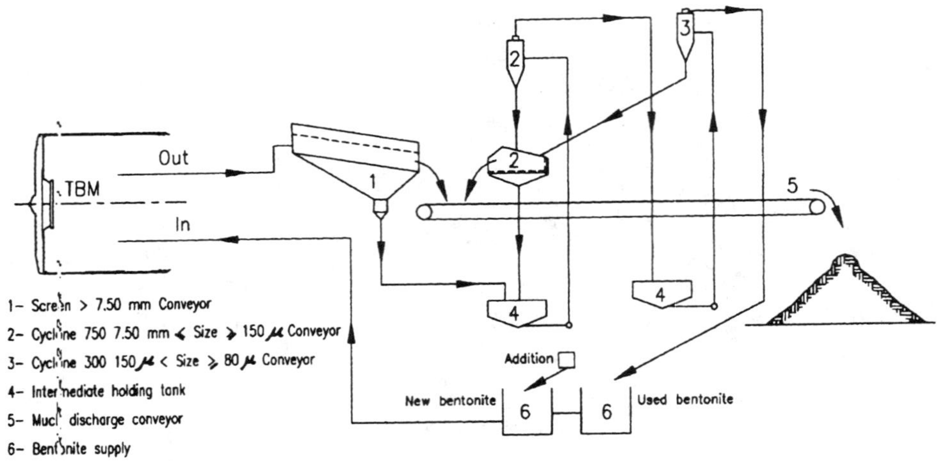

Figure 6. Slurry Treatment Plant Schematic Flow Chart

Lower deck: Slurry pump station, telescopic slurry lines for negotiating curves, slurry density measuring device, mortar pump control system, compressed air tanks.

Trailer 3. Upper deck: Electrical distribution cabinets, transformer station, cable drum.

Lower deck: Workshop and storage area, air compressor, sliding rail fastenings, suspension and guidance for telescopic slurry and feed lines.

Trailer 4. Upper deck: Crew container, crane for changing slurry pipes, ventilation hose storage.

Lower deck: Pipe storage for slurry and feed lines, mechanized telescopic pipe retraction, end suspension for long telescopic pipes.

The TBM, including all four trailers, was 65m in length with a total weight of 8045 KN (820T).

The TBM equipment was initially assembled with the assistance of Herrenknecht staff from the factory under subcontract to the Civil Works Joint Venture. In addition, this arrangement was implemented again when the TBM's required their first disassembly, transport, and reassemble at a new start up site. After the initial subcontract disassembly, all other disassembly/reassembly activities were done using the Contractor's staff.

Slurry Treatment Plant

The treatment plant was designed for slurry processing at a rate of 1500 m^3/hr at a density between 1.0 and 1.2g/cm^3. It was designed to handle a maximum of 15 percent of the solids > 7mm in size, 99 percent between 7mm and 80 microns, and 15 percent smaller than 80 microns. The treatment process included several steps of solids removal, starting with passage of the slurry over a vibrator screen to remove particles > 5mm in size. The next step passed the slurry through six cyclone desanding units removing particles down to

Table 3. Bored Tunnel Summary

LOT NO.	LOCATION	LENGTH (m)	TBM NO.	NO. OF ANNEX STRUCT	NO. OF GALLERY CONNECT	START* DATE	COMPL* DATE	DURATION CALENDAR DAYS
12	KHA TO STT	795	1	2	2	06/07/94	20/10/94	105
14	STT TO REF	743	1	1	2	25/11/94	13/02/95	80
16	REF TO MAS	888	2	1	2	08/09/94	01/03/95	174
18	MAS TO MUB	998	2	2	2	03/04/94	23/06/95	81
20	MUB TO ATT	740	2	1	2	21/10/95	11/01/96	82
22	ATT TO ABD	986	1	1	1	21/07/95	09/10/95	79
24	ABD TO SAD	683	1	1	1	03/04/96	21/05/96	49

* Day/Month/Year

80-100 microns. Finally, the slurry passed through two units of eight cyclones each for silt removal down to 50 microns in size. A flow chart for the MS treatment plant is shown in Figure 6.

TBM Operations Staff

When both TBM's were operating simultaneously, the Contractor had a support staff of approximately 280 people, including 43 expatriate managers and supervisors. When the tunnel boring schedule required only one machine operating at a time, the total staff was reduced to approximately 190 people, including 26 expatriates. The typical TBM operation and support staff for one TBM on any particular day or shift was as follows:

Operations Underground:	22 local;	3 expatriates
Operations Topside:	8 local;	no expatriates
Treatment Plant:	3 local;	1 expatriate
Supervision:	2 local;	10 expatriates

In addition, there were two expatriate mechanics plus 12 locals on day shift and six on night shift to support all TBM related operations. The summary also does not include staff at the on-site batch plant that supplied tail void grout, nor does it include staff involved in segment production.

The success of the tunnel boring operations was strongly dependent upon having knowledgeable and experienced expatriate supervision familiar with the state-of-the-art equipment and techniques being used. Use of the same equipment by inexperienced staff would not likely have achieved the same successes in ground control and excavation progress.

Support Services

The tunnel boring operation was supported during boring by electrical power, tunnel lighting and ventilation, industrial water, industrial air, and construction trackwork for a fleet of diesel locomotives and associated vehicles for hauling personnel, materials, and equipment to the TBM. Electrical power was supplied by an on-site unit of five 1360 KVA diesel generators for each TBM so that tunnel boring operations would not be susceptible to, or otherwise interrupted by, unscheduled outages and voltage fluctuations. Surface support facilities included a 25 ton capacity gantry crane for lowering of precast concrete lining segments and other materials to the construction train. The TBM components were installed and removed by a 220 ton capacity gantry crane. Other top-side facilities included offices, material storage areas, and mechanical, electrical, and welding shops.

Table 4. Bored Tunnel Progress (M)

LOT NO.	BEST DAY	AVG.** DAY	BEST SHIFT	AVG.** SHIFT	BEST WEEK	AVG.** WEEK	DELAY DAYS	BEST MONTH	AVG. MONTH
12	25.5	8.6/8.1	10.5	2.9/2.7	100.5	48.7	3	289.5	213
14	27.0	9.9/4.8	10.5	4.8/3.3	150.0	70.8/59.4	25	349.5	306/248
16	27.0	10.3/5.6	10.5	3.5/1.9	135.5	58.0/32.9	70	378.5	240/148
18	34.5	17.5/14.3	13.5	5.8/4.8	168.0	81.8/76.7	6	537.0	388/360
20	33.0	20.9/11.3	13.5	7.5/3.6	205.5	125.3/65.7	34	565.5	543/291
22	27.0	13.4	12.0	4.5	138.0	80.8	0	414.0	350
24	34.5	19.5	13.5	6.5	169.5	117.0	0	367.5	422
TBM 1	34.5	12.9/11.5	13.5	4.7/4.3	169.5	79.3/76.5	28	414.0	323/308
TBM 2	34.5	16.2/10.4	13.5	5.6/3.4	205.5	88.4/58.4	110	565.5	390/266
OVERALL	34.5	14.3/11.0	13.5	5.1/3.9	205.5	83.2/68.7	138	565.5	352/290

** XXX/YYY = WORKED DAYS/ACTUAL (CALENDAR DAYS)

Note: TBM No. 1= Lots 12, 14, 22, 24: Best Month = 355m Avg.
TBM No. 2= Lots 16, 18, 20: Best Month = 493m Avg.

TUNNEL CONSTRUCTION

Progress Rates

Normal TBM operations were carried out 24 hours per day, six days per week. As can be seen in Table 3, boring of the first section of tunnel on Phase 1A (Lot 12) started on 6 July 1994. Completion of the last section of tunnel in Phase 1B (Lot 24) was on 21 May 1996. The completion date was 50 calendar days ahead of the contractual completion date of 10 July 1996. Productivity varied considerably for the different tunnel drives. Details are shown in Table 4.

It can be seen in Table 4 that progress was relatively good, despite delays encountered for various reasons. Average boring rates even considering the unanticipated delays, were 11.0m/day. This actual boring rate exceeded the assumed Contract boring rate of 7.6m/day (average for the seven tunnel drives) by about 45 percent. If considering the average production rate during actual boring of 14.3m/day, the assumed baseline average rate was exceeded by 88 percent.

Alignment Control

The specifications as amended during contract negotiations allowed tunnel construction tolerances of 75mm misalignment from the theoretical centerline, and 50mm out-of-roundness measured on the radius, for a total radial alignment tolerance from the theoretical centerline of 125mm.

The Contractor used a computerized guidance system (CAP) on the TBM supported by conventional laser survey to bring survey controls from the ground surface into the tunnel and following the TBM as it advanced. The CAP system, developed by Campenon Bernard SGE, is a computerized system used as a navigation and steering aid. The program calculates the best trajectory and the corresponding sequence of rings required to correct for deviations from the theoretical centerline. This data is then input into the steering unit which controls the shoves automatically. Smooth automatic steering and optimization of the TBM's mechanical functions saves cost to the Contractor and gives better quality results to the owner.

Using the designed taper of the standard tunnel lining ring, the theoretical layout of tunnel lining rings was initially evaluated to establish the required key position to achieve the theoretical alignment. As the TBM advanced, the attitude of the shield in three dimensional space was constantly monitored and deviations from the theoretical centerline computed. Using this data, an estimate was made of the actual position of the last ring relative to the theoretical position. The computer guidance program then developed the required thrust jack pressures and elongations to force the TBM back to it's theoretical position, over the next four or five lining rings. In addition, the orientation of the key segment for the next lining ring was calculated, considering the actual orientation of the previously erected ring and the correction measures implemented on the thrust jacks for the next shove. This procedure was continuously repeated throughout each tunnel drive with good results, with the theoretical alignment tolerance exceeded for only about 3 percent of the total number of lining rings installed.

Even for these out of tolerance rings, the actual ring shape deviation was sufficiently small that the out-of- tolerance on alignment could be absorbed by the remaining unused shape tolerance such that the total allowable tolerance of 125mm was never violated. A summary of as-built alignment data is included in Figure 7. The measured average vertical deviation of the actual centerline position from the theoretical centerline was 22mm and the average horizontal deviation was 19mm.

Segment Damage Statistics

Segment damage experienced during construction of the bored tunnels was mainly of four types: (1) tension cracks in the segments of completed rings while shoving for the next ring, (2) spalling of concrete due to localized concentrated stresses on the counter key segment when forcing the key into an opening that was too small, (3) minor spalls and chips during handling and erection, and (4) less frequent but still important, spalling due to contact between the erected lining ring and the tailskin. These are discussed individually in the following paragraphs.

Cracks: This type of damage was the most frequent. The cracked segments averaged about 2.6 percent of the total, and ranged from about 0.5 to 4.6 percent for the different drives as shown in Table 5. It can be seen in Table 5 that the cracking frequency was higher for TBM # 1 than for TBM # 2, indicating a difference in the performance of the machine and/or the skill of the crews. In addition, it can be seen that the crack frequency was influenced by the "learning curve", with a decrease in frequency with succeeding tunnel drives for both machines.

Cracking problems generally increased on tight horizontal curves due to the non-uniform jack forces required to force the shield around the curve, and the forced eccentricity between the tunnel centerline and the shield centerline. Exceptions to this are in Lot 12 which had the highest crack frequency with no tight curves, and in Lot 20 which had a relatively low crack frequency, but had several tight horizontal curves.

There appeared to be three factors which contributed to the control of cracking. First was the ability of the TBM crews to maintain ring plane, so that there were no high points on the previous ring against which a new ring was erected. Although most rings were erected without any packing materials between rings, when ring planarity checks indicated non-plane conditions, thin plywood, fiberboard, or ductile plastic sheeting was installed at appro-

Figure 7. Summary Alignment Statistics As-Built

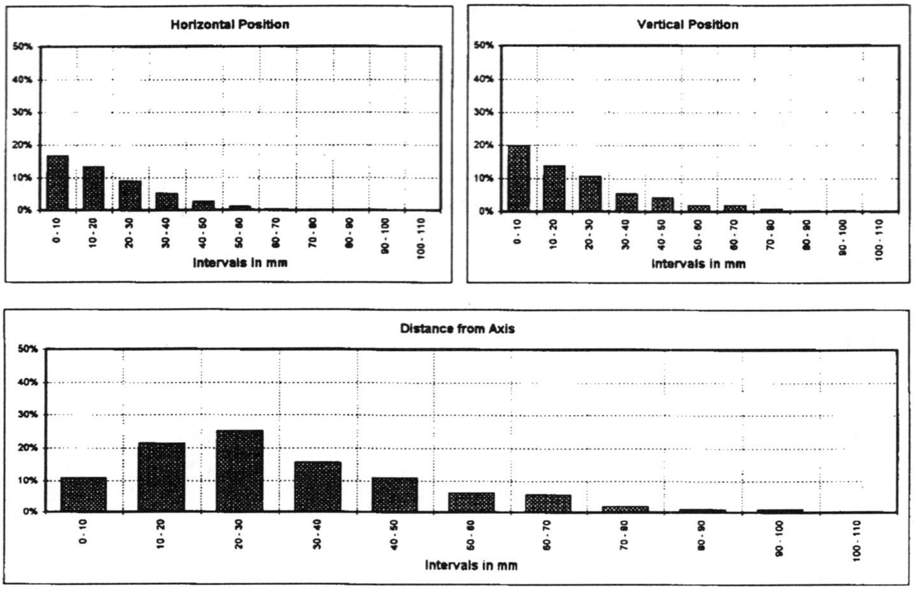

priate locations and in appropriate thickness to correct the out-of-plane conditions. The second factor contributing to cracking appeared to be eccentric loading from the thrust jack load distribution pads. Cracking appeared to be slightly more frequent when these pads were not "centered" in the leading face of the segment. the characteristic and efficiency of the TBM guidance system described previously which provided a relatively uniform load distribution on the rings, which kept the center of thrust in a 5mm circle, as well as basing all guidance calculations on the plane position of the previously erected ring.

Keyway Spalls: Segment damage due to keyway spalls were less frequent, only occurring 31 out of 3886 rings, or 0.8 percent of the completed rings. The damage was easily preventable by maintaining the correct keyway segment opening width before key insertion. This type of damage usually occurred only when the erection crews were in a hurry to make another ring, and neglected to provide the proper clearance required by the approved ring erection procedures. When this occurred the damaged segment was removed and replaced if possible, but if not, a Non Compliance Notice was issued and the damaged segment was repaired in-situ according to approved procedures.

Minor Chips and Spalls: This damage in handling usually occurred either at the segment edge if knocked against the TBM frame during handling or at the centering hole when the pickup device was slightly off center. All damages of this type were repaired after the ring was erected.

Iron-Bound Spalls: Occasionally, the Contractor was unable to keep the erected lining ring centered within the tailskin. In these cases, in extreme circumstances, the lining came

Table 5. Segment Damage Summary

SEGMENT DAMAGE PARAMETER	DAMAGE SUMMARY BY LOT NUMBER							
	12	14	16	18	20	22	24	OVERALL
TBM NUMBER	1-1	1-2	2-1	2-2	2-3	1-3	1-4	-
NO. HORIZONTAL CURVES	2	1	2	5	4	3	NONE	17
MINIMUM CURVE RADIUS (m)	1501	2001	2001	221	201	201	-	201
NO. ERECTED RINGS	530	495	592	665	493	656	455	3886
NO. ERECTED SEGMENTS	4240	3960	4736	5320	3944	5248	3640	31008
NO. CRACKED SEGMENTS	194	131	121	122	48	160	20	796
TOTAL NO. OF CRACKS	274	235	152	48	160	20	32	1189
% CRACKED TO TOTAL SEGMENTS	4.6	3.3	2.6	2.3	1.2	3.0	0.5	2.6
MAX. CRACKS IN ONE SEGMENT	5 X 1	5 X 2	4 X 2	7 X 1	4 X 2	4 X 2	3 X 4	7 X 1
NO. CRACKS/NO. CRACK SEGMENTS	1.40	1.79	1.25	1.46	1.63	1.50	1.60	1.49
NO. CRACKS/TOTAL SEGMENTS	0.06	0.06	0.03	0.03	0.02	0.05	0.009	0.038
NO. SEGMENTS WITH SPALLS	82	18	147	4	0	25	5	281
% SPALLED TO TOTAL SEGMENT	1.93	0.45	3.10	0.07	0	0.48	0.13	0.9
NO. CRACKS REQUIRING REPAIR	4	0	0	0	11	0	0	15

into contact with the tailskin (iron-bound) and sometimes caused spalling of the leading edge of the lining segments at the point of contact. This occurred on only 14 of 3886 rings, or 0.4 percent of the completed rings.

<u>Production Damage:</u> Segment damage during the casting and handling process at the precast factory was minimal, with only 0.3 percent of the segments rejected due to damage or poor quality.

Tail Void Grout

The lime based mortar discussed previously was the primary tail void grout used in bored tunneling. However, for the last few rings erected at a break-out to a station, an "active" cement based grout was used to provide a more rapid setting time. In isolated cases where the completed tunnel was not watertight using the conventional grout materials, polyurethane foam was injected to supplement the standard grout materials. The theoretical injection volume to fill the annular void was 7.24m^3 per ring. This was generally exceeded, with injection volumes as high as 26m^3. Occasionally, less than the theoretical volume was injection, as low as 4m^3. Overall, the actual injected volume exceeded the theoretical by about 9 percent.

The "excessive" grout injection was attributed to bleeding as discussed previously, use of higher injection pressures to minimize ground displacements, slight overcut on curves to allow the shield to negotiate the curve easier, but which increases the thickness of the annular void, and possibly allowed the grout to flow past the shield into the cutting chamber. The low grout volumes were attributable to slight closure of the annular void before completion of grouting or breakdown of the pumps. The highest grout volumes were more difficult to explain, but were likely due to flow of the grout forward past the shield to mix with the bentonite at the face. This was speculated on several occasions by dramatic changes of bentonite properties due to suspected contamination by mortar, and checked by monitoring fluid levels in the cutting chamber while the TBM was stopped, but mortar being injected. It could conceivably have been a malfunction of the measuring equipment. Checks of surface settlements at these locations did not show any evidence of over excavation.

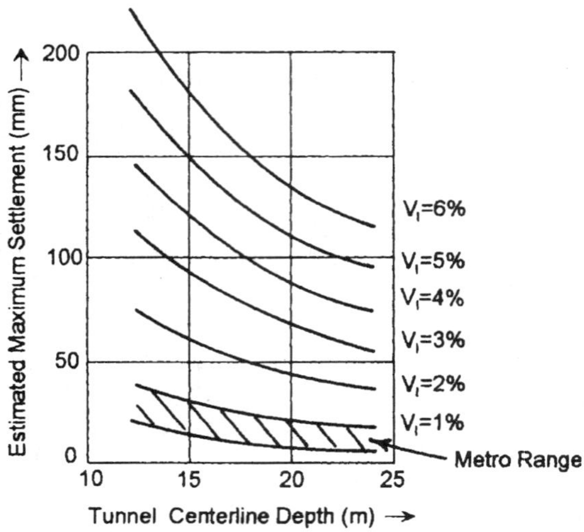

Figure 8. Shield Tunnel Settlement

During driving of the tunnel, the mechanism of measuring grout volumes was to count the number of strokes of the pump and multiply it by the volume of each stroke. However, there were occasional problems with the counters giving unreliable data in which case the volumes were estimated by monitoring the level of mortar in the mixing tank on the TBM. These minor problems introduced some degree of uncertainty in the grout volume data. Tail void filling was reasonably complete and effective, demonstrated by the observed small surface settlements.

Table 6. Surface Settlemtn Summary

PARAMETER	VALUE BY TUNNEL LOT NO.							
	12	14	16	18	20***	22	24	OVERALL
TUNNEL LENGTH (M)	795	743	888	998	740	986	683	5833
NUMBER OF ERPS	18	32	27	43	29	27	33	209
MAX δ (MM)**	54	36	73	50	8	14	15	73
MIN δ (MM)	5	10	10	0	1	4	2	0
AVERAGE δ (MM)**	16	16	23	17	5	9	6	13

**EXCLUDES ISOLATED COLLAPSE INCIDENTS

*** HIGHER THAN USUAL SLURRY PRESSURES USED TO PROTECT OVERLYING 800MM WATER MAIN PARALLELING THE TUNNEL ALIGNMENT

Surface Settlements

In an urban area, settlements from tunneling are extremely important, particularly when dealing with old, sensitive structures in close proximity to the tunnel route. This, combined with generally shallow foundations led to the incorporation of the continuous grouting system through the tail of the shield, as noted previously. Bored tunneling with the "automatic grouting system" was very successful at maintaining ground and ground water control with only a few exceptions. Considering the size of the tunnel excavation (9.43m), and the often shallow overburden adjacent to the stations (as little as 7.3m - i.e. < one diameter), settlements were very small.

Bored tunneling on the Greater Cairo Wastewater System (Flint and Foreman, 1991, and Flint 1992) on the east bank of the Nile, using both EPB (Earth Pressure Balance) and bentonite slurry shield machines recorded significant surface settlements for smaller and generally shallower tunnels. Actual ground loss values ranged from about 3 to 6 percent. If similar ground losses had been experienced on the larger metro tunnels, ground settlements would have ranged from approximately 30 to 130mm as seen in Figure 8. However, the constant and efficient supporting of the tunnel face and the automated grouting system kept ground loss generally less than one percent, which reduced settlements to about 30mm or less (see Figure 8). These were of tremendous benefit in minimizing damage to adjacent structures as a result of construction settlement.

Actual measured settlements for the Metro tunnel construction are summarized in Table 6. It can be seen in Table 6 that "average" settlements ranged from about 5-25mm for the individual tunnel drives, averaging about 13mm overall. It can be seen in Figure 8, that settlements of this magnitude correspond to ground loss volumes between about 0.5 and 1.0 percent.

These low values of settlement were attributable to two key factors: (1) The production of a bentonite slurry which formed a good, supple filter cake at the excavation face under a slurry pressure sufficient to minimize ground movement ahead of the face without excessive risk of blow outs.; (2) The TBM provision to grout the tail void through the tailskin of the shield as the TBM advanced, providing sufficient pressures and volumes to minimize the amount of closure of the tail void as the erected lining came out of the tailskin.

It should be noted however, that there were a few exceptions to the generally good TBM performance at controlling the ground and ground water. These exceptions usually occurred either as the cutterhead was leaving the treated ground at a break-in, or when the cutting chamber was under compressed air for maintenance and/or inspection of the cutterhead. At six locations, face stability was lost and a chimney to the ground surface with corresponding surface subsidence occurred.

Watertightness

Watertightness of the completed tunnels was very good, generally exceeding the specification requirements. At isolated locations, supplemental grouting through the segments was required to produce dry joints between the lining segments. This usually occurred where minor lining displacements occurred due to the disturbances to the ground and the stress distribution changes at the connection between the bored tunnel and the annex connection galleries, or at the first (or last) lining connection at the station diaphragm wall. Correction of these "leaks" was a relatively minor effort. Caulking, although provided for in the technical specification, was not necessary to achieve the required watertightness performance. Areas requiring supplemental treatment to control seepage are as noted below:

Lot 12	6 rings =	1.13 %		Lot 20	5 rings =	1.01%
Lot 14	4 rings =	0.81 %		Lot 22	3 rings =	0.45%
Lot 16	3 rings =	0.51%		Lot 24	None =	0.00%
Lot 18	None =	0.00%		Overall	21 rings =	0.54%

Bentonite Consumption

Bentonite consumption in these fine to medium sands of low permeability was low. Approximately 17Kg of dry bentonite per in-situ cubic meter of excavated material was the gross average consumption on the project for the seven bored tunnel drives.

TBM Performance

The two Herrenknecht TBMs and supporting MS slurry treatment plants exhibited very good dependability. The majority of delays of any significance were attributable to non TBM related activities. TBM performance data is summarized in Figure 9.

The performance of tunnel boring machines depend firstly upon equipment setup and dependability. The global setting from treatment plant to gantries or tower cranes, locomotives and finally TBM proved to be excellent. Workmanship and organization helped to prevent unnecessary stoppages by systematic checks and maintenance programs. From the diagrams (Fig. 9) it can be noted that both TBMs had their first drives slightly slower than the later drives, which was due to the learning curve and is not surprising.

The fact that globally, the performance did not really improve that much over time when compared to the overall rates for the duration of the project, shows that good choices of equipment were made from the early stages of planning, and that sufficient experienced manpower was available. It shows as well that once local staff were properly trained, similar or even better rates were achieved with fewer expatriates. It should be noted that if non-TBM related delays are excluded, and only actual working days are considered, TBM availability ranged from 79 to 96 percent for the seven tunnel drives, averaging 87 percent for TBM #1,93 percent for TBM #2, and 90 percent overall. Actual machine utilization averaged 67 percent for TBM #1,72 percent for TBM #2, and 62 percent overall.

PROBLEMS AND SOLUTIONS

Despite the remarkably good performance of the tunnel boring machine and the successes at minimizing settlement and maintaining good alignment control, several construction requirements presented unusual problems requiring innovative solutions. These are described briefly following.

Crossing Beneath Existing Tunnel

In Lot 16 north of Masarra Station, the existing 2.20m diameter (ID) Boulac Spine

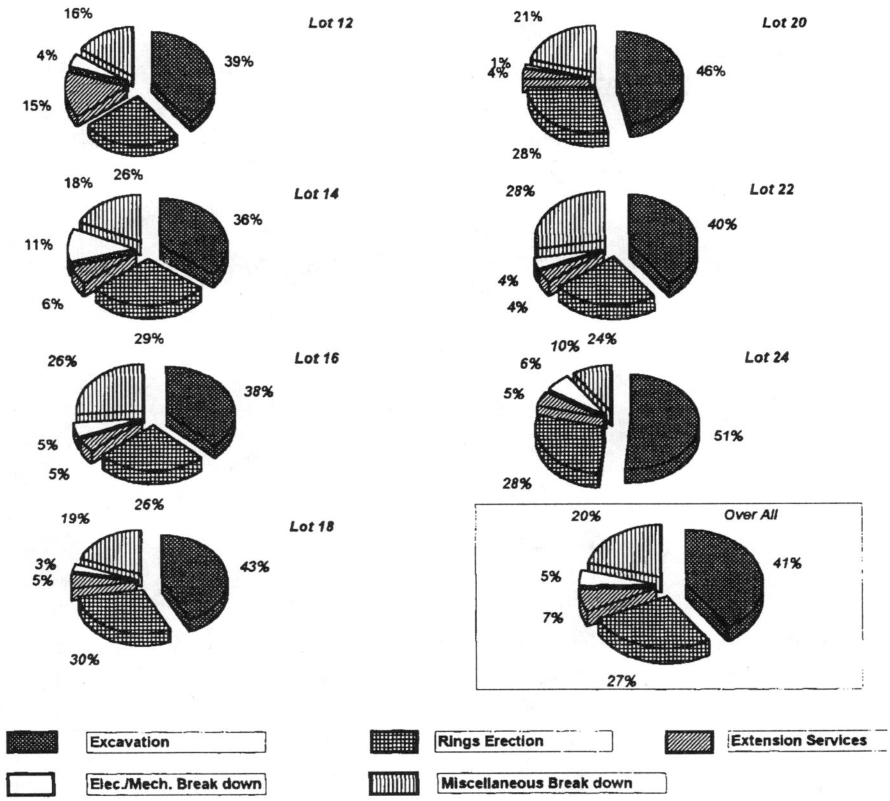

Figure 9. TBM Performance Statistics

Tunnel of the Cairo Wastewater Organization (CWO) crossed over the Metro tunnel alignment with a clearance in between of 5.0m. Although the existing tunnel had been strengthened with an internal steel lining when originally built, in anticipation of the future Metro tunnel beneath it, the potential settlements from Metro construction would still have presented a risk to the tunnel structural integrity. Consequently, the Metro Contractor, in conjunction with NAT and CWO/Ambric (the CWO consultant) developed an underpinning scheme of ground treatment by mineral grout to consolidate the ground surrounding the existing tunnel before the Metro tunnel passed beneath. Upon completion of this grouting, instrumentation was installed to monitor the settlements of the existing tunnel during Metro construction. The Metro construction was successfully completed between 6 and 9 November 1994 with only 9mm of measured settlement on the Boulac Spine Tunnel.

ENR Mainline Tracks and Offices-Lot 18

Over a 100m section in Lot 18, the Metro tunnel passed beneath seven mainline tracks of Egyptian National Railway (ENR) with a vertical clearance of 10m. In addition, 25m of this section of tunnel crossed beneath the ENR headquarters offices. In this critical zone, the bentonite slurry pressure at the face was increased slightly as well as a slight increase in the

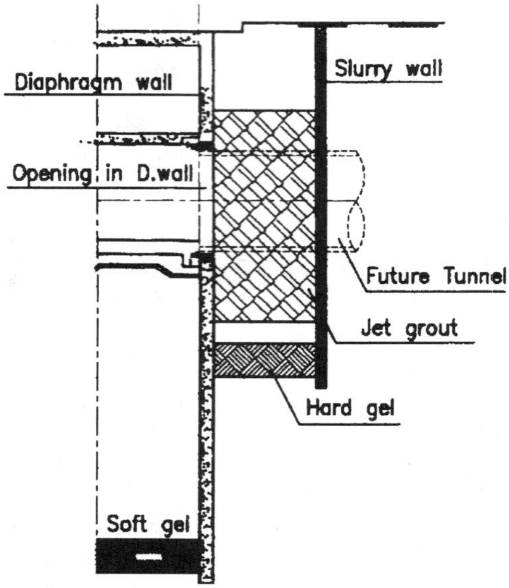

Figure 10. Composite Break-in

grout injection pressure in order to further reduce potential ground surface settlement. The tunnel passed safely beneath the ENR tracks with only 7mm of settlements, and beneath the structures, with only 4mm of settlement. There was no observable distress to any of the facilities.

Future Tunnel Crossing in Lot 20 & 22

At these locations, the Metro tunnel construction preceded CWO construction. The

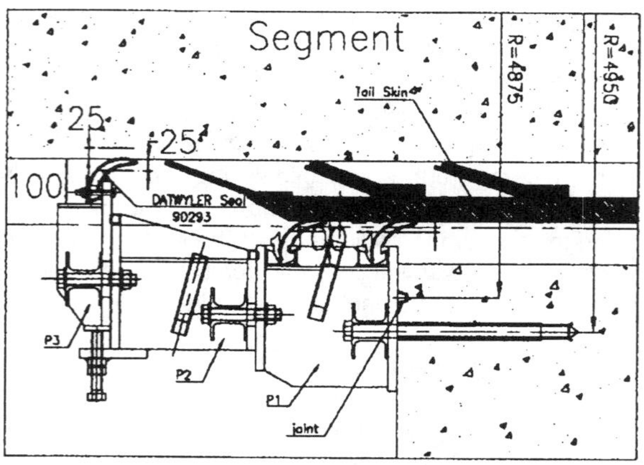

Figure 11. Break-in Seal Arrangement

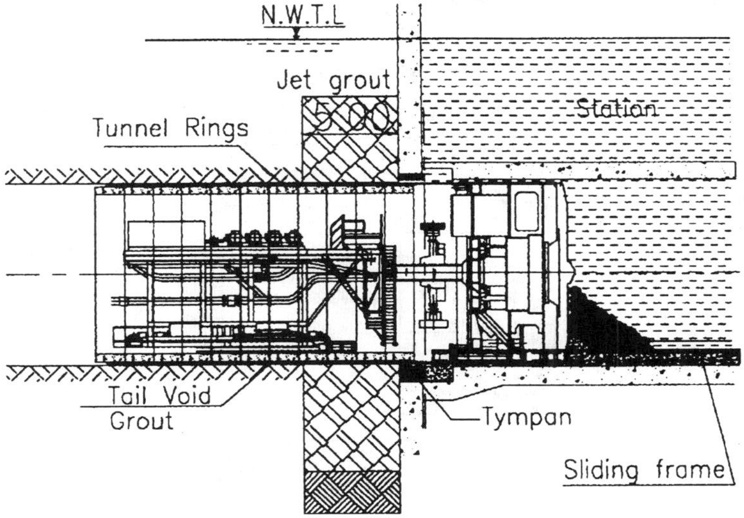

Figure 12. Groundwater Balance Concept

future CWO construction included an 0.80m diameter micro tunnel crossing over the Metro tunnel with a vertical clearance of 3.50m (Lot 20), and a 2.00m diameter bored tunnel using EPB equipment (Lot 22). Prior to Metro construction, joint surveys were made by Metro and CWO staff to establish baseline conditions prior to any tunnel construction. Elevations at reference points along the Metro tunnel centerline and on adjacent structures were agreed upon initially, and then re-surveyed following the Metro TBM passing the site. The effects of Metro tunnel construction in Lot 20 was a maximum of 6mm of settlement of the ground surface overlying the tunnel, and 3mm of maximum settlement of adjacent structures. In Lot 22, the Metro construction produced 10mm of settlement on adjacent structures, and the CWO crossing added 6mm more settlement.

Break-In & Break-Out

Ground treatment for the break-in and break-out of the TBM at structures varied considerably from site to site as a function of space available, equipment available, time available, and previous experience with the success or failure of a particular concept. The final concepts of break-in and break-out were based on the premise that the safest and lowest risk option should include ground treatment of sufficient length to allow several permanent lining rings to be grouted in before the cutterhead broke into either virgin ground (break-in) or into the station structure (break-out). The final concept is shown in Figure 10. For further details, the reader is referred to Richards et. al., 1996.

In addition to ground treatment, supplemental seals were required inside the station structure to control possible infiltrating water before the first few permanent rings were installed at break-in. The seal used successfully on the Cairo Metro (Figure 11) included two circular rubber seals (P-1) placed against the perimeter of the shield body. These were mounted in a circular frame around the tunnel eye, which was itself sealed against the struc-

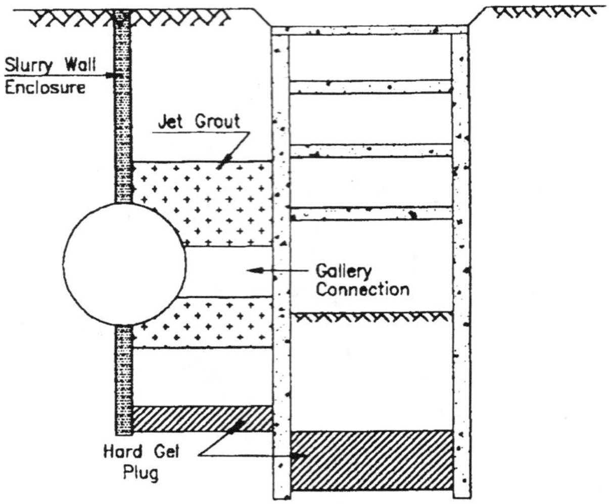

Figure 13. Annex Connection Schematic

ture wall. As the shield advanced forward, with the completed lining left in place, the lining being of smaller diameter than the shield body required installation of a smaller diameter circumferential seal (P-3) behind the tailskin of the shield. For additional protection against ground water leaFiFkage past the two mechanical (rubber) seals, grout such as polyurethane or cement mortar was systematically injected into the frame (P-2) connecting the P-1 and P-3 seals. As the TBM advanced forward, the tunnel lining rings were grouted into place to provide a permanent seal.

At three of the break-out locations, the schedule, space, or equipment availability did not permit development of the full ground treatment as outlined above. Consequently, a modified concept with a shorter length of ground treatment was developed in which the ground treatment was supplemented by flooding of the entry structure to improve ground water control. This "ground water balance" concept is shown in Figure 12. Further details are given by Segnana et. al (1997) and Richards et. al. (1996).

Boulders

In the Lot 22 tunnel drive, approximately 500m of the 986m drive encountered a layer of cobbles and boulders that significantly hindered TBM progress. This layer was later revealed to be approximately 1m in thickness. The boulders affected progress in two ways. First, the boulders had a tendency to block the slurry pumps and/or piping system. The efforts to stop the TBM and clear the blockages took a cumulative total of approximately 120 hours. The boulders creating the blockages were generally on the order of 20-25cm in smallest dimension, and oblong in shape, composed of either limestone or petrified wood (suspected palm trees). The second effect, and more serious, was the breakage of cutter teeth, which first became evident when they were discharged at the slurry treatment plant. After observation of this phenomenon, a compressed air intervention was undertaken to inspect the cutterhead and assess the damage. This inspection revealed that 40 percent of the

Figure 14. Typical TBM Breakout

teeth and teeth holders were either damaged or missing, resulting in severe damage to the cutterhead as a result of abrasion in the area of the missing teeth. The broken teeth were replaced and new teeth put in the opening where the roller discs had previously been removed. These efforts required an additional 50 hours. The drive was successfully completed, and cutterhead repair fully executed while the TBM was stopped at Abdeen Station.

Annex Structures

At the approximate center of each tunnel drive, generally coincident with the low point to the drive, an annex structure was located for purposes of ventilation/dewatering and/or electrical cables. These structures were composed of a shaft constructed by conventional diaphragm walls with a grouted plug at the bottom, connected to the bored tunnel with a gallery hand mined from the shaft to the completed tunnel through a block of treated ground. For eight of the nine annex connections, ground treatment for gallery construction was completed from the ground surface prior to passing of the TBM. The final design concept was a block of jet grout surrounded by a slurry wall and underlain by a hard gel plug as shown in Figure 13. Details are given by Campo et. al 1997. In total, twelve gallery connections to the bored tunnel were completed successfully, with only 2mm displacement measured on the previously completed tunnel lining rings.

CONCLUSIONS

From the previous discussions it is evident that except for a few isolated incidents, the bored tunneling by bentonite slurry shields proved to be very successful at maintaining ground control to minimize surface settlements while at the same time achieving admirable advance rates. The tunnels were driven within the allowable alignment tolerance with isolated exceptions. They have proven to be essentially watertight, with only a few leaks requiring supplemented treatment to seal off the leak. The TBMs proved to be very reliable and dependable and provided a completed tunnel with minimal cracking of the segments, none of which experienced any water leakage through the cracks. A photograph of a typical break-out is

Figure 15. Completed Tunnel

shown in Figure 14 and the completed tunnel is shown in Figure 15.

ACKNOWLEDGEMENTS

The authors acknowledge the following organizations and individuals for their support in preparing this paper:

- Parsons Brinckerhoff International - Cairo, Egypt, for support in prepar ing the paper.
- National Authority for Tunnels - Cairo, Egypt, for permission to use selected data.
- Campenon Bernard SGE, and Bouygues; both of Cairo, Egypt and Paris, France.
- Karen Richards for final preparation of the manuscript.
- David Campo, Ahmed Zakeria, and Khalid Farid for assistance in preparation of figures.

REFERENCES

Abdel Salam, M.E., 1991, "Cairo Metro Network", Proc. Conf. Tunnelling in Congested Cities, Cairo, pp. 3-20.

Abdel Salam, M.E., 1992, "Cairo Metro Network Line 2", Proc. Cong. Current Experiences in Tunnelling, Cairo pp. 1-15.

Burchell, A.J. and F.W.S. Ellis, 1987, "Recent Developments in Tunnelling as Applied inthe Construction of the Singapore MRT", Proc. Intl. Conf. on Modern Techniques in Construction, Singapore, pp., 466-484.

Campo. D.W., D. P. Richards and M.Coudry, 1997, "A Review of Grouting on Line 2 of the Greater Cairo Metro", Proc. RETC, Las Vegas, SME, Littleton, CO.

Flint, G.R., and W. Foreman, 1991, "Bentonite Tunnelling for the Greater Cairo Wastewater Project,", Proc. Conf. on Tunneling in Congested Cities, Cairo, pp. 231-252.

Flint, G.R., 1992, "Tunneling Using Earth Pressure Balance Machines for the Boulac Spine Sewers for the Greater Cairo Wastewater Project", Proc. Cong. Current Experiences in Tunneling, Cairo, pp. 299-325.

Hamza, M., 1993, Geotechnical Investigation and Recommended Geotechnical Properties - Cairo Metro Line 2, Phase 1, Various Reports.

Herrenknecht, M. and B. Maidl, 1994, "Transferring the European Experience Using Mixshields for the Employment in Cairo", Proc. ITA Conf. Tunnelling and Ground Conditions, Cairo, pp. 333-337, A. A. Balkema, Rotterdam.

Maidl, B., M. Herrenknecht and L. Anheuser, 1996, Mechanical Shield Tunnelling, Ernst & Sohn, Berlin.

Richards, D.P. and A.J. Burchell, 1994 "Engineering Geology consideration in the Planning, Design, and Construction of the Cairo Metro", Proc. 7th IAEG Congress, Lisbon, pp., 4223-4231, A. A. Balkema, Rotterdam.

Richards, D.P., A.J. Burchell, D.W. Campo and P. Ramond, 1996, "Review of Break-in and Break-out Concepts for Tunnel Boring Shields in Saturated Soft Ground", Proc. ITA Conf. North American Tunneling 96, Wash. D.C., pp. 451-460, A.A. Balkema, Rotterdam.

Shata, A.A., 1988, "Geology of Cairo, Egypt", Bulletin of the Association of Engineering Geologists, Vol. XXV, No. 2, pp. 149-183.

Segnana, J., P. Ramond, B. DeLorenzi and D.P. Richards, 1997, The Ground Water Balance Concept for Tunnel Break-outs, under preparation.

14

Soft Ground Case Histories

Chairman
D. Heinz
Kenny Construction Co.
Wheeling, IL

Chapter 45

AN INNOVATIVE GROUND SUPPORT SOLUTION USING "SOFT-CENTERED" JET GROUTING

Rafael C. Castro,[1] Gregg Sherry,[2] John Overton,[3] and David A. Sutter[4]
[1]Senior Engineer, Haley & Aldrich, Inc., San Francisco, CA; [2]Vice President, Haley & Aldrich, Inc., Denver, CO; [3]Project Manager, Tutor-Saliba-Perini Joint Venture, San Francisco, CA; [4]Project Manager, San Francisco Public Transportation Commission, San Francisco, CA

ABSTRACT

This paper presents an innovative ground support solution developed to construct a tunnel cross-passage for the MUNI Metro Turnback (MMT) Project. A "soft center" jet grouting approach was designed to provide solid ground support while maintaining a soft center to allow for hand-mining of a 2.44 m (8 ft.) wide, 3.66 m (12 ft.) tall and 4.6 m (15 ft.) long cross-passage through soft Bay Mud. Prohibitive costs and schedule impacts associated with mining under compressed air (at the time being used for main tunnel construction) were the catalysts to search for alternative methods for cross-passage construction. Limited surface disruption, relatively low costs, ease of subsequent excavation, nominal schedule impacts and proven local success of jet grouting were key factors in selecting and designing this method of ground support.

INTRODUCTION

The MMT Project consisted of extending the San Francisco MUNI Metro System from the existing Embarcadero Station below Market Street to surface rail on the newly reconstructed South Embarcadero Roadway. The Owner and Project Manager was the San Francisco Public Transportation Commission and the prime Designer and Construction Manager was the Bechtel Corporation. The project was constructed by a joint venture of Tutor-Saliba-Perini.

The project's main components, shown on Figure 1, consisted of an underground turnback facility for switching and storage of MUNI rail cars, twin tunnels connecting the turnback facility with Embarcadero Station, a cross-passage between the twin tunnels and a ventilation structure. The turnback facility and ventilation structure are located just south of Justin Herman Plaza and were constructed using cut-and-cover construction methods. The tunnels were excavated under compressed air with an open face tunneling shield.

The tunnels were approximately 18 feet in diameter and 850 feet long. The tunnel inverts were approximately 40 feet below the ground surface and the distance between centerlines was approximately 32 feet. The initial liner also served as the final liner and consisted of 2.5 feet wide steel liner plates (a total of six segments per ring).

The cross-passage was located below Justin Herman Plaza approximately 220 feet north of the turnback section as shown on Figure 1. The length of the passage between tunnel springlines was fifteen feet. Finished dimensions of the passage were four feet wide by seven feet tall.

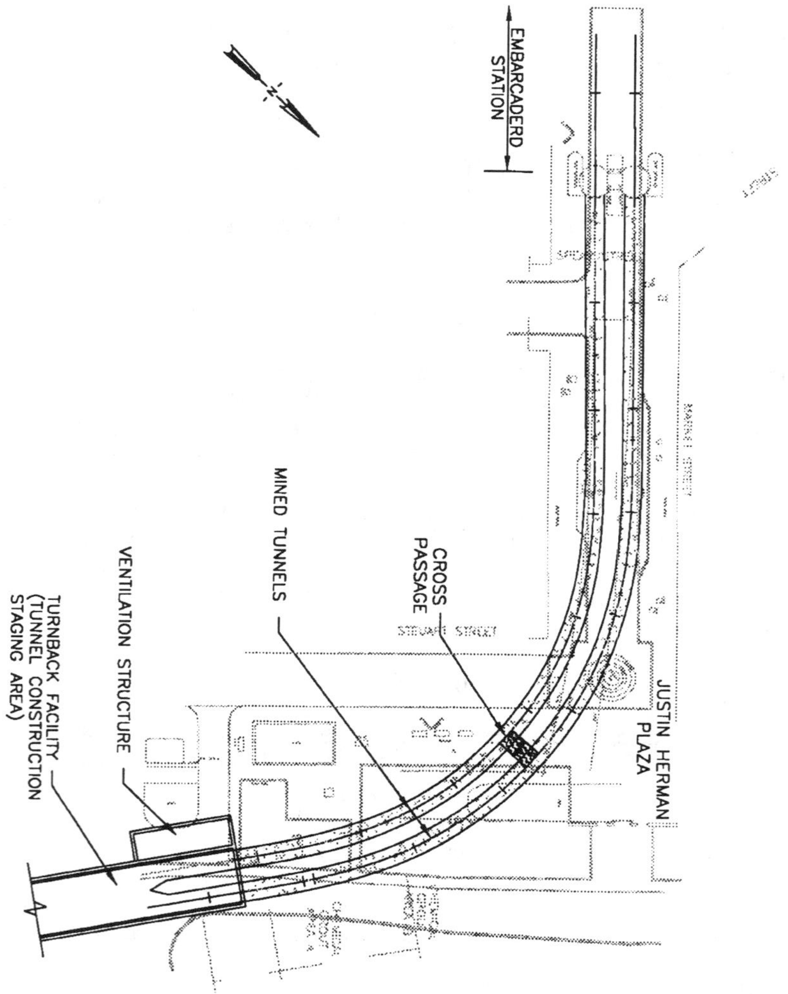

Figure 1.

An innovative ground support solution using "soft-centered" jet grouting was developed to construct the cross-passage. This paper presents a description of subsurface conditions, the catalyst to search for alternative ground support methods, considerations for selecting and designing a jet grout ground support, and the methods used and observations made during cross-passage construction.

SUBSURFACE CONDITIONS

Subsurface conditions varied along the project alignment, generally consisting of a sandy fill overlying a layer of recent Bay Mud which extended below the tunnel inverts. A mix face of the two or a full face of either was encountered while mining the main tunnels. The recent Bay Mud in this area can be described as a soft, gray-black, fat clay, with an undrained shear strength of approximately 28.7 Kpa (600 PSF). The Sand Fill consisted mainly of clean, poorly graded, loose fine sand. The water table was approximately 2.44 m (8 ft.) below the ground surface.

Timber piles were encountered along the tunnel alignments at an average rate of approximately 1 pile for every 0.61 m (2 ft.)and often in clusters of 2 to 5 piles. Pile diameters ranged from 25.4 cm (10 in.) to as high as 91.4 cm (36 in.) and primarily extended vertically through the full face.

In the vicinity of the cross-passage, soil borings and probing from the main tunnels indicated the Sand Fill / Bay Mud interface was approximately 1 m (3.2 ft.) above the crown of the cross-passage. A full face of Bay Mud was anticipated for cross-passage excavation. In addition, probing revealed timber piles would also be present along the cross-passage alignment.

CATALYST FOR CHANGE

The original plan for ground control during mining of the cross-passage was to use compressed air. The use of compressed air for ground control would require main tunnel excavation to temporarily halt while cross-passage excavation was completed. A general construction schedule for this scheme is shown on Figure 2a. The tunnel construction and demobilization of the compressed air plant and staging area were on the critical path. In this scheme, the time required to mine the cross-passage would be added to the overall tunneling schedule and therefore the need for the air plant and staging area would extend the duration of the project.

An alternative method to construct the cross-passage under "free air" would have less of an impact to the construction schedule and ultimately cost less.

SELECTION OF JET GROUT ALTERNATIVE

In addition to providing adequate ground control, the method of construction of the cross-passage would have to fulfill an additional criteria - main tunnel excavation could not be interrupted. Two alternatives were considered. First, conventional excavation from the surface exposing the two main tunnels, constructing the cross-passage and backfilling. The second alternative was to improve the soft Bay Mud between the two tunnels and mine the cross-passage under "free air".

The drawback of the conventional excavation method was the surface disruptions at Justin Herman Plaza. Approximately six weeks of excavation, hauling of excavated soils and cast-in-pace construction would pose major disruptions and was unacceptable. This option, however, did meet the criteria that it would not impact the second tunnel excavation.

There were several benefits to improvement of the Bay Mud and mining the cross-passage under "free air". Grouting from the surface could be completed with relatively minimal surface disruptions, and, as shown on Figure 2b, without stopping main tunnel excavation. Subsequent excavation of the cross-passage from the first tunnel could be completed concurrent with demobilization of the tunneling shield from the second tunnel. Another favorable aspect was that jet grouting was recently used successfully for similar soil conditions at the nearby Islais Creek Transport Tunnel Contract E, therefore grouting results could be predicted. Also, grouting equipment was locally available. This local experience coupled with the advantage of removing the cross-passage excavation from the critical path were the basis for selecting the jet grout soil improvement alternative.

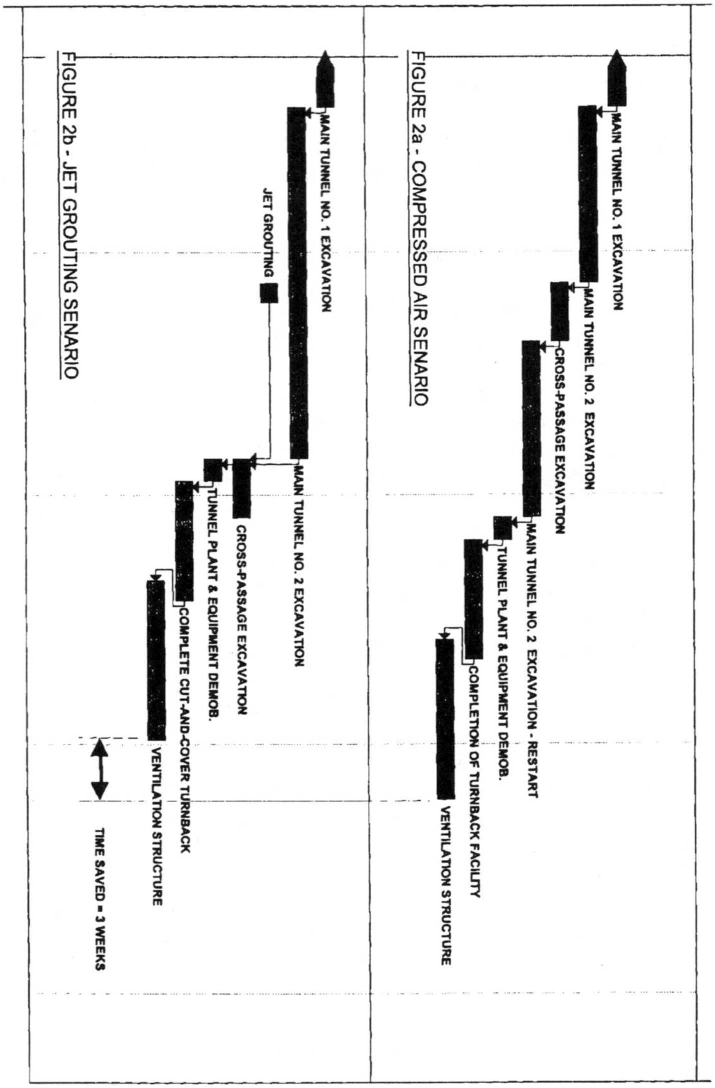

Figure 2.

JET GROUTING DESIGN CONSIDERATIONS

General

The "three-fluid" system used for grouting consists of injecting, at high pressures, a combination of water, cement grout and air into the soil mass. This is done through a special probe which is rotated as it is withdrawn at rotation and withdrawal rates selected to achieve the column diameters and improved soil strengths desired for specific applications. The injection process creates a mixture of soil and cement grout. Design considerations were impacts to existing and new structures, configuration of the improved soil mass and layout of probe holes. Provided below is a brief description of the design considerations, cement grout mix, injection pressures and withdrawal and rotation rates used for this project.

Design Considerations

Impacts to Existing and New Structures

Fluids injected into the ground at high pressures can have detrimental effects on nearby utilities and have the potential to heave the ground surface. Locations of utilities were well documented and not near the cross-passage. The cross-passage was located below a sidewalk in Justin Herman Plaza with surface features including park benches, trees and a MUNI bus stop enclosure. These surface features were not considered critical with respect to their ability to undergo some vertical movement. Impact to existing surface and subsurface structures was considered to be minimal. Prior to grouting, a vertical survey of the area was conducted for restoration of original grades if significant changes occurred.

New construction in this area consisted only of the two tunnels. Grouting at high pressures adjacent to the tunnel liners was a concern, particularly for the tunnel which was under compressed air. The tunnel liners were constructed of steel skin plates 15.9 mm (5/8 in.) thick, stiffened by T-type sections, end plates and gusset plates of similar thickness [Sutter, et. al. 1995]. Liner plates were bolted together by 2.54 cm (1 in.) bolts and sealed by two hydroswelling seals between end flanges. Deformation of the tunnel liner and breaking the seals created by the hydroswelling rubber seals were the main concerns. An observational approach was planned with grouting beginning adjacent to the "free air" tunnel. Continuous contact between the grouting crew on the surface and monitoring crew in the tunnels was required. Monitoring crews in the tunnel surveyed the liner for movement (convergence/ divergence) and visually inspected the liner joints for grout intrusion. If movement or grout intrusion were detected, modifications would be made prior to grouting adjacent to the compressed air tunnel. The observational approach would continue throughout grouting operations.

Configuration of Improved Soil Mass

In defining a grouted soil mass cross-sectional area, two aspects of the grouted area were considered. The first was the ability of the grout along the excavation perimeter to support the surrounding soil loads and second was the ease of mining.

Because of the flexibility and ease with which grout injection can be started and stopped at different depths, a "soft center" approach to establish the cross-sectional area was developed. This would allow for relatively easy hand mining of ungrouted Bay Mud while having 2000 to 3500 kPa (300 to 500 psi) grouted walls, roof and floor slab or a "box" shape around the excavation limits. Vertical and horizontal soil loads were used to design the roof and wall thickness respectively. The Contractor's design modeled the box as a continuous beam with hinges located at the corners to determine displacements and maximum moments for each element of the "box". Each element was then analyzed for rupture. Figure 3 shows the cross-sectional area and element thickness and widths resulting from the analyses performed. The "soft center" approach provides both support and ease of mining. The grouting methods used to construct this "box" are described in detail in the next section, Cross-Passage Construction.

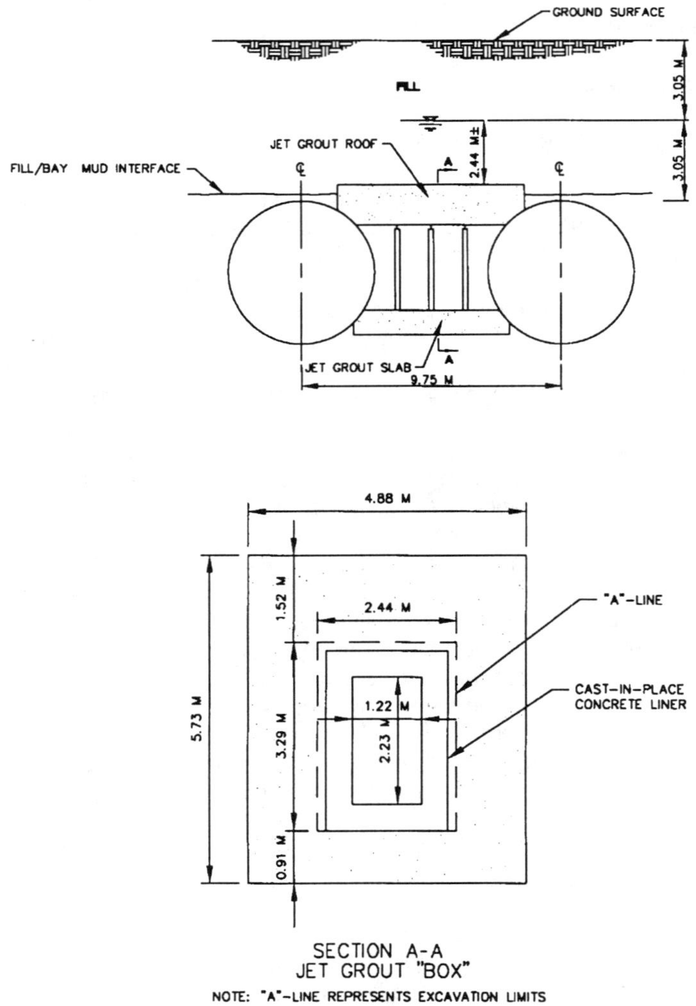

Figure 3.

Jet Grout Column Layout Considerations

In addition to laying out grout columns to provide the cross-sectional configuration described above, two other factors were considered. First, the effects of timber piles on grout column diameters and, second, intrusion of water along the main tunnels' steel liners.

In general, an overlapping pattern of grout columns was desired. Without considering the effect of timber piles, columns could have been spaced roughly on 1.82 m (6 ft.) centers. The uncertainty introduced by the presence of timber piles was the potential for ungrouted shadowed areas adjacent to piles. In other words, a piece of the pie would be missing from the grout column. Column spacing was reduced to 1.3 m (4.3 ft.) in an attempt to mitigate potential grout shadowing through increasing the column overlap. In the areas adjacent to the main tunnels, an additional row of grout columns was designed to stop water intrusion along the liners.

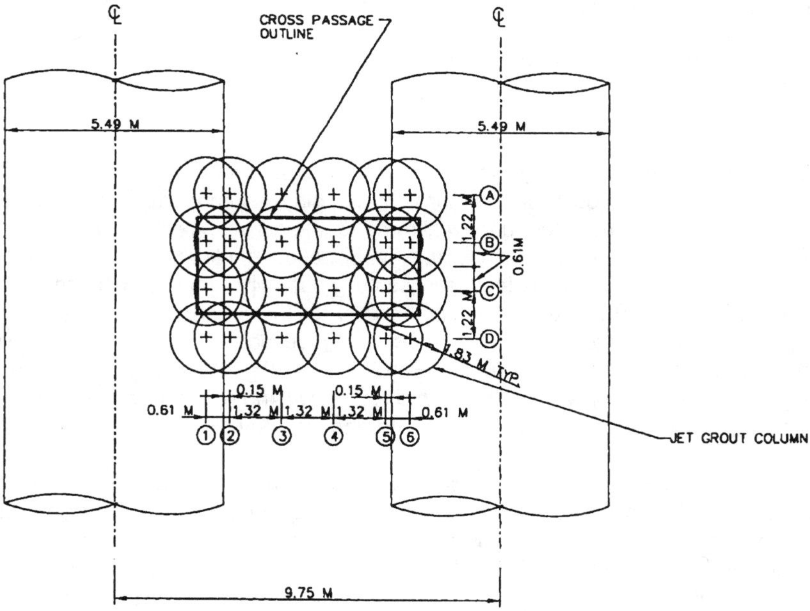

Figure 4.

As shown on Figure 4, a total of 24 holes were used to achieve the grouted soil mass configuration described in this section and the previous section.

Grout Mix, Injection Pressures and Injection Rates

The successful grouting methods recently used for tunneling and results of a full scale test program performed for the Islais Creek project provided the basis for selection of grouting method and grout mix design. Test program results indicated six foot diameter or greater columns with compressive strengths of 2000 to 3500 kPa (300 to 500 psi) were attainable in

Bay Muds with an undrained shear strength of 28.7 kPa (600 psf) using a "three-fluid" system. As suggested by its name, this system injects three different media into the soil - cement grout, water and air. Injection pressures of 400 bar for the cement grout and water and 9 bar for air were used. The water to cement ratio of the cement grout was 0.90. [Pelegrino et al, 1995] Probe withdrawal rate was 12 cm per minute and rotation rate was 10 revolutions per minute. [Adams, 1996] Each one of these variables including the type of soil to be grouted contribute to the resulting diameter, strength and homogenialty of the grouted columns.

CROSS-PASSAGE CONSTRUCTION

General

Grouting and excavation of the cross-passage was accomplished concurrently with main tunnel construction and nominal impact to the project's critical path schedule. Jet grouting was completed from the surface and the cross-passage mined by hand in "free air".

Construction Methods and Considerations

Jet Grouting

Jet grouting was performed from the surface in an area of downtown San Francisco which is a hub for popular outdoor summer activities such as concerts, plays and general gatherings. There was a one week window allowed for the grouting operations. In addition to attaining the desired grouted soil mass results, mobilization, containment of spoil and clean up were key factors which would define a successful grouting program. Due to the amount of spoil return from the grout holes (approximately 20 to 30 percent of soil grouted) to the surface a continuous effort to clean up the return was necessary. The Contractor used water sprayed from hoses to break up chunks of Bay Mud and direct flow of the spoil toward a trash pump. The trash pump would then pump to an awaiting empty ready-mix and stockpiled on site to allow it to dry prior to final hauling and disposal.

The staging area for the grouting operation was approximately 13.7 m (45 ft.) wide by 54 m (175 ft.) long. It's configuration was dictated by landscaped features of the park. Equipment in this staging area included, a series of mud tanks, a cement mixer and agitator, a large grout pump capable of pumping at 1000 bar pressure, drill rig and the clean up equipment described above. Concrete Jersey Barriers (K-Rails) and sand bags were placed at the perimeter of the staging area to contain the spoils. A 2.44 m (8 ft.) fence with plastic tarp was erected on top of the barriers to keep grouting spoil from splashing outside of the area.

Grouting proceeded in an alternating hole pattern so that a recently grouted hole would not be affected by grouting an adjacent hole. This restriction limited the number of grout holes completed per day to 5 or 6. As mentioned in the design considerations section of this paper, grouting began adjacent to the "free air" tunnel. Observations were made during grouting of the first hole from within the adjacent tunnel to determine if the tunnel liner would deform or if the liner seals would be breached. Neither occurred.

Referring to Figure 4, the holes are designated with the letters A, B, C, and D, 1 through 6. These designations indicate the following process used for grouting and obtaining the "soft center" desired. Columns designated as A2 through A5 and D2 through D5 are columns used to form the walls of the "box". These columns were created by grouting from the bottom of the "box" to the top of the "box". There were eight wall columns. The columns designated as B2 through B5 and C2 through C5 are columns that formed the roof and floor slabs. These columns were grouted first from the bottom of the "box" for 0.91 m (3 ft.) then grouting would stop (creating a floor slab), the probe would then be withdrawn without injecting grout, water or air to the bottom of the roof slab where grouting would continue for 1.5 m (5 ft.). The columns along rows 1 and 6 formed the water cut-off portion of the grout mass adjacent to the main tunnels. For these columns, the probe was lowered until it touched the liner then withdrawn while grouting to approximately the top of the "box".

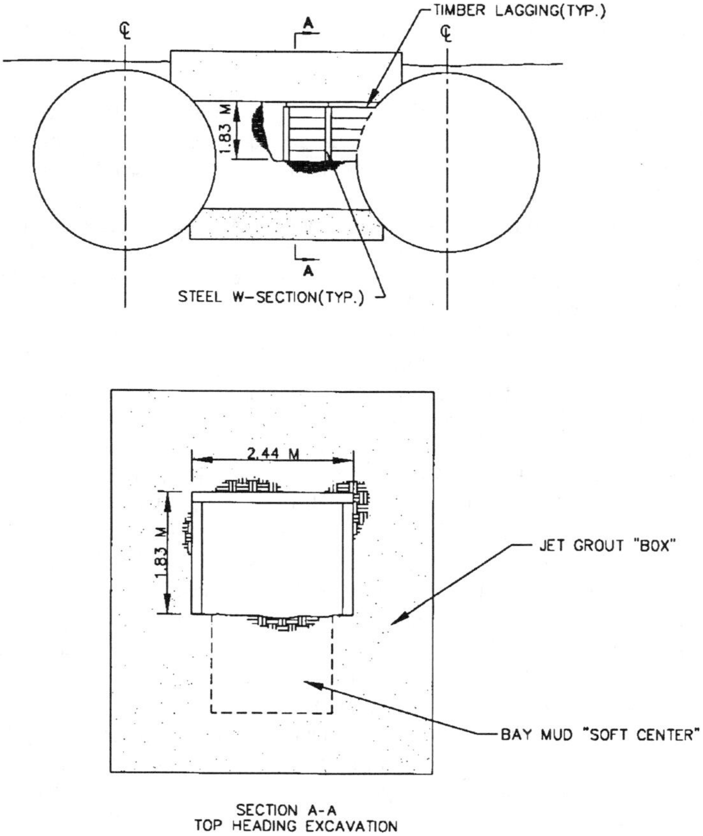

Figure 5.

Cross-Passage Excavation

Because of the 3.65 m (12 ft.) height of excavation, a top heading was mined first and then the bottom heading. Each heading was 1.82 m (6 ft.) tall by 2.44 m (8 ft.) wide. Figure 5 shows the general overview of the excavation process. The Bay Mud was excavated with hand held pneumatic spades. The mining crew generally consisted of two miners excavating the face and two miners mucking out behind them onto an electric conveyor and eventually into a muck car. Ten centimeter (4 inch) steel posts and caps were erected with wooden lagging spanning their 1.22 m (4 ft.) center to center spacing. Steel beams and lagging were installed primarily to control possible spalling. Plywood was then secured to the wood lagging and a cement grout pumped behind the lagging to fill any voids created by overexcavation. The cross-passage excavation was completed and readied for forms, rebar and concrete for the final structural concrete.

OBSERVATIONS AND CONCLUSIONS

Observations during cross-passage construction are summarized below.

- Surface settlements due to excavation were nominal, however heave of the sidewalk due to grouting was on the order of one to four inches. Had their been nearby utilities or surface structures sensitive to heave, procedures during grouting could have been taken to reduce the amount of heave.
- The resultant grouted soil mass was homogeneously grouted and strength appeared to be fairly consistent. In areas where grouted soil had to be excavated, hand held pneumatic spaders were fitted with spikes. As expected, excavation of the grouted soil was more difficult than the Bay Mud.
- The tighter grout column pattern resulted in only minor "shadowing" by timber piles (six timber piles were encountered).
- Water infiltration into the excavation either around the timber piles or along the edge of the main tunnel liners was minor. At no time did the water pose a threat to the stability of the excavation.
- During excavation, the headings could have been and in some cases were advanced greater than the 1.22 (4 ft.) required for the next steel set. In these cases the grouted roof element was stable - no spalling occurred.
- A strong ammonia smell let off by the grouted soil was at times intolerable. Ventilation - even for a 1.6 m (15 ft.) long tunnel was necessary.
- Cross-passage excavation rates were better than expected. Top and bottom heading excavation and installation of steel and lagging was completed in three weeks. This can be attributed to having the ground support (jet grouted "box") in place prior to excavating. Also, a cleaner excavated perimeter allowed for quicker steel and lagging installation. The time savings shown on Figure 2 (approximately three weeks) does not reflect additional savings since mining under compressed air would have required the use of breasting boards and more careful mining techniques. The additional time saved was likely on the order of one week. A total time saved off the critical path was therefore one month.

ACKNOWLEDGMENT

Overall, excavation of the cross-passage was completed as planned and on schedule. The success of the cross-passage construction was a result of the project's Contractor, Construction Managers and Owner working together to implement and improve the use of new technology. A specialty contractor with local experience in jet grouting was important to meet the project goals. Nicholson-Rodio provided this experience and expertise.

REFERENCES

Sutter, D., Frobenious, P., Wu, C., 1995, "Tunnel design of MUNI Metro Turnback Project", Proceedings, Rapid Excavation and Tunneling Conference San Francisco, California

Sutter, D., Sherry, G., Connolly, J., Wu, C., 1996, "Tunnel Construction of MUNI Metro Turnback Project", North American Tunneling, Washington, DC.

Pellegrino, G., Adams, D., 1996, "The Use of Jet Grouting to Improve Soft Clays for Open Face Tunneling", North American Tunneling, Washington, DC.

Adams, D., Jacobs Associates, 1997, San Francisco, CA USA, Telephone Interview, January 1997

Chapter 46

WEST SEATTLE TUNNEL

Martin Oatman[1] and Lawrence Lenahan[2]
[1]Construction Manager, KCDNR; [2]Vice President, McNally Tunneling Corporation

ABSTRACT

The West Seattle Tunnel is part of a regional sewer conveyance system located in Seattle Washington. The tunnel was constructed for the Waste Water Treatment Division of the King County Department Of Natural Resources, from June of 1995 through the first quarter of 1997. The Geologic setting for this soft-ground tunnel consists of glacial, interglacial, glacio-lacustrine, and alluvial soils. This paper, prepared jointly by the owner and contractor, will describe project-related site geology, engineering considerations, construction, and dispute resolution/avoidance practices.

INTRODUCTION

King County has a program to upgrade primary sewer treatment facilities within its jurisdiction. Faced with providing secondary sewer treatment for sewage generated in the West Seattle drainage basin, the County determined the most cost effective method of treatment involved conveyance of raw sewage from West Seattle to its regional treatment plant at West Point. Conveyance of sewage from its Primary Treatment Plant on Alki Head to its West Point Treatment plant required installation of about 6705.6 m (22,000 ft) of new pipeline including the West Seattle Tunnel.

Working with its consultants, the County selected a tunnel alignment beneath West Seattle. The tunnel will convey sewage across Alki Head to a new pump station constructed at Harbor Avenue and Hanford Street. Under present day conditions, the maximum flow capacity for the tunnel is 19 million gallons per day (MGD). A 122 cm (48 in) diameter pipe constructed at a slope of 0.0092 ft/ft has sufficient capacity to carry present maximum flow. Long range forecasts indicate 96 MGD will be the ultimate sewage flow collected from West Seattle area. The design initially called for a 238.8 cm (94 in) diameter tunnel to carry the ultimate flow. No regular access to the tunnel was anticipated or incorporated into the tunnel design. The actual as constructed size of the tunnel is 3.38 m (133 in).

LOCATION

The 3128.2 m (10,263 ft) West Seattle Tunnel route begins on the east side of Alki Point, near the intersection of Harbor Avenue and Hanford Street. Proceeding westerly, the tunnel winds through a short horizontal curve to align it with the SW Hanford Street right-of-way. The alignment follows SW Hanford Street for over 1615.4 m (5,300 ft) at depths of over 91.4 m (300 ft) until it reaches Schmitz Park. Beneath Schmitz Park the tunnel passes through two horizontal curves with equal radii of 243.8 m (800 ft) and lengths of approximately 237.7 m (780 ft). This exaggerated "S" curve aligns the tunnel with the SW Spokane Street right-of-way. Following the Spokane Street alignment the tunnel reaches its terminus at the west portal in the intersection of Spokane Street and 61st Avenue southwest.

SITE GEOLOGY

Geotechnical Investigations

In a two-phase program, Converse Consultants Northwest, a geotechnical engineering firm under the direction of prime consultant HDR, completed geotechnical investigations for the tunnel. Program elements involved a feasibility study for a deep tunnel and preferred portal locations. Eighteen borings were completed on the tunnel alignment. The borings had an average depth of 79.2 m (260 ft) and a maximum depth of over 128.0 m (420 ft). Interpretation of subsurface investigation data showed that the tunnel would pass through a series of glacially consolidated soil units deposited from 15,000 to 250,000 years ago. A Geophysical investigation identified bedrock depth in areas where bedrock was suspected near the tunnel invert. Groundwater pump tests were performed at the two portal locations. Bore hole slug tests were completed on several of the deep tunnel borings to evaluate aquifer properties and identify dewatering requirements for the alignment.

Tunnel Geology

The tunnel passed through an ancient landslide zone from station [Tunnel stationing begins at 10+00 for clarity and convention. Subtract 1000 ft from stationing to determine distance from east portal.] 10+00 to about station 11+60. Following the tunnelman's ground classification system, tunnel zone soils were classified as squeezing to fast raveling then cohesive-flowing to flowing as tunnel excavation moves from east to west. The youngest soil unit is a fine grained glacio-lacustrine clay (Lawton Clay), which was encountered from the east portal for about 243.8 m (800 ft) , where a transition to fine grained, pre-Vashon age interglacial deposits occured at about station 18+00.

From station 18+00 to station 43+00, the tunnel passed through a glacio-lacustrine (pre-Vashon) fine-grained unit. This glacially consolidated soil consists predominately of hard, moist, silty clay, clayey-silt, non-plastic silt, very dense silty sand, and fine to medium sand. The tunnel zone is classified as raveling with a transition to cohesive-flowing to flowing in this portion of the alignment. A variety of face conditions ranging from clayey-silt to sand and gravelly-sand or a combination of soil layers were encountered.

The majority of the tunnel from station 43+00 to the west portal is pre-Vashon medium and coarse grained deposits consisting of very dense, wet, fine to coarse sand, and sand and gravel mixtures. The meandering nature of the depositional systems that is believed to have laid down the pre-Vashon units, resulted in a high degree of interbedding. The tunnel zone is classified as flowing throughout this section of the tunnel.

ENGINEERING CONSIDERATIONS

Using interpretation of subsurface conditions observed in the geotechnical explorations, a program was developed to identify engineering and construction risks for the West Seattle Tunnel. Engineering and construction risks were defined as; safety, groundwater, flowing ground, methane, boulders, and the disturbed slide material near the east portal. The risks were evaluated, then plans and specifications were developed to take the risks into account. The owner assigned personnel with tunnel construction experience recently obtained on its downtown transit and sewer system soft ground tunnels to work with its consultants on the design. Current industry practice and experience indicated that geotechnical base line conditions were required for bidding, construction, and dispute resolution. A Geotechnical Design Summary Report (GDSR) was prepared and incorporated into the construction documents.

Contaminated Ground Water

Groundwater and soil samples at the east portal site indicated that contaminated groundwater would not be generated from construction at the portal site. Groundwater samples from wells within the hydrogeologic zone of influence of the east portal shaft produced both short and long term concerns. In the short term, contaminated ground water posed a potential threat to the health of construction workers. In the long term, there were concerns about the potential for spreading a plume of contaminated groundwater from a former landfill into the permanent pump station site at the east portal. Several engineering solutions were evaluated, including an aquifer recharge system, water tight shoring, and slurry cut-off wall. A cement/bentonite slurry cut-off wall was selected for its ability to address short term and long term concerns. The slurry trench was constructed before portal excavation. The slurry mix [6% bentonite, 25% cement, 0.14% Spercene, water-cement ratio: 0.23 (proportions based on slurry mis weight.] had a compressive strength of about 40 psi and a hydraulic conductivity of 4 x 10^{-7} cm/sec. It was placed in a trench 12.9 m long x 15.2 m deep x 0.91m wide.

Groundwater In Tunneling

Groundwater was observed in soil boring holes above the tunnel invert elevation throughout the length of the alignment. For all practical purposes this proved to be the case during tunnel construction. Measured groundwater levels indicated a maximum hydrostatic head on the tunnel invert of about 18.3 m (60 ft) within the fine-grained, low permeable Lawton Clay deposits near the east portal. As expected groundwater was not a significant factor in advancing the heading through the Lawton Clay deposits. The hydrostatic head on the tunnel invert decreased as the tunnel moved westward. In the western two-thirds of the tunnel, hydrostatic heads range from 2.4 m to 4.3 m (8 to 14 ft) above the invert. In this section of tunnel, saturated coarse to fine-grained soils presented a greater challenge to face control and heading advance.

Flowing Ground

Laboratory testing of soil samples from the tunnel zone in the western two-thirds of the alignment showed less than 10 percent fines (material passing the US number 200 sieve). The lack of fine material together with the measured groundwater heads of between 2.4 m to 4.3 m (8 to 14 ft) above the invert suggested that flowing ground conditions would be an important factor in advancing the heading in the 914.4 m (3000 ft) of tunnel leading up to the west portal. Flowing ground conditions were encountered as reported in the GDSR.

McNally chose a Lovat TBM over several competing manufacturers. No tunnel boring machine is perfectly suited to every condition expected, but this one certainly came close. The tunnel boring machine was equipped with injection ports for soil conditioning ahead of the cutting wheel. Using this machine feature, McNally successfully advanced the heading with soil conditioners [Morrison Mud products: Tunl-Kwik T-50 coagulant & Tunl Kwik Foam] injected ahead of the cutting wheel and into the muck plenum. Although the TBM was supplied with a screw conveyor, it was never operated with the screw conveyor in place. Muck volumes were closely monitored at the heading by the owner's inspection staff (see Figure 1.0 - Muck Volume and Annulus Grout). Theoretical muck volumes were adjusted to compensate for the different soil conditions encountered along the alignment. McNally injected compensating grout and regrouted as needed to fill ground loss voids. Surface settlement monitoring did not reveal settlement above the specified maximum of 1.27 cm (0.5 in).

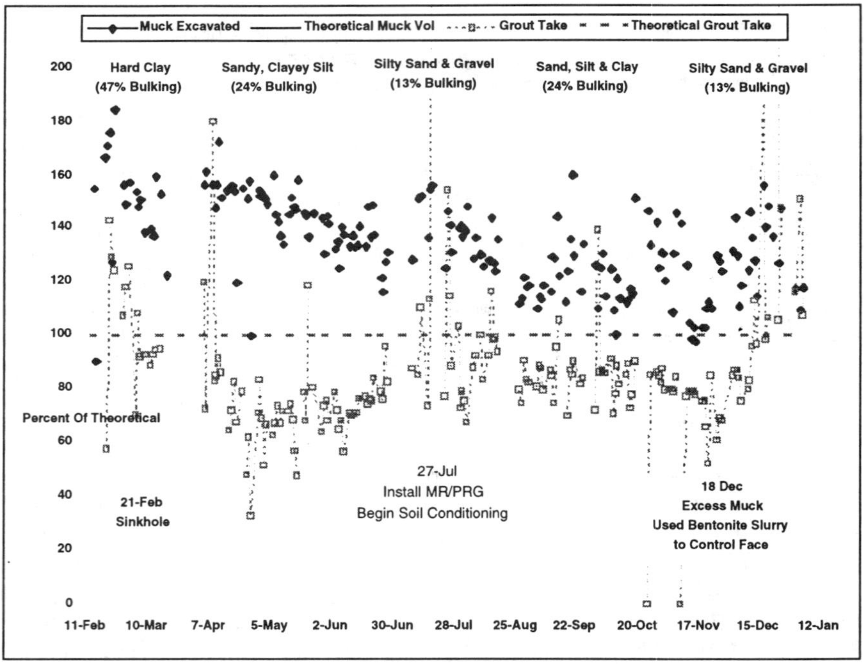

Figure 1.0. Muck Volume and Annulus Grout

Methane

Measurable quantities of methane were detected within the pre-Vashon deposits in two methane wells installed above the groundwater table and four groundwater wells along the tunnel alignment. These measurements, as well as the presence of peat and other organic matter in soil samples, suggested that methane gas could be present in the soil and groundwater within the tunnel zone. Due to the presence of methane, the tunnel was classified as potentially gassy. The specifications addressed the presence of Methane within the tunnel zone and established monitoring and air quality standards for tunnel construction. McNally implemented a comprehensive tunnel safety program which included training in the use of self rescue and related tunnel safety apparatus. Although methane was detected during tunneling, advance planning negated its effects on the work.

Boulders

Boulders were encountered in the pre-Vashon granular deposits at two boring locations. Statistical analysis and experience from other tunnel projects constructed in the soil units identified by the project geotechnical investigation, suggested that boulders could be a significant factor in tunnel heading advance. The gravelly pre-Vashon deposit encountered in the western half of the tunnel alignment was deposited by a meandering river or stream system. As rivers and streams meandered, they cut into older deposits to rework, concentrate, and redeposited the coarse-grained fraction from the older sediments. The lowest portions of channel deposits typically contain gravel, cobbles and possibly boulders. Incised channels of boulder-bearing gravel were thought to be present along the alignment within the coarse grained pre-Vashon deposits expected in the western portion of the tunnel. As expected, boulders were encountered during construction. The TBM cutting wheel was fitted with discs and drag teeth, replaceable from inside the TBM as specified. Boulders were broken by the cutting tools and readily removed from the muck plenum after passing through cutting wheel flood doors.

Slide Plane

The tunnel passes through an ancient slide zone in its first 76.2 m (250 ft) from the east portal. Tunnel liner was designed to resist a tensile force of 614 kips through this zone. This requirement ensures tunnel integrity during a seismically induced slide at the east portal. Such an event would otherwise produce a catastrophic failure of the tunnel. To protect permanent structures on the east portal site, a permanent hillside retaining wall was constructed as part of the east portal shaft installation. This retaining wall features sloping horizontal drains at two elevations located 10.7 m (35 ft) apart, near the top and bottom of the wall. These drains are intended to relieve hydrostatic pressure to reduce loads on the retaining wall. In addition to the horizontal drains, plans called for a surface water drainage corridor.

Tunnel Boring Machine

The importance of specifying and selecting the correct tunnel construction equipment for a project cannot be overstated. In the low bid competitive atmosphere of public works projects, cost at bid time often drives the choice of tunnel mining equipment. One obvious effect of a cost driven Tunnel Boring Machine (TBM) choice is that the importance of the TBM to project success may be overlooked or inappropriately considered. Lack of critical attention to tunnel construction equipment has been at the root of many abandoned tunnel projects or projects with poor performance records. From the owner's perspective leaving the choice of tunnel boring equipment entirely in the hands of the construction contractor is simply too risky in complex geologic settings with deep cover and long tunnel drive lengths.

As previously stated, geotechnical investigation revealed a variable soft ground geologic setting along the tunnel alignment. With the tunnel invert below the water table, and over 70% of the tunnel classified as "Flowing Ground", the choice of tunnel construction equipment was of utmost importance to the success of the project. The length of the tunnel drive, and the depth of cover were important in deliberations over the type of tunneling equipment and whether it should be new or used. After prolonged debate and evaluation, the design team determined that project risks could be greatly reduced by the use of a new TBM designed for the project. Specifications were developed around a new Earth Pressure Balance Tunnel Boring Machine (EPBM).

A literature search revealed that machine tool wear could be a significant problem in the geology expected along the alignment. TBM specifications indicated a complete cutting tool replacement at a minimum once every 304.8 m (1000 ft) of tunnel length. Cutting tool replacement did not occur at the rate indicated in the specifications. But tool wear became a matter of concern to the project team when excessive wear lead to the loss of integrity in the forward shield.

Main bearing problems were considered a high risk area too. Project specifications required a spare main bearing and its insitu replacement. A spare bearing was provided, but never used. Whether it was a strange coincidence or not, bearing seal grease became contaminated with water and foreign particles. The knowledge that a bearing would be immediately available partially relieved project team worry over the presence of contaminated bearing seal grease.

Although these specification requirements certainly added to the cost of the project, the County believes that a far greater risk of increased costs exists from a failed tunnel drive. Tunnel construction histories are filled with examples of such failed drives and the unanticipated costs which greatly exceed the cost of adequately addressing ground conditions in tunnel equipment specifications.

Tunnel Size

A pipe diameter of 2.1 m (7 ft) is sufficient to handle the ultimate flow capacity from the Alki drainage basin. For short drive lengths, a 2.1 m (7 ft) pipe could have been jacked using readily available bore and jack technology. But pipe jacking was ruled out because of depth and length of the pipeline. Although it is feasible to construct a 2.1 m (7 ft) diameter tunnel, the geology and 3128.2 m (10,263 ft) drive length, dictated careful consideration of the risks associated with constructing a small diameter tunnel with a single access point. The most economical tunnel size would be one that provided sufficient space for safe and efficient construction. A study was completed to determine the minimum cost effective and intrinsically safe tunnel diameter for the project.

The first concern was safety during tunnel construction. Adequate working space is crucial to safe and efficient construction. Cramped working conditions would be hot and fatiguing to workers. Tired workers would be susceptible to accidents. A potentially gassy classification necessitated adequate ventilation, which would be problematic with surface drop shafts out of the question. As tunnel excavation advances, efficient muck and material transport become are essential to a successful tunnel drive. Clearly, muck disposal and liner costs grow with increasing tunnel size. However, it is not as obvious that both length and depth of the tunnel are very important considerations in determining an efficient and cost effective tunnel size. Detailed cost estimates showed that a 2.1 m (7 ft) diameter tunnel would cost about the same to construct as a tunnel of 2.9 m (9.5 ft) inside diameter for the tunnel design depth and length. Careful evaluation of tunnel equipment and working space available in the 2.9 m (9.5 ft) tunnel supported a decision to establish it as the minimum safe size for the project. All thirteen of the bids submitted for the project included a tunnel with a finished inside diameter of 3.05 m (10 ft).

Tunnel Liner

Although it would have been feasible to construct about 914.4 m (3,000 ft) of the tunnel with a two pass system employing steel ribs and lagging, the design team believed it would be difficult to persuade tunnel construction contractors to change to a water tight system if such became a necessity. Geotechnical investigation suggested that the tunnel would require a water tight initial liner for about 2133.6 m (7,000 ft) of tunnel. To avoid confusion during bidding and construction, the design team opted for water tight initial liner throughout the tunnel length. Three construction options permitted bidders to determine whether they would use single or two pass liner systems. The options included; two pass liner with cast in place concrete final liner, two pass liner with pipe and grouted annulus, or single pass precast bolted gasketed liner. All thirteen bidders proposed a single pass precast bolted gasketed concrete liner.

BIDDING AND AWARD

Bidding Method

The soft ground geology along the West Seattle Tunnel alignment is varied with a potential for mixed face conditions, hydrostatic pressure, and methane gas. These conditions increase risks in safety, claims, delays, and added costs at completion. The County and its consultants recognized these risks and crafted bid and technical specification documents reflecting a rational approach to risk allocation.

Although the contract is a lump sum type, the bid was structured with 8 bid items focused to address critical or risky construction elements. These bid items served the purpose of calling the bidders' attention to the risky work, and to make clear the County's intent and methods of payment. A TBM bid item included a newly manufactured Earth Pressure Balance Tunnel Boring specified for the project; and a boulder removal bid item identified the method and extent of payment for this activity.

Boulders are common in the tunnel alignment geology. One can only estimate how many boulders might be encountered, or large the boulders will be. Traditionally, tunnel construction contracts have been written with a boulder size and quantity threshold to trigger payment for removal, or alternately attempt to put all the burden on the contractor. These methods have not been entirely successful allocating risk or reducing claims for additional compensation. The approach taken for boulder removal on the West Seattle tunnel project was to recognize the high possibility of boulder encounters and to compensate the tunnel construction contractor for removing boulders. For bidding purposes, 700 hrs [Unit price based on 700 hrs of composite crew time to remove boulders, including labor, equipment, materials, and indirect costs.] was used as the time baseline to establish costs for crews involved in boulder removal, equipment and plant standby, and overhead. One obvious risk in structuring the bid item for boulders in units of time is that a bidder might load the bid item with a high price, then not move aggressively to remove boulders. To minimize the potential of an unbalanced bid, the boulder removal price was included in the total bid price. A bidder inflating the cost of removal would jeopardize their potential to be the successful low bidder.

The unit price for boulder removal proved to be a successful method of payment for the project. There were no claims for additional compensation involving boulder removal. While the unit price method of payment insured that the contractor received adequate compensation for boulder removal, the owner was protected from excessive billings and claims.

Bid Opening and Notice to Proceed

Bids for the West Seattle Tunnel were opened on June 23, 1994. There were 13 bids received ranging from $20 million-US to $28 million-US. The three low bidders were within $650,000 of each other. The third low bidder mounted an unsuccessful challenge to the award of the contract to McNally. This unsuccessful bid challenge delayed award of the contract to McNally Tunneling by nearly eight months.

CONSTRUCTION

Pre Award Activities

The eight month time period from bid opening until Notice to Proceed presented challenges for both King County and McNally Tunneling. The County's schedule for conveyance system construction had to be assessed to determine whether the bid protest delay required a new construction sequence for any of the remaining five construction contracts. The County delayed bidding of its Pump Station contract because Pump Station site preparation was part of the tunnel construction contract. McNally was concerned with the potential outcome of the protest and how The County might handle the situation in the current industry climate of rebidding and legal wrangling. Although the bid protest delay had its adverse consequences, McNally was able to use the time to the project benefit. McNally could not issue any purchase orders, subcontracts, assign personnel, or commit any assets without a contract. However, during the bidding process it became evident that purchase, design, approval, and manufacture of a new TBM and the same process for a lining to match, were critical path items. The bid protest delay afforded McNally the opportunity to get a productive start on the selection and design of the TBM and the liner. This advance work helped to get the project off to a quicker start.

Advance Planning

Nine machine manufacturers and two segmental lining manufacturers submitted quotes at bid time. McNally asked all of the manufacturers to provide their past and current experience history of performance on projects similar to the West Seattle Tunnel. McNally staff made contacts on tunnel projects world wide. It became evident that contractors working in Europe and Asia were dealing with comparable geology, and also successfully using some of the cutting-edge technology and tools required by the contract TBM specifications. McNally sorted through the projects and determined that technology and techniques employed in Europe should be studied for application to the West Seattle Tunnel Project.

McNally staff visited all major European tunnel boring machine manufacturing facilities where machines similar to the one specified for the project were in fabrication. Buchan liner manufacturing locations in the UK were also visited. Numerous visits were made to job sites of different contractors throughout the European continent. It was apparent that contractors were successfully using a Lovat TBM in combination with a Buchan liner to advance tunnel headings in geology similar to that expected for the West Seattle Tunnel. Nowhere was this more evident from a cultural standpoint than at the Mangfall Project outside of Munich, Germany. On this project, a German company, Bilfinger Berger, selected a Canadian manufactured Lovat TBM, a liner designed by an English firm, C.V. Buchan, then constructed the project on German soil.

The benefits of advance planning became evident to all parties involved as the project progressed. It should be noted that during this time frame, tunnel equipment demand picked up significantly world wide. This coupled with the earthquake in Kobe Japan, a major world manufacturing area, caused lead time for TBM manufacture to increase four to eight months from what it had been at bid time.

Schedule

The contract established a duration of 731 calendar days to complete the project. Construction and land use permits were obtained by the County before construction. Key submittals were identified in the bid documents and time frames set for their preparation and delivery. McNally and County staff worked cooperatively in the development of a draft schedule. The draft schedule was presented to key subcontractors for input and concurrence. The schedule development process was streamlined and it resulted establishment of a realistic schedule within the first few weeks of the project. McNally and the County kept separate project as-built schedules whose differences were reconciled on a monthly basis.

Approximately fifteen months after starting work on the project, McNally was contractually obligated to share the east portal shaft with a contractor constructing a pump station. McNally also had to complete work at the west portal in time to allow a third contractor access to the area for pipeline construction. These constraints were not without challenges. Unfortunately, events transpired that delayed access to areas shared by McNally with the pump station contractor. The County determined that the best course of action was to delay the pump station contract so McNally could finish its work. The pump station contractor, McNally, and the County worked cooperatively to reschedule work, change sequence and make the best of a situation that was outside anyone's direct control.

There were many regulatory agencies involved in project oversight. Some of these agencies had approval authority over important aspects of the project such as excavation and shoring on public property. The County, McNally and engineering consultants worked closely to ensure timely and complete submittal reviews. This made it possible in many cases to send a submittal through the system once. The joint effort enabled McNally subcontractors to start work quickly on critical path tasks such as portal construction.

McNally faced and overcame many obstacles in completing the project. Some of those obstacles, especially mechanical problems and late tunnel equipment delivery, adversely affected schedule performance. McNally worked two production shifts and an equipment maintenance shift. Crews frequently performed maintenance work on weekends. Tunneling started on February 14, 1996 and progressed at a rate of about 13.7 m/day (45 ft/day). Holing through was milestoned for November of 1996, but as of January 9, 1997 about 305 m (1,000 ft) of tunnel remained to be constructed.

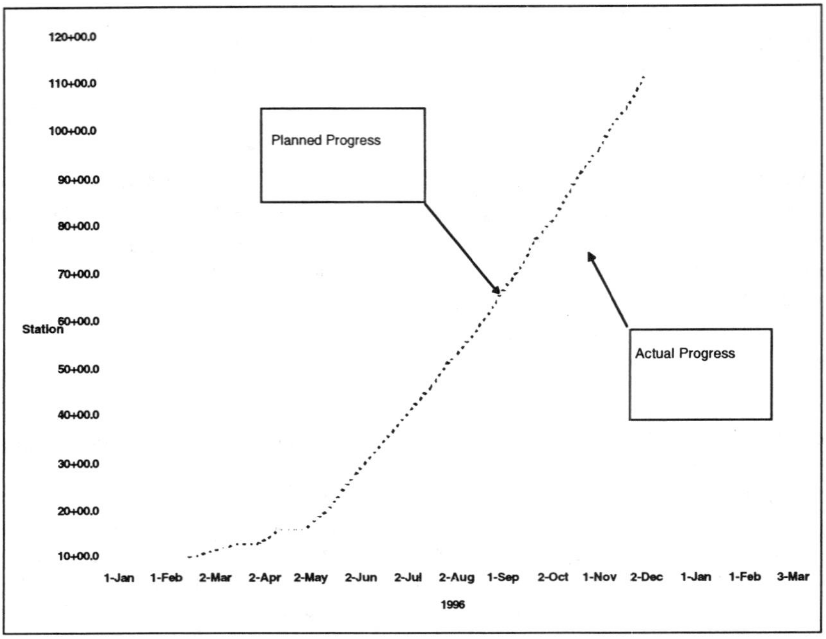

Figure 2.0. Tunnel Heading Advance Rate

Shaft Construction

The small site footprint posed a challenge to the design team. It was clear that the tunnel contractor would need more space than could be obtained on the County's property. The County leased additional space, and plans called for the construction of temporary detour and use of existing street surface area for construction staging. The detour was the first item constructed, followed by the previously noted slurry cut-off wall. McNally subcontracted shaft construction to Kulchin-Condon Associates/American Pile Driving (KCA/AP). Their contract included shaft excavation, spoils haulage, and excavation support systems installation. East portal shaft construction involved both permanent and temporary excavation support systems. About 622.9 m^3 (22,000 y^3) of spoils were removed for disposal off site. Permanent supports included forty eight, 19.8 m (65 ft) long, composite steel and concrete cylinder piles. One hundred and twenty, permanent, double corrosion protected soil tie-back anchors were installed in six rows. The permantent tie-back anchors were about 45.7 m (150 ft) long, and had lock-loads in the 200 kip range. Temporary excavation support was provided by a contractor-designed soldier pile timber lagging retaining wall with soil tie-back anchors. Following a modest soil tie-back anchor test program, KCA/AP started work on the east portal in August of 1995. The work progressed in stages until it was completed and ready for TBM launch platform construction by the end of November, 1995.

Tunneling Setup and Initial Drive

As outlined above, a purchase order was issued to Lovat Tunnel Equipment, Inc. upon contract award to design and manufacture Model ME157SE SERIES 15600 TBM. The contract requirements were quite onerous as pointed out by all manufacturers. For brevity, only some of the main TBM features will be noted here.

The TBM has an outside diameter of 398.8 cm (157 in.). It had to perform in open or EPB modes. In the EPB mode it was fitted with a 61 cm (24 in) axial screw conveyor. It was successfully used in open mode with a muck ring, pressure relieving gates, and a 76.2 cm (30 in) belt conveyor. This was made possible with use of soil modifying agents injected ahead of the cutting face and into the muck plenum. A triple roller main bearing, flood control doors, back loading ripper teeth with interchangeable disk cutters, and two face injection ports were incorporated into the machine. The machine has a total horse power rating of 1280 hp with 1000 hp dedicated to the cutting heading. The machine developed a maximum thrust of 1750 ton. Buchan participated jointly with Lovat in the design and fabrication of a hydraulic segment erector arm for the machine.

North American Segment Company (NASCO) was licensed under Buchan Ltd. to produce precast concrete liner segments. The segments were designed and manufactured to exacting standards specified by two full sections of the contract documents covering 17 pages. These specification included a statistical sampling plan to assure final product quality. A six piece, tapered trapezoidal, bolted, gasketed ring, 106.7 cm (42 in) long, 337.8 cm (133 in) ID, 383.5 CM (151 IN) OD was chosen as the optimum segment system. The molds were produced in the UK and shipped to Seattle. The segments were manufactured in Seattle and shipped to the job site by truck. NASCO began segment manufacturing in November of 1995 and completed nearly 30,000 segments by July of 1996. These segments out performed all expectations during tunnel construction. The segments followed the machine and sealed the ground throughout the entire alignment of the tunnel.

The TBM left Toronto, Ontario, Canada, December 15, 1995 by rail across Canada and arrived at a railroad siding in Seattle January 27, 1996. It was off loaded and trucked to the job site where it was set in the east shaft February 2, 1996. The eye of the portal was cut and the machine turned under on February 15, 1996. The failure of a 125 HP elect ric motor on the propulsion pump on February 18, 1996, before the machine was fully turned under ground, was a prelude to coming mechanical problems. Although a new machine allows for specific characteristics to be incorporated into the original design and manufacture, it must also pass through an initial break-in period to sort out and correct flaws. The project schedule allowed for an initial break-in and crew training period, but the time allotted proved to be insufficient.

The East shaft was 30.5 m (100 ft) across at the tunnel eye. This allowed for easy trailing assembly as the TBM was advanced through the first 76.2 m (300 ft) of the tunnel. Contact during the initial drive was accomplished with grout produced on site, then pumped into place from a grout car with an integral Moyno pump. The trailing assembly was fully completed and the TBM had advanced 152.4 m (500 ft) before the permanent shaft setup could be prepared for full production.

Both the County and McNally agreed that the initial drive could be made without the screw conveyor due to the competent ground conditions encountered during east portal shaft excavation. The limited site space, combined with the inability to set up the permanent muck handling facility, led to the decision to proceed without the screw. Control of the TBM during the initial drive proved difficult. It was discovered that grout had accumulated on the outside of the TBM. The grout was removed by hand mining around the TBM. Subsequently, grouting procedures were modified to prevent buildup of grout on the TBM exterior, and the drive continued without a repeat of the grout adherence problem.

Main Drive

The east shaft setup was complete and the full production of tunnel excavation began May 21, 1996. The shaft area required numerous operations to take place efficiently to adequately supply material to the heading and remove muck. Grout was batched and loaded into the grout car. Segments and their accessories were loaded onto segment cars. Muck cars were dumped and the train configured in the proper order. Each train was sized to sustain the effort of each segment ring advance shove. A California switch was installed in the tunnel to expedite muck train turn-around time.

Upon arrival at the heading by the train, excavation by the TBM commenced while the segments were unloaded and prepared for installation. At the same time, grout injection took place through previously installed liner segments. After the TBM completed a 106.7 cm (42 in) advance, the train exited while segmental ring erection took place. Rail and utility installation were concurrent with other operations. A 76.2 cm (30 in) bagline was used for ventilation at the heading. Production was conducted on two shifts. The third shift was used for survey, patching of segments and supply when needed. Due to the high maintenance requirements that developed during excavation, the third shift also allowed for a permanent maintenance crew dedicated to the TBM.

Many different ground conditions were encountered during excavation. Numerous combinations of TBM features were employed to advance the tunnel heading. Recognizing the conditions and adapting the TBM to changing conditions took its toll in lost time. McNally used soil conditioners injected through the cutting face to control ground that otherwise would have required the use of the screw conveyor. Although the screw conveyor was available, it was not used. At the time of this writing, 304.8 m (1000 ft) of tunnel remains to be excavated, and all parties believe that the excavation will be completed successfully without the use of the screw conveyer. Production to date is as follows; average daily rate 13.7 m (45 ft); best daily production 32.9 m (108); best weekly production 168.5 m (553 ft); best monthly production 595.3 m (1953 ft).

Problems

No job is without its problems and this one was no exception. It should be reiterated that this project from inception was known to be technically challenging. The site, the geology, and the equipment added to the project challenges and brought their share of problems to the project team.

As noted above, during the eight months of delay caused by the bid protest, world tunnel equipment demands increased rapidly, creating a large backlog and long lead times for the manufacture of tunneling equipment. Prior to the arrival of the Seattle TBM, demand was so high that a TBM was actually air lifted from Toronto to England. Strained manufacturing capacity extended the time required to complete the TBM for the project. Instead of shipping in late October the machine was actually ready for shipment in early December.

Shipping was to take 14 calendar days by train as experienced previously by Lovat. As noted above, the machine arrived January 27, 1996. Shipping delays occured because of holidays and winter weather conditions crossing the Canadian Rocky Mountains. The delayed arrival of the TBM in Seattle placed the project behind schedule and the time was never recovered.

Grade control was a constant concern due to the one-pass liner approach, which did not allow for second lining correction. Sewage flow requirements and the nearly level grade 0.0004 m/m (0.0014 ft/ft) exacerbated alignment control problems. Skilled operators with abilities matching the sophistication of the machinery are difficult to find in many geographic markets. This also contributed to the difficulty maintaining control of line and grade. The County and McNally worked jointly to minimize the cause, effect, and correction of the grade control problems. To date, only 36.6 m (120 ft) of tunnel requires remedial action to correct out of tolerance construction. With about 304.8 m (1000 ft) remaining to complete the drive, the project team is confident that no further remedial action will be required.

Mechanical problems presented a constant challenge. All of the equipment saw periods of heavy usage and presented both expected and unexpected maintenance problems. A new TBM should be considered a test unit requiring a test operator and test mechanic until its capabilities are proven in field applications. Manufacturing QC and involvement of experienced operating personnel during TBM design and fabrication are imperatives of successful TBM deployment during tunneling. Much time, effort, and funds were expended to eventually overcome many machine inadequacies during a time when the TBM was expected to be operational and in use. The decision whether to use a new or used TBM must be carefully assessed. The TBM used on this project is state of the art. It has many capabilities and some weaknesses. Unlike at the beginning of the project, these capabilities and weaknesses are now known, because unlike at the beginning of the job, it is a used machine.

DISPUTES AND CLAIMS

The potential for disputes and claims was a major concern for the design team. Differing site conditions, delay claims, and construction-related problems were potentially problematic. Very often these are the things that cause disputes in underground construction. The design team felt strongly that the underground construction industry was moving in the correct direction in the use of dispute avoidance and resolution techniques. The bid documents included a Geotechnical Design Summary Report (GDSR); a formalized Partnering Agreement; clear disputes resolution procedures including a Dispute Review Board; and bid escrow documents. Good intentions cannot prevent disputes on fast moving, complex, and evolving construction projects. Obviously, disputes can be the source of disintegration of relations between the contracting parties. Disintegrated relations often lead to loss of focus on getting the job done. Bid escrow documents, Partnering, and formal disputes resolution procedures help maintain focus on the work instead of disputes.

Partnering Agreement

From the beginning it was clearly stated and understood that the partnering agreement was not a replacement for the contract. County and contractor staff took the partnership concept seriously and used the process to defuse communication problems between the two organizations. A questionnaire was circulated at quarterly intervals to determine key trouble spots. The process was successfully employed to maintain constructive dialogue between the contracting parties, the engineer, and other important stakeholders. An independent facilitator [Ann Smith of Smith-Culp Consulting, Seattle, Washington] used the questionnaire to prepare agendas for working sessions between the various parties. The facilitator moderated the working sessions to keep them focused and productive. These sessions were used to confront issues such as deteriorating communications, schedule losses, and technical challenges. Problems were openly discussed in working sessions. This allowed the stakeholders to clearly state their needs, resulting in better communication. Problem solving teams consisting of key personnel from the County, McNally and engineering staff tackled the most difficult problems in a cooperative, professional atmosphere. These problem solving teams determined and implemented the best solution for all parties. The authors believe the Partnering Agreement was beneficial to the contracting parties and that the agreement helped keep the stakeholders focused on getting their jobs done. We didn't always agree, but when we disagreed we used the dispute resolution process to resolve our differences.

Dispute Review Board

Project specifications stipulated the use of a formalized dispute resolution process including a Dispute Review Board (DRB). Senior owner and contractor staff believed very strongly that the project had a better chance of succeeding if disputes could be avoided, or resolved impartially and in a timely manner. A Dispute Review Board was constituted for the project. The DRB met on a quarterly basis. Their expenses were jointly covered by the contracting parties who entered into a third party agreement with the Board members [Ed Peterson (Chair), Russel Clough, Paul Eller].

Although owner and contractor staff took great pride in their ability to resolve disputes, there was one issue where agreeing to disagree was the best that could be accomplished. The dispute grew over a claim for additional compensation resulting from a work stoppage to remove grout from the TBM exterior and reconstruct a section of tunnel invert. Between July and November of 1996, the project team made several attempts to resolve the dispute at the project level. The parties were able to agree on cost facts, but could not assign responsibility. As the project advanced it became apparent that an agreement could not be reached and the dispute was handed to the DRB. The DRB convened a special meeting to hear the positions of both sides. Their deliberation and findings resulted in a 50% split in the costs for the disputed work. Both parties agreed with DRB's determination and the County issued a change order for its share of the costs. The authors believe that the project benefited from following the specified dispute resolution procedures.

CONCLUSIONS

Completing a soft ground tunnel like the West Seattle Tunnel provides many engineering and construction challenges. These challenges are not without risks. The authors believe the project was successful because the people who prepared the design, built the project, and managed the work came together to accomplish common goals. There were disputes and problems, but these were overcome because the people maintained professional attitudes and openly discussed problems. One of the greatest successes of the project was demonstrating that industry practices to reduces disputes, share risks, and accept responsibility very definitely contributed to the kind of project that people can look at with pride and a sense of accomplishment. The authors are proud of the project and the accomplishments of the people who designed it, built it, and managed its construction.

ACKNOWLEDGMENTS

The authors wish to acknowledge the following individuals for their contributions to the project. Without the efforts of these people the project successes never would have happened:

- The staff of McNally Tunneling Corporation
- Brad Cowles, Resident Engineer, KCDNR
- The KCDNR Construction Management team
- Leon Maday, Project Manager, KCDNR
- Al Larson, Project Engineer, HDR Engineers
- HDR Engineers and subconsultants
- Mike Byers, Geotechnical Engineer, Pac-Rim
- Paul Godlewski, Manager of Geotechnical Engineering Support, Shannon & Wilson

Chapter 47

IMPACTS OF GLACIAL TILL FILLED VALLEYS ON MINING OF THE UPPER SCIOTO WEST INTERCEPTOR SEWER, COLUMBUS, OHIO

David A. Day,[1] Paul H. Marshall,[2] Charles H. Arnold, III,[3] and C. Timothy Fallara[4]

[1]Technical Project Manager, Dodson -Lindblom Associates Inc.; [2]Project Manager, KM&M, Joint Venture; [3]Tunnel Consultant, Jenny Engineering Corporation; [4]Project Manager, Division of Sewerage and Drainage, City of Columbus

ABSTRACT

The Upper Scioto West Interceptor Sewer (USWIS), Parts 1 and 2, consists of approximately 7,128 meters (23,385 ft) of 328 cm (129 in.) diameter tunnel. The project extends from the City of Dublin to Griggs Dam in the City of Columbus. The majority of the tunnel was designed for excavation in rock by TBM. The remaining portions span pre-glaciation formed valleys in-filled with glacially-derived sediments. Tunnel construction in one of the glacial till filled sections includes excavation by drilling and controlled blasting. The other section will be constructed by ground consolidation followed by excavation by TBM.

This paper describes the methods employed to mine the glacial till filled valleys.

PROJECT BACKGROUND

The original design of the Upper Scioto West Interceptor Sewer was completed in 1987. The original design included approximately 1,829 m (6,000 ft) of open cut sewer in developed residential areas along the west bank of the Scioto River and through a City of Columbus Recreational Park and approximately 7,315 m (24,000 ft) of tunneled sewer ranging in depth from 14 to 23 meters (45 to 75 ft). As originally designed the tunnel would intersect four glacial till filled valleys.

Following the completion of the design, a Value Engineering (V.E.) Review was conducted. One recommendation of the Value Engineering Team was to increase the depth of the tunnel approximately 8 meters (25 ft) and extend the tunnel through the open cut section. The basis for the recommendation was to place the tunnel excavation further into bedrock and avoid the glacial till filled valleys and the associated soft ground mining. In addition, the elimination of the open cut segment was favored to remove the adverse impacts to the river front, residential areas and the City's park.

The V.E. recommendation was partially accepted by the design team. Due to elevation connection constraints at the ends of the project, the profile could be lowered only 5.5 m (18 ft). At this depth, the tunnel would encounter only one of the valleys. Following the V.E. Review the project was redesigned at the lower profile and bid.

The construction of the USWIS is of critical importance to the community due to the

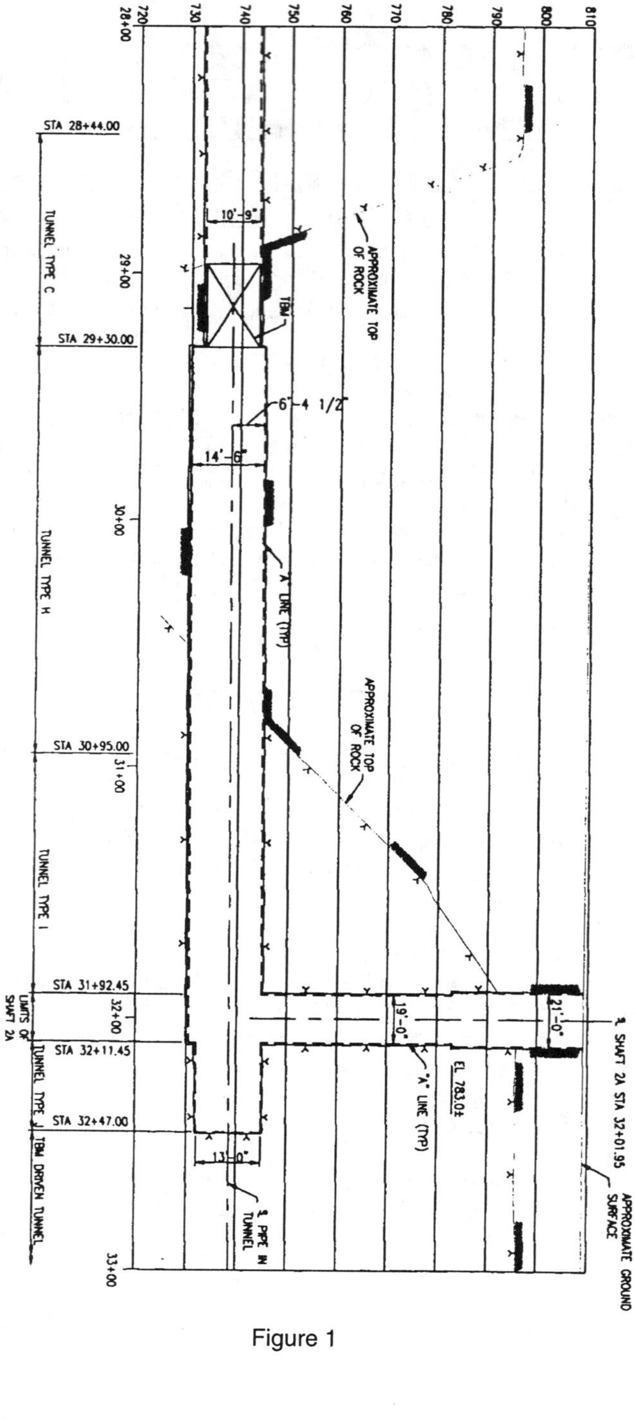

720
730
740
750
760
770
780
790
800
810
28+00
29+00
30+00
31+00
32+00
33+00
STA 28+44.00
STA 29+30.00
STA 30+95.00
STA 31+92.45
STA 32+11.45
STA 32+47.00
TUNNEL TYPE C
TUNNEL TYPE H
TUNNEL TYPE I
LIMITS OF SHAFT 2A
TUNNEL TYPE J
TBM DRIVEN TUNNEL
10'-9"
TBM
APPROXIMATE TOP OF ROCK
6'-4 1/2"
14'-6"
"A" LINE (TYP)
APPROXIMATE TOP OF ROCK
19'-0"
21'-0"
"A" LINE (TYP)
EL 783.0±
13'-0"
℄ PIPE IN TUNNEL
℄ SHAFT 2A STA 32+01.95
APPROXIMATE GROUND SURFACE

Figure 1

extensive development of the northwestern Franklin County, southeastern Union County, and southwestern Delaware County. The area tributary to the USWIS encompasses approximately 12,383 hectares (30,600 acres). The primary function of the tunnel is to convey sewage flows from the tributary area to the east side of the Scioto River where it connects to the existing East Scioto Interceptor Sewer. Once in service, the USWIS will enable the abandonment of eight major pump stations.

Part 1 of the project extends from Station 0+00 to Station 102+79 with the tunnel beginning at station 6+00 (3,134 m or 10,279 ft). The tunnel portion of Part 1 was expected to lie entirely within Columbus limestone. An inverted sewer siphon was constructed from Station 0+00 to Station 6+00 to convey flows under the Scioto River to the existing East Scioto Interceptor Sewer. The Part 1 Contract also included the construction of four access shafts and hydraulic connection structures.

Part 2 of the project begins at Station 102+79 and extends to Station 240+04 (4,184 m or 13,725 ft). The entire Part 2 Contract was designed to be tunneled. The majority of the Part 2 tunnel was expected to lie within Columbus limestone. The Part 2 contract also included the valley not eliminated by the lowering of the tunnel. This section extends from Station 121+09 to Station 126+00 (150 m or 491 ft) and was designed for drilling and controlled blasting and soft ground hand excavation. The Part 2 Contract also included the construction of six access shafts and hydraulic connection structures.

PREDICTED GEOLOGICAL CONDITIONS

Subsurface Exploration Program

An extensive subsurface exploration program was completed during the design of the project to define the geological conditions. The program consisted of site reconnaissance, approximately 84 vertical and angled boreholes, borehole logging, water pressure testing of the boreholes, and laboratory testing of the soil samples and rock cores. A Geotechnical Design Summary Report (GDSR) was prepared by Jenny Engineering Corporation to summarize the geotechnical data collected during the program and present engineering interpretations of the anticipated soils, rock and groundwater conditions. The GDSR was included in the Contract Documents for the Project, completing the fair contracting practices facilities of Disputes Review Board and Escrow Bid Documention.

Expected Site Geologic Conditions

The topography within the project area represents a moderately undulating, gently sloping ground surface. The ground surface ranges from approximately 231 to 256 meters (760 to 840 ft) above mean sea elevation.

Soil sampling indicated that glacial and alluvial soils cover the bedrock and range from 0.5 to 23 meters (1.5 to 75 ft) in thickness.

Based on the geologic data obtained from the exploration program the bedrock within the project area contains the following rock types (listed in order from upper to lower layers):

1. Delaware limestone.
2. Delaware limestone with chert nodules.

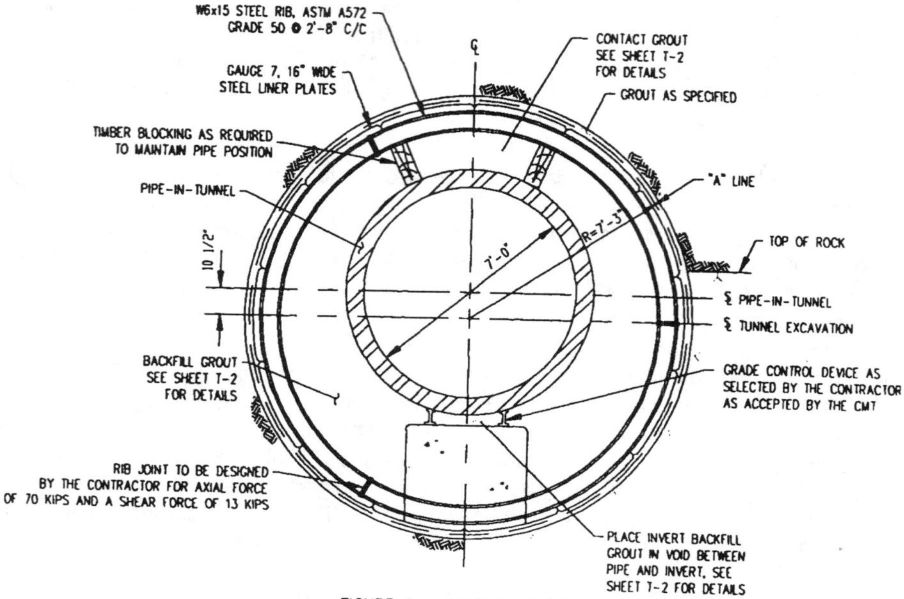

FIGURE 2 - TYPE H TUNNEL

MINED BY HAND MINING METHODS.
IN SOFT GROUND MIXED FACE AND ROCK

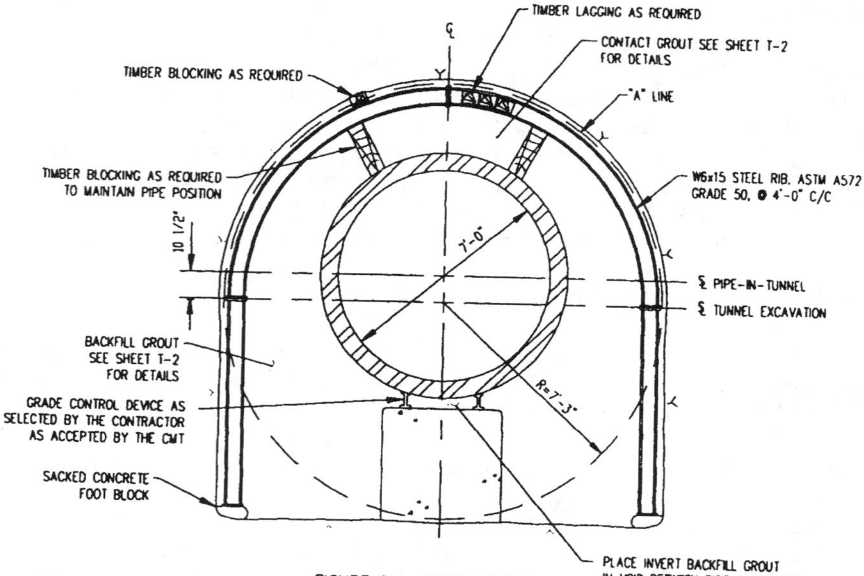

FIGURE 3 - TYPE I TUNNEL

MINED BY DRILL AND CONTROLLED
BLASTING AND HAND MINING METHODS

3. Columbus limestone.
4. A chert layer within the Columbus limestone.
5. Columbus limestone with chert nodules.
6. Dolomite.

The majority of the tunnel was anticipated to be within the layers of Columbus limestone, chert, and Columbus limestone with chert nodules. The Columbus limestone is composed of calcareous and magnesium deposits. The formation is approximately 30 meters (100 ft) thick with the lower 12 meters (40 ft) consisting of a brown magnesian limestone with small sections of chert and pockets of calcite crystals. The Rock Quality Designations (RQDs) of the rock ranged from 40% to 94% and averaged at approximately 57%. The tunnel zone was expected to have several high angle open and soil-filled fractures and broken zones. The unconfined compressive strength (qu) values of rock cores ranged from approximately 34 to 90 mpa (5,000 to 13,000 psi) with an average value of 59 MPa (8,500 psi).

Glacial Till-Filled Valleys.

Four glacial till filled valleys which run perpendicular to the tunnel alignment were identified in the GDSR. Vertical borings were concentrated in these areas to help define the depth and length of the valleys and to determine if the valleys would impact construction methods. Based on the borings, only one of the valleys was expected to extend into the tunnel zone. This again, was located in Part 2 of the project at approximately station 121+09.

TUNNEL DESIGN

The diameter of the excavated tunnel opening was defined by the hydraulic requirements for the finished sewer. The tunnel excavation was designed for a 328 cm (129 in.) diameter. The depth of the tunnel ranges from approximately 18 to 27 meters (60 to 90 ft). The Contract required that TBM driven tunnel be initially supported with systems ranging from steel channel straps secured with a 2-dowel or 4-dowel pattern on 1.2 meter (4 ft) centers (Type A and B) to steel ribs spaced on 1.2 meter centers (4 ft) with timber lagging (Type C) depending on the ground conditions. The starter and tail tunnels were designed for Type D tunnel which consisted of a 4.3 meter (14 ft) horseshoe tunnel with 6-#7 rock dowels, 3 meters (10 ft) in length, spaced radially on 0.6 meter centers (2 ft) and 1.2 meters (4 ft) longitudinally. The approximately 150 meters (491 ft) of soft ground area required Type E and F tunnel. The initial support for Type E tunnel consists of circular or horseshoe shaped W5x16 steel ribs on 1.2 meter (4 ft) centers and 5 gauge, 61 cm (24 in.) steel, gasketed liner plates. Type F initial support consists of circular shaped W5x16 steel ribs on 1.2 meter (4 ft) centers and 5 gauge, 61 cm (24 in.) steel gasketed liner plates.

Final lining of the tunnel consists of pre-cast reinforced concrete pipe with imbedded pvc liner. The annulus between the excavation and pipe was specified to be backfilled with cement, sanded grout to be placed in multiple lifts utilizing slicklines and grout holes.

TUNNEL CONSTRUCTION

The Contract for Parts 1 and 2 of the USWIS was awarded to KM&M, A Joint Venture (Kassouf Company, Murray Hill Construction Company, and Mole Constructors Inc.) with "Notice to Proceed" given on September 21, 1996.

Photo 1. Steel Pipe Center Support with Screw Jacks

Following the construction of Shaft 1, a tail tunnel and starter tunnel, the tunnel mining of the USWIS, Part 1, was initiated on May 24, 1996. The mining proceeded upgrade from Shaft 1 with a fully shielded Lovat tunnel boring machine (TBM) modified with a unidirectional, 43 cm (17 in.) disc cutter rock head. During the initial several thousand feet of this drive, numerous occurances of clay filled pockets were encountered; some as extensive as 11 meters (35 ft.) in length and covering approximately 50% of the face. Fortunately the fully shielded TBM which thrusts off a 360° gripper shield was capable of advance through these areas.

The Unexpected

As the TBM mining progressed, the rock quality rapidly deteriorated at approximately Station 29+00. As the TBM advanced, mixed face conditions were encountered and a full

Photo 2. Type H Tunnel Installation

Photo 3. Unearthing of TBM

face of soil was encountered at Station 29+19 on July 8, 1996. Behind the gripper shield the ground began to ravel requiring that modified Type C initial supports (ribs spaced on two foot centers) be installed with skin tight timber lagging.

KM&M promptly notified the city of Columbus of the formation and proposed to take several borings to define the extent of the tunnel zone which would be affected by the soft ground. The borings were also taken to allow the sampling and classification of the material within the formation. The City's Construction Management Team (CMT) acknowledged that a "differing site condition" was encountered and authorized KM&M to complete the proposed borings.

The first boring (B-1) was taken at station 29+43. The boring extended to a depth of 25 meters (82 ft), more than 1.5 meters (5 ft) below the bottom of the proposed tunnel excavation, without encountering bedrock. A boring (B-2) was then taken at station 31+47 which encountered bedrock at approximately 9 meters (30 ft), or more than 12 meters (39 ft) above the tunnel invert. Boring B-2 was extended to a depth of 24 meters (79 ft) with rock cores taken from 9 to 24 meters (30 ft. to 79 ft). Borings were then taken at station 30+23 (B-4) and station 30+66 (B-5) which did not encounter bedrock in the full 25 meters (82 ft) of the boring. A boring at station 30+90 encountered rock at 19 meters (62 ft) or approximately 3 meters (10 ft) above the tunnel crown. Boring B-7 was completed at station 29+90 to provide additional soil sampling within the karst formation. Based on the soils sampled from the borings it appeared that the feature encountered was actually a karst which was below glacial till. The feature was predicted to impact the tunnel zone for approximately 49 meters (160 ft).

The CMT and KM&M reviewed the boring information and concluded that it was not feasible to mine the soil within the karst with the TBM outfitted for medium rock excavation. The potential to retract the TBM back to Shaft 2 (station 20+00) and hand mine the clay was examined. However, retracting the TBM would require the removal of supports behind the TBM gripper ring, potentially impacting the stability of the tunnel excavation and resulting in a safety hazard unacceptable to the CMT.

Photo 4. TBM Advancing on Skid Beam Through Hand Mined Tunnel

Based on the soil type anticipated (stiff to running), the impacts to Dublin Road, the impacts to the project schedule and the associated costs, consolidation grouting was not accepted as a viable method to construct the tunnel within the karst feature.

The CMT and KM&M next considered constructing a temporary shaft near the problem area. This solution would allow the soft ground to be hand mined. The TBM would be "walked" forward to competent rock and relaunched from a newly constructed starter tunnel. Two sites for the temporary shaft were examined.

The first site reviewed was directly in front of the TBM within Dublin Road pavement. Construction of the shaft within Dublin Road pavement would require either the closure of Dublin Road or the construction of a temporary bypass roadway.

Photo 5. Fabricated "Shoe"

The second site considered for the construction of a temporary shaft was an area immediately to the north of the karst feature (at approximately station 32+00). The area is outside of the paved roadway allowing Dublin Road to remain open while the temporary shaft and recovery tunnel was constructed. This alternative would require a realignment of the tunnel. After reviewing schedule and cost estimates of the two alternative sites, it was determined to realign the tunnel, and locate the shaft in the area north of the karst feature and hand mine down station to the TBM (see Figure 1).

KM&M held several meetings with residents of condominiums adjacent to the shaft site to inform the residents of the required work. Site restoration plans were prepared based on input from the residents. Efforts were made to lessen the impacts of the construction to the residents. The site fencing was designed to provide sufficient site distance for motorists entering Dublin Road from the condominium roadway. Sound absorbing blankets were placed on the site fencing to minimize noise.

The CMT and KM&M held several meetings to determine the best design for the temporary shaft and recovery tunnel. Shaft and tunnel geometrics were discussed to ensure that available equipment could be used. Construction materials were evaluated so that specified materials were already on site or readily available.

Based on the meetings, the temporary shaft was designed for a 6.4 meter (21 ft) excavation diameter in the overburden. The supports were designed for a combination of 5 gauge, 61 cm (24 in.) steel, gasketed liner plates and W6x20 steel ribs spaced on 1.2 meter (4 ft) centers. The excavation diameter of the shaft in the rock was 5.8 meters (19 ft). The supports in the rock section included 2.4 meters (8 ft) long, #10 rock dowels spaced 1.2 meters (4 ft) vertically with a combination of welded wire fabric (6x6 W2.9xw2.9) and 10 cm (4 in.) of shotcrete.

The recovery tunnel from station 31+92 to station 30+95 (30 m or 97 ft) was expected to be mined in rock (boring B-6, station 30+90 encountered rock 3 m (10 ft) above the tunnel crown). This section of the recovery tunnel was designed as a 4.4 m (14.5 ft) horseshoe section (Type I) and included W6x15 steel ribs on 1.2 m (4 ft) centers with skeletal timber lagging (see Figure 3). From station 30+95 to station 29–19 the recovery tunnel was expected to be mined in a mixed face of rock/soil and a full face of soil. This section of the tunnel was designed as a 4.4 meter (14.5 ft) circular section (Type H) and included W6x15 steel ribs on 81.3 cm (32 in.) centers with 7 gauge, 40.6 cm (16 in.) wide steel, gasketed liner plates with grout fill between the liner plates and the limits of the excavation (see Figure 2).

The relaunch tunnel was designed to extend from station 32+11 to station 32+47 as a 4 meter (13 ft) horseshoe section (Type J) and included W6x15 steel ribs on 1.2 m (4 ft) centers with timber lagging. A 3.4 meter (11 ft) diameter cast-in-place concrete tube was included in the Type J tunnel to provide a circular section for the gripper shield to relaunch the TBM.

On August 23, 1996 sitework for the recovery shaft was initiated. Several large mature trees were relocated in keeping with the restoration plans developed in cooperation with the adjacent residents. Site fencing with privacy slats and sound dampening blankets was installed. New Jersey style concrete barriers were placed between Dublin Road and the site fencing. Pre-blast surveys of structures near the shaft site were completed while the sitework

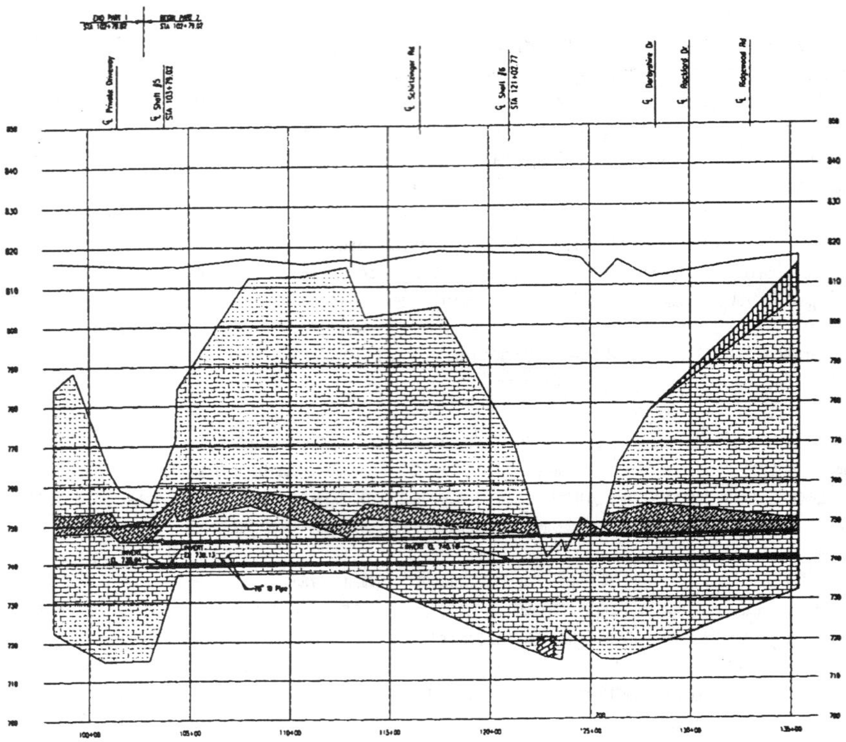

Figure 4. Buried Valley, Jet Grout Zone

was being completed.

Shaft excavation was initiated on September 4, 1996. The excavation within soil was completed on September 20, 1996. The depth to competent rock was approximately 11 meters (36 ft). The shaft excavation within rock was completed on October 5, 1996. Following the completion of the shaft excavation, drill and controlled blast mining (Type I and Type J) was started for the recovery tunnel and the relaunch tunnel. An alternating heading method was used. The mining of the relaunch tunnel was completed on October 17, 1996. The Type I tunnel excavation and initial support was continued until was completed on November 2, 1996 at station 30+83.

Because of the expectation that raveling and possibly running ground could be encountered, KM&M carefully chose the excavation and initial support methods for the Type H tunnel. The main concern emanated from the magnitude of the diameter of this soft ground heading under a major traffic artery. The excavation initiated with the installation of a 40.6 cm (16 in.) diameter steel pipe by conventional bore and jack methods at springline. Approximately 42.7 m (140 ft) of the pipe was installed in the center of the Type H tunnel section to provide a concentric bearing point for screw jack supports to hold the liner plates in place until bench and heading excavation was complete and a full circle rib could be installed. A platform for the auger was constructed on 2 muck cars. The platform was secured in place by welding steel I-beams to the muck cars and to rock dowels installed into the floor of the Type I tunnel. The installation of the pipe was completed to the face of the TBM on November 9, 1996.

Following the installation of the springline casing, a top heading and bench method was utilized to excavate the Type H tunnel with the steel pipe being utilized as a center support. The top half of the heading was excavated forward to a depth which allowed a row of arch plates to be installed in the top heading. Screw jacks were immediately extended from the pipe to the plates providing temporary support until the bottom section of plates and ribs could be installed.

Generally, the top heading was advanced to a length required to accomodate approximately 8 plates before the bench was excavated. Arch plates were installed in the bottom of the heading as the bench excavation progressed. Ribs were installed when the full circular section of liner plates were complete. The steel pipe was simply cut off in sections as the excavation progressed. Photos 1 and 2 depict the Type H tunnel in various stages. This process was repeated until the soft ground hand mining was completed on January 6, 1997. Photo 3 depicts the unearthing of the TBM.

After the completion of the recovery tunnel, the invert was cleaned and a W24x161 skid beam was installed on the floor of the tunnel. The skid beam was imbedded in a concrete cradle formed to the curvature of the TBM. Steel cables were also imbedded in the concrete to provide a means to secure the final pre-cast pipe liner in place during the placement of grout backfill. The skidbeam was used to advance the TBM into the relaunch tunnel. (Photo 4 depicts the TBM being advanced on the skid beam. Three inch diameter holes were cut in the web of the skid beam on approximately 1.2 m (4 ft) centers and a "shoe" (Photo 5) was fabricated with a 7.6 cm (3 in.) diameter peg to provide a backstop for the TBM to push against.

The TBM was advanced into the temporary shaft on January 10, 1997. Following maintenance to the TBM, the TBM was advanced into and relaunched from the starter tunnel on January 11, 1997.

The Expected

A major glacial till filled valley was discovered during the geotechnical investigation for the Upper Scioto West Interceptor from Stations 121+09 through 126+00 a distance of 150 meters (491 ft). This valley (see Figure 4) was filled with materials ranging from broken, weathered limestone at the interface, overlain with sands, gravels and silts and capped with a very dense clay. The depth of this valley is well below the invert of the tunnel excavation, which was about 24.4 meters (80 ft) from the surface. As a result of the presence of this buried valley and the expected difficulty that a TBM designed to mine the limestone present throughout the vast majority of the tunnel would encounter, the designers specified two different primary linings in this area in anticipation of a drill and blast method in conjunction with soft ground hand mining techniques to mine this area. A dewatering system was also specified to facilitate this operation. The original designed primary linings consisted of 4.1 meters (13.5 ft) diameter 5 gauge linerplates with internal W5 x 16 steel ribs on 1.2 meter (4 ft) centers. It was anticipated that this hand driven excavation would be performed prior to the TBM advancing into this area, and the TBM would then have been walked through on a bottom skid beam.

Soon after award of the Contract and notice to proceed, KM&M Joint Venture began to investigate other possible means and methods to solve this challenging geologic circumstance. They retained Lachel & Associates, Inc. Geologists, Engineers to perform an additional geotechnical study of this immediate area including 6 soil borings. Subsequent to the Lachel report, consideration was given to ground freezing, compressed air, and ultimately jet grouting was selected to modify the ground to an appropriate medium for the TBM to mine with the disc cutter head.

Jet grouting is a technique which originated in Japan in 1965, it has however just recently become prevalent in North America. The ASCE Geotechnical Engineering Division Committee on Grouting (1980) defined jet grouting as a "technique utilizing a special drill bit with horizontal and vertical high speed water jets to excavate alluvial soils and produce hard impervious columns by pumping grout through the horizontal nozzles that jets and mixes with foundation material as the drill bit is withdrawn." Combinations of high pressure water, grout and air are used to tailor the system to the geology encountered. The air is used to "air-lift" material up the drill stem to the surface which has been cut by the water and displaced by the cement and water slurry. Typical pressures for the water and grout are in the 400 to 500 bar range.

Three specialist firms were contacted to render a proposal and Nicholson Construction Company a Rodio Group Company was selected to perform the necessary ground modification.

The design criteria of the grout block was as follows:

- 5.5 MPa (800psi) ultimate compressive strength at 28 days.
- Dimensions: 97.5 meters (320 ft) in length, approximately 6.1 meters (20 ft) wide and 1.8 meters (6 ft) above and below the tunnel excavation of 3.4 meters (11 ft) in diameter.

A test section was to be performed and then unearthed to verify the jet grout columns.

The test section was installed in comparable geology, however at a much shallower depth to allow the recovery of the columns. Several columns were created using various speeds of rotation, pace of retraction, pressure changes and variance in mix design. Upon hydration, the columns were removed and the best combination of techniques to achieve design results was selected. The test column selected was 1.8 meters (6 ft) to 2.1 meters (7 ft) in diameter, had the appearance of conglomerate and had adequate compressive strength.

Following the test column program, the CMT accepted the proposal by KM&M to employ jet grouting followed by TBM excavation in lieu of the designed tunnel (Type E and F tunnel) . Acceptance of the jet grout technique allowed for the initial support design to be revised for this area. The support system accepted with the jet grout program consists of a modified Type C initial support (steel ribs spaced on 1.2 meter centers (4 ft) with timber lagging and geotextile fabric).

The area to be jet grouted is immediately north of a Shaft 6 which provides a convenient site for the setup of the grout supply machinery. Some of the equipment is shown in Photo 6.

Four rows of columns were installed on 1.5 meter (5 ft) centers and a total of 65 rows

Photo 6. Jet Grout Batch Plant

(260 columns) were placed throughout this tunnel alignment area. Difficulty was encountered by Nicholson when it was determined that the columns were not developing as anticipated. The extremely porous soils at the interface between the weathered dolomite and the soft overlaying soils in conjunction with the very fractured and jointed bedrock would not allow the development of the columns. Additional testing, varying the procedures and introducing accelerating agents proved to be the correct solution. Many of the columns had to be re-jetted to assure proper ground modification so that the design requirements would be accomplished. At the time of this printing, the tunnel had not advanced into the jet grouted area.

SUMMARY

Through the cooperative efforts of KM&M, the City of Columbus, and the City's Construction Management Team, two separate processes have been employed to mine difficult ground conditions encountered on the Upper Scioto West Interceptor Sewer. An unexpected soft ground formation required the utilization of a heading and bench mining technique which proved to be a safe, efficient method to excavate the medium encountered. An expected difficult soft ground area which was designed to be excavated by convential hand mining techniques is being modified by a jet grout ground consolidation method followed by the excavation by TBM. Jet grouting appears to have definite wide spread applications in the tunneling industry to modify many types of difficult ground to make it appropriate for Tunnel Boring Machine excavation.

AKNOWLEDGEMENTS

Ralph Dodero, KM&M, is the project superintendant.
Roger C. Au, H. R. Gray & Associates, is the Project Coordinator.

REFERENCES

ASCE Geotechnical Engineering Division Commitee on Grouting, 1980.

Chapter 48

A 10-YEAR REVIEW OF MICROTUNNELING IN NORTH AMERICA

Glenn M. Boyce[1] and Timothy R. Coss[2]
[1]Senior Professional Associate, Parsons Brinckerhoff, San Francisco, CA;
[2]President, Microtunneling, Inc., Boulder, CO

ABSTRACT

Microtunneling first appeared with some regularity in the United States in 1987. Until then, only Japan and Germany utilized this method of construction. Its high costs initially limited its application to special situations. The last ten years have seen the use of new pipe materials, new machines, new cutterheads, new techniques, and an average growth of 30 percent each year. This paper documents the growth and changes that have occurred with microtunneling. Based on current industry practices and advances from abroad, this paper postulates what the next ten years may hold.

INTRODUCTION

In North America, microtunneling is defined as a sophisticated, remote-controlled, guided, non-personnel entry tunneling method with the capability of installing underground conduit at relatively accurate lines and grades within 25 mm (1 inch), plus or minus. The remote-controlled tunneling machine can be either an earth-pressured or a slurry-type machine and is advanced through the ground by hydraulic jacks pushing a string of pipe. Commonly, the pipe is a carrier pipe and no secondary casing is required. Pipe is jacked to fill and support the circular opening formed behind the microtunneling machine, unlike conventional tunneling techniques which assemble liner within a tunnel shield. The remote-control aspect of microtunneling makes it unique.

This paper discusses the growth and changes that have occurred with microtunneling in North America during the last ten years. The current state of the industry is documented. Trends for the next ten years are discussed.

MICROTUNNELING IN THE 1970S AND 1980S

Japan's development of microtunneling technology began as a result of an increasing public demand to control water pollution and improve the environment in the 1960s. The Japanese government responded to this demand by developing an aggressive policy and program to cleanup the environment. Successive five-year plans were developed and financed by the government. The initial stages of the clean-up program emphasized sewage treatment facilities and large-diameter pipe systems. By the early 1970s, the emphasis was diverted to the installation of small-diameter sewer systems. The need arose to develop methods which minimized surface disruptions of congested streets (Iseley, 1990).

The first machines were manufactured by Komsuta in 1975. Iseki Poly-Tech Inc. of Japan followed with a mechanical earth pressure counter-balance (MEPCB) machine in 1976. This machine had a large diameter man-controlled shield. Iseki realized that future demand would be for smaller diameter machines with remote-control and earth pressure balance capabilities. In 1979, a miniature version of the MEPCB machine, the Telemole, was manufactured. Iseki continued development. In 1981, it introduced the Crunchingmole able to advance through soft ground containing boulders.

In the 1980s, Germany took the same technology and began producing its version of microtunneling machines. The broadly uniform soil conditions found in Japan and Germany were important factors in the early development work.

THE NORTH AMERICA EXPERIENCE

Microtunneling first arrived in the United States in 1984. An Iseki machine was used to install a 1800-mm (72-inch) diameter 190-m (615-foot) long steel pipe under I-95 in Miami, Florida. The machines and methods were considered too expensive and sophisticated and the geology in the United States too variable for routine use. However in 1987, the City of Houston began to expand its existing sewer system in an effort to meet Federal Clean Water Regulations. Houston's geology consisting of unconsolidated sediments, a flat topography, and a need for massive expansion of the sewer system created ideal conditions to stimulate the use of microtunneling (Merritt, Crisci, and Klein, 1991). BRH-Garver realized the potential benefits of the method and was able to under bid the open trench contractors. BRH-Garver's successful bid for the River Oaks Projects helped bring microtunneling to the forefront in Houston (Tubbs, 1993). Once citizens found out their streets did not have to be dug up, microtunneling became the preferred method of sewer installation in Houston. Between 1987 and 1991, over 21,000 meters (68,000 feet) of microtunnels and 17,900 meters (58,000 feet) of jacked pipe were installed. The success of the contractors, the method, and the reduced capital costs of equipment have made microtunneling the leading method of sanitary sewer construction in the Houston area. Microtunneling has reached a point in Houston where trenching is no longer price competitive.

Since the Houston experience, microtunneling has been repeatedly used elsewhere. Early projects were completed in Dallas and Austin Texas, New York City, and Florida. City and county public works departments and water, power, and cable companies throughout North America are beginning to look at microtunneling and other trenchless techniques as ways to modernize and expand their existing systems.

There have been over 250 North American microtunneling case histories reported in the literature as of January 1997. In most of the early cases, contractors opted to use microtunneling to install sewer lines where projects were designed to be open trenched. In other cases, microtunneling has been used in special applications. These applications include crossings under airport runways, railroads, or highways; intake structures; or initial support for larger tunnels.

Texas has over 70 projects and is clearly the center for microtunneling. California, Florida, New York, and Hawaii are beginning to produce their share of projects. Table 1 shows the percentage of microtunnels installed by state as tabulated by the Trenchless Technology Center (TTC) in Louisiana (Atalah and Hadala, 1996). The increased use of microtunneling in these states is an indication of the way microtunneling has grown in more populated states.

The amount of microtunneling completed with each successive year has increased. Figure 1 shows a breakdown of the amount of microtunnels design and constructed by year. The 1980s showed steady work based solely on work in Houston. In the 1990s, the amount of work has increased to new heights each year in part because microtunneling is being required by designs.

Table 1. Percentage of Microtunnels Completed by State (Atalah and Hadala, 1996)

State	Installation (m)	% of Total Installation
Texas	61,050	48.7%
California	12,600	10.1%
New York	8,500	6.8%
Hawaii	6,400	5.1%
Washington	4,010	3.2%
New Jersey	3,900	3.1%
Florida	2,680	2.2%

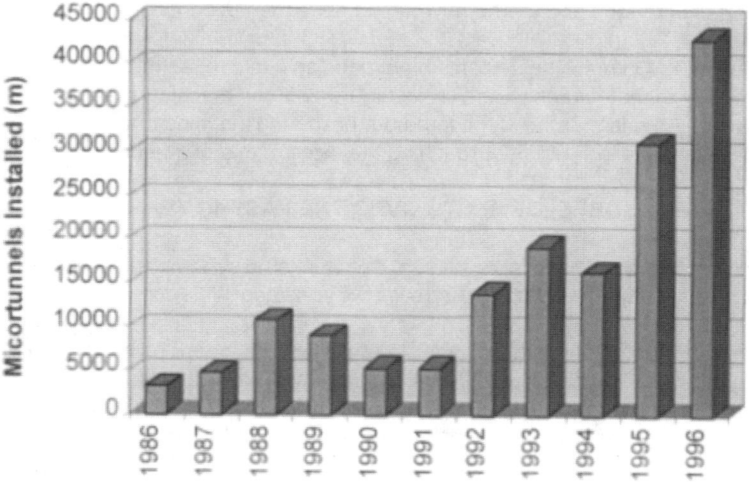

Figure 1. Footage of Microtunnels Designed or Construction Per Year

THE CURRENT STATE OF THE INDUSTRY

Microtunneling installations vary between 350 to 2200 mm (14 to 88 inches). At larger diameters, conventional tunneling methods are used. In North America, the smaller diameters are installed using other trenchless methods. Diameters of 450 mm (18 inches) are the smallest microtunnels installed in North America.

Microtunnels are typically installed with drive lengths vary between 30 and 300 meters (100 to 1,000 feet). Distances shorter than 30 meters (100 feet) utilitize other trenchless methods. It is not cost effective to mobilize microtunneling equipment for short distances. The range of drive lengths is a function of pipe diameter. Figure 2 shows the typical range of drive lengths for different pipe diameters. The ranges have been broken up between poor ground conditions and favorable ground conditions. Poor ground conditions include cobbly, bouldery, or mixed conditions. Favorable ground conditions included uniform ground types like sand, silts, or clay. The upper range represents the limits currently reached by microtunnels. The ultimate drive length a contractor will successfully complete will depend on a combination of many factors including: jacking capacity, cutterhead, lubrication, overcut, pipe materials, and actual ground conditions.

The maximum possible drive lengths are also controlled by the pipe diameter. Pipe materials' ability to withstand jacking forces controls drive lengths for microtunnels with diameters between 300 to 600 mm (12 to 24 inch). For the smaller diameters, drive lengths will approach 100 to 160 meters (300 to 500 feet). Slurry system's pumping capacity and the inability to install intermediate jacking stations (IJS) limit drive lengths for machines in the 600 to 900 mm (24 to 36 inch) range. These machines approach drive lengths of 160 to 200 meters (500 to 600 feet). The machines 900 mm (36 inch) diameter and larger have increased power, more room for the slurry pumps, and can use intermediate jacking stations. The drive lengths can immediately approach 300 meters (1,000 feet). The maximum drive lengths plateaus at around 400 meters (1,300 feet) for the larger diameters because of limitations on jacking systems, jacking forces, and lose of production. At long lengths, it becomes easier to stop and start a new tunnel drive then to continue.

As of 1996, 36 machine systems have been sold to contractors in North America. The breakdown by manufacturer is shown in Figure 3. Four machine manufacturers are still active in North America. Some of the larger tunnel boring machine manufacturers are beginning to produce smaller TBMs approaching the sizes of microtunneling.

DEVELOPMENTS OVER THE LAST 10 YEARS

Many changes that have taken place in microtunneling systems and associated equipment over the last few years. Equipment supplied even five years ago differs from a system supplied today.

A microtunneling system is made up of many component parts, including hydraulic, electric, electronic, and mechanical. The technology in these industries has developed so microtunneling machine manufacturers have taken advantage of this new technology and the equipment has improved. Microtunneling systems today are more flexible, powerful, reliable and productive than the early systems. Some of the developments which have occurred are discussed in the following.

Survey Techniques

Time and electronics have carried survey techniques to greater distance over the past years and should go even further. Some issues are correct laser placement for machine guidance and ground anomaly survey techniques.

Tunnel laser placement is critical to a microtunnel's completing a drive within 25 mm

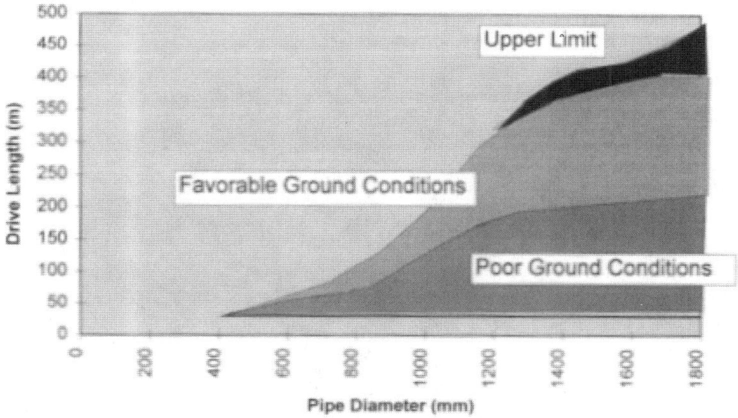

Figure 2. Typical Drive Lengths for Different Pipe Diameters

(1 inch) of line and grade. Using a horizontal string and plumb line for short drives to control line and grade often performs well. Longer drives demand exact laser placement. Laser set ups have successfully been accomplished by the use of rotating vertical lasers. The set up places the rotating laser top side in a "crows nest", obtaining its alignment from PK's spaced as much as 300 meters (1,000 feet) from the jacking pit. This near perfect alignment is transferred down the shaft in the rotation of the laser beam. The tunnel laser is then set and maintaining line and grade to 25 mm (1 inch) over 465 meters (1500 feet) have been attained. Self leveling lasers provide reaches of 465 meters (1500 feet) plus with a usable spot diameter of 40 mm (1.7 inches). These techniques have

The most exciting innovation is the surveying of subsurface anomalies with the use of horizontal bore hole sonics (Miller, 1996). A horizontal directional drill is used to place a 100 mm (4 inch) diameter HDPE pipeline along the pipe alignment. An electronic transmitter and receiver are pulled from one end of the HDPE to the other. Reflected signals from buried anomalies are recorded. This method has proven valuable on several situations within the past year. It has saved considerable pain prior to being stopped by obstructions during the actual drive.

This area has shown considerable growth and remain the one whose growth in the years to come is needed most. Innovative minds coupled with the electronics and computer industries expected advances should make it easier to know where we are going with a reasonable assumption of getting there.

Guidance and Steering Systems

Laser guidance and steering systems have come a long way since their inception. Major manufacturers have either manual or automatic steering systems. Most are digital type smart boxes and some use TV cameras and all work remarkably well. Automatic steering systems are an integral part of the machine's control system. Systems come with individual bells and whistles and control line and grade to within 25 mm (1 inch). A machine on automatic without a trained operator is like an aircraft without a pilot. The automated sys-

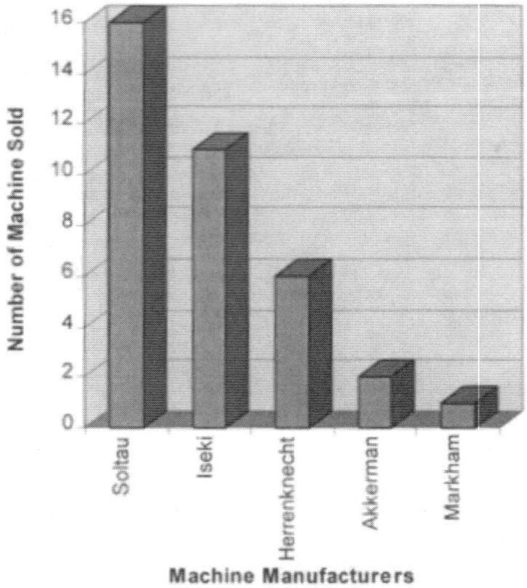

Figure 3. Microtunneling Machines Currently Owned in North America

tems take mundane tasks away from the technician but do not replace him. Data logging is an important part of the system's capability and is covered in the following.

These systems come at a premium. The future holds increased automation of the system. Hopefully in the years ahead, these systems will remain user friendly but at a much reduced cost.

Cutterheads

One area that has improved over the years is cutterhead designs. The improvements contribute to the advances realized by microtunneling in recent years. A few years ago, the cutterhead design concept had a recessed face, as shown in Figure 4. The idea was to allow the ground to enter into the face. The head of the machine was equipped with a cone crusher to crush all particles to a size which could be carried by the auger or slurry system. The rule of thumb that a microtunneling machine could excavate boulders one-third the machine diameter was solely related to this type of cutterhead. The cutterhead was used primarily in soils.

In the last few years, the philosophy of cutterhead design has changed. Heads with three to five arms are often used, see Figure 5. The head cuts the material and directs the ground into the machine head. Some of the larger particles will pass between the arms and are crushed. Particles too large remain in front of the machine and are attacked by the cutterhead. The cutters, chisel points, or picks repeatedly hit the obstructions reducing its size so it can pass into the machine crusher.

High pressure jets were introduced to microtunneling in 1994. The jets are protected by the mass of the cutterhead. The jets are used to improve performance notably in clays. Fixed jets focus the water stream to within the machine's cutterhead assembly. One manufacturer takes this a step further by incorporating this rotary jet cutting action on the cutterhead by using a rotary union on the main shaft. This action is more efficient and ensures a better cutting action within the throat of the machine.

Figure 4. Microtunneling Machine with the Recessed Cutterhead

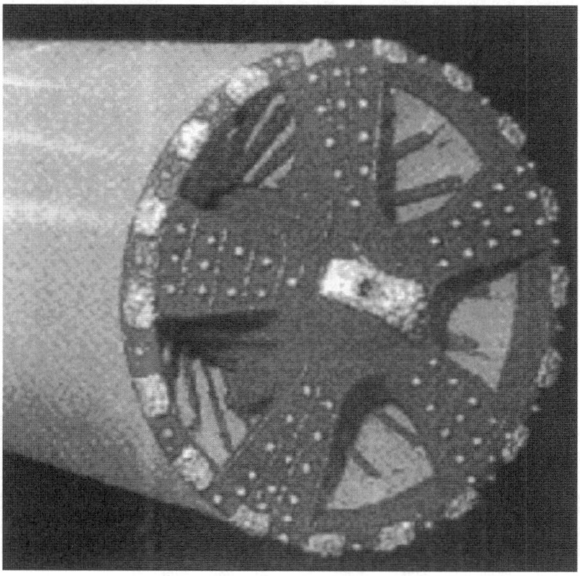

Figure 5. Microtunneling Machine with the Five-arm Cutterhead

The most recent advancement has been the development of the mini-disc cutters (Friant, Ozdemir, and Ronnkvist, 1994). The disc cutters are 100 to 125 mm (4 to 5 inches) in size. They are now mounted on a dome-shaped cutterhead, see Figure 6. The cutters extend ahead of the soil slots. If rock is encountered, the cutters begin to chip the rock into particle sizes which pass into the slots for further crushing. In soil conditions, the cutters penetrate into the ground and the soil slots, mounted with chisels, excavate though the soil. These latest development now allow microtunneling through rock and mixed face conditions.

Machine Drive Motors and Torque

Drive motor size, torque, and speed have increased over the years. The increases have allowed machines to develop more power to excavate through rock and mixed face conditions. Discussion of motors and torque between rival equipment suppliers will keep one amused for quite a while. Horse power, torque, electric drive and hydraulic drive are some of the buzz words to initiate such an interchange. It becomes complicated for those who know the difference and difficult to explain, that is why there seems to be so much artistic liberty in the literature.

Something to think about while being bombarded. One machine can have 271,000 Nm (200,000 ft/lbs) of torque utilizing a 11 KW (15 HP) motor. Another machine can have the same torque with a 110 KW (150 HP) motor. The question should be at what speed does the machine have with what torque? A variable speed cutter head makes a machine more advantageous over one of fixed speed. A machine with variable speed has no advantage if it does not maintain a reasonable torque curve over its entire speed range.

Figure 6. Microtunneling Machine with the Dome-Shaped Cutterhead for Excavating in Rock and Mixed Conditions

The electric drives torque, if a fixed speed machine, should be stated at its running drive rotation speed. Variable speed cutter heads should have torque stated at half and full speed. The specification should calculate the motor at 100 percent of its name plate rating.

A better determination by the end user can be made as to the machines ability to negotiate specific ground conditions. A more common method of measuring torque is not in the best interest of some suppliers so the future is cloudy in this regard.

In-Line Pumps

With the event of longer jacking distances, faster production rates and larger machine diameters, booster and intermediate pumps have become a necessity. These pumps are not new to the industry and have been around for the past ten years. The manufacturing of pumps has improved, providing better capacity and reliability.

Booster pumps provide a added boost for slurry from the charge pump and to the discharge pump. Maintaining critical minimum velocities prevent solids separation in the slurry line. Pump size and horse power coupled with the optimum line size ensure the success of the slurry circuit. All pumps need to be considered and sized with the other pumps of the system.

We have found over the years the sizes of slurry system pumps have increased. They are being speed controlled and are an important part of most projects. The in-line pumps are giving way to the booster pumps type. Booster and the in-line pumps perform the same job while doing it a little differently.

The system consists of:

- A charge pump supplies slurry to the microtunnel operation and is located on top of the ground adjacent to the slurry separation tanks. A 56 KW (75 HP) charge pump with a variable speed controller ensures its flexibility and is recommended on machine sizes of 1200 mm (48 inch) and larger.
- The in-line pump is located within the pipe string at various distances from the machine, there can be more than one in-line pump, and sizes can vary from 3.5 to 22 KW (5 to 30 HP).
- The booster pump is normally a larger pump, from 30 to 45 KW (40 to 60 HP) and is located behind the tunneling shield. The booster pump is driven by a variable speed motor controller and provides seamless operation in conjunction with the charge and discharge pump.
- The discharge pump located at the shaft bottom completes the circuit and is in the 37 to 55 KW (50 to 75 HP) range.
- The slurry line sizes are the most neglected part of the slurry circuit. Improper sizing of a 100 to 125 mm (4 to 5 inches) slurry line can require twice the horse power to achieve the same effect.

Ventilation

Ventilation when microtunneling first started was left to contractors who had no experience with it. Tunnels were placed in the ground with no ventilation at all because no one entered the microtunnel. The microtunnel environment becomes stagnate and hot. In the early days, no one attributed problems with alignment control to a lack of proper ventilation. Time to focus on the problem has led to the use of efficient blowers that are able to provide positive and negative flows to the heading. Positive flow furnishes air pressure through an air line located within the pipe string terminating at the microtunnel machine. Negative flow utilizes vacuum at the blower to draw air through the line that is ending at the heading.

Efficient blowers in the industry today supply 400 to 800 cfm utilizing small air lines 100 to 125 mm (4 to 5 inches) in diameter. The small air lines enable there insertion within the pipe string without creating a size obstacle. The proper balance of line size and blower pressure ratings can create enough negative air flow to vacuum water from the heading in down hill drives. This saves the use of a special pump and return line for this purpose. Water separation packages are needed for this use.

The use of blowers today are usually called out in the specifications. There use compiles with OSHA regulations for 200 cfm per person, within the tunnel. Laser image quality is also greatly enhanced with clean air preventing laser distortion.

Today we enjoy adequate equipment to perform the job at hand. Future needs are not the equipment but rather educating the industry.

Jacking Forces

A lot of discussion has developed over jacking forces. Jacking forces are the main concern particularly for the smaller pipe diameters because the pipe wall thicknesses are thinner and the pipes have lower strengths. Jacking forces can get very high when certain conditions are allowed to occur. These conditions include:

- Installing the pipe string with multiple steering corrections causing a plowing effect
- Using pipe with a rough exterior surface increasing the coefficient of friction
- Allowing the ground to collapse around the pipe string
- Using pipe than is not round
- Trying to jack long distances
- Using only one jacking system.

Jacking forces can be theoretically calculated, but many factors contribute to the ultimate value. High jacking forces can be reduced by using a combination of smooth round pipe materials, lubrication, overcuts, intermediate jacking stations, and keeping the pipe string moving through the ground. A lot of data on jacking forces has been collected from microtunneling case histories (Coller, Staheli, Bennett, and Post, 1996). Estimates of the range of possible values have been developed. This information is useful for designing jacking systems with sufficient capacity.

Intermediate Jacking Stations

Intermediate Jacking Stations (IJS) have been around since before microtunneling was performed in 1984. Their use is to extend the useful jacking capacity of the pipe string and the main jacking station. A project in Staten Island, New York by E. E. Cruz and Company saw the use of four 800-ton IJSs plus 800 tons from the main jacks, a total of 3,000 tons. This effort saw the installation of 482 meters (1560 feet) of 800 ton rated Hobas pipe. The actual jacking tonnage used was 2250 tons.

Actual projects using IJS have not been common because shorter microtunneling distances are attempted. We are seeing an increase in the jacking distances being attempted today and in poor ground conditions. The increase in jacking distances are in part to the reduced costs related to increased tunnel lengths. There is a point of diminishing returns as the tunnel length is increased with the use of IJSs. IJSs are costly for each one placed in the pipe string and productions are greatly impacted. A 800 to 1,200 ton push with one or two IJSs installed is probably the limit of cost effectiveness. Longer pushes for various reasons may make the use of many IJSs practical.

The state of the art as it relates to IJS hardware is effective today, although experience industry wide is lacking. Proper use of IJS is related to the stations placement within the pipe. Improper estimating of jacking loads could cause too many stations being placed or not enough to do the job. Over or under estimating their use can lead to failure or costing too much. Pipe manufacturers joint details often indicate a lack of experience with IJS and the average contractor does not have the ability to correct it. More in the way of education is needed.

The control of these units electrically where more than one station can be used at once is an area that needs to be defined. The use of hydraulic power via hoses from the operation room utilizing quick disconnects becomes complicated because of the volume of hydraulic oil and pressure consumed for multiple station operation.

IJS technology is where it needs to be but the knowledge to make it happen is lacking. The future needs to be reserved to educate the industry.

Lubrication Systems

Pipe string lubrication has been a requisite since the first microtunnel was performed in 1984. The use of lubricants with a slurry type tunneling machine is noticeably different from that of the dry tunneling method.

The use of bentonite as a lubricant has and will probably remain a constant within the industry. Polymers have been introduced and can be used alone or in conjunction with bentonite. Chemical additives to prevent clays from sticking to the microtunneling machine, prevent clay from swelling and balling are important to the process.

The proper geotechnical investigation helps the contractor determine what can be expected from the soils and help select the correct lubricant. This alone will not insure success, placement techniques are critical in this process. Lubricant mixing techniques, pump flow rates, pump type, pump pressures, lubricant supply line configuration, placement within the pipe string through out are some of the factors affecting the systems efficiency.

We have come a long way in the past ten years. We think the future will provide even better lubricants. What is needed is a better handle on how lubricants react in varying ground conditions as they relate to predicting jacking loads.

Electrical Systems

Electrical systems were left to the contractor in the early days. A evolution in this regard has been slow but as of this writing the manufacturers have come a long way. A contractor of old would decide how he "thought" things should be connected without a lot of knowledge of how everything interconnected electrically.

Today some operation control rooms look and perform like futuristic rocket ships. Distribution panels provide places to plug in all of the many individual circuits required to operate a microtunneling set-up. Ground fault protection supplements the panels circuit breaker protection. Electrical cables have evolved a great deal. The older cables would trip out and shut off the machine if it even got damp. Today the cables although not water proof can withstand hose down with a pressure washer. Variable speed AC drives have enabled better control of slurry systems. The computer age has reached the microtunneling operation room. Programmable controllers centralize the control of all system elements. Modular, plug-in components have provided greater control parameters, ease of maintenance at a lower cost.

We can only imagine the future of the electrical portion of the microtunneling process. Cost to build these systems will likely come down with the amount of control and sophistication to increase. Computers and modular components will lead the way. The repair and trouble shooting of these sophisticated control systems will place the need for a higher caliper technician on the site.

Data Acquisition Systems

Microtunneling systems have become electrically very sophisticated. Computer interfaces with system components provide better control and also the opportunity to record this data. Computer data are in digital format ensuring very accurate control of the microtunneling process. This same digital information is very easy to store in the memory of the same computer used to access it.

Data logging is the retrieval of information at a future date to evaluate system performance. This information can be evaluated in the event of a problem or to just study the operation. The information can be generated in the form of a report. The format is usually in the form of a hard copy.

Some information that can be retrieved includes:

- Jacking length
- Jacking pressure
- Rate of speed
- x, y, or z deviation
- Roll
- Pitch
- Slurry flows
- Valve positions
- Cutter head speed
- Cutter head torque
- Steering cylinders.

These separate functions will vary slightly depending of the manufacturer. The future will no doubt provide more information with better report generation at a lesser cost. Computer advancements will lead the way. The industry is in a good position today with the technology available.

Soil Separation Systems

Separation plants have become common place within the microtunneling community in the past several years. The system comprises of several components that need to work in harmony with one another and often time in conjunction with polymers.

Accurate soil borings are critical for the proper planning of the separation process. Since this mechanical separation cuts out specific particle sizes with screens and/or hydrocyclones it is important to know these sizes prior to setting up the separation plant. Particle determination to 1 micron are needed.

Settlement - Separation by the settling of solids within settlement tanks has its origins from the beginning of the microtunneling technique. Cuttings from the excavation of the microtunneling machine are carried away by the slurry system. If the solids are large or heavy enough, then once deposited in the separation tanks they will fall out of suspension. A variable to this method is the actual catchment time of the particles in the tanks, it needs the time to fall out. Several, large volume tanks are required. The spoils are than removed using a backhoe or clam bucket. The method at best is messy, but works well with larger particle sizes of say 100 to 200 microns and up.

Shale Shakers - Shakers are the first component of the separation process. Typically, they are placed on the settlement container as the slurry comes to ground level from the microtunneling process. The unit is basically a vibrating frame surrounding a screen of varying mesh sizes. The slurry flow on top of the vibrating screen causes particles larger than the mesh of the screen to be retained by the screen. The movement of the screen carries the particles away from the slurry and drops them on the ground or in a tank for disposal.

De-sanding Cones - Usually 250 to 300 mm (10 to 12 inches), single or double configurations provide added cleaning of the fluids prior to the mud cleaner. This process prevents plugging and ensures efficient operation of the hydro-cyclones.

Mud Cleaners - A mud cleaner incorporates hydro-cyclones over a fine mesh vibrating screen. The cyclones usually 75 to 100 mm (3 to 4 inches) depending on its setup configuration can give cuts of 10 to 20 microns.

Polymers - The addition of polymers to the slurry aid in the process by helping to drop very small particles that can not be cut out mechanically.

Excellent separation equipment is available today. It would be hard to conceive that the equipment will get a lot better in the years to come. Education of the industry is in the process with approximately ten separation plants on construction sites today. The industry needs to be educated to make a science of the process rather than a guessing game. The use of mechanical separation equipment and the use of polymers should become common place in the years to come.

Pipe Materials

Numerous pipe materials are available. Early installations used steel casing for highway, railroad, and runway crossings. Concrete was used because of its low cost. Direct burial concrete pipe is not required to be round and ASTM standards were not very defining for microtunneling.

Today, pipe choices have never been greater, with choices leading quality. Pipe materials have come a long way with since we started microtunneling in the 1980s. Below is a summary of the many pipe choices available (Boyce and Gray, 1996).

Fiberglass Reinforced Pipe (FRP) - Hobas Pipe has lead the microtunneling market with approximately 68,000 meters (220,000 feet) installed to date. The Houston based manufacturer has been a pioneer in the field. The fiber glass pipe provides a smooth interior and exterior surfaces. This smoothness reduces jacking loads and enables higher interior flow rates per diameter inch. The pipe can be designed as both as a gravity and force main pipe. Quality is very high.

Steel Pipe - This pipe was the first microtunnel pipe to be jacked in Florida in 1984. Today it is estimated that 20,000 meters (65,000 feet) of steel pipe has been installed. A large boost to this footage has been the introduction of Permalok Pipe. Permalok is a flush ID and OD pipe that snaps together quite easily. The advantages are the normal welding times are eliminated thus increasing productivity. Permalok has introduced a pressure joint that should be welcomed warmly within the industry.

Reinforced Concrete Pipe (RCP) - Concrete pipe was first used in 1986, with approximately 54,000 meters (175,000 feet) installed to date. It is normally easy to find all over North America. The pipe comes with a draw back, being the wide variance in quality. Several suppliers do a fantastic job while some have a lot to learn.

PVC Lined Concrete Pipe - This pipe is new to the industry and has come on quickly. The pipe is manufactured by Pipe Form of Mendota Heights, MN. The core of the pipe is concrete with a PVC wrap on the inside and outside, while the pipe ends are ground square for dimensional accuracy. This product should find a niche in the market and do well in the future.

Polymer Reinforced Concrete Pipe - With a trade name of Polycrete, the pipe is a high strength polymer based concrete mix providing a thinner wall pipe with the same jacking strengths as a conventional concrete pipe. Its polymer composition providing a smooth interior and exterior surface with good corrosion resistance. The pipe is new to North America and promises to do well in the future.

Vitrified Clay Pipe (VCP) - Clay pipe with three manufacturers in North America will probably over take over concrete pipe footage in the not to distant future. The pipe is made in diameters to 1200 mm (48 inches) with pipe lengths from 1.2 to 2.5 meters (4 to 8 feet) in length. Its corrosion and chemical resistance coupled with its fine quality standards have helped make it the fastest growing pipe in the industry in the past three years.

THE NEXT 10 YEARS

What does the next 10 years hold?

Jacking on Curves

Curved microtunneling has yet to be implemented in North America. Curved tunnels of much larger size have been done successfully for years. Curved microtunneling has been implemented in Europe, the electronics on the market should ensure success once tried in the North America. Experience will probably be slow to realize in this area. The issue of joint tightness also has to be addressed for curves with tight radii. Tapered pipe segments which roll to form straight sections and curved sections still need to be developed and then manufactured at a low cost.

Jacking Distances

Jacking distances today are being stretched faster than the industry has experience with. The North American record is 482 meters (1560 feet), was extremely difficult and came with a great price tag. There are times where longer distances will justify the cost and risk but for the most part 300- to 370-meter (1,000- to 1,200-foot) drives remain within the realm of being practical. The costs for intermediate jacking stations, electric modifications, pump considerations, laser limitations, hydraulic constraints and much reduced production times make drives longer than 370 meters (1,200 feet) only for special circumstances.

A push of over a kilometer was accomplished in Europe in 1996 at enormous costs. Site specific evaluations must always be evaluated. Longer drives must be weighted carefully with technical issues. A second shaft is often times a better choice.

Tunnel Systems

Currently we push pipe through the ground using the cutterhead to cut and excavate the earth materials. The jacking distance is a function of the jacking forces that can be applied to the pipe and equipment. The future holds not only pushing, but pushing and pulling the pipe through the ground. This process was initially proposed by Markham of England (Bristow, 1994). Markham's system uses prefabricated pipe sections to push the machine through the ground. When the machine reaches the receiving pit, the machine is removed and the product pipe laid in the receiving pit. By pushing the pipe and pulling the prefabricated pipe section, the pipe is pulled into place. The system was successfully demonstrated as part of the CPAR program in Vicksburg, Mississippi (Bennett and Staheli, 1995) and with a project in Boston. The system is too time consuming for regular use. The system requires two passes and increases installation time. While the system will not likely be used in the future, the concept of pushing and pulling will.

Microtunneling is not the only method of installing small-diameter pipelines. Horizontal directional drilling (HDD) is another method, which could be used with microtunneling. Right now, HDD involves drilling a small-diameter (50 to 200 mm) pilot hole. The alignment of the pilot hole can be controlled horizontally and vertically. Depending on the final diameter of the pipe, the pilot hole is reamed to larger diameters and the pipe pulled into place. HDD installations are approaching 1700 meters (5,100 feet) in length. The future holds drilling a pilot hole with a HDD rig. The drill string will be attached to the microtunneling machine and in combination with the jacks, the pipe is pushed and pulled into place. The microtunneling machine in essence is a forward reamer, excavating and collecting the ground as the pipe is installed. This approach will allow microtunnels to be installed in lengths of hundreds to thousands of meters.

Pipe Materials

The next generation of pipe materials appears to be the use of polyethylene (PE) and polyvinyl chloride (PVC). The Japanese have begun to use the materials with increased regularity. At an technical exchange seminar held in conjunction with No-Dig '96, Japanese engineers noted 40 percent of pipelines are installed with PE or PVC. This number included concrete lined pipe, but is an indication of a future trend. The key aspect of PE and PVC is the pipe is flexible and does not have the strength to be jacked directly into place. To install this flexible pipe, a load carrying sleeve is required. The sleeve can be a rigid pipe surrounding the flexible pipe. This type of installation results with the both pipes remaining in the ground. The second approach is to use an internal mandrel to carry the jacking forces during construction. Upon installation, the mandrel is removed and the flexible pipe is left in the ground. This type of installation was successfully demonstrated as part of the CPAR test program in Vicksburg, Mississippi (Najafi, Iseley, Pumphrey, and Nishida, 1993).

Layout and Designs

Microtunneling is an effective means of installing utilities underground. The process is developing acceptance. Microtunneling designs are beginning to develop. Design documents will begin to become better and less risky. To this end, engineers are needed to produce the design documents necessary to produce competitive bids, at the lowest overall cost, and with the lowest acceptable risk for the owner as well as the contractor.

To accomplish these objectives, owners, engineers, and contractors must begin to fully understand the microtunneling process. They must be able to establish the proper layouts of plan, profile, and shaft locations. They must anticipate the potential problems and hazards by documenting the geotechnical conditions, and establishing a geotechnical baseline. Everyone also learn to recognize when conditions are not right for microtunneling.

CONCLUSIONS

Microtunneling has grown over the last ten years. The growth has seen microtunneling machines able to excavate through not only soils, but now cobbly ground, mixed conditions, and rock. The support systems have also improved. Alignment control systems have developed from pipe lasers and cameras to high intensity lasers, photocells, fuzzy logic, and Window-driven consoles. The pipe materials have improved, becoming stronger, rounder, and more environmental inert. New pipe materials are always available every few years. The lubricates and lubrication systems are improving with new polymers and slurries. All of these improvements are helping to increase the successful installation of microtunneled pipelines. The future holds a lot of promise. Jacking on curves, with flexible pipe, for longer distances with automated tunneling systems is just on the horizon.

REFERENCES

Atalah, A. and P. Hadala, 1996. "Microtunneling Database for the USA and Canada from 1984 to 1995, " Proc. of Pipeline Crossings 1996, ASCE, June, pp 332-339.

Bennett, D. and K. Staheli, 1995. "Case Study: Retrievability and Reaming Capabilities Tested in Third WES Microtunneling Test," No-Dig Engineering, Trenchless Technology, Vol. 2, No. 1, March , pp 8-13.

Boyce, G.M. and W.W.S. Gray, 1996. "A User's Guide of Pipe," Proc. of International No-Dig '96, NASTT, New Orleans, April, Paper 6A-2, pp. 520-532.

Bristow N., 1994. "Microtunneling Using an Upsizing Reamer," Proc. of North American No-Dig '94, NASTT, Dallas, April, Paper E1.

Coller, P.J., K. Staheli, D. Bennett, and R. Post, 1996. "A Review of Jacking forces by Both Theoretical and Empirical Methods as Compared with 20 Years of Practical Experience," Proc. of International No-Dig '96, NASTT, New Orleans, April, Paper 2A-1, pp. 126-150.

Friant, J.E., L. Ozdemir, and E. Ronnkvist, 1994. "Mini-Cutter Technology - The Answer to a Truly Mobile Excavator," Proc. of North American Tunneling '94 Conference and Exhibition, Denver, June.

Iseley, D.T., 1990. "Trenchless Excavation Construction: Microtunneling - The United States Experience," Proc. of the International Symposium on Unique Underground Structures, USBR, Vol. 2, Chapter 92, June.

Merritt, B.K., A. Crisci, and G.H. Klein, 1991. "Houston Pipe Jacking - Large and Small," Proc. of Rapid Excavation and Tunneling Conference, SME, Chapter 22, June, pp. 391-407.

Miller, R.J., 1996. "Hazard Recognition in Trenchless Technology," No-Dig Engineering, Trenchless Technology, Vol. 3, No. 6, November/December, pp. 13-15.

Najafi, M., D.T. Iseley, N.D. Pumphrey, and H. Nishida, 1993. "Details of Kidoh/Iseki/Carlon Research Project Conducted at TTC," Proc. of Trenchless Technology: An Advanced Technical Seminar, Trenchless Technology Center, Louisiana Tech University, January, pp. 399-432.

Tubbs, C.L., 1993. "Microtunneling: A Houston Contractor's Perspective," Proc. of Trenchless Technology: An Advanced Technical Seminar, Trenchless Technology Center, Louisiana Tech University, January, pp. 249-257.

15

Water & Gas

Chairman
W. Mariucci
Kiewit Construction Co.
Omaha, NE

Chapter 49

NIAGARA FALLS WATER INTAKE TUNNEL CITY OF NIAGARA FALLS, NEW YORK

C.A. Wiggins
Jenny Engineering Corp.

ABSTRACT

The new Niagara Falls Intake Tunnel, mined in the highly fractured Lockport Dolomite, encountered prolific inflows of water contained in rock seams directly linked to the Niagara River. These inflows were pressurized and, more significantly, contained high concentrations of dissolved H_2S gas. The contractors attempts to control the water/gas inflows included cement and chemical grouting from inside the tunnel. Ultimately, improving the rate of air exchange in the tunnel by increasing the fan capacity, in conjunction with a pattern down-hole pre-grouting program along the tunnel alignment allowed the contractor to control the concentration of H_2S gas and complete the tunnel.

INTRODUCTION

The original water intake tunnel, built in the 1930's and extended in the 1950's, is located just west of the location of the new Water Treatment Plant. This reinforced concrete lined tunnel had to be decommissioned due to migration of contaminants from the nearby petrochemical manufacturing plants into the tunnel opening.

The new intake tunnel and water treatment facilities are located on an undeveloped area on the northern bank of the Niagara River, bordered to the south by the Robert Moses Parkway and the Niagara River, to the north by Buffalo Ave., and to the east by the North Grand Island Bridge carrying I-190 over the river. Much of the land in this area was reclaimed by dumping fill derived from the nearby downstream construction of the massive Intake tunnels in the 1960's supplying water to the New York Power Authority Robert Moses Power Project.

The new water treatment facilities will provide potable water for the City of Niagara Falls. The water will be taken from mid-stream of the Tonawanda Channel of the Niagara River through an Intake Structure constructed as a part of the tunnel project. The reinforced concrete Intake structure is anchored to the bedrock underlying the river in approximately 7.6 m (25 ft) of water. Designed to protrude above the maximum water level, the above water part of the Intake Structure was fitted with Navigational aids to safeguard river traffic. The land shaft, which was used as the work shaft from which the tunnel was mined, is located approximately 243.8 m (800 ft) north of the north bank of the Niagara River. This would become the wet well for the intake tanks housed at the start of the treatment process in the adjacent treatment facilities.

Other work associated with this project included the buildings housing all the water treatment processes, mechanical equipment and administrative offices, and a water tower nearby. These facilities were administered under separate contracts.

Design Engineers for this project were O'Brien and Gere Eng., Inc. and Parsons Brinkerhoff Gore and Storrie, Inc. provided Architectural services for the treatment facilities and Geotechnical Engineering Design services for the water tower and the tunnel. O'Brien Kreitzberg had the contract to provide Construction Management services for the owners, the City of Niagara Falls. Jenny Engineering were sub-contracted to O'Brien Krietzberg to provide supervision and inspection services for the construction on the Intake tunnel. Stern and Wheeler, LLC were also sub-contracted to O'Brien Kreitzberg to assist with the inspection of the mechanical equipment installation for the water treatment facilities. Pigott Contractors Inc. of Canada, who became Walter & SCI, Const. (USA), Inc. during the project, was the Contractor for the water treatment plant facilities. Construction of the Intake tunnel facilities was performed by Herbert F. Darling, Inc. of Williamsville, New York. A local contractor, Sevenson Construction Co., got the contract to construct the water tower.

GEOLOGICAL ISSUES

The tunnel alignment is located within the Lockport Dolomite which makes up the bed rock in the area for a depth of approximately 54.9 m (180 ft). The soil cover above the bedrock varied from a maximum value of approximately 13.4 m (44 ft) at the land shaft to approximately 9.1 to 10.7 m (30 to 35 ft) under the river. The overburden encountered at the land shaft consists of miscellaneous fill ,varved soils and glacial till overlying the Lock Dolomite. At the Intake structure the first two strata were replaced by alluvial deposits overlying the glacial till.

The Lockport Dolomite was reported as containing vertical to near vertical joints spaced 5.8 cm (2 in) to 3.1 m (10 ft) with horizontal bedding ranging from a few centimeters to 2.4 m (8 ft) and well distributed fractures. The joints along the bedding planes form the dominant feature in the rock mass. In addition the Engineers reported the potential for rock squeeze due to mainly historical evidence of high insitu horizontal stresses. This never became an issue for the tunnel construction.

The geotechnical reports cited historical precedence to forewarn the contractor of the potential for encountering methane and hydrogen sulfide gasses. Because the boring program did not reveal the presence of these gasses no measurements were available to quantify the expected quantities of these. Excerpts from construction of the existing tunnels made reference to the presence of hydrogen sulfide gas in the construction excavations borne by water inflows.

Thin bedding planes, which were identified as the primary water bearing openings, interconnected with vertical open joints. Ground water levels in the rock is presumed to be directly linked to the Niagara river by means of these interconnecting openings which continue deep into the rock mass. The Engineers predicted significant inflows during construction, based both on the results on permeability tests, where actual flows were anticipated to be even higher than the test results were indicating, and on historical precedence.

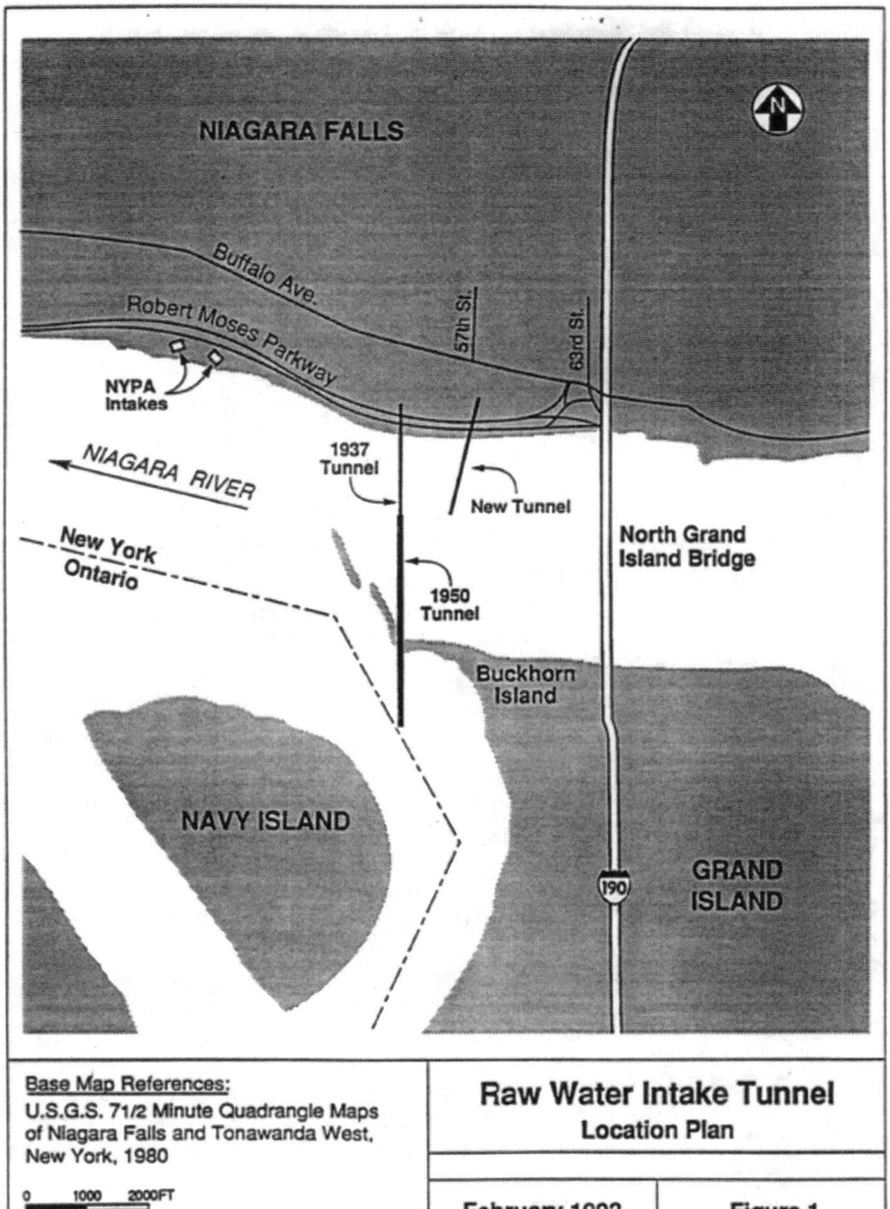

Figure 1. (Reproduced from Geotechnical Design Summary Report)
(With permission of City of Niagara Falls)

This paper is primarily concerned with the methods used by the contractor to control water inflows, which were directly associated with gaseous conditions which developed in the tunnel.

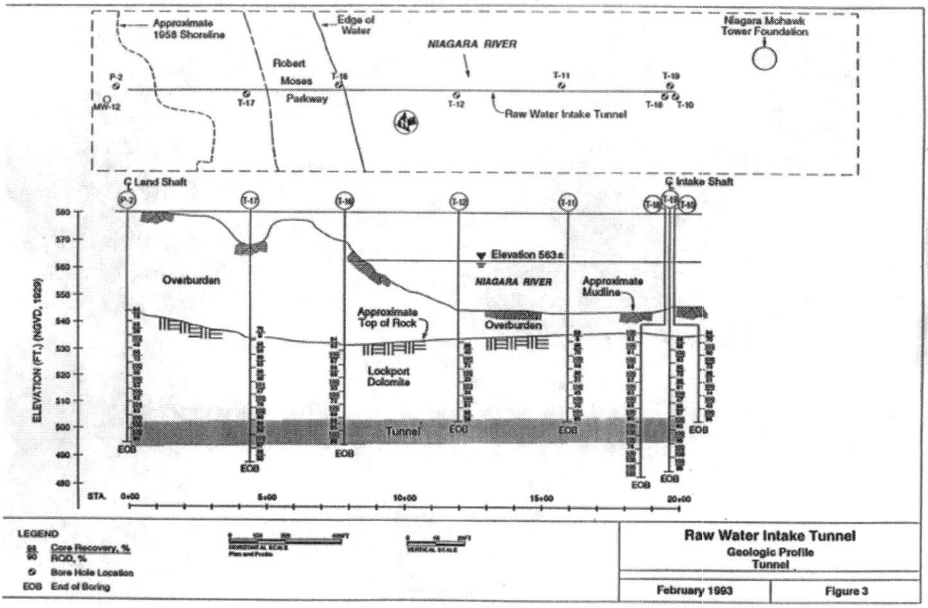

Figure 2. (Reproduced from Geotechnical Design Summary Report)
(With permission of City of Niagara Falls)

MINING AND INITIAL GROUND SUPPORT

All drilling except for the relief holes were accomplished using pneumatic hand drills mounted on jack legs. Relief holes in the land shaft were drilled using an Air Track, and in the tunnel using a "center" drill mounted on a contractor made drilling platform on rails. After pregrouting of the rock mass, the overburden at the land shaft was removed using a large backhoe to excavate to bedrock. A pre-assembled steel-liner shaft support was set on this base and the surrounding excavation backfilled. At the Intake shaft the rock/sheet piling interface was similarly pre-grouted to prevent excessive water inflows and potential rock uplift during excavation and the subsequent dewatering. The overburden here was removed using a clam shell and finished using a large backhoe on a barge.

The rock portions of the land shaft, the tunnel and the intake shaft connecting the intake structure to the tunnel were all excavated using drill and blast techniques.

From the top of the bedrock the invert of the 6.1 m (20 ft) diameter land shaft was advanced using generally 60 to 70 drill holes per set with a 1.5 m (5 ft) pull. As the shaft progressed 10 split set friction rock dowels spaced at 1.5 m (5 ft) vertically were installed. Loose rock was scaled before each shift. At the completion of the blasting program in the land shaft welded wire fabric was fixed to the exposed rock walls to safeguard the shaft invert workspace from rock falls due to spalling.

The mining program in the 3.1 m (10 ft) diameter horseshoe tunnel used on average 35 holes per set with 2.4 m (8 ft) pulls to advance the tunnel heading. Three split set friction dowels, spaced at 1.2 m (4 ft) intervals with welded wire fabric, comprised the initial pattern ground support installed as mining progressed.

The 4.6 m (15 ft) diameter intake shaft was constructed by raise boring from the tunnel opening. Initially a 1.5 m (5 ft) diameter shaft was mined through to the cofferdam using 10 drill holes per set with 1.8 m (6 ft) pulls. Drilling was accomplished from a platform fabricated from pipe sections and clamps. Working from above, within the cofferdam, the shaft was increased to the desired 4.6 m (15 ft) diameter by a series of slash shots.

FINAL LININGS

Final linings in the land shaft, the intake shaft and the tunnel were all Prestressed Concrete Cylinder Pipe (PCCP) sections. Finished diameters were respectively 3.7 m (12 ft), 2.4 m (8 ft) and 1.8 m (6 ft). A short 1.4 m (4.5 ft) diameter section of PCCP joined the wet well to the low lift pump station. Normal weight concrete was used to backfill the pipe sections in both shafts. A lightweight foamed concrete mix was used to backfill the pipe sections in the tunnel.

WATER AND GAS CONTROL

Water was first encountered during the sinking of the land shaft at which time it also became apparent that the Hydrogen Sulfide gas was present. During the shaft excavation, and for approximately the first 76.2 m (250 ft) of tunnel , the rock remained relatively dry with some water seepage from rock seams. A fairly constant background H_2S level of 4 ppm approximately was established.

HEADING?

Increasingly significant inflows were encountered during drilling and installation of the split set friction dowels. With increasing frequency the split set dowel would form a ready conduit for the inflow of water into the tunnel opening. It became apparent by monitoring gas readings that high levels of H_2S gas was concentrated at the points of water inflow where readings would peak locally into the hundreds ppm, particularly after a shut down of the venting system

At this stage water inflows were primarily located at the dowel locations. Efforts to control these relatively modest flows included installing packers either in the bolt holes, at the end of leaking bolts and/or grouting these with polyurethane foam. Occasionally the flow would be panned locally to run down the tunnel wall which would cause the gas to be released and be vented by the air flow. The water would drain down the tunnel invert to a sump at the land shaft from which it was pumped to a settling basin. Pumping rates at this time where of the order of 3.5 l/s (55 gals/min).

Local Panning

101 M (330 FT.) TO 305 M (1000 FT.)

At approximately 100.6 m (330 ft) into the tunnel the first inflow of water from the heading was encountered. Following this the frequency and intensity of the encountered water both in the face and in the crown increased. Additionally the inflow of water, particularly in the crown, was no longer primarily limited to the dowel locations but would increasingly flow through seams in the rock mass. The occurrence of these inflows began to significantly impact negatively on the mining progress.

Water Inflow from Packer Valves in Heading

Face with Numerous Packers Ready for Grouting

With the first inflow in the heading the contractor initiated a cement based cut-off grouting program to control the water inflows. Cut-off grouting was performed in packers installed in drilled dowel holes encountering heavy flows, in bolts installed with lighter flows, or where flows started or increased after bolt installation.

Additionally grout holes purposely located to seal off the estimated water flow path behind the working face would be drilled. This was initially successful but increasingly became less so as grouting a new location would either initiate the failure of an old location, or the flows became so heavy that the grout became less effective. Modifications to the water cement ratio had a noticeable but limited usefulness as did additives such as Slick-pak II and Intra-Plast "N" used to aid in inhibiting dispersion of the grout. The volume of water flows rendered these approaches ineffective eventually.

The 40 KW (50 hp) fan primarily blowing through a 91.4 cm (36 in) diameter vent line became increasingly unable to maintain average readings of H_2S below the OASHA mandated criteria of 10 ppm and 15 ppm for varying exposures. This was also impacting on the mining progress rates. Pumping rates at this time were on the order of 10.1 l/s (160 gals/min).

305 M (1000 FT.) TO 396 M (1300 FT.)

With the inflows of water, and the consequent increases in the H_2S levels in the tunnel, becoming more pervasive the contractor took on the services of a local grouting subcontractor experienced in the use of chemical grouting techniques to control ground water problems in construction projects in the area. Using a poly-urethane foam the team from Vara-Science, Inc. realized a modest level of success in stemming the flow in some of the worst areas. One such area between approximately 274.3 m and 304.8 m (900 ft and 1000 ft), which was dubbed the rain forest, where the water constantly poured down with consequent high concentrations of H_2S gas prior to grouting. This gas would be vented down the tunnel where it would accumulate at the bottom of the work shaft occasionally producing conditions in which the contractor was not comfortable having his work force.

However the success of the grouting program proved to be temporary. As a result of

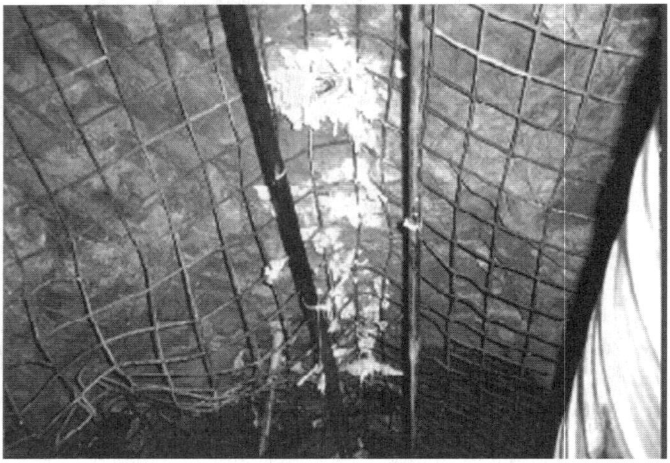

Results of Chemical Grouting

stress relief, creep, and possibly aided by rock movements generated by the blasting, seams that had been sealed by the chemical grouting effort were reopening. The contractor continued the use of normal cement grouting in the face using a colloidal mixer. Additionally mixes were modified to include type III cement, bentonite and flyash to improve the cut-off performance.

Eventually, in spite of these efforts, at approximately 396.2 m (1300 ft) into the tunnel, with the background H_2S readings generally approaching 20 ppm and higher, the contractor shut down mining until a method of controlling the gas satisfactory to all parties could be formulated.

Cement Grouting Using Rail Mounted Colloidal Mixer

Pre-emptive Down-hole Grouting

396 M (1300 FT.) TO COMPLETION

During the approximately two weeks that mining operations were suspended various methodologies and procedures to overcoming the gas problem were discussed. Many ideas were considered including providing breathing apparatus for those working in the tunnel. Ultimately it was decided to focus on increasing the rate of air exchange in the tunnel and inhibiting the dissolution of the H_2S gas from the water into the tunnel air. To this end the contractor increased the rate of air flow by increasing the static head pushing the air into the tunnel through a larger, additional, 91.4 cm (36 in) ventline. Two 60 KW (75 hp) fan units were mounted "piggy-back" venting directly to the working face through the new 91.4 cm (36 in) diameter line. Two additional 40 KW (50 hp) fans vented to the worst offending area for water/gas inflows through the existing, relocated 61 cm (24 in) vent line. Areas where the worst water/gas inflows occurred were panned continuously with sheets of plywood to the invert to prevent out-gassing as the water drained to collecting sumps. From these the water was pumped directly to the larger sump at the land shaft for removal.

The most significant modification introduced by the contractor was the use of the barge, interrupting its operations at the intake shaft, to initiate a program of pattern grouting. Starting at the current heading location a series of 6.4 cm (2.5 in) grout locations at 7.6 m (25 ft) intervals were grouted. Through the overburden a guide tube attached to an H-pile was driven to the top of rock. From here the grout hole was drilled through the guide tube to approximately 1.5 m (5 ft) below the tunnel invert. Using inflatable packers a stage grouting operation at 3.1 to 4.6 m (10 to 15 ft) intervals pumping at 414 kPa (60 psi) or to refusal was performed until all levels of the rock were grouted. Grout used include type III cement, bentonite, flyash and the use of the colloidal mixer.

As the resumed mining progressed, it became clear that the incidence of new inflows had dramatically decreased. Where new inflows were encountered these tended to be stemmed relatively easily by application of cut-off grout within the tunnel. With no increase in the gas-bearing water inflows, the improved air exchange maintained a background H_2S level of below 10 ppm with localized peaks. The contractor was thus able to complete mining without any further major disruptions.

Once the tunnel was mined through to the cofferdam, it was feared that the air flow through the open system would not maintain the H_2S levels at workable levels. To ensure the air flow would be controlled in one direction, two jet nozzles were installed at the work shaft invert to force air flow from the land shaft to the intake shaft. During installation of the PCCP lining the work was occasionally interrupted as a result of H_2S gas emissions from the water in the tunnel in the more confined space. During grouting the gas was controlled by piping the grout-displaced water through drain ports in the PCCP into an 20 cm (8 in) diameter pvc pipe along the pipe invert to the land shaft to be pumped to the settling basin.

CONCLUSIONS

For obvious reasons encountering significant water during a tunnel drive is a situation that the contractor would rather not have to deal with. At best it can make life difficult and at worst it can be very dangerous.

In the case of this contract, while the water inflow was significant it was never unmanageable. It was the high concentrations of H_2S gas carried into the tunnel by the water which caused complications on this project.

On the face of it the grouting program specified, i.e.. probing and grouting ahead of the face would probably have worked in most cases. However the incidence of numerous open vertical joints interconnecting with the even more numerous thinly laminated horizontal bedding joints in areas rendered this approach inefficient. In hindsight, it was simply not possible to angle grout holes sufficiently to intersect enough of these interconnecting water ways from inside the tunnel to make the grouting effort effective. The use of chemical grouting highlighted a potential shortcoming of the normally effective application of chemical grouting when used in conjunction with a drill and blast program.

Ultimately a simple application of down-hole pre-grouting proved to be an extremely efficient approach. This was clearly, after the fact, the preferred method for sealing the Lockport Dolomite with its complicated interconnecting joint patterns and given that the most significant waterbearing feature was the open horizontal bedding planes.

Finally the cement cut-off grout was in general less sensitive to the effects of minute rock movements resulting from stress relief, creep and/or blasting. Once applied it was not usual for a cement grouted area to open up. The chemical grouting was effective in sealing the finer joints in the rock mass and did little to seal off larger openings in the rock. This made the chemically grouted seals highly susceptible to leaking with very slight movements of these fine cracks. In contrast, with access to the larger rock joints provided by vertical grouting holes, the cement grouting program was able to seal off a large enough region around the tunnel that was not sensitive to minute rock movements.

BIBLIOGRAPHY

February, 1993, "Geotechnical Design Summary Report - Raw Water Intake Tunnel, New Drinking Water Treatment Plant" by Parsons Brinckerhoff Gore Storie, Inc. and O'Brien & Gere Engineers, Inc..

Chapter 50

PERFORMANCE OF GROUTING PROGRAM IN CAMBRIDGE ARGILLITE AT INTER-ISLAND TUNNEL

Troy L. Page,[1] David R. Crouthamel,[2] and Francis L. Fong[3]
[1]S.A. Healy Construction Co., McCook, IL; [2]Stone & Webster Engineering Co., Boston, MA; [3]Sverdrup Civil, Inc., Boston, MA

ABSTRACT

This paper focuses on the results and the performance of the grouting program as a means to control groundwater infiltration during the excavation of the 7.7 km (25,160 lf.), 3.5 m (11 ft. - 7 in.) diameter Inter-Island Tunnel beneath Boston Harbor in Massachusetts. Two major operations of grouting were performed - water cutoff grouting behind the tunnel heading after excavation and face grouting ahead of the Tunnel Boring Machine (TBM) operation. An array of products from cementitious to chemical grouts and various grouting procedures and equipment were employed to evaluate their effectiveness in controlling the water inflows. Evaluation of the overall performance of each of the grouting rations, overall costs and comparison of face grouting versus post excavation grouting is made.

INTRODUCTION

Boston Harbor's size, history, natural resources, and recreational value make it unique among major American seaports. However, the continuous dumping of sewerage and industrial waste into the Harbor since the colonial times poses substantial environmental threats to the coastal cities and communities around Boston. In September 1985, the Federal District Court ruled that the discharge of "primary" treated effluent into the Boston Harbor was unlawful, and ordered the Commonwealth to provide full "secondary" treatment of all wastewater discharges. A newly created State agency, the Massachusetts Water Resources Authority (MWRA), was given the mandate to conceive, design and construct the facilities to comply with the requirements of the court order, and to reclaim the Harbor to its previous glory.

The decision was made to construct a new primary and secondary water treatment plant at Deer Island, and the Inter-Island Tunnel which would convey untreated wastewater from the Boston's south shore at Nut Island to Deer Island. Once treated, sewerage will be discharged via a 15-km (10-mile) long outfall tunnel extending from Deer Island beneath Massachusetts Bay. Figure 1 illustrates the Boston Harbor and its vicinity and the alignment of the Outfall and Inter-Island Tunnels.

INTER-ISLAND TUNNEL

The Inter-Island tunnel system consists of shafts at both Deer and Nut Islands and an intermediate drop shaft at Long Island. The tunnel is 7.7 km (25,160 lf.) in length and 3.5 m

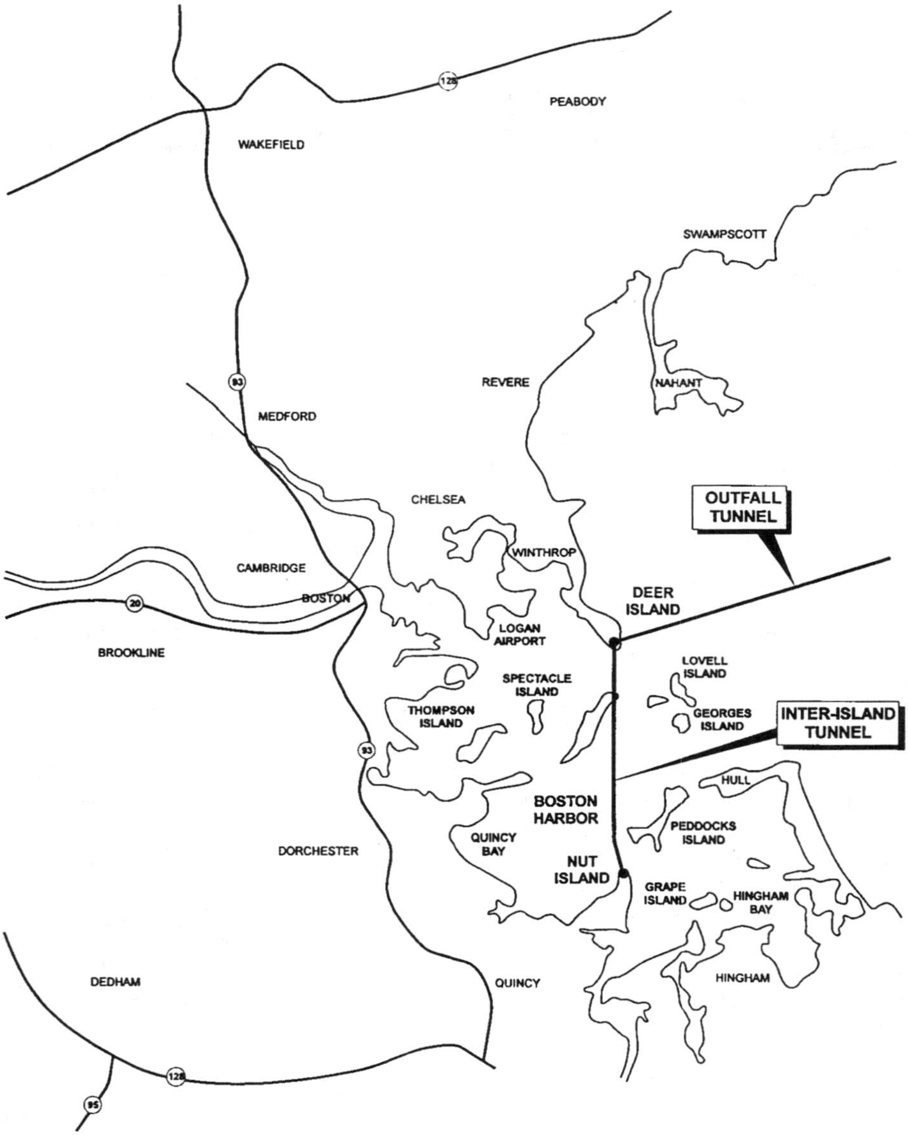

Figure 1. Site Location Map

(11 ft. - 7 in.) in finished diameter. The tunnel elevation is between 82 m (270 feet) and 63 m (200 feet) below sea level at the Deer and Nut Island shaft, respectively, and is embedded entirely in bedrock. The estimated thickness of bedrock above the crown of the tunnel varies from 25 m to 55 m (85 feet to 180 feet). [1]

The tunnel lies in the eastern portion of the topographic and structural depression known as the Boston Basin. The bedrock in the basin contains a series of inter-layered sedimentary rocks intruded by igneous rocks. Structurally the bedrock is folded and faulted longitudinally in a east-west direction. Secondary faults are also present treading approximately north-south, roughly parallel to the alignment of the tunnel.

The Cambridge Formation, located at the top of the basin, is characterized by laminated bedding with alternating layers of light gray, sandy, and dark gray clayey argillite. Severe alternation of the argillite, known as kaolinization, which results in a soft, whitish rock or even clayey, occurs in random areas within the Formation. Igneous intrusions, mainly diabases, are also abundant. The intrusions occurred along fracture planes as dikes that cut across other bedrock units. The Inter-Island Tunnel is constructed entirely within the Cambridge Formation.

The design of the Inter-Island Tunnel project was performed by Sverdrup Civil, Inc. in association with Jacobs Associates, Goldberg-Zoino & Associates and Delon Hampton & Associates. The construction contract was awarded in 1990 to a joint venture of S.A. Healy Construction Co. and Modern Continental Construction Co. at a contract price of $62 million. The Construction Manager was ICF Kaiser Engineers, Inc. in association with Stone & Webster Engineering Co.

TUNNELING CONDITIONS

Excavation of the tunnel was commenced in April 1993 and was completed in November 1995. The Tunnel Boring Machine (TBM) was manufactured by Robbins, Model No. 146-210, which has been used successfully in the past in the Calumet Sewer Tunnel and Dupage Project in Chicago and Sultan River Hydro project in the State of Washington. This TBM is a conventional hard rock main beam machine equipped with finger shield support. The excavated diameter of the tunnel was approximately 4.2 m (13 ft. - 8 in.).

The bedrock conditions along the tunnel alignment varied from the very competent, massive argillite with minimal rock bolt support to the blocky and jointed diabase intrusive rock zones which required full circle steel rib support with lagging. An extensive zone of kaolinized argillite, approximately 150 m (500 lf.) in length along the tunnel alignment, was also encountered. The presence of the soft, soil-like material within the kaolinized zone had posed significant excavation difficulties for the hard rock open type TBM. Hand mining was performed in a limited stretch of the kaolinized zone when it was determined that TBM operation would be ineffective.

GROUNDWATER CONDITIONS

Since intact rock material is relatively impervious, inter-granular seepage is not a significant source of inflow during tunnel excavation. Groundwater infiltration into tunnel excavation is via joints, cracks, shear zones, and faults. These discontinuities may either crowd together within a stretch of tunnel or appear as a singular feature.

For the Inter-Island tunnel, the majority of the groundwater seeped through fissures and small cracks in the rock mass in a more or less uniform manner. Groundwater seepage appeared as drops, drips and continuous leaks along the excavated, unlined tunnel wall.

The seepage conditions within any given 15 m (50 feet) stretch of the tunnel typically varied from completely dry to 750 liter per minute (200 gallon per minute). Singular occurrence such as an open fault with significant water inflow was rare. The maximum inflow from a singular source, which was usually located at the diabase intrusions, was no more than 190 liter per minute (50 gallon per minute). The groundwater infiltration in the kaolinized zone was minimal.

The total quantity of inflow at the completion of the tunnel excavation was approximately 0.33 m^3/sec (5,300 gpm). Generally speaking, this was a sustainable inflow which did not decay appreciatively during the duration of the construction. The actual inflow might have been even greater if not due to the extensive post- and pre-grouting performed during the tunnel excavation. The distribution of groundwater infiltration along the tunnel alignment is illustrated in Figure 2.

Although the tunnel was located some 60 to 80 m below the sea level, only in several locations groundwater inflows exhibited 0.6 to 1.2 kN/m^2 (10 to 20 psi) pressure in the excavated, unlined tunnel. However, limited measurements ahead of the TBM heading indicated the groundwater pressure was significantly higher at 2.4 kN/m^2 (40 psi).

It is speculated that the sources of inflow were primarily seawater in the Boston Harbor and, to a lesser extend, freshwater aquifers located near the shafts. The salt content in the groundwater discharge from the tunnel was measured 15 to 20 mg/l, and was less than the 30 mg/l salinity typically for seawater taken from the Boston Harbor.

CONTRACT GROUTING REQUIREMENTS

Three major types of grouting were specified in the Contract Documents for controlling groundwater inflows: face grouting, water cutoff grouting and consolidation grouting.

Face grouting (pre-excavation grouting) should be performed ahead of the tunnel excavation such that the inflowing water did not become an impediment to the advance of the TBM. It was required when the inflow at the face exceeded 190 lpm (50 gpm), and was considered complete when the inflow was reduced to 60 lpm (15 gpm). Water cutoff grouting (post-excavation grouting) would be performed, if necessary, to reduce the overall infiltration into the excavated tunnel before the placement of the final concrete lining. It was required if the total tunnel inflow exceeded 0.25 m^3/sec (3,200 gpm). Consolidation grouting program was designed to consolidate the rock mass behind the final lining for the purpose of reducing final groundwater leakage to acceptable limits.

All three types of grouting were to be performed under pressure to assist the movement of grout in the rock mass. The injection pressure (gauge pressure) typically varied from 12 to 120 kN/m^2 (20 to 200 psi) but did not exceed 240 kN/m^2 (400 psi). Contract Documents specified the use of high early strength Portland Type III cement with a water/cement ratio between 4:1 and 0.5:1 by volume [2]. Other grout materials included a micro-fine cementitious grout, polyurethane chemical grouts and various admixtures.

This paper focuses on the results and performance of the face grouting and water cutoff grouting programs and the overall cost effectiveness in controlling the water inflows.

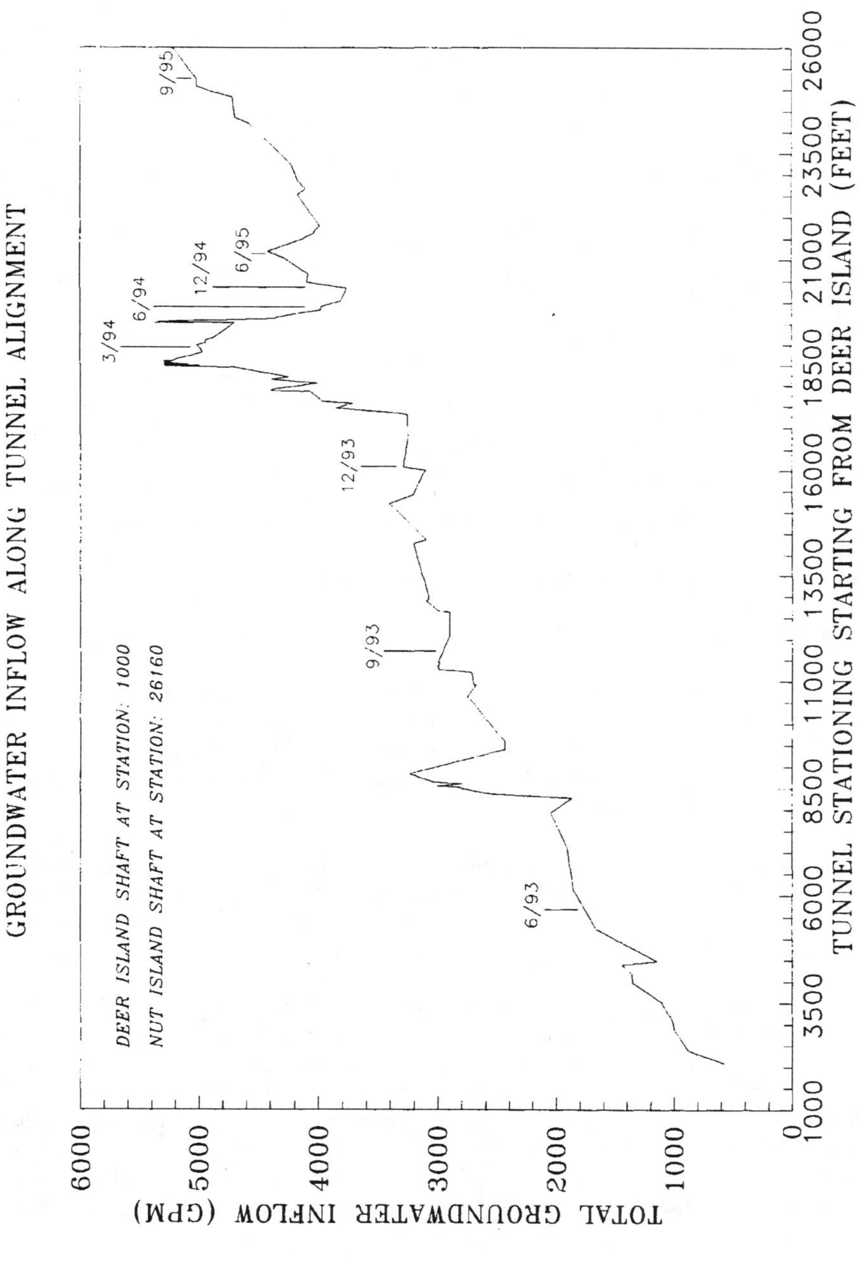

Figure 2. Distribution of Groundwater Infiltration Along Tunnel Alignment

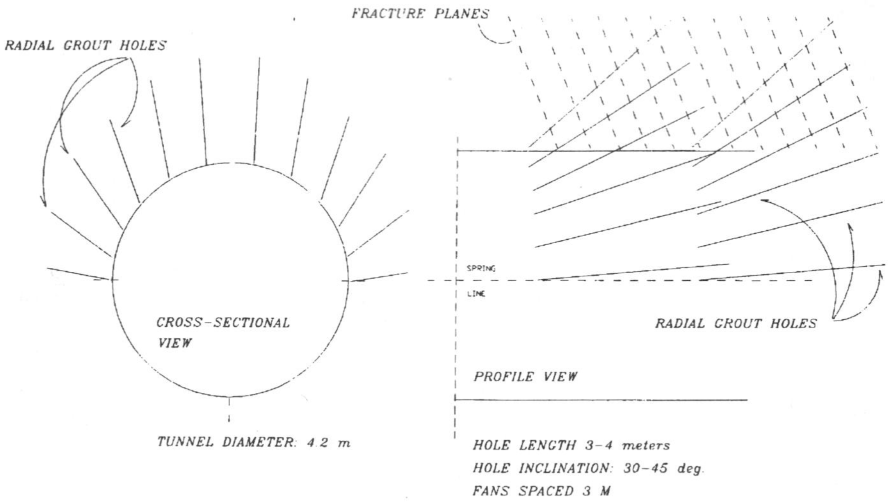

Figure 3. Typical Radial Fan Grout Hole Pattern for Post Excavation Grouting

POST EXCAVATION GROUTING PROGRAM

Means and Method

Post excavation grouting in the unlined rock tunnel was performed well behind the TBM heading typically to reduce overall inflow either in wet zones, stretches of the tunnel of high permeability and inflows or isolated water bearing zones such as a single rock joint or seam. Leakage in the Inter-Island tunnel was typically the result of open rock joints or fractured zones which tended to have higher inflows.

Depending upon the grout material used, the quantity of water inflows and structural orientation of the exposed joint pattern, grout holes were usually advanced along the water bearing seams or joints with a minimum thickness of 30 cm (1 ft). Typical drill patterns are described in Figure 3.

The primary drilling equipment to advance individual grout holes included a jack leg. A single boom Atlas Copco drill jumbo (Figure 4) was also utilized for drilling angular grout holes in the radial fan pattern. However, it's performance was crippled by it's relatively large size in comparison to the 4.2 m tunnel diameter, (limited angle, interference with rail muck haulage), lack of operating experience within the available labor pool and reduced reliability in salt water conditions. Rail muck haulage was also impacted since all rail traffic had to pass beneath the drill jumbo and grouting equipment. Drilling of radial fans with a jack leg was avoided due to the lack of any inclination possible by drilling from the invert.

Grouting was performed off an elevated work platform which housed the grout pump, colloidal mixer, and a minimal storage of materials (Figure 5). The work deck straddled the muck haulage rails, allowing traffic to pass beneath with only minor hindrance to the rail traffic. The work deck was transportable with air operated winches and a folding carriage which traveled on the rails. Bagged materials were delivered to the jumbo via rail. Crew size ranged from 3 to 4 and was split between pumping grout and drilling new holes.

The grout pumps used were either Moyno worm type pumps or double piston pumps. The Moyno pumps typically worked well in controlling low pressure injections less than 1 MPa (140 psi) with high injection flow rates whereas the piston pumps worked best in low injection flow rates at high injection pressures greater than 1 MPa (140 psi).

Grout Materials and Mix Designs

Cementitious grouts consisted primarily of Type III Portland cement. However, Type I/II Portland cements and micro-fine cements were also used when appropriate. The water/cement ratio for the injection of the cementitious grouts was based upon the thickness and extent of the water bearing structure, potential of cross communication with the injection hole and adjacent grout holes, magnitude of water inflow, depth of the injection hole and the existing water head. The water/cement ratio typically used for both Type I/II/III or micro-fine cement grouts was 0.5. Admixtures consisted primarily of a thixotropic stabilizer which limited washout and enhanced grout communication. The thixotropic stabilizer was particularly effective in reducing grout leakage in areas of high inflow and return. Other materials used were granular bentonites to increase the plasticity and vermiculites to reduce leakage and plug open fractures. The thixotropic stabilizer behaved in similar fashion and was likely to result in filter blockage of the injection hole.

Polyurethane based chemical grouts were used, on an experimental basis, in areas where high volume and velocity water flows were encountered. Both hydrophobic and hydrophillic polyurethanes were used. The material costs were equivalent to the micro-fine based cement grouts. Regular use of polyurethanes grouts required specialized equipment for injection and additional safety precautions for safe handling. Furthermore, the storage and application of the material were more problematic due to it's sensitivity to the lower ambient temperatures in the tunnel, which increased the reaction time and reduced the material's effectiveness.

PRE-EXCAVATION GROUTING PROGRAM

Means and Method

Pre-excavation grouting was required to reduce water inflows which impacted TBM operations. It was performed ahead of the TBM excavation in areas of anticipated higher permeability and poor ground conditions. Probe drilling was used as the means to evaluate the actual conditions ahead and was advanced from 23 m (75 ft) to 37 m (120 ft) ahead of the TBM cutterhead. In areas where probe drilling was continuous , the holes were overlapped 4.5 m (15 ft).

Grouting ahead of the TBM was performed using the probe holes as grout injection holes. Generally, pre-excavation grouting was required for inflows at the face or through the probe holes which exceeded 190 lpm (50 gpm). The typical layout and orientation for the probe/grout hole drilling are shown in Figure 6.

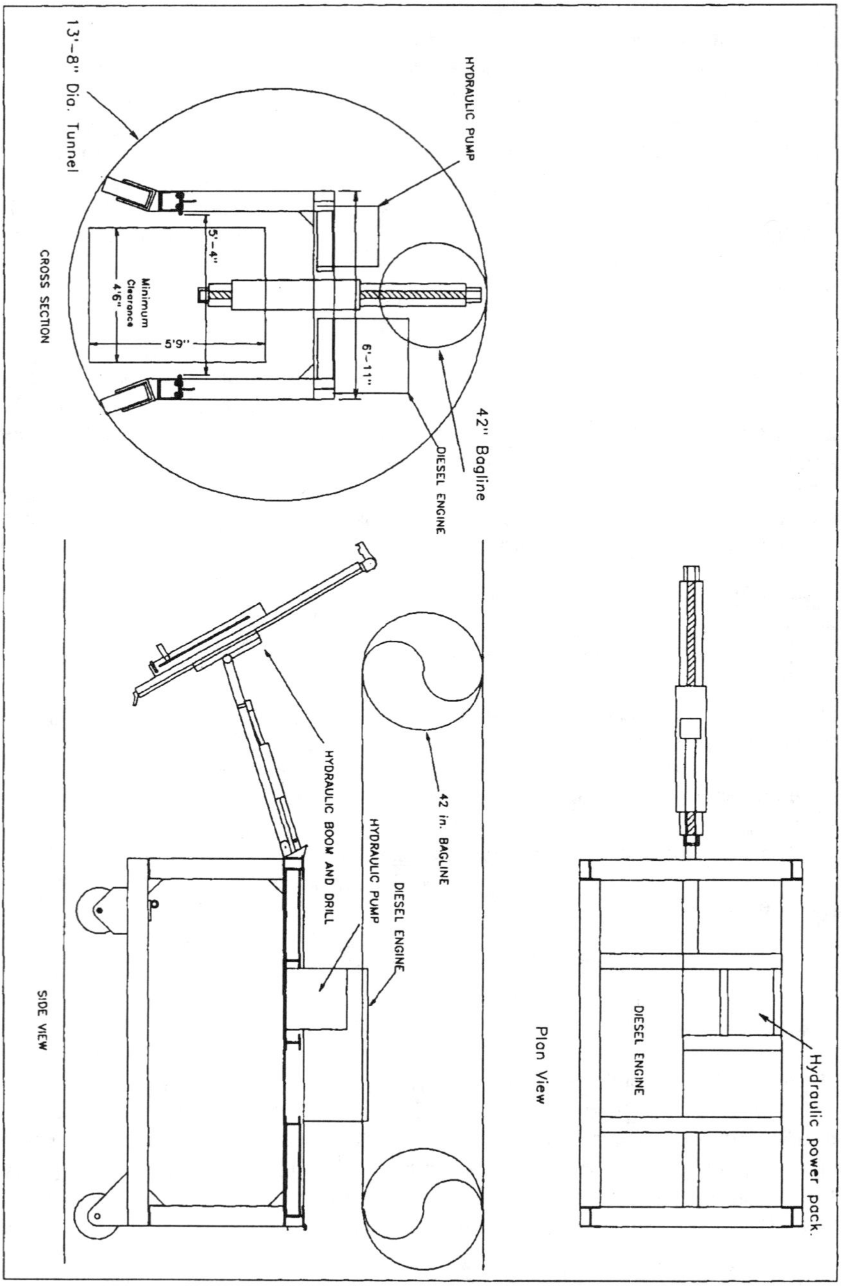

Figure 4. Atlas Copco Drill Jumbo for Post Excavation Grouting

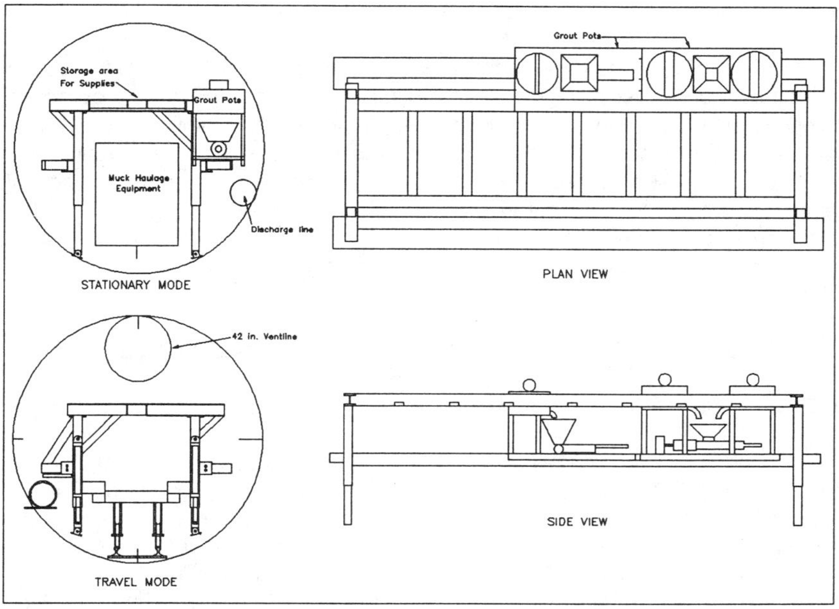

Figure 5. Grout Jumbo for Post Excavation Grouting

Probe or grout hole drilling ahead of the TBM was performed by a Tamrock hard rock drill located at the tunnel springline. Drilling of a secondary probe hole was made possible with the use of a compact Boyle's Brother's Bazooka drill mounted on the gripper carriage. The drills were located at a distance less than 1 m (3.25 ft) behind the cutterhead rear support at a typical inclination of 15° off the tunnel trend. The drill holes were 4.5 cm (1.75 inch) in diameter.

Probe/grout hole locations were constrained by the interference of the TBM cutterhead and the drilling equipment. The configuration of the project TBM was based on a dated Robbins design which prevented drilling access directly through the cutterhead.. Therefore, due to the limited space constraints, the availability of drilling equipment capable of advancing a deep probe hole beyond the cutterhead was limited. Significant modifications had not been made to the TBM since it was believed that only a limited amount of pre-excavation grouting would be necessary.

A double piston grout pump was used for grout delivery ahead of the TBM. This pump was selected due to it's higher pressure capability and limited pressure differential during pumping compared to a single piston pump. Figure 7 describes the configuration of the equipment used for pre-excavation grouting in the heading. The equipment includes a colloidal mixer transferring grout materials into a standby mixer for storage and then to the double piston grout pump. All of the equipment were outfitted on a flat car for quick delivery and removal from the heading. In general when cutoff pressures reached 180 to 240 kN/m^2 (300 to 400 psi) grouting was terminated. Typical embedment of the packer in the probe hole was 1.5 m (5 ft).

Grout Materials and Mix Designs

Because of its reduced fine content and the greater penetrability, micro-fine cementitious grout was best suited for high pressure grouting ahead of the TBM. The purposes of the grouting were to reduce the groundwater inflow and also to consolidate and enhance the structural integrity of the rock mass. Grout mix was initially set at 3:1 water/cement ratio (by weight) and decreased as necessary to 0.5:1 until the cutoff pressure was achieved. A dispersant was typically pre-mixed by the manufacturer in the bagged cement. Chemical grouts such as polyurethanes were not used because significant consolidation of the rock mass could not be achieved and there was limited control over leakage and travel of the grout material.

COMPARISON OF GROUT MATERIALS AND COSTS

Post Excavation Grouting

Grouting behind the TBM was performed for a total duration of approximately 8 months. During that time 23,107 man-hours were expended to grout approximately 1,650 meters of tunnel (5,415 linear feet). A total of 5,802 meters (19,037 ft) of holes were drilled and an estimated 3,645 liters per minute (963 gallons per minute, gpm) of water inflow was stopped. In summary, the post excavation grouting program achieved an overall performance of 6.3 man-hours per liter (24 man-hours per gallon) stopped. The man-hours included herein were direct labor only. Indirect labor hours such as supporting crew in manning the shafts were not included.

The results of the post excavation grouting could easily be evaluated by taking intermittent water flow readings at the invert of the tunnel. From these readings, it could be determined if any reductions in water flow had occurred as a result of the grouting operation. However, the performance and cost comparison of the post excavation grouting program varied significantly according to the grout materials used. Each type of material will be discussed below. All prices included herein are in 1994 dollars.

<u>Type I/II/III Portland Cement</u> Type I/II/III Portland cement was priced at an average of US$ 0.027 per kg (US$0.06 per lb.), and was delivered in pallets of 42.6 kg (94 lb.) bags. The total quantity of water inflow was estimated to have been reduced by approximately 2,396 lpm (633 gpm) as a result of the post-excavation grouting by Type I/II/III Portland cement. The total quantity of cement used was 147.8 metric tons (163 short tons), or 0.06 metric tons per liter (0.26 short tons per gallon) of water cutoff. A total of 18,700 man-hours or 7.8 man-hours per liter (29.5 man-hours per gallon) were expended to achieve this water stoppage.

<u>Micro-fine Cement</u> The micro-fine cement was priced at an average of US$ 0.318 per kg (US$0.70 per lb.), and was delivered in pallets of 13.6 kilogram (30 lb.) bags. A total reduction in groundwater inflow of approximately 530 liters per minute (140 gpm) could be attributed to the performance of the micro-fine cement grouting. The quantity of micro-fine cement used was 26.5 metric tons (29.17 short tons) or 0.05 metric tons per liter (0.21 short tons per gallon) of water cutoff. A total of 2,300 man-hours or 4.33 man-hours per liter (16.4 man-hours per gallon) were expended to achieve this water stoppage.

<u>Polyurethane Based Chemical Grouts</u> Some experimentation with chemical polyure-

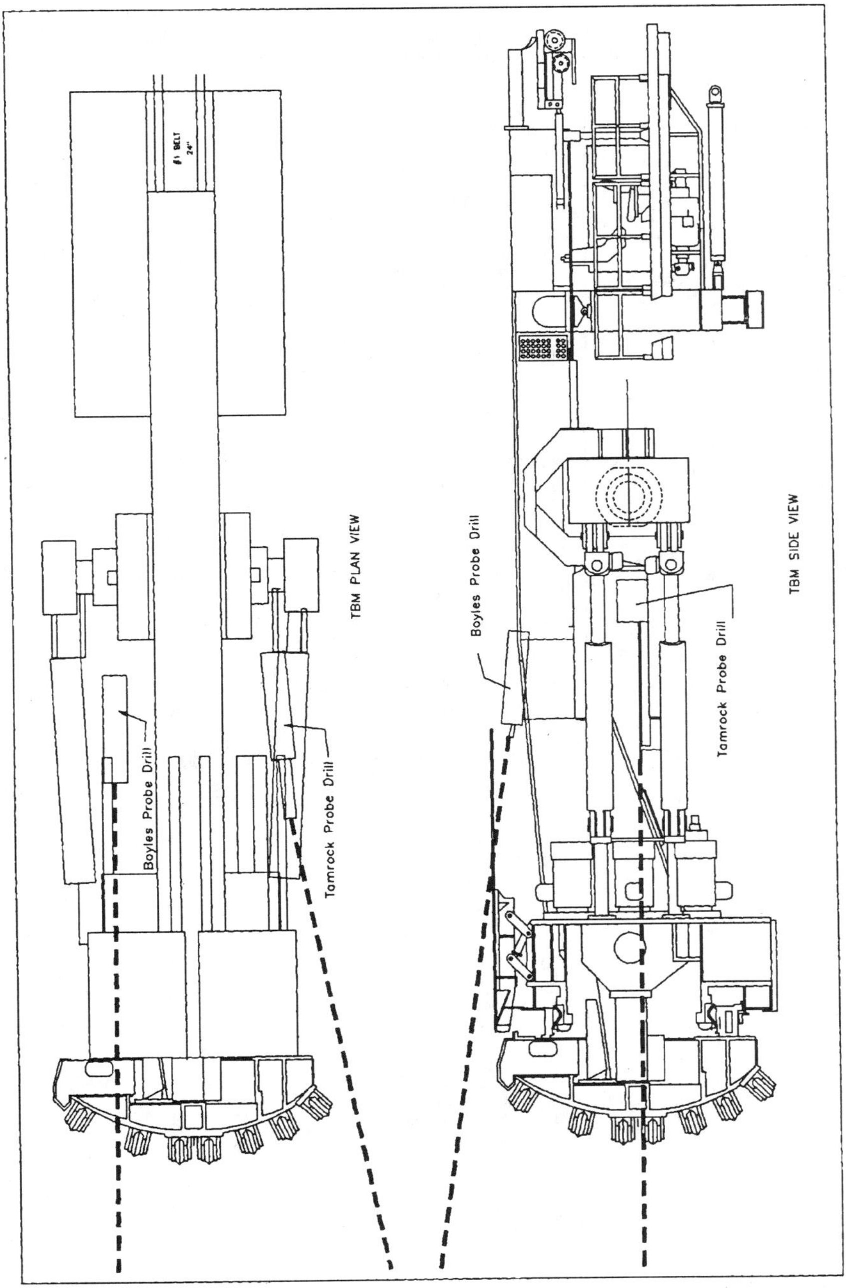

Figure 6. Layout for Grout/Probe Hold Drilling Ahead of the TBM, for Pre-Excavation Grouting

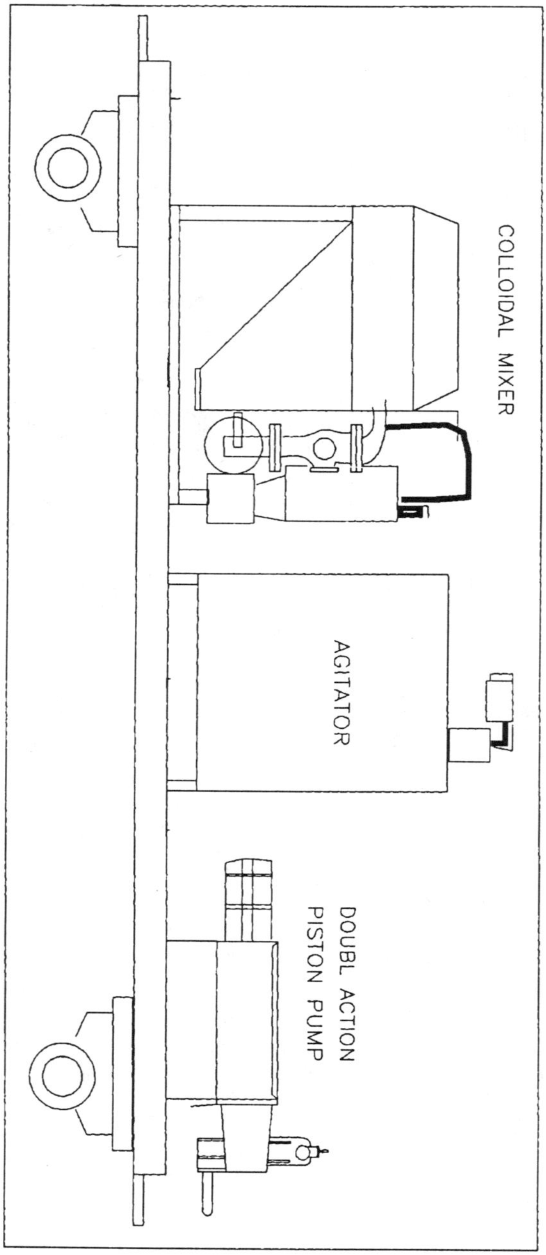

Figure 7. Equipment Layout for Pre-excavation Grouting in the TBM Heading

thane grouts were performed. These grouts solidified rapidly in contact with water and were used primarily to stop concentrated inflows of higher velocity. Both hydrophobic and hydrophillic polyurethanes were used. The cost of the polyurethanes ranged from US$ 9.60 per kg (US$ 21.16 per lb) to US$ 10.25 per kg (US$ 22.59 per lb) and was delivered in 20.4 kg pails (45 lbs) or 224.5 kg drums (495 lbs). Water cutoff of approximately 719 liters per minute (190 gpm) could be attributed to the application of polyurethanes. The total quantity of polyurethane used was 5,560 kilograms (12,254 pounds) or 7.73 kilograms per liter (64.5 pounds per gallon) of water cutoff. A total of 2,107 man-hours or 2.9 man-hours per liter (11.1 man-hours per gallon) were expended to achieve this water stoppage.

Grout Additives Various grout additives for Type II and III cement grouts were also used during the 8-month excavation grouting program and a thixotropic stabilizer was found to be the most effective. The thixotropic stabilizer was used in an attempt to minimize leakage and washout of the grout and to enhance communication of the grout through the rock mass. The cost of this material was US$ 1.6 per kg (US$ 3.52 per lb) and was delivered in 13.6 kg (30 lb) bags. Approximately 8,816 kilograms (4,000 pounds) of the thixotropic stabilizer was used. Water cutoff and man-hours associated with the addition of the thixotropic were not distinguishable, therefore are included in the Type II and Type III Portland cement quantities.

A sodium silicate solution was also tried in an attempt to flash set the Portland cement, sealing off the water bearing seams. The cost of this material was US$ 1.22 per liter (US$ 4.64 per gallon,) and was delivered in 208 liter (55 gallon) drums. Approximately 4,164 liters (1,100 gallons) of the sodium silicate solution was used. Water cutoff and man-hours associated with the addition of the sodium silicate solution were not distinguishable, so they are included in the Type II and Type III Portland cement quantities.

Pre-Excavation Grouting

Pre-excavation grouting was performed intermittently for a duration of approximately 2 months within the same time frame during the post-excavation grouting. Additional laborers were typically required due to the limited work room at the TBM heading. Approximately 3,802 man-hours were expended to cutoff an estimated 715 liters per minute (189 gpm) of groundwater inflow at an overall performance rating of 5.4 man-hours per liter (20.5 man-hours per gallon). Since pre-excavation grouting was performed at or prior to the passage of the TBM, the actual quantity of groundwater inflow that would have otherwise seeped into the tunnel was not known. Therefore, the total quantity of water cutoff for pre-excavation grouting was based on estimates from probe hole inflow data. The material used was primarily micro-fine cement, but Type II and Type III cements were also included.

Table 1. Comparison of Materials for Post-Excavation Grouting

Material Used	Unit Cost (US$/kg)	Total Water Cutoff (liters/min)	Material Used vs. Water Cutoff (Metric ton/liter)	Man-hour vs. Water Cutoff (man-hours/liter)
Type I/II/III Cement	0.027	2,396	0.06	7.8
Micro-fine Cement	0.318	530	0.05	4.3
Polyurethanes	9.93	190	0.01	2.9

Table 2. Comparison of Materials for Pre-excavation Grouting

Material Used	Unit Cost (US$/kg)	Total Water Cutoff (liters/min)	Material Used vs. Water Cutoff (Metric ton/liter)	Man-hour vs. Water Cutoff (man-hours/liter)
Type II/III Cement	0.027	151	0.06	6.0
Micro-fine Cement	0.318	564	0.03	5.1

Type II/III Portland Cement The cost of the Type II and Type III cement is described above. An estimated water cutoff of approximately 151 liters per minute (40 gpm) could be attributed to the use of the Type II and Type III Portland cement during the pre-excavation grouting. A total of 9.4 metric tons (10.3 short tons) or 0.06 metric tons per liter (0.26 short tons per gallon) of Type II and Type III Portland cement was used. In addition, a total of 916 man-hours was expended. Therefore, an average of 6.0 man-hours was consumed to cutoff one liter of groundwater inflow (22.9 man-hours per gallon).

Micro-fine Cements The cost of the micro-fine cement is described as above. An estimated water cutoff of approximately 564 liters per minute (149 gpm) could be attributed to the use of the micro-fine cement during the pre-excavation grouting. A total of 17.2 metric tons (18.9 short tons) or 0.03 metric tons per liter (0.13 short tons per gallon) of micro-fine cement was used. A total of 2,886 man-hours, or 5.1 man-hours per liter (19.4 man-hours per gallon) was to achieve this water cutoff.

COMPARISON OF PRE- VS. POST-EXCAVATION GROUTING

Total Groundwater Cutoff

An estimated quantity of 4,360 liter per minute (1,152 gpm) of groundwater inflow was cutoff by the two grouting programs. This was roughly equal to 22 percent of the maximum quantity of inflow encountered during the tunnel excavation. Approximately 16 percent of the groundwater inflow was eliminated by pre-excavation grouting program. These grouting results do not necessarily indicate that water cutoff by post excavation grouting is more effective. The Contractor simply spent more time and focused their efforts performing the post-excavation grouting.

Assuming the average labor rate (wage plus benefit) in the Boston underground construction market is $50 per man-hour, the estimated total cost (labor and material) varied from US$250 to $300 per liter of groundwater seepage cutoff. This cost of the cement based grout(s) was usually less than 1 percent of the total cost in performing either pre- or post excavation grouting. For the more expensive chemical based grouts, material cost was approximately 40 percent of the total amount spent. However, according to Table 1, the chemical or micro-fine cement grouts did appear to be more effective in sealing off groundwater leakage to justify their additional expenses.

Performance Comparison

Based on the analyses above, the performance of the pre-excavation and post-excavation grouting programs varied from 5 to 6 man-hours and 4 to 8 man-hours expended per liter of groundwater cutoff for the Inter-Island Tunnel. This seems to suggest the cost efficiency of the two programs were somewhat comparable.

This observation was perhaps not surprising. Theoretically, grouting ahead of the TBM, or pre-excavation grouting, permits pre-sealing of any water bearing seams in the rock in advance of the excavation. Success of this type of application was evidenced by the presence of grout materials in rock seams after the TBM advanced. However, due to the limited drilling capabilities ahead of the TBM on this project, it was difficult to insure that the entire cross sectional area of the tunnel would be "pre-grouted". The micro-fine cement was used in an attempt to enhance cross communication and to increase the effective area to be grouted. The greater travel distance of the micro-fine cement was noted by leakage observed some 21 meters (70 ft) behind the location of the point of injection in the tunnel. Data presented in Table 2 also confirms that the micro-fine cement grouting was 100 percent more effective in stopping water inflow from the heading. Approximately 0.03 metric ton of micro-fine grout was used to remove one liter of inflow, as compared to the consumption rate of 0.06 metric ton of regular cement(s) per liter of inflow.

Difficulties Encountered - Pre-Excavation Grouting

One of the more difficult decisions had to be made in the field during the construction of the Inter-Island Tunnel was whether pre-excavation grouting was justified. One might argue that pre-excavation grouting is necessary only when excessive water inflow was observed through probe holes in front of the TBM. Exactly how much inflows that might be cutoff by pre-excavation grouting was only an estimate, because it could not be determined with confidence that whether the flow was sustainable or whether it would bleed off during the grouting operation. Therefore, to shut down the TBM operation for an extended period of time for grouting might not be practical since the outcome of the grouting program was less than certain.

For Inter-Island Tunnel, the performance, as measured in terms of total man-hours expended and material used per liter of groundwater cutoff, for both the pre- and post-excavation grouting program were roughly similar. This seems to suggest that there was no measurable cost disadvantage in performing the pre-excavation grouting program. However, TBM production must be halted during pre-excavation grouting which can increase the amount of man-hours per liter if the heading crew is placed on standby during grouting operations.

Difficulties Encountered - Post-Excavation Grouting

When the water bearing features were located above the water line at the invert, the impact of the post excavation grouting program was readily noted. However, during several occasions groundwater seepage was seen to be "chased" along seams around the tunnel perimeter and down into the invert. It was questionable whether the seepage was cutoff or whether it simply emptied itself into the existing discharge at the tunnel invert. Grouting directly into the invert was not practical due to the negative impact on the tunnel traffic.

Several times during post excavation grouting, groundwater was observed moving upstation and downstation along the tunnel alignment, not only along visible seams, but through fissured sections of the argillite. At a highly fissured section of the tunnel, grout holes in a radial fan pattern were drilled and grouted with micro-fine cement. Holes were advanced as close to the invert as possible to allow almost complete circumferential grouting. As a result, groundwater was displaced and finally emerged at the end of a 35 meter

(115 foot) section where more competent argillite was encountered. The resulting increase in water pressure at this section made it far more difficult to seal the final 9 meters (30 feet) of the unlined tunnel.

Due to the fissured nature of the rock mass and the absentee of any major singular water yielding feature in the tunnel, it was often difficult to pin point specific areas in the tunnel for extensive post excavation grout treatment. For the Inter-Island Tunnel, groundwater seepage tended to be more or less distributed uniformly along the tunnel alignment. Therefore, it may be concluded that post-excavation grouting program appears to be less effective since the seepage might have been displaced to other locations in a fissured rock mass.

CONCLUSIONS AND RECOMMENDATIONS

Grouting is by no means a very precise science. The bedrock and groundwater conditions will dictate grouting procedures, equipment and materials to be used at each application.

Based on the results of the grouting program for the Inter-Island tunnel project, it appears that pre-excavation grouting was marginally more cost effective than post-excavation grouting in cutting off groundwater inflow in a fissured rock mass. There was a better chance of stopping leakage in the rock mass in situ rather than after the ground was relaxed and disturbed by the TBM. Pre-excavation grouting provided more confinement and allowed the grout to travel further from the injection hole and penetrate more of the water bearing formation. After the passage of the TBM, post excavation grouting would appear to be less effective since leakage could occur at any location along the exposed areas.

However, when the cost of indirect labor due to TBM standby time was included in evaluating the cost efficiency and performance of the two grouting programs, pre-excavation grouting would be significantly more expensive.

Pre-Excavation Grouting

If pre-excavation grouting were warranted, more than one hole extending in front of the TBM should be used. This should insure a better distribution of grout ahead of the TBM. The use of micro-fine cement would enhance grout travel throughout the entire cross sectional area and maximized grout penetration. However, because of the relatively high material cost of the micro-fine grout, water pressure tests should be performed prior to grouting to determine if a less expensive material say Type II/III cements could be used more effectively. In addition, there may be added cost efficiencies by better organization of the work task and utilization of the grout crew.

Post-Excavation Grouting

In post-excavation grouting, means and methods to be used would be dictated by actual ground conditions. The least expensive and most commonly utilized material used on this project was the Type I/II/III Portland cements. Generally speaking, Portland Type I/II/III cements worked well in the relatively open fracture argillite. However, fine cracks were better suited for use with the micro-fine cements where penetrability was critical. Micro-fine's greater penetrability in tight features required the use of fewer holes to be drilled and connections to be made. An open fracture could lead to excessive waste and leakage if micro-fine grout is used. The polyurethanes were more useful at locations of isolated high flow, relatively open features. Sodium silicate solution could also be used under similar conditions to the polyurethanes, but could not be relied upon for producing a durable grout and required additional holes and skill for drilling connecting holes.

Greater success might have been achieved if the invert had been dewatered while grouting, and the invert grouted off. However, invert grouting operations would not have been compatible with the rail based muck haulage and might be impractical with the high amount of water flow in the invert. Patterned grout holes along the tunnel alignment worked well in highly fissured argillite where cross communication between bedding planes were frequent.

A significant amount of time and money could be spent experimenting with the grouting procedures and/or materials to determine what would work best for the given ground conditions. A post excavation grouting program might keep the overall grouting costs down until familiarization with the materials, equipment and ground conditions was achieved, before pre-excavation grouting was performed at the heading where TBM production would be halted.

ACKNOWLEDGEMENTS

The authors would like to acknowledge Peter Petrofsky of Jacobs Associates and Ken Chin of the Massachusetts Water Resources Authority.

REFERENCES

[1]Sverdrup Civil, Inc., "Geotechnical Design Summary Report - Contract Documents for Inter-Island Tunnel", October 1990.
[2]Sverdrup Civil, Inc., "Technical Specifications - Contract Documents for Inter-Island Tunnel". October 1990.

Chapter 51

GASOLINE EXPLOSION IN EAST LANSING SEWER TUNNEL

Barry R. Doyle,[1] Steven W. Hunt,[2] and Jane A. Kettler[3]
[1]Consulting Geotechnical Engineer, Cheboygan, Michigan; [2]Principal, STS Consultants, Ltd., Milwaukee, WI; [3]Project Chemist, STS Consultants, Ltd., Milwaukee, WI

ABSTRACT

A sewer tunnel being constructed through an urban commercial area was to pass the site of a former gasoline service station. The site was known to have contained leaking underground storage tanks. An environmental assessment of soils on the alignment was performed, and measures were implemented to handle expected contaminants. Despite these precautions, an explosion occurred in the heading when liquid gasoline was encountered. Four workers were injured. This paper presents the findings of an investigation into the cause of the explosion, and examines how concerted efforts by the parties involved failed to prevent it. The level of guidance provided by OSHA Code provisions for tunneling in gassy conditions is discussed.

INTRODUCTION

The subject project is part of the Combined Sewer Overflow Abatement Program undertaken by the City of East Lansing, Michigan. The project included 2.5 km of 3-m finished diameter tunnel, and 0.5 km of 1.5-m finished diameter tunnel, both in soils. The tunnel alignment runs through an urban commercial area, much of it beneath city streets.

The alignment of the 3-m finished diameter tunnel passes property formerly occupied by a gasoline service station. The property had been identified as a leaking underground storage tank (LUST) site. An environmental investigation revealed gasoline contaminants on the alignment. Recommendations were made in the investigative report for handling and disposal of contaminated soil and groundwater. The contractor made preparations in response to these recommendations. Despite what was known, an explosion occurred when gasoline was encountered in the heading. Four workers were injured.

The City of East Lansing Fire Department asked STS Consultants to determine the cause of the explosion. This investigation began three weeks after the explosion. We examined the tunnel, reviewed written statements by workers, reviewed geotechnical and environmental reports that had been prepared prior to the explosion, and reviewed drilling and environmental test data obtained by others shortly after the explosion. Subsequent to our investigation, additional information was obtained from excavation of a recovery shaft 4 m ahead of the TBM. We have not considered project management issues. Nothing stated in this paper should be interpreted to infer responsibility among the parties involved.

Gasoline vapors in a tunnel atmosphere behave like other hazardous gases. This incident provides an opportunity to consider the current U.S. OSHA Code provisions for tunneling in ground containing hazardous gases. This paper will show that the Code provisions, by themselves, are an insufficient guide to safe construction.

PRE-EXPLOSION SUBSURFACE INVESTIGATIONS AND CONSTRUCTION METHODS

GEOTECHNICAL INVESTIGATIONS

Subsurface conditions along the alignment were described in a geotechnical report included in the contract documents. Preparation of the geotechnical report began in 1978-79, when the project was initially designed. When the design was finalized in 1994, a current evaluation of the original report was appended. No new borings were drilled in the vicinity of the LUST site as part of the update. The geotechnical report included subsurface profiles depicting soil layers found in individual borings. Soil conditions were not interpolated between borings. Soils were identified as 'glacial'.

According to the geotechnical report, in the vicinity of the LUST site the tunnel would be excavated in interbedded layers of granular and cohesive soils. The cohesive soils were generally described as sandy clay, silty clay, and clayey silt. The granular soils were generally described as sand to sandy silt, and sandy gravel. Bedrock was weathered limestone and shale, about 15 m below ground surface. Groundwater monitoring wells at the soil-rock contact showed water levels from about springline to crown. Perched water was present in granular soils above. Geotechnical investigation borings located in the vicinity of the LUST site are shown in Figure 1 (g1 to g3).

The contractor drilled a boring in the vicinity of the LUST site as one of a series of groundwater monitoring wells along the alignment. Conditions encountered were consistent with the geotechnical report. Petroleum odors were not noticed. The boring location is shown in Figure 1 (c8).

ENVIRONMENTAL INVESTIGATIONS

Part of the tunnel alignment follows Grand River Avenue, a busy boulevard. The southeast corner of the intersection of Grand River Avenue and Bogue Street had been occupied by a gasoline service station from 1949 to 1993. An environmental investigation on the service station property in 1988 showed that gasoline that had leaked from the underground storage system had escaped the tank cavity. The service station subsequently closed and the business was converted into a convenience store. All underground storage tanks were removed in 1993. No free product was observed during tank removal. The relationship of the service station property to the tunnel alignment is shown in Figure 1.

In 1994, a remedial investigation was performed on the former service station property to determine the extent of contaminated ground. Boring locations for this investigation are shown in Figure 1 (r1 to r7). One boring was drilled 13 m to rock (r1), the remainder were drilled about 8 to 11 m deep. Contaminated soils were found to a maximum depth of 10 m. High concentrations of contaminants in some samples suggested that liquid gasoline was present. High concentrations of contaminants in borings on the property boundaries led investigators to conclude that contaminants had probably migrated offsite, to the north, west, and southwest. They were unable to determine how much gasoline had been released.

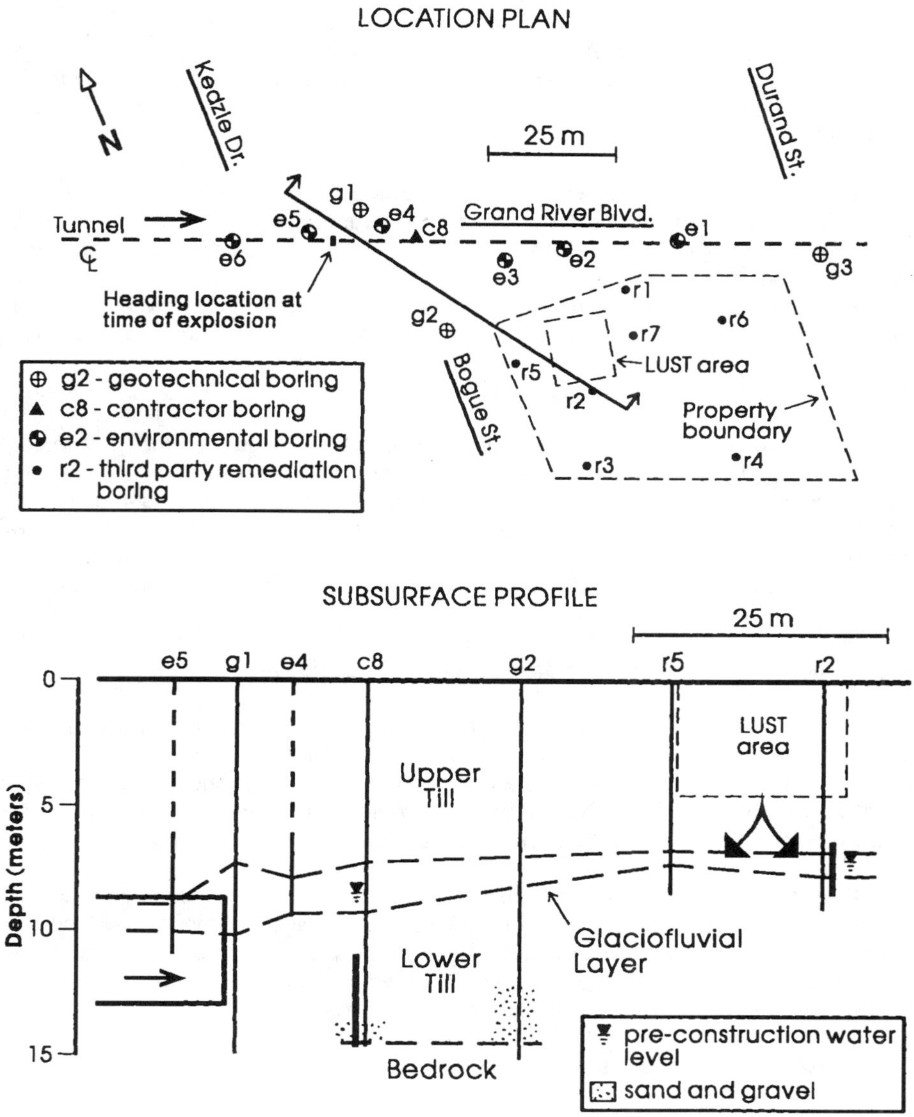

Figure 1. Subsurface conditions in vicinity of explosion. All borings drilled prior to explosion. Fill not shown. Till layers predominantly clayey with granular inclusions. Glaciofluvial layer predominantly fine sand. Lower till contains sand and gravel near top of rock.

In 1995, while tunneling was under way, the designer ordered an environmental investigation adjacent the LUST site, to determine the extent of contaminants on the alignment and to evaluate their effects on health and safety during tunneling. Six borings were drilled, as shown in Figure 1 (e1 to e6). Borings were mistakenly drilled to only about 10 m, 3 m above the excavated invert. Petroleum contaminants were found in four borings (e1 to e4), to full depth in two of these (e2 and e3).

The investigators recommended that groundwater entering the tunnel from the contaminated zone be treated prior to discharge into sewers, that contaminated soils be disposed of in an appropriately classified landfill, that tunnel workers be trained in hazardous waste operations, and that the Site Health and Safety Plan (HASP) be adhered to regarding personal protective equipment and air monitoring. The HASP did not provide specific instructions on tunneling through ground contaminated with high concentrations of gasoline.

TUNNELING METHODS

The tunnel was being excavated with a Lovat TBM. Soil entered the TBM through adjustable outer doors, into a wheel chamber. From there it was conveyed on a belt, through an open muck door in an inner bulkhead, to muck cars. The muck door measured about 0.7 m square, and was positioned in about the center of the bulkhead. Initial ground support was provided by steel ribs and wood lagging, erected inside the tail can of the TBM. Ribs were set at 1.2-m centers. The TBM pulled a trailing gear that extended about 35 m behind the tail can.

Workers employed in the heading included the TBM operator, four men erecting the lining, a conveyor operator, a locomotive operator, and a foreman.

The tunnel was ventilated using an overlapping ventilation system. The main fan was located above a shaft about 600 m from the face. The main fan blew air through a 0.5-m diameter rigid fiberglass duct, which extended past the end of the trailing gear. A smaller fan and a 0.3-m diameter rigid fiberglass duct, referred to here as the forward fan and duct, were mounted on the trailing gear. The fan was at the trailing end of the gear, and the duct extended 30 m forward.

The tunnel atmosphere was monitored with combustible gas monitors, set to warn at 10% and alarm at 20% of the lower explosive limit (LEL) for methane. One monitor was fixed to the TBM with its sensor head located above the muck door. It was set to cut electrical power to TBM circuits at 20% LEL. Several handheld monitors were distributed among the workers to supplement the fixed monitor. All of these monitors were products in common use in underground construction.

Muck and materials were hauled with a diesel locomotive, three muck cars, and two flat cars.

The tunnel operation had not been declared “potentially gassy” or “gassy” under provisions of the OSHA Code.

ENCOUNTERS WITH GASOLINE

SECOND SHIFT

The second shift was excavating their seventh set when strong gasoline fumes rapidly filled the heading. The heading was about 45 m from the LUST area, and 5 m short of the station where contaminants were anticipated. The time was around 8:00 P.M. A handheld gas monitor indicated 5% LEL inside the wheel chamber. The foreman reported conditions to the second shift superintendent. Excavation was stopped. The crew extended the main ventilation duct to the rib expander. They also placed a small fan attached to a 0.15 m diameter flexible duct in the TBM, and extended the duct into the wheel chamber. The main fan, the forward fan, and the small fan in the TBM were all blowing. The fumes subsided.

Tunneling recommenced, and while excavating the eighth set, the fumes grew more intense. Crew members reported burning eyes, throat irritation, and a foul taste. A gas monitor indicated 2% LEL outside the wheel chamber. The crew were not wearing the protective clothing required to handle contaminated soils, so the foreman decided to end the shift early. The set was completed, and the crew left the tunnel about 10:00 P.M. The ventilation system remained running. The second shift superintendent left a note for the first shift, informing them that contaminated ground had been encountered, and that gas monitors had indicated 5% LEL. He ordered that a 10-m length of 0.3-m diameter flexible duct be taken to the heading to extend the forward ventilation duct into the TBM. He also reported the situation to the project manager.

FIRST SHIFT

The first shift entered the tunnel about 5:15 A.M., seven hours after the second shift had departed. Shift supervisors were aware of conditions in the heading. The crew wore the protective clothing required for handling contaminated soils. On arriving at the heading, handheld monitors were used to check for combustible gas. None was detected. One worker reported smelling gasoline. Other workers reported no odor or were unsure.

The crew had brought in the 10-m length of flexible duct. They used it to extend the forward ventilation duct to 1.4 m from the muck door. The small fan and duct that had been placed in the TBM by the second shift were removed. The forward fan was reversed to exhaust air from the heading. The arrangement is shown in Figure 2. The TBM operator reported checking the forward fan and finding it operating properly.

Excavation began. Gasoline fumes were noticed soon after the wheel began to turn. By the second set, workers reported watering and burning eyes, and intense odor. Sometime while the second set was being erected, the locomotive operator complained of the fumes being discharged in his vicinity by the forward fan. In response to his complaint, the forward fan was shut off. This fan was turned back on about the time that excavation for the third set began.

In addition to the fixed gas monitor, there were four handheld monitors in the heading. The TBM operator and foreman each had one, as did the conveyor operator and the locomotive operator. Readings of 3 to 5% LEL were reported among the fixed and the two handheld monitors in the heading.

As excavation for the third set began, the TBM operator was in his station on the right side. The foreman was opposite him on the left. Two workers who erected the upper half of the initial support ring were positioned on catwalks halfway up the left and right sides of the tailcan. Two workers were nearby in the invert. The conveyor operator was at his station at

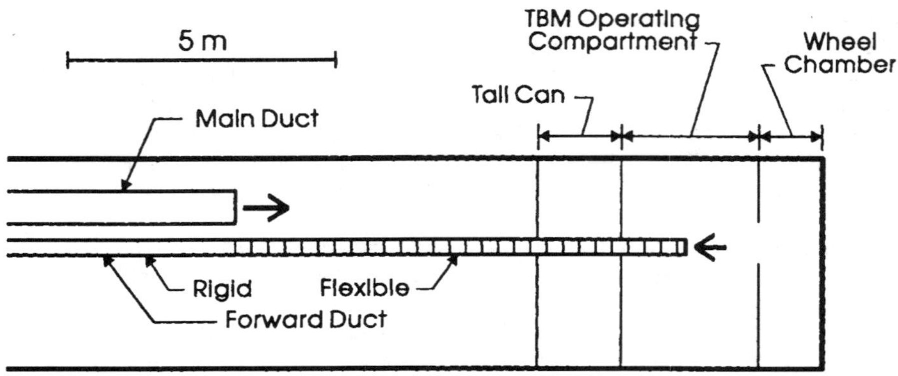

Figure 2. Elevation view of ventilation arrangement in forward area of heading at time of explosion. Bends in flexible duct are not shown.

the tail pulley, near the end of the trailing gear. The locomotive operator was in his cab nearby. The gasoline fumes became more intense. The TBM operator reported that about 1 minute into the shove the cutter head struck a boulder. At that moment, a bright flash of flame came through the muck door and shot across the top of the tunnel.

The force of the explosion knocked the TBM operator and foreman down in their stations, and blew both workers off the catwalks into the invert. These four were burned on the face and neck, two were burned on their arms, two suffered eye injuries, and one a closed head injury. The two workers stationed in the invert were blown 3 to 6 m down the tunnel, but were not injured. The conveyor operator and locomotive operator were struck by the force of the explosion but also were not injured. A small fire started in the heading, apparently fueled by foam plastic packing rods and leaked hydraulic oil.

After the explosion the TBM continued running. A worker cut the power at the transformer on the trailing gear. The explosion was reported to the shaft. Three workers remained briefly to extinguish the fire, and then followed the rest of the crew out. Evacuation took about 15 minutes.

INVESTIGATIVE FINDINGS

From our investigation we concluded that the immediate cause of the explosion was combustion of gasoline vapors inside the wheel chamber of the TBM, initiated by a friction spark from the cutterhead striking a boulder. The vapors were apparently generated from liquid gasoline that entered the wheel chamber through granular soils. Beyond this immediate cause, we attempted to identify technical issues that were not adequately addressed during design and construction, that were important to control of the tunnel atmosphere. Some of these issues were contributing factors to the explosion.

GEOTECHNICAL ISSUES

Soil and groundwater conditions encountered in the tunnel up to the time of the explosion had been consistent with what had been described in the geotechnical report. The

Table 1. Effects of exposure to Gasoline Vapors.

Concentration (ppm)	Exposure Time	Effects
0.25	-	Approximate threshold for odor.
500	1 h	Moderate irritation to the eyes.
900	1 h	Irritation to nose and throat.
1000	15 min - 1 h	Drowsiness, dizziness, nausea, and numbness.
5 000 - 16 000	5 min	Irregular heartbeat leading to cardiac arrest.

SOURCE: Adapted from Sittig, 1991; and Weaver, 1992.
NOTE: Volume concentration, in percent, = ppm x 10^{-4}.

tunnel had encountered mixed glacial soils, occasional cobbles and boulders, and little groundwater. At the time of our investigation nearly all of the face was visible from inside the wheel chamber, except for a small portion of the invert. The face consisted of hard sandy silty clay, some gravel. A 1-m diameter granitic boulder with fresh scrape marks was exposed near the crown. There was no visible groundwater seepage from the face, though the TBM operator stated that it had been wet near the crown. There was no evidence of petroleum products inside the TBM or the wheel chamber, except for a slight odor. However, during a tunnel reentry by others, hours after the explosion, a rainbow sheen on water on the tunnel invert was observed. The rate of groundwater discharge from the heading was visually estimated to be about 25 L min^{-1}. Some of this inflow apparently came up through the invert, in back of the tail can.

Our examination of the logs of borings drilled in the vicinity of the explosion revealed details of the subsurface stratigraphy that were not previously noted. Our interpretation of the geologic profile is shown in Figure 1. The soil column consists of two layers of till, separated by a layer of glaciofluvial deposits that represent an interglacial period. The upper till is predominantly clayey, but contains layers of sandy and silty soils. The glaciofluvial layer consists of narrowly graded coarse to fine sand, to silty fine sand. The glaciofluvial layer was not evident from the wheel chamber. The lower till is predominantly clayey, but more granular than the upper till. Sand and gravel was found near the top of rock in several borings.

The glaciofluvial deposits and the granular till near the bedrock surface provide potential migration routes for gasoline. The glaciofluvial layer was present in nearly all the borings shown on the plan in Figure 1. The granular till was less apparent because few borings were drilled to bedrock. A more complete assessment of potential contaminant migration routes would likely have raised the level of concern for liquid gasoline entering the tunnel. In our opinion, the absence of a detailed geologic profile for the vicinity of the LUST site was a contributing factor in the explosion.

ENVIRONMENTAL ISSUES

The Nature of Gasoline

Gasolines are blends of from 50 to 150 predominantly C_5 to C_{10} hydrocarbon compounds (Cole, 1994), with non-hydrocarbon additives to affect the performance of internal combustion engines. Relative amounts of constituents vary among the grades and brands of gasoline. Despite their varying compositions, the bulk chemical properties of gasolines are essentially the same.

Gasoline may pose a toxicological hazard. Exposure to low concentrations of vapors can irritate the eyes and respiratory tract. Inhalation of higher concentrations can cause more serious symptoms and may lead to cardiac arrest. The effects of gasoline vapors on the body are shown in Table 1. The effects of gasoline vapors are so irritating that they usually force one to leave the vicinity before the vapor concentration reaches the level of a life threatening toxicological hazard. Ingestion of gasoline compounds, via contaminated groundwater, is also a health threat. Some of the most toxic constituents of gasoline are benzene, ethylbenzene, toluene, and the xylenes (BETX compounds). They constitute 15 to 25% of regular gasolines (Baugh and Lovegreen, 1989). BETX compounds are among the aromatic group of hydrocarbons, the group responsible for the distinctive odor of gasoline.

The most serious hazard from gasoline in underground space is explosion of vapors. The flash point of gasoline lies between -43 and -30_C (Cole, 1994). At temperatures above the flash point, liquid gasoline can generate a combustible concentration of vapors. The lower flammability limit of gasoline vapors is 1.4% volume concentration, and the upper flammability limit is 7.6%. Compared to methane, combustible concentrations of gasoline vapors ignite at lower temperatures, and release about the same amount of energy on ignition.

When gasoline spills into soils, some of its constituent compounds may partition from the liquid into other phases. The four phases of gasoline are: (1) free liquid within soil pore space, (2) vapors within soil pore space, (3) solute in groundwater, and (4) adsorbate on soil surfaces. The most volatile constituents preferentially enter the vapor phase, and the most soluble constituents enter the solute phase. These constituents are generally light, simply constructed compounds. Heavier constituents are represented in the adsorbate phase. The liquid, vapor, and solute phases are mobile. The adsorbate phase is immobile.

Over time, liquid phase gasoline becomes depleted in light constituents. After partitioning out of the liquid into the vapor and solute phases, the light constituents escape to the atmosphere, are carried away in groundwater, or are consumed by bacteria. The overall process is referred to as weathering. Weathering diminishes the volatility of gasoline, but does not eliminate the toxic or explosion hazards from vapors. Fresh gasoline has an equilibrium vapor concentration of about 34%. A vapor concentration of about 5% for weathered gasoline, reported in one LUST study (Johnson, et al., 1990), is still within the explosive range.

Evaluating the Explosion Hazard

All of the four phases of gasoline may contribute vapors to the tunnel atmosphere. Rapid vapor dispersion from the liquid, solute, and adsorbate phases is aided by the agitation that occurs when soil and groundwater enter the tunnel. The relative importance of each phase in contributing vapors to the tunnel atmosphere can be determined by considering each phase separately. We made this determination for the East Lansing tunnel.

We calculated that less than 1 liter of liquid gasoline was enough to generate an explosive concentration of vapors inside the wheel chamber of the TBM. This estimate was made using the ideal gas law ($pV = nRT$), a molecular weight for weathered gasoline of 111 g mol^{-1} (Johnson et al., 1990), and a density for gasoline of 0.78 g ml^{-1} (Cheremisinoff, 1992). The void space inside the wheel chamber at the time of our investigation was 7 m^3, the chamber being three-quarters empty of soil. This amount of liquid gasoline may have seeped into the wheel chamber, or been carried within the pore space of a small volume of soil.

The solubility of gasoline hydrocarbons in groundwater is about 50 to 100 ppm, at 20°C (Cole, 1994). We calculated that an inflow of 9 m^3 of groundwater containing 50 ppm dissolved gasoline would be required to generate an explosive concentration of vapors inside the wheel chamber. This assumes that all solute comes out of solution upon entering the wheel. Considering the low rate of groundwater inflow to the tunnel, the solute phase does not appear to be significant in this case.

We calculated that approximately 2 m^3 of soil pore space containing gasoline vapors at a partial pressure of 0.05 atm (5% vapor concentration) would have to enter the wheel chamber to generate an explosive concentration. Considering that most soils at tunnel depth were saturated clays, this phase does not appear to be significant in this case.

The adsorbate phase may have contributed vapors to the wheel chamber, but the significance of this phase is not clear. Adsorption would have been limited by low soil organic matter content and water-wet soil particles. Detailed information on adsorption characteristics of soils has been presented by others (Lyman, Reidy, and Levy, 1992).

Contaminants on the Alignment

The risk of encountering liquid gasoline in the tunnel was not made apparent by the pre-explosion environmental investigation, though the approach taken was typical for identifying gasoline contaminants. The lower 6 m of each borehole was sampled continuously. Soil samples were screened in the field for volatile hydrocarbons using a photoionization detector (PID). The PID is capable of detecting low concentrations of volatiles. Samples showing high PID readings were tested in the laboratory for BETX compounds. BETX compounds, in addition to being toxic, are some of the most mobile of gasoline constituents. They are common target compounds in gasoline contaminant analysis. Groundwater samples were collected from two borings (e4 and e6) where water was found in sufficient quantity, and tested for BETX compounds.

Evidence of contamination was found in borings e1 to e4. PID readings of 7000 and 8000 ppm were obtained from soil samples from boring e3, and a reading of 700 ppm was obtained from e2. PID field screening data in this range indicate that liquid gasoline may be present. A soil sample from boring e3 showed a total BETX concentration of 156 ppm, also suggestive of liquid gasoline. Other soil and groundwater samples tested showed total BETX concentrations of 2 ppm or less, not indicative of liquid phase gasoline.

As a reasonable worst case interpretation, the environmental data showed that small quantities of liquid phase gasoline may be present in granular soils above tunnel springline. Small quantities of liquid gasoline constitute a vapor hazard. The investigative report did not address the issue of gasoline vapor hazards developing in the tunnel. In our opinion, this omission was a contributing factor to the explosion.

Excavation of the TBM recovery shaft provided another possible explanation for how gasoline may have entered the wheel chamber when concentrations on the alignment appeared low at most sampling locations. Gasoline odors were noted while excavating through the glaciofluvial deposits, though odors dissipated as the excavation entered the lower till. Serious problems developed as the shaft bottom advanced into granular till near tunnel invert elevation. Groundwater and liquid gasoline began flowing from the granular till. The amount of liquid gasoline pumped from the shaft is unknown, but by the time the shaft was completed an estimated 680 L of gasoline had leaked into the tunnel invert. Gasoline was apparently drawn toward the shaft excavation, through granular soils in the till, as the groundwater table was drawn down in the immediate vicinity. This same process may have occurred previously when the tunnel approached the contaminated area.

We offer the following suggestions for evaluating the hazards of encountering gasoline during tunneling. Close coordination of geotechnical and environmental investigative efforts may be the best approach to define potential contaminant migration routes. The possibility of encountering liquid phase gasoline should be carefully considered. Installation of monitoring wells would allow long term monitoring of pore fluids. The effects that the excavation may have on the groundwater table, and gasoline floating on the groundwater table, should be considered. BETX concentrations alone do not provide an accurate indication of gasoline constituent concentrations, because BETX compounds are the among the most volatile constituents and may be lost during sample handling. Test methods that may provide a more complete assessment of gasoline constituent concentrations include the total petroleum hydrocarbon (TPH) scan or the gasoline-range organics (GRO) scan.

The environmental investigation performed on the East Lansing project did not fulfil the expectations of other project participants. Insight as to how this may have occurred is provided by considering the normal focus of environmental investigations. These investigations often address the potential health threat from contaminated groundwater, the potential for contaminant migration to other sites, and regulated disposal of contaminated soil and groundwater. In their experience, exposure to the contaminant almost always occurs at the ground surface, where vapor hazards may not be a major issue. Environmental professionals frequently represent owners of contaminated sites, and so may not be inclined toward a worst-case interpretation of the data. The tunnel designer cannot expect the environmental professional to be familiar with the dynamics of tunnel excavation, in terms of what occurs in the ground surrounding the excavation, and how this effects the tunnel atmosphere.

CONTRACTOR'S MEANS AND METHODS

Ventilation System

We could not directly evaluate how the ventilation system had been performing at the time of the explosion. At the time of our investigation, the main fan was operating, but additional air was being blown into the tunnel from a new ventilation shaft located a short distance behind the trailing gear. In addition, the forward fan was off.

Before the explosion, when gasoline fumes became apparent in the heading, the tunnel crew improved the performance of the ventilation system. The main duct was advanced, and at the time of the explosion terminated about 6 m from the tail can. Airflow from the main duct could be felt at the back of the TBM. A few minutes prior to the explosion, the contractor's field engineer measured an airflow velocity of 18 m min^{-1} (twice the OSHA minimum) behind the end of the trailing gear, confirming the adequacy of airflow from the main fan and duct.

The forward fan was exhausting air from the heading, correctly operating in reverse to the main fan. The forward duct overlapped the main duct by 7.5 tunnel diameters, exceeding the minimum recommended overlap distance of two tunnel diameters. Air flow direction in the two ventilation ducts, and duct overlap, were appropriate to prevent recirculation in the heading (Kissell, 1996).

We could not establish whether or not the forward fan and duct were moving air at the minimum velocity of 9 m min^{-1} required by OSHA, at the time of the explosion. Calculations indicate that adding the flexible duct raised friction losses enough to cause the fan to operate near the aerodynamic stall zone. Air flow rate from a fan operating near stall cannot be reliably calculated. The TBM operator stated that he found the system to be operating properly, which suggests that it was exhausting air, but of an unknown quantity. Addition of the flexible duct may have caused a significant reduction in airflow, and demonstrates why it is important that all changes to the ventilation system be engineered beforehand.

Evidence indicates that vapors did not reach explosive concentrations outside the wheel chamber. We found almost no evidence of burning on ragged edges of rags and paper scraps found in the TBM operating compartment. Gas monitors did not register combustible concentrations of gas. The TBM continued to run after the explosion. The force of the explosion, while injurious to workmen, caused no apparent damage to equipment. From this and other evidence noted herein we concluded that the quantity of airflow from the forward fan and duct, though possibly below required capacity, was not a contributing factor in the explosion.

A workman's response to the locomotive operator's complaints about fumes, by shutting down the forward fan, indicates that he did not understand the critical importance of the ventilation system in maintaining control of the tunnel atmosphere. There was no evidence that this action was a contributing factor in the explosion.

The ventilation system was failing to keep up with the entry of gasoline into the heading, and vapors reached concentrations in excess of levels allowed by OSHA. Maximum exposure levels are 300 ppm for eight hours, and 500 ppm for 15 minutes. Worker's reports of physical discomfort suggest concentrations of approximately 500 to 900 ppm, as shown in Table 1. Effects of vapors on workers indicate that explosive levels had not been reached outside the wheel chamber. The LEL of gasoline vapors is about 1.4%, or 14,000 ppm, a fatally toxic concentration (Sittig, 1991).

Supervision did not recognize that conditions were not what had been anticipated. The crew continued working, relying on construction methods and health and safety procedures not suited to the conditions encountered. In our opinion, the failure to recognize that conditions had changed was a contributing factor to the explosion.

Atmospheric Monitoring System

The operation of the combustible gas monitors had been duly checked prior to the explosion, in anticipation of encountering contaminated ground. The fixed monitor was checked the day before the explosion, and the handheld monitors five days before.

However, there were no atmospheric monitoring devices in the heading capable of accurately measuring gasoline vapors at low but toxic concentrations. A monitoring device such as a PID would have been appropriate. The sense of smell is a reliable indicator of the presence of gasoline vapors, but is inaccurate as to concentration.

The monitors in use were of the catalytic type. They burn combustible gases in the atmosphere on a heated filament, and the device responds electronically to the heat generated. Gasoline vapors have different combustion characteristics than methane. The handheld monitors, calibrated for methane but in the presence of gasoline vapors, indicated a percent LEL about one-half the true value for gasoline vapors. This response characteristic was known by workers prior to the explosion. The fixed monitor would have reacted in a similar manner. Multiplication factors for that particular unit could not be obtained because the instrument model number was obscured.

The gas monitors were not appropriate for identifying hazardous concentrations of gasoline vapors. This did not appear to have been a contributing factor in the explosion. No readings greater than 5% LEL (methane) were reported. Doubling these readings to account for gasoline vapors does not exceed the 20% LEL alarm and shutdown level.

Electrical System

We examined the electrical system in the heading to identify potential ignition sources, not immediately ruling out the possibility of an ignition source outside the wheel chamber. When the TBM was manufactured, electrical switchgear and motors had been enclosed in explosion-proof boxes. The TBM had not been maintained in an explosion-proof condition. Covers for explosion-proof boxes were either missing or were held on with less than the required number of bolts. Electrical gear not designed for hazardous locations had been added to the TBM, including outlet boxes and a portable light. Numerous pieces of electrical gear not designed for use in hazardous atmospheres were located in back of the TBM. These included the transformer, the forward ventilation fan, and lights.

The electrical system provided numerous potential ignition sources. The condition of the electrical system in the TBM was particularly undesirable. However, considering other evidence, the electrical system did not appear to have been a contributing factor in the explosion.

EFFECTIVENESS OF THE OSHA CODE ON CONTROLLING THE TUNNEL ATMOSPHERE

The U.S. Occupational Safety and Health Administration (OSHA) Code for underground construction was revised in 1989 to better address, among other things, tunneling in ground containing hazardous gas (OSHA, 1990). The revisions represent a step forward in developing a safe standard of practice for controlling the tunnel atmosphere. However, deficiencies in the Code make it an unreliable guide to dealing with gas. The East Lansing tunnel explosion provides an opportunity to point out some of these deficiencies.

OSHA Code provisions for controlling the tunnel atmosphere can be divided into three categories: (1) general provisions that apply to all tunnels, (2) provisions for tunnels classified as "potentially gassy", and (3) provisions for tunnels classified as "gassy." Code provisions for classifying construction operations as "potentially gassy" or "gassy" are summarized as follows:

Potentially gassy operations

(i) air monitoring discloses 10% or more of the LEL for flammable gases in any underground work area for more than a 24 hour period;

(ii) the history of the geographical area or geological formation indicates that 10% or more of the LEL for flammable gases is likely to be encountered.

Gassy operations

(i) air monitoring discloses 10% or more of the LEL for flammable gases in any underground work area for more than 3 consecutive days;

(ii) there has been an ignition of flammable gases emanating from the strata that indicates the presence of such gases.

These provisions for classifying operations are not appropriate for making a risk assessment. Regarding provisions (i) under both classifications; air monitoring data are not a reliable indicator of the amount of gas in the ground. They are more a measure of the relative capacity of the ventilation system with respect to gas encountered. Of two tunnels in the same gassy ground, the better ventilated tunnel will exhibit the lesser amount of gas, though the risk of encountering gas in either tunnel is the same. The 24-hour and 3-day time factors under these provisions belie the fact that dangerous levels of contaminant gas can accumulate in a tunnel heading in a much shorter period, as in the East Lansing tunnel.

Provision (ii) of the "potentially gassy" classification can mislead if the history of an area has not revealed the presence of gas. A gas problem may be shrouded by (1) lack of tunneling experience in the area, (2) prior use of tunneling methods that impede gas entry, such as compressed air or slurry shield, or (3) good ventilation practice or luck on the part of previous contractors.

Provision (ii) of the "gassy" classification shares the same deficiencies as provisions (i). An ignition confirms the presence of combustible gas, and a failed atmospheric control system, but says little else.

The OSHA provisions for classifying tunnel operations resemble those used by the U.S. Mine Safety and Health Administration (MSHA) to control methane in metal and non-metal underground mines (MSHA, 1987). While underground mining codes and literature contain a wealth of information on atmospheric control, the mining approach to hazard assessment, with its emphasis on observations made during excavation, is inappropriate for civil works tunnels. Entry of even a small amount of gas in a confined tunnel heading is a safety concern, and reliable means of controlling the inflow should be in place at the start of excavation.

The specific demands that the OSHA Code makes on contractors operating under "potentially gassy" or "gassy" classifications do not rise to the occasion. The provisions for "potentially gassy" operations require little additional investment over unclassified operations. The ventilation system must be powered by an acceptable electrical system and be constructed of fire resistant materials. The air monitoring requirements are more detailed. But these provisions may not be enough to control the tunnel atmosphere. Additional control measures to consider include ensuring turbulent mixing of air at the face; 24-hour per day ventilation; automatically de-energizing all electrical equipment that may be exposed to a combustible concentration of gas, not just that in the TBM; and eliminating all electrical ignition sources in the TBM, where high concentrations of contaminant gas are most likely. The East Lansing explosion raises a concern regarding the need to ventilate an unsealed wheel chamber. Ventilating the chamber would more quickly transport contaminated air to gas monitors outside, shortening the response time of alarms and automatic electrical shut-down systems.

The OSHA Code provisions for "gassy" operations are similar to those for "potentially gassy" operations, with an additional requirement that all equipment be approved for use in hazardous locations. For a mechanized operation this can mean a substantial investment. The alternative is to apply labor intensive methods with limited equipment, the approach taken to complete the East Lansing tunnel. But eliminating electrical ignition sources will not ensure safety in combustible atmospheres, as the East Lansing explosion demonstrates.

Attention to detail in every aspect of controlling the tunnel atmosphere is important to success. This is not reflected in the OSHA Code. Consider our findings in the East Lansing tunnel investigation. The ventilation system, for the most part, had a high capacity and had been extended far forward. But friction losses in the forward duct weakened the system close to the face, where gas inflow was most likely to occur. The forward fan was not designed for use in hazardous atmospheres, as allowed in unclassified operations. But had gas alarms sounded, the fan would have had to have been shut down, a difficult decision to make while facing a gas inflow. The number of gas monitors in the heading was more than adequate, but none were designed to address the unique toxic hazard posed by the gas encountered. The condition of the electrical system in the TBM conformed with Code provisions for unclassified operations, but provided numerous ignition sources in an area where gas inflow was most likely to occur. Actions taken by workmen, for the most part, demonstrated caution and skill, but some mistakes were made. These operational deficiencies and others like them are not unique to the East Lansing project. A series of deficiencies, seemingly so minor as to be easily overlooked, can accommodate a serious accident.

A risk assessment made prior to construction provides a sound basis for designing a reliable system to control the tunnel atmosphere. A risk assessment should include an explanation of the hazardous properties of the gas, the manner in which gas occurs in the ground, and the manner in which it may enter an excavation. An accurate risk assessment may demand close coordination between geotechnical and environmental investigations. Prior tunneling experience in an area is an important consideration, but should not be relied upon. Conditions encountered during tunneling should, of course, be continually evaluated against conditions anticipated. Beyond promoting safe construction, the risk assessment will help to eliminate delays and extra costs associated with unanticipated gas inflows. When the East Lansing explosion occurred, 140 m of the tunnel remained to be excavated. It was completed by hand excavating a 2-m diameter jacked pipe. The project was delayed 200 days, and the cost of that length of tunnel rose from $400,000 to $4,000,000. Further investigation along the alignment of the smaller, 1.5-m finished diameter tunnel, not yet excavated, revealed more petroleum contaminants from other LUST sites. This tunnel was deleted from the contract, pending redesign.

In constructing the tunnel, the best approach to controlling the tunnel atmosphere may be to apply simple measures that are well-engineered and regularly evaluated. Control measures implemented should reflect the level of risk. Redundant control measures are desirable as a means of addressing unforeseen conditions, mechanical breakdowns, and other unforeseeable events. Redundancy should not be relied upon to make up for deficient control measures. A reliable system will be capable of handling any amount of gas and any manner of entry that can reasonably be anticipated, with some accommodation for unforeseen conditions. Gas entry is not precluded by unclassified status, and differing site conditions can occur with respect to gas, just as they occur with respect to other geologic features. Most important, the contractor should not count on an opportunity to modify his system, once construction is under way.

CONCLUDING REMARKS

The factors that we determined to have contributed to the East Lansing tunnel explosion are: (1) lack of a clear understanding of subsurface conditions, even though boring information was available, (2) failure to identify the explosion hazard, even though an environmental investigation aimed at hazard evaluation was performed, (3) lack of a clear understanding of the hazardous properties of the contaminant, consequently failing to recognize that conditions had changed from what was expected and continuing the work while relying on methods not suited to the changed conditions.

These factors that contributed to the explosion form a chain that bind design and construction. Despite awareness of gasoline contaminants near the tunnel alignment, each party to the project failed in their effort to address the hazard. Each party's effort was focused well within their particular experience. As a result, each party's effort showed a lack of awareness of important issues outside their area of expertise, and was ultimately flawed. A more collective effort almost certainly would have produced a comprehensive perspective of the hazard, and been successful.

A goal that has been proposed to advance safety and health in the workplace is to apply engineering principles to the total human-machine-environment system (Tarrants, 1992). This concept seems an appropriate approach to the problem of controlling the tunnel atmosphere. Developing this approach may require raising our standard of practice, through the efforts of both designers and contractors. An accurate assessment of the subsurface environment and of the risks associated with gas forms the basis for implementing reliable control measures. The equipment necessary to control the tunnel atmosphere is for the most part readily available, but too often is not integrated into a reliable system. Worker training to maintain the integrity of control systems and to ensure appropriate response in an emergency is critically important. The OSHA Code is not an adequate guide to completing this work. Control of the tunnel atmosphere need not become a central issue in most tunnel projects. But because the consequences of failure can be catastrophic, it is an issue that demands sound engineering solutions.

In closing, the authors would like to acknowledge Jay Dee Contractors, Inc., of Livonia, Michigan, for their active support in preparing this paper. Their willingness to share this experience in the interest of promoting safety in the tunneling industry is commendable. We also acknowledge the contributions of Tom Kroeger, Principal Hydrogeologist, STS, and Roger Ilsley of R I Geotechnical, who reviewed the paper.

REFERENCES

Baugh, A.L., and Lovegreen, J.R., 1989, "Differentiation of Crude Oil and Refined Petroleum Products in Soil," Petroleum Contaminated Soils, Vol. 3, Kostecki, P.T., and Calabrese, E.J., eds., Lewis Publishers, Chelsea, Michigan.

Cheremisinoff, P.N., 1992, A Guide to Underground Storage Tanks: Evaluation, Site Assessment, and Remediation, Prentice Hall, Englewood Cliffs, New Jersey.

Cole, G.M., 1994, Assessment and Remediation of Petroleum Contaminated Sites, Lewis Publishers, Boca Raton, Florida.

Johnson, P.C., et al., 1990, "A Practical Approach to the Design, Operation, and Monitoring of In Situ Soil-Venting Systems," Ground Water Monitoring Review, Vol. 10, pp. 159-178.

Kissell, F.N., 1996, How to Control Air Contaminants During Tunnel Construction, Information Circular 9439, U.S. Department of the Interior, Bureau of Mines.

Lyman, W.J., Reidy, P.J., Levy, B., 1992, Mobility and Degradation of Organic Contaminants in Subsurface Environments, C.K. Smoley, Chelsea, Michigan.

MSHA, 1987, 30 CFR, Parts 0-199, Mineral Resources, U.S. Department of Labor, Mine Safety and Health Administration, Government Printing Office, Washington, D.C.

OSHA, 1990, 29 CFR, Part 1926, Safety and Health Regulations for Construction, Section 800, U.S. Department of Labor, Occupational Safety and Health Administration, Government Printing Office, Washington, D.C.

Sittig, M., 1991, Handbook of Toxic and Hazardous Chemicals and Carcinogens, 3rd ed., Noyes Publications, Park Ridge, New Jersey, pp. 850-852.

Tarrants, W.E., 1992, "Emerging Trends in Safety and Health," forward in Accident Prevention Manual for Business & Industry: Engineering and Technology, 10th ed., National Safety Council, Itaska, Illinois.

Weaver, N.K., M.D., 1992, "Gasoline," Hazardous Materials Toxicology: Clinical Principles of Environmental Health, Sullivan, J.B., Jr., M.D., and Krieger, G.R., M.D., eds., Williams & Wilkins, Baltimore.

Chapter 52

PRESSURE BALANCE SHIELD PIPE JACKING TO AVOID CONTAMINATION

Marco D. Boscardin,[1] R. Lee Wooten,[1] and James M. Taylor[2]
[1]GEI Consultants, Inc., [2]Fay Spafford & Thorndike, Inc.

ABSTRACT

An earth pressure balance (EPB) tunnel boring machine (TBM) and pipe jacking were used to construct a 430-meter-long segment of sanitary sewer force main. The force main was constructed by first jacking a 2045-mm-outside-diameter, reinforced concrete cylinder pipe (RCCP) casing through alluvial sands and gravels. Later, a 1220-mm-diameter, ductile iron carrier pipe was installed within the casing pipe. The main concern affecting the construction method selection was the elimination of the need for dewatering, which could cause groundwater flow from deeper, contaminated sands up across a silt aquitard into cleaner, shallower sands. In addition, the method selected reduced the volume of soil to be handled and, consequentially, the potential amount of contaminated soil to be managed. Items discussed in the paper include provisions for handling boulders and controlling ground movements as excavation and jacking occurred approximately 300 mm beneath an active, 4800 kPa, 254-mm-diameter natural gas transmission pipeline.

INTRODUCTION

A section of sewer force main, 430 meters long and 1220 mm in diameter, was constructed through an area where ground water contamination had been detected. Field investigations revealed that the upper sand and gravel layer in the area of the force main construction was not contaminated, but ground water contamination was detected in the lower sands and gravels. A silt layer that appears to separate the upper and lower sand and gravel layers was observed in most borings. A concern existed that conventional trenched construction methods involving dewatering could create gradients and flows from the lower sand layer to the upper sand layer and potentially a spread of the contamination. Several alternative construction methods were considered, and earth pressure balance pipe jacking was selected to meet these challenges.

BACKGROUND

The force main is located under University Road in the northwest corner of the Town of Canton, Massachusetts, and is part of the Massachusetts Water Resources Authority (MWRA) New Neponset Valley Force Main project, which involved the construction of approximately 5,650 meters of 1220-mm-diameter force main connecting a new pump station to the existing New Neponset Valley Sewer. The alignment beneath University Road is within a 9-meter-wide sewer easement that also contains a previously constructed 1400-mm by 1500-

mm concrete sewer comprising the existing New Neponset Valley Sewer, Fig. 1. The force main is located approximately 3 to 4.6 meters east of the existing sewer and about 1 to 1.2 meters deeper.

At the north end of University Road, the force main comes within about 20 meters of two sets of MBTA commuter rail tracks, makes a right angle turn to the east, and traverses under the railroad tracks. After passing beneath the railroad tracks, the alignment turns northeast and passes through the Dedham Street highway embankment. These crossings, about 50 and 65 meters, respectively, were also made using pipe jacking methods.

For approximately 135 meters, the alignment passes within the wellhead protection area for Canton Well No. 2, where regulations required that the carrier pipe be placed within a casing pipe. The force main also crosses 300 mm below an active 254-mm-diameter, 4800 kPa natural gas transmission pipeline that remained in service under full pressure during construction.

SUBSURFACE CONDITIONS

The project area is within the Neponset River flood plain, where the naturally deposited surficial soils are mapped as generally consisting of fine-grained alluvial sand, silty sand, fine gravel and silt. Peat and organic silt were reported in the upper natural strata during the earlier sewer construction in the area. Fill placed during construction of the concrete sewer and University Road comprises the upper soils in the project area.

Sixteen borings were drilled along the University Road alignment during the subsurface investigations. The initial geologic investigations consisted of three borings ranging from 6.4 to 7.9 meters deep. An additional thirteen borings ranging from 5.2 to 18.1 meters deep were drilled later to investigate contamination detected. The borings were about 30 meters apart along most of the alignment under University Road. However, at the south end of the project, there were only two borings that were about 150 meters apart. A subsurface profile developed using the boring data is presented in Fig. 2 and described below.

- Fill materials were encountered from the ground surface to approximately El. 44.2 (meters) in the borings. Along the southern, unpaved 122 meters of University Road, the ground surface is at about El. 46, and along the paved portion of University Road, the ground surface is approximately El. 47.2. Sand and gravel with cobbles and boulders comprise the fill.
- A 0.3- to 1-meter-thick layer of peat is generally encountered below the fill, but at the north end of University Road, 1.2 to 1.5 meters of peat is present. The base of the peat is typically at El. 43.6 or higher.
- An upper sand layer ranging from 1.5 to more than 3 meters thick is found below the peat. The upper sand consists of medium dense to very dense, fine to medium-grained sand, often with a trace of fine gravel.
- Below the upper sands, a layer of medium dense to very dense, non-plastic to low plasticity silt to fine sandy silt is present at most locations. Where present, the silt layer ranges from 1 to more than 2.4 meters thick. The silt layer was absent in borings in a 100-meter-long section of the alignment centered at the boundary of the paved and unpaved portions of University Road.
- Below the silt layer, a lower sand layer is found and consists of medium dense to very dense, very fine to coarse sand with a trace of gravel. The lower sand is generally fine at

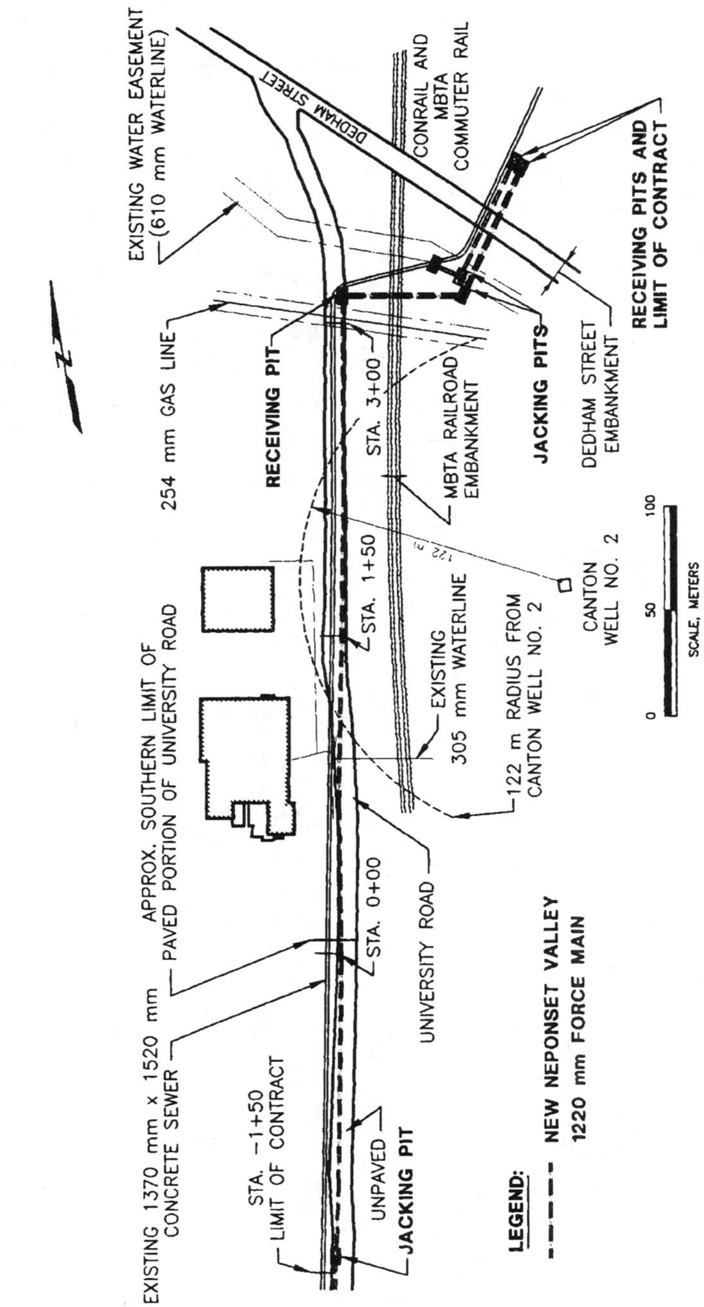

Figure 1. New Neponset Valley Force Main Site Location Plan

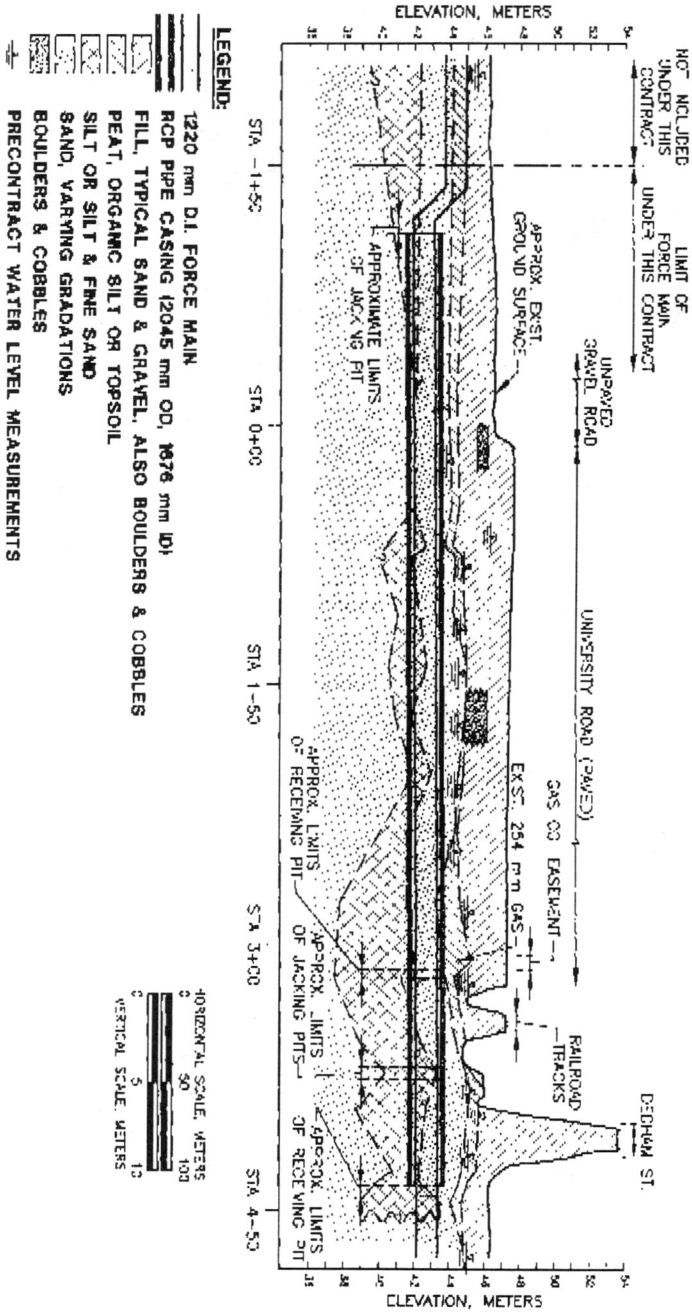

Figure 2. New Neponset Valley Force Main — Profile

the top of the layer and becomes coarser with depth.

- Ground water levels in the borings and monitoring wells were in the range of El. 44.3 to 45.1. It was evident prior to construction that the invert of the force main or casing pipe would be about 1.8 to 3 meters below the ground water level and would require dewatering if constructed using conventional open excavation with trench box support.

ALTERNATIVES FOR FORCE MAIN CONSTRUCTION

During design, a variety of alternatives for construction of the force main were examined:

- Conventional open excavation with trench box support,
- Open excavation using a trench box and subaqueous construction,
- Open excavation with sloped sides and subaqueous construction,
- Open excavation with temporary sheeting,
- Open excavation with permanent sheeting and tremie concrete base mat,
- Conventional open face or auger methods of pipe jacking,
- Slurry or earth pressure balance methods of pipe jacking, and
- Open excavation using ground freezing for support and ground water control.

Initially, the technical merits of each alternative were evaluated relative to the amount of dewatering required, amount of excavation in the event that contaminated soils were encountered, impact on future construction, impact on use of the area, and constructibility. The method of construction selected had to meet the following technical criteria:

- Eliminate or minimize the need for dewatering and the potential for introducing hydraulic connection and ground water flow between the upper and lower aquifers,
- Minimize the potential volumes of contaminated soils and ground water to be managed,
- Minimize the potential disruption of the use of and access to properties along University Road,
- Minimize the permanent impact on the hydrologic regime and future construction along the force main alignment, and
- Be constructible.

Those alternatives considered to be technically satisfactory were then compared on the basis of estimated relative cost. Based on the initial evaluation of the technical merits for the alternatives described, only slurry or earth pressure balance pipe jacking methods and open excavation using ground freezing were retained for further evaluation. The ground freezing approach was then eliminated based on estimates indicating that an approach using ground freezing would cost several times that of an approach using a pressure balance TBM and pipe jacking. In addition, the surface disruption and the duration of construction were judged to be significantly greater with a ground freezing approach in this case.

DESIGN

The preliminary design of the force main was based on conventional open cut methods of construction. Switching to a pressure balance pipe jacking construction method required several changes. To improve face control and reduce ground surface settlements, and to reduce the potential for the pressurized face causing a blowout or objectional heaving of the ground surface, the alignment of the force main was placed deeper so that there was more soil cover. If the crown of the jacked pipe was less than about 2 pipe diameters below the ground surface, surface settlement/heave would tend to increase and face control would be more difficult. The alignment of the force main was lowered so that the crown of the jacked pipe was approximately El. 43.3, about 4 meters below the surface of the paved portion of University Road and about 3 meters below the surface of the unpaved portion.

Lowering the force main alignment placed the jacking operations below the level of the fill and reduced the potential for encountering cobbles and boulders, which were reported above El. 44.2. The crown of the jacked pipe was below the bottom of the fill detected in the borings and generally in sand, silty sand, fine gravel and silt. Although boulders were not expected in the face of the pipe jacking excavation, it was specified that the TBM be able to handle boulders up to 30 percent of the diameter of the TBM. In cases where boulders or other obstructions larger than that size block progress of the TBM, a sheeted or trench box supported pit filled with water could be excavated at the face of the TBM to remove the obstruction because the pipe is shallow and in a relatively open area. After removal of the obstruction, the pit would be filled with a sand and gravel fill and jacking could then continue.

A second change was to design the entire jacked pipe length (430 meters) of the force main as a 1220-mm-diameter ductile iron carrier pipe within a 1676-mm- inside-diameter (2045-mm-OD) RCCP casing pipe to meet Massachusetts Department of Environmental Protection requirements for public water supply well head protection. Jacking two different size pipes was not considered feasible due to equipment mobilization costs and the goal of locating the jacking pit outside the area where contamination had been detected. The receiving pit was located at the north end of the run where the force main turns to the east to pass under the railroad tracks. Contamination had been reported in the area of the receiving pit. Dewatering outside the jacking and receiving pits was not permitted, and provisions for grouting or sealing were required in the event that the pits leaked and ground water levels outside the pits dropped.

A slurry pressure balance TBM was specified for the construction. The slurry method was selected because the bentonite slurry and associated filter cake that would form at the cutting face would support the face, control the tendency for the sands to run, and control groundwater infiltration by serving as a binder, and preventing runs and flows. Control of the slurry pressure was required to avoid the potential for a blowout under the low cover conditions. To reduce the potential for overstress of the jacked pipe, intermediate jacking stations were required. Estimated settlements over the pipeline centerline were in the range of 30 mm, and the associated settlement trough was estimated to be about 6 meters wide.

CONSTRUCTION

The contractor proposed that an EPB TBM be used for the pipe jacking work instead of a slurry pressure balance TBM. The TBM proposed was a 2045-mm-diameter Lovat Model M77 equipped with flood gates at the face and pressure relieving gates. The pressure relieving gates control removal of muck from the muck chamber and maintain a pressure on the face. In addition, a binding medium, either bentonite or polymer based, could be injected into the muck chamber to control the face, ground water infiltration, or running sands if needed.

The contractor's reasons for wanting to use the EPB TBM included less risk of a blow-out through the shallow ground cover with the more passive EPB support, compared to actively pumping slurry under pressure, smaller area requirements for muck handling and processing staging, and easier muck disposal because the muck was not mixed with slurry.

The TBM had a bi-directional, high torque cutter head mounted in an articulated shield with three sections having a total length of about 7 meters. Articulation of the 2-meter-long front or steering section of the shield was controlled by hydraulic rams. Line, grade, and rotation of the TBM were monitored using a laser and twin target arrangement. Four, 83-ton hydraulic rams were available to thrust the TBM forward independent of the pipe string. The pressure relieving gate doors of the TBM would permit passage of 360-mm-diameter boulders. Carbide ripper teeth were set on the spokes, and hardened picks were set along the perimeter of the cutter head, to create an aggressive toothing arrangement so that larger boulders encountered might be broken up or cut through. The gage cutters were set so that a 2108-mm-diameter opening was cut.

The main jacking station at the jacking pit consisted of four, 1800-kN-capacity, 3-meter stroke hydraulic cylinders and was designed to jack up to 6 meters of 1676-mm-ID RCCP at a time using three strokes of the cylinders. Each section of RCCP was 3 meters long and had a rated thrust capacity of 8900 kN. The first intermediate jacking station (IJS) was located about 60 meters behind the TBM. Additional IJSs were provided at about 90 meter intervals thereafter. Each IJS had a stroke of 267 mm and consisted of 10 hydraulic rams, with a total rated capacity of 5600 kN.

Bentonite lubricant was injected through ports located directly behind the TBM tail shield. The volume of lubricant injected was limited to 100 to 110 percent of the theoretical volume of the annulus between the pipe and the wall of the excavation cut by the TBM. To avoid heaving of the ground surface under the shallow cover conditions, volumetric control was used rather than pressure control. Secondary injections were used as needed to reduce jacking loads or to control ground surface settlements. The contractor proposed and used a lubrication/grout mix with a retarded setup time. This eliminated the need for replacing the bentonite lubricant with a grout after jacking was completed.

As muck was discharged through the pressure relieving gates, it was removed from the heading area via a conveyor system and deposited in rail-mounted muck cars about 10 meters behind the face. The contractor's estimate for the jacking time was 35 working days. The jacking took 60 days to complete due to start up and mechanical delays.

GAS MAIN CROSSING

At the location of the force main/gas main crossing, several protective measures were considered. These included increasing the clearance between the TBM and the gas main, installing a split shell carrier pipe placed around the gas main to isolate it from jacking related ground movements, depressurizing the gas main during the crossing, and instrumenting the TBM approach and crossing with borros settlement points to monitor ground movements and provide feedback for control of the TBM. Because depressurization of the gas main was not allowed, and because the clearance could not be increased, the gas transmission company exposed the gas main crossing using a vacuum excavated pit to a depth just below the gas main to permit monitoring of the gas pipeline as the TBM and jacking passed under it. The pit allowed visual confirmation of the clearance between the gas main and the TBM, monitoring of the gas main for leaks and distress, and prevented heaving pressures due to the TBM operation from affecting the gas main. The use of the earth pressure balance TBM, instead of a slurry pressure balance TBM, allowed the TBM to make the crossing with little to no cover.

GROUND SURFACE SETTLEMENTS

The borros point monitoring program along the path of the TBM indicated that at the gas line, ground heave and settlements were typically less than 25 mm. However, more significant surface settlements and sinkholes were observed over about one half to one third of the force main alignment following rainfall events.

Between Stations 1+50 (meters) and 3+15, damage in the form of settlements, sinkholes, and pavement cracking occurred. Numerous pavement settlements, sinkholes, and cracks were observed on the east side of University Road. Typically, settlements and sinkholes were located within about 1.2 meters of the pavement edge and directly over the force main alignment from about Station 1+50 to 3+15. Pavement drops up to 450 mm deep were observed in the sinkholes. Many cracks in longitudinal and transverse directions were evident over much of the west side of University Road between approximately Stations 1+50 and 3+15. Spalling of the pavement topcoat was observed in two locations.

Modifications were made to the TBM to reduce the potential for overexcavation-related ground losses prior to the subsequent TBM installed sewer crossing under a major highway for another part of the contract. The modifications included: reducing the TBM overcut; supporting the annular space around the pipe with a thicker grout; improving monitoring of tunneling operations; and minimizing or eliminating weekend shutdowns to avoid potential startup difficulties.

CONCLUSIONS

The installation of the force main using an earth pressure balance TBM combined with pipe jacking permitted the successful construction of the force main without dewatering. There was a concern that dewatering during construction could cause hydraulic gradients and consequently a spread of contamination from a lower sand stratum to an upper sand stratum. The method was also used successfully to pass about 300 mm under an active, 4800 kPa natural gas transmission line without incident. Ground movements in shallow cover cases are highly dependent on the details of the TBM cutting profile, operation methods, and control. Despite the ground surface settlements and pavement cracking that developed, earth pressure balance TBM pipe jacking was an appropriate method for this set of circumstances.

ACKNOWLEDGMENTS

The Massachusetts Water Resources Authority is the owner of the project. Fay, Spofford, and Thorndike, Inc. is the designer with GEI Consultants, Inc. as the geotechnical consultant. Caliacco Corporation is the contractor with Laxfield Corporation as the pipe jacking subcontractor.

1997 RETC EXHIBITORS

(as at 4/21/97)

Akkerman Equipment, Inc.
Brownsdale, MN

American Commercial, Inc.
Louisville, KY

Atlas Copco Robbins, Inc.
Kent, WA

Boart Longyear-Percussive Products Div.
Lively, ON

Boretec, Inc.
Solon, OH

Chemgrout, Inc.
La Grange Park, IL

Cleco Manufacturing, Inc.
Wallingford, CT

Construction & Tunneling Services, Inc.
Kent, WA

Dynex/Rivett, Inc.
Pewaukee, WI

Engelhard-Mearl Cellular Concrete
Roselle Park, NJ

Fosroc, Inc. Mining & Tunneling Division
Georgetown, KY

Geokon, Inc.
Lebanon, NH

Goodman Equipment Corporation
Bedford Park, IL

Hany Grouting Systems
Louisville, KY

Hayward Baker, Inc.
Odenton, MD

Herrenknecht Corporation
Greenville, SC

Jacobs Associates
San Francisco, CA

Long-Airdox Company
Oak Hill, WV

Lovat Tunnel Equipment, Inc.
Etobicoke, ON

Master Builders
Beachwood, OH

Maxon Industries, Inc.
Milwaukee, WI

National Mine Service Company
Indiana, PA

Nicholson Construction Company
Bridgeville, PA

Precision Centerline, Inc. (DYWIDAG)
Suwanee, GA

Schauenburg Flexadux Corp.
Clifton, CO

Schwing America, Inc.
White Bear Lake,MN

Shotcrete Technologies, Inc.
Idaho Springs, CO

Slope Indicator, Co.
Bothell, WA

Tunnels & Tunnelling International
London, England

Williams Form Engineering Corp.
Portland, OR

Woodward-Clyde (Consultants)
Concord, CA

World Tunnelling
London, England

Zeni Drilling Company
Morgantown, WV

Index

1997 RETC Proceedings

D

E

F

O

P

Q

R

U

V

W

Y

Z